Moses Fayngold

Special Relativity and How it Works

Related Titles

Liebscher, D.-E.

The Geometry of Time

ISBN: 978-3-527-40567-1

Moses Fayngold

Special Relativity and How it Works

WILEY-VCH Verlag GmbH & Co. KGaA

The Author

Moses Fayngold
New Jersey Inst. of Technology
Department of Physics
Newark, New Jersey
USA

Cover and Illustrations
Roland Wengenmayr

Library of Congress Card No.: applied for

British Library Cataloguing-in-Publication Data
A catalogue record for this book is available from the British Library.

Bibliographic information published by the Deutsche Nationalbibliothek
Die Deutsche Nationalbibliothek lists this publication in the Deutsche Nationalbibliografie; detailed bibliographic data are available in the Internet at http://dnb.d-nb.de.

Typesetting Thomson Digital, Noida, India
Printing betz-druck GmbH, Darmstadt
Binding Litges & Dopf GmbH, Heppenheim

Printed in the Federal Republic of Germany
Printed on acid-free paper

ISBN 978-3-527-40607-4

To the memory of Mary, Rachel, and Ezra.

Subtle is the Lord. . .
Einstein

Contents

Special Relativity and How it Works. Moses Fayngold

ISBN: 978-3-527-40607-4

Preface

This book has been inspired by a positive feedback to my previous book [1]. It includes a few revised sections of [1], but most of the book presents totally new material. In addition, I have included problems in the end of almost each chapter. I tried to make them so diverse as to cover the most essential parts of the theoretical material. As a result, the book can be used either as part of modern physics courses in college curricula, or as a main/complementary textbook for a separate course on special relativity, or else as a reference/textbook for distant learning courses offered online.

I tried to maintain, wherever possible, the two-layered structure similar to that used in [2, 3], with the lower level using an intuitive approach and the higher level presenting a more rigorous treatment appropriate for graduate and senior undergraduate students.

The book greatly expands the material in [1] on relationship between superluminal motions and the principle of causality. In particular, I have included some important results from quantum information theory related to the problem of superluminal signaling [4–9]. The reader will also find here the discussion of a new feature precluding the tachyons – hypothetical superluminal particles [10, 11] – from causality violation, and the description of some latest results on superluminal group velocities in quantum tunneling [12–14]. The connection between the front velocity and the high-frequency limit of the phase velocity (the Sommerfeld–Brillouin theorem) is discussed together with the basic properties of signal transfer [15, 16] and the role of causality in determining optical characteristics of a medium (Kramers–Kronig theorem) [17, 18].

There is a widespread misconception that special relativity is restricted exclusively to inertial (uniform) motions and is not suited to describe accelerated systems and phenomena therein. This is the same as saying that arithmetic is restricted only to integers and cannot treat fractions. One can find statements to this effect about special relativity even in some recently published monographs on relativity and cosmology. Such statements made by respected authors reflect the flaws in college physics curricula. One of the goals of this book is to dispel the myth that accelerated motions cannot be treated in the framework of special relativity. The reader will find

Special Relativity and How it Works. Moses Fayngold

ISBN: 978-3-527-40607-4

a standard treatment of accelerated motion in Chapter 5, devoted especially to *relativistic dynamics* of a point mass. In the last section of this chapter, relativistic motion in Coulomb's field and precession of elliptical orbits are described. This precession can be considered as an embryo of the famous effect known in general relativity – the precession of planetary orbits, widely believed to be a specifically general-relativistic effect. A few sections in Chapter 7 describe subtle phenomena associated with accelerated motion of extended bodies and the whole Chapter 8 is devoted to motions in rotating reference frames, including the famous experiments with atomic clocks flown around the Earth [19, 20]. Section 11.4 discusses accelerated superluminal motions. A special chapter on applied relativity covers the operation of particle accelerators and relativistic jet propulsion.

The book contains new examples of some of the most paradoxical-seeming aspects of the theory, such as the Ehrenfest paradox and the "conveyer belt" paradox.

In Chapters 4 and 5, the basic kinematic and dynamic characteristics of a physical system are introduced as vectors or tensors in pseudo-Euclidean vector space (space-time). This allows one to present mechanics and electromagnetism in tensor notations, which explicitly shows the Lorentz invariance of the laws of Nature. The readers will find here a thorough discussion of the mass–energy relation. The general theory is applied in the end to the above-mentioned motion of a charged particle in Coulomb's field.

Many of the existing texts on Special Relativity can be divided into two categories: one highly popularized, the other highly abstract. The former conveys some basic concepts, but does not give professional knowledge necessary for analyzing actual relativistic phenomena or solving concrete problems, let alone active working in the field. The latter is only accessible to the most advanced and diligent readers with a solid mathematical background.

On the other hand, there are many good college textbooks on Modern Physics, which fill the gap between the above two extremes (see, for instance, [21–23]). But they mostly draw upon a limited pool of traditional material, often combining Special Theory of Relativity and Quantum Physics in a one-semester course of Modern Physics. I have also taught such a course, and my experience (shared by many of my colleagues) shows that this is not sufficient to give the students a solid background in either field. In addition, most available texts are restricted to very special examples of point-like particles moving with relativistic speeds. As it comes to extended bodies, they are mentioned only once – just to illustrate the relativistic length-contraction effect, and again, only for a special case of uniform motion. As a result, most students complete the course with a rather vague and severely restricted knowledge of what relativity *is*. They perceive it as a theory describing a remote realm of some exotic events having little if anything to do with the real world around us. As to general public, many people even with higher education regard Einstein's theory merely as an intellectual curiosity, rather than the experimentally proved and self-consistent description of the world.

I tried to include the features that show not only what relativity *is*, but also *how relativity works*. The book shows the intimate connections between the fundamental principles of relativity and some everyday life phenomena that everybody is familiar with. These connections are so strong, and the Special Relativity is ingrained so deeply into fabric of the world, that the moment it would be cancelled by some divine providence, all of us and the entire world as we know it would disappear. The study of these connections can give the reader a better understanding of some basic phenomena of life, such as the existence of light; provide him/her with a working knowledge for solving non-standard problems, understanding physical restrictions for signal propagation in communication lines, principles of design and functioning of particle accelerators, relativistic rockets, etc.

The chapter on Electromagnetism and Optics has been written in the spirit of John Bell's famous paper "*How to Teach Special Relativity*" [24]. But the approach used is different. Namely, instead of showing how special relativity follows from electromagnetic phenomena, it is shown how all Electromagnetism *follows from relativity* with only one additional element – Coulomb's law. Accordingly, the students will learn how to view magnetic field as a purely relativistic effect resulting from relative motion between the observer and an electric charge. This will provide them with a deeper understanding of the statement that electric and magnetic field are manifestations of a more general entity – the electromagnetic field. Many think that this statement is true only in dynamics, when either of the fields can be created by changing the other field. The examples in the new chapter show that this statement has even deeper roots, because it works also in stationary situations and uniform fields. Generally, either field can emerge from its counterpart by switching between reference frames. This is a *manifestation of the relativity of fields.* Understanding of this deeper aspect may be especially important for those who want to specialize in high-energy physics, where the concept of unification of all known types of forces plays a fundamental role.

The transformation rules for electric and magnetic fields are shown to follow from the basic properties of space and time already found in the previous chapters. Hopefully, the used approach can help the students achieve a better understanding of both relativity and electromagnetism.

Some chapters include *discussions* aimed at gaining a broader view of a problem, or seeing it from another angle. A discussion, which may even include controversial views, can help the reader gain a deeper insight into a studied topic. I believe that exposing the reader, and especially the student, to such discussions may enhance his/her imagination and critical thinking necessary for a better understanding of the world.

I benefited from many colleagues in the process of working on this book – too many to mention all of them here. I am grateful to many readers and reviewers of my first book for their comments, questions, and suggestions. My special thanks to Edmund Immergut for his support of this project and even suggesting the title of the book; Edward Parilis for his encouragement and insightful remarks; David Green for stimulating discussions; Boris Bolotowski for interesting discussions of the visual image of moving bodies; my elder son Albert for his help in the initial design and drafting of the front cover; my younger son Vadim for his invaluable technical help. I also want to thank Michael Ibison and all participants of the Austin Forum (May 2006) for creative and stimulating atmosphere in discussing the new approaches and latest developments in relativity, quantum mechanics, and cosmology, which greatly inspired my work.

I enjoyed working with Roland Wengenmayr, who brought in the same skill and creativity in his illustrations as in my first book. I want to thank the Project Editor Esther Doerring for her infinite patience in dealing with my numerous delays on the latest stage.

I am deeply grateful to my wife Sophie for her unconditional support.

New York, May 2006 *Moses Fayngold*

1 Introduction

The job of science is to enable the inquiring mind to feel at home in a mysterious universe.

Lewis Carroll Epstein, *Relativity Visualized*

1.1 The Meaning of Relativity

The theory of relativity (special and general) is one of the cornerstones of modern physics. Its basic element is the principle of relativity. The word "relativity" here reflects only one, although very important, aspect of this principle: certain physical characteristics of a system are relative, in the sense that a numerical value of such characteristic measured by one observer may be different from the value measured by another observer moving with respect to the first one. The second aspect, inseparable from the first one, is that all laws of Nature are independent of the observer's motion. This statement reflects the "absolute" aspect of the principle of relativity, namely, that the *physical laws are the same for all observers*. And the two aspects are inseparable because one directly follows from the other. Indeed, the relativity of motion ("the states of rest and motion do not have the absolute meaning") follows immediately from absoluteness of natural laws (they are the same regardless of the state of motion of an observer).

We will start here with the relativity aspect. And a good starting point may be the discussion of such familiar characteristics of motion as velocity. Even a person with only rudimentary education can easily understand that velocity is a relative characteristic. If you are riding on a train and see another passenger passing from the rear of the train car to her seat in the front, you could estimate her velocity as about 2 miles/h. But an observer outside the train may estimate her velocity as 42 miles/h, owing to the additional 40 miles/h made by the train.[1)] The velocity of an object acquires exact meaning only when we specify relative to what it is measured. In this respect, it is a "flexible" characteristic. An object that is perceived by a ground-based observer to be moving is at rest to another observer moving together with this object. A third observer, moving in the same direction, but faster than the second one, will see the

1) We will see later that this simple addition of velocities is only an approximation to the more general rule.

Special Relativity and How it Works. Moses Fayngold

ISBN: 978-3-527-40607-4

same object moving in the opposite direction. We will call such quantities as velocity "observer-dependent," or relative.

Not all physical quantities are relative, however. Some of them are observer-independent, or absolute. Here is a simple example: if a car with three passengers has a velocity 45 miles/h, then the fact of it having this velocity is of a quite different category than the fact of it having three passengers inside. The latter is absolute because it is true for anyone regardless of one's state of motion. The former is relative because it is only true for those standing on the ground. But it is false, say, for a driver in another car moving along the same straight road. The driver will agree with you on the number of passengers in the first car but disagree on its velocity. He may hold that the first car has zero velocity because it has always been at the same distance from him.

Who is right – you or the second driver? Both are. And there is no contradiction here, because each observer relates what he sees to his own "reference frame."

Moreover, even one and the same observer can measure different velocities of the same object, depending on the observer's state of motion. A policeman in a car, using radar for measuring speeds of moving objects, will record two different values for the velocity of a vehicle, if he measures this velocity first time when his own car just stands on the road and the second time when his car is moving. We emphasize that nothing happens to the observed vehicle, it remains in the same state of motion with constant speed on a straight highway. And yet the value of this speed as registered by the radar is different for the two cases.

We thus see that the value of a speed does not by itself tell us anything. It only becomes meaningful if you specify *relative to what* this speed is measured. This is what we mean by saying that speed (more general, velocity) is a relative physical quantity.

Understanding the relative nature of some physical quantities (and absolute nature of some others) is the first step to acquiring the main ideas of special relativity.

Let us start with the widespread public perception of the theory of relativity: "Einstein has proved that everything is relative. Even time is relative."

One of these statements is true and profoundly deep; the other one is totally misleading.

The true statement is: *time is relative*. The realization of relative nature of time was a revolutionary breakthrough in our understanding of the world.

The wrong statement in the above "popular" account of relativity is that *everything* is relative. We already know that, for instance, the number of passengers in a car (or the chemical composition of a certain material) is not relative. One of the most important principles in relativity is that, together with natural laws, *certain physical quantities are absolute (invariant)*. One of such invariable quantities is the speed of light in vacuum. Also, a certain combination of time and distance turns out to be invariant. We will discuss these absolute characteristics in the next chapters. They are so important that we might as well call the theory of relativity the theory of absoluteness. It all depends on which aspect of the theory we want to emphasize.

We will now discuss in more details the relativity aspect, but keep in mind that, as emphasized above, its essence is the absolute status of the laws of Nature.

Let us first recall the classical principle of relativity in mechanics. Suppose you are inside a train car that moves uniformly along a straight track. If the motion is smooth enough, then, unless you look out of the window, you cannot tell whether the train moves or is at rest on the track. For instance, if you drop a book, it will fall straight down with acceleration, as it would do on the stationary platform. It will hit the floor near your feet, as it would do on the platform. If you play billiards, the balls will move, and collide, and bounce off in precisely the same manner as they do on the platform. And all other experiments will be indistinguishable from those on the platform. There is no way to tell, whether you are moving or not, by performing mechanical tests. This means that the states of rest and uniform motion are equivalent for mechanical phenomena. There is no intrinsic, fundamental difference between them. This general statement was formulated by Galileo and it came to be known as his principle of relativity. According to this principle, the statement "My train is moving" has no absolute meaning. Of course, you can find out that it is moving, the moment you look out of the window. But the moment you do it, you start referring all your observations to the platform. You then can say: "My car is moving relative to the platform." Platform constitutes your reference frame in this case. But you may as well refer all your data to the car you are in. Then the car itself will be your reference frame, and you may say: "My car is at rest, while the platform is moving relative to it." Now, pit the last two quoted statements against each other. They seem to be in contradiction, but they are not, because they refer to different reference frames. Each statement is meaningful and correct, once you specify the corresponding frame of reference.

We see that the concept of reference frame plays a very important role in our description of natural phenomena. We can even reformulate the principle of relativity in terms of reference frames. To broaden the pool of examples (and make the further discussion more rigorous!), we will now switch from jittering trains, and from spinning Earth with its gravity, far into deep space. A better, and more modern, realization of a suitable reference frame would be a nonrotating spaceship with its engines off, coasting far away from Earth or other lumps of matter. Suppose that initially the ship just hangs in space, motionless with respect to distant stars. You may find this an ideal place to check the basic laws of mechanics. You perform corresponding experiments and find all of them confirmed to even higher precision than those on Earth.

If you release a book, it will not go down; there is no such thing as "up" or "down" in your spaceship, because there is no gravity in it. The book will just hang in air close by you. If you give it an instantaneous push, it will start moving in the direction of the push. Inasmuch as you can neglect air resistance, the book will keep on moving in a straight line with constant speed, until it collides with another object. This is a manifestation of Newton's first law of motion – the famous law of inertia. Then you experiment with different objects, applying to them various forces or combination of forces. You measure the forces, the objects' masses, and their response to the forces. In all cases, the results invariably confirm Newton's second law – the net force accelerates an object in the direction of the force, and the magnitude of the acceleration is such that its product by the mass of the object equals the force. This explains why the released book does not go down – in the absence of gravity it does not know where "down" is. With no gravity, and possible other forces balanced, the net force on the book and thereby its acceleration is zero. Then you push against the wall of your compartment and immediately find yourself being pushed back by the wall and flying away from it. This is a manifestation of Newton's third law: forces always come in pairs; to every action there is always equal and opposite reaction.

Let us now stop for a while and make a proper definition. Call a system where the law of inertia holds, an inertial system or inertial reference frame. Then you can say that your ship represents an inertial system. So does the background of distant stars relative to which the ship is resting.

Suppose now that you fall asleep and during your sleep the engines are turned on. The spaceship is propelled up to a certain velocity, after which the engines are turned off. You are still asleep, but the ship is now in a totally different state of motion. It has acquired a velocity relative to the background of stars, and it keeps on coasting with this velocity due to inertia. The magnitude of this velocity may be arbitrary. But even if it is nearly as large as that of light, it will not by itself affect in any way the course of events in the ship. After you have woken up and checked if everything is functioning properly, you do not find anything unusual. All your tests give the same results as before. The law of inertia and other laws hold as they had done before. Your ship therefore represents an inertial reference frame as it had been before. Unless you look outside and watch the "sky" or measure the spectra of different stars, you would not know that your ship is now in a *different state of motion* than it had originally been. The reference frame associated with the ship is therefore also different from the previous one. But, according to our definition, it remains inertial.

What conclusions can we draw from this? First, any system moving uniformly relative to an inertial reference frame is also an inertial reference frame. Second, all the inertial reference frames are equivalent with respect to all laws of mechanics. The laws are the same in all of them. The last statement is the classical (Galilean) principle of relativity expressed in terms of the inertial reference frames.

The classical principle of relativity is very deep. It seems to run against our intuition. In this era of computers and space exploration, I have come across a few students in my undergraduate physics class who argued that if a passenger in a uniformly moving subway car dropped an apple, the apple would not fall straight down, but rather would go somewhat backwards. They reasoned that while the apple is falling down, the car is being pulled forward from under it, which causes the apple to hit the floor closer to the rear of the car. This argument, which overtly invokes the platform as a fundamental reference frame, overlooks one crucial detail: before being dropped, the apple in the passenger's hand was moving forward together with the car. This preexisting component of the motion persists in the falling apple due to inertia and exactly cancels the effect described by the student, so that the apple as seen by an observer in the car will go down strictly along its vertical path (Figure 1.1). This conclusion is confirmed by innumerable observations of falling objects in moving cars. It is a remarkable psychological phenomenon that sometimes not even such strong evidence as direct observation can overrule the influence of a more ancient tradition of thought. About a century and a half ago, when the first railways and trains appeared, some people were afraid to ride in them because of their great speed. The same story repeated at the emergence of aviation. Many people were afraid to board a plane not only because of the altitude of flight, but also because of its

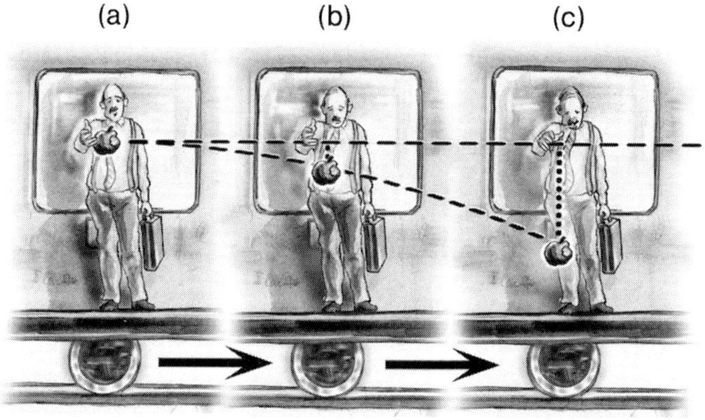

Figure 1.1 The fall of an apple in a moving car as observed from the platform: (a), (b), and (c) are the three consecutive snapshots of the process. The passenger sees the apple fall vertically, while it traces out a parabola relative to the platform. The shape of trajectory turns out to be a relative property of motion.

great speed. Apart from the fear of a *collision* at high speed, it might have been the fear of the speed itself. Many believed that something terrible would happen to them at such a speed. It took a great deal of time and new experience to realize that speed itself, no matter how great, does not cause any disturbance in the regular patterns of natural events so far as velocity remains constant. It is the *change* of velocity (acceleration) during braking, collision, or turn that can be felt and manifests itself inside of a moving system. If you are in a car that is slowing down, you can immediately tell this by the appearance of a force that pushes you forward. Likewise, if the car accelerates, everything inside experiences a force in the backward direction. It is precisely because of these forces that I wanted you to fall asleep during the acceleration of the spaceship. Otherwise, you would have immediately noticed the appearance of a new force and known that your ship was changing its state of motion, which I did not want you to do.

A remarkable thing about this new force is that it does not fit into the classical definition of a real force. It appears to be real because you can observe and measure it; you have to apply a real force to balance it; when unbalanced, it causes acceleration, as does any real force; it is equal to the product of a body's mass and acceleration, as is any real unbalanced force. In this respect, it obeys Newton's second law. Yet, it appears to be fictitious if you ask a question: Who (or what) exerts this force? Where does it come from? Then you realize that it, unlike all other forces in Nature, does not have a physical source. It does not obey Newton's third law, because *it is not a part* of an action–reaction pair. You cannot find and single out a material object producing this force, not even if you search out the whole universe. Unless, of course, you prefer to consider the whole universe becoming its source when the universe is accelerated past your frame of reference.

The new force has been called the inertial force – for a good reason. First, it is always proportional to the mass of a body it is applied to – and mass is the measure of

Figure 1.2 A chandelier in an accelerated car. To Alice, the tension force in the deflected chain acquires a horizontal component causing the chandelier to accelerate at the same rate as the car. Tom explains the deflection of the chandelier as the result of the inertial force. This force balances the horizontal component of the chain's tension.

the body's inertia. In this respect, it is similar to the force of gravity. Second, its origin can be easily traced to a manifestation of inertia. Imagine two students, Alice and Tom. They both observe the same phenomenon from two different reference frames. Tom is inside a car of a train that has just started to accelerate, while Alice is on the platform. Alice's reference frame is, to a very good approximation, inertial, while Tom's one is not. Tom looks at a chandelier suspended from the car's ceiling. He notices that the chandelier deflects backward during acceleration. He attributes it to a fictitious force associated with the accelerating universe. Alice sees the chandelier from the platform through the car's window (Figure 1.2), but she interprets what she sees quite differently. "Well," she says, "this is just what should be expected from the Newton's laws of motion. The unbalanced forces are exerted on the car by the rails and, may be, by the adjacent cars, causing the car to accelerate. However, the chandelier, which hangs from a chain, does not immediately experience these new forces. Therefore, it retains its original state of motion, according to the law of inertia, which holds in my reference frame. At the start, the chandelier accelerates back relative to the car only because the car accelerates forward relative to the platform. This transitional process lasts until the deflected chain exerts sufficient horizontal force on the chandelier." "Finally, Alice concludes, this force will accelerate the chandelier relative to the platform at the rate of the car, and there will be no relative acceleration between the car and chandelier." All the forces are accounted for in Alice's reference frame. In Tom's frame of reference, the force of inertia that keeps the chandelier with the chain off the vertical is felt everywhere throughout the car but cannot be accounted for. This state of affairs tells Tom that his car is accelerating.

Figure 1.3 Water in an accelerated fish tank. The rear wall of the tank rushes upon the water, raising its adjacent surface, while the front wall accelerates away from the water, giving it an extra room in front, which causes the water there to sink. To Tom, tilt of the water surface is caused by the inertial force. The tilted chain of the chandelier makes the right angle with the tilted water surface.

Tom has also taken the pain of bringing along an aquarium with fish in it. When the train starts accelerating, both Tom and Alice see the water in the aquarium bulge at the rear edge and subside at the front edge, so that its surface forms an incline (Figure 1.3). Alice interprets it by noticing that the rear wall of the aquarium drives the adjacent layers of water against the front layers, which tend to retain their initial velocity. This causes the rear layers to rise; in contrast, the front layers sink because the front wall of the fish tank accelerates away from them. Thus, the water surface tilts.

Tom does not see any accelerated motions within his car, but he feels the horizontal force pushing him toward the rear. "Aha," Tom says, "this force seems to be everywhere indeed. It pushes me and the chandelier back, and now I see it doing the same to water. It is quite similar to the gravity force, but it is horizontal and seems to have no source. Its combination with the Earth-caused gravity gives the net force tilted with respect to the vertical line." Being as good a student as Alice, Tom knows that the water surface always tends to adjust itself so as to be perpendicular to the net force acting on it. Since the latter is now tilted toward the vertical, the water surface in the aquarium becomes tilted to the horizontal by the same angle. The only trouble is that there is no physical body responsible for the horizontal component of the net force. "This indicates," Tom concludes, "that horizontal component is a fictitious inertial force caused by acceleration of my car."

In a similar way, one can detect a rotational motion, because the parts of a rotating body accelerate toward its center. We call it centripetal acceleration. For instance, we could tell that the Earth is rotating even if the sky were always cloudy so that we would be unable to see the Sun, Moon, or stars. That is, we could not "look out of the window." But we do not have to. Many mechanical phenomena on Earth betray its rotation. The Earth is slightly bulged along the equator and flattened at the poles. A free-falling body does not precisely fall along the vertical line (unless you experiment at one of the geographical poles). A pendulum does not swing all the time in one plane. Many rivers tend to turn their flow. Thus, in the Northern Hemisphere, rivers are more likely to have their right banks steep and precipitous and the left ones shallow. In the one-way railways, the right rails are being worn out faster than the left ones because the rims of the trains' wheels are pressed mostly against the right rail. In the Southern Hemisphere, the situation is the opposite. It is easier to launch a satellite in the east direction than in the north or south, let alone west direction. All these phenomena are manifestations of the inertial forces.

We will illustrate the origin of these forces for a simplified model of a train moving radially on a rotating disk. Suppose that the train is moving down a radial track toward the center of the disk and you observe this motion from an inertial stationary platform (Figure 1.4). At any moment the instantaneous velocity of the train relative to the platform has two components: radial toward the center and transverse, which is due to the local rotational velocity of the disk. The peripheral parts of the disk have larger rotational velocity than the central parts. As the train moves toward the center, it tends, following the law of inertia holding on the platform, to retain the larger

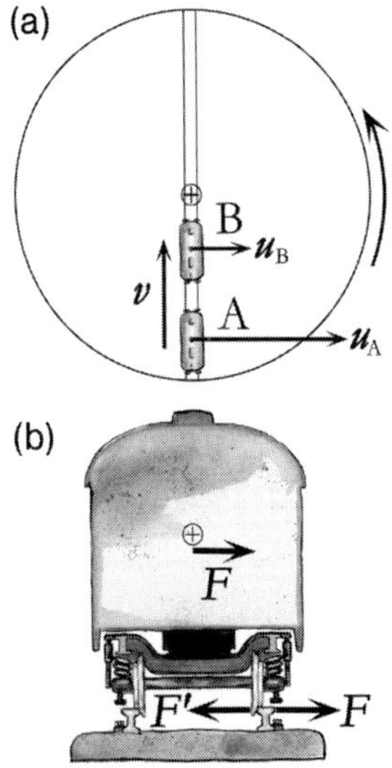

Figure 1.4 Schematic representation of the inertial forces acting on a moving car in a rotating reference frame. (a) View from above. The train moves from A to B with the speed v. Because of inertia, the train tends to transport its original rotational velocity u_A from A to B. Since u_A is greater than u_B, the train experiences transverse inertial force F. (b) View from behind. The force F is balanced by force F'.

rotational velocity "inherited" from the peripheral parts of the disk. This would immediately cause the derailment onto the right side of the track, had it not been for the wheels' rims that hold the train on the rails. The same effect causes the overall asymmetry between the left and the right banks of the rivers. We thus see that these phenomena are, in fact, manifestations of the inertia. Their common feature is that they permeate all the space throughout an accelerated system and cannot be attributed to an action of a specific physical body. Because of them, the Earth can be considered as an inertial system only to a certain approximation. Careful observation reveals the Earth's rotation without anyone having ever to look up in the sky.

All these examples show that inertial systems in classical physics form a very special class of moving systems. The world when looked upon from such a system looks simpler because there are no inertial forces. You can consider any inertial system as stationary by choosing it to be your reference frame without bringing along any inertial forces. There is no intrinsic physical difference between the states of rest and uniform motion. All other types of motion are absolute in a sense that Nature provides us with the criterion that distinguishes one such motion from all the others. We can also relate all observational data to an accelerated system and consider it motionless. But there are intrinsic physical phenomena (inertial forces) that reveal its motion relative to an inertial reference frame. Not only can we detect this motion without "looking out of the window," but we can also determine precisely all its characteristics, including the magnitude and direction of acceleration, the rate of rotation, and the direction of the rotational axis.

We thus arrive at the conclusion that Nature distinguishes between inertial and accelerated motions. It does not at all mean that the theory cannot describe accelerated motions. It can, and we will see the examples of such description further in the book. The special theory of relativity can even be formulated in arbitrary accelerated and therefore noninertial reference frames [19]. But the description of motion in such systems is far less straightforward, to a large extent because of the appearance of the inertial forces. The general theory of relativity reveals deep connections between the inertial forces in an accelerated system and gravitation. We will in this book be concerned mostly with special relativity and make a very brief outline of basic ideas of general relativity in the last chapter.

Discussion

Here we want to mention another misconception of relativity of time among the general public. It can be expressed as an extension of a passerby's remark quoted above: "Einstein has proved that time is relative. One minute or one million years makes no difference. So relax and take it easy."

While the first statement here is true, the second one is a good example of what can happen to an idea when one takes it to extremes and on the way changes its meaning. In the given case, the result is what philosophers of science call "relativistic nihilism." The corresponding view is not even wrong. It is utterly meaningless. The relativity of time refers to the duration of a process observed or measured from *different* reference frames. By contrast, the last statement in the quotation is about duration of a process in *the same* reference frame. And in this case saying that there is no difference between one minute and one million years is equivalent to saying that there is no difference between one cent or one million dollars in your bank account. If this passerby is rich, but someday (God forbid!) loses his millions, with only one cent left on him, let him then use his own good advice – relax and take it easy.

Another important point relevant to the above discussion is the difference between a reference frame and a coordinate system. A reference frame is a physical object (usually – but not always – a sufficiently rigid body) to which we refer our measurements and observations. It may be a car, a plane, a spaceship, a planet (say, our Earth), a galaxy, or even a set of galaxies.

A coordinate system is a way we specify a position by assigning to it a set of numbers (coordinates). We can use an infinite variety of *different* coordinate systems associated with the *same* reference frame. The most familiar are the Cartesian, spherical, and cylindrical coordinate systems. The most commonly used is a Cartesian system – a system of three mutually perpendicular directions x, y, z, taken as reference directions. Geometrically, these can be represented as a triad of unit vectors $\hat{\mathbf{x}}$, $\hat{\mathbf{y}}$, $\hat{\mathbf{z}}$, and a point in space is specified by the components (orthogonal projections) of its position vector onto the corresponding directions.

We can use an infinite variety of such triads. They are all distinct and at the same time can be obtained one from another by appropriate rotations and/or reflections.

A spherical coordinate system is determined by a radial distance of a particle from a reference point taken as the origin and its two angular coordinates – the polar and azimuthal angles (similar to latitude and longitude in geography). In this case, we also have an infinite variety of different systems of spherical coordinates – all associated with the same reference frame. They differ from one another by their position of origin, orientation of the polar axis, and the choice of the reference plain for azimuthal angle (similar to the choice of Greenwich meridian).

We see from these examples that a reference frame and a coordinate system are *different concepts*. And, in particular, the former does not specify the latter. We will often use in this book both terms, and it is important to distinguish between them. We will frequently abbreviate a reference frame as RF.

1.2
Weirdness of Light

The special theory of relativity has emerged from studies of electromagnetic phenomena. One of them is motion of light.

Let us extend our discussion of motions of physical bodies to situations involving light. Previously, we had come to the conclusion that one can catch up with any object. Does this statement include light? This question was torturing a high school student Albert Einstein more than a century ago and eventually brought him to special relativity. What we have just learned about velocity prompts immediately a positive answer to the question. Velocity is a relative quantity – it depends on reference frame. It can be changed by merely changing the reference frame. For instance, if an object is moving relative to Earth with a speed v, we can change this speed by boarding a vehicle moving in the same direction with a speed V. Then the speed of the object relative to us will be

$$v' = v - V. \tag{1.1}$$

We can change v' by "playing" with the vehicle – accelerating or decelerating it. For instance, reversing the speed of the vehicle would result in changing the sign of V in the above equation, and accordingly, would greatly increase the relative speed of the object without touching it. If we want to catch up with the object, we need to bring its relative velocity down to zero. We can do it by accelerating the vehicle to the speed $V = v$.

Because this works for objects like bullets, planes, or baseballs, people naturally believed that it should work for light as well. It is true that we never saw light at rest before. But, as an old Arabic saying has it, "if a mountain does not go to Mohammed, then Mohammed goes to the mountain." If we cannot stop the light on Earth, then we have to board a spaceship capable of moving relative to Earth as fast as light does and use this "vehicle" to transport us in the direction of light. Let c be the speed of light relative to Earth and V be the speed of a spaceship also relative to Earth. If Equation (1.1) is universal, then we can apply it to this situation and expect that the speed c' of light relative to the spaceship will decrease by the amount V:

$$c' = c - V. \tag{1.2}$$

Suppose that the rocket boosters accelerate the spaceship, its velocity V increases, and c' decreases. When V becomes equal to c, the speed c' becomes zero. In other words, light stops relative to us, that is, we have caught up with light. The same principle that has helped us "stop" the object in (1.1) at $V = v$, works here to help us catch the light. The law (1.1) of addition of velocities says that it is possible.

But there immediately follows an interesting conclusion. We know that the Earth can to a good approximation be considered as an inertial reference frame, and all inertial reference frames, according to mechanics, are equivalent. Einstein thought that this principle could be extended beyond mechanics to include all natural phenomena. If this is true, then whatever we can observe in one inertial system, can as well be observed in any other inertial system. If light can be stopped relative to at least one spaceship, then it can be brought to rest relative to any other inertial system, including Earth. In physics, if Mohammed can come to a mountain, the mountain can come to Mohammed. To stop light relative to the spaceship, we need to accelerate the ship up to the speed of light. To stop light relative to Earth, we may, for example, put a laser gun onto this ship and fire it backwards. Then the laser pulse, while leaving the ship with velocity c relative to it, will have zero velocity with respect to Earth. We then will witness a miraculous phenomenon of stopped light.

I can imagine an abstract from a science fiction story exploiting such a possibility, something running like this:

Mary stretched her arm cautiously and took the light into her hand.
She felt its quivering wave-like texture, which was constantly

changing in shape, brightness, and color. Its warm gleam has gradually penetrated her skin and permeated all her body, filling it with an ecstatic thrill. She suddenly felt a divine joy, as though a new glorious life was being conceived in her.

But, alas! Beautiful and tempting as it may seem, our conclusion that freely traveling light can be stopped relative to Earth or whatever else, is not confirmed by observation. It stands in flat contradiction with all known experiments involving light. As it had already been established before Einstein's birth, light is electromagnetic waves. The theory of electromagnetic phenomena, developed by J.C. Maxwell, shows a remarkable agreement with experiments. And both theory and experiments show quite counterintuitive and mysterious behavior of light: not only is it impossible to catch up with light, it is impossible even to change its speed in vacuum by a slightest degree – no matter what spaceship we board and in what direction or how fast it moves.

We have arrived at a deep puzzle. Light does not obey the law of addition of velocities expressed by Equation (1.1). The equation appears to be as fundamental as it is simple. And yet there must be something fundamentally wrong about it.

"Wait a minute!" the reader may say. "Equation (1.1) is based on a vast amount of precise experiments. It is therefore absolutely reliable, and it says that. . ."

"What it says is true for planes, bullets, planets, and all the objects moving much slower than light. But it is not true for light," – I answer.

"Well, look here: the speed of light as measured in experiments on Earth is about 300 000 km/s. Suppose a spaceship passes by me with the velocity of 200 000 km/s and I fire the laser pulse at the same moment in the same direction. Then one second later the laser pulse will be 300 000 km away from me, whereas the spaceship will be 200 000 km away. Is it correct?"

"Absolutely."

"Well, then it must be equally true that the distance between the spaceship and the pulse will be 100 000 km, which means that the laser pulse makes 100 000 km in one second relative to the spaceship. It is quite obvious!"

"Apparently obvious, but not true."

"How can that be?"

"This is a good question. The answer to it gives one the basic idea of what relativity is about. You will find the detailed explanations in the next chapter. It starts with the analysis of one of the most known experiments that have demonstrated mysterious behavior of light mentioned above. But to understand it better, let us first recall a simple problem from an Introductory Course of College Physics."

1.3 A Steamer in the Stream

The following is a textbook problem in nonrelativistic mechanics; however, its solution may be essential for understanding one of the experimental foundations of special relativity.

The problem is this: A steamer has a speed of u km/h relative to water. It starts at point A on the bank of the river with the stream velocity v km/h. It moves downstream to the point B on the same bank at a distance L from A, immediately turns back and moves upstream. How long will it take to make round-trip from A to B and back?

The solution is pretty simple. In the case of still water, the answer would be

$$t_0 = 2\frac{L}{u}. \tag{1.3}$$

Now, take account of the stream. If the steamer makes u km/h relative to water and the stream makes v km/h relative to the bank, then the steamer's velocity relative to the bank is $(u+v)$ km/h when downstream and $(u-v)$ km/h when upstream. We are interested in the resulting time, which is determined by the ratios of the distance to velocities. We must therefore use the speed averaged *over time*. The total time consists of two parts: one (t_{AB}), which is needed to move from A to B, and the other (t_{BA}) to move back from B to A. The time t_{BA} is always greater than t_{AB}, since the net velocity of the steamer is less during this time. Thus, the net velocity of the steamer is greater than u during the shorter time and less than u by the same amount during the longer time. Therefore, its average over the whole time is less than u. As a result, the total time itself must be greater than t_0. It must become ever greater as v gets closer to u. This result becomes self-evident when $v = u$. Then the steamer after turning back is carried down by the stream at the same rate as it makes its journey upstream. So it will just remain at rest relative to the bank at B and will never return to A. This is the same as to say that it will return to A in the infinite future, that is, the total time is infinite.

What if v becomes greater than u, that is, the stream is faster than the steamer? Then the steamer after the turn is even unable to remain at B; the stream will drag it down, taking it ever farther away from its destination. We can formally describe this situation by ascribing negative sign to the total time t.

Let us now solve the problem quantitatively. The time it takes to go from A to B and then from B to A is, respectively,

$$t_{AB} = \frac{L}{u+v}, \quad t_{BA} = \frac{L}{u-v}. \tag{1.4}$$

So the total time

$$t_{\downarrow\uparrow} = t_{AB} + t_{BA} = \frac{L}{u+v} + \frac{L}{u-v} = \frac{t_0}{1-(v^2/u^2)}, \tag{1.5}$$

where t_0 is the would-be time in the still water, given by Equation (1.3).

If we plot the dependence (1.5) of time versus stream velocity, we get the graph shown in Figure 1.5.

Equation (1.5) describes in one line all that was written over the whole page and, moreover, it provides us with the exact numerical answer for each possible situation. The graph in Figure 1.5 describes all possible situations visually. You see that for all $v < u$ the time t is greater than t_0, it becomes infinite at $v = u$, and negative at all $v > u$. When v is very small relative to u, Equation (1.5) gives $t_{\downarrow\uparrow} \approx t_0$. This is natural, since

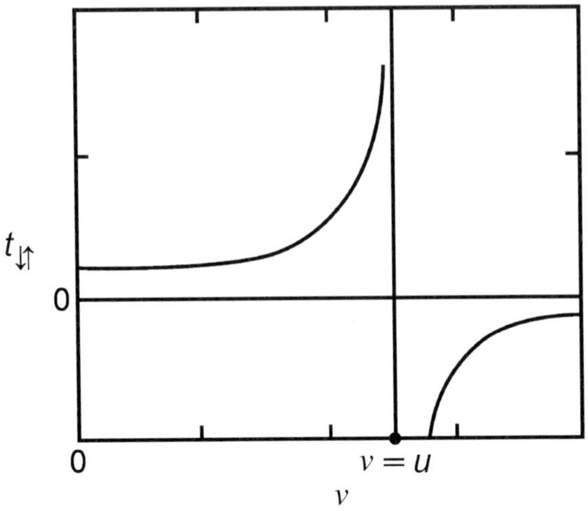

Figure 1.5 The dependence of round-trip time $t_{\downarrow\uparrow}$ on speed v.

for small v the impact of the stream is negligible, and we recover the result (1.3) obtained for the lake.

Now, consider another case. The river is L km wide. The same steamer has to cross it from A to B right opposite A on another bank and then come back, so the total distance to swim relative to the banks is again $2L$. How long will it take to do this?

The only thing we have to know to get the answer is the speed of the steamer u' in the direction AB right across the river. The steamer must head all the time a bit upstream relative to this direction to compensate for the drift caused by the stream. If during the crossing time the steamer has drifted l km downstream, then to get to B, it must head to a point B′, l km upstream of B. Thus, its velocity relative to water is u and directed along AB′, the velocity of the stream is v and directed along B′B, and the resulting sought-for velocity of the steamer relative to the banks is directed along AB. These three velocities form the right triangle (Figure 1.6), and therefore

$$u' = \sqrt{u^2 - v^2} = u\sqrt{1 - (v^2/u^2)}. \tag{1.6}$$

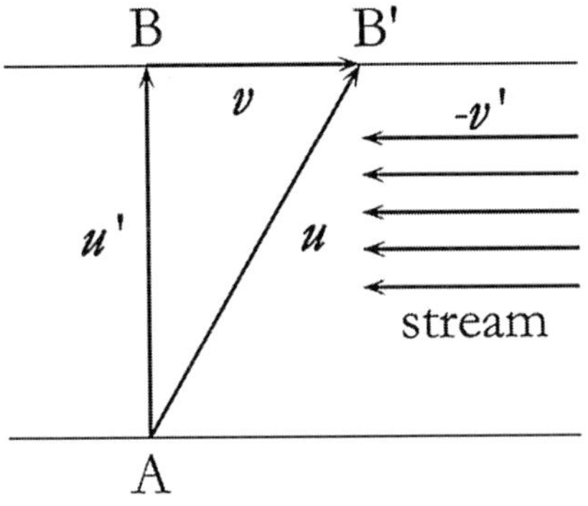

Figure 1.6

So our final answer for the total time back and forth between A and B is

$$t_{\perp} = \frac{2L}{u'} = \frac{2L}{u\sqrt{1-(v^2/u^2)}} = \frac{t_0}{\sqrt{1-(v^2/u^2)}}. \tag{1.7}$$

Note that Equations (1.6) and (1.7) give a meaningful result only when $v < u$ (a side of the right triangle is shorter than hypotenuse). Then, according to (1.7), time $t_{\perp}$ is also greater than t_0, but it is less than $t_{\downarrow\uparrow}$. Thus, one can write

$$t_0 < t_{\perp} < t_{\downarrow\uparrow}. \tag{1.8}$$

If $v > u$, the triangle in Figure 1.6 cannot be formed. The steamer's drift per unit time exceeds its velocity u and the steamer would not be able to reach the point B, let alone return to A. This circumstance is reflected in the mathematical structure of Equations (1.6) and (1.7), which yield imaginary numbers when $v > u$. They say that there is in this case no physical solution that would satisfy the conditions of the problem.

Now, what is the link between this problem and the experiment with light mentioned above? Take the running waves on the water surface instead of the steamer and you turn the mechanical problem into the hydrodynamic one. Then take the sound waves in air during the wind instead of the steamer on the water stream and you get the same problem in fluid dynamics. And as the last step, consider the light that propagates in a moving transparent medium in transverse and longitudinal directions and here you are with the optical problem that is identical to the initial mechanical one.

This is why I started the book with this *introductory physics* problem. Its mathematical description is exactly the same as that of the problem ahead. But keep in mind that the treatment of this "sample problem" is nonrelativistic! This is totally OK for a steamer or for the sound in air; the corresponding errors in results obtained are quite negligible in these cases, so we can use them safely. But when we apply the same to the case of light, the result will be totally erroneous. It was precisely this wide discrepancy between the nonrelativistic prediction and actual observation that became one of the crucial scientific evidence in favor of Einstein's theory.

Problems

1.1 A steamer has a speed u relative to water. It started from A, went downstream, reached B on the same bank at distance L, immediately turned back and returned to A. The recorded round-trip time of this travel was t. Find the speed V of the stream in terms of u, L, and t.

1.2 With the same data as in the previous problem, find the speed of the stream for the case, when the points A and B lie exactly opposite each other across the river.

1.3 For the same data as in Problem 1.1, find the average speed of the steamer over time t and over distance $2L$. Which average is greater?

2
Light and Relativity

2.1
The Michelson Experiment

In the history of the study of the world, one can trace a tendency to explain the greatest possible number of phenomena using the smallest number of basic principles. In the eighteenth and nineteenth centuries, it seamed that the solution of this task was not far off. That period witnessed a spectacular flourish of Newtonian mechanics. Using its basic concepts, scientists made astonishing progress in astronomy, navigation, technology, Earth studies, and so on. Later on, the advance of the molecular-kinetic theory helped describe the huge field of thermodynamic phenomena in the language of mechanics.

This engendered a hypothesis that *all* natural phenomena can be reduced to mechanics; that is, one can construct an entirely mechanical picture of the world – a picture based on the laws of Newton and on the corresponding concepts of absolute time and space. Consequently, physicists sought to integrate electromagnetic phenomena and particularly the propagation of light into mechanical theory.

By that time it had been proved that light propagation is a wave process for which the phenomena of interference and diffraction, common for all waves, could be observed. And since all waves known in mechanics could propagate only in some medium with elastic properties, it seamed reasonable to assume that light waves are also mechanical oscillations of some elastic medium, which penetrates all physical objects and fills all space in the universe. This hypothetical medium was called ether.

The ether hypothesis leads to a number of inferences, whose examination may confirm or refute the hypothesis itself. In this section, we will consider one such inference, whose analysis has played an important role in the history of science.

Let us assume that the space is filled with ether. Then, since the Earth is traveling through the ether, an earthly observer may expect to discover an "ether wind." The speed of light in the ether as measured by the earthly observer may in this case depend on direction. If the wind has a speed v relative to the Earth, the observer would expect to measure for the speed of light $c_{\uparrow} = c + v$ in the direction of the wind and

Special Relativity and How it Works. Moses Fayngold

ISBN: 978-3-527-40607-4

$c_{\downarrow} = c - v$ in the opposite direction. And what is the speed of light in the transverse direction? In order for light to move perpendicular to the wind, it is necessary to compensate for the lateral "drift," which means that the light's velocity relative to the ether must have a longitudinal component against the wind, equal to v. However, the total velocity of light relative to the ether is equal to c. Therefore, according to our results in the previous chapter, the transverse component must be equal to $c_{\perp} = \sqrt{c^2 - v^2}$ (Figure 1.6 with $u = c$ and $u' = c_{\perp}$). If our reasoning is correct, the speed of light relative to the Earth must be anisotropic (i.e., it would depend upon the direction) due to the Earth's motion through the ether. Conversely, an observation of such anisotropy would enable us to detect this motion and to find its speed. In other words, optical phenomena would reveal a fundamental difference between a moving reference frame (RF) and a "privileged" frame attached to the ether. This would mean that the relativity principle formulated by Galileo for mechanical phenomena is invalid for optical phenomena and so we would be able to distinguish the state of uniform motion in a straight line from the state of "absolute rest."

The prominent physicist–experimenter Michelson, later accompanied by Morley, had tried to actually discover this effect in a series of experiments. The idea of these experiments was very simple and based on the interference of light waves. For instance, consider two rays with the same oscillation frequency ν, which have been obtained by splitting a beam from a small light source. The splitting of the beam occurs in a glass plate P, which partially transmits and partially reflects light. At a certain position of the beam splitter, the reflected and transmitted parts of the light wave propagate in two mutually perpendicular directions and then come back, after reflection in the mirrors A and B (Figure 2.1a). Because the split beams have taken different ways, they may accordingly have spent different times traveling along their respective paths. As a result, their oscillations will have a certain phase shift with respect to one another when they recombine. The phase shift can be

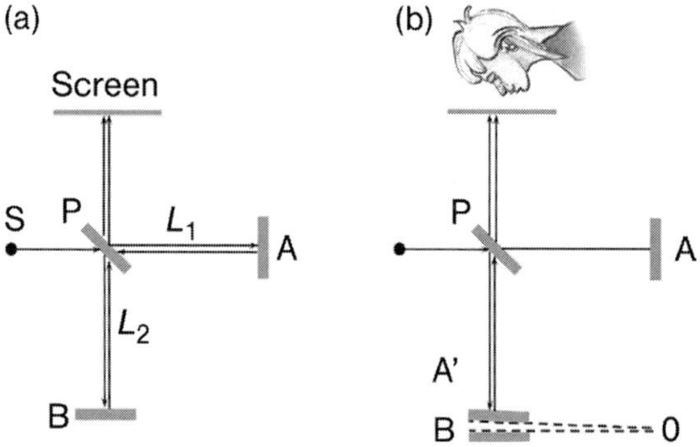

Figure 2.1 (a) Schematic representation of the Michelson interferometer. S denotes source of light, and A and B denote mirrors. (b) An equivalent air wedge A′OB produced by an angular misalignment of mirrors A and B.

determined as a ratio of the relative time lag to the oscillation period T, multiplied by 2π. If the two waves of the same frequency and the same individual light intensity I_0 meet with a phase difference $\Delta\phi$ at a certain point, the net intensity at this point will be

$$I = 2I_0(1 + \cos\Delta\phi). \tag{2.1}$$

For the waves oscillating in synchrony, we have $\Delta\phi = 0$, and the waves reinforce each other, producing the net intensity equal to four individual intensities (constructive interference). When the wave oscillations are totally out of phase ($\Delta\phi = 180°$), the waves cancel each other out, giving zero net intensity at corresponding point. In this case, light combined with light produces darkness (destructive interference).

Generally, the phase shift $\Delta\phi$ is different for different points on the screen. Consider, for instance, an interferometer with its mirrors not ideally perpendicular to each other. Interference in this case is similar to that on a wedge-shaped layer of air between two interfaces. Imagine your eye placed at the screen (Figure 2.1b). Then you will see simultaneously the mirror B and the image A′ of the mirror A. If the mirrors are not ideally perpendicular, then the image A′ is not parallel to B and the interference is equivalent to that on an air wedge BOA′. It is clearly seen from Figure 2.1b that the farther from the edge, the greater the path difference between the interfering beams, and accordingly the phase shift $\Delta\phi$. Thus, the phase shift is a function of distance y between the observation point and the image of the edge on the screen: $\Delta\phi = \Delta\phi(y)$. As you sweep across the screen, you will pass places with different phase shifts between the combining waves and accordingly different light intensity. The screen will display a pattern of bright and dark fringes (i.e., alternating regions of high and low intensity). Such a pattern will be observed even when the "arms" of the interferometer (the distances between the center of the beam splitter and the centers of the mirrors) are the same: $L_1 = L_2 = L$.

Let us consider this case and calculate an *additional* phase difference caused by a possible time lag owing to the hypothetical ether wind. Suppose that the wind "blows" along one of the arms of the interferometer. We can treat this problem in *total analogy with our treatment of the "steamer in the stream"* in Chapter 1. The light here will play the role of the steamer and the ether wind will be the stream. Then, by the same reasoning, the time required for the light to travel there and back along the "longitudinal" arm should be equal:

$$t_{\downarrow\uparrow} = \frac{L}{c+v} + \frac{L}{c-v} = 2\frac{Lc}{c^2 - v^2} \approx 2\frac{L}{c}(1 + \beta^2), \tag{2.2}$$

where β is the ratio v/c (which in this case is much smaller than 1).

The round-trip time in the transverse direction is determined by the above-mentioned "transverse" speed $c_\perp$ and equals

$$t_\perp = \frac{2L}{c_\perp} = \frac{2L}{\sqrt{c^2 - v^2}} = \frac{2L}{c\sqrt{1-\beta^2}} \approx \frac{2L}{c}\left(1 + \frac{1}{2}\beta^2\right). \tag{2.3}$$

In the last two equations, we also wrote the approximations to the exact expressions to the accuracy of the second order of β. Thus, the time lag between these two waves will be

$$\Delta t = t_{\downarrow\uparrow} - t_{\perp} = \frac{L}{c}\beta^2. \tag{2.4}$$

The corresponding phase shift, according to the above definition, is

$$\Delta\phi_e = 2\pi\frac{L\beta^2}{cT} = 2\pi\frac{L\beta^2}{\lambda}. \tag{2.5}$$

Here $\lambda = cT$ is the wavelength of light (the distance traveled in one period).

As we see from (2.5), the contribution from the ether wind depends only on the wavelength, the arm length, and the speed of the Earth relative to ether. Therefore, it must be to a high accuracy the same for all points on the screen. Thus, the possible influence of the ether wind can be described as a constant (2.5), added to the phase $\Delta\phi(y)$ in Equation (2.1). If a constant is added to a phase in the sine or cosine function, the graph of this function will just shift along the y-axis. Therefore, with the ether wind, the observed interference pattern on the screen would be shifted relative to its position in the absence of the wind.

Suppose now that we have turned the whole device by 90°, so that the beam, which was parallel to the "wind," now travels in the transverse direction, and vice versa. Then, the wave that previously had arrived late at a given point will now arrive earlier; in other words, the time lag will change sign. This must result in the shift of the observed interference pattern corresponding to the change in phase difference by $2\Delta\phi_e$. Thus, if there is no ether wind, the turning of the device will not affect the interference pattern. If the wind exists and affects the speed of light, the interference pattern will shift with the turning of the device. It was this shift that Michelson and Morley wanted to observe in their experiments.

In order to observe the effect, the pattern on the screen must shift a distance comparable with the fringe spacing; that is, the additional phase shift $\Delta\phi_e$ due to the expected "ether wind" must be comparable with 2π. According to (2.5), this requires an experimental setup in which the distance L is on the order of λ/β^2. For the wavelengths of visible light and the speed of ether wind comparable with the speed of Earth's motion around the Sun, the length of the travel path of light in the device must be no less than 100 m. Therefore, the light in the Michelson interferometer was made to travel many times back and forth along either of the two paths before recombining to make the interference pattern on the screen [2]. The whole setup was state of the art by the time (1881–1887) the experiments were carried out.

The experiments conducted along this scheme and repeated many times thereafter with ever-increasing accuracy did not produce the expected result. The ether wind and thereby the motion of Earth could not be detected. This can be considered as evidence that motion of a reference frame does not affect the speed of light.

A plethora of studies have been devoted to the analysis of the Michelson experiment. In some of them, the authors tried to retain the concept of ether. To account for the negative results of the Michelson experiment, they had to assume that the ether wind is precluded from being observed by some countereffect. For instance, the change of direction of the ether wind relative to the device could deform the interferometer's arms in such a way as to compensate for the change of the interference pattern. As a result, no effect would be observed. Precisely such an explanation was proposed by physicists H.A. Lorentz and G.F. FitzGerald.

Lorentz and FitzGerald had assumed that any system moving at a speed v relative to the ether contracts in the direction of motion by the amount $(1 - v^2/c^2)^{1/2}$. Such a contraction explains the negative result of the Michelson experiment. Indeed, if we multiply the longitudinal size in formula (2.2) by the above factor, the time $t_{\downarrow\uparrow}$ will become equal to $t_{\perp}$, which means that the light's traveling time for both rays and, accordingly, the interference pattern, will no longer depend on the interferometer's orientation. Such explanation is logically consistent, but it is unduly complicated. It implies the necessity of a few independent postulates:

1. The ether does exist (and in addition, it must possess a number of very special and hardly compatible properties, and each of them must also be postulated).
2. The motion of any system through the ether contracts the system in the longitudinal direction.
3. This contraction is such as to compensate all observable manifestations of the ether wind.

In addition to its complexity, the described scheme is faulty in two respects. First, its primary substance (ether), whose existence it postulates, does not reveal itself in the observed phenomena (the scheme itself has been designed to account for this fact). Second, it leads to a number of subsequent difficulties and complications. Therefore, it could not have become a foundation for a physical theory.

All these difficulties were eliminated in Einstein's special theory of relativity, which does not in any way mention ether. In the basis of the theory lies Einstein's principle of relativity, according to which *all* natural laws and thus *all* physical phenomena (and not only mechanical ones) look similar in all inertial reference frames. In other words, all inertial systems are totally equivalent.

This principle easily explains why no indications of the Earth's motion were detected in the Michelson experiment. Since the Earth's orbital motion is inertial with a high degree of accuracy on any small segment of its orbit, it cannot affect the outcome of any laboratory experiment.

Thus, Einstein's principle of relativity makes the negative result of the Michelson experiment obvious from the very beginning. An interesting historical fact is that Einstein himself was probably unaware of the Michelson experiment when he published his famous first article on the theory of relativity. It does not mean, however, that such an experiment was unnecessary. Regardless of whether it was or was not known to Einstein at that time, the Michelson experiment is one of the cornerstones in the experimental basis of the theory of relativity. Its result has greatly facilitated the acceptance of this theory and helped to quickly apprehend its striking revelations about the basic properties of time and space. And this is what comprises the historical role of the Michelson experiment.

2.2
The Speed of Light and the Principle of Relativity

Let us now try to interpret the results of the above-mentioned experiments with light. These results contradict our intuition based on observing motions much slower than light. Our experience expressed in Equation (1.1) shows that the velocities of such motions just add together. In particular, this equation accurately describes a well-known fact that if a surfer reaches the same speed as a running ocean wave by just riding it, then the speed of the wave relative to the surfer is zero.

But what can be done with an oceanic (or sound) wave cannot be done with light. The experiments did not support the viewpoint that light waves are just perturbations in a specific medium (ether) permeating the whole space. And with no scientific evidence, it makes no sense to speak about such medium. Therefore, we accept the viewpoint that space does not contain any light-carrying substance (ether), in which light could spread as the sound in the air or waves in the ocean. A light wave can exist "all by itself" in a free space and only in motion. A notion of "still" or even "slow" light waves in an empty space contradicts both the electromagnetic theory and the experiment. Light always moves with the same universal speed. We cannot tell whether we are on a stationary platform or in a uniformly moving car or else in a speeding spaceship with engines off, by measuring the speed of light: in either case the result is the same. Nor can we tell uniform motion from rest by observing any other electromagnetic phenomena. These phenomena, as well as the mechanical ones, are "insensitive" to a state of uniform motion of the observer. Einstein accepted this statement as part of a universal principle he had formulated (Einstein's principle of relativity) – that *all* natural phenomena look the same in all inertial reference frames. In other words, Nature possesses a deep symmetry that is manifest in the equivalence of all inertial systems. All observed phenomena confirm this conclusion.

A skeptical reader could object: "Excuse me, but this conclusion seems ridiculous. I can understand that the invariance of the speed of light, difficult as it is to grasp, indicates that all inertial reference frames are equivalent. However, a speed of an object such as a stone or bullet is *not* invariant and yet you say that this is also a manifestation of the same principle of relativity. How can that be?"

The answer to this is that the speed of a stone may vary even in *one* reference frame, depending on initial conditions or applied forces. Therefore, any difference in such speed measured by different observers reflects only the difference in the initial conditions, not the difference in laws of Nature. For instance, the falling item in Figure 1.1 has no initial velocity in horizontal direction as seen from the train and has initial horizontal velocity equal to that of the train as seen from the ground. Therefore, it falls straight down relative to the train and traces out a parabola relative to the platform. But it might as well start moving without initial horizontal component if dropped by the person on the platform, in which case it would fall straight down relative to the platform. Or, it might start moving in the train car with initial horizontal component if pushed horizontally by the passenger, in which case it would move there in a parabola, as it does on the platform under similar conditions. Thus, if we have two identical systems in two different inertial reference frames K and

K′, and both systems start from identical initial conditions, they perform identical motions. And in either frame, the speed of corresponding mass can vary within the same range – from zero to the speed approaching that of light. This is more rigorous formulation of the principle of relativity for systems like stones or planes.

Light, on the other hand, can move only with one fixed speed in one reference frame. The principle of relativity in this case requires that this fixed speed remains the same in any other inertial reference frame, regardless of the initial conditions.

But here the same thoughtful reader may ask another question: "OK, this explanation is logically consistent, *if* we accept that the speed of light, unlike speeds of most other objects, is the fixed quantity. But how can it be that light, which moves in the same space and time as do objects like cars, bullets, and planets, does not obey the law (1.1) of addition of velocities that applies to these objects?"

This question, as I noticed in Chapter 1, is crucial for understanding relativity.

Let us trace the origin of the law of addition of velocities. Consider two inertial reference frames K and K′. Let K′ move relative to K in the x-direction with a speed V, and the origins of both systems coincide at the moment $t=0$. Consider an object M at a later (nonzero) moment t. By this moment, the origin of system K′ will have traveled a distance Vt (Figure 2.2). Therefore, the x-coordinate of the object in K at this moment will differ from its x'-coordinate in K′ by this distance:

$$\left.\begin{aligned} x &= x' + Vt \\ t &= t' \end{aligned}\right\}. \tag{2.6}$$

The second of the Equations (2.6) expresses the obvious fact that time is the same in both systems. Relations (2.6) between the space and time coordinates of an event observed in two different reference frames are known as Galilean transformations. The law of addition of velocities follows directly from these transformations. The speed of the object in K is $v = \mathrm{d}x/\mathrm{d}t$. Its speed in K′ is $v' = \mathrm{d}x'/\mathrm{d}t'$. Since $t = t'$, we have

$$v = \frac{\mathrm{d}x}{\mathrm{d}t} = \frac{\mathrm{d}x'}{\mathrm{d}t} + V = \frac{\mathrm{d}x'}{\mathrm{d}t'} + V = v' + V. \tag{2.7}$$

This is the law of addition (1.1). For $v = c$, we recover Equation (1.2) as a special case.

But, since Equation (2.7) does not hold for light, it must be generally wrong, even though it describes accurately the slow motions. But how can it be wrong if it follows directly from the most fundamental properties (2.6) of space and time? There can be only one explanation: the "fundamental" properties (2.6) that we had considered as

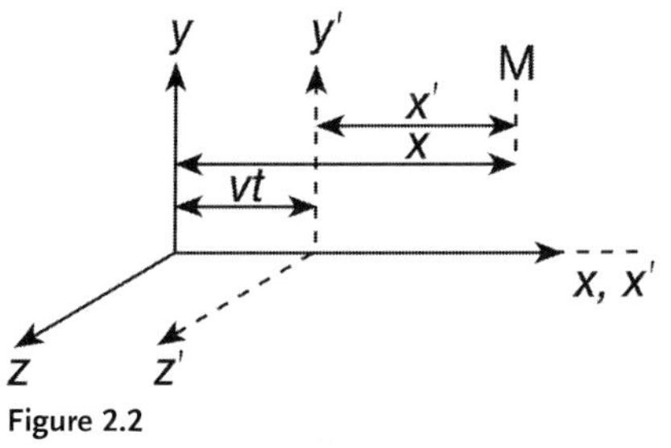

Figure 2.2

self-evident must themselves be generally wrong and need critical revision. That was Einstein's brilliant idea, which became a starting point of his theory of relativity.

2.3
"Obvious" Does Not Always Mean "True!"

When we enter the domain of speeds comparable with the speed of light, we must generalize the law of addition of velocities in such a way that one formula would describe both – the simple addition of low-speed motions and the "weird" behavior of light. And to do this, we will analyze in more detail the initial premises on which the law of velocities' addition is based.

Consider the following situation: a spaceship (system K′) moves at a speed V relative to an inertial system K, assumed to be stationary. An object moves inside the spaceship from its rear to its front (i.e., parallel to the spaceship's velocity) at a speed v'. The speed v of the object relative to K is then given by the "obvious" formula (2.7).

Let us now scrutinize the definition of speed used in the previous section. The object's speed v' relative to the spaceship is $v' = \Delta x'/\Delta t'$, where $\Delta x'$ is the length of the spaceship and $\Delta t'$ is the time it takes for the object to travel this length. Thus, Equation (2.7) means that

$$v = V + \frac{\Delta x'}{\Delta t'}. \tag{2.8}$$

Scrutinize the meaning of all the terms in this equation. The first two terms, the speed v of the object and V of the spaceship relative to K, are measured using rulers and clocks, which belong to system K and *do not participate in the spaceship's motion.* The last term, the speed of the object relative to the spaceship, is measured by the spaceship's crew using *the rulers and the clocks they find on the spaceship.* Of course, the rulers and clocks in K and K′ are identical in the sense that they have been constructed in the same way (the possibility of their structures being identical is guaranteed by the fact that the laws of Nature are identical in both systems, i.e., by Einstein's principle of relativity). However, the two systems of rulers and clocks are *moving* relative to each other and we do not know beforehand how this will affect the result of their *direct comparison with one another.* That is why it is utterly wrong to measure the two terms on the right of (2.8), which contribute to the net speed v, in the units belonging to *different* reference systems. The motion of system K′ relative to K may affect its rulers and clocks, and vice versa.

The correct formula, corresponding precisely to the definition of velocity of an object in K, is the following:

$$v = V + \frac{\Delta x}{\Delta t}, \tag{2.9}$$

where Δx is the length of the spaceship measured *in units of system K* and Δt is the corresponding time (i.e., the time it takes for the object to move from the rear to the front of the spaceship) measured *using the clocks of system K.*

The correct formula (2.9) can be reduced to (2.8) only if we make two *additional assumptions*:

1. The distance $\Delta x'$ in K′ (in our case, the length of the spaceship measured by its own rulers) is transferred without any change to system K (i.e., $\Delta x' = \Delta x$).
2. The duration $\Delta t'$ of a process (in our case, the time that the object spends in motion) in system K′ is the same as its duration in system K (i.e., $\Delta t' = \Delta t$).

In other words, the sizes of objects (or distances between objects) and durations of processes (or time intervals between events) had been assumed to be absolute regardless of the state of motion of the system to which we attach our clocks and scales. The absoluteness of distances and the invariance of time in all reference systems must result in simple addition (2.7) of velocities. But, since the "simple addition" law, when applied to light, clashes with the experiment, it must be generally wrong. Therefore, the assumption that space and time are absolute also must be wrong. We have already emphasized that the belief in absoluteness of space and time has been "born" in the world of low speeds. But the speed of light is not low! It follows that the concepts of absolute time and space, upon which Equation (2.7) was based, must be changed in such a way as to obtain a description of the world, which would hold for *any* motions, slow or fast.

2.4
Light Determines Simultaneity

It is quite natural that light, whose "weird" behavior has prompted us to revise the concepts of time and space, is itself suggesting the direction of such a revision. In fact, not only does it suggest it, but also points unambiguously to the only possible solution.

Light propagates in the same physical space where other objects are moving. However, while the speeds of most objects can change, particularly after transition into another reference frame, the magnitude of the velocity of light remains constant. The properties of time and space must be reconciled with this fundamental fact.

The invariance of the speed of light suggests, as the above analysis shows, that the time interval $\Delta t'$ between two events of system K′ is generally different from the time interval Δt between the same events in system K, that is, $\Delta t \neq \Delta t'$.

In particular, this means that if $\Delta t = 0$ for two spatially separated events in K (i.e., both events occur simultaneously in K), then $\Delta t'$ may be different from zero, and these same events will not be simultaneous in system K′. It is at this point where the most fundamental break with Newtonian concepts lies.

The classical notion of absolute simultaneity is based upon the intuitive idea that time is something universal and is the same at any moment for all points in space and in any reference frame. Space itself is perceived as the locus of all points (or, more precisely, "events"), "snapped" at some moment of time.

But what does it mean – one and the same moment of time for two points A and B a way apart?

Let two clocks with identical structure be placed at the points of interest. We call two events occurring at these points simultaneous if the clocks A and B show the same time readings at the corresponding moments. But this definition is based on an unspoken assumption that both clocks had been started at the same time. It follows that the simultaneity of the two given events at A and B depends upon the definition of simultaneity of another pair of events (the starts of clocks A and B). Since a concept cannot be defined in terms of itself, it is necessary to find some other definition.

The concept of simultaneity for spatially separated events (and thereby the mere idea of space "at a given moment") can only have a clear physical meaning (i.e., be based on a realizable experimental procedure) if there exists a universal and reliable way to compare the events happening at different places. Light provides us with such possibility! The process of propagation of light (or, more generally, electromagnetic interactions) is precisely what makes it possible to connect the time "there" with the time "here." Being a universal "messenger" between different regions of space, light makes it possible to judge the simultaneity of spatially separated events. The experimental fact that the speed of light is independent of reference frame allows to define an electromagnetic procedure for clocks' synchronization, which is *uniform for all inertial systems.* The clocks A and B at distance x from one another and at rest *in a given reference frame* are synchronized if the light signal emitted from A at the moment t_A arrives at B at the moment $t_B = t_A + x/c$. It follows from this definition that the two light signals from a flash at a moment t_C at the point C right in the middle of the segment AB reach the ends A and B simultaneously:

$$t_A = t_B = t_C + \frac{1}{2}\frac{x}{c}. \tag{2.10}$$

If we reverse this procedure, we will come to Einstein's definition of simultaneity: two events at points A and B are simultaneous if the light signals from these events meet exactly in the middle between A and B.

Because of the invariance of the speed of light, the conclusion about *relativity* of simultaneity follows immediately from Einstein's definition. Let us consider again the spaceship from the previous section, assuming that its walls are transparent and that a detector of light signals is positioned in the middle of the spaceship. This detector does not respond to a signal coming from only one direction or to signals arriving from the opposite directions at different moments. If, however, the detector is lit from both directions *simultaneously,* a wiring device switches on and the detector explodes. A similar detector is put at the point C of the "stationary" system K (Figure 2.3a). Suppose that precisely at the moment when both detectors were coincident ($t_C = t'_C = 0$) we marked the instantaneous positions of the end points A and B of the spaceship in system K. The phrase "precisely at the moment" has now a clear physical meaning owing to the definition of simultaneity: it means that if two flashes of light occur at points A and B at this moment, then the emitted signals will meet exactly in the middle of the segment AB, that is, at the point C, where our detector is located, and the latter, being lit simultaneously from the opposite directions, will explode. Since the spaceship is transparent, we can also observe the

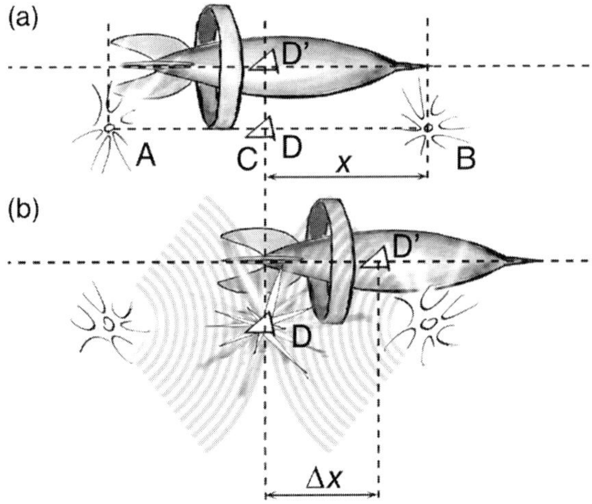

Figure 2.3 A thought experiment illustrating the relativity of simultaneity. (a) The initial moment: two flashes at the end points of a moving spaceship are simultaneous in K. (b) The final moment: the photons from the flashes explode detector D, but not detector D′.

course of events in the spaceship while remaining outside (Figure 2.3b). We will watch the spaceship's detector moving toward one signal and running away from the other while it goes past the point C. By the moment when both signals meet at C, exploding our detector, the detector on the spaceship will reach some other point C′ and remain intact because it will be lit by only one (oncoming) signal. Thus, in the spaceship's system, the signals will meet not at its center but at some other point and so the detector will not explode. Relative to the spaceship, however, both signals move with the same speed c as they do in system K. So, how is it possible that they do not meet at the middle of the spaceship? Or, more precisely, why does the signal from the front travel a longer distance than that from the rear? There is only one plausible answer: because it was emitted earlier! To put it another way, in the spaceship's system, the flashes were *not* simultaneous. The flash in the front occurred earlier than the flash in the rear.

Let us calculate how much earlier. Suppose $\Delta x'$ is the distance between the center of the spaceship and the point C′ where the signals meet, *measured relative to the spaceship*. We will call it the proper distance. To play it safe, we will avoid the statement that $\Delta x'$ is equal to $\Delta x = CC'$ with CC′ being the distance measured in system K (later we will see that such a precaution is justified). On the contrary, since the segment $\Delta x'$ is moving together with the spaceship at a speed V relative to K, its length $\Delta x = CC'$, measured in system K, might differ from its length $\Delta x'$ measured in system K′ by a factor $\gamma(V)$, which depends on V:

$$\Delta x = \gamma^{-1}(V)\Delta x'. \tag{2.11}$$

In system K, however, the distance Δx is $\Delta x = Vt$, where t is the time interval between the flashes at A and B and the detector's explosion at C. If we denote the distance AC = CB (i.e., half the spaceship's length in system K) as x, then $t = x/c$, and therefore

$$\Delta x = \frac{V}{c}x. \tag{2.12}$$

Excluding Δx from (2.11) and (2.12), we find

$$\Delta x' = \gamma(V)\frac{V}{c}x. \tag{2.13}$$

Thus, we have found that the signal coming from the front of the spaceship has traveled a distance longer by $2\Delta x' = 2\gamma(V)(V/c)x$ than the signal coming from the rear. This means that it was emitted earlier by the time interval $2\Delta t' = 2\Delta x'/c$. Suppose now that a clock has been attached to each of the two detectors and that both clocks read zero time ($t_C = t'_C = 0$) at the moment when they were coincident. We will then obtain the following result for the moments of two flashes at A and B:

$$\left.\begin{array}{ll}\text{In system K}: & t_A = t_B = 0 \\ \text{In system K}': & t'_A = \dfrac{\Delta x'}{c} = \gamma(V)\dfrac{V}{c^2}x; \quad t'_B = -t'_A\end{array}\right\}. \tag{2.14}$$

We can put it this way: when the flash occurred at A, the spaceship's clock located at that point reads the time $t'_A = \Delta x'/c$ and when the flash occurred at B, the corresponding clock reads $t'_B = -\Delta x'/c$ *if both clocks on the spaceship had previously been synchronized in their reference frame* according to Einstein's definition of simultaneity.

We want to emphasize the importance of this conclusion. We are discussing natural phenomena. A pair of spatially separated events is being considered. And it turns out that these events are simultaneous in one reference frame but nonsimultaneous in another. This means that *simultaneity is relative.* Its relativity is due to the fact that the speed of light is invariable. If light obeyed the simple law of velocities' addition, the light signal would travel faster from the front to the rear than from the rear to the front in the reference frame of the spaceship. This would account for the fact that the two signals do not meet in the center of the spaceship. The flashes would remain simultaneous. In that case, however, the laws for electromagnetic phenomena (e.g., the speed of light propagation) and the corresponding procedures used to define simultaneity would not be the same for all inertial systems. There would be only one "privileged" system of reference, where the speed of light would be the same in all directions. The clocks of all other systems would be set according to the clocks of this "absolutely still" system, which would bring us back to Newtonian concept of absolute time.

But light does not leave us such a possibility, because it moves with one fixed speed in all inertial systems, rendering them all equivalent. Thus, Einstein's principle of relativity, together with invariance of the speed of light, implies the relativity of time.

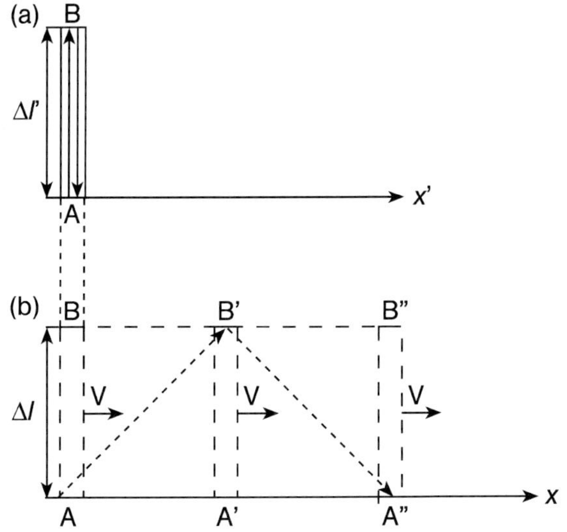

Figure 2.4 The light pulse in a vertical cylinder that is moving horizontally. (a) In the rest frame of the cylinder (system K′); (b) in system K.

2.5
Light, Times, and Distances

The relativity of time causes the relativity of distances and time intervals: these quantities are different in different reference systems.

Let us consider a vertical cylinder of length $\Delta l'$ with mirrored butt ends; a light signal is traveling back and forth periodically inside the cylinder (Figure 2.4a). In system K′ attached to the cylinder, the time interval between two successive arrivals of the signal to a chosen end is equal to

$$\Delta t' = 2\frac{\Delta l'}{c}. \tag{2.15}$$

The interval $\Delta t'$ can be called an eigen (proper) period of the signal's motion. Now, we can analyze the whole process in system K, in which the cylinder moves horizontally at the speed V. What is the time of this process in system K? Denote this time as Δt. In system K, light participates simultaneously in two motions: in the vertical direction (along the cylinder's axis) and in the horizontal direction (together with the cylinder). As a result, during the period Δt of one "oscillation" up and down, the signal will travel the distance $V\Delta t$ in the horizontal direction and so its trajectory will become a broken line AB′A″ (Figure 2.4b).

The length $l_{AB'A''}$ of the element AB′A″ is equal to

$$l_{AB'A''} = 2\sqrt{\Delta l'^2 + \left(\frac{1}{2}V\Delta t\right)^2} = \sqrt{c^2\Delta t'^2 + V^2\Delta t^2}. \tag{2.16}$$

It is greater than $2\Delta l'$. At the same time, the *speed* of light along the broken line in system K must remain equal to c. To travel a greater distance at the same speed takes a longer time. Indeed, putting $l_{AB'A''} = c\Delta t$ for the element's length in Equation (2.16), we will obtain the following relation between Δt and $\Delta t'$:

$$c\Delta t = \sqrt{c^2 \Delta t'^2 + V^2 \Delta t^2}. \tag{2.17}$$

It follows

$$\Delta t' = \Delta t \sqrt{1 - \frac{V^2}{c^2}}. \tag{2.18}$$

As we can see from (2.18), the period of the same process – one complete oscillation of the light signal inside the cylinder – is different in different systems and is smallest in a system where the cylinder is at rest.

There is another way to see it. In system K light moves along AB′ with the speed c, while moving horizontally with a speed V. We can see from Figure 2.4 that the vertical component of its motion must be

$$v = \sqrt{c^2 - V^2} = c\sqrt{1 - \frac{V^2}{c^2}}. \tag{2.19}$$

The motion of light along the cylinder is slower than c when the cylinder is moving and its period Δt is accordingly greater than $\Delta t'$ by the same factor, which is the essence of Equation (2.18).

Further, we have suggested (Equation (2.11)) that the "longitudinal" size of an object, that is, its length in the direction of the velocity of its relative motion, may be relative, too. To find the law for the length transformation, we will slightly modify our experiment by directing the axis of the cylinder *along* its relative velocity (Figure 2.5).

Obviously, in system K′, this operation will not affect the cylinder's length $\Delta l'$ (the size of an object in its rest system does not change when the object is rotated). Accordingly, the period of motion (Equation (2.15)) of the light signal will remain the

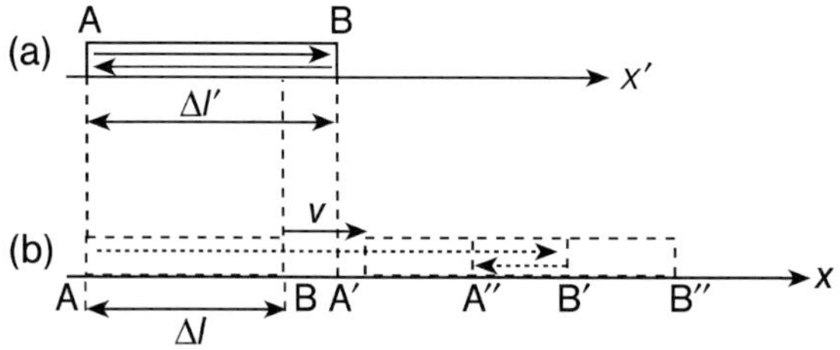

Figure 2.5 The same as in Figure 2.4, but now the cylinder is horizontal. (a) In the rest frame of the cylinder (system K′); (b) in system K. AB is the initial position of the cylinder (the pulse starts at A); A′B′ is its intermediate position (the pulse reaches the front at B′); A″B″ is its final position (the reflected pulse returns to the rear at A″).

same. But if the time interval $\Delta t'$ between two events (the emission and return of the signal) at one point (point A of the cylinder) in system K′ does not depend on orientation of the cylinder, then the corresponding time interval Δt between those same events considered from system K would not change either. Therefore, Equation (2.18) must also hold for the cylinder in the horizontal position. Using the relation (2.18) between Δt and $\Delta x'$, we obtain

$$\Delta t = \frac{\Delta t'}{\sqrt{1-\dfrac{V^2}{c^2}}} = \frac{2\Delta l'}{c}\left(1-\frac{V^2}{c^2}\right)^{-1/2}. \tag{2.20}$$

Now let us express the time interval Δt in terms of the "longitudinal" length Δl of the cylinder in system K. In this system, the light now travels in a moving horizontal "corridor" of length Δl, catching up with the mirror B, which runs away from it at a speed V. How long does it take for light to catch up with the front end B? Denote this time interval as Δt_1. The distance traveled by the light pulse from point A to point B′ where it catches up with the front of the cylinder is $c\Delta t_1$. The same distance can be expressed in terms of the length of the cylinder as $V\Delta t_1 + \Delta l$ (Figure 2.5). Thus, we have $c\Delta t_1 = V\Delta t_1 + \Delta l$, so that $\Delta t_1 = \Delta l/(c-V)$. After the reflection from mirror B, the light returns to point A (rear of the cylinder), which moves *toward* it at speed V. The time Δt_2 it takes the reflected signal to return to this point can be found in a similar way and is equal to

$$\Delta t_2 = \Delta l/(c+V). \tag{2.21}$$

The total time Δt the signal spends between its departure and return to the same end of the cylinder is

$$\Delta t = \Delta t_1 + \Delta t_2 = \frac{\Delta l}{c-V} + \frac{\Delta l}{c+V} = \frac{2\Delta l}{c}\left(1-\frac{V^2}{c^2}\right)^{-1}. \tag{2.22}$$

It turns out to have been calculated in the same way as the corresponding time for the "horizontal" beam in the Michelson experiment. However, we must keep in mind that here the quantities $c-V$ and $c+V$ are the rates of change of the distance between the signal and the butt ends of the cylinder *in system K*, and they by no means represent the speeds of the signal *relative to the cylinder*, that is, its speed in system K′. Similarly, the distance Δl is the length of the cylinder in *system K*, and, as we will now show, it is indeed different from its length measured in its rest frame.

Comparing (2.20) and (2.22), we obtain

$$2\frac{\Delta l}{c}\left(1-\frac{V^2}{c^2}\right)^{-1} = 2\frac{\Delta l'}{c}\left(1-\frac{V^2}{c^2}\right)^{-1/2}, \tag{2.23}$$

that is,

$$\Delta l = \Delta l'\sqrt{1-\frac{V^2}{c^2}}. \tag{2.24}$$

According to (2.24), the length of a moving segment is smaller than its length in its rest frame by a factor of $(1 - V^2/c^2)^{-1/2}$. In other words, the sizes of moving objects are contracted in the direction of motion. This effect is called the Lorentz contraction. But it has a quite different meaning from the contraction introduced by Lorentz in connection with the Michelson experiment. Lorentz assumed that the longitudinal contraction appears *only* when an object is moving relative to the ether, which serves as a universal system of reference. Consequently, a segment that is stationary relative to the ether possesses the greatest length. In reality, the length contraction is observed for any object moving relative to *any* inertial system. And the segment has the greatest length in its own system of rest, which may be moving relative to a given inertial system at an arbitrary speed v. The ratio

$$\frac{\Delta l}{\Delta l'} = \gamma^{-1}(V) \equiv \sqrt{1 - \frac{V^2}{c^2}} \tag{2.25}$$

is precisely the same proportionality coefficient (2.11) between the length of a moving segment and its "rest" length that has been introduced in the previous section. Now, analyzing our thought experiments with light, we have found the exact value of this coefficient. Because we will come across this coefficient quite often, we will give it a special name, the Lorentz factor, and stick to our symbol $\gamma(V)$, remembering that the explicit form (2.25) of the function $\gamma(V)$ is known to us. Using this symbol, we can rewrite the essential formulas (2.18) and (2.24) for time and length transformation:

$$\Delta t = \gamma(V)\Delta t', \tag{2.26}$$

$$\Delta l = \gamma^{-1}(V)\Delta l'. \tag{2.27}$$

It is essential for the correct use of these formulas that we clearly understand the physical meaning of the related quantities: $\Delta t'$ is the time between two different events taking place at *one and the same point* of system K′; Δt is the time between *the same* events in K, where these events are being observed at different points of space. Similarly, $\Delta l'$ is the length of a segment in its rest system K′; Δl is its length in K that slides along the segment with a speed V. Since in system K the segment is moving, the coordinates of its end points must be recorded at *one and the same moment in* K; the quantity Δl represents the distance in K between these two instantaneous positions. As a consequence of the relative nature of simultaneity, the recordings of instantaneous positions of the end points, performed simultaneously in system K, are not simultaneous in system K′.

2.6
The Lorentz Transformations

The examples considered in the previous sections demonstrate how the invariance of the speed of light leads to the relativity of time and space. Here we will derive the equations, which provide a complete description of the fundamental properties of time and space. These equations are called the Lorentz transformations. They show how the coordinates of any arbitrary event (the Cartesian coordinates of a point where the event has occurred and the corresponding moment of time) get transformed after transition from one inertial system (reference frame) to another.

Let axes x, y, z of system K be parallel to the axes x', y', z' of system K$'$, moving in the direction x at a speed V relative to K (Figure 2.2). Let the origins O and O$'$ of both systems coincide at the moment $t = t' = 0$ on their clocks there. This can always be achieved by a proper choice of the initial moments of time in both systems.

Let at this moment a flash of light at the origin produce a diverging spherical wave. *Because of the invariance of the speed of light, this wave will be spherical in both reference frames.* Let us express this fact in mathematical terms.

By the moment t in system K the wave front will form a spherical surface of radius $r = ct$ centered at the origin; this surface is described by the equation

$$x^2 + y^2 + z^2 - c^2t^2 = 0. \tag{2.28}$$

Similarly, we conclude that the space and time coordinates of the expanding wave front in system K$'$ must also satisfy the equations of the spherical surface centered at the origin of K$'$ and having a radius $r' = ct'$:

$$x'^2 + y'^2 + z'^2 - c^2t'^2 = 0. \tag{2.29}$$

Equating the left parts of these two equations, we obtain

$$c^2t^2 - x^2 - y^2 - z^2 = c^2t'^2 - x'^2 - y'^2 - z'^2 = 0. \tag{2.30}$$

The expression

$$s \equiv \sqrt{c^2t^2 - x^2 - y^2 - z^2} \tag{2.31}$$

is called the space–time interval. Its mathematical form suggests that it can be considered as a distance between two points in an abstract four-dimensional set, one of the dimensions being time. As we will see later, this is indeed the case. It generalizes the spatial distance between two events.[1]

If the two events are arbitrarily close to one another, then the interval between them can be defined as

$$ds^2 \equiv c^2dt^2 - dx^2 - dy^2 - dz^2. \tag{2.32}$$

In another inertial RF we will have

$$ds'^2 \equiv c^2dt'^2 - dx'^2 - dy'^2 - dz'^2. \tag{2.33}$$

Note that an interval ds between two events can be zero, d$s = 0$, in two quite different ways. First, when the two events coincide, that is, they happen at the same

1) The "spatial" part of the interval (2.31) is $-r^2$, where r is the length of the vector with coordinates (x, y, z).

point and at the same time (in this case $dr = c\,dt = 0$). Second, when the events are (or can be) connected by a light signal. The events themselves can be separated by huge distances and eons of time (in this case both r and ct can be very large, and $(ct)^2 - r^2 = 0$ but $ct \neq 0$). Both possibilities are described by conditions (2.8)–(2.10).

According to (2.30), if $ds = 0$ in one RF, it is also zero in another RF. In this case, the interval is invariant with respect to change of reference frame. It seems natural to assume that this important property can be extended onto nonzero intervals as well. We can show it by the following argument [20].

Consider a pair of two very close events from two different inertial RF. The coordinate system used in either frame must be "smooth" or continuous in the sense that close events are assigned close sets of coordinates [3]. Then the values ds and ds' of an interval between two close events must be of the same order of magnitude in both systems. It follows that ds and ds' must be proportional to one another:

$$ds' = \alpha\, ds. \tag{2.34}$$

Because of the isotropy of space, the proportionality coefficient α can depend only on *magnitude* of the relative velocity V between K and K′. Also, since all points and moments of time are equivalent, α cannot depend on $\boldsymbol{r}$ or t. Thus, $\alpha = \alpha(V)$. This means that the change in the *direction* of relative motion between the two frames does not change α. Therefore, as well as (2.34), we can write

$$ds = \alpha\, ds'. \tag{2.35}$$

Indeed, the velocity of K relative to K′ is negative of the velocity of K′ relative to K, and this cannot change α. Combining (2.34) and (2.35), we find that $\alpha^2 = 1$, that is, $\alpha = \pm 1$. Now, $\alpha(V)$ must be a continuous function of V, and therefore either always 1 or always -1 (otherwise it should go through intermediate values between 1 and -1, which are forbidden). By the same token, if $V \to 0$, either of Equations (2.34) and (2.35) reduces to identity $ds' = ds$, which can be considered as a special case with $\alpha = 1$. Once we find at least one case for which $\alpha = 1$, it fixes the same value of α for all possible cases. Thus, we come to a fundamental conclusion that always

$$ds' = ds. \tag{2.36}$$

And the same must therefore be true for any interval of finite value between any two events:

$$s' = s. \tag{2.37}$$

If the origins of the two systems coincide, Equation (2.37) also holds for an interval between an arbitrary event and the origin. Interval is an absolute (invariant) characteristic of any two events, which is independent of RF or any observers.

If we use identical synchronized clocks, as well as Cartesian coordinates with identical length units in all inertial systems of reference, then the mathematical form of this expression (the combination of squares of all four coordinates, each square taken with a definite sign) will be maintained. This property of the space–time interval is called covariance.

Now, we must find the relations between coordinates (t, x, y, z) and (t', x', y', z'), which will satisfy the requirement that the interval be covariant. First of all, these relations, that is, functions $t(t', x', y', z')$, $x(t', x', y', z')$, $y(t', x', y', z')$, $z(t', x', y', z')$, have to be linear:

$$\left.\begin{aligned} ct &= a_{00}ct' + a_{01}x' + a_{02}y' + a_{03}z' \\ x &= a_{10}ct' + a_{11}x' + a_{12}y' + a_{13}z' \\ y &= a_{20}ct' + a_{21}x' + a_{22}y' + a_{23}z' \\ z &= a_{30}ct' + a_{31}x' + a_{32}y' + a_{33}z' \end{aligned}\right\}. \tag{2.38}$$

For those familiar with the formalism of linear algebra, this can be represented as a matrix equation

$$\begin{pmatrix} ct \\ x \\ y \\ z \end{pmatrix} = \begin{pmatrix} a_{00} & a_{01} & a_{02} & a_{03} \\ a_{10} & a_{11} & a_{12} & a_{13} \\ a_{20} & a_{21} & a_{22} & a_{23} \\ a_{30} & a_{31} & a_{32} & a_{33} \end{pmatrix} \begin{pmatrix} ct' \\ x' \\ y' \\ z' \end{pmatrix}. \tag{2.39}$$

Equations (2.38) and (2.39) follow from the fact that only linear relations between coordinates can ensure the covariance of the expression (2.37) for the interval. If instead of (2.38) we put into expression (2.37) any other function of the primed coordinates, we will not obtain the sum of squares of these coordinates. Further, the differentials of coordinates must be independent of position of the origin: the time interval $dt = t_2 - t_1$ between two events E_1 and E_2 does not depend on our choice of the zero moment. Similarly, the distance $dx = x_2 - x_1$ between the x-coordinates of these events does not depend on our choice of the reference point, and the same is true for the two other spatial coordinates. This property is called *translational invariance*. The linear dependence (2.38) automatically guarantees the translational invariance of the interval in all inertial reference frames. In addition, the linear function satisfies the requirement that its inverse function is also linear. This requirement must hold here because the equivalence of systems K and K′ implies that the inverse relation between (t', x', y', z') and (t, x, y, z) has the same mathematical form as the direct one.

These mathematical conditions have a simple physical meaning: the linearity of the relation $(ct, x, y, z) \Leftrightarrow (ct', x', y', z')$ expresses both the equivalence of all points of space and moments of time, and the equivalence of inertial systems K and K′.

Thus, the physics of the considered phenomena dictates the linearity of the transformation between the coordinates of an event observed in different inertial systems. In our case, when the velocity $\boldsymbol{v}$ is parallel to the x-axis, the transverse coordinates must have the same values in both systems:

$$y = y'; \quad z = z'. \tag{2.40}$$

The reason is that the speed of the relative motion of systems K and K′ is zero in directions y and z. Under conditions (2.40), the matrix equation (2.39) simplifies to

$$\begin{pmatrix} ct \\ x \\ y \\ z \end{pmatrix} = \begin{pmatrix} a_{00} & a_{01} & 0 & 0 \\ a_{10} & a_{11} & 0 & 0 \\ 0 & 0 & 1 & 0 \\ 0 & 0 & 0 & 1 \end{pmatrix} \begin{pmatrix} ct' \\ x' \\ y' \\ z' \end{pmatrix}, \tag{2.41}$$

that is,

$$\left.\begin{array}{l} ct = a_{00}ct' + a_{01}x' \\ x = a_{10}ct' + a_{11}x' \\ y = y' \\ z = z' \end{array}\right\}. \tag{2.42}$$

But the point O′ moves at a speed V relative to K, so by the moment t its position in K is given by the coordinate $x = Vt$. Assuming $x' = 0$ and putting into (2.42)$x = Vt$, we have

$$\left.\begin{array}{l} ct = a_{00}ct' \\ Vt = a_{10}ct' \end{array}\right\}. \tag{2.43}$$

It follows

$$\frac{a_{10}}{a_{00}} = \frac{V}{c}. \tag{2.44}$$

Thus, we have the following matrix equation for the transformation of x and t:

$$\begin{pmatrix} ct \\ x \end{pmatrix} = \begin{pmatrix} a_{00} & a_{01} \\ \frac{V}{c} a_{00} & a_{11} \end{pmatrix} \begin{pmatrix} ct' \\ x' \end{pmatrix}. \tag{2.45}$$

To find the remaining three unknowns, we substitute linear expressions (2.45) into the equation

$$c^2t^2 - x^2 = c^2t'^2 - x'^2 \tag{2.46}$$

and obtain

$$(a_{00}ct' + a_{01}x')^2 - (a_{00}Vt' - a_{11}x')^2 = (ct')^2 - x'^2. \tag{2.47}$$

Now we demand that the expression on the left might be reduced to the form $c^2t'^2 - x'^2$. This is possible only if

$$\left.\begin{aligned} a_{00}^2\left(1-\frac{V^2}{c^2}\right) &\equiv a_{00}^2\gamma^{-2}(V) = 1 \\ a_{01} - \frac{V}{c}a_{11} &= 0 \\ a_{01}^2 - a_{11}^2 &= -1 \end{aligned}\right\}. \tag{2.48}$$

We have a system of three equations for three unknowns. The solution is simple and gives

$$a_{00} = a_{11} = \gamma(V); \quad a_{01} = \frac{V}{c}\gamma(V). \tag{2.49}$$

Taking into account (2.44), we finally get

$$\begin{pmatrix} ct \\ x \end{pmatrix} = \gamma(V)\begin{pmatrix} 1 & \frac{V}{c} \\ \frac{V}{c} & 1 \end{pmatrix}\begin{pmatrix} ct' \\ x' \end{pmatrix} \tag{2.50}$$

or

$$\left.\begin{aligned} ct &= \gamma(V)\left(ct' + \frac{V}{c}x'\right) \\ x &= \gamma(V)(x' + Vt') \end{aligned}\right\}. \tag{2.51}$$

This is the Lorentz transformation. Changing the sign of V, we can obtain the inverse transformation:

$$\left.\begin{aligned} ct' &= \gamma(V)\left(ct - \frac{V}{c}x\right) \\ x' &= \gamma(V)(x - Vt) \end{aligned}\right\}. \tag{2.52}$$

If we substitute these transformations into (2.38), it will become an identity, which means that the expression (2.39) for the interval is covariant with respect to Lorentz transformations. In a special case when the speed V is much smaller than c, the ratio V/c is negligible, the Lorentz factor approaches 1, and (2.52) reduces to Galilean transformations (2.6).

Finally, we can write an expression for the Lorentz transformation in the matrix form for the case of relative motion along the x-direction, but including explicitly all four dimensions:

$$\begin{pmatrix} x^0 \\ x^1 \\ x^2 \\ x^3 \end{pmatrix} = \mathcal{L} \begin{pmatrix} x'^0 \\ x'^1 \\ x'^2 \\ x'^3 \end{pmatrix}. \tag{2.53}$$

Here the superscripts 0, 1, 2, and 3 stand for ct, x, y, and z respectively, and the same for primed variables. Do not confuse these superscripts with powers. The transformation "coefficient" $\mathcal{L}$ is, of course, a matrix:

$$\mathcal{L} = \begin{pmatrix} \gamma(V) & \beta\gamma(V) & 0 & 0 \\ \beta\gamma(V) & \gamma(V) & 0 & 0 \\ 0 & 0 & 1 & 0 \\ 0 & 0 & 0 & 1 \end{pmatrix}. \tag{2.54}$$

We will see later that from the mathematical viewpoint this matrix performs the rotation in the (ct, x)-plane from one coordinate system to the other in the four-dimensional vector space.

2.7
The Relativity of Simultaneity

Your "now" is not my "now."

Charles Lamb, 1817

Using the Lorentz transformation in its general and explicit form, we can arrive at the results that are already familiar to us and, in particular, the conclusion about relativity of time. This relativity can be not only observed but also described quantitatively, with the help of the following thought experiment. Let us arrange equally spaced identical clocks along the x-axis in system K, assuming the clocks to be synchronized in K according to the previously described Einstein's definition of simultaneity. Let us place identical clocks at equal distances from each other along the x'-axis of K′. As previously described, the expression "identical" means that the clocks of system K′ are constructed in this system from the materials and under conditions identical to the materials and conditions in system K; the possibility of such identical conditions is guaranteed by Einstein's relativity principle (independence of all laws of Nature from a state of inertial motion). All clocks of system K′ are also synchronized in this system using the light signals according to Einstein's procedure. In the final result, we have two linear sets of identical clocks synchronized in their respective systems and moving relative to one another at a speed V.

Let us define the initial moment so that when the origins of both systems coincide ($x = x' = 0$), the corresponding local clocks there would read the same time $t = t' = 0$. Consider other points of space *in system K* at this moment. The phrase "at this moment" means that all the K-clocks placed at these points read the same time $t = 0$. All events that occur at this moment form a set that is simultaneous in K (What we call "space" is just an infinite and continuous set of different but simultaneous events!). In system K′, however, these same events *do not* form a simultaneous set, and thus the system's clocks have different readings at different points. To ascertain this fact, we use the Lorentz transformation. Substituting $t=0$ into (2.52), we obtain

$$t' = -\gamma(V)\frac{V}{c^2}x, \tag{2.55}$$

that is, the already familiar result (2.14).

We see that the spatially separated events occurring all at one moment $t=0$ in K occur at different moments in K′, depending on location of a K′-clock. Namely, a K′-clock to the right of the origin of K′ (at $x' > 0$) reads an earlier time than does the K′-clock currently passing by the origin. A clock to the left of the origin (at $x' < 0$) reads a later time than the K′-clock at the origin. According to (2.55), the discrepancy between readings of "moving" and "stationary" clocks is proportional to their distance from the origin (Figure 2.6).

Generally, if we observe a set of moving clocks (set K′) lined up along the direction of motion, then at any moment of time in K, for any pair of clocks in the moving set, the leading clock reads an earlier time than the trailing clock, and the time difference between their readings is

$$\Delta t' = -\gamma(V)\frac{V}{c^2}\Delta x, \tag{2.56}$$

where Δx is the instant distance between the two clocks, as measured in K.

Note that this result is obtained for a set of clocks *synchronized* in their rest system K′. Thus, we can make the apparently paradoxical, but actually true and very important, statement: the time discrepancy $\Delta t'$ (2.56) between any two clocks in a moving train K′, which are separated by a distance $\Delta x' = \gamma(V)\Delta x$ is, in fact, the indication that all the clocks in that train are synchronized. To make it look more physical, imagine an episode from a *Star Treck* serial: we have met in deep space an abandoned spaceship from an alien civilization. We want to learn as much as possible about the alien ship and the reason why the crew abandoned it. But the ship is zipping by very fast and we do not have time and resources to adjust our motion accordingly. We can make a few snapshots, however, of its interior through

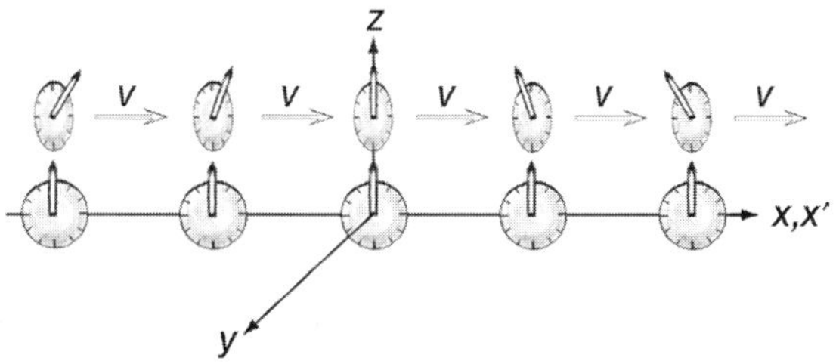

Figure 2.6

its transparent shell. After processing the images with account taken for different time needed for light to come to the camera from different points of the ship, we find its clocks still ticking, but any of the two clocks along the line of the ship's motion reading different time. The question arises: were the clocks properly synchronized? Maybe they were not, and this might have caused some trouble, which had forced the crew to abandon the ship? We know already that different readings on different moving clocks at one moment of *our* time is, by itself, not sufficient to judge upon their synchronization: one moment of our time is not necessarily one moment of their time, and vice versa. So what is the criterion for right synchronization?

We process further the data on the images and find that for any two clocks along the line of motion the discrepancy in their reading is proportional to their spacing; that is, the ratio of discrepancy to spacing is constant and has the correct sign:

$$\frac{\Delta t'}{\Delta x} = -\gamma(V)\frac{V}{c^2} \quad \text{or} \quad \frac{\Delta t'}{\Delta x'} = -\frac{V}{c^2}, \tag{2.57}$$

where $\Delta x'$ is the proper distance between the two respective clocks in K′. This is equivalent to Equation (2.56), indicating that the clocks of the ship were correctly synchronized.

Exercise

As an expert in relativity, you have received a letter with a proposal aimed at setting universal procedure that would produce sets of synchronized clocks for all inertial reference frames. The basic idea of the proposal is to synchronize a row of clocks in one, arbitrarily chosen RF, and then boost this row as one single whole into another RF. To see how it works, it is sufficient to consider only two identical synchronized clocks C_1 and C_2 in a reference frame K. Now, imagine that we act on these clocks simultaneously in K with two equal forces along the line C_1C_2. Then, both clocks will synchronously accelerate until they wind up moving with the same velocity V relative to K. Remaining at rest with respect to one another, they now belong to another inertial reference frame K′. The beauty of this idea, according to its author, lies in the fact that the outcome does not depend on whether the boost affects the rate of ticking or not. Since the force on each clock is the same, regardless of its effect on the ticking rate, the readings of the clocks after the boost will be the same. Thus, we can provide all RF with clones of clocks synchronized originally in only one RF. What will your conclusion be?

Solution

The procedure as such is realizable, but has little if anything to do with synchronization. Immediately after the boost, the clocks C_1 and C_2, while belonging now to another RF K′ and showing *the same* time when observed from K, are *not* synchronized in K′. For them to be synchronized in K′, they have to be desynchronized in K, namely, the difference between their readings recorded at one moment of time in K must satisfy the criterion (2.57).

It should be emphasized that the conclusions we obtained are quite general in their character, since they do not depend on any assumptions about the nature or internal structure of the clocks. A clock can be based on some periodic process such as mechanical or electromagnetic oscillations, or radioactive decay of some nuclei. It does not matter what physical phenomenon provides the basis for functioning of a given clock. In any case, the result will be the same. This allows us to ignore the properties of specific clocks or physical processes and focus on the properties of time itself. Let us repeat the essential point of our previous discussion: spatially separated events that are simultaneous in one system of reference are generally not simultaneous in another one. An exception occurs when the given events are taking place in a plane perpendicular to the direction of the systems' relative motion.

All relativistic phenomena are, in fact, the consequences of this fundamental property of time, which itself originates from the existence of the invariant speed c. We will find out shortly how relativity of time affects the appearance of a process or an object in different reference frames.

2.8 A Proper Length and a Proper Time

Consider a point (x', y', z') in system K′. Suppose, the clock records the moments t'_1 and t'_2 of the two successive events at this point. The difference $\Delta t' = t'_2 - t'_1$ is the time interval between these events in system K′. The time interval between the two events occurring at the same point is called the proper time. What time between these two events will be measured in another reference system K?

Using the Lorentz transformation (2.51), we have for a fixed value of $x'_1 = x'_2 = x'$:

$$\left.\begin{aligned} ct_1 &= \gamma(V)\left(ct'_1 + \frac{V}{c}x'\right) \\ ct_2 &= \gamma(V)\left(ct'_2 + \frac{V}{c}x'\right) \end{aligned}\right\}. \tag{2.58}$$

It follows

$$\Delta t = t_2 - t_1 = \gamma(V)(t'_2 - t'_1) = \gamma(V)\Delta t'. \tag{2.59}$$

Since $\gamma(V)$ is always greater than 1, we find that $\Delta t > \Delta t'$, that is, in system K, where the considered events are observed at *different* points of space, they are separated by a longer time interval than in system K′. A process localized at one point in some reference frame lasts the shortest when it is observed from this reference frame. Any change of an object's location owing to its motion is accompanied by slowing down of its evolution. A moving clock "ticks" slower than a stationary one by a factor of $\gamma(V)$.

Let t'_1 and t'_2 be the moments of birth and decay of a particle, for instance, a μ-meson, in its proper system. Then the time interval $\Delta t' = t'_2 - t'_1$ is its proper lifetime. For a μ-meson this time, inferred from the numerous experimental data, is about 2×10^{-6} s. But in system K, in which the μ-meson is moving at a speed v, its birth and decay occur at different points; that is, because of the particle's motion, its life is "spread" in space. Accordingly, its life span in the new system is $\gamma(v)$ times longer than its "proper" lifetime. If v is large enough (e.g., $v \to c$), then Δt will be much greater than $\Delta t'$, and the particle will cover a vast distance before its decay. Precisely this phenomenon had been discovered for μ-mesons created in the upper atmosphere (at altitudes of about 100 km) under the influence of cosmic rays. The greatest distance a particle can travel in 2×10^{-6} s is $\Delta r = c\Delta t' \cong 2 \times 10^{-6}\,\mathrm{s} \times 3 \times \mathrm{m/s} = 600\,\mathrm{m}$. Therefore, if time were absolute, μ-mesons would decay not far from the point of their creation – practically at the same altitude of about 100 km above the Earth's surface. However, these particles can be detected at the sea level, which means that they manage to travel through the whole thickness of the Earth's atmosphere during their lifetime. This fact reveals the relativity of time in its full sway: since in the Earth's reference system the μ-meson moves at a subluminal speed, its lifetime in this system greatly exceeds its "proper" lifetime. This allows it to travel such a long distance.

Now, let us consider the consequences of the Lorentz transformation for space. A rod of length

$$\Delta x' = x'_2 - x'_1 \tag{2.60}$$

is positioned parallel to the x'-axis in system K′. What is the rod's length in K? As was previously mentioned, to measure the length, we have to mark the instantaneous coordinates x_1 and x_2 of the edges of the rod at one and the same moment t in *system K*. The clearing between the marks gives us the wanted length:

$$\Delta x = x_2 - x_1. \tag{2.61}$$

According to Lorentz transformations, we have for the moment t:

$$x'_1 = \gamma(V)(x_1 - Vt), \quad x'_2 = \gamma(V)(x_2 - Vt),$$

so that

$$\Delta x' = x'_2 - x'_1 = \gamma(V)(x_2 - x_1) = \gamma(V)\Delta x, \tag{2.62}$$

or

$$\Delta x = \gamma^{-1}(V)\Delta x', \tag{2.63}$$

in full agreement with (2.27). We have obtained an already familiar result: the length of the rod is the greatest in a system where the rod is at rest (the proper length). In a system where the rod is moving while being parallel to its velocity, its length diminishes by a factor of $\gamma(V)$.

The described effect of the length contraction allows one to explain the result of the experiment with atmospheric μ-mesons from the viewpoint of a fictitious observer traveling together with these particles, that is, at a speed $V = v$. Relative to this observer, the μ-mesons are motionless, whereas the Earth is rushing toward them at a speed v. The observer measures the lifetime of μ-mesons in his system of reference and, naturally, finds it equal to their *proper* lifetime 2×10^{-6} s. In such a small time interval, the Earth, even at a speed close to c, will travel less than 600 m toward him. How can it be then that the μ-mesons born in the upper layers of atmosphere, that is, about 100 km from the Earth, can be detected near the Earth's surface? The explanation lies in the effect of Lorentz contraction of the Earth in the direction of relative motion. The 100 km length that we have mentioned is the *proper* thickness of the atmosphere. In the system of μ-mesons moving toward the Earth, both the Earth and its atmosphere are flattened because of the length contraction. The atmosphere's thickness of 100 km is contracted down to about 600 m – just enough to pass by the observer in as small a time as 10^{-6} s.

Thus, the entire picture of the process becomes intrinsically consistent. Time and space transform in such a way that even though different inertial observers will measure different values for these quantities, they all register the same result: a μ-meson is created in the upper layers of the atmosphere and decays at the sea level or even in the depths of the ocean. The relativity of time and space is consistent with the covariance of dynamic laws.

The described experiment with μ-mesons demonstrates the unity of Nature. Our statement about the relativity of time and space was deduced from studies of electromagnetic interactions. Meanwhile, interactions that lead to decay of μ-mesons are not electromagnetic. They have a different nature and are called weak interactions. But the time dilation of μ-meson decay follows the same Lorentz transformation rules that we have obtained analyzing the properties of light. Different kinds of interactions turn out to have common properties.

This suggests that at a deeper level all kinds of interactions will prove to be manifestations of yet unknown universal interaction just like the electric and magnetic fields, which initially appeared fundamentally different, have been proven to be merely special cases of the electromagnetic field. Now, we have a developed theory that unifies electromagnetic and weak interactions. According to this theory, both interactions, despite all the differences between them (we have just pointed out that they are different!) are manifestations of a more fundamental *electroweak* interaction. The theory had predicted hitherto unknown phenomena, which have now received an experimental confirmation.

In the last years of his life, Einstein tried unsuccessfully to find a theory that would unify all observable physical phenomena ("the theory of everything"). It is only now at a more advanced level of our knowledge that the approach to the long coveted "Great Unification" of all theories appears to be discernible.

Thus, Einstein's principle of relativity that had been postulated on the basis of mechanics and electromagnetism more than a century ago still remains a powerful tool in our search for understanding the workings of Nature.

Problems

2.1 A rod of 1 m proper length is observed from a reference frame moving along the rod's length. The experimenter in this frame measures the length of the same rod and finds it to be only 0.5 m. How fast is the observer moving?

2.2 A police car is moving at 75 miles/h. What is percent difference between its proper length and contracted length? (A percent difference is the ratio of the amount of contraction to the proper length, times 100%.)

2.3 Imagine that you have boarded a spaceship in a *Star Treck* serial and set out to a distant star at the periphery of our Galaxy, 50 000 light years away from us (1 light year is a distance unit used in astronomy, it is a distance traveled by light in 1 year). Nobody expects you to be alive by the time your spaceship reaches its destination. However, to your own surprise, you do reach it within the span of your lifetime, having aged by only 25 years, while all your contemporaries on Earth have long been dead. How fast was your spaceship moving? Give the percent difference between the speed of your spaceship and the speed of light.

2.4 A relativistic spaceship from a *Star Treck* serial is moving "horizontally" past the Earth at 99% of the speed of light and the water rises in the ship's swimming pool at 5 m/s.
   (a) Find the tilt of the water surface with respect to the horizontal, as observed from Earth.
   (b) Reverse the conditions: let now the spaceship "crawl" at 5 m/s and the light wave move up in the ship's atmosphere at 99% of the speed of light in vacuum. Find now the tilt of the propagating wave front with respect to the horizontal.
   (c) Let the spaceship and the light wave inside move (in mutually perpendicular directions, as before) each at 99% of the speed of light in vacuum. Find the tilt of the wave front in this case.

2.5 Two events occur in one and the same place in the laboratory reference frame and are separated in time by 3 s. These events are also observed from a rocket flying by.
   (a) What is the spatial distance between these events in the reference frame of the rocket, if the time interval between the events is 5 s by the rocket's clocks?
   (b) What is the speed of the rocket?

2.6 Imagine that you have a very long track with synchronized clocks along it. There is a very long train moving to the right down the track at $v = 300$ m/s. At the moment when the central clock on the train passes by your central clock, both clocks read the zero time. The same time is read by your clock x m to the right at the place where the front of the train is observed now. But the corresponding synchronized clock on the train's front is reading a slightly earlier time, 10^{-9} s before the zero moment. What is the distance x?

2.7 A transparent spaceship with proper length 360 m passes by you. At the moment $t = 0$ of *your* time, the clock in the middle of the spaceship also reads $t' = 0$. However, another clock on the ship, positioned at its rear, reads $t' = 2.7 \times 10^{-8}$ s at this moment.
   (a) How fast is the ship moving?
   (b) What is the ship's length in your reference frame?

2.8 As a chief inspector of the Galactic Positioning System, you are passing by the planet Tarlan at $v = 0.95c$ and observe a pair of clocks there along the course of your spaceship. The distance between the clocks in your RF is 70 km and the clock that you pass first reads earlier time than the other clock, so that the time discrepancy between these clocks at the same moment of your time is 6.3×10^{-4} s. Are these clocks properly synchronized?

2.9 Prove that the interval between any two events is invariant with respect to Lorentz transformations, that is, its value is the same in all inertial reference frames.

2.10 Synchronized clocks are positioned at regular intervals, a 1.5 million km apart, along a straight line. When the clock next to you reads 12 noon:
(a) What time do you *see* on the 40th clock down the line?
(b) What time do you *observe* on that clock? (Observation means obtaining objective physical data based upon use of scientific instruments and scientific reasoning. (Section 7.1))

2.11 Derive the Lorentz contraction effect from the invariance of an interval.

2.12 Find
(a) the expression for the Lorentz transformation matrix $\mathcal{LLL}$ from K′ to K when K′ is sliding down the *y*-axis of K; down the *x*-axis of K;
(b) in either case find the matrix of inverse transformation (from K to K′)
(c) prove by direct multiplication that $\mathcal{L}\mathcal{L}^{-1} = \mathcal{L}^{-1}\mathcal{L} = I$, where I is the unitary matrix.

2.13 Show how the classical (Newtonian) low of addition of velocities follows from Galilean transformations.

3
The Velocities' Play

3.1
The Addition of Collinear Velocities

We now have sufficient background to find a universal rule of velocity transformation. It would relate the velocities of an object measured in two different reference frames (RFs).

Let us start with simple examples showing that velocities cannot simply add up in relativity as they do in Newtonian mechanics.

The first example involves light or anything moving with a speed of light in vacuum. Since this speed is invariant, it cannot just add up with a speed of another reference frame.

The second example: consider two events A and B happening simultaneously in a reference frame K. We can say that they are connected by an object moving, say, from A to B with an infinite speed $v = \infty$. Consider the same events from another reference frame K′, moving relative to K with a speed V along the line connecting the events. What is the speed of the object connecting these events in K′? According to Newton,

$$v' = v - V = \infty - V = \infty. \tag{3.1}$$

If we take any finite number and add it to or subtract it from the infinity, we will again get infinity. It is the infinite speed that plays the role of the invariant speed in Newtonian mechanics.

In relativity, the velocities cannot add according to (3.1), because the two events that are simultaneous in K are not simultaneous in K′. They can therefore be connected by an object, moving with a certain finite velocity $v' = \Delta l'/\Delta t'$, where $\Delta l'$ and $\Delta t'$ are, respectively, the spatial separation and the time interval between A and B in system K′. Obviously, the finite speed v' cannot be obtained from infinite speed by adding or subtracting finite velocity V. Thus, relativity of simultaneity automatically cancels the Newtonian law of the addition of velocities.

The most general expression for the result of superposition of two motions (along the same line) with velocities V and v' has the form

$$v = \eta(V, v')(V + v'). \tag{3.2}$$

Special Relativity and How it Works. Moses Fayngold

ISBN: 978-3-527-40607-4

The function $\eta(V, v')$ must be such that for $V \ll c$, $v' \ll c$ the value of $\eta(V, v')$ approaches 1. On the contrary, the invariant speed c added with a speed V must again give c. In other words, for $v' = c$, there must be

$$v = \eta(V, c)(V + c) = c. \tag{3.3}$$

It follows that

$$\eta(V, c) = \frac{c}{V + c} = \frac{1}{1 + (V/c)}. \tag{3.4}$$

The symmetry with regard to the interchange between V and v' demands that

$$\eta(c, v') = \frac{1}{1 + (v'/c)}. \tag{3.5}$$

To satisfy both requirements, $\eta(V, v')$ must have the form

$$\eta(c, v') = \frac{1}{1 + (Vv'/c^2)}. \tag{3.6}$$

Thus, the result of addition of two collinear velocities, which complies with the postulate that c is constant, is given by

$$v = \frac{V + v'}{1 + (Vv'/c^2)}. \tag{3.7}$$

Note the following fact: In deriving this equation, we did not use the Lorentz transformation explicitly. Our reasoning was based only on the general requirement that the expression has to be symmetrical with regard to the velocities being added and that if one of the velocities equals c, the result of their addition must also be equal to c. However, as we have seen in Chapter 1, this latter requirement leads immediately to the Lorentz transformation for spatial and time coordinates. That is why the use of this condition in derivation of (3.7) is equivalent to an implicit use of the Lorentz transformation itself.

Let us derive formula (3.7) in another way, where the Lorentz transformation is used explicitly. With that purpose, we will take the correct formula (2.9) and express the length Δx in system K in terms of the proper length

$$\Delta x = \gamma^{-1}(V)\Delta x', \tag{3.8}$$

and the time of the object's motion, in terms of the corresponding quantities $\Delta x'$ and $\Delta t'$, according to (2.45):

$$\Delta t = \gamma(V)\left(\Delta t' + \frac{V}{c^2}\Delta x'\right). \tag{3.9}$$

The reader should remember that Δx is the length in K of a moving segment whose ends' coordinates are being measured simultaneously and Δt is the time in K of motion of some particles along this segment. The beginning and the end of this motion are being observed both in K and K′ at different points in space, namely, at the beginning and the end of the segment Δx. This is why both $\Delta t'$ and $\Delta x'$ are present in the expression (3.9) for Δt. The time $\Delta t'$ in (3.9) is not a proper time of some process in K′, since the particle in question is moving in both reference frames.

Finally, we compose the ratio

$$\frac{\Delta x}{\Delta t} = \gamma^{-2}(V)\frac{\Delta x'}{\Delta t' + (V/c^2)\Delta x'} = \gamma^{-2}(V)\frac{v'}{1 + (Vv'/c^2)} \tag{3.10}$$

and add the resulting expression to V:

$$v = V + \frac{\Delta x}{\Delta t} = V + \left(1 - \frac{V^2}{c^2}\right)\frac{v'}{1 + (Vv'/c^2)} = \frac{V + v'}{1 + (Vv'/c^2)}. \tag{3.11}$$

The result we obtained is in agreement with (3.7).

Example 3.1

A production team is shooting a movie *The Wild Cosmos*. At a certain moment, the main hero (Tom) sitting in the middle of the moving spaceship has to fire from his two guns simultaneously in two opposite directions – one to the front and the other to the rear of the ship. Both of Tom's guns are identical, with the proper length of their bullets 10 cm. Alice records the scene from her platform. The speed of the ship relative to Alice is $V = 0.85c$, the muzzle speed of each bullet relative to its respective gun (recoil neglected) is $v' = 0.9c$.

What are the Lorentz-contracted lengths of the bullets as recorded by Alice?

Solution

Alice expects both bullets to be Lorentz contracted, but with different Lorentz factors because, having the same speed relative to Tom, the bullets have different speeds relative to her, owing to the motion of Tom's ship. She denotes the speed of the bullet moving toward the rear of the ship as v_- and its length as l_-. Similarly, the speed of the bullet moving toward the front is denoted as v_+ and the corresponding length as l_+. Then she expresses the bullet's lengths in terms of their respective speeds:

$$l_- = \frac{l_0}{\gamma(v_-)}, \qquad l_+ = \frac{l_0}{\gamma(v_+)}. \tag{3.12}$$

Now the problem reduces to finding the corresponding Lorentz factors in terms of the given speeds v' and V. Using Equation (3.11) with v' for the bullet fired to the front of the spaceship and $-v'$ for the bullet fired to its rear, Alice writes

$$v_+ = \frac{V + v'}{1 + (Vv'/c^2)}, \qquad v_- = \frac{V - v'}{1 - (Vv'/c^2)}. \tag{3.13}$$

The equations tell her that the bullet fired in the direction of the ship's motion is moving relative to her nearly as fast as light, whereas the other bullet must be moving significantly slower. Numerically, she gets $v_+ \cong 0.9915c$ and $v_- \cong 0.213c$. Her next impulse is to put these numbers into (3.12) and be done with it; but being curious, she decides to look at what happens if she puts there instead the general analytical expressions (3.13). Overcoming her dread of algebra, she does it and after a minute of manipulation comes up with the expressions

$$\gamma(v_+) = \gamma(V)\gamma(v')\left(1 + \frac{Vv'}{c^2}\right), \qquad \gamma(v_-) = \gamma(V)\gamma(v')\left(1 - \frac{Vv'}{c^2}\right). \tag{3.14}$$

Now, it is a straightforward matter to obtain the analytical expressions for the sought-for lengths of the bullets:

$$l_+ = \frac{l_0}{\gamma(V)\gamma(v')(1 + (Vv'/c^2))}, \qquad l_- = \frac{l_0}{\gamma(V)\gamma(v')(1 - (Vv'/c^2))}. \tag{3.15}$$

Plugging in the numbers, Alice obtains $l_+ \cong 1.3$ cm, $l_- \cong 9.7$ cm. The bullet fired forward is much shorter than the identical bullet fired back. The reason is that the former is moving much faster than the latter. The analytical expression for the ratio of the corresponding Lorentz-contracted lengths is

$$\frac{l_+}{l_-} = \frac{\gamma(v_-)}{\gamma(v_+)} = \frac{1 - (Vv'/c^2)}{1 + (Vv'/c^2)}. \tag{3.16}$$

As a by-product of her work, Alice has obtained an important result (3.14). Thanks to her, we now know that if

$$v = \frac{V \pm v'}{1 \pm (Vv'/c^2)},$$

then

$$\gamma(v) = \gamma(V)\gamma(v')\left(1 \pm \frac{Vv'}{c^2}\right). \tag{3.17}$$

3.2
The Addition of Arbitrarily Directed Velocities

The Lorentz transformation also allows us to obtain the general law of addition of velocities with arbitrary directions.

Suppose again that the system K′ is moving relative to K at a speed V along the x-axis. Consider an object in a state of relative motion at a velocity $\mathbf{v}'$ with respect to K′. The nature of this object has no importance whatsoever: it can be a massive particle, a photon, or just a mathematical point with a given law of motion. The components of velocity $\mathbf{v}'$ are

$$v'_x = \frac{dx'}{dt'}, \qquad v'_y = \frac{dy'}{dt'}, \qquad v'_z = \frac{dz'}{dt'}. \tag{3.18}$$

Now, what is the velocity of the same object as observed from the system K? The answer follows almost automatically from the Lorentz transformation. According to the definition of velocity, one has in system K

$$v_x = \frac{dx}{dt}, \qquad v_y = \frac{dy}{dt}, \qquad v_z = \frac{dz}{dt}. \tag{3.19}$$

Here, the infinitely small time interval dt and corresponding increments of coordinates dx, dy, dz separate two events – the passing of the object through two close points. For an observer in K′, the same events are separated by increments dt', dx', dy', dz'. Now all we have to do is to express the increments in K in terms of the corresponding increments in K′. Taking differential of Lorentz transformation (2.51), we have

$$dx = \gamma(V)(dx' + V dt'), \qquad dy = dy', \qquad dz = dz',$$
$$dt = \gamma(V)\left(dt' + \frac{V}{c^2}dx'\right). \tag{3.20}$$

It follows immediately that the sought-for components of velocity are

$$v_x = \frac{dx}{dt} = \frac{dx' + V dt'}{dt' + (V/c^2)dx'} = \frac{V + v'_x}{1 + (Vv'_x/c^2)}, \qquad v_y = \frac{v'_y}{\gamma(V)(1 + (Vv'_y/c^2))},$$
$$v_y = \frac{v'_y}{\gamma(V)(1 + (Vv'_y/c^2))}. \tag{3.21}$$

We have expressed the velocity components in the stationary system in terms of the velocity components in the moving system.

If we regard K′ as stationary and K as moving, we can obtain the reciprocal transformation by just swapping primed and unprimed components and changing the sign of the relative velocity:

$$v'_x = \frac{V - v_x}{1 - (Vv_x/c^2)}, \qquad v'_y = \frac{v_y}{\gamma(V)(1 - (Vv_y/c^2))},$$
$$v'_y = \frac{v_y}{\gamma(V)(1 - (Vv_y/c^2))}. \tag{3.22}$$

(The reader can obtain the same result by solving (3.21) for the primed components of the velocity $\mathbf{v}'$.)

We have thus arrived at the sought-for general formulas for velocity transformation, which describe both the simple addition of slow (relative to light) motions and the "weird" behavior of light.

Discussion

For those who like to discover the intimate connections between the apparently unrelated phenomena, it may be interesting to notice a similarity of the addition of collinear velocities with geometrical addition of the angles. Consider two angles α' and δ in a plane and their sum

$$\alpha = \alpha' + \delta. \tag{3.23}$$

What does it have to do with addition of collinear velocities?

Compare the first expression of Equations (3.21) and (3.22) with the known geometric identity for the tangent of the sum of two angles:

$$\tan \alpha \equiv \frac{\tan \alpha' + \tan \delta}{1 - \tan \alpha' \tan \delta}. \tag{3.24}$$

We see immediately the similarity in the algebraic structure of the two expressions. This similarity suggests that the collinear velocities figuring, say, in (3.21) can be represented by the corresponding angles according to the rule

$$\frac{v}{c} = \tan \alpha, \qquad \frac{v'}{c} = \tan \alpha', \qquad \frac{V}{c} = \tan \delta. \tag{3.25}$$

If we put this into the first expression of Equation (3.21), we obtain the expression which, although very similar, is not identical to (3.24). The difference is in the sign of the denominator. But even so, the analogy is striking.

We can extend the analogy even farther if we switch from trigonometric to hyperbolic functions of an angle. Write instead of (3.25),

$$\frac{v}{c} = \tanh \alpha, \qquad \frac{v'}{c} = \tanh \alpha', \qquad \frac{V}{c} = \tanh \delta. \tag{3.26}$$

In this case, the first expression of Equation (3.21) will give

$$\tanh \alpha \equiv \frac{\tanh \alpha' + \tanh \delta}{1 + \tanh \alpha' \tanh \delta}. \tag{3.24a}$$

This is identical to the formula for hyperbolic tangent of the sum of two angles. Thus, if we make a substitution (3.26), the relativistic formula for addition of collinear velocities corresponds to the addition of the angles representing these velocities. We can therefore expect that Lorentz transformation between two different RFs with relative velocity $\mathbf{V} || \hat{\mathbf{x}}$, and the corresponding velocity transformation, is mathematically equivalent to rotation through the angle

$$\delta = \operatorname{arctanh} \frac{V}{c} \tag{3.25a}$$

in the (ct, x)-plane of an abstract vector space. We will discuss the nature of this space and its relation to real space and time in the next chapter.

3.3 The Velocities' Play

Let us now "play" a little with the derived formulas to see how they work in different situations.

First, we want to make sure that for the motions slow with respect to light, our equations reduce to simple addition of velocities. Setting in (3.21) $V \ll c, v'_x \ll c$, we obtain

$$v_x \approx V + v'_x, \qquad v_y \approx v'_y, \qquad v_z \approx v'_z. \tag{3.27}$$

The theory of relativity has not overturned the previous theory; it has only revealed its approximate character and clearly charted the domain of its applicability. One only has to keep in mind that the borders of this domain are themselves relative – they depend on the accuracy of our measurements. Increasing this accuracy, we can, in principle, notice the approximate character of relations (3.27) even at very small velocities. We will discuss one such situation in detail in Section 8.2.

The new and unexpected (from the viewpoint of Newtonian mechanics) phenomena become significant when the velocities involved are comparable to c.

Consider an object moving in system K′ with velocity $\mathbf{v}' \perp \mathbf{V}$, that is, $v'_x = 0$. Then, the exact equations (3.21) yield

$$v_x = V, \qquad v_y = \gamma^{-1}(V)v'_y, \qquad v_z = \gamma^{-1}(V)v'_z. \tag{3.28}$$

We see that the transverse components v_y, v_z are diminished in system K by a factor of $\gamma(V)$ in total accord with the effect of time dilation.

Let now $v'_y = c, v'_z = 0$; this corresponds to the experiment with the light pulse within a vertical cylinder, moving horizontally, which has been considered in Section 2.5. According to (2.19), the speed of the pulse along the cylinder as measured in system K is equal to

$$v_y = \gamma^{-1}(V)c, \tag{3.29}$$

which is by a factor of $\gamma(V)$ less than c. But it does not follow from here that the light pulse moves slower in system K. Indeed, formula (3.29) is not yet the *whole* result of transformations (3.21), but only one part of it, related to the transverse component of the velocity. Let us now take into account its second part – the emergence of the longitudinal component in system K:

$$v_x = V, \tag{3.30}$$

which is caused by the motion of the cylinder itself. Owing to this component, the velocity vector turns out to have been rotated, so that in system K the velocity $\mathbf{v}$ is not perpendicular to $\mathbf{V}$.

In our case, when the object moving inside the cylinder is a photon, this longitudinal component gives the precisely right contribution to the magnitude of the rotated velocity vector to compensate for the slowdown of the transverse motion, so that the resulting speed is

$$v = \sqrt{V^2 + v_y'^2} = \sqrt{V^2 + c^2\left(1 - \frac{V^2}{c^2}\right)} = c. \tag{3.31}$$

Thus, the light velocity vector is oriented differently in system K than in system K′, but its magnitude remains equal to c.

Now, we are going to consider a few cases when the velocity of an object is parallel to the relative velocity of the system K and K′.

1. Let system K′ be a spaceship moving relative to K with a speed V close to the speed of light. There is a particle accelerator working inside the spaceship, producing high-energy particles moving from the rear to the front with speed v' relative to the spaceship. This speed is also close to that of light, so that we can write

$$V = c - \Delta V, \qquad v' = c - \Delta v', \tag{3.32}$$

where

$$\Delta V \ll c, \qquad \Delta v' \ll c. \tag{3.33}$$

What is the particles' speed relative to system K?

According to the "obvious" prerelativistic formula (1.1), we would have

$$v = V + v' = 2c - \Delta V - \Delta v', \tag{3.34}$$

that is, the particles' speed in K must be nearly twice the speed of light. In reality, however, the particles' speed in K is described by Equation (2.9), leading to (3.7), so that we have

$$v = \frac{V + v'}{1 + (Vv'/c^2)} = \frac{2c - \Delta V - \Delta v'}{2 - ((\Delta V + \Delta v')/c) + (\Delta V \Delta v'/c^2)}$$
$$= \frac{c}{1 + (\Delta V \Delta v'/c(2c - \Delta V - \Delta v'))}. \tag{3.35}$$

Taking into account the conditions (3.33), we can, to a high accuracy, approximate this exact equation by

$$v \cong c\left(1 - \frac{1}{2}\frac{\Delta V \Delta v'}{c^2}\right). \tag{3.36}$$

We see that v remains less than c.

2. Imagine now two relativistic spaceships moving away from each other. They represent two different inertial systems K_1 and K_2, and their speeds v_1 and v_2 are measured relative to one and the same system K. Because in this case we use the same time t when measuring both speeds, the simple addition formula

$$\frac{d(x_2 - x_1)}{dt} = v_1 + v_2 \tag{3.37}$$

retains a certain physical meaning. It shows the rate of change of the distance between the spaceships in system K. For instance, if the spaceships fly asunder in system K at a speed $0.9c$ each, then the separation between them, as seen by an observer in K, increases at a rate of $1.8c$. This does not in any way contradict the theory of relativity because the $1.8c$ is *not* the speed of their relative motion. To determine their relative speed, one has to use the measuring devices of *one of the spaceships*, that is, to transfer to the rest frame of this spaceship and then measure the speed of the other spaceship. In this case, one will obtain

$$v = \frac{v_1 + v_2}{1 + (v_1 v_2/c^2)} = \frac{1.8c}{1 + (0.9)^2} \cong 0.995c; \tag{3.38}$$

that is, v remains less than c again.

Thus, the number $1.8c$ in this example emerges as an intermediate result of the algebraic calculation. It has a physical meaning of the rate of separation change between the two objects in the "alien" reference frame K in which they are both moving. This number is *not* the experimental result of the direct measurement of the speed of one of the objects relative to the other.

One could ask: "I see two objects flying asunder each with the speed $0.9c$ relative to me, don't I therefore see them flying asunder with the relative speed $1.8c$?"

The answer to this would be: "No. They do fly apart with the speed $1.8c$. But this is not the speed of their *relative* motion. The relative speed is observed in the rest frame of one of the objects."

This example shows the difference between the "speed" $v_1 + v_2$ and the relative speed v.

3. The situation is again the same as in case 1, only the objects in question are photons because instead of particle accelerator, the crew of the spaceship is now using a laser. Then $v' = c$. Equation (1.1) would give for the photon speed in K the value $v = c + V$; in reality, however, we will have

$$v = \frac{V + c}{1 + (Vc/c^2)} = c. \tag{3.39}$$

Here, we do not even need any special assumptions about the value of V (whether it is much smaller than c or close to c). At *any* V, the relativistic law of addition of velocities gives for the speed of a photon the same value c. The equation works as a simple and at the same time as an intricate machine; no matter what speed is entered into the machine together with the "c," there always comes out "c" in the output. The same result follows if the speed "c" of a photon is determined first in a stationary system K and then one looks at its speed in the spaceship dashing after the photon with the speed V. Here we come back to the question from which Einstein had started his musings that led him to the theory of relativity. And the theory gives an immediate answer: the spaceship's system K′ is equivalent to K, and one observes in it the same laws of Nature and the same rule of addition of velocities. One only has to change V to $-V$ because the system K, where the photon speed c has been measured, is moving relative to K′, in the opposite direction. We obtain

$$v' = \frac{c - V}{1 - (Vc/c^2)} = c. \tag{3.40}$$

No matter how hard the spaceship is accelerating, the photon will run away from it with the same speed c.

We considered these examples, perhaps even with too minute details, to emphasize once again the basic fact lying at the core of the theory of relativity: the invariance of the speed of light. We will discuss this invariance in more detail in Section 9.2.

3.4 Observable Manifestations

We consider here two observable manifestations of the relativistic addition law for velocities.

3.4.1 The Fizeau Experiment

As far back as 1851, a French physicist Fizeau carried out an interesting experiment whose result can be explained by the relativistic addition law for velocities.

Suppose we have a transparent refracting material with the index of refraction n. The speed of light in this material is $u = c/n$. Now, what happens if the material itself is boosted with respect to the laboratory at a speed V? There were two different answers to this question, both based on the ether hypothesis. According to this hypothesis, the ether permeated the whole space and its density is greater within bodies with $n > 1$, which explains the slowdown of light within such a body.

According to this picture, if the body is moving, the results would be different depending of ether's response to this motion. There may be two extreme cases:

(a) If the ether filling the body remains in place and *only the location* of its increased density changes together with the body's location because of its motion, then this motion will not affect the propagation speed of light and we will measure the same speed $v' = u = c/n$ in the moving body as in the stationary body.

(b) If the ether *itself* gets completely involved into body's motion, so that the ether inside the body is moving together with the body with the speed V relative to the laboratory, then the speed of this motion will add to the speed of light c/n relative to the body and the resulting speed will be

$$v' = V + u = V + \frac{c}{n}. \tag{3.41}$$

The experiment carried out by Fizeau was designed to distinguish between these two possibilities. The simplified version of this experiment is shown in Figure 3.1. The role of a moving body was played by flowing water. The monochromatic light from a small source S was directed by the lens L_1 along the two parallel tubes. The water in one tube was flowing with the speed V in the direction of propagation of light,

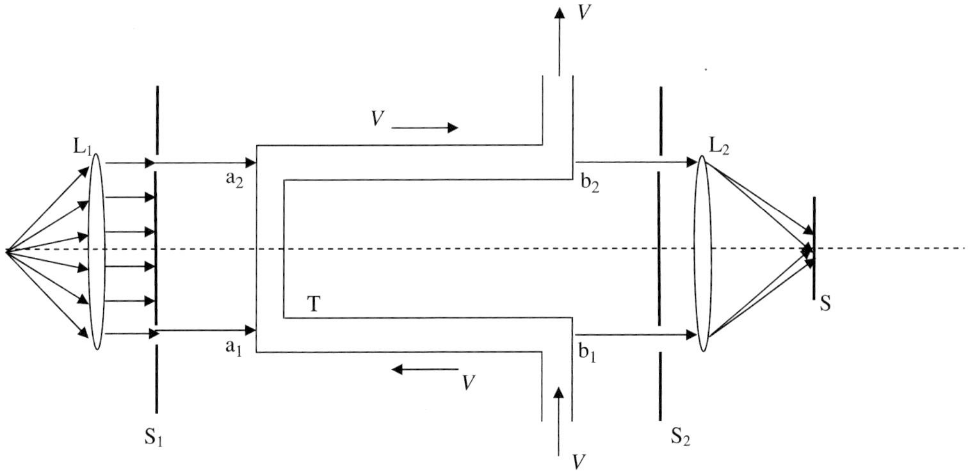

Figure 3.1 Schematic of the Fizeau experiment. L_1, L_2: lenses; S_1, S_2: perforated screens; T: the tube with flowing water; S: the image screen.

whereas the water in the second tube was flowing against the direction of propagation [18]. Applying the Newtonian addition law, the speed of light in the respective tubes relative to the laboratory would be given by

$$v'_{1,2} = \frac{c}{n} \pm V. \tag{3.42}$$

The windows b_1 and b_2 in the tube are transparent, as are the windows a_1 and a_2. Therefore, the split light beams from both tubes exit and are recombined on the screen OO′ after passing through the second lens L_2. Since the recombined beams come from the same source, they are to a high accuracy coherent and can interfere. The brightness of the image on the screen depends on their optical path difference, much in the same way as the local brightness on the screen in the Michelson experiment, and is described by the corresponding Equation (2.1), where I_0 now is the light intensity of each individual beam and $\Delta\phi$ is the phase difference between them.

If l is the length of the tube, then the analysis similar to that in Section 2.1 yields the corresponding phase difference

$$\Delta\phi = \omega(t_2 - t_1) = \omega l\left(\frac{1}{v'_2} - \frac{1}{v'_1}\right). \tag{3.43}$$

Equation (3.43) gives different results depending on mechanism. In case (a), both speeds are equal and the phase difference is zero. In case (b), Equation (3.43) together with (3.42) gives

$$\Delta\phi = 2lV\frac{\omega}{c^2}n^2. \tag{3.44}$$

Actual experimental result observed by Fizeau was

$$\Delta\phi = 2lV\frac{\omega}{c^2}(n^2 - 1). \tag{3.45}$$

It corresponded to neither of the two possibilities (a) or (b), but rather to a somewhat exotic situation between these two extremes. Namely, it indicated that the speed of light relative to the laboratory was the sum of its speed in the medium and the medium's speed V taken with a coefficient α. In other words, the (Newtonian) speeds of light in the respective tubes must be

$$v'_{1,2} = \frac{c}{n} \pm \alpha V, \quad \text{with} \quad \alpha = 1 - \frac{1}{n^2}. \tag{3.46}$$

They are different from the analogous speeds (3.42) corresponding to the cases (a) or (b). This showed that neither of these extreme assumptions was correct.

To explain the experimental results, Fresnel suggested the hypothesis of *partially dragged* ether. According to Fresnel, the ether neither remained stationary within a moving body ($\tilde{V} = 0$) nor set in motion with the speed of the body ($\tilde{V} = V$), but was rather only partially involved with the speed that depended on refraction index and was between the two extremes:

$$\tilde{V} = V(n) = \left(1 - \frac{1}{n^2}\right)V \equiv \alpha(n)V. \tag{3.47}$$

The coefficient

$$\alpha = \alpha(n) \equiv 1 - \frac{1}{n^2} \tag{3.48}$$

was named the drag coefficient.

As we know, at that time light was considered as waves in ether considered as an elastic medium with density ρ and the Young's modulus N. This made it possible to explain the Fizeau experiment within the framework of the ether hypothesis by playing with one of these parameters. Fresnel proposed a very compelling explanation. He made an assumption that any material condenses the ether contained in it, but does not affect its Young's modulus. If ρ is the ether density in the absence of all matter, its density $\tilde{\rho}$ in a material with refraction index n is greater than ρ, but its Young's modulus N is the same as in the free space. Applying the theory of elastic waves to ether gives the expression relating the speed of light to the introduced mechanical characteristics of ether:

$$c = \sqrt{\frac{N}{\rho}} \quad \text{(free space)}, \qquad \tilde{c} = \sqrt{\frac{N}{\tilde{\rho}}} \quad \text{(transparent medium)}. \tag{3.49}$$

Therefore, the observed refraction index of the material depends only on ratio of the ether densities in this material and in free space:

$$n = \frac{c}{\tilde{c}} = \sqrt{\frac{\tilde{\rho}}{\rho}}. \tag{3.50}$$

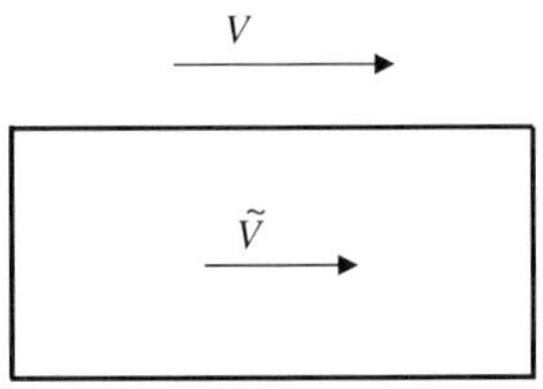

Figure 3.2 V – the velocity of the etherflow outside the glass; $\tilde{V}$ – the velocity of the etherflow inside the glass.

Now, let a bulk of this material (say, a glass plate) move in ether at a speed V. This means that ether *outside* the glass flows past it with the speed $-V$. However, the ether *within* the glass must flow relative to it with a *different* speed $\tilde{V}$ due to difference in density (Figure 3.2). The ether flux must be continuous, that is, the incoming flux density ρV on the outer side of the plate's surface S must be equal to the outgoing flux density $\tilde{\rho}\tilde{V}$ on the inner side of S:

$$\rho V = \tilde{\rho}\tilde{V}. \tag{3.51}$$

Combining this with (3.50) gives

$$\tilde{V} = V\frac{\rho}{\tilde{\rho}} = \frac{V}{n^2}. \tag{3.52}$$

Thus, the condensed ether within the material flows with the speed $\tilde{V} < V$ and the Newtonian difference in the speed of the ether within and outside is

$$\delta V = V - \tilde{V} = \left(1 - \frac{1}{n^2}\right)V. \tag{3.53}$$

But this difference is the speed (3.47) of the condensed ether relative to the laboratory! Adding it to the speed c/n of light in glass, we obtain the result (3.46) of the Fizeau experiment. The expression $1 - n^{-2} \equiv \alpha(n)$ is Fresnel's famous drag coefficient. It means that ether is dragged by the moving glass ($\tilde{V} \neq V$) but only partially, not completely ($\tilde{V} \neq 0$).

Thus, Fresnel's refined assumption of the ether partially dragged by a moving body allowed him to explain the results of Fizeau experiment.

According to relativity, there is no ether. Then the question arises: can relativity explain the results of Fizeau's experiment? And, if yes, then how does it do it? The answer to both questions is: it does it automatically as the result of relativistic addition of velocities. Indeed, applying Equation (3.11) with $v' = c/n$ gives for $V \ll c$

$$\begin{aligned} v' &= \frac{(c/n) \pm V}{1 \pm (V/cn)} \approx \left(\frac{c}{n} \pm V\right)\left(1 \mp \frac{V}{cn}\right) \\ &= \pm\left(1 - \frac{1}{n^2}\right)V + c\left(\frac{1}{n} \mp \frac{V^2}{c^2}\right) \approx \frac{c}{n} \pm \left(1 - \frac{1}{n^2}\right)V. \end{aligned} \tag{3.54}$$

That's it! There is no need for the ether hypothesis, let alone the additional, although intuitively clear, assumptions about its behavior.[1)]

But a skeptic reader may still ask, why do we need a sophisticated new theory to explain an experiment that already has an intuitively clear explanation in the framework of the old theory?

To answer this question, let us first ask another question: in the framework of the old theory, what is the speed of light in a moving piece of glass for an observer sitting on this piece? The glass is moving relative to ether, but is stationary relative to the observer. The answer is simple: the light propagates with the speed $\tilde{c}$ in condensed ether within the glass, but this condensed ether, according to Fresnel (Equation (3.52)), is moving within the glass with the speed $\tilde{V} = -V/n^2$.

For the case when motions of the glass and the light are collinear, this gives

$$\tilde{c}' = \tilde{c} - \frac{V}{n^2} \neq \tilde{c}. \tag{3.55}$$

Thus, the speed of light in glass moving in ether is different from its speed in glass stationary in ether. This difference makes the two cases experimentally distinct. In other words, we would have the means to distinguish one inertial RF from another and, moreover, could find a privileged RF stationary in ether. But this contradicts the experiment, which shows no such differences!

This is the answer to the original question.

3.4.2 Aberration of Light

A particle in K′ is moving with velocity

$$\mathbf{v}' = v'_x \hat{\mathbf{x}} + v'_y \hat{\mathbf{y}}. \tag{3.56}$$

System K′ is moving relative to K in the x-direction at a speed V. Find the particle's velocity in K.

Applying the rules (3.21) to this case, we obtain

$$\mathbf{v} = \frac{V + v'_x}{1 + (Vv'_x/c^2)} \hat{\mathbf{x}} + \frac{v'_y}{\gamma(V)(1 + Vv'_x/c^2)} \hat{\mathbf{y}}. \tag{3.57}$$

It follows for the angle α between $\mathbf{v}$ and $\hat{\mathbf{x}}$:

$$\tan\alpha = \gamma^{-1}(V) \frac{v'_y}{V + v'_x}. \tag{3.58}$$

Similarly, we introduce the angle α' between v' and $\hat{\mathbf{x}}' = \hat{\mathbf{x}}$ in K′. In terms of this angle, the previous equation takes the form

$$\tan\alpha = \frac{v' \sin\alpha'}{\gamma(V)(V + v' \cos\alpha')}. \tag{3.59}$$

1) For simplicity, we have neglected dependence of the light-speed in a medium on frequency due to dispersion. Taking into account this effect brings in additional small term into Equation (3.54), which also can be obtained from the relativistic addition law for velocity.

This equation determines the change in the direction of the velocity vector when we switch from one RF to another. It has an important application in a special case when a particle in question is a photon. In this case, $v = v' = c$ and Equation (3.62) reduces to

$$\tan\alpha = \sqrt{1-\beta^2}\,\frac{\sin\alpha'}{\beta+\cos\alpha'}, \qquad \beta \equiv \frac{V}{c}. \tag{3.60}$$

It is easy to express the dependence (3.60) between the angles in terms of $\sin\alpha$ and $\cos\alpha$:

$$\sin\alpha = \sqrt{1-\beta^2}\,\frac{\sin\alpha'}{1+\beta\cos\alpha'}, \qquad \cos\alpha = \frac{\beta+\cos\alpha'}{1+\beta\cos\alpha'}. \tag{3.61}$$

These equations describe the effect well known in astronomy as aberration of light. It is the difference in the direction of light rays observed from two different RFs [28].

Let us consider two cases of this effect. First, suppose that a distant source of light (e.g., a certain star) shines right overhead for an observer in K′. The light from this source streams vertically down along the $\hat{y}'$-direction, so the angle α' is $90\,^\circ$. The same star is observed by an astronomer in another system K moving relative to K′ with a speed $\mathbf{V} \perp \hat{y}'$. Without doing any calculations, we can make a qualitative prediction that the light will not fall vertically in the astronomer's RF, so the angle α will not be $90\,^\circ$. This is similar to a familiar situation on a rainy day: suppose the rain is falling vertically down onto the Earth. However, if you are driving a car, the same rain is tilted relative to the car. If you are not in the car, then the rain is vertical when you are stationary, but it is tilted relative to you when you are running. In the latter case, you have to accordingly tilt your umbrella to remain dry. Now, let us calculate the necessary tilt for the case of light. In this case, it will be more convenient to express the effect in terms of the angles θ, θ' that the respective velocities make with the *vertical* direction. All we need is to express each angle in the above equations through its complementary angle. We can use any of the above three equations. Let us use the first one. We will get

$$\cot\theta = \sqrt{1-\beta^2}\,\frac{\cos\theta'}{\beta+\sin\theta'}, \qquad \beta \equiv \frac{V}{c}. \tag{3.62}$$

Since the light is falling vertically down in K′, the angle $\theta' = 0$. Putting this into (3.62), we obtain

$$\tan\theta = \beta\gamma(\beta). \tag{3.63}$$

Try to remember this result. We will return to it later in Section 6.13 from a somewhat different perspective.

In the second case, we suppose angle α to be arbitrary, but the relative speed V small, that is, $\beta \ll 1$. This also represents a relevant astronomical situation, with K′ attached to the background of distant stars and K instantly comoving with our planet in its orbital motion around the Sun. Since α is arbitrary, the expected result must give the deviation of light rays

$$\Delta\alpha = \alpha' - \alpha \tag{3.64}$$

for *any* location of a respective star in the sky. The sought-for deviation angle (or angle of aberration) must, under given conditions, be small, that is, $\Delta\alpha \ll 1$.

It is now convenient to use the first expression of Equation (3.61). At small β, it takes the form (to the first-order accuracy in β)

$$\sin\alpha = \sin\alpha'(1 - \beta\cos\alpha'), \tag{3.65a}$$

or

$$\sin\alpha - \sin\alpha' = -\beta\sin\alpha'\cos\alpha'. \tag{3.65b}$$

Since α and α' are in this case very close to each other, we can, to the same accuracy, treat the difference on the left as the differential of $\sin\alpha$. This gives

$$\Delta\alpha = \beta\sin\alpha'. \tag{3.66}$$

This result correctly describes the well-known astronomical data for the aberration of light.

Problems

3.1 You have fired a laser pulse in the positive x-direction. A relativistic spaceship is rushing in the same direction at 99% of the speed of light, trying to catch up with the pulse. Show that the pulse is rushing away from the ship at the same speed of light as it is from you.

3.2 Two spaceships are moving toward each other, each with the speed 99% of c in your reference frame.

(a) Find the rate of change of separation between the ships in this frame.
(b) Find the relative speed of the ships with respect to one another.

3.3 In a Star War game, an alien spaceship moving at $0.75c$ relative to the center of our Milky Way galaxy (system K) pursues our spaceship that only goes $0.5c$ with respect to K. The aliens fire a rocket whose velocity relative to their gun is $0.33c$. Does the rocket reach its target

(a) according to Galileo;
(b) according to Einstein? (D. Griffiths)

3.4 A space traveler moving along the x-direction relative to Earth fires a laser pulse. The laser barrel makes $60°$ with the positive x-direction in the rest frame of the spaceship.

(a) What is the angle between the barrel and the positive x-direction for an Earth-based observer?
(b) At what angle, according to this observer, does the *laser pulse* move with respect to the x-direction?
(c) Draw a free-body diagram showing the orientation of the laser barrel and the trajectory of the laser pulse in respective reference frames.

3.5 A source in the middle of Tom's ship moving at a speed V emits two photons – one in the forward direction and the other in the backward direction. The proper length of the ship is L. There are two detectors – one at the front and the other at the rear of the ship. The clocks on the ship and the platform are set so that Alice's clock at the origin and Tom's clock in the middle of the ship both read zero time when they pass by one another.

(a) When, by Alice's clock, will the first and the second detector fire? (Find the answer to this question by two different methods.)
(b) Find the simplest equation relating these two moments to one another.
(c) How far from the center of platform will the second photon be when the first detector fires?
(d) What is the instant distance (in Alice's frame) between Tom and this photon?

3.6 The conditions are the same as in the previous problem, but now, instead of the photons, the source fires two bullets flying in the two opposite directions with the same speed v' relative to the spaceship.

(a) Find the moments of Alice's time when the bullets will hit, respectively, the rear and the front of the ship.
(b) Write the expressions for the ratio and the product of these moments.
(c) Find the positions of the rear and the front at these moments.
(d) Find the instant distances, Δx_1 and Δx_2, between each bullet and Tom, measured on the platform (instant distances mean at the same moment of the platform time). What is the ratio of these distances?
(e) Find the limit of the answers to (a)–(d) at $v' \to c$. How do the limits in cases (a) and (b) relate to the corresponding answers in the previous problem?

3.7 Consider a particle uniformly moving with velocity $\mathbf{v}$ relative to an inertial reference frame K. Its equation of motion is $\mathbf{r} = \mathbf{v}t$. Derive the relativistic velocity addition rules for $\mathbf{v}$ by applying Lorentz transformations to $\mathbf{r}$ and t.

3.8 Consider two simultaneous events A and B separated by a distance x in a reference frame K. You can think about these events as connected by a fictitious infinitely fast particle. Imagine another observer in a reference frame K′ moving with a speed V along the line connecting the events. Our intuitive expectation would be that, since infinity is an overwhelmingly strong thing, the speed of the above fictitious particle would be infinite in any other reference frame as well.

(a) What is the actual speed of this particle in K′ according to Newton?
(b) According to Einstein? Obtain the answers to both questions (a) and (b) by two different methods.
(c) What is the speed of this particle in another reference frame K″ moving in a direction *perpendicular* to the line connecting the events?

3.9 A particle in an inertial reference frame K′ is moving in the plane xz with a speed v' at an angle θ' with the z-axis. The frame K′ itself is moving with a speed V along the x-axis of system K. As always, we assume the respective axes of both systems (x and x', y and y', z and z') to be parallel.

(a) What is the speed of this particle in the reference frame K?

(b) What is the angle between the velocity vector $\mathbf{v}$ and the z-axis in K?

3.10 Replace the particle in the previous problem by a photon and derive the relationship between the angles θ and θ' for this case. You will obtain the description of the well-known effect – the aberration of light.

3.11 Suppose that an analogue of the Fizeau experiment is carried out with plasma instead of water. Using the refraction index for plasma $n < 1$, find the Fresnel's drag coefficient. Interpret the results.

3.12 A particle is moving at a speed v along the y-axis of an inertial system K.

(a) Find the speed v' of this particle in the inertial system K′ moving with velocity $\mathbf{V}$ relative to K along its x-axis.

(b) Find the corresponding Lorentz factor $\gamma(v')$ in terms of v and V.

4
Space-Time

From this moment on, the notions of time by itself and space by itself will sink into mere shadows, and only their combination in a 4-dimensional entity – space-time – will become an absolute reality.
Hermann Minkowski

In the previous chapters, we were interpreting the Lorentz transformations in terms of their unusual but physically observable manifestations, such as time dilation and relativistic length contraction. In this respect, we were studying its physical face. Here, we examine the Lorentz transformations from another perspective, showing their geometrical face. We will realize that this geometrical aspect forms a basis for a powerful generalization made by Minkowski – unification of space and time into a continuous four-dimensional set – space-time.

4.1
Minkowski's World

I see the Past, Present, and Future existing all at once before me.
William Blake

Our study of relationship between space and time can be made geometrically clear by using a remarkable construction – the so-called space-time diagrams. These diagrams were introduced by an outstanding mathematician, Hermann Minkowski, who had been the young Einstein's instructor in the Zurich Polytechnicum. After Einstein had published his theory of relativity, Minkowski allegedly said, "To tell the truth, I did not expect this from Einstein." Einstein, on his part, also allegedly remarked after the publication of Minkowski's work: "After the mathematicians had taken care of my theory, I no longer can understand a bean in it." He joked, of course. The geometrical approach introduced by Minkowski was an important contribution to the theory. It leads to a clear and elegant description of the world, and was later used by Einstein himself to extend his theory to include gravitation. Now, this approach is universally used in both special and general theories of relativity.

Special Relativity and How it Works. Moses Fayngold

ISBN: 978-3-527-40607-4

A space-time diagram is in its essence just a graph of a displacement versus time for a given motion. The basic element in this construction is a concept of an *event*. An event is any phenomenon so fleeting and occupying so small a region that it can be considered as instantaneous and pointlike. It therefore can be characterized by one moment of time t and three spatial coordinates (x, y, z). Thus, to any possible event there corresponds a set of four numbers (t, x, y, z), and vice versa, each set of four numbers (t, x, y, z) specifies an event in space and time. For instance, a set of numbers (43 s, 2 m, –54.6 m, 0.33 m) labels an event that happened at 43 s a.m. today, at a point with coordinates 2 m, –54.6 m, and 0.33 m along the x-, y-, and z-directions, respectively. All possible events that have ever happened, are happening, and are going to happen, form a continuous four-dimensional set – a combination of three spatial and one temporal dimensions. This set was called the space-time, or the Minkowski world [2,29–32].

In order for all four coordinates of Minkowski's world to have common dimension, it is convenient to use ct rather than t as a time coordinate. It means that instead of measuring time in seconds, we measure it in equivalent units of length – so-called light seconds (1 light second (LS) is the distance traveled by light in 1 s). Accordingly, we are going to express spatial distances in these new units of length. We thus get one common unit for all four dimensions of our world.

As a first illustration of a space-time diagram, consider a light signal moving in the positive x-direction: $x = ct$ (Figure 4.1). This motion is represented graphically by a straight line OS that is a bisector of the angle between the axes x and ct (it is convenient to direct ct-axis upwards). The light signal moving in the opposite direction ($x = -ct$) will be represented graphically by a straight line OS′. We can consider either of the two lines as a trajectory of the light pulse in a plane (x, ct). We call such a trajectory *the world line*. Thus, OS is a world line of a light signal moving in the positive x-direction and OS′ is the world line of such signal moving in the negative x-direction.

The notion of a world line is different from the notion of an interval introduced in Section 2.6. An interval is only a number determined by two events (two points in space-time). A world line, as is clear from the term, is a path (which can be an arbitrarily curved line) connecting these points. The interval between any two events in space-time is the pseudo-Euclidean distance between them, that is, the "kinematic length" of the straight path connecting them.

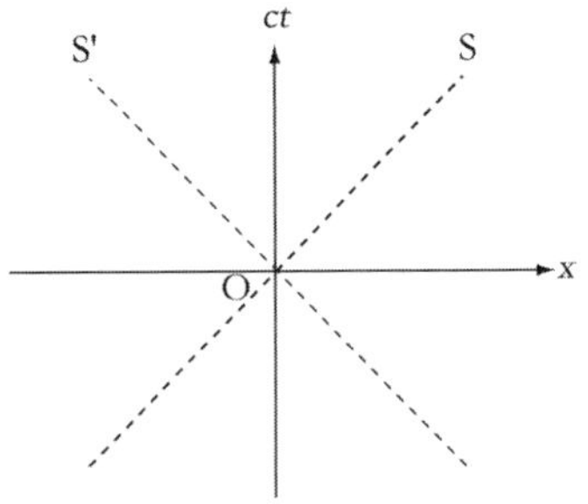

Figure 4.1

Our next example will be a rehash of Section 2.6, but from a somewhat different perspective.

Consider two events: the first is the emission of a light pulse from a laser gun at a point (x_1, y_1, z_1) at a moment t_1 and the second one is the absorption of this signal by a detector at point (x_2, y_2, z_2) at a moment t_2. The spatial distance between the gun and the detector is

$$r_{12} = \sqrt{(x_2 - x_1)^2 + (y_2 - y_1)^2 + (z_2 - z_1)^2} = c(t_2 - t_1). \tag{4.1}$$

We can write this relation between the distance r_{12} and time $t_{12} \equiv t_2 - t_1$ in the form

$$c^2(t_2 - t_1)^2 - (x_2 - x_1)^2 - (y_2 - y_1)^2 - (z_2 - z_1)^2 = 0. \tag{4.2}$$

Denote the expression on the left-hand side of this equation as s_{12}^2. Then the equation for the light pulse can be written as

$$s_{12}^2 = 0. \tag{4.3}$$

The quantity s_{12} is an already familiar four-dimensional interval (or just the interval) between the events 1 and 2 (recall Equation (2.31) in Section 2.6). We see that the interval for a light signal in vacuum is equal to zero. Accordingly, we will call it the zero, or the null interval.

Suppose the same process is being observed from another inertial reference frame K′. The space and time coordinates for the same two events (emission and absorption of the signal) in the system K′ will be $(ct_1', x_1', y_1', z_1')$ and $(ct_2', x_2', y_2', z_2')$, respectively. The distance r_{12}' between the laser gun and the detector and the time interval $t_{12}' \equiv t_2' - t_1'$ between the emission and absorption of the signal will also be different from those measured in K. But because of the invariance of the speed of light, there must be $r_{12}'/t_{12}' = c$, or $c^2(t_{12}')^2 - (r_{12}')^2 = 0$. Therefore, if an observer in the system K′ also considers the quantity $(s_{12}')^2 \equiv c^2(t_{12}')^2 - (r_{12}')^2$, he can write

$$(s_{12}')^2 = s_{12}^2 = 0. \tag{4.4}$$

The zero value for a null interval turns out to be a universal property independent of the reference frame.

Thus far, the invariance of the four-dimensional interval under four-dimensional rotations (Lorentz transformations) was established in Equation (4.4) for the null intervals only, as a direct consequence of the invariance of the speed of light in vacuum. This result, however, turns out to be much more general. Consider two *arbitrary* events (t_1, x_1, y_1, z_1) and (t_2, x_2, y_2, z_2), which are not necessarily connected by the light signal (they do not lie on the light's world line), and introduce the corresponding interval

$$s_{12}^2 = c^2(t_2 - t_1)^2 - (x_2 - x_1)^2 - (y_2 - y_1)^2 - (z_2 - z_1)^2. \tag{4.5}$$

Applying the Lorentz transformations (2.45), the reader can prove that a nonzero interval also turns out to be independent of choice of a reference frame (Problem 4.1):

$$s_{12} = s_{12}'. \tag{4.6}$$

Thus, the Lorentz transformations derived in Section 2.6 ensure the invariance of *any* interval. Accordingly, this ensures that not only the length of a world line of something propagating with the invariant speed but also the length of *any* world line for *all* possible processes is invariant for all observers. This reflects the fundamental physical properties of time and space as a whole.

Indeed, look at the result (4.6) from a geometrical viewpoint. In a three-dimensional space, the distance between two points (the length of the segment connecting them) does not depend on orientation of the axes (x, y, z). When we switch to another system of axes (x', y', z'), each point is being assigned a new set of coordinates. The new coordinates of a point are linear combinations of the old ones with the coefficients depending on the angles between the old and new axes; however, the length of a segment does not change.

The behavior of a four-dimensional interval displays striking similarity to this property of a segment: as we switch to another reference system K′, the new coordinates of an event are expressed in terms of the old ones by linear equations (Lorentz transformations), whose coefficients depend on relative velocity between the two systems; however, the interval itself does not change. We can, therefore, consider the interval s_{12} as the "distance" between the corresponding points in a four-dimensional set called space-time; the Lorentz transformations from this viewpoint can be thought of as a rotation from one four-dimensional system of axes to another one.

However, the discovered analogy with ordinary three-dimensional space is not complete. Unlike the distances in ordinary space, the squares of the temporal and spatial coordinates in the interval (4.5) enter with the *opposite* signs. This fact is the manifestation of the fundamental physical difference between time and space. To emphasize this difference, their combination is sometimes called $(1+3)$-space, rather than four-dimensional space. It is a four-dimensional space, where one dimension (time) is physically different from the other three. Therefore, the term "space-time" seems to be most appropriate. From a physicist's viewpoint, the temporal dimension is fundamentally different from the spatial ones, because the corresponding experimental procedures for measuring them are quite different. We measure time between the events A and B by using some periodic process and measure how many periods (cycles) fit between A and B. We measure distance by applying the meterstick along the line connecting A and B and determining how many such metersticks fit into this line.

One can hear sometimes a counterargument that a distance between two objects can be measured by measuring the time it takes a light pulse to travel between them, that is, the length measurement can be reduced to time measurement. This is true, and was, in fact, implemented in a few known experiments (e.g., laser location of the Moon). But this is an *indirect* measurement, based on additional information we had obtained about the connection between space and time, such as the invariance of the speed of light in deep space. Using such indirect measurements can mask but does not eliminate the difference between space and time. It is all the more remarkable that, with this difference, both can be combined into one entity, space-time.

As mentioned above, the physical difference between the two constituents of this entity is manifest in the different signs, with which the squares of the corresponding coordinates enter the expression (4.5) for 4-interval. This leads to a peculiar property: the value of s_{12}^2, unlike the square of the ordinary distance, can be negative; and it can be zero even when the interval connects two *different* events. As a result, for any chosen point (i.e., an event t_1, $\mathbf{r}_1$) in Minkowski's world, the set of all other events can be divided into three different regions.

(1) The region where

$$s_{12}^2 > 0, \tag{4.7}$$

that is, $c^2(t_2 - t_1)^2 > r_{12}^2$ (the temporal component of the interval between events 1 and 2 is greater than the spatial component). We call such intervals the *timelike* intervals. In particular, when $\mathbf{r}_1 = \mathbf{r}_2$, that is, $\mathbf{r}_{12} = 0$ (both events occur in the same place), the 4-interval reduces to just the time interval between the events (a proper time!), multiplied by c.

(2) The region where

$$s_{12}^2 < 0, \tag{4.8}$$

that is, $c^2(t_2 - t_1)^2 < r_{12}^2$ (the spatial component of the interval s_{12} is greater than the temporal component). We call such intervals *spacelike*. In particular, when $t_1 = t_2$ (the two events are simultaneous), the interval is determined only by the distance between the events. We see from (4.1) and (4.5) that the spatial "contribution" to s_{12}^2 is the negative square of the distance r_{12}.

(3) The region where

$$s_{12}^2 = 0 \tag{4.9}$$

(the temporal and spatial components in the interval have equal weights). We call such intervals the *zero* or else the *null intervals* (the term "zero" or "null" is self-evident from definition (4.4)). They are also called the *isotropic* intervals. The term "isotropic" stems from ordinary three-dimensional geometry, where a vector $\mathbf{r}_{12} = 0$ reduces to a single point and therefore cannot be characterized by any specific direction. In this respect, it is isotropic. The reader must be careful about this part of the analogy, because a four-dimensional zero interval does not necessarily reduce to a single point! It can connect two *different* points. Such a pair of points represents two different events in space-time, which can be linked to one another by an agent moving with the invariant speed (e.g., a light signal). Accordingly, the corresponding interval is also called the *lightlike* interval. Thus, all four names (the *zero, null, isotropic,* and *lightlike*) denote the different "faces" of the same kind of interval. The zero length of this kind of interval reflects the fact that the squares of its temporal and spatial parts, being equal in magnitude, enter the interval with the opposite signs.

Because of the invariance of the interval, its belonging to one of the three different types is an absolute characteristic, that is, it does not depend on choice of a reference frame.

Representing the physical events as points of the four-dimensional space-time allows us to visualize the relations between different events by using space-time diagrams. We cannot adequately image all four dimensions of Minkowski's world on a sheet of paper having only two dimensions. We will, therefore, plot on the graph only one spatial dimension x and the time dimension ct. Physically, it means that we consider only the set of events happening on one straight line (the x-axis). Looking at it slightly differently, we may say that we consider the history of this axis.

Pick up one of the events in this history (call it event O) as a reference event. It will serve as an origin for measuring time ct and the spatial coordinate x. Draw ct-axis through this point up perpendicularly to the x-axis (Figure 4.1). Let us take the directions indicated by arrows in Figure 4.1 as positive. Now, all the events that happen at the "origin" of the line x after the event O are represented by the points on the ct-axis above point O; all the events preceding the event O belong to the part of the ct-axis below this point. We say that the ct-axis forms *the world line* of the point $x=0$ (i.e., a set of all events successively happening at this point).

On the contrary, all the points of x-axis at the moment $t=0$ are represented by the x-axis itself. We may call it a world line of all events on x that are simultaneous with the event O, or just simultaneity line in K. Alternatively, we may call it a world line of a hypothetical superluminal particle that traces out all the line x at one moment $ct=0$ (it must therefore move with an infinite speed).

Consider now an event at the point $x=0$ at the moment $ct=1$ LS. It will be represented by a point O′ on the ct-axis. The world line of all the events on x, which are simultaneous with the event O′ in K, is the line parallel to x and passing through point O′. This is just the x-axis itself taken at the moment 1 LS of its history. We can also say that it is a world line of a fictitious superluminal particle zipped along the x-axis with an infinite speed at the moment 1 LS after the zero time. Applying this to all other moments of time, we see that all the (ct–x)-plane is a continuous set of straight lines parallel to x and representing all the x-axis at different moments of time. In other words, it is the *worldsheet* traced out by the x-axis through space-time. Alternatively, we can say that the (ct–x)-plane is a continuous set of world lines of all points on the x-axis (Figure 4.2).

If we attempt to take into the picture the y-axis (perpendicular to both ct and x), then the world lines of all the photons moving in the xy plane and passing through the origin (event O) form the generatrices of a conical surface with the open angle 90° (Figure 4.3). This surface is called the *light cone*. Because there is also the z-axis,

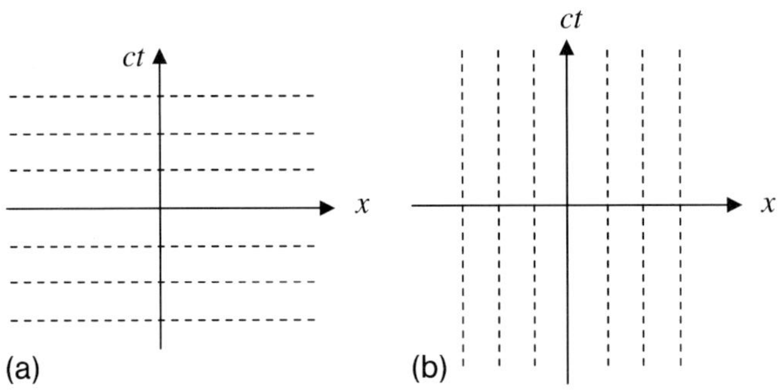

Figure 4.2

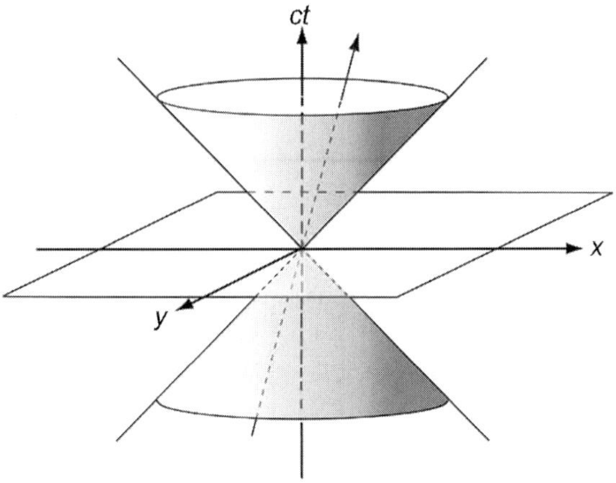

Figure 4.3

the (ct, x, y) light cone forms a sort of "hypersurface" of the four-dimensional Minkowski's world. If you want to include the z-direction to make the complete description, you have to consider all the photons moving in three-dimensional space (x, y, z) and passing through the origin at the zero moment. Their world lines will form a three-dimensional conical surface in four-dimensional space-time. Mathematicians call this type of surface "hypersurface." We can try to somehow imagine it, but we cannot depict it in real space.

From the geometrical viewpoint, any world line in space-time is just a graph of motion of some particle or process. It forms a trajectory in space-time. We should not confuse it with purely spatial trajectory of the given motion. For example, the *space-time* trajectory (the world line) of the stationary particle at point $x = 0$ is the ct-axis. The purely *spatial* trajectory of the same point is the point $x = 0$ itself. The spatial trajectory is in this case reduced to a point in space (because the particle is stationary!). The spatial trajectories of the photons moving through the origin in the plane (x, y) all lie in this plane, whereas their world lines all lie on the surface of the light cone in Figure 4.3.

Let us now plot on our diagram all possible types of intervals connecting the event O with other events. We will then see that the three types of intervals indicated above fill in three different regions of Minkowski's space-time.

Indeed, all timelike intervals (of the type (4.7)) fall inside the light cone. The spacelike intervals (of the type (4.8)) lie outside the light cone. And finally, all zero intervals (of the type (4.9)) lie along generatrices of the light cone itself. Thus, three different types of intervals analytically distinguished by the criterion (4.7)–(4.9) correspond to their locations in geometrically different domains. This geometrical difference, in turn, corresponds to fundamental physical difference between them. We can see it with full clarity if we write down the expression (4.5) for an interval in terms of the speed of corresponding object or process connecting its end points

$$v = \frac{|\mathbf{r}_2 - \mathbf{r}_1|}{t_2 - t_1} = \frac{r_{12}}{t_2 - t_1}. \tag{4.10}$$

We will have

$$s_{12}^2 = c^2(t_2 - t_1)^2 - r_{12}^2 = c^2(t_2 - t_1)^2\left(1 - \frac{v^2}{c^2}\right) = c^2(t_2 - t_1)^2\gamma^{-2}(v). \quad (4.11)$$

It follows that for the timelike intervals $\gamma^2(v) > 0$ and the speed $v < c$; for the null intervals $\gamma(v) = \infty$ and $v = c$; finally, for spacelike intervals, $\gamma^2(v) < 0$, which corresponds to $v > c$. The intervals of the first two types lie inside or on the surface of the light cone in Figure 4.3. They connect the events, which can be either the cause or the effect of the event O. Consider, for example, an event N within the light cone below O (i.e., $t(O') < t(O)$, Figure 4.3). It can be connected with O by a line O′O that might be a world line of a subluminal particle and thus affect the event O. In other words, the event O can in this case be caused by event O′. Since causal relation between two events is the same for all observers, all of them will agree that event O′ (possible cause) precedes the event O (possible effect). Therefore, any point O′ within the lower fold of the light cone is said to be in the *absolute past* of point O. Similarly, any point O″ within the upper hold of the light cone is in *absolute future* of point O.

In contrast, the intervals of the type (4.8) (the spacelike intervals) lie outside the light cone; they may connect events only by a process characterized by superluminal speed $v > c$. As we will see later, this type of a process cannot be used for signaling; therefore, there is no causal connection between the corresponding events. The absence of such connection is absolute for all observers. Therefore, an event P outside the light cone with vortex at O is said to be *absolutely remote* from O. One and the same event absolutely remote from O can be observed as occurring earlier or later than the event O, depending on a chosen reference frame. In particular, we can find a system of reference in which the event occurs simultaneously with O. The concept of "earlier–later" is relative for such pairs of events.

We will illustrate this relativity graphically in the next section.

Example

Let us see, how the concept of interval can be used to derive the formula for proper time. Consider two successive events A and B on a body moving with constant speed V relative to a reference frame K. The proper time between the events is the difference $\Delta t' = t'_B - t'_A$ between the moments t'_A and t'_B of these events read by a clock K′ attached to the moving body. Since in K′ both events occur at the same place, we have $x'_{AB} = x'_B - x'_A = 0$, and the interval between the events in this reference frame is reduced to $s'^2_{AB} = c^2(\Delta t')^2$. The clocks in the stationary reference frame K read the moments t_A and t_B for the events A and B, respectively, so that the time interval between the events is $\Delta t = t_B - t_A$. In this frame, the events are spatially separated by the distance $x_{AB} = V\Delta t$. The four-dimensional interval between the events expressed in terms of the coordinates of system K is $(s_{AB})^2 = c^2(\Delta t)^2 - V^2(\Delta t)^2$. Because of the invariance of the interval, we have $c^2(\Delta t')^2 = (c^2 - V^2)(\Delta t)^2$ or

$$\Delta t' = \gamma^{-1}(V)\Delta t. \quad (4.12)$$

This is the result (2.18) obtained in different way in Section 2.5: the moving clock ticks slower than a stationary one.

Now consider the Lorentz contraction effect. We have a rod of proper length l_0 (system K′) moving longitudinally with a speed v relative to K. Find its length in K.

We know that measuring the length of a moving body requires recording the *instant* positions of its front and rear in K. The distance between the obtained markings is the length l of the body in K. Since the time interval between the recordings in K is zero, the corresponding expression for the interval reduces to

$$\Delta s^2 = -l^2 \tag{4.13}$$

On the other hand, because of the invariance of the interval, we have $(\Delta s')^2 = \Delta s^2$, so that

$$(\Delta s')^2 = c^2(\Delta t')^2 - l_0^2 = -l^2. \tag{4.14}$$

Here, $\Delta t'$ is the time interval in K′ between the markings. Now, the time difference in K′ between the two events simultaneous in K can be obtained from Equation (2.55)

$$\Delta t' = -\gamma(v)\frac{v}{c^2}\Delta x = -\gamma(v)\frac{v}{c^2}l. \tag{4.15}$$

Putting this into (4.14) yields

$$\gamma^2(v)\frac{v^2}{c^2}l^2 - l_0^2 = -l^2, \tag{4.16}$$

or

$$l = \frac{l_0}{\gamma(v)}. \tag{4.17}$$

4.2 Rotations in Space-Time (Representation in Euclidean Plane)

Here, we will use the approach developed in the previous section for geometrical representation of *two* different RF. We are going to picture the basis axes of the two systems in one graph using Minkowski's diagrams. We assume, as before, that both systems have their x, x'-axes, respectively, lined up with the direction of their relative motion (say, direction of motion of K′ relative to K). The local clocks at the origins of the respective systems read zero time at the moment of their instant coincidence.

We have already noticed that the time axis of an inertial system is coincident with its origin's world line. Let now a real particle move uniformly along the x-axis with a speed v and pass the point $x=0$ at the moment $t=0$, so that

$$x = vt = \frac{v}{c}\cdot ct. \tag{4.18}$$

The particle is stationary in a reference frame K′ comoving together with it. If the particle is at the origin of this reference frame, Equation (4.18) describes the world

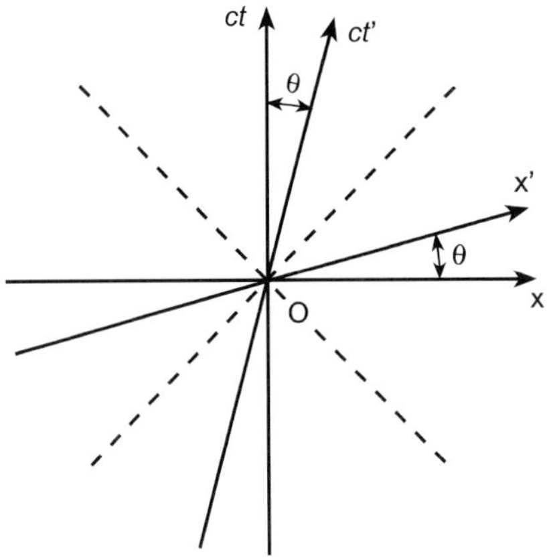

Figure 4.4

line of the origin and thereby determines the time axis ct' of the associated reference frame K′ *as viewed from K* (Figure 4.4). According to (4.18), the time axis ct' of system K′, when depicted in K, forms with ct an angle

$$\theta = \arctan\frac{v}{c}. \tag{4.19}$$

For positive v, this line is in the first and third quadrants of the (ct, x)-plane. For a particle moving in the negative x-direction we have $v < 0$, and the corresponding world line passes in the second and fourth quadrants of the plane (ct, x). If this particle is a photon, then $v = c$ and $\theta = \pm 45^\circ$. We see again that the world lines of photons moving in the positive or negative x-direction form mutually perpendicular bisectors of the angles between the coordinate axes.

Since the world line of the origin O′ represents the time axis of corresponding system K′, all the events happening at O′ after the moment $t' = 0$ are represented by the points on ct' above the origin; all the events that occur there before this moment are on the lower part of the line ct'.

Let us now turn to the x'-axis. How is it represented in our diagram? From the symmetry, we expect it to run symmetrical to ct' with respect to the photons' world line. This expectation is based on two facts: first, the photon's world line is the bisectrix of the coordinate angle (ct, Ox) in K; second, all the inertial FRs are equivalent. It follows that the same photon's world line has to be the bisectrix of the coordinate angle (ct', Ox') in K′.

It is easy to see that this is, indeed, the case. Earlier, you remember, we defined a spatial coordinate axis as the set of all events on this axis that are simultaneous with the event O. Because now we are interested in the x'-axis of moving system K′, we must consider the events on this axis that are simultaneous with the event O *by clocks of system K′*. All these events are characterized by one moment $t' = 0$ in readings of all K′-clocks. The set of these events determines the line of simultaneity in K′ with event $(0, 0)$. We can also say that this is a world line of a particle moving along the x-direction with an infinite speed $v' = \infty$ in K′. Setting $t' = 0$ in Equation (2.52), we find that in K

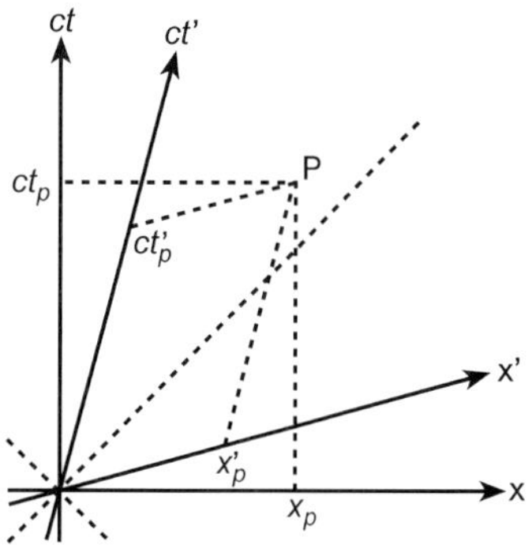

Figure 4.5

all these events occur at *different* moments depending on their coordinate x:

$$ct = \frac{v}{c}x. \tag{4.20}$$

But this is just the equation for the line passing through the origin and making with the x-axis the same angle as in (4.19). This line represents the x'-axis of system K′ from the viewpoint of system K (Figure 4.5). We can also say that it represents the world line of the just-mentioned fictitious superluminal particle. Since it is tilted with respect to x, we see that its speed in K is finite, although it remains, of course, greater than c.

Note that it is *not* an actual position of the x'-axis in space (the space in our example is only one dimensional). But it is an actual position in space-time of the set of all events on OX′, which are simultaneous in K′.

We see that the axes ct' and x' turn out to have been rotated through the same angle relative to ct and x', respectively (Figure 4.4). However, in contrast with the usual geometry of purely spatial rotations, the rotations here are in the *opposite senses* (this is yet another manifestation of the physical difference between space and time). As a result, the coordinate system (ct', x') seems skewed (deformed) from the viewpoint of system K (in exactly the same way the system (x, ct) would look distorted from the viewpoint of system K′). Precisely because of such a "deformation," the photon's world line remains, as it should, the bisector of the angle between the ct' and x' axes, that is, the speed of light in K′ is also equal to c. But now the photon's world line in Figure 4.4 makes the angle $\alpha = 45^\circ - \theta < 45^\circ$ with the axes ct', x'. This is due to the fact that these axes are plotted in the system K, which is "alien" for them. At $v \to c$ the angle θ approaches 45° and α approaches zero, that is, the axes ct' and x' approach the photon's world line; but ct'-axis always remains inside, while x' remains outside of the light cone.

If the relative speed v were negative (the system K′ were moving to the left of K), the axes ct', x' would make an obtuse angle. They can be considered as obtained from ct, x by rotating *away* from each other through the same angle θ.

Consider an event P (Figure 4.5). Its coordinates (ct_P, x_P) in K are the normal (orthogonal) projections of point P on the axes ct, x, and coordinates (ct'_P, x'_P) in K′ (when depicted on the *same* plane) are the oblique projections of this point on the axes ct', x'.

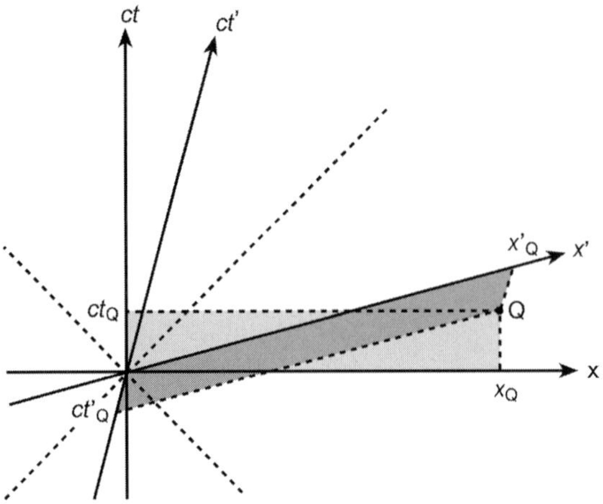

Figure 4.6

We can now consider the relation "earlier–later" between different events graphically, from the viewpoint of the systems K and K′. Let, for example, the event P lie inside the light cone with the vertex at O. It means that the events O and P admit causal connection. If $ct_P > 0$ (P lies in the upper fold of the light cone), then event P can result as an effect (the consequence) of event O. Projecting point P onto the axis ct', we see that at any possible tilt of this axis, projection ct'_P lies in the upper semiplane, that is, $ct'_P > 0$. The event P occurs after event O for all possible observers. As mentioned before, it is in the *absolute future* with respect to O.

If the point P lies in the lower fold of the light cone, then event P can be the cause of the event O and, accordingly, it will be observed *before* O in any reference frame. In this case, it is in the *absolute past* with respect to O.

Consider now another point Q lying *outside* the light cone (Figure 4.6). Let this point lie above the x-axis, that is, its projection $ct_Q > 0$. Physically, it means that Q occurs later than O in the inertial system K. But the projection ct'_Q of the very same point onto the ct'-axis can be negative (this will be the case for any system with the x'-axis passing above Q, as seen from Figure 4.6). Physically, it means that in corresponding inertial system K′, the event Q occurs earlier than O. The relation earlier–later is not invariant for pairs of events like O and Q. For such events, one can always find a reference frame for which their time ordering is switched to the opposite. And there exists such a system, for which both events of the pair are simultaneous ($ct_Q = 0$). The x'-axis of this system passes right through Q. Accordingly, the interval between the two events in this system is determined entirely by their spatial separation r'_{OQ}. According to the definition in the previous section, all the events outside the light cone can be called *absolutely remote* with respect to the event O. They are connected with O by spacelike intervals.

It is easy to see that any communications between the event O and an event outside the light cone would involve signals moving faster than light in vacuum. We will discuss later the relation between such signals and a fundamental physical principle of causality.

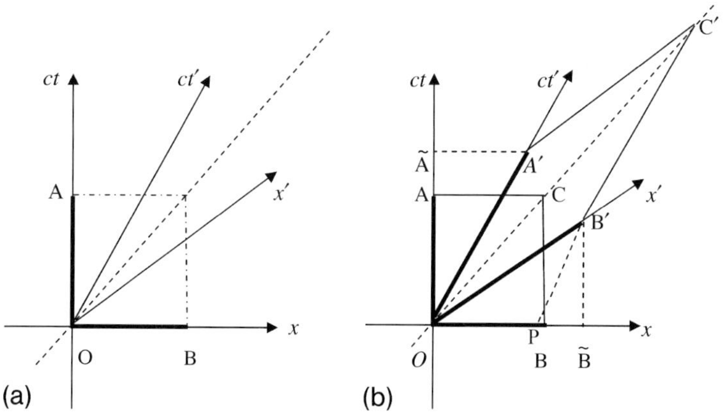

Figure 4.7

So far we have discussed only the angular relations between the respective axes of the two systems. Our next question is about the *metric* relations between them, that is, the ratio of the corresponding segments representing the equal time or space intervals. Suppose the segment OA of length τ along the *ct*-axis represents one temporal unit in K, that is, for instance, value of *ct* for $t = 1$ s (1 LS, Figure 4.7a). Suppose now that an observer in K′ plots the same unit on his axis *ct*′. What is the length $OA' = \tau'$ of the corresponding segment representing one temporal unit in K′, *on the same graph*, where mutually perpendicular axes *ct* and *x* represent time and distances *observed in K*? In other words, where on the *ct*′-axis (as drawn in K) should we put the mark A′ representing 1 light second by the clock C′ at the origin of system K′?

A similar question can be asked about the segment OB along the *x*′-axis, which represents the same unit length, in our case, 1 light second of distance. What length $OB' = \tau''$ along the *x*′ will represent this unit?

These would be trivial questions in Euclidean geometry of the 3-space – a rotation in a Euclidean plane changes only the orientation of a figure, not its shape or size. Therefore, OA and OB would remain orthogonal and equal, that is, $\tau' = \tau'' = \tau$. But in the pseudo-Euclidean geometry of space-time, where the basis axes are turned *toward* each other under the corresponding rotation, the outcome is not so straightforward. After a little thought, however, we can make a qualitative prediction on the basis of two assumptions: first, the symmetry between both dimensions; and second, the invariance of certain geometrical characteristics. The first assumption suggests that rotation, even if changing the sides OA and OB of the initial square, must change them by the same factor, so that τ' and τ'', while no longer equal to τ, will be equal to each other. In other words, it will take a square to a rhombus. Now, consider the second assumption – the invariance. On a Euclidean plane, the length of a segment and the area of a region remain invariant under rotation. This must remain true under rotations in the pseudo-Euclidean plane in space-time, and we know that the kinematic "length" of a segment (the 2D interval *s*) does, indeed, remain invariant under rotations. The same must be true for the area and, therefore, we may assume that the Euclidean area τ^2 of the original square with the side τ on the axes ct and x must remain unchanged under rotation. But, by the first assumption, the pseudo-Euclidean rotation takes the square to the rhombus. Combining the two, we see that the rotation takes the square to the rhombus of the same area. In the depicted case, corresponding to the motion of K′ in the *positive* x-direction, this rhombus has the acute angle between ct' and x'. Our final qualitative prediction is, that, in order to maintain the area of this rhombus, its sides (representing each *the same* length τ in K′) must be *greater* than τ.

Now, let us see how relativity works in ensuring this. Imagine Tom sitting at the origin of K′ and handling the local clock there. Precisely when his C′-clock reads 1 s after the zero time, this reading is recorded by Alice in K. Since the moving clock ticks slower by a factor of $\gamma(v)$, its 1 s corresponds to γ seconds on the local clock in K instantly coincident with C′. Thus, the position A′ must be such that its normal projection $O\tilde{A}$ onto the ct-axis is equal to $\gamma\tau$ (Figure 4.7b). Since the angle between the corresponding axes is θ, we have

$$\tau' = \frac{\gamma\tau}{\cos\theta} \equiv \gamma\tau\sqrt{1+\tan^2\theta}. \qquad (4.21)$$

Putting here $\tan\theta = v/c \equiv \beta$, we obtain from (4.19)

$$\tau' = \tau\sqrt{\frac{1+\beta^2}{1-\beta^2}}. \qquad (4.22)$$

As we had expected, τ' turns out to be, indeed, greater than τ. But here we have already its quantitative evaluation.

Let us now turn to the side OB′. By definition, OB′ represents the "meterstick" (1 LS long) oriented along the x'-axis of system K′. The point B′ represents the instant position of its leading edge at the moment $t' = 0$ of Tom's time (which corresponds to a certain *nonzero* moment of Alice's time). The line parallel to ct' and passing through B′ is the world line of this edge. The intersection P of this line with the x-axis is an oblique projection of B′ onto this axis. Physically, the event P is the marking by Alice of the instant position of the stick's leading edge at the zero moment of Alice's time. Since at this very moment the trailing edge is passing the origin O, the distance OP is the Lorentz-contracted length of the stick in K. Since the proper length of the stick is τ, its Lorentz-contracted length is $\gamma^{-1}\tau$. Now, consider the triangle OPB′. The angles POB′ and OPB′ are, respectively, θ and $\pi/2+\theta$ by the initial conditions. Then, denoting $OB' = \tau''$ and applying the sine theorem, we can write

$$\frac{\tau''}{\sin((\pi/2)+\theta)} = \frac{\gamma^{-1}\tau}{\sin((\pi/2)-2\theta)},$$

or

$$\tau'' = \gamma^{-1}\tau\frac{\cos\theta}{\cos 2\theta} = \gamma^{-1}\tau\frac{\cos\theta}{\cos^2\theta - \sin^2\theta}. \qquad (4.23)$$

Expressing again the sine and cosine of an angle in terms of its tangent and using $\tan\theta = \beta$, we obtain

$$\tau'' = \tau' = \tau\sqrt{\frac{1+\beta^2}{1-\beta^2}}. \tag{4.24}$$

As one would expect, it is the same as in (4.22). Thus, the laws of relativity do ensure the assumed conservation of symmetry with respect to a photon's world line. Since this line serves as a diagonal of a square, as in Figure 4.7, then the rotation in the (ct, x) plane transforms the square to a rhombus with its diagonal along the same line. Let us call this diagonal longitudinal. Then, the other diagonal, which is perpendicular to the photons' world line, can be called transverse.

Now, it remains to prove that the area $\Delta A'$ of this rhombus is equal to the area $\Delta A = \tau^2$ of the initial square.

We have

$$\Delta A' = (\tau')^2 \sin 2\varphi, \tag{4.25}$$

where $\varphi = (\pi/4) - \theta$ is the angle between a side of the rhombus and the photon's world line.

Putting this into (4.25) gives

$$\Delta A' = (\tau')^2 \cos 2\theta = (\tau')^2\left(\cos^2\theta - \sin^2\theta\right) = (\tau')^2\frac{1-\tan^2\theta}{1+\tan^2\theta} = (\tau')^2\frac{1-\beta^2}{1+\beta^2}. \tag{4.26}$$

Combining this with (4.24), we obtain

$$\Delta A' = \tau^2 = \Delta A. \tag{4.27}$$

Thus, the laws of relativity also ensure the area conservation under rotations. Generally, we can conclude that if a reference frame K is represented graphically by mutually perpendicular axes ct, x, y, z, then the Lorentz transformation to another RF K′ moving along the x-axis takes a square OACB in the (ct, x)-plane to the rhombus OA′C′B′ with the same area, the same directions of its diagonals, and the sides given by (4.24).

Exercise 4.1

What is the length $O\tilde{A}'$ of the oblique projection of segment OA′ onto the ct-axis?

A good way to answer this question is to consult with the K′ observer. By reciprocity, this observer will say that the segment OA′ representing his temporal unit τ is the normal projection of $O\tilde{A}$ onto the ct'-axis (Figure 4.8). He knows that point $\tilde{A}$ is the reading of our clock sitting at the origin of K. He also knows that this clock is ticking γ times slower than his clocks, because it is moving with respect to him with a speed v. Since the moment $O\tilde{A}'$ corresponds to the unit τ of his time, he concludes that $O\tilde{A}'$ must be equal to $\gamma^{-1}\tau$. We accept this conclusion as well.

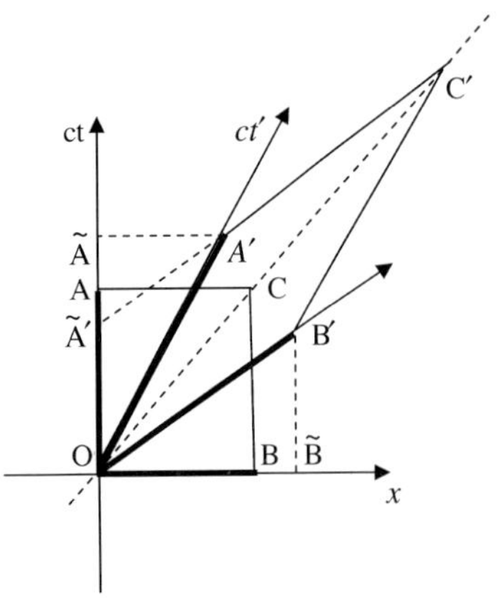

Figure 4.8

An interesting corollary of this result is that the side of our square OACB turns out to be the geometric mean of $O\tilde{A}$ and $O\tilde{A}'$ (Figure 4.8)

Exercise 4.2

Suppose we launch a continuous set of spaceships simultaneously from the origin O of K, all along the x-axis, but with different speeds, so that speed v becomes a continuous variable ranging within $-c < v < c$. Accordingly, we obtain a continuous set of RF, each represented by its respective rhombus, so that each rhombus can be characterized by this variable v. Find the locus of the end points of the transverse diagonals of these rhombuses.

Consider an event represented by the upper end point of such a diagonal. As shown in Figure 4.7, the ct-coordinate of this point is $\gamma\tau$ and the corresponding x-coordinate is $\tau' \sin\theta$. Using $\tan\theta = \beta$ and Equation (4.22), we obtain two equations

$$\begin{aligned} ct &= \gamma\tau = \frac{\tau}{\sqrt{1-\beta^2}}, \\ x &= \tau\sqrt{\frac{1+\beta^2}{1-\beta^2}}\frac{\beta}{\sqrt{1+\beta^2}} = \beta\gamma\tau. \end{aligned} \tag{4.28}$$

Elimination of β from these equations gives

$$(ct)^2 - x^2 = \tau^2. \tag{4.29}$$

This is the equation of a hyperbola with its symmetry axis along the ct-axis (Figure 4.9).

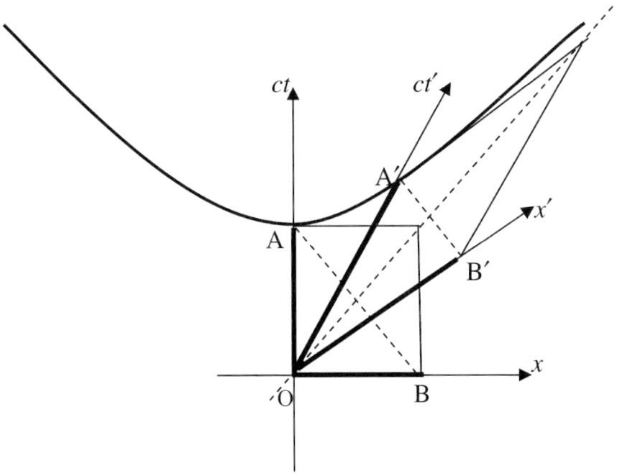

Figure 4.9

4.3 Lorentz Transformations as Rotations

The simple case we have considered so far involved two inertial RF with their respective axes parallel to each other, in relative motion parallel to the x, x'-direction. The corresponding Lorentz transformation involves two variables, x and ct, transformed into x' and ct'. This transformation is linear and conserves the interval

$$s'^2 = c^2t'^2 - x'^2 = c^2t^2 - x^2 = s^2. \tag{4.30}$$

Were it not for the minus sign, with which the square of x (x') enters the equation, we would say that $s = s'$ is the distance of the corresponding point from the origin. We know from geometry that a transformation in a two-dimensional (2D) space, which involves both coordinates, is linear and conserves the distances, is rotation. It was therefore natural to suppose, as we in fact did in the previous section, that (4.30) represents a "distance" in a special kind of 2D space, and the Lorentz transformations are rotations in this space. Here we will express this analogy analytically.

The indicated analogy is close to, but not exactly compatible with, rotations in a Euclidean plane. Consider two Cartesian coordinate systems x, y and x', y' rotated through an angle θ with respect to one another (Figure 4.10). The reader can easily verify (Problem 4.2) that the coordinates of a point P in the respective systems are related by

$$\begin{aligned} x' &= (\cos\theta)x + (\sin\theta)y, \\ y' &= (-\sin\theta)x + (\cos\theta)y \end{aligned} \tag{4.31}$$

or, in the matrix form,

$$\begin{pmatrix} x' \\ y' \end{pmatrix} = \begin{pmatrix} \cos\theta & \sin\theta \\ -\sin\theta & \cos\theta \end{pmatrix} \begin{pmatrix} x \\ y \end{pmatrix}. \tag{4.32}$$

Comparing this with the Lorentz transformation (2.52) or (2.54) does not seem to show any impressive analogy. The coefficients in the Lorentz transformations are the

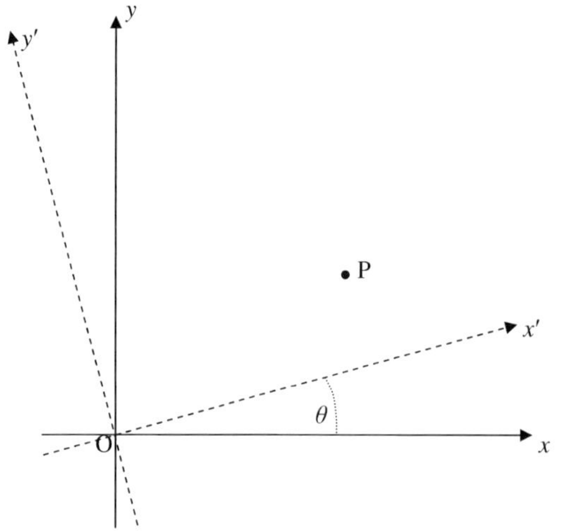

Figure 4.10

irrational functions of relative velocity between two inertial RF, whereas the coefficients in the 2D rotations are quite different (trigonometric) functions and depend on a different variable – rotation angle θ.

But a closer look reveals something interesting. If we express sine and cosine functions in (4.32) in terms of tangent of the same angle, we obtain

$$\begin{pmatrix} x' \\ y' \end{pmatrix} = \frac{1}{\sqrt{1+\tan^2\theta}} \begin{pmatrix} 1 & \tan\theta \\ -\tan\theta & 1 \end{pmatrix} \begin{pmatrix} x \\ y \end{pmatrix}. \tag{4.33}$$

Further, let us set

$$\tan\theta \equiv \frac{v}{c} \equiv \beta. \tag{4.34}$$

Then

$$\begin{pmatrix} x' \\ y' \end{pmatrix} = \frac{1}{\sqrt{1+\beta^2}} \begin{pmatrix} 1 & \beta \\ -\beta & 1 \end{pmatrix} \begin{pmatrix} x \\ y \end{pmatrix}. \tag{4.35}$$

In order to compare with the Lorentz transformation in the form (2.52), let us redenote in (4.35) $y \to ct$ and change the ordering of the variables; instead of (x, ct) we will write (ct, x) just as we do with time and space coordinates. Equation (4.35) will then take the form

$$\begin{pmatrix} ct' \\ x' \end{pmatrix} = \frac{1}{\sqrt{1+\beta^2}} \begin{pmatrix} 1 & -\beta \\ \beta & 1 \end{pmatrix} \begin{pmatrix} ct \\ x \end{pmatrix} \tag{4.36}$$

Compare this with the Lorentz transformation in the form (2.52):

$$\begin{pmatrix} ct' \\ x' \end{pmatrix} = \frac{1}{\sqrt{1-\beta^2}} \begin{pmatrix} 1 & -\beta \\ -\beta & 1 \end{pmatrix} \begin{pmatrix} ct \\ x \end{pmatrix}. \tag{4.37}$$

Now we can see a considerable similarity. The Lorentz transformation (4.37) for ct, x, relating them to the coordinates ct', x' of the system moving with velocity V relative to K along the x-axis, is similar to rotation in the plane ct, x through the angle related to velocity by Equation (4.34).

Still, the analogy, while being essential, is not complete. The difference is that the transformation matrix is symmetrical for the Lorentz transformation but antisymmetrical for Euclidean rotation. This difference can be eliminated if, instead of rotating both axes in *the same* direction, (true rotation) we rotate them in the *opposite* directions. For instance, when x is rotated counterclockwise, ct should be rotated clockwise. The resulting coordinate system (ct', x') will now be skewed rather than rotated; but from the mathematical viewpoint, the coordinate transformation will look like that of the rotation, with the only exception that the sign difference in nondiagonal elements of the transformation matrix will disappear. In all other respects, the transformation retains the properties specific for rotation – it remains linear and conserves the radial distances. Therefore, from a mathematician's viewpoint, it can be considered as a "generalized rotation." Let us accept this view, for its mathematical elegance and consistency.

Finally, we have the minus sign in the Lorentz factor in (4.37), whereas it is plus in (4.36). This difference reflects the fact that (4.35) describes a rotation in the Euclidean plane, preserving the form

$$r^2 = x^2 + y^2 = x'^2 + y'^2 = r'^2, \tag{4.38}$$

whereas the Lorentz transformation preserves the form (4.30).

The invariance of the form of the type (4.30) is characteristic of the so-called *pseudo-Euclidean* or *hyperbolic* (or else, Lorentzian) geometry. The corresponding rotation preserving the form (4.30) can be expressed in terms of the *hyperbolic* functions of the rotational angle, rather than trigonometric functions $\sin\varphi$ and $\cos\varphi$:

$$\mathrm{sh}\,\phi \equiv \frac{1}{2}(\mathrm{e}^{\phi} - \mathrm{e}^{-\phi}), \qquad \mathrm{ch}\,\phi \equiv \frac{1}{2}(\mathrm{e}^{\phi} + \mathrm{e}^{-\phi}), \qquad \mathrm{th}\,\phi \equiv \frac{\mathrm{sh}\,\phi}{\mathrm{ch}\,\phi} = \frac{\mathrm{e}^{\phi} - \mathrm{e}^{-\phi}}{\mathrm{e}^{\phi} + \mathrm{e}^{-\phi}}. \tag{4.39}$$

From the definition (4.39) of these functions it is easy to prove the following identities:

$$\mathrm{ch}^2\,\phi - \mathrm{sh}^2\,\phi \equiv 1, \qquad \mathrm{sh}\,\phi \equiv \frac{\mathrm{th}\,\phi}{\sqrt{1-\mathrm{th}^2\,\phi}}, \qquad \mathrm{ch}\,\phi = \frac{1}{\sqrt{1-\mathrm{th}^2\,\phi}}. \tag{4.40}$$

The second important difference between the rotations in a Euclidean and a pseudo-Euclidean plane is that, contrary to (4.35), the transformation matrix for hyperbolic rotation is symmetrical:

$$\left.\begin{aligned} x &= x'\mathrm{ch}\,\phi + y'\mathrm{sh}\,\phi, \\ y &= x'\mathrm{sh}\,\phi + y'\mathrm{ch}\,\phi \end{aligned}\right\} \quad \text{or} \quad \begin{pmatrix} y \\ x \end{pmatrix} = \begin{pmatrix} \mathrm{ch}\,\phi & \mathrm{sh}\,\varphi \\ \mathrm{sh}\,\phi & \mathrm{ch}\,\phi \end{pmatrix} \begin{pmatrix} y' \\ x' \end{pmatrix}. \tag{4.41}$$

Suppose again that the (ct-axis) of the chosen subset (ct, x) of space-time corresponds to the y-axis in the hyperbolic plane (y, x), that is, $y \leftrightarrow ct$. Then, using the properties (4.40), it is easy to show that the Lorentz transformation is indeed equivalent to rotation in the pseudo-Euclidean plane (ct, x), with the rotational angle

$$\phi = \operatorname{arcth} \beta. \tag{4.42}$$

Accordingly, the parameter β of the Lorentz transformation is a hyperbolic function of φ considered as an angle of rotation

$$\beta = \operatorname{th} \phi. \tag{4.42a}$$

This is consistent with Equations (3.25) and (3.26) obtained in Section 3.2 for the addition of parallel velocities.

All this seems now self-consistent and beautiful, but still rather abstract. We can make it more intuitive. In the previous section, we depicted the relation between the two different RF in the Euclidean plane (what else could we do on a sheet of paper?). Accordingly, the angle θ in (4.19) is not the best variable that can be used to describe rotation in a *pseudo*-Euclidean plane. We can see it from the following argument. Suppose that Tom rushes past Alice in a spaceship whose speed V is comparable to c. Then its temporal ct' and spatial x' axes depicted on a sheet of paper used to represent Alice's RF look as the sides of a rhombus (Figure 4.9). According to the results of the previous section, the angle θ between their respective temporal axes is

$$\theta = \operatorname{Arctan}\frac{V}{c} \to \operatorname{Arctan} 1 = 45^\circ \text{ as } V \to C. \tag{4.43}$$

Let a third observer, Peter, slide relative to Tom in the same direction with a speed V' also close to c. Then, by the same argument, Peter's ct''-axis will make an angle θ' close to 45° with *Tom's* ct'-axis.

In Euclidean geometry, if we rotate an axis through an angle θ and then rotate again through an angle θ' in the same plane, the resulting angle between the initial and final directions of the axis (in our case, the angle between the axes ct and ct'') will be the sum $\theta + \theta'$ of the individual angles. Accordingly, if we try to depict graphically the position of Peter's ct''-axis on *Alice's* (ct, x)-plane, it should make an angle close to 90° with the ct-axis. But it will not! We know that Peter's velocity relative to Alice

$$V'' = \frac{V + V'}{1 + (VV'/c^2)} < c. \tag{4.44}$$

It remains less than c. Accordingly, the ct''-axis cannot make an angle with ct, which would exceed 45°. In other words, it cannot go beyond the photon's world line. As the invariant speed c is the unattainable barrier for all massive objects, its representative – the photon's world line – is unattainable position for the ct'-axes of any RF made of massive objects. Since rotation is a continuous operation, the c-barrier has to be attained before being passed. Once the former is impossible, the latter becomes impossible. We see that a succession of rotations, which can amount to an arbitrary large angle, is represented by an angle θ that is always less than 45° on the Alice's (Euclidean) plane and, in any event, is less than the sum of individual angles. In other words, the angle we have used in the previous section is an additive variable under Euclidean rotations, but not an additive variable under space-time rotations!

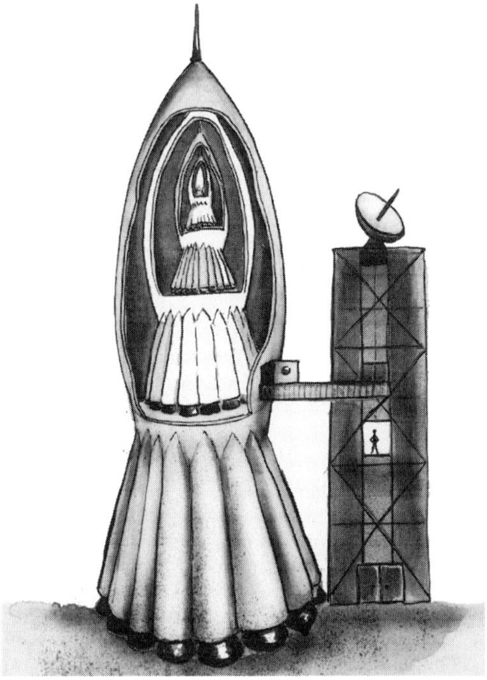

Figure 4.11 Mr. X uses an elevator to board his rocket containing a set of ever smaller rockets inside of one another. His invention is a prototype of a well-known multistage rocket used in space exploration. Such a rocket is essentially a set of few rockets of decreasing size atop of one another, and the astronauts are from the very beginning positioned inside of the spaceship on the top rocket.

The question arises how to find an appropriate variable that would be additive under space-time rotations?

Imagine an extravagant gentleman named Mr. X. He has practically unlimited financial resources, but very limited understanding of science. He thinks that the theory of relativity is just an intellectual curiosity with no roots in the real world and bashfully believes that he can reach and even supersede the speed of light by creating a sufficient amount of fast rockets and boosting them in a certain sequence using each previous rocket as a launching pad. He does acquire a set of such rockets of various sizes but all equally fast (making each $0.9c$ with respect to its respective launching pad) and installs them one inside the other, as in a Russian "matryoshka" (Figure 4.11). He starts from system K where he shoots a laser pulse and jumps into the big rocket containing all the rest. After the boost in the direction of the pulse he measures the speed of the pulse and still finds it equal to c. He then jumps into the second rocket and launches it in the same direction in the hope that this time the pulse would make at least $0.1c$ relative to that rocket. Instead, he sees again the pulse moving at $v = c$, as if nothing happened. Slightly embarrassed, he jumps into the third rocket, but after the boost sees again the pulse moving at the speed c. He gets mad and jumps from one rocket to another until he uses the last of them – with the same outcome. He radios back to Alice and Tom, who remain in K, asking them to tell what they observe. They answer that with each new boost, his new resulting velocity, instead of being the sum of the previous boost velocities, is greatly devaluated because of the factor $(1+(VV'/c^2))^{-1}$ in the velocity addition law. Then they make their own calculation.

They represent Mr. X's every jump from one rocket to another geometrically as rotation through an angle $\Delta\theta_i$, with i ranging from 1 to N, where N is the total number of jumps. In Euclidean geometry, a succession of rotations through individual angles $\Delta\theta_i$ is equivalent to one rotation through an angle

$$\theta_N = \sum_{i=1}^{N} \Delta\theta_i. \tag{4.45}$$

We express this property by saying that such operations form a group, and each individual rotation, as well as their sum, is an element of this group.

But, when viewed by Alice and Tom from K, the rocket-jumper will remain subluminal no matter how many jumps he performs. To them, the resulting angle is not equal to their sum (4.45). They try to find instead of θ another variable ϕ such that when the resulting speed β is expressed in terms of ϕ, the function $\beta(\phi)$ approaches 1, the angle ϕ goes to infinity rather than to 45° on the Alice's graph. The way to such variable clarifies when we compare it with traditional (Euclidean) relationship

$$\beta = \tan\theta = \frac{1}{i}\frac{e^{i\theta} - e^{-i\theta}}{e^{i\theta} + e^{-i\theta}}. \tag{4.46}$$

We immediately notice that if we drop the i's on the right of (4.46) and, accordingly, change notations $\theta \to \phi$, we will get

$$\beta = \frac{e^{\phi} - e^{-\phi}}{e^{\phi} + e^{-\phi}} \Rightarrow \begin{cases} 1, & \phi \to \infty, \\ 0, & \phi \to 0, \\ -1, & \phi \to -\infty. \end{cases} \tag{4.47}$$

This is precisely what we wanted. And the expression in the middle of (4.47) is our old friend – the hyperbolic tangent function of ϕ. The inverse of this function gives us the thought-for variable

$$\phi = \operatorname{arcth}\frac{V}{c}. \tag{4.48}$$

This is identical to (4.42) and consistent with our previous result (3.26).

Thus, when Mr. X jumps from one rocket to another with a relative speed V, he, possibly to his own surprise, performs a rotation in space-time through an angle determined by Equation (4.48). This new angle is an additive characteristic of relative motion. It can be considered as a very relevant parameter whenever we deal with succession of boosts. In this case, the result can be represented by the parameter, which is the sum of the individual parameters associated with the respective boosts. In other words, it behaves in a succession of space-time rotations through consecutive angles $\Delta\phi_i$ according to the law

$$\phi_N = \sum_{i=1}^{N} \Delta\phi_i, \tag{4.49}$$

just as the angle θ does in (4.45) in the succession of Euclidean rotations in space.

The corresponding parameter ϕ is called the rapidity since it can be used as an *additive* characteristic of how rapid the resulting motion is after a succession of boosts.

4.4
What is Horizontal and What is Vertical?

Any rotation in 3-space x, y, z can be represented as a combination of three rotations in the planes xy, yz, and zx, respectively. Similarly, a rotation in 4-space can be decomposed into six rotations in six respective mutually perpendicular planes. In particular, a rotation in space-time can be represented as rotations in the planes xy, yz, zx, xt, ty, zt. The first three correspond to regular spatial rotations within *the same* RF. The last three rotations, involving the temporal dimension, correspond to transitions between *different* RF. Thus, as we have learned, the xt-rotation describes the relative motion of two RF with their respective axes x and x' lined up with the direction of their relative motion (Figure 2.2). Similarly, the t, y-rotation describes a transition from frame K to another RF moving along the y-direction and having its y'-axis parallel to y. Changing $y \to z$ in the above statement, we will obtain the true statement for space-time rotations in the zt-plane.

With this in mind, we want now to consider a more general case of *combined* space-time rotations in at least two different planes involving the temporal axis. This would correspond to an object or system moving in an *arbitrary* direction with respect to a fixed inertial reference frame K. If we attach a RF to this object, its origin would have a velocity **v** with at least two nonzero components (say, v_x, v_y) along the respective axes of frame K (Figure 4.12). Study of this situation shows some new interesting aspects of space-time.

Let us invite a character from book [1] (Mr. O'Bryan) for a double mission. First, we want him to board an aircraft carrier (frame K'_x) moving along the x-axis of the stationary RF K associated with the seashore. He is to watch a rising elevator in the carrier, with a passenger, a fish tank on the table, and a chandelier inside (Figure 4.13), and record his observations. This would constitute mission 1.

Second (mission 2), we want him to rise in a huge blimp (frame K'_y) at the same rate as the elevator in the aircraft carrier and watch the interior of a car sliding down the rails along the length of the blimp. The interior and the proper shape of the car are identical to that of the elevator in the first mission. The rail speed of the car along the x-axis is equal to the speed of the aircraft carrier. In other words, both – the elevator in the aircraft carrier and the car in the blimp have the same velocity **V** relative to the seashore (frame K). Therefore, they both constitute a part of the same inertial reference frame K″, moving with velocity **V** relative to K. And he should again

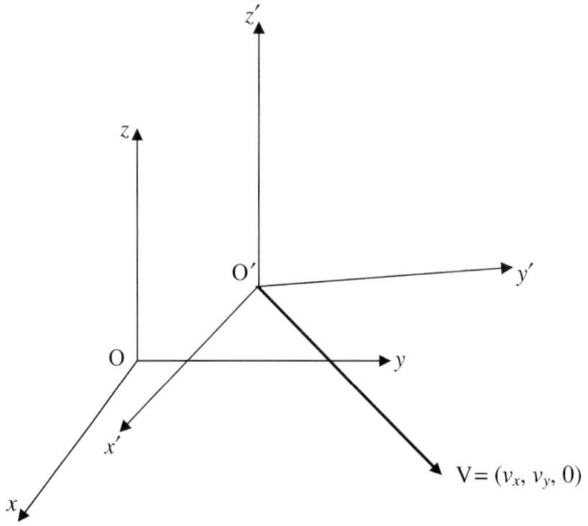

Figure 4.12 The origin O′ of an inertia! RF K′ is moving relative to K with velocity *V* parallel to the *x–y* plane. Note that the orientations of the primed (horizontal) axes may not be mutually orthogonal in K even if they are mutually orthogonal in K′ (or vice versa)!

record the data, now from his rising outpost (both the elevator and the aircraft-carrier are transparent and their interiors are open for observation from outside). After completing both missions, Mr. O'Bryan is assigned to compare his results with one another and with the observations carried out on the seashore.

When in frame K'_x associated with the uniformly moving aircraft carrier, Mr. O'Bryan notes that the water surface in the fish tank inside of the rising elevator remains horizontal, as it does when the ship is anchored. The same is true about all other horizontal surfaces in the rising elevator, so we will in further discussion use the water surface not only in its own right, but also as a representative of all other horizontal surfaces in this elevator (Figure 4.14a).

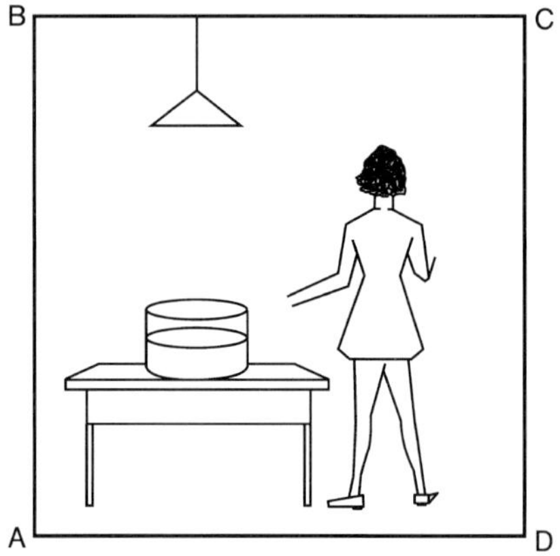

Figure 4.13

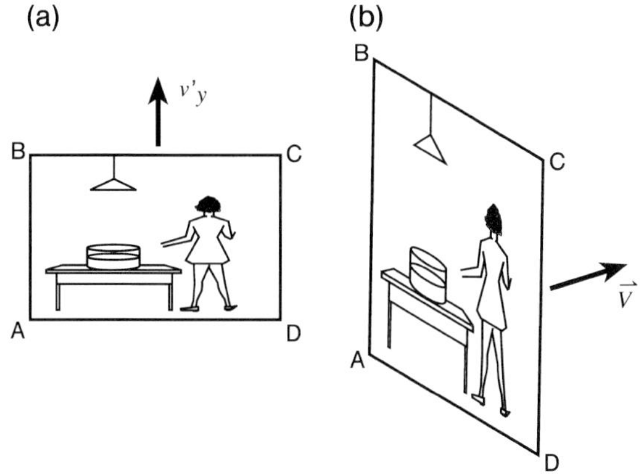

Figure 4.14

"What a sound manifestation of the principle of relativity," Mr. O'Bryan murmurs. "A ship's resident cannot tell whether the ship is moving or anchored, by observing the water surface in the rising elevator."

At this very time an observer on the seashore also watches the same process and notices an unusual phenomenon: the surface of the water rising with the elevator in the moving ship is tilted to the horizon, so that the water at the rear edge of the tank is higher than that at the front edge (Figure 4.14b). The faster the ship's motion, the steeper the tilt. It even occurs to the observer (who also happens to be a good engineer) that this phenomenon could be used for measuring the ship's speed. He writes the Lorentz transformation on a page in his notebook:

$$\Delta t' = \gamma(v_x)\left(\Delta t - \frac{v_x}{c^2}\Delta x\right),$$

and sets there $\Delta t = 0$ for the two simultaneous events at the opposite edges A and D of the elevator.

"What a sound manifestation of the relativity of time," the engineer thinks. "Two simultaneous events in my RF, a distance Δx apart along the ship's motion, are not simultaneous in the moving ship. They are separated by a time $\Delta t' = -\gamma(v_x)v_x\Delta x/c^2$, if v_x is the ship's speed. The points A and D that I mark *now* at the front and the rear edges of the elevator floor, are marked with the different moments of time by the ship's synchronized clocks. *Now* the ship's clock at D lags behind the ship's clock at A by the amount $\Delta t'$. Therefore, if the elevator's floor rises at a rate $v'_y = dy'/dt'$ by the ship's clock, its level at A must be higher than at D by the amount $\Delta y = \Delta y' = v'_y\Delta t'$, since it had more time to rise. Using the above expression for $\Delta t'$, one would get for Δy:

$$\Delta y = \gamma(v_x)\frac{v_x v'_y}{c^2}\Delta x = \gamma^2(v_x)\frac{v_x v_y}{c^2}\Delta x, \qquad (4.50)$$

where v_y is the rate of the elevator's rise in the shore reference frame. Thus, the elevator's floor, as well as the water surface in its tank as observed by me, must be inclined to *my* horizontal by the angle α_x:

$$\tan\alpha_x = \frac{\Delta y}{\Delta x} = \gamma^2(v_x)\frac{v_x v_y}{c^2} = \gamma(v_x)\frac{v_x v'_y}{c^2}. \tag{4.51}$$

If I know v'_y," the engineer concludes, "then measuring α_x (provided my instruments are sensitive enough) will give me a quantitative measure for the ship's velocity v_y through Equation (4.51)."

Now, let us read and interpret this equation. First, we must stress the difference from the apparently similar tilt of axis x' as depicted in Figure 4.4. Both tilts are intimately related to one another, but have different physical meaning. The x'-axis in Figure 4.4 represents the line of simultaneity in a moving system K′, which is a line *in space-time* (the set of events simultaneous in K′). Physically, the events themselves are all *on one line* – the x'-axis of K′ is sliding along the x-axis of K, so that both axes are coincident at any time. It is only when we unfold the events using also *temporal* dimension that x' acquires the tilt in the (ct', x') plane.

In contrast to that case, here we have the *real tilted position* of rising surface, a position *in space* as seen by the K-observer.

Equation (4.51) says that there is no tilt when $v_x = 0$. Makes sense, doesn't it? When the ship is anchored, the water surface in the rising elevator is horizontal at any moment for either Mr. O'Bryan or the engineer. If $v'_y = 0$ (the elevator is stationary in K'_x), there is no tilt either, which also makes sense. The water surface in the *stationary* (relative to the aircraft carrier) elevator is horizontal for both observers no matter how fast the ship moves so far as its motion remains uniform. It is only when the ship and the elevator in it are both moving that the effect is being observed by the engineer. When $v'_y < 0$ (the elevator in the ship is going down), the sign of $\tan\alpha$ is negative, which means that the water level as observed by the engineer, would be higher at the front edge than at the rear. Thus, the surface of the same pool of water is horizontal in one inertial reference frame (the ship) and may be inclined in another one (e.g., the seashore). The property of being horizontal turns out to be relative even for the two observers in the same locality, because of the relativity of time.

Some of the readers may be tempted to think that the phenomenon is of the same nature as one observed by Alice and Tom (recall the experiment with the aquarium in Figure 1.3). But the water tilt in that experiment was caused exclusively by the *acceleration*, not by the velocity. And this makes all the difference. The water tilt in the "aquarium effect" is determined by the ratio a/g, where a is the acceleration of Tom's reference frame and g is the acceleration caused by gravity. The effect is easy to observe since the acceleration a comparable to g is easy to achieve. And it was observed for even the still water (the aquarium did not rise, $v'_y = 0$!) by *both* – Alice and Tom. It is therefore of the effects that we would call absolute. It reveals the acceleration of a system to an insider. In contrast, the "elevator-in-the-ship" effect is observed only by the engineer when the elevator rises or descends ($v'_y \neq 0$!) and the ship moves *without* acceleration at a speed v_x. We have no g as a scaling constant here. Instead, a new constant – that of c – enters the picture. The water surface tilt is in this case determined by *two* ratios – v_x/c and v_y/c. In real conditions, both of them are small and, accordingly, the surface tilt α would be ridiculously small. For example, if the aircraft carrier rushes at the speed of 360 km/h (as a small plane!) and the elevator in it rises at a rate of 1 m/s, Equation (4.51) gives $\alpha_x \approx 1.1 \times 10^{-13}$. We see that we have greatly overestimated the precision of the engineer's devices in our thought experiment. But the equations we use are correct. For other conditions, they can give, as we will see further, quite noticeable values of α_x. And, what is most important, it is not associated with the liquidity of water. As was emphasized above, the water level was used as a representative of all the surfaces that are horizontal in the elevator. We could have applied the same argument to the bottom of the tank or to the surface of the desk on which it is resting, or to the floor of the rising elevator, and obtain the same result. All horizontal surfaces in K'_x, liquid or solid, are tilted in K if $v'_y \neq 0$, and the angle of tilt is given by Equation (4.51). This result describes a new relativistic effect that is seen and registered differently in different inertial frames: a rising (or sinking) horizontal line parallel to x in the frame K'_x is not parallel to x in frame K (Figure 4.14b).

On the other hand, the directions determined by Mr. O'Bryan as vertical in his frame K'_x remain vertical in K. For instance, as seen in the same figure, the chandelier hanging vertically and the elevator's passenger standing on the floor are vertical for both Mr. O'Bryan and the engineer. This observation is especially evident for the elevator's walls, since they are at any moment coincident with the walls of the elevator's shaft, which is not moving in K'_x. This follows directly from the fact that simultaneity is the same for all events in a plane perpendicular to the direction of relative motion of the two RF (in our case, K and K'_x).

As you think of the obtained result, it may seem very strange or even downright wrong. Indeed, a K-observer could focus entirely on the elevator and totally ignore the ship. It then seems to be evident that both components – v_x and v_y – of the elevator's velocity **v** must be on the same footing. Therefore, in the framework of this approach, we would be tempted to think that a line's tilt should: (1) be observed for a vertical line as well as for a horizontal one and (2) be described by an expression symmetrical with respect to both components of **v**.

Both these assumptions are at odds with our results. Figure 4.14b shows vertical directions unchanged, and the expression (4.51) for changing the horizontal directions is not symmetric with respect to the components of **v** due to the factor $\gamma(v_x)$.

To get a deeper insight into this phenomenon, let us now reassign Mr. O'Bryan to mission 2. Instead of accompanying the aircraft carrier, that is, moving along x with a speed v_x in K, he will now accompany the rising blimp; that is, he will move along the y-axis with a speed v_y in K. What happens in this case? Well, under this new condition, the blimp's car (which is an exact clone of the aircraft carrier's elevator) now remains on the same level with him and is only moving past him in the horizontal direction. Mr. O'Bryan watches the chandelier hanging freely from the suspension point. "What a sound manifestation of the relativity principle," he murmurs. "You cannot tell whether your blimp is ascending or just sitting on the shore by observing the passing chandelier; in both cases, the chain holding the chandelier remains vertical." (Again, there is nothing special in the hanging chandelier, and we will use it as a representative of *all* verticals in the blimp.)

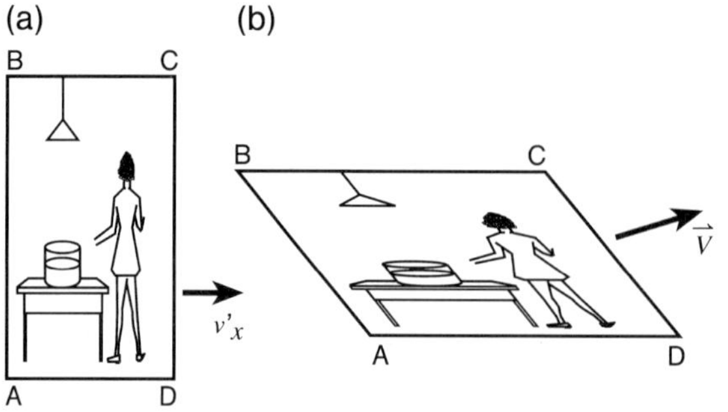

Figure 4.15

At this very time the engineer, who indeed is sitting on the shore, notices a very strange thing: the chandelier's chain, as well as all other lines parallel to it, are tilted with respect to *his* vertical direction. And the reason for this is that for him, the car is moving with a tilted velocity **v** rather than just passing by horizontally. Applying the same reasoning as in the previous case, we can together with the engineer derive the expression for the corresponding tilt angle α_y (Problem 4.5):

$$\tan \alpha_y = \gamma(v_y) \frac{v'_y v_x}{c^2} = \gamma^2(v_y) \frac{v_x v_y}{c^2}. \tag{4.52}$$

However, all the directions that are horizontal in K'_y and K″ are horizontal in K as well. If we depict these results in a figure, we will obtain something close to Figure 4.15b.

But this leads us to even more bizarre conclusion: the result observed by the engineer, it turns out, depends on the option chosen by Mr. O'Bryan to accompany the object (elevator or car): to follow the horizontal or the vertical component of its motion. Clearly, our observation should only depend on the experimental conditions specifying the motion of the object relative to K (e.g., its velocity **V** in K), not on the way we represent this motion using other observers. The fact that it is not so in the considered situation can have only one explanation: the relative velocity between two RFs is *not all* information that is necessary for complete description of a system.

To see what else is essential, let us look at the situation in a slightly different way – not as a thought experiment but as an operational procedure. We have two different RFs – K and K″. The origin of K″ is moving relative to K with velocity **v**. This (together with the origin's instant position) completely defines K″ as a *reference frame* but not as a *coordinate system*. We can use an infinite variety of different coordinate systems within one RF. Consider an important class of such systems used throughout this book – the Cartesian coordinate systems. To define one such system, we must specify its orientation in space, that is, to determine orientations of its axes with respect to the fixed axes of the system K. This requires three additional variables (degrees of freedom), say, two angles specifying orientation of the z''-axis and the third angle specifying additional rotation of the system K″ about z'' (i.e., rotation in $x''y''$-plane). If we take into account these additional variables, we can reformulate the above results quantitatively.

So consider again the whole situation as an operational procedure. Suppose we observe the launch of a spaceship from rest to velocity **v** with components v_x and v_y. We can do it in many different ways out of which the three simplest are the following:

1. We can first launch it along the x-direction to the speed v_x and then, from the comoving system K'_x, we can launch it vertically along the y'-direction up to the speed $v'_y = \gamma(v_x)v_y$ (in this case, according to the law of velocity addition (3.21), we will obtain the correct value of the v_y component in K). The result will be the desired velocity **v**, but all the lines parallel to x in K'_x and K″ will be tilted to x in K, with the tilt described by Equation (4.54). This is equivalent to Mr. O'Bryan's first mission.

2. Alternatively, we can first launch the ship along the y-direction up to the speed v_y and then from the accompanying system K'_y, perform the additional launch in the x'-direction with the speed $v'_x = \gamma(v_y)\, v_x$ (in this case, according to the law of velocity addition (3.21), we will obtain the correct value of the v_x component in K). The result will be again the desired velocity **v**, but now all the lines parallel to y in K'_y and K″ will be tilted to y in K, with the tilt described by Equation (4.52). This corresponds to Mr. O'Bryan's second mission.

3. We can launch the ship directly along the direction **v** with the speed v (this is the most simple way). In this case, both kinds of lines – those parallel to x'' and those parallel to y'' in K″ – will be tilted to the respective axes in K. And the amount of the tilt can be obtained by just considering a square with a unit side in its rest frame K″, which had been launched from K in one single boost with velocity **v**. In this case, we can simply consider the Lorentz-contracted shape of this square as observed in K. The Lorentz contraction will be observed along the direction **v**. In the simplest possible case, when v makes 45° with the horizontal, it will be Lorentz contracted along its diagonal by a factor $\gamma(v)$ (Figure 4.16). Using

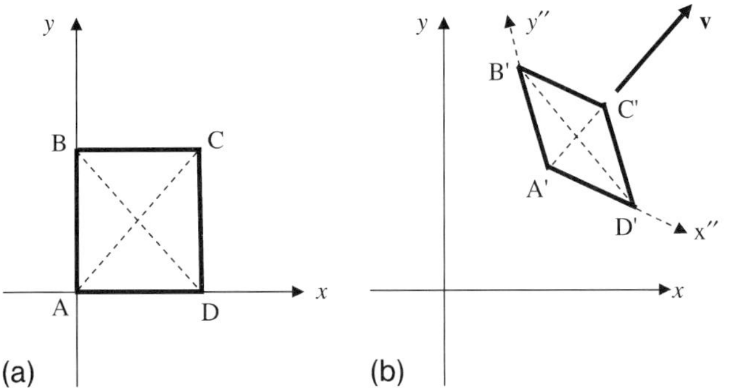

Figure 4.16 A square figure (a) before the launch and (b) after the launch along AC with a speed V.

this fact, we can easily find the tilt of both sides of the square as observed in K (Problem 4.6):

$$\tan \alpha_x = \tan \alpha_y = \frac{\gamma(v) - 1}{\gamma(v) + 1}. \tag{4.53}$$

In all three cases, we have different outcomes. The directions of the coordinate axes of a boosted system observed from a fixed reference frame K depend not only on the boost's final velocity **v** but also on the boost's history. The obtained results just show that in all three cases we will, indeed, obtain the same reference frame K″, but its coordinate axes will be different depending on the details of the boost.

We can illustrate this conclusion from slightly different perspective. Consider again the simplified situation of a square ABCD launched from K directly along its diagonal AC to the speed v. In other words, the square is transferred from frame K *directly* to another frame K″ moving with velocity **v** relative to K, with v parallel to AC. The orientation of each side of the square is preserved during the launch.

Let us now reverse the argument and ask what would Mr. O'Bryan observe in this case, if he boards the same reference frame K'_x as in the case of the aircraft carrier? By definition, the frame K'_x is moving relative to K with the speed v_x equal to the x-component of **V**, so in this frame the square will only move along the direction parallel to y. The question now can be formulated as follows: can this square represent the elevator within the aircraft carrier considered in the beginning of this section? Recall that, according to Mr. O'Bryan's observation (Figure 4.14a), the elevator in that case had a rectangular shape. The engineer in K had observed it as a parallelogram with tilted top and bottom sides (Figure 4.14b).

Now the engineer observes it as a rhombus A′BC′D in Figure 4.16b. Already from this fact alone, we can predict qualitatively that Mr. O'Bryan will also observe a shape different from the rectangle in the case of aircraft carrier. We want now to find this shape.

Let us pick a point (x_0, y_0) on the segment A′D. Since this segment is tilted in the coordinate system x, y by the angle determined by (4.53), it forms part of the tilted straight line described by the equation

$$y - y_0 = \frac{1 - \gamma}{1 + \gamma}(x - x_0). \tag{4.54}$$

But the whole segment, together with the chosen point, is moving in K, so that the coordinates of this point are the linear functions of time:

$$x_0(t) = v_x t; \quad y_0(t) = v_y t \tag{4.55}$$

(we have chosen the point in question so that at the zero moment it passes through the origin). Also, in our special case we have

$$v_x = v_y = \frac{v}{\sqrt{2}}. \tag{4.56}$$

Using this information, we can after simple algebra (Problem 4.8) bring Equation (4.55) to the form

$$(\gamma - 1)x + (\gamma + 1)y = \sqrt{2}\gamma vt. \tag{4.57}$$

As a simple check, consider the case $v = 0$. In this case, $\gamma = 1$, and Equation (4.60) then gives $y = 0$. This is precisely what one would expect, as the square in this case remains stationary in K, with its bottom side horizontal.

Now, we are prepared to find the orientation of this line (the bottom side of the moving square) as observed by Mr. O'Bryan. All we need is to express the coordinates x, y, t in terms of the corresponding coordinates of system K'_x.

Write the Lorentz transformation

$$\begin{aligned} t &= \gamma\left(t' + \frac{v_x}{c^2}x'\right), \\ x &= \gamma(x' + v_x t'), \\ y &= y'. \end{aligned} \tag{4.58}$$

Note that this time the Lorentz factor is a *function of* v_x ($\gamma = \gamma(v_x)$), since this is the relative velocity between K and K'_x. Putting these expressions into (4.57) and using (4.56), we obtain

$$(1 - \gamma)x' + (1 + \gamma)y' = \frac{1}{\sqrt{2}}\gamma(1 + \gamma)vt'. \tag{4.59}$$

As we had suspected, this is *not* a horizontal line. Only in the extreme special case when $v_x = v/\sqrt{2} = 0$, that is, when Mr. O'Bryan is sitting together with the engineer in K and observing the stationary elevator, do they both see the segment as horizontal. Indeed, in this case $\gamma = 1$, and Equation (4.69) reduces to $y' = 0$. Otherwise, Mr. O'Bryan, while observing the elevator moving along the vertical line in his frame of reference, will see its floor, ceiling, desk, and the water surface in the fish tank all tilted with respect to the horizontal. The angle of tilt can be easily obtained from (4.59):

$$\tan \alpha'_x = \frac{\gamma(v/\sqrt{2}) - 1}{\gamma(v/\sqrt{2}) + 1}. \tag{4.60}$$

The same equation obtains for the angle α'_y between the side A′B and the y'-axis (Problem 4.7). Note that this time, according to Equation (4.60), the angle α'_x is plotted counterclockwise from the positive direction of axis x' and the angle α'_y is plotted clockwise from the positive direction of the axis y'. Combining the derived information, we obtain the shape of the rising (or falling) square (as observed by Mr. O'Bryan), shown in Figure 4.17. It is not a square, and generally, not even a rhombus. It is a parallelogram! But, in contrast with the rhombus observed in *the same* experiment by the engineer from K, the parallelogram in K'_x results from the Lorentz contraction along the diagonal BD. We conclude that the square ABCD in this boost cannot represent the elevator in the two-stage boost described above.

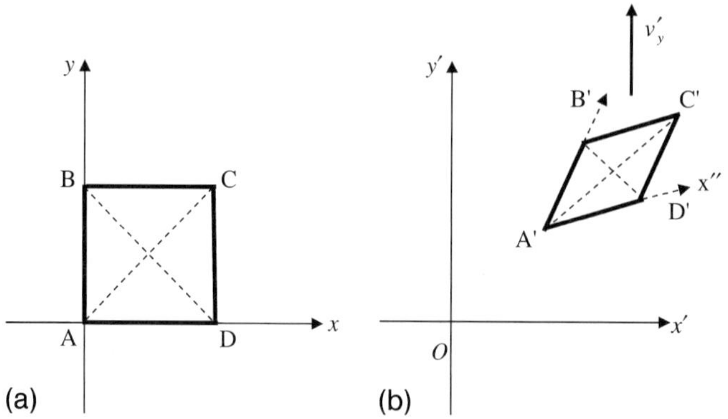

Figure 4.17 (a) The same square plate as in Figure 4.16, launched following the same procedure (as a single boost along V), but now (b) it is observed by Mr. O'Bryan from the aircraft carrier.

By symmetry, we can obtain the same conclusions for the case when Mr. O'Bryan follows the vertical component of the square's motion.

An interesting question is what specific mechanism can be used to ensure that the spatial orientation of a segment transferred between two different RFs is indeed preserved? The law of the conservation of angular momentum provides one such possibility. If we have a spinning object (a gyroscope), then, according to this law, the direction of its rotational axis is conserved provided no torques are acting on it. Therefore, if we consider the rotational axis of such a gyro as a transferred segment, and see to it that the forces needed for the boost are applied only to the center of mass of the gyro so that they produce no torque, then it will be a physical realization of a segment with conserved direction in any comoving RF. Applying this to our situation, we can have all sides of the square rigidly attached to their respective gyroscopes and the accelerating forces acting only on their centers of mass. As a result, we will have in K″ the same square with horizontal side AD and vertical side AB. However, we have just found that these very sides will, as observed in K, be slanted with respect to the corresponding directions in K through the angle (4.53) (Figure 4.16b).

Now we can summarize and explain these remarkable results, using the same object (the square) in the following way. In the two-stage boost (case 1), each side of the square is transported parallel to itself from K first to K'_x and then from K'_x to K″. In this case, the intermediate RF (frame K'_x) is moving along x in K. Accordingly, the side AB remains vertical in K, since all the events on it that are simultaneous in K'_x are simultaneous in K. The side AD, on the contrary, is slanted in K, since all the events on it that are simultaneous in K are not simultaneous in K'_x, and vice versa. A similar argument shows that in another two-stage boost (case 2), when the square is transported parallel to itself first to K'_y and then from K'_y to K″, sides AB and CD of the square will be slanted with respect to vertical direction in K, and sides AD and BC will remain horizontal in K. Finally, the direct parallel transport from K to K″

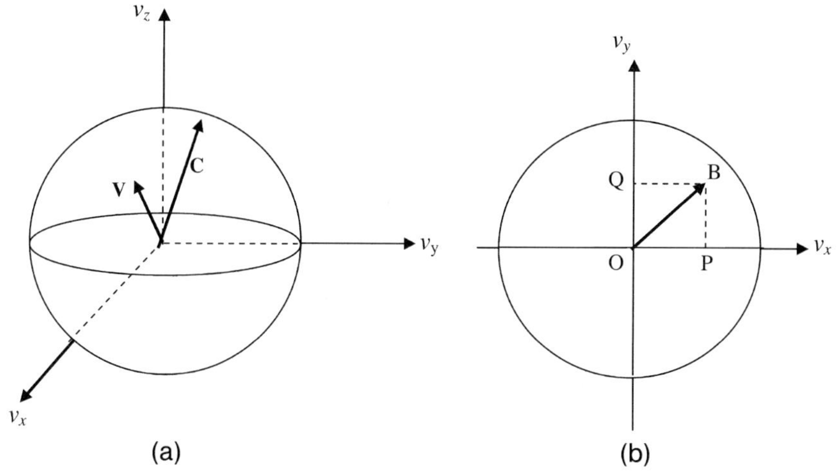

Figure 4.18

results in the tilt of all sides of the square, thus making it a rhombus as observed in K.

Thus, the choice of intermediate RF (the history of the boost) is crucial for the resulting shape of the launched object observed from the initial RF.

These results can also be described in terms of the velocity vector space.

Consider the 3-velocity vector space, on whose mutually perpendicular axes there are plotted the components v_x, v_y, v_z. In this space, all velocities accessible for the known physical objects are represented by points within or on the spherical surface of radius $v = c$, that is (Figure 4.18a),

$$v_x^2 + v_y^2 + v_z^2 \leq c^2. \tag{4.61}$$

To simplify the discussion, set $v_z = 0$. In this case, the continuous set of available velocities reduces to the circle of radius c in the v_x, v_y plane and its interior:

$$v_x^2 + v_y^2 \leq c^2. \tag{4.62}$$

Select a point $\mathbf{v} = (v_x, v_y)$ (Figure 4.17b). Just as any point in the configuration space (the subset x, y, z of space-time) can be reached from the origin by infinite number of different paths, so can the point $\mathbf{v}$ in the velocity space. The first operational procedure considered above (starting with the boost $K \rightarrow K'_x$ along the x and then performing the appropriate boost $K'_x \rightarrow K''$) corresponds to the path OPB in the velocity space shown in Figure 4.17b. The second procedure (starting from the boost $K \rightarrow K'_y$ along the y and then performing the appropriate boost $K'_y \rightarrow K''$) corresponds to the path OQB in this space. And the third (the most simple) procedure (the direct boost $K \rightarrow K'$) corresponds to the straight path OB in the velocity space. As we have seen, each path ends up with different orientation of the resulting coordinate axes x'', y'', z'' as observed in the frame K. In other words, the resulting orientation depends on path in the velocity space. Generally, *an axis turns out to have undergone a specific rotation in the process of a boost*, which depends on a chosen succession of boosts leading to the same final boost. Note that this kind of rotation occurs *without any torque* on a segment or line, representing a coordinate axis. In the process of the boosts leading from K to K″, each boosted segment *retains its orientation* in any intermediate comoving system (examine from this viewpoint the systems K'_x or K'_y in Figures 4.13a and 4.14a). And yet a segment as observed in a stationary frame K may wind up being rotated through a certain angle after the boost! This is another remarkable result of the relativity of time. How can we understand the underlying physical mechanism leading to this result in terms of space-time geometry?

Recall that each boost, even with *purely spatial* directions fixed, is a rotation in space-time, involving temporal and the spatial axes along the direction of the boost. Therefore, even though there are no intermediate rotations *in space*, there is a succession of *space-time* rotations in the process of multistage boosts. For instance, the succession of boosts OP and PB, represented by path OPB in Figure 4.18, is equivalent to rotation of the corresponding RF first in plane (ct, x) and then in plane (ct', y'). Denote these two rotations as $\hat{R}_x$ and $\hat{R}_y$, respectively. Then their succession can be written as $\hat{R}_y\hat{R}_x$. The corresponding expression can be called the group product of the two operations of rotation and should be read from right to left: we apply first the operation $\hat{R}_x$ and then $\hat{R}_y$.

The succession of boosts OQB, on the contrary, is equivalent to rotation of the initial RF first in plane (ct, y) and then in plane (ct', x'). Using our notations, we can write the succession of both rotations as $\hat{R}_x\hat{R}_y$. This is another group product of the same operations but performed in the reversed order: now we apply the operation $\hat{R}_y$ first and then $\hat{R}_x$. As is well known, the group of rotations is not commutative – the product of the two rotations is generally different for different ordering:

$$\hat{R}_x\hat{R}_y \neq \hat{R}_y\hat{R}_x. \tag{4.63}$$

Here is a rather simple relevant illustration from Euclidean geometry: take a figure in the xy-plane and rotate it, say, counterclockwise through 90° first in the xz-plane and then in the same sense in the $z'y'$-plane. Using our notations, the result of these two consecutive rotations can be written as $\hat{R}_y\hat{R}_x$ (Figure 4.19a). In another trial, do the same rotations of the figure first in the yz-plane and then in the $z'x'$-plane. The succession of the two rotations can be written as $\hat{R}_x\hat{R}_y$ and gives the result shown in Figure 4.19b. The results are different.

A similar situation is observed when we replace the z-axis with the ct-axis. Physically, such replacement means that instead of purely spatial rotations we now perform the corresponding boosts; but since, according to the theory of relativity, each boost is a rotation in space-time, and two rotations are generally not commutative, we obtain different results for the above-considered successions of boosts. For each succession, we wind up with the same final RF K″ (uniquely defined by its velocity **V** with respect to K), but the Cartesian coordinate systems in this RF turn out to be different. They are spatially rotated with respect to orthonormal triad $(\hat{\mathbf{x}}, \hat{\mathbf{y}}, \hat{\mathbf{z}})$ of frame K (from which they have originated in the above-described processes of consecutive boosts) and (as observed from K) with respect to one another.

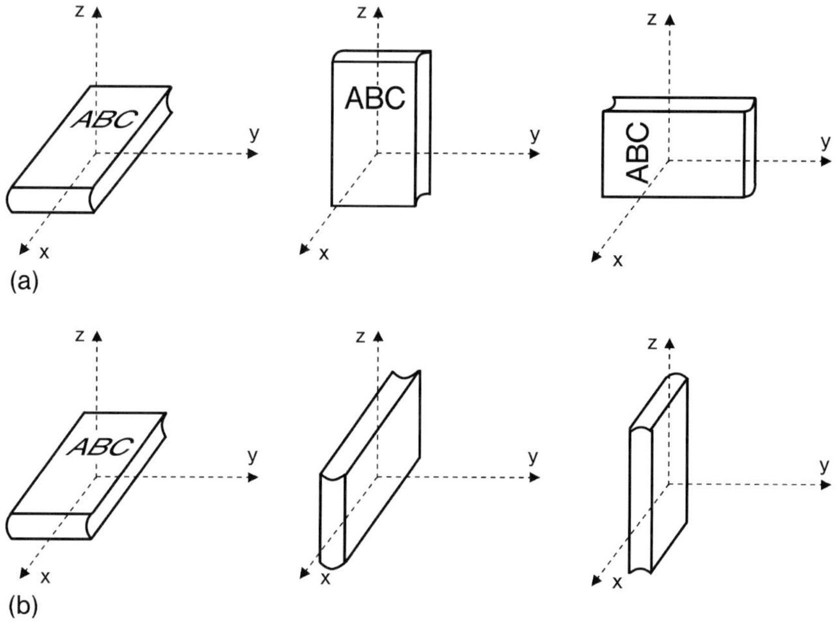

Figure 4.19

Thus, due to relativity of simultaneity, any launch of a spaceship, or generally, any transition between two different RFs is equivalent to the corresponding rotation in space-time. This, in turn, leads to the corresponding rotation of a spatial segment through a certain angle, even though its orientation is preserved in each intermediate boost. And the most simple illustration is provided by the direct boost $\mathrm{K} \to \mathrm{K}''$ in Figure 4.16. Considering the two sides of a square in this figure as a dyad $\hat{\mathbf{x}}, \hat{\mathbf{y}}$, we see that its both vectors $\hat{\mathbf{x}}', \hat{\mathbf{y}}'$ after the boost $\mathrm{K} \to \mathrm{K}''$ are turned with respect to $\hat{\mathbf{x}}, \hat{\mathbf{y}}$, even though their directions were maintained during the boost.

So, when we speak of the tilt of a *boosted* segment with respect to its earlier direction in a fixed frame, we mean the segment with preserved orientation. Because of the relativity of time, the spatial orientation of a segment is a relative characteristic. This property has profound consequences. One of them will be considered in the next section. The other one will be discussed in Section 8.5.

4.5 General Lorentz Transformations

We have so far considered a very special case of the Lorentz transformations, when the respective axes of two different RFs were parallel to one another and the relative velocity of the two systems was parallel to x and x'. In other words, the velocity vector of relative motion had only one nonzero component in both frames.

Here, we want to generalize this to the case of *arbitrary* direction of this velocity. The material of the previous section shows that we have to be careful in doing this, since, for instance, the sides of a fixed orthonormal dyad x', y' of frame K′ are generally *not parallel* to the respective sides x, y of frame K, even though their spatial directions have been preserved during transition $\mathrm{K} \to \mathrm{K}'$.

Instead of trying to figure out how to incorporate this property of the mutual orientation of the coordinate axes, we can get around this difficulty by applying the approach suggested by Moeller [31, 32]. Namely, we can start with the *special* Lorentz transformations, in which case the respective axes of both frames – K and K′ – remain parallel to one another, as shown in Figure 2.2. Having written down the transformation rules (2.54) for this case, we then perform a *purely spatial rotation* of coordinate system identical for both frames. This will automatically take the vector **V** of relative velocity to an arbitrary orientation with respect to the coordinate axes of either frame. Let us perform this procedure step by step.

1. Since we are going to consider rotations in 3-vector space, redenote the coordinates x, y, z as the components of position vector **r** of an event:

$$x \to r_x, \qquad y \to r_y, \qquad z \to r_z, \tag{4.64}$$

and similarly for the primed coordinates. Then do the same for relative velocity **V**:

$$\mathbf{V} = (v_x, v_y, v_z) \xrightarrow[v_y = v_z = 0]{} (V, 0, 0). \tag{4.65}$$

2. Write down Equation (2.54), using the new notations, as

$$\begin{aligned} t' &= \gamma(V)\left(t - \frac{v_x}{c^2} r_x\right); \\ r'_x &= \gamma(V)(r_x - v_x t); \\ r'_y &= r_y; \quad r'_z = r_z. \end{aligned} \tag{4.66}$$

3. Make the second equation in this system look more similar to the last two by adding and subtracting r_x on the right side:

$$r'_x = r_x + \gamma(V)(r_x - v_x t) - r_x. \tag{4.67}$$

4. Using the fact that, in the considered special case, $v_x = V$, rewrite this as

$$r'_x = r_x + v_x \gamma(V)\left(\frac{v_x r_x}{V^2} - t\right) - v_x \frac{v_x r_x}{V^2} = r_x + v_x \left[(\gamma(V) - 1)\frac{v_x r_x}{V^2} - \gamma(V) t\right]. \tag{4.68}$$

5. In the same special case we have, according to (4.65),

$$v_x r_x = v_x r_x + v_y r_y + v_z r_z = \mathbf{V} \cdot \mathbf{r},$$

so that we can rewrite (4.69) as

$$r'_x = r_x + v_x\left[(\gamma(V)-1)\frac{\mathbf{V}\cdot\mathbf{r}}{V^2} - \gamma(V)t\right]. \tag{4.69}$$

6. In view of (4.65), this equation automatically describes the last two equations of (4.66) as well, if we replace the subscript "x" by "y" or "z". Therefore, we can write the whole Lorentz transformation (4.66) in the vector form

$$t' = \gamma(V)\left(t - \frac{\mathbf{V}\cdot\mathbf{r}}{c^2}\right), \tag{4.70a}$$

$$\mathbf{r}' = \mathbf{r} + \mathbf{V}\left[(\gamma(V)-1)\frac{\mathbf{V}\cdot\mathbf{r}}{V^2} - \gamma(V)t\right]. \tag{4.70b}$$

Similarly, we can derive the inverse transformations:

$$t = \gamma(V)\left(t' - \frac{\mathbf{V}'\cdot\mathbf{r}'}{c^2}\right), \tag{4.71a}$$

$$\mathbf{r} = \mathbf{r}' + \mathbf{V}'\left[(\gamma(V)-1)\frac{\mathbf{V}'\cdot\mathbf{r}'}{V^2} - \gamma(V)t'\right]. \tag{4.71b}$$

Here, $\mathbf{V}' = -\mathbf{V}$ is the velocity of K with respect to K′.

So far, this is simply a general notation for a special case (4.65). But it is written in the form that is invariant under 3-rotations in space. Therefore, we can now subject the triads $(\hat{\mathbf{x}}, \hat{\mathbf{y}}, \hat{\mathbf{z}})$ and $(\hat{\mathbf{x}}', \hat{\mathbf{y}}', \hat{\mathbf{z}}')$ to the same spatial rotation in frames K and K′, respectively, without changing the equations. Since the triads' rotations are identical, the vectors $\mathbf{V}$, $\mathbf{r}$, on the one hand and $\mathbf{V}'$, $\mathbf{r}'$ on the other, undergo the same transformation, so that Equations (4.70) and (4.71) remain valid. But now they describe a far more general case. Indeed, after we have rotated, say, the triad $(\hat{\mathbf{x}}, \hat{\mathbf{y}}, \hat{\mathbf{z}})$, the relative velocity $\mathbf{V}$ of the two systems is no longer along its $\hat{\mathbf{x}}$-direction. It is now *arbitrarily oriented* with respect to the new axes. The same is true for the triad $(\hat{\mathbf{x}}', \hat{\mathbf{y}}', \hat{\mathbf{z}}')$, so that we can write

$$\mathbf{V} = -\mathbf{V}' = (v_x, v_y, v_z), \tag{4.72}$$

where the specific values of the component of velocity depend on performed rotation. Finally, returning again to the initial notations in terms of coordinates, we can write

$$\begin{aligned}
t' &= \gamma(V)\left[t - c^{-2}(v_x x + v_y y + v_z z)\right],\\
x' &= x - v_x\left[\gamma(V)t - V^{-2}(\gamma(V)-1)(v_x x + v_y y + v_z z)\right],\\
y' &= y - v_y\left[\gamma(V)t - V^{-2}(\gamma(V)-1)(v_x x + v_y y + v_z z)\right],\\
z' &= z - v_z\left[\gamma(V)t - V^{-2}(\gamma(V)-1)(v_x x + v_y y + v_z z)\right].
\end{aligned} \tag{4.73}$$

The inverse transformations are obtained directly by swapping the primes and changing sign in $\mathbf{V}'$.

The derived equations perform the Lorentz transformation between the components of an event considered from two different RFs moving relative to one another with an arbitrarily directed velocity, but with their respective axes parallel to one another. Their parallelism is preserved since they undergo the identical rotations from their initial orientations in which the respective axes were parallel. But, as we have already mentioned, based on our previous experience with Mr. O'Bryan and the engineer, this does not necessarily mean that we will actually observe the respective axes of the two frames parallel to one another. In most cases, they will not be. This can be seen from the following argument. Consider two identical stationary coincident dyads $(\hat{\mathbf{x}}, \hat{\mathbf{y}})$. Suppose that one of the dyads is launched from its initial position along its bisector with a speed V. This results in two dyads – one $(\hat{\mathbf{x}}, \hat{\mathbf{y}})$(initial) and the other $(\hat{\mathbf{x}}', \hat{\mathbf{y}}')$ moving with velocity $\mathbf{V}$ that makes 45° with the x-direction. Identifying the sides of each dyad with two sides of the unit square, we reiterate the special case discussed in the second part of the previous section. On the other hand, the two dyads can now be considered as two different RFs, whose relative velocity is not parallel to any of their reference directions. Suppose that these respective directions remain parallel to one another, that is,

$$\hat{\mathbf{x}} \| \hat{\mathbf{x}}' \quad \text{and} \quad \hat{\mathbf{y}} \| \hat{\mathbf{y}}'. \tag{4.74}$$

In this case, the moving dyad would be observed as undistorted square as it had been before the boost. This would mean one of the two things: either it has not undergone the Lorentz contraction or its shape has not been preserved during the boost, so that now the *proper* shape of this dyad in its own rest frame is no longer a square but a rhombus extended along the direction of relative motion. In other words, the moving dyad in this case no longer constitutes a Cartesian coordinate system. The first option contradicts the laws of Nature and the second one contradicts the condition that the proper shape of an object be preserved during the boost. This contradiction shows that our assumption (4.74) cannot be true.

Reversing the argument, we can say that a dyad $(\hat{\mathbf{x}}', \hat{\mathbf{y}}')$, with its sides parallel to $(\hat{\mathbf{x}}, \hat{\mathbf{y}})$ and moving along the bisector of the first quadrant, will be observed as the sides of the moving square in Figure 4.16b, – when in motion along the diagonal they become the sides of a rhombus.

These conclusions, of course, hold for triads as well. Generally, two triads brought into a state of motion relative to one another in a boost preserving their parallelism will have their axes nonparallel after the boost. This counterintuitive property of objects in space-time is just another manifestation of the relativity of simultaneity. We will see other unusual examples in Chapter 7.

We can obtain the expressions for even more general transformation if we subject one of the RF to additional rotation or each of them to different rotations. In this case, we will have two triads with arbitrary direction of their relative motion and arbitrary orientations of their respective axes. A detailed description of the most general Lorentz transformation can be found in Refs [31, 32].

4.6 The Path Integrals

Consider a moving system K′. As we know, all the clocks in this system are ticking slower compared to the stationary clocks in K. It may appear that there is a paradox here. Indeed, an observer sitting on the moving body in system K′ can say that his clocks are stationary, while our clocks in K are all moving, and therefore *they* must tick slower than his clocks. Thus, a clock in K′ runs slower from the viewpoint of system K and vice versa.

But as we think of it more carefully, we can see that there is no contradiction here, because the statements of the two observers are about two different procedures. The observer in K compares *one* moving clock K′ with *two different* clocks K_1 and K_2 of his reference frame: first time when K′ passes by K_1 and the second time when K′ passes by K_2. This procedure shows that the difference between the readings of K′ at these two coincidences is *less* than the difference between the corresponding readings of K_1 and K_2.

The observer in K′ compares *one* moving clock K with *two different* clocks K'_1 and K'_2 of his reference frame: first time when K passes by K'_1 and the second time when K passes by K'_2. This procedure shows that the difference between the readings of K at these two coincidences is *less* than the difference between the corresponding readings of K'_1 and K'_2.

You can easily see that in these two procedures there figure two *different pairs* of events and therefore there is no contradiction between the statements of the observers. Each measuring procedure is intrinsically asymmetric: whenever you want to measure time read by an inertially moving clock, you have to compare this clock with *more than one* of the synchronized clocks of your system. And it is always the clock that is being compared with two (or more) different clocks of another reference frame that turns out to run slower. A more detailed discussion of this "paradox" can be found in Ref. [1].

But if you persist in your critical inquiry, you may think, hey, what if I, at some intermediate moment, send the moving clock K′ back in its track or just make it move in a closed curve (closed in space, not in space-time!) to compare it again with *the same* stationary clock K as before (Figure 4.20). Since my reference frame is inertial, the conclusion about time dilation of any moving clock applies to it. Therefore, since the clock K′ keeps on moving, albeit now not uniformly, it must read less time interval than my stationary clock K, between their two consecutive meetings. On the other hand, the observer sitting on that moving clock will claim that it is *he* who is stationary and therefore it is *my* clock K that must show the lesser time. And this time it appears to be a true logical contradiction because it refers to the same pair of events happening at the same point of space in either system!

We will discuss this paradox in more detail in Section 7.6. But here suffice it to say that now the two observers are not equivalent. One of them (K′) is moving *with acceleration*, and even though he may legitimately claim that he is stationary if he refers the observations to his frame, the relativistic conclusions that hold in the *inertial* reference frames do not necessarily apply to his frame. Therefore, this time

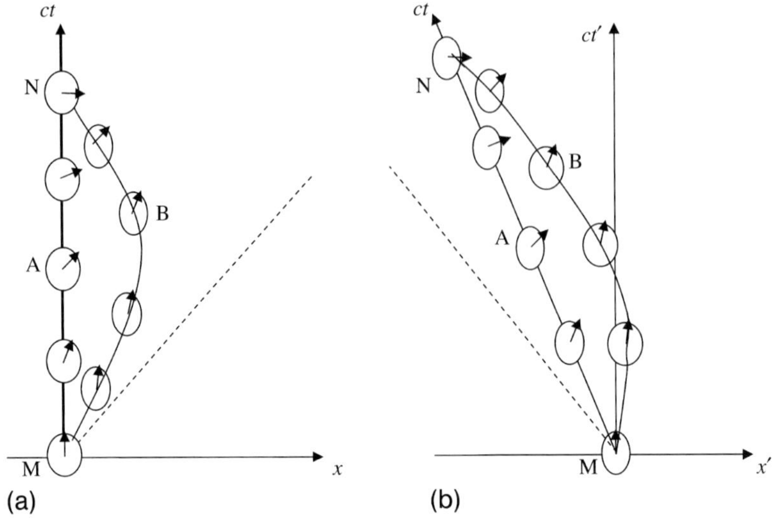

Figure 4.20 The world lines of the two identical clocks A and B that are coincident at event M and then at event N. (a) As observed from an inertial RF K in which the clock A is stationary (the moving clock B represents a noninertial RF attached to it); (b) as observed from another inertial RF K′ in which the clock A is moving with a speed V along the common x-, x'-direction. Note that here, in contrast with Figure 4.4, the systems of axes (ct, x) and (ct', x') are drawn each in its respective RF, that is, (ct, x) in K and (ct', x') in K′, so that $ct \perp x$ and $ct' \perp x'$.

only K-observer is right; as to the K′-observer, he is wrong if he expects the K-clock to read less time between the events than his clock K′. The clock K′ will read the lesser time for *both* observers.

Thus, the reading of a clock is a relativistic invariant. If a clock K reads seven units of time between two events, M and N, it will be seven units for the observer sitting on this clock and for *any* other observer watching the readings of this clock K. If another clock K′ passing by K first when the event M happens there and then when the event N happens, reads only three units of time between these events, it will be three units for all observers watching the clock K′.

The reason why we mentioned this at this earlier stage of the book is to show how the concept of proper time between two events can be related to the line integral between them.

Let us visualize the above thought experiment using the concept of the world line in the Minkowski's world. Denote the two events in question as M and N. Then we will see that the physical difference between the motions of two clocks is reflected geometrically in the shapes of their respective world lines. Indeed, the world line of clock K is straight as a laser beam, whereas the world line of K′ is curved (Figure 4.20). Consider two infinitesimally close points on a world line. According to the first example in the end of Section 4.1, *in the comoving system* the time interval between the corresponding events is the 4-displacement ds between them divided by c. Together with ds, it is an invariant of Lorentz transformations – it is a proper time of the clock moving along the corresponding line. Then the net proper time between the events M and N is given by the line integral

$$\tau = \frac{1}{c}\int_M^N ds = \int_M^N \frac{dt}{\gamma(v)}. \tag{4.75}$$

This is a special case of what is known as *path integrals* in space-time.

Since each incremental 4-displacement is Lorentz invariant, their sum (in the limit – the integral (4.75)) taken along the path MN is also Lorentz invariant:

$$\tau = \int_M^N \frac{dt}{\gamma(v)} = \text{inv}. \tag{4.76}$$

On the other hand, two *different* paths in space-time connecting M and N have different lengths in one RF. We can illustrate this in a simple example with a famous experiment – laser location of the Moon (1973). The researchers shot a very short laser pulse in the direction of the Moon and were able to record the arrival of the pulse reflected from a specially designed reflector left on the Moon by the astronauts in the preceding expedition.

Consider the space-time diagram of this experiment. It consists of two world lines between the points M and N. Point M corresponds to the event of the pulse emission and N corresponds to the event of detection of the pulse returned after reflection from the reflector on the Moon's surface. The world line MN represents the "motion" of the lab with the research team, laser and detector of the reflected pulse, through space-time; the world line MRN represents the pulse itself (Figure 4.21). The latter world line is broken in two halves each corresponding to the motion of the pulse

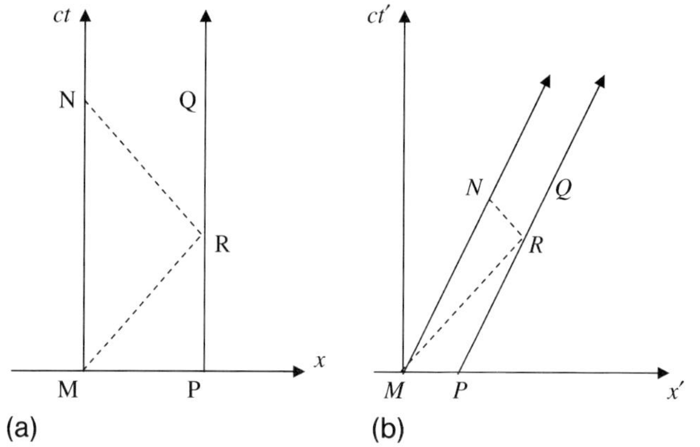

Figure 4.21 A diagram of the laser location of the Moon (the curvature of the Moon's world line can be neglected here). (a) In the (nonrotating) RF attached to the center of the Earth; (b) in the RF associated with the center of the Sun. Both RFs are to high accuracy inertial within a short time of the experiment (~3 s) between the emission and the return of the laser pulse; in either RF we have $\tau(\text{MN}) = \text{const} > 0 = \tau(\text{MRN})$. MN and PQ are the world lines of the Earth and Moon, respectively; MRN is the world line of the laser pulse. The tilt of the world lines MN and PQ, and accordingly the asymmetry in the right triangle MRN in (b), is greatly exaggerated.

respectively toward (MR) and away (RN) from the Moon. Whatever the shape, we have two different paths connecting points M and N.

The length of line MN is just the proper time of the experiment, that is, the Lab time between the emission and the absorption of the pulse. The length of line MRN is, of course, zero, since the pulse has traveled with the invariant speed. Thus, the value of the integral (4.76) is individual for each individual path connecting the points M and N. We can express this by writing

$$S = c\tau = \int_M^N ds = S\{x(t)\} = \text{inv.} \tag{4.77}$$

The expression $S\{x(t)\}$ emphasizes that we have a "function of function" – the value of S depends not only on the end points, but also on the shape of a curve connecting them.

If the line connecting M and N is curved (the rest frame of the clock K′ is noninertial), then the speed v in (4.75) and (4.76) is a function of time. If the line is straight, then $v = \text{const}$. Here is the crucial point: the value of time τ depends, as just shown, on choice of a line connecting the two events; but once the choice is made, for each line the value of the integral, being the invariant, does not depend on choice of reference frame used for calculation. For instance, if you board a vehicle comoving with the clock K (straight world line), then the clock is stationary and the same straight world line MN will be vertical in this frame. We will see later (Sections 7.8 and 9.3) that a stationary inertial clock always reads longer time than a moving clock between the same events. This means that the integral (4.75) between two given points M and N has the maximal possible value when it is taken along the straight line between these events. That is, out of all lines connecting two events (two different points in space-time), the straight line has the greatest, not the least, length.

4.7 4-Vectors and Tensors

We can now develop the relativistic kinematics of a particle. The most natural way to do this would be to use the concept of Minkowski's world. By analogy with the four-dimensional vector of an event $(ct, \mathbf{r}) = (ct, x, y, z)$, we introduce a general concept of a four-dimensional vector A as any quantity characterized by four components, which transform as the four components (coordinates) of an event in Minkowski's world. These transformations, as we know, leave invariant the form

$$A^2 = A_t^2 - A_x^2 - A_y^2 - A_z^2. \tag{4.78}$$

We will call this form the square of the length of vector A and the number $|A|$ measuring this length its norm. In most texts, the norm is denoted just A, without the sign of the absolute value. We will also use this convention. Following the same analogy with a 4-displacement vector s, the temporal component A_t is sometimes called timelike and the components A_x, A_y, A_z of the spatial part $\mathbf{A}$ are called spacelike components. In essence, Equation (4.78) is the Pythagorean theorem in four-dimensional pseudo-Euclidean (Lorentzian) geometry.

Some people feel uncomfortable seeing the squares of the spatial components entering expression (4.78) with the sign opposite to that of the temporal component. This property, common to that of the interval (4-displacement) in space-time, is often referred to as "negative metric signature" [29]. We can define it formally as a product of signs of the coefficients in A^2 in (4.78):

$$(1)\,(-1)\,(-1)\,(-1) = -1\,. \tag{4.79}$$

There is nothing to be done about it – the negative metric signature is the intrinsic characteristic of the world. But it is still possible to write expression (4.78) in a more convenient form with all the terms having the same sign, if we introduce two different "sorts" of component. We will do it in two steps. First, let us replace the subscript "t" by the superscript "0", and subscripts "x", "y", "z" by superscripts 1, 2, 3, respectively. Expression (4.78) will take the form

$$A^2 = (A^0)^2 - (A^1)^2 - (A^2)^2 - (A^3)^2 = (A^0)^2 - \sum_{\alpha=1}^{3} (A^\alpha)^2. \tag{4.80}$$

(The superscripts in parentheses here are indexes, *not* powers!) All four components so defined are components of the first sort. They are called contravariant components and, according to their definition, they transform under rotations in space-time as the coordinates of an event in Minkowski's world:

$$A^0 = \gamma(V)\left(A'^0 + \frac{V}{c}A'^1\right), \qquad A^1 = \gamma(V)\left(A'^1 + \frac{V}{c}A'^0\right), \qquad A^2 = A'^2,$$
$$A^3 = A'^3. \tag{4.81}$$

(Caution again: from now on the reader should use judgment to distinguish between powers and superscripts denoting contravariant components of a vector.)

Next step: we introduce components of the second sort with *subscript* indexes, defining them as

$$A_0 = A^0, \qquad A_1 = -A^1, \qquad A_2 = -A^2, \qquad A_3 = -A^3. \tag{4.82}$$

Components of the second sort are called covariant components. It is easy to see that the Lorentz transformations of a 4-vector expressed in terms of its covariant components is different (in signs) from the same transformation of its contravariant components. Indeed, using (4.82) to express (4.81) in terms of the covariant components, we obtain

$$A_0 = \gamma(V)\left(A'_0 - \frac{V}{c}A'_1\right), \qquad A_1 = \gamma(V)\left(A'_1 - \frac{V}{c}A'_0\right), \qquad A_2 = A'_2,$$
$$A_3 = A'_3. \tag{4.83}$$

With definition (4.82), we can now rewrite Equation (4.78) as

$$A^2 = A_0A^0 + A_1A^1 + A_2A^2 + A_3A^3 = \sum_{j=0}^{3} A_jA^j. \tag{4.84}$$

The reader can now prove (Problem 4.1) that expression (4.84) remains invariant under Lorentz transformations.

Consider a subset of the LT affecting only the three spatial axes. Just as three components x^1, x^2, x^3 form a regular 3-vector **r**, the three spatial components of A^j form a regular 3-vector **A** under purely spatial rotations within the same reference frame. Similarly, the temporal component A^0 is, under the same transformations, a 3-scalar, as is the time under coordinate transformations within the same reference frame. This fact is reflected in frequently used notations for a 4-vector in terms of its temporal and spatial parts:

$$A = A^j = (A^0, \mathbf{A}) \tag{4.85a}$$

and

$$A = A_j = (A_0, -\mathbf{A}) \tag{4.85b}$$

for the contravariant or covariant components of a vector, respectively. Equation (4.84) can be written as

$$A^2 = (A^0)^2 - \mathbf{A}^2 = (A_0)^2 - \mathbf{A}^2, \tag{4.86}$$

which is, of course, equivalent to (4.84). However, (4.84) looks more familiar, since all the products enter the sum with the same sign. You can consider (4.84) as a dot product of a 4-vector by itself, expressed in terms of its contravariant and covariant components.

Here and afterwards we will use a Latin index (e.g., "j") to denote *all* four components of a 4-vector A and the Greek index α to denote only spatial components of the vector. Accordingly, "j" will range from 0 to 3 ($j = 0, 1, 2, 3$) as in expression (4.84), and "α" will range from 1 to 3 ($\alpha = 1, 2, 3$) as in expression (4.80).

We can further simplify the notations, if we use Einstein's summation rule, according to which summation is assumed over each index appearing twice in an expression (such indices are called "dumb indices"). In the expression on the right of (4.84) we see the index "j" figuring twice, therefore Einstein's summation rule allows to rewrite it as

$$A^2 = A_j A^j. \tag{4.87}$$

In particular, we can now represent the interval in the previous sections, using both, covariant and contravariant coordinates, as

$$ds^2 = dx_j dx^j. \tag{4.88}$$

A more general invariant is a dot product of two *different* 4-vecors, A and B. We will encounter such situations later. Using only one sort of components, we will have

$$AB = A_0 B_0 - A_1 B_1 - A_2 B_2 - A_3 B_3 = A_0 B_0 - A_\alpha B_\alpha \tag{4.89}$$

in covariant components, or

$$AB = A^0 B^0 - A^1 B^1 - A^2 B^2 - A^3 B^3 = A^0 B^0 - A^\alpha B^\alpha \tag{4.90}$$

in contravariant components. Using both sorts (one sort for A and the other sort for B) in the same expression, we will have

$$AB = A_0 B^0 + A_1 B^1 + A_2 B^2 + A_3 B^3 = A_j B^j. \tag{4.91}$$

Of course, it does not matter which sort of components for which vector you will use.

Note that the rule (4.82) for *spatial* covariant components arises entirely from the fact that all spatial components form only a subspace of the 4-vector space. Although the metric signature of space-time is negative, within the Euclidean 3-space alone it is positive:

$$\mathbf{r}^2 = (x^1)^2 + (x^2)^2 + (x^3)^2 = x_1^2 + x_2^2 + x_3^2. \tag{4.92}$$

Therefore, if we use Cartesian coordinates, there is no need to distinguish between contra- and covariant components of a 3-vector. These components, however, may become distinct in coordinates other than Cartesian. Consider, for instance, a rectilinear but not rectangular coordinate system with x- and y-axes making an angle φ with one another (Figure 4.21). A 2-vector $\mathbf{A}$ has in this system two contravariant components A^1 and A^2, which are just its *oblique* projections onto the axes x and y, respectively. Using the cosine theorem, we find for the norm of this vector

$$A^2 = (A^1)^2 + (A^2)^2 + 2A^1 A^2 \cos\varphi. \tag{4.93}$$

Rewriting this as

$$\begin{aligned} A^2 = (A^1)^2 + A^1 A^2 \cos\phi + A^1 A^2 \cos\phi + (A^2)^2 = A^1(A^1 + A^2\cos\varphi) \\ + A^2(A^2 + A^1\cos\varphi), \end{aligned} \tag{4.93a}$$

we notice that the expressions in the parentheses on the right are the *orthogonal projections* of $\mathbf{A}$ onto the corresponding axes. Denote them as A_1 and A_2:

$$A_1 = A^1 + (\cos\varphi)A^2, \qquad A_2 = A^2 + (\cos\varphi)A^1 \tag{4.94}$$

or, in the matrix form

$$\begin{pmatrix} A_1 \\ A_2 \end{pmatrix} = \begin{pmatrix} 1 & \cos\varphi \\ \cos\varphi & 1 \end{pmatrix} \begin{pmatrix} A^1 \\ A^2 \end{pmatrix}. \tag{4.94a}$$

Then we can write

$$A^2 = A^1 A_1 + A^2 A_2. \tag{4.95}$$

This expression looks somewhat like Pythagorean theorem, even though the triangle formed by $\mathbf{A}$ and its components A^1 and A^2, is not a right triangle. This is, again, achieved at the cost of introducing, in addition to (A^1, A^2), the second set of components (A_1, A_2) and using this set, together with the first one, in the same expression. The set (A_1, A_2) is the set of covariant components, and their geometrical meaning is clear from Figure 4.22.

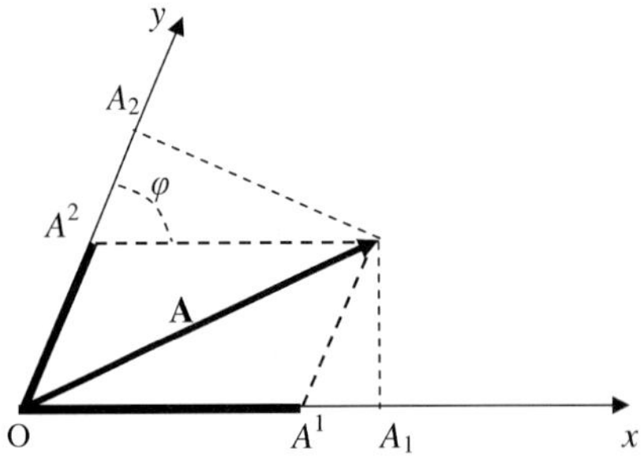

Figure 4.22 Contravariant (OA^1, OA^2) and covariant ($0A_1$, OA_2) coordinates of a vector in a nonrectangular coordinate system.

Equation (4.95) looks more simple than the equivalent Equation (4.93). But this kind of simplicity comes at a price. Apart from dealing with two sorts of components, you can see immediately the inconvenience of being always alert to the difference between the square of the vector on the left of (4.93) and (4.95) and one of its contravariant components A^2 on the right. You can hide this inconvenience by using the Einstein summation rule and writing $A^2 = A_\alpha A^\alpha$, or in the case of all four dimensions involved, $A^2 = A_j A^j$, but it can always pop up again when you go to a specific calculation. So beware!

The examples (4.82) and (4.94) are only two special cases. They show that, generally, the covariant components of a vector are the *linear superpositions* of its contravariant components (and vice versa!). We can express this by writing

$$A_j = g_{jk} A^k, \tag{4.96}$$

where g_{jk} are the expansion coefficients. Here j is *not* a dumb index, because it appears only once in either side of the equation. However, k is a dumb index and according to Einstein's rule, the summation is assumed in (4.96) over all k ranging from 0 to 3. Since the remaining free index j takes on four values $j = 0, 1, 2, 3$, Equation (4.96) is actually a system of four equations determining all four covariant components in terms of their contravariant counterparts. Now, recall the previous chapter. Just as a set of four numbers A_j can be represented as a column matrix and a set of the coefficients in the Lorentz transformation can be represented as a square matrix $\mathcal{L}$, the set of coefficients g_{jk} can be represented also as a square matrix:

$$g_{jk} = \begin{pmatrix} g_{00} & g_{01} & g_{02} & g_{03} \\ g_{10} & g_{11} & g_{12} & g_{13} \\ g_{20} & g_{21} & g_{22} & g_{23} \\ g_{30} & g_{31} & g_{32} & g_{33} \end{pmatrix}. \tag{4.97}$$

Therefore, we can rewrite the four linear equations (4.96) as one matrix equation:

$$\begin{pmatrix} A_0 \\ A_1 \\ A_2 \\ A_3 \end{pmatrix} = \begin{pmatrix} g_{00} & g_{01} & g_{02} & g_{03} \\ g_{10} & g_{11} & g_{12} & g_{13} \\ g_{20} & g_{21} & g_{22} & g_{23} \\ g_{30} & g_{31} & g_{32} & g_{33} \end{pmatrix} \begin{pmatrix} A^0 \\ A^1 \\ A^2 \\ A^3 \end{pmatrix}. \tag{4.96a}$$

As we see from the matrix g_{jk}, the first index in g_{jk} numbers rows and the second index numbers columns. Generally, we have 16 elements g_{jk}. The set of these coefficients forms the important geometric entity called the *metric tensor*. Since it depends on two indices, it is called the second rank tensor. From this broader viewpoint, a vector is a first rank tensor, since it depends on only one index; similarly, a scalar is a zero-rank tensor.

In contrast to matrix $\mathcal{L}$ of Lorentz transformation, which relates the same sort of components of a 4-vector in two *different* reference frames, the g-matrix in (4.96) and (4.97) or in (4.94a) relates two different sets of the components of a vector in *the same* reference frame. In this respect, the $\mathcal{L}$-matrix acts as the machine converting the components of a vector in one RF into its components of the same sort in another RF; the g-matrix acts as a machine converting one sort of components into the other sort within the same RF. In Equations (4.96) or (4.96a), it converts contravariant components into the covariant components, by lowering the corresponding indices. Thus, the metric tensor g_{jk} works as the lowering machine: it takes four contravariant components as an input and returns four covariant components (of the same vector) as the output. In the above two examples, the components of the metric tensor are

$$g_{jk} = \begin{pmatrix} 1 & 0 & 0 & 0 \\ 0 & -1 & 0 & 0 \\ 0 & 0 & -1 & 0 \\ 0 & 0 & 0 & -1 \end{pmatrix} \tag{4.98}$$

(case (4.81)), and

$$g_{jk} = \begin{pmatrix} 1 & 0 & 0 & 0 \\ 0 & -1 & -\cos\varphi & 0 \\ 0 & -\cos\varphi & -1 & 0 \\ 0 & 0 & 0 & -1 \end{pmatrix}. \tag{4.99}$$

(case (4.83)). In both cases, the minuses in the spatial part of the tensor are introduced to meet the requirement that expression (4.84) or (4.91) be in all terms the *sum* of the products of the covariant and contravariant components. In the three-dimensional subspace of the space-time, the minuses would be excessive and we can use the original expression (4.94) relating covariant and contravariant components of a 3-vector.

(As an exercise, the reader can attempt to write the inverse operation converting covariant components into the contravariant ones, by using symmetry considerations (Problem 4.11).)

Note also that if we define the signature as (4.79), then it is, as seen from (4.98), just the determinant of the g-matrix.

As we see from these examples, the elements of the metric tensor depend on coordinate system. Therefore, apart from being a conversion machine between the contra- and covariant components of a vector, they also store information about the nature of the coordinate system used (Problem 4.3). It seems, therefore, natural to apply all this to an element of the 4-interval in an arbitrary coordinate system, by writing, as in (4.96)

$$dx_j = g_{jk} dx^k. \tag{4.100}$$

Then (4.88) can be written as

$$ds^2 = g_{jk} dx^j dx^k. \tag{4.101}$$

Thus, the double sum (4.101) with a 4-displacement vector gives the square of magnitude of this vector. This is the scalar (or dot) product of the displacement vector with itself. In the same way we can use g_{ik} to express the square of magnitude of *any* vector as its scalar product with itself:

$$A^2 = A_i A^i = g_{ik} A^i A^k. \tag{4.101a}$$

Similarly, we can use it to form the dot product of any two vectors A and B:

$$AB = A_i B^i = A^i B_i = g_{ik} A^i B^k. \tag{4.101b}$$

In this respect, the tensor g_{ik} acts as a machine with two inputs for two vectors A and B and one output, producing the scalar product of these vectors.

Generally (in curved coordinates), the elements g_{jk} may be themselves functions of coordinates:

$$g_{jk} = g_{jk}(x^0, x^1, x^2, x^3). \tag{4.102}$$

For instance, in polar coordinates r, φ in a 2D Euclidean plane the length element is $ds^2 = dr^2 + r^2 d\varphi^2$. This means that the metric tensor is of the form

$$g_{\alpha\beta} = \begin{pmatrix} 1 & 0 \\ 0 & r \end{pmatrix}, \tag{4.102a}$$

and one of its matrix elements is a function of r.

This example illustrates another important aspect of the metric tensor – it acts as the storage of information about the coordinates used and even about the geometry of the space-time itself. It is especially important in general relativity (Chapter 13).

Note that both indices on the right in (4.101) appear twice. Therefore, the summation has to be taken over either of them. In other words, the expression on the right is the *double* sum. Using this fact, one can show that the metric tensor is symmetrical.

The form of expression (4.96) helps illuminate another important question – about the transformation properties of a tensor. The answer to this question can be obtained by asking another question: what happens if we, instead of g_{jk}, write in (4.96) the products

$$\eta_{jk} = P_j Q_k, \tag{4.103}$$

where P_j and Q_k are the covariant components of two arbitrary 4-vectors P and Q? We will have, instead of (4.96), the expression of the form

$$A_j = P_j Q_k A^k. \tag{4.104}$$

By virtue of Equation (4.91), the sum $Q_k A^k$ on the right is the dot product of the two vectors and, as such, is a scalar. Call it κ. Then the resulting expression on the right can be written as κP_j and since the product of a vector and a scalar is again a vector, this expression is, in its transformation properties, a covariant component of a 4-vector;

this is no surprise since it *must* have the same transformation properties as A_j on the left. But the tensor g_{jk} in the original expressions (4.96) and (4.96a) also produces a covariant j-component of a vector from its contravariant components – it does the same job as the set P_jQ_k. *Therefore it must transform as the set* (4.103) *does*. In other words, we can define an arbitrary second-rank tensor as any set of elements η_{jk} that transforms as the set of products P_jQ_k.

In this example, both 4-vectors P and Q were represented by their covariant components. Accordingly, a tensor η_{jk} with the subscript indices is a covariant second rank tensor. But, since both sorts are equivalent, we could as well produce a second-rank *contravariant* tensor η^{jk} as the set of products of the contravariant components of the same two vectors. Similarly, we could define a *mixed* second-rank tensor as a set of products P_jQ^k. But in the latter case we must, generally, distinguish which index is superscript and which one is subscript. In other words, $\eta_j{}^k$ and $\eta^j{}_k$ are, generally, two different mixed tensors. The geometrical reason for this is very simple: generally, $P_jQ^k \neq P^jQ_k$ (even when the indices j and k are equal). The metric tensor g_{jk} itself, being a second-rank tensor, can come in three varieties: covariant (g_{jk}), contravariant (g^{jk}), and mixed ($g_j{}^k$).

For a tensor, the connection between any two different sorts of component is determined by the same rule (4.82): in Cartesian coordinates, the lowering or raising the temporal index (0) does not change the corresponding component; the lowering or raising a spatial index ($\alpha = 1, 2, 3$) changes the sign of the corresponding component (Problem 4.16).

Different components of a tensor behave differently under a given type of transformation. Consider, for instance, a second-rank contravariant tensor η^{jk}. Under three-dimensional transformations affecting only space, the corresponding nine components $\eta^{\alpha\beta}$ form three-dimensional tensor. It is easy to see that the components $\eta^{0\alpha}$ and $\eta^{\alpha 0}$ behave under pure spatial transformations (say, 3-rotations) as 3-vectors and the component η^{00} is a 3-scalar.

A tensor η^{jk} is called symmetric if $\eta^{jk} = \eta^{kj}$ and antisymmetric if $\eta^{jk} = -\eta^{kj}$. It is easy to show that all the diagonal elements of an antisymmetric tensor are zeros (Problem 4.20). Also, for the mixed components of a symmetric tensor we can prove a property $\eta^j{}_k = \eta_k{}^j$ (Problem 4.21). In such cases, we can just write η^j_k, that is, place the indices in a column.

An important operation on a second-rank tensor is "tracing." It consists of two stages. First, if the tensor is given as contravariant, we lower one of the indices (does not matter which one), thus forming a mixed tensor. For instance,

$$\eta_j{}^k = g_{jl}\eta^{lk}. \tag{4.105}$$

(If we start with the covariant components, we raise one of the indices.)

Next, we sum all the diagonal elements of the mixed tensor:

$$\eta \equiv \eta_j{}^j. \tag{4.106}$$

This sum of the diagonal elements is called the *trace* of a tensor, hence the name of the whole operation. Evidently, the trace is a scalar. From this viewpoint, the dot product of the two vectors considered above (Equation (4.90)) is just the formation of a tensor $\eta_j{}^k = A_j B^k$ and then finding its trace.

An interesting special case is a symmetric second-rank tensor with components

$$\delta_j^k = \begin{cases} 1, & \text{if} \quad k = j; \\ 0, & \text{if} \quad k \neq j. \end{cases} \tag{4.107}$$

A tensor thus defined has the same components in all coordinate systems. This is the famous Kronecker delta. Its trace is, of course, equal to 4. However, the trace of any antisymmetric second-rank tensor, being the sum of its diagonal elements, is zero.

Up to this point, talking about the transformations of the coordinate systems, we implicitly assumed only rotations and translations. There is another important kind of transformation called inversion or mirror reflection [33]. We can look at it as either mirror reflection of an object under study, with coordinate axes fixed, or mirror reflection of the coordinate axes, with the object fixed. We are dealing with this type of transformations every day. Each time when you are looking in the mirror, you see your inverted self. An interesting thing about the complete spatial reflection is that it constitutes an independent type of transformation only in spaces of odd dimensionality. Let us consider a few simple examples, starting with one dimension. One-dimensional space is just a line (for simplicity assumed straight). A directed segment on this line represents a 1D vector **A**. Its mirror reflection will represent another vector $\mathbf{A}' = -\mathbf{A}$ (Figure 4.23). There is no way you could produce this vector from **A** by any transformation other then mirror reflection. An apparent objection that you could do it by rotating **A** through 180° does not apply here: there are no rotations in one-dimensional space! Now, let us add another dimension. We will get a two-dimensional surface for simplicity assumed to be a plane with the axes x and y. In this case, the reflected vector $\mathbf{A}'$ alone could as well be obtained from **A** by an appropriate rotation in the (xy) plane (Figure 4.24a). But a reflected two-dimensional figure, say, a duplet of vectors **A**, **B**, if reflected only in one dimension, for instance, in the x-dimension, cannot be obtained by any rotation in this plane (Figure 4.24b). If, however, we reflect in both dimensions, that is, perform a *complete inversion*, then the result can be obtained by rotation through 180° (Figure 4.24c). Therefore, the reflection of a 2D figure in only one direction is an independent operation, while the complete inversion in a 2D space is not an independent operation. Similar reasoning shows that complete inversion in a 3D space and a reflection of a 3D object in only one direction are both irreproducible by any other type of transformation.

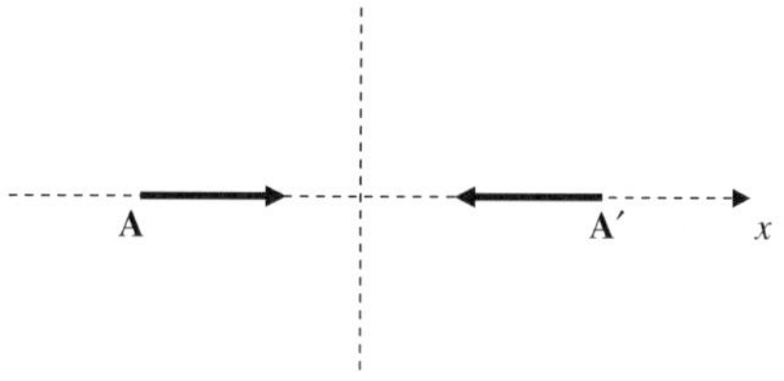

Figure 4.23

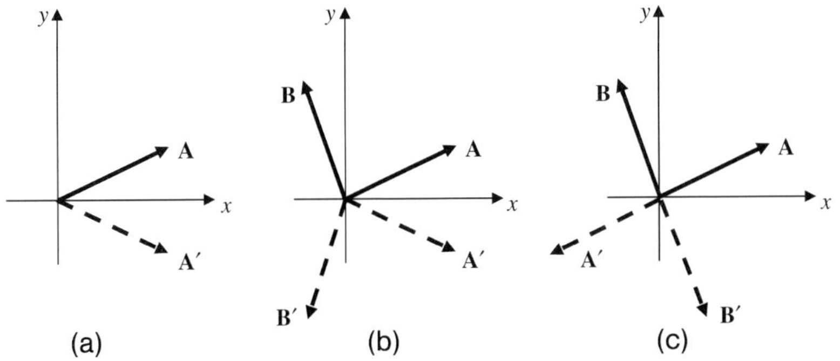

Figure 4.24 (a) In 2D space the mirror reflection **A**′ of a vector **A** can also be obtained by an appropriate rotation in the (x, y)-plane; (b) the mirror reflection of a doublet (**A**, **B**) produces a reflected doublet (**A**′, **B**′), which cannot be obtained by any rotation in the (x, y)-plane; (c) a complete mirror reflection in two dimensions (i.e., reflection with respect to x and then with respect to y-direction) produces the result that can also be obtained by rotation in the (x, y)-plane through 180°.

Thus, for three dimensions, the inversion again becomes an irreducible transformation, forming separate class independent from rotations. Indeed, if you try to obtain your inverted (mirror-reflected) self by rotating through 180°, you will find yourself only reoriented in space, but in all other respects in the same internal state as before. If you are right-handed, you remain right-handed; if your heart is in your left side, it remains in the left side. Your mirror-reflected self, in contrast, is left-handed and has the heart in the right side. The general conclusion is that in an N-dimensional space, the reflection in odd number of the basis axes constitutes an independent operation, while the reflection in even number of such axes can be reproduced by a suitable rotation. As a corollary, it follows that the complete inversion in a space of odd dimensionality is always an independent operation. Also, it is easy to see that in such a space, a complete inversion is equivalent to reflection in only one dimension plus a suitable rotation.

At this point we realize an important fact: in a broader class of transformations including mirror reflections, all vectors form two different types – genuine vectors and pretenders. A genuine (or polar) vector is one exhibiting the above-described behavior under inversions. All 3-vectors originating from displacement $\mathbf{r}$ (velocity $\dot{\mathbf{r}}$, acceleration $\ddot{\mathbf{r}}$, momentum $m\,\dot{\mathbf{r}}$, Newton's force $m\,\ddot{\mathbf{r}}$) are the polar vectors. Analytically, a polar vector $\mathbf{V}$ is defined by the property that

$$\mathbf{V} \rightarrow -\mathbf{V}(V_\alpha \rightarrow -V_\alpha) \tag{4.108}$$

under inversions. Pretenders (more politely, pseudovectors, or axial vectors) are trying to look and behave like regular vectors, and they succeed in all vector transformations other than inversions. Under inversions, however, they fail to change in sign. Such behavior seems rather strange. You can visualize the difference in behavior of the two types of vector in the following way. Take a vector, put a mirror in front of its tip and look at the mirror image. The image of a polar vector, as

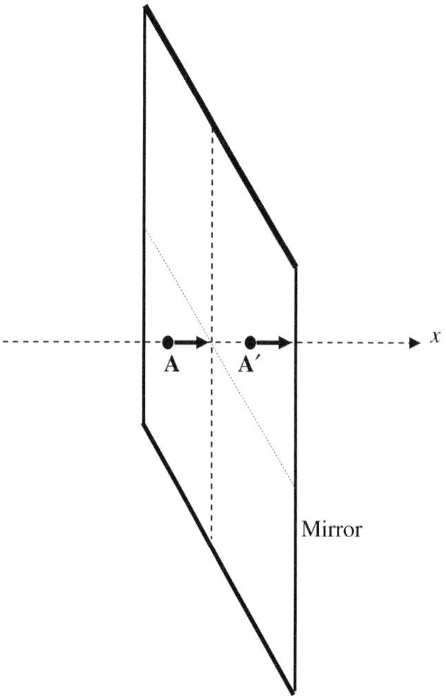

Figure 4.25 The mirror image of a *pseudovector* A. It is not inversed under mirror reflection. Such a behavior justifies the name "pseudo" given to this type of a vector. Note that the position vector of the *origin* of **A** (represented by the dot) is reflected normally (inversed), thus displaying the feature of a true vector.

described above, will look as that of any decent physical object (go back to Figure 4.23): it will point in the direction opposite to the original. If you throw an item in the positive x-direction (perpendicular to the plane of the mirror), your mirror image throws it in the negative x-direction. In contrast, the image of a pseudovector will be identical to the original. If a pseudovector points in the positive x-direction, so does its mirror image (Figure 4.25). It does not respond to reflection!

If you prefer to represent the inversion as reflection of the *basis axes* rather than the reflection of an object itself, that is, if you perform the coordinate transformation

$$x^{\alpha} \rightarrow x'^{\alpha} = -x^{\alpha}, \tag{4.109}$$

then the components of a polar vector will accordingly change the sign as well. This time they change the sign because of change of representing basis, not because of change of the vector itself.

The behavior of a pseudovector, on the contrary, will look even more weird than in mirror representation. Its components will *not* change the sign, which means that the vector itself, represented as a directed segment, must reverse its actual direction together with the axes (Figure 4.26). Since when a physical object turns around only because somebody turned the representing basis? Well, a pseudovector does

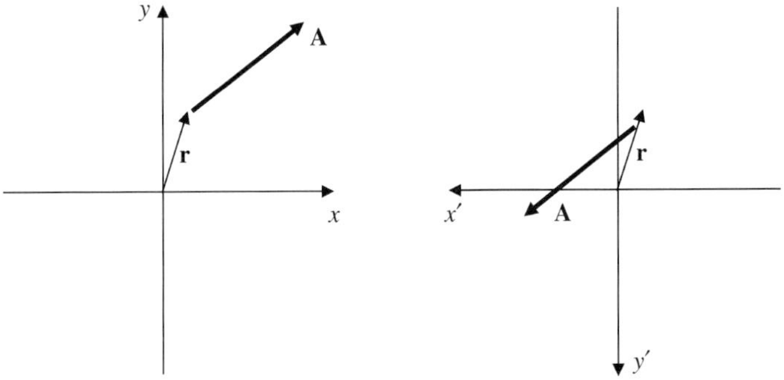

Figure 4.26 A pseudo-vector **A** under inversion of the coordinate system. (a) in the initial coordinate system x, y; (b) the same pseudo-vector in the inverted system x', y'. Note that a true vector **r** (the position vector of the origin of **A**) is not affected by the inversion of the coordinate system.

precisely this and the word "pseudo" is quite appropriate in describing this kind of behavior.

An example of a pseudovector in Euclidean 3-space is a cross product of two polar vectors

$$\tilde{\mathbf{V}} = \mathbf{A} \times \mathbf{B}. \tag{4.110}$$

In coordinate representation, this can be written in an elegant form using the antisymmetric third-rank unitary pseudotensor $\varepsilon_{\alpha\beta\gamma}$. This tensor is defined as a set of components that change the sign at the interchange of any two indexes. Therefore, only the components with all three indexes different are nonzero. For these components we set $\varepsilon_{123} = 1$; the rest are equal to 1 or -1, depending on even or odd number of transpositions takes 123 to $\alpha\beta\gamma$. Using this tensor, we can write for any component of $\tilde{V}$:

$$\tilde{V}_\alpha = \varepsilon_{\alpha\beta\gamma} A^\beta B^\gamma. \tag{4.111}$$

If we perform the spatial inversion in the expression (4.110) or (4.111), then in (4.110) vector $\tilde{\mathbf{V}}$ will remain unchanged, whereas in (4.111) its components change the sign.

Indeed, in the "mirror representation," both vectors **A** and **B**, represented as directed segments in (4.110), will flip over when mirror-reflected. Accordingly, there will appear the product of two minuses in (4.110), leaving the result on the left unchanged. This is precisely how a pseudovector behaves – its direct geometrical representation is insensitive to reflection.

The two examples of such vector are familiar to you from the introductory physics. These are the torque and angular momentum:

$$\tau = \mathbf{r} \times \mathbf{f}, \qquad \mathbf{L} = \mathbf{r} \times \mathbf{p}. \tag{4.112}$$

There should be no confusion between a polar and axial vector, on the one hand, and contravariant and covariant components of a vector, on the other. Polar and axial vectors are more than two different vectors – they are two different types of vector. Contravariant and covariant components are different types of components of *the same* vector, which may be either polar or axial vector.

What happens if we form a three-dimensional dot product of a polar and an axial vector or a 3D dot product of two axial vectors? Clearly, the result will be a scalar under all transformations but inversions. Under inversions, the first product will change the sign. It would look like a white sphere becoming black (and vice versa) when reflected by a mirror. Such an entity, remaining invariant under all transformations but inversions, and changing sign under inversions, is called a pseudoscalar. Thus, a dot product of a polar and an axial vector is a pseudoscalar.

The second kind of product, when both vectors are pseudovectors, is a true scalar again.

Now, you remember that a second-rank 4-tensor can be represented as a set of 16 elements being all the possible products of components of two 4-vectors (recall (4.103)). This can be used in visualizing an antisymmetric second-rank 4-tensor as

$$\eta_{jk} = P_j Q_k - P_k Q_j \tag{4.113}$$

and the same for contravariant components of the tensor.

This allows us to represent an axial 3-vector as a specific combination of the spatial components of an antisymmetric second-rank tensor.

Spatial components of an antisymmetric 4-tensor η^{jk} form an antisymmetric 3-tensor under spatial transformations; according to what was said above, its components can be expressed as an axial 3-vector. The components $\eta^{0\alpha}$ form, under the same transformations, a three-dimensional polar vector. In a matrix form

$$\eta^{jk} = \begin{pmatrix} 0 & B_1 & B_2 & B_3 \\ -B_1 & 0 & -A_3 & A_2 \\ -B_2 & A_3 & 0 & -A_1 \\ -B_3 & -A_2 & A_1 & 0 \end{pmatrix}. \tag{4.114}$$

Here **B** and **A** are, with respect to spatial transformations, a polar and an axial vector, respectively.

So far, we have considered linear vector algebra in 4-space with pseudo-Euclidean metric. Now we can consider the basic elements of vector calculus in the same space. The basic element in vector calculus is a differential del-operator, which in Cartesian coordinates has the form

$$\vec{\nabla} \equiv \frac{\partial}{\partial x}\hat{\mathbf{x}} + \frac{\partial}{\partial y}\hat{\mathbf{y}} + \frac{\partial}{\partial z}\hat{\mathbf{z}}, \tag{4.115}$$

or using Einstein's summation rule

$$\vec{\nabla} = \frac{\partial}{\partial x^{\alpha}}\mathbf{x}^{\alpha}. \tag{4.116}$$

Be careful in reading these expressions. They may appear to look like a dot product of two vectors and, therefore, one may be tempted to interpret the expression on the left as a scalar. This would be wrong, because the elements $\hat{\mathbf{x}}^{\alpha}$ on the right are three

($\alpha = 1, 2, 3$) different unit vectors (basis vectors), not the three different components of the same vector. Accordingly, $\vec{\nabla}$ is, as we know, a vector operator, which is emphasized here by the arrow on the top. The partial derivatives $\partial/\partial x^{\alpha}$ are the components of this vector along the respective directions $\vec{\hat{\mathbf{x}}}^{\alpha}$. If we apply the del-operator to a scalar function $\Phi(\mathbf{r})$, we obtain a vector function $\nabla\Phi(\mathbf{r})$– the gradient of $\Phi(\mathbf{r})$. If we apply it to a vector function $\mathbf{A}(\mathbf{r})$ as in dot product, we obtain a scalar function $\vec{\nabla}\cdot \mathbf{A}(\mathbf{r})$ – the divergence of $\mathbf{A}(\mathbf{r})$; finally, if we apply it to the same vector function $\mathbf{A}(\mathbf{r})$ as a cross product, we obtain another vector function $\vec{\nabla}\times \mathbf{A}(\mathbf{r})$ – the curl of $\mathbf{A}(\mathbf{r})$. Similarly, we can consider the dot product of $\vec{\nabla}$ with itself to obtain ∇^2 – the famous Laplacean operator.

To extend all this onto the four-dimensional pseudo-Euclidean space, we add, together with the new temporal dimension $x^0 = ct$, the corresponding differential operation – partial derivative with respect to time $\partial/\partial x^0 = \partial/c\,\partial t$. Note that such derivative has been used in vector calculus and classical field theory as long as the spatial derivatives have. What makes it a new relativistic generalization is the realization of the fact that temporal derivative acts on the same footing as the spatial ones and behaves under 4-rotations as a component of a 4-vector. We can thus form a 4-vector differential operator (4-gradient)

$$\Box \equiv \left(\frac{\partial}{\partial x^0}, \vec{\nabla}\right). \tag{4.117}$$

Now, the moment we expand our operations onto this broader stage, there comes the distinction between the two sorts of a vector even for Cartesian coordinates in 3-space. The question arises that if we apply the 4-vector operator (4.117) to a scalar function $\Psi(\mathbf{r}, t)$, we obtain the 4-gradient

$$\Box\Psi = \frac{\partial\Psi}{\partial x^0}\hat{\mathbf{x}}^0 + \frac{\partial\Psi}{\partial x^1}\hat{\mathbf{x}}^1 + \frac{\partial\Psi}{\partial x^2}\hat{\mathbf{x}}^2 + \frac{\partial\Psi}{\partial x^3}\hat{\mathbf{x}}^3 = \frac{\partial\Psi}{\partial x^j}\hat{\mathbf{x}}^j, \tag{4.118}$$

whose components are the four partial derivatives of the scalar Ψ. They must form a 4-vector, but the question is: Are they the contravariant or covariant components of this vector? The correct answer is suggested by the structure of expression (4.118) – it contains the derivatives with respect to coordinates x^{α}, which are the *contravariant* components of the 4-vector of an event in space-time. Having a contravariant component in a denominator of a fraction must make the fraction itself a covariant component. But this argument, while being cogent, is not rigorous. To make it more rigorous, consider the perfect differential of a function $\Psi(\mathbf{r}, t)$:

$$d\Psi = \frac{\partial\Psi}{\partial x^j}dx^j. \tag{4.119}$$

The perfect differential of a scalar is also a scalar. On the other hand, the expression on the right is a scalar (dot) product of two 4-vectors. In Equation (4.119), the components of one of the vectors (dx^j) are contravariant. Therefore, the components of the other vector ($\partial\Psi/\partial x^j$) must be covariant.

Now it is straightforward to see that the dot product of the □-operator with a 4-vector **A**(**r**) will give a 4-divergence:

$$\Box A = \frac{\partial A^j}{\partial x^j}. \tag{4.120}$$

Similarly, we can generalize the curl operation to four dimensions by composing a set of products

$$\eta_j^k \equiv \frac{\partial}{\partial x^j} A^k \tag{4.121}$$

and then forming a set an their antisymmetric combinations

$$\Xi_j^k \equiv \frac{\partial A^k}{\partial x^j} - \frac{\partial A^j}{\partial x^k}. \tag{4.122}$$

This will produce an antisymmetric second-rank 4-tensor, whose spatial nondiagonal elements are the components of curl **A**. Finally, taking the dot product of the 4-vector □-operator with itself, we will obtain $\Box^2$ – the famous D'Alambertian.

(Warning: the notations in (4.117)–(4.122) are not conventional – the square symbol on the left is universally used to denote D'Alambertian. But the notations used here seem to be more natural.)

Problems

4.1 Using the Lorentz transformations (4.4) and (4.6), prove that
   (a) the kinematic length of a 4-vector A remains invariant for observers in different inertial RF;
   (b) the dot product of two different 4-vectors is also invariant under the Lorentz transformations.

4.2 For a two-dimensional rotation in the x, y plane derive the transformation (4.32) between the old (x, y) and the new (x', y') coordinates of a point A.

4.3 Consider a transformation from Cartesian coordinate system on a 2D-plane to a nonrectangular coordinate system described in Section 4.7 with angle φ between coordinate axes and find the new coordinates x', y' for the two cases:
   (a) The coordinates x', y' of the point A are its *normal* projections onto the new axes.
   (b) The coordinates $\tilde{x}'$, $\tilde{y}'$ of the point A are its *oblique* projections onto these axes. Compare the results with one another and with the transformation (4.32) under "pure" rotation in the previous problem.

4.4 A satellite is first launched vertically up beyond the atmosphere and then horizontally to acquire a necessary speed for orbiting the Earth at a given distance. Is this identical to the operation 2 (the second mission of Mr. O'Bryan) Section 4.4? Explain.

4.5 Derive Equation (4.52).

4.6 Derive Equation (4.53) for the corresponding conditions $v_x = v_y$.

4.7 Find the tilts α_x and α_y for a boost along a direction **V** specified by an angle θ the vector **V** makes with vertical in K.

4.8 Derive Equation (4.57).

4.9 Consider a square ABCD shown in Figure 4.15a and boosted along the direction of its diagonal AC. After the boost, this square, as observed from its initial rest frame K, is a rhombus shown in Figure 4.15b. Its bottom side A′D makes an angle alpha$_x$ with the horizontal in K, determined by Equation (4.57). Find the angle α'_y between A′B and the y'-axis, as observed by Mr. O'Bryan in the reference frame K'_x described in Section 4.4.

4.10 In Figure 4.8 consider the triangle $OA'\tilde{A}'$ and show that $(OA)^2 = O\tilde{A} \cdot O\tilde{A}'$ using Equation (4.22) and the sine theorem.

4.11 Derive the expression inverse to (4.96a), that is, express the components A^j in terms of A_k.

4.12 Prove that the three components $\eta^{0\alpha}$ of the second-rank tensor η^{jk} form a 3-vector, and η^{00} is a scalar, under purely spatial coordinate transformations.

4.13 Find the determinant of matrix (4.99).

4.14 (a) A relativistic spaceship from a Star Trek serial is moving "horizontally" past the Earth at 99% of the speed of light and the water rises in the ship's swimming pool at 5 m/s.Find the tilt of the water surface with respect to the horizontal, as observed from Earth.

(b) Reverse the conditions: let now the spaceship "crawl" at 5 m/s and the light wave move up in the ship's dense atmosphere at 99% of the speed of light in vacuum.Find now the tilt of the propagating wave front with respect to the horizontal.

(c) Let the spaceship and the light wave inside move (in mutually perpendicular directions, as before) each at 99% of the speed of light in vacuum. Find the tilt of the wave front in this case.

4.15 Show that the metric tensor is symmetrical, that is, $g_{mn} = g_{nm}$.

4.16 What is the difference between the statements $\eta_j{}^k \neq \eta^j{}_k$ and $\eta_j{}^k \neq \eta_k{}^j$?

4.17 Suppose you have a tensor $\eta^{0\alpha}$ in Cartesian coordinates. Show that $\eta^{0\alpha} = \eta_0{}^\alpha$, but $\eta^{0\alpha} = -\eta^0_\alpha$.

4.18 Show that for a metric tensor $g^i{}_k = g_i{}^k$.

4.19 Show that
(a) the components $\eta^{0\alpha}$ of a tensor η^{jk} behave as a 3-vector under purely spatial rotations;

(b) the component η^{00} behaves as a scalar (remains invariant) under such rotations.

4.20 An antisymmetric second-rank tensor is defined as $\eta_{mn}=-\eta_{nm}$. What are the diagonal elements of such a tensor?

4.21 Show that for the mixed components of a symmetric second-rank tensor η^{mn} we have $\eta^{m}{}_{n} = \eta_{n}{}^{m}$.

4.22 Show that

(a) there are only three independent spatial ($\alpha=1, 2, 3$) components of the angular momentum tensor defined by Equation (4.111) or (4.112);

(b) these components transform as a 3-vector under spatial rotations.

4.23 How many independent ("mutually perpendicular") 3D subspaces are contained within Minkowski's space-time? Explain.

4.24 Suppose we observe the objects that can freely pass though one another without deviation or absorption. In this respect, they behave much like photons, which move practically without scattering on one another. But our objects, unlike photons, can move with different speeds in the same space. Let us call them "ghosts."Now, three such ghosts move at constant (but different) speeds across a flat field [26].

(a) We know that any three objects can form three different pairs. Suppose we observe that two of these pairs pass through each other at different times. Can we say with certainty whether the third pair will pass (or passed at some earlier time) through each other?

(b) If instead of three, we now have four ghosts, they can form six pairs. Five of these pairs pass through each other at different times. Can we say with certainty whether the sixth pair also passes through each other? Prove your answers to (a) and (b) and explain the similarity (or difference) between them.

4.25 Prove that the metric tensor is symmetric, that is, $g_{jk}=g_{kj}$.

4.26 Assuming that you are given the metric tensor g_{jk}, find its contravariant counterpart g^{jk} (*Hint*: Consider the differentials dx_j in (4.100) as given, and the differentials dx^j as unknown).

4.27 Consider two arbitrary two-dimensional vectors **P** and **Q** in a plane with axes as depicted in Figure 4.22, and form the second-rank two-dimensional tensors $\eta_{\alpha\beta}$, $\eta^{\alpha\beta}$, η_{α}^{β}, and η_{β}^{α}.

(a) Is the tensor $\eta_{\alpha\beta}$ symmetrical?

(b) Is the tensor $\eta^{\alpha\beta}$ symmetrical?

(c) Determine by direct comparison whether $\eta_1{}^1$ is equal to $\eta^1{}_1$, or not.

4.28 Find the expression for the metric tensor in

(a) spherical coordinates;

(b) cylindrical coordinates
by writing the spatial part dr^2 of the square of the interval in the corresponding coordinates and comparing the result with the general expression (4.101).

4.29 Using the expression (4.103) and property (4.105), find the relations between
(a) η_{00} and η^{00};
(b) η_{01} and η^{01};
(c) η_{11} and η^{11};
(d) $\eta^0{}_0$ and η^{00};
(e) $\eta^1{}_0$ and η^{01};
(f) $\eta^0{}_1$ and η^{01};
(g) $\eta^1{}_1$ and η^{01}.

4.30 Show that all diagonal elements of an antisymmetric tensor η^{jk} are zeros.

4.31 Prove that if $\eta^{jk} = \eta^{kj}$, then $\eta^j{}_k = \eta_k{}^j$.

4.32 A non-Cartesian coordinate system has its x- and y-directions making 60° with one another. A displacement vector **A** with a norm of $2m$ lies in the x–y plane and makes 30° with the x-direction. Find
(a) the contravariant components of **A**;
(b) the covariant components of **A**;
(c) prove that the expression for the norm of this vector in terms of its components of both sorts gives exactly $2m$.

5
Relativity at Work I: Mechanics

After learning the concept of a 4-vector in space-time, as well as basic operations on 4-vectors (such as dot product), we can now turn to direct physical examples of 4-vectors other than the 4-displacement *s*. The concept of 4-displacement can be used as a fundamental building block to form all other important physical characteristics of a system, such as velocity, acceleration, linear and angular momentum, and so on. Considered in space-time, they are all either 4-vectors or 4-tensors of a higher rank. Introducing 4-vectors and tensors, which represent important characteristics of motion, will allow us to reformulate mechanics in relativistically invariant form.

5.1
4-Velocity

> *"Does time stand still or does it flow?*
> *If it is in a state of flux, what is its speed?*
> *And since we measure speed by the ratio of traveling distance*
> *to the periodic motion of the clock,*
> *how are we supposed to measure the velocity of time?*
> *By time itself?"*
>
> Paul Hartal

We will first consider four-dimensional velocity (or just 4-velocity) of a particle. Suppose we picked up two close points on the particle's world line: $P_1 = (ct_1, x_1, y_1, z_1)$ and $P_2 = (ct_2, x_2, y_2, z_2)$. They are connected by a small interval ds that can be specified by its projections ($c\,dt$, dx, dy, dz) onto coordinate axes, where $dt = t_2 - t_1$, $dx = x_2 - x_1$, $dy = y_2 - y_1$, $dz = z_2 - z_1$. In Minkowski's world, these projections form the components of a four-dimensional displacement with the temporal part $c\,dt$ and the spatial part $d\mathbf{r} = (dx, dy, dz)$.

In Newtonian mechanics, the motion of a particle from P_1 to P_2 is characterized by velocity $\mathbf{v} = d\mathbf{r}/dt$. The velocity $\mathbf{v}$ is a three-dimensional vector in space. We obtain it by dividing the 3-displacement $d\mathbf{r}$ by the corresponding time interval dt, which is invariant under *three-dimensional* rotations and Galilean transformations. If we want

Special Relativity and How it Works. Moses Fayngold

ISBN: 978-3-527-40607-4

to obtain a *four-dimensional* analogue of velocity in Minkowski's world, we must take a 4-displacement ($c\,dt$, $\mathrm{d}\mathbf{r}$) and divide it by a quantity that is related to the motion of a particle between the points P_1 and P_2 and at the same time is invariant under *four-dimensional* rotations, that is, under Lorentz transformations. Such a quantity is the interval ds! Therefore, we can define 4-velocity as

$$u = (u_t, u_x, u_y, u_z) = \left(\frac{c\mathrm{d}t}{\mathrm{d}s}, \frac{\mathrm{d}x}{\mathrm{d}s}, \frac{\mathrm{d}y}{\mathrm{d}s}, \frac{\mathrm{d}z}{\mathrm{d}s}\right). \tag{5.1}$$

The 4-velocity thus defined turns out to be dimensionless, as is, for instance, the ratio $\beta \equiv v/c$. It also follows from this definition that the square of any 4-velocity is equal to unity:

$$u_t^2 - u_x^2 - u_y^2 - u_z^2 = \frac{c^2\mathrm{d}t^2 - \mathrm{d}x^2 - \mathrm{d}y^2 - \mathrm{d}z^2}{\mathrm{d}s^2} = 1. \tag{5.2}$$

Using the two types of coordinates introduced in Section 4.7, this can be written as

$$u_j u^j = 1. \tag{5.2a}$$

This means that in a four-dimensional space on whose axes are plotted the components of a 4-velocity vector, the tip of this vector always remains on the surface of a "sphere" of unit radius, centered about the origin. Mathematicians call such a sphere a pseudosphere. The word "pseudo" here reflects the fact that unlike the conventional sphere of unit radius in 3-space

$$x^2 + y^2 + z^2 = 1, \tag{5.3}$$

the square of the "temporal" component of the sphere (5.2) in the 4-velocity space enters the expression with the sign opposite to that of the squares of the "spatial" components. In this respect, it is similar to the expression for the square of an interval, but with the important distinction that all the vectors of 4-velocity have one fixed length $u = 1$. Geometrically, 4-velocity is a unit vector, tangent at each point to the world line of a particle.

Now, as you look at expression (5.2), you may wonder how can such a vector of constant "length" characterize the real physical velocity of a particle, which can, apart from varying in direction, also vary over continuous range of *magnitudes*?

This becomes clear if we turn to the *components* of the 4-velocity and express them in terms of the ordinary 3-velocity **v**. To this end, we write the interval in the form

$$\mathrm{d}s^2 = c^2\mathrm{d}t^2 - \mathrm{d}x^2 - \mathrm{d}y^2 - \mathrm{d}z^2 = c^2\mathrm{d}t^2 - \mathrm{d}r^2 = c^2\mathrm{d}t^2\left(1 - \frac{v^2}{c^2}\right) = \gamma^{-2}(v)c^2\mathrm{d}t^2. \tag{5.4}$$

Then we find using (5.1)

$$u_t = \gamma(v), \qquad u_x = \frac{v_x}{c}\gamma(v), \qquad u_y = \frac{v_y}{c}\gamma(v), \qquad u_z = \frac{v_z}{c}\gamma(v) \tag{5.5a}$$

or

$$u = \left(\gamma(v), \frac{\mathbf{v}}{c}\gamma(v)\right). \tag{5.5b}$$

We see that even though the square of a 4-velocity is always equal to 1 for any possible motion of a particle, its components are variables depending on **v**. For instance, a stationary particle ($\mathbf{v} = 0$) has the Lorentz factor $\gamma(v) = 1$. The temporal and spatial components of its 4-velocity are, respectively, (1, 0, 0, 0). The length of the vector is determined by $u^2 = 1 - 0 - 0 - 0 = 1$, so that $u = 1$. This is somewhat a counterintuitive result, but we have to keep in mind that u is the *4-velocity*, not 3-velocity, so it is nonzero even when **v** (its spatial part) is zero. This is because even resting objects "flow" with time, that is, move along the temporal dimension of space-time. Accordingly, the 4-velocity vector of a resting body is all lined up along the *ct*-axis – a quite natural result for a particle at rest. For a physical particle moving at an arbitrary speed $v < c$, we have

$$u^2 = \gamma^2(v) - \frac{v^2}{c^2}\gamma^2(v) \equiv 1. \tag{5.6}$$

As we could expect from the general definition (5.1), the result is 1 regardless of the particle's speed.

For a photon, the temporal and at least one of the spatial components of its 4-velocity are infinite. They can combine into a vector of unit length only because the squares of the temporal and spatial components add with the opposite signs. The infinite values for the *components* of the 4-velocity of a photon emphasize the special role of the speed of light and its unattainability for the particles with nonzero rest mass.

Discussion

Here, we will consider a few important features in the notion of 4-velocity.

For any particle other than a photon or a graviton, the magnitude of its 4-velocity can be determined either directly from the general definition (5.2) (i.e., *before or even without* calculating the components), or indirectly by first calculating the components (5.5) and then adding up their squares according to (5.6). The result will be the same. However, if we apply the second (indirect) procedure to a photon ($v = c$, $\gamma(c) = \infty$), the expression (5.6) will give us the difference between the two actual infinities: $du^2 = \infty - \infty$. We can hope that the result will turn out to be finite because one of the above infinities *subtracts* from the other, but nobody in his/her right mind would ever try to figure out the exact outcome of such an operation with explicit infinities. In this case, we can use a limiting procedure known in calculus – for instance, writing the finite expression for some $v < c$ and considering the result at the limit $v \to c$. Already in the first stage, you will see that the final result turns out to be totally independent of v. This enables you to conclude that the result holds for $v = c$ even without considering the limit $v \to c$.

But you can do even better – start directly from the general definition (5.1). It allows you to immediately find u^2 by just writing $u^2 = ds^2/ds^2 \equiv 1$ for *any* object of the universe, *without* even calculating the components.

It follows that in terms of their 4-velocity, all objects in the world, for instance, a bullet, a snail, a stationary rock sitting in place for millions of years, and a photon, move through space-time with the same speed – the speed of light!

The realization of the fact that *all* objects in the world flow through *space-time* with the same invariant speed $u = c$ (or, in our units, $u = 1$) is one of the most beautiful statements of the theory of relativity.

Well, if this is so, what is then the difference between the photon and the motionless rock in this respect? The answer is: only in the *tilt* of their respective world lines to the time axis.

Now we can take our quest for the meaning of 4-velocity one step further. All the previous examples correspond to the motion of known particles. According to the basic definition (5.1) with the interval in the form (5.4), for these particles all u^0 come out positive. If we use the vertical axis for the u^0 component of 4-velocity, plotting it up for $u^0 > 0$, and two horizontal axes for its x- and y-components (u^1 and u^2), as we had done for the 4-displacements in Minkowski's space, then the locus of points of all possible 4-velocities of such particles forms the upper fold of a two-folded hyperboloid of revolution (Figure 5.1a). What about the lower fold of this hyperboloid? Clearly, it should correspond to negative u^0. Here, we came to a point where the mathematical or geometrical symmetry of an abstract construction may hint at some hidden physical symmetry. The reason why all the u^0 components are positive is that all the time intervals we had considered in the expression $u^0 = dx^0/ds = (c\,dt/c|dt|)\gamma(v)$ were positive. This corresponds to the "regular" world, with everything moving "forward" in time (from past to future). Negative u^0 could arise from this expression only if we allow negative time intervals $dt < 0$. Physically, this would correspond to motions "backward" in time (from future to past). Actually, the relativistic equations have solutions describing this kind of motion. Usually, such solutions are dropped as "nonphysical," but actually they can be incorporated into the picture of the world on an equal footing with the "regular" solutions describing the evolution forward in time [34–36]. They can also be

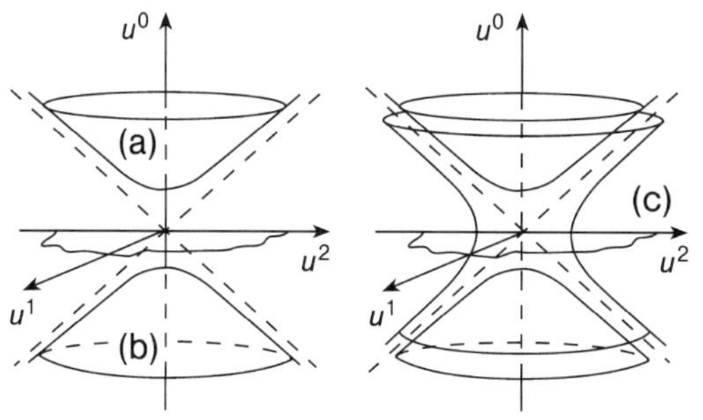

Figure 5.1

interpreted as states with negative energy $\mathcal{E} < 0$. The more complete physical symmetry thus achieved matches the geometric symmetry of Figure 5.1 including both folds – a and b.

But here we can say again that the symmetry is still not totally complete. The algebraic structure of the 4-velocity space defined in (5.5) allows one to include also the single-folded hyperboloid of revolution (Figure 5.1c). What kind of motion does it correspond to? The answer is obvious: since the fold c lies outside the light cone, it must describe superluminal motions, or hypothetical objects propagating faster than light. Such motions were extensively discussed in many sources (e.g., [1,5]) and some more examples will be considered in Chapter 11 of this book. For now suffice it to note that, regardless of whether superluminal objects exist or not, definitions (5.1) and (5.2) attribute a 4-velocity vector to *any* two events P_1 and P_2 in space-time, including those separated by a spacelike interval. In the latter case, you can formally think of a fictitious superluminal particle starting at P_1 when the event 1 happens there and arriving at P_2 when the event 2 happens there.

If you apply straightforwardly definition (5.1) to such a particle, you will get imaginary numbers for some or all components of u. Geometrically, this means that the components of the 4-velocities of superluminal motions along spacelike intervals are not mapped onto any real domain in the 4-velocity space. The straightforward version (5.1) of the definition of the 4-velocity in a sense leaves out the superluminal motions. For u^2 we still get $u^2 = 1$. This is because, even though the square of the interval is negative for superluminal objects, the algebraic structure of expression (5.2) implies that the square in the denominator is in this case also negative. This makes u^2 identically equal to 1 even when $v > c$! The statement that the square of the 4-velocity (more accurate, its dot product with itself) *always* equals to 1 is accepted in most textbooks without much comment. This is perfectly OK, since it does not lead to any contradictions or inconsistencies in the description of the world. Aesthetically, however, the definition in which two events connected by a spacelike interval are represented by a spacelike 4-velocity (and with all its components real!) seems more preferable. Such definition will follow from (5.1) if we consider ds as the *modulus* of 4-displacement, as mentioned in the beginning of this section. In this case, the denominator of (5.2) will always be nonnegative, while the numerator may be positive, zero, or negative for timelike, lightlike, and spacelike intervals, respectively. Applying this to any event absolutely remote from that at the origin in Figure 5.1, we find for the corresponding 4-velocity that $u^2 = -1$. All such 4-velocity vectors point outside the light cone in Figure 5.1c, and have their tips on the fold c. The advantage of this interpretation is that it maps the interior of the light cone of Minkowski's space-time onto the inner folds a and b of the 4-velocity space and the exterior of the light cone of Minkowski's space-time onto the outer fold c of the 4-velocity space. Its disadvantage is that now u becomes imaginary for $v > c$ and, accordingly, u^2 for the photons turns out to be double-valued: it is $+1$ or -1 depending on whether we approach the speed of light from one or the other side of the light barrier (graphically, this corresponds to approaching the light cone by sliding away from the origin over the inner fold or over the outer fold in Figure 5.1c). But, on the other hand, this may be an indication that the invariant speed c is the barrier not only for all subliminal motions but also for all hypothetical superluminal motions – it may be a two-sided barrier! Which definition to choose is a matter of taste; in any case, as it comes to the *magnitude* (modulus) of u, both interpretations of definition (5.1) produce the same outcome $|u| = 1$ for all entries without exception – be they luminal, subliminal, or superluminal. But the second interpretation explicitly suggests that both logically possible worlds, subliminal and superluminal, can in a certain sense be considered on the equal footing. This brings in an additional symmetry between these domains of space-time.

On the other hand, we must carefully distinguish between Minkowski's space-time and the 4-velocity space. As we know, the intervals in Minkowski's space-time may have *any* length; the tips of 4-displacement vectors that can be drawn from the origin fill in the whole space-time, thus forming the continuous unbounded four-dimensional set. This set can be divided into three different domains – spacelike ($s^2 < 0$), null or lightlike ($s^2 = 0$), and timelike ($s^2 > 0$).

In contrast, the tips of all possible 4-velocity vectors that can be drawn from the origin of the 4-velocity space fill in only a three-dimensional subset of this space. Figure 5.1 shows an incomplete representation of this situation. It is, by necessity, restricted to only two "spacelike" variables – u_x and u_y (the "space" of the third one is occupied by timelike variable u_t). As we have seen from the figure, the subset forms the sheets of hyperboloid of revolution about the temporal axis u_t. The top sheet represents motions as they occur in Nature – forward in time (from past to future); the bottom sheet represents hypothetical motions backward in time (from future to past); the outer sheet represents motions along the spacelike intervals. Of course, the whole thing should be completed by adding the third "spacelike" coordinate u_z, which must be also perpendicular to u_t, as well as to u_x and u_y. This would extend an incomplete two-dimensional subset in the figure to a complete three-dimensional subset of the 4-velocity space.

Thus, while the tips of all possible displacements fill all space-time of Minkowski's world, the tips of all possible 4-velocity vectors reside only on three-dimensional "hypersurface" of the abstract 4-velocity space. As you can see from the figure, this hypersurface is a three-folded hyperboloid of revolution about the u^0-axis. It does *not* include the "light cone" (the surface, for which $u^2 = 0$)! In contrast with the lightlike 4-displacements in Figure 4.2 (Section 4.1), whose norm is zero, there is no such thing as a lightlike 4-velocity vector with the zero norm!

5.2
4-Acceleration

By analogy with definition of 4-velocity, we can consider the second derivative of coordinates

$$w^j = \frac{\mathrm{d}u^j}{\mathrm{d}s} = \frac{\mathrm{d}^2 x^j}{\mathrm{d}s^2}. \tag{5.7}$$

This is 4-acceleration. There is an important universal relation between 4-velocity and 4-acceleration. We can grasp it intuitively from the familiar classical analogy. As a point is moving, its velocity vector changes in both – magnitude and direction. The rate of change of the tip of the velocity vector generates another vector – 3-acceleration **a**. Generally, this vector can have any magnitude and any direction. But in the special case, when 3-velocity only changes in direction, not in the magnitude (e.g., in a circular motion), the acceleration is perpendicular to the velocity. Now, what is only the special case for 3-velocity and 3-acceleration, is a fundamental law – the only possible relation – for 4-velocity and 4-acceleration. Since the magnitude of 4-velocity is a fixed universal quantity, it can only change in direction and, accordingly, the 4-acceleration can be only perpendicular to it.

We can get the formal proof by differentiating u^2 and using definition (5.7):

$$\frac{\mathrm{d}u^2}{\mathrm{d}s} = 2u \cdot \frac{\mathrm{d}u}{\mathrm{d}s} = 2u \cdot w. \tag{5.8}$$

But the derivative on the left-hand side of (5.8) is zero since u^2 is a constant equal to 1. It follows then

$$u \cdot w = u_j w^j = 0. \tag{5.9}$$

This is a mathematical definition of orthogonality between two vectors.

Discussion

In view of the second interpretation of the definition of 4-velocity discussed in the previous section, the result (5.9) requires more careful consideration for objects moving with the invariant speed. These objects reside on the "razor blade" between subluminal and superluminal worlds. Graphically, they form the light cone in space-time (Figure 4.1) and the infinitely remote region of the 4-velocity space, where the inner folds of the 4-velocity vectors merge with the outer fold (Figure 5.1). According to the second interpretation,

$$u^2 = \begin{cases} 1, & \text{on the inner folds,} \\ -1, & \text{on the outer fold.} \end{cases} \tag{5.10}$$

In other words, u^2 has a sharp discontinuity on the light cone in space-time. It follows then that Equation (5.8) can yield singularity on this cone. Formally, the magnitude of 4-acceleration becomes infinite in such a case.

This, however, should not be considered as an inconsistency, because it has a reasonable physical interpretation. First, as just mentioned, the singularity occurs only in the infinitely remote regions of the 4-velocity space. Second, and more important, it only occurs when the two end points of displacement ds lie *on the opposite sides* of the light cone, which physically would correspond to the transport of an object between subluminal and superluminal worlds, that is, its passing through the light barrier. No surprise then that the theory gives the infinite (and thereby physically unattainable) acceleration for such a case.

5.3
4-Momentum

Using the concept of 4-velocity, we can formulate the relativistic mechanics of a point mass.

In Newtonian mechanics, the product of the particle's velocity by its mass gives the momentum

$$\mathbf{P} = m\mathbf{v}. \tag{5.11}$$

In relativistic mechanics, the analogous product of the rest mass by the 4-velocity and by c gives the 4-momentum:

$$P_j = m_0 u_j c. \tag{5.12}$$

As we have arranged in the previous section, the index j takes on the values 0, 1, 2, 3, which stand, respectively, for the ct, x, y, and z component of a 4-vector. Also, as mentioned before, when considering only three spatial dimensions, we will often use the Greek letter α ranging through 1, 2, 3. If they are used as superscripts, they will represent contravariant components, otherwise – covariant components. As we know, in Cartesian coordinates, the covariant and contravariant components are indistinguishable, except for the sign. Therefore, wherever it is convenient, we can represent spatial projection $\mathbf{p}$ of a 4-vector p by its covariant components, writing, for example, p_α ($\alpha = 1, 2, 3$) instead of p^α. Using covariant components, we avoid the confusion with the powers.

All spatial components of the 4-momentum have a simple physical meaning. They can be written as products of corresponding components of 3-velocity by the common coefficient m:

$$P_\alpha = m_0 \gamma(v) v_\alpha = m v_\alpha \quad \text{or} \quad \mathbf{P} = m\mathbf{v}, \tag{5.13}$$

where

$$m(v) = m_0 \gamma(v). \tag{5.14}$$

Expression (5.13) for relativistic momentum looks exactly like classical momentum if we identify the factor m multiplying $\mathbf{v}$ with the mass of a moving body. This factor is called the relativistic mass. We see that the relativistic mass depends on speed of the particle. When the speed is small, m is practically indistinguishable from the particle's rest mass m_0. This is the situation that we have been used to in nonrelativistic physics. However, at high enough speeds the mass increases. As the speed approaches c, mass m becomes infinitely large.

Historically, physicists had suspected that the mass of a particle should increase with its speed even before the advent of the theory of relativity. Apart from theoretical arguments following from electrodynamics [34], there was accumulating experimental evidence that mass depends on speed. For example, an electron moving with velocity $\mathbf{v}$ in a region with a homogeneous magnetic field $\mathbf{B}$ experiences, as we will find in the next chapter, the so-called Lorentz force

$$\mathbf{f} = q_e(\mathbf{v} \times \mathbf{B}), \tag{5.15}$$

where q_e is the electron's charge. The force **f**, being a cross product of the two vectors, is perpendicular to either of them. Therefore, it plays the role of centripetal force bending the charge's trajectory. If the electron's initial velocity was perpendicular to **B**, its trajectory will be bent into a circle. We can easily find its radius R by equating the centripetal force on a particle with the electron's mass m_e to the force (5.15)

$$m_e \frac{v^2}{R} = q_e v B. \tag{5.16}$$

Here, we have dropped the sign of the cross product because $\mathbf{f} \perp \mathbf{B}$, and we are interested only in the magnitudes of the resulting vectors. We obtain

$$R = \frac{m_e}{q_e} \frac{v}{B}. \tag{5.17}$$

From (5.16), there also follows the expression for angular frequency of the electron's orbit:

$$\omega = \frac{v}{R} = \frac{q_e}{m_e} B. \tag{5.17a}$$

The radius here is proportional to the speed of the particle and inversely proportional to the strength of the magnetic field. According to this result, if we increase v and B by the same factor, the radius R should remain the same. Experiments, however, showed that it becomes increasingly difficult to keep the particle in the same orbit even if we adjust the field B to increased speed as required by (5.17). In order to keep R constant, we have to increase B *more* than we increase v. Since the charge definitely does not depend on v, the only possible explanation of this effect is that the particle's mass increases with its speed. An object in motion must be more inert than at rest! And both the experimental data and the theoretical considerations showed that the mass dependence on speed must be of the form $m(v) = m_0\gamma(v)$ – precisely the result obtained so effortlessly by forming the expression (5.12) for the relativistic 4-momentum. The dependence of mass on speed and its specific form (5.14) are intimately connected with the fact that expression (5.12) forms a 4-vector.

Here we have reached an important landmark. Having realized that the spatial part of 4-momentum can be interpreted as a 3-vector of regular momentum with the mass depending on speed according to (5.14), we now come to the question: what is the physical meaning of the temporal part P_0?

The temporal component of the 4-momentum is given by

$$P_0 = m_0 c\gamma(v) = mc. \tag{5.18}$$

(The subscript "0" in P stands for the "*ct*" component of momentum. Do not confuse it with the similar subscript in m used to indicate the *rest* mass.)

Before doing any math, we can come up with a tentative answer about the meaning of (5.18) by analogy. In Minkowski's world, two counterparts of reality – space and time – are combined into space-time. Out of four components x^j of a space-time interval, the component x^0 represents time. Similarly, we can expect that in relativistic mechanics the two conserving characteristics of motion – momentum **P** and energy $\mathcal{E}$ – must combine into 4-momentum in the four-dimensional momentum space.

The reason for such analogy is that **p**, like **r**, is a 3-vector, and $\mathcal{E}$, like t, is a scalar under 3-rotations. If this is true, then the component P_0 must represent the energy of the system.

We can confirm this expectation by the following argument. Denote $v/c \equiv \beta$ and expand the mass (5.14) into the Taylor series:

$$m = m_0(1-\beta^2)^{-1/2} = m_0 + \frac{1}{2}m_0\beta^2 + \frac{3}{8}m_0\beta^4 + \cdots.$$

Then multiply by c^2:

$$mc^2 = m_0c^2 + \frac{1}{2}m_0v^2 + \frac{3}{8}m_0\frac{v^4}{c^2} + \cdots. \tag{5.19}$$

The second term on the right is the familiar (nonrelativistic) expression for kinetic energy. Therefore, the following terms must be the relativistic corrections and the whole sum of all the terms depending on speed must represent *relativistic* kinetic energy K of a body moving with a speed v. Thus, we can write

$$mc^2 = m_0c^2 + K. \tag{5.20}$$

Since energy can be added only to energy, the first term on the right must also represent energy, and the same therefore can be said about the term on the left. What kind of energy are these? Now, we have only one possible interpretation: since m_0 is the rest mass of the system, the product m_0c^2 must be its rest energy; similarly, the product on the left must be the total relativistic energy of the moving system. Naturally, the total turns out to be the sum of the rest energy and the kinetic energy. Denoting this total as $\mathcal{E}$, we come to the most famous equation in physics – the mass–energy relationship first suggested by Einstein:

$$\mathcal{E} = mc^2. \tag{5.21}$$

We will discuss this relation and its implications, as well as the meaning of the rest energy, in the following sections.

Now, combining (5.18) with (5.15), we can write

$$P_0 = \frac{\mathcal{E}}{c}. \tag{5.22}$$

Just as we had expected, the temporal component of 4-momentum has the physical meaning of the total energy of the particle divided by c.

These expressions can be written in the form showing the similarity between time–space on the one hand and energy–momentum, on the other. Using (5.12), (5.18), and (5.22), we obtain direct relation between $\mathcal{E}$ and **P**:

$$\frac{\mathcal{E}^2}{c^2} - P_x^2 - P_y^2 - P_z^2 = \frac{\mathcal{E}^2}{c^2} - \mathbf{P}^2 = m_0^2c^2. \tag{5.23}$$

Solving this for $\mathcal{E}$, we obtain

$$\mathcal{E} = \pm\sqrt{m_0^2c^4 + P^2c^2}. \tag{5.24}$$

Thus, the theory predicts two possible signs for the energy of a system with given rest mass and momentum. We will delay the discussion of the very interesting implications of this fact and right now will only choose the positive sign as physically possible.

The algebraic structure of the left-hand side of Equation (5.23) is the same as the structure of the expression for the interval in Minkowski's world, or the 4-velocity in the velocity space. Each one of the values of $\mathcal{E}/c$ and $\mathbf{P}$ is different for different observers (that is in different reference frames), but, because the rest mass of a particle is constant, the difference of their squares is invariant. Therefore, the values $\mathcal{E}/c$, P_x, P_y, P_z can be considered as coordinates of a point in an abstract four-dimensional space similar to Minkowski's space-time (the momentum space). The coordinate $\mathcal{E}/c$ in this space is similar to the time coordinate ct, and P_x, P_y, P_z are similar to the spatial coordinates x, y, z. It follows that the quantities $\mathcal{E}/c$ and $\mathbf{P}$ must transform in the same way as do ct and $\mathbf{r}$ when one switches between two inertial frames moving with relative velocity V:

$$\begin{aligned}
&\mathcal{E}' = \gamma(V)(\mathcal{E} - VP_x),\\
&P_x' = \gamma(V)\left(P_x - \frac{V}{c^2}\mathcal{E}\right), \qquad P_y' = P_y, \quad P_z' = P_z.
\end{aligned} \tag{5.25}$$

Just as the Lorentz transformation of the time and space coordinates has geometrical interpretation as rotation in Minkowski's space-time, the same transformation applied to the components of a 4-momentum (Equation (5.25)) can be considered geometrically as rotation in four-dimensional momentum space.

By combining time and space into space-time, relativity also combines energy and momentum into more general entity – 4-momentum. In contrast to Newtonian physics, there are no two separate conservation laws – one for the energy of a system and one for its momentum. There is one law of conservation of the 4-momentum of the system.

If we know the energy and momentum of a given particle, we can find its velocity. The general expression for the velocity of the particle in terms of its energy and momentum is the same as in Newtonian mechanics (Problem 5.11):

$$v_\alpha = \frac{\partial \mathcal{E}}{\partial p^\alpha}, \qquad \alpha = 1, 2, 3. \tag{5.26}$$

For the velocity magnitude we can write

$$v = \frac{d\mathcal{E}}{dP}. \tag{5.27}$$

Differentiating Equation (5.24) with respect to P, we get

$$v = \frac{d\mathcal{E}}{dP} = \frac{P}{\mathcal{E}}c^2 = \sqrt{1 - \frac{m_0^2 c^4}{\mathcal{E}^2}} \le c. \tag{5.28}$$

Because the total energy of a free particle is always larger than (or equal to) its rest energy, the expression under the square root is positive and less than 1, and the particle's speed automatically comes out always less than c. For a photon, which does not have a rest mass because it never rests ($m_0 = 0$), Equation (5.28) gives $v = c$.

Note that we could derive the relation (5.28) between the velocity, momentum, and energy without using (5.27). Instead, we can write directly from (5.11)

$$\mathbf{P} = m\mathbf{v} = \frac{\mathcal{E}}{c^2}\mathbf{v}. \tag{5.29}$$

This is a more general (vectorial) form of (5.28). As an example, let us apply it to a photon in vacuum. In this case, $\mathbf{v} = \mathbf{c}$, and

$$\mathbf{P} = \frac{\mathcal{E}}{c}\hat{\mathbf{n}}, \tag{5.30}$$

where $\hat{\mathbf{n}}$ is the unit vector in the direction of the motion of the photon. According to quantum mechanics, the energy of a photon is uniquely determined by its frequency ω as

$$\mathcal{E} = \hbar\omega, \tag{5.31}$$

where $\hbar$ is the Plank's constant. Therefore, a photon with frequency ω has the momentum

$$\mathbf{P} = \frac{\hbar\omega}{c}\hat{\mathbf{n}}. \tag{5.32}$$

Here we treat the photon as a particle, but it propagates as a wave. If you divide the speed u of the wave by its frequency f, you will obtain the wavelength, $u/f = \lambda$. In our case $u = c$ and $\omega = 2\pi f$, so

$$\omega/c = 2\pi/\lambda, \tag{5.33}$$

where λ is the wavelength associated with the given photon. The ratio (5.33) is called the wave- or propagation number and is denoted as k. Physically, k is the number of waves fitting into 2π units of lengths used for measuring distances. For instance, if we measure the wavelengths in meters, then k is the number of waves within 2π meters. The product $k\hat{\mathbf{n}}$ is a vector of length k in the direction of propagation of the wave. It is called the wave vector and is an important characteristic of the wave motion. Using this concept, we can express the momentum of the photon in terms of its wave vector by a very simple equation:

$$\mathbf{P} = \hbar\mathbf{k}. \tag{5.34}$$

Equations (5.31) and (5.34) expressing the energy and momentum of a photon (considered as a particle) in terms of its frequency and wave vector relate the wave properties of light to its corpuscular properties. They were first introduced into physics by Einstein [28], who was inspired by Plank's discovery of quantization of the energy levels of molecular or atomic oscillators [29]. This work of Einstein was based on considerations having little, if anything, to do with relativity as such. And yet, the expressions for the frequency and the wave number in terms of the energy and momentum are intrinsically relativistic, which is no surprise, since they were originally obtained exclusively for the photons. Indeed, taking a close look at (5.31) and (5.34), we note that ω and $\mathbf{k}$ can be considered as the temporal and spatial components of a 4-vector! We will discuss later in Chapter 10 the profound significance of this fact.

5.4 Relativistic Angular Momentum

In the previous sections, we have considered some of the most important relativistic characteristics of a physical system. Their distinctive feature is that they are either scalars (invariants, e.g., the rest mass) or vectors under the Lorentz transformations. In the language of algebra, they are, respectively, zero and the first-rank tensors.

Here, we consider another important (and very familiar!) characteristic of a system – its angular momentum with respect to a fixed center:

$$\mathbf{L} = \mathbf{r} \times \mathbf{p}, \tag{5.35}$$

or, in Cartesian coordinates,

$$L_x = yp_z - zp_y, \qquad L_y = zp_x - xp_z, \qquad L_z = xp_y - yp_z. \tag{5.36}$$

In three spatial dimensions, we have only three components and they transform as a pseudovector under 3-rotations of a reference frame (RF) or/and mirror reflections. Actually, since their components are formed from the products of the components of the two vectors, they form a second-rank tensor. But this tensor is antisymmetric and therefore out of its $3\times 3 = 9$ elements, three diagonal elements are zeros, and out of the remaining six, only three are independent owing to the antisymmetry condition $L_{\alpha\beta} = -L_{\beta\alpha}$.

Consider now the extension of this definition to space-time. Suppose we have a moving particle, which may be passing by or orbiting around a fixed point. The particle is characterized by two 4-vectors – its 4-displacement $x = (x^i)$ and 4-momentum $p = (p^j)$. Following the algorithm described in Section 4.7 we can now form a set of 16 elements $Q^{ij} = x^i p^j$. By definition, this set is a second-rank tensor. As a next step, we define a set

$$L^{ij} \equiv Q^{ij} - Q^{ji} = x^i p^j - x^j p^i. \tag{5.37}$$

This is an antisymmetric second-rank tensor defined in the whole space-time and we can accordingly represent it by a 4×4 matrix.

Getting back to more traditional notations $(1, 2, 3) \to (x, y, z)$, we immediately recognize in the subset $L^{\alpha\beta}$ of this matrix the familiar components of the angular momentum given by (5.36):

$$L^{12} = L_z, \qquad L^{13} = -L_y, \qquad L^{23} = L_x. \tag{5.38}$$

Thus, the matrix representing the relativistic angular momentum tensor can be written as

$$L = \begin{pmatrix} 0 & -L^{01} & -L^{02} & -L^{03} \\ -L^{01} & 0 & L_z & L_y \\ -L^{02} & L_z & 0 & L_x \\ -L^{03} & L_y & -L_x & 0 \end{pmatrix}. \tag{5.39}$$

The three remaining nonzero components $L^{0\alpha}$ form a 3-vector under rotations of the coordinate system in 3-space. Indeed, using again the traditional notations, we have

$$L^{0\alpha} = \left(ct \cdot p_x - \frac{\varepsilon}{c} x;\ ct \cdot p_y - \frac{\varepsilon}{c} y;\ ct \cdot p_z - \frac{\varepsilon}{c} z \right). \tag{5.40}$$

But ct and ε are the constants under spatial rotations, so that each element in the row matrix (5.40) is a linear superposition of the corresponding components of vectors **r** and **p** and thus itself is a component of a 3-vector:

$$\tilde{\mathbf{L}} \equiv ct\mathbf{p} - \frac{\varepsilon}{c}\mathbf{r}. \tag{5.41}$$

What are transformation rules for the whole matrix L under the 4-rotations? As before, we will restrict ourselves to a special case of Lorentz transformation between systems K and K′ with their respective axes parallel to one another and x, x' along the relative velocity **V** of K′ with respect to K. We then will have

$$L_x = yp_z - zp_y = y'p'_z - z'p'_y = L'_x. \tag{5.42}$$

Thus, the component of the angular momentum parallel to the direction of relative motion does not change under the Lorentz transformation. Note the difference: for spatial displacement **r** and momentum **p** it is the *perpendicular* components that remain constant under such transformation.

The behavior of the longitudinal component of the angular momentum has a very clear physical explanation. Suppose that the particle's orbit in K is a closed curve. We can consider the orbiting particle as a clock with one ticking corresponding to one complete cycle of its orbital motion. For a K′-observer, this clock is ticking slower by a factor $\gamma(V)$ due to time dilation, that is, the particle is orbiting more sluggishly. But its relativistic mass increases by the same factor, so that its transverse momentum, and thereby longitudinal angular momentum, remains the same.

Now, what is the behavior of the transverse components of **L**?

We have

$$\begin{aligned} L_y = zp_x - xp_z &= \gamma(V)\left[z'\left(p'_x + \frac{V}{c^2}\varepsilon'\right) - (x' + Vt')p'_z\right] \\ &= \gamma(V)\left[L'_y + \frac{V}{c}\left(\frac{\varepsilon'}{c}z' - ct'p'_z\right)\right]. \end{aligned}$$

Combining this with (5.41) gives

$$L_y = \gamma(V)\left(L'_y + \frac{V}{c}\tilde{L}'_z\right). \tag{5.43}$$

Similarly,

$$L_z = \gamma(V)\left(L'_z + \frac{V}{c}\tilde{L}'_y\right). \tag{5.44}$$

Introducing, as before, the subscripts "$_{||}$" and "$_{\perp}$" for the longitudinal and transverse components, respectively, we can write all three equations (5.42)–(5.44) as

$$L_{||} = L'_{||}; \quad \mathbf{L}_{\perp} = \gamma(V)\left[\mathbf{L}'_{\perp} + \frac{V}{c}(\tilde{\mathbf{L}}'_{\perp} \times \hat{\mathbf{n}})\right], \tag{5.45}$$

where $\hat{\mathbf{n}}$ is the unit vector along $\mathbf{V}$.

We see that the elements of the angular momentum tensor in one RF are, as they should, the linear combinations of the elements of this tensor in another RF. In particular, the elements $\mathbf{L}$ in K are linear combinations of $\mathbf{L}'$ and $\tilde{\mathbf{L}}'$ measured in K′.

Generally, a system will include N particles. If the particles do not interact with each other, the net angular momentum tensor of the system is the sum of the individual angular momenta L_n^{ij}, $n = 1, 2, 3, \ldots, N$. Using the same notations as in (5.41), we can write

$$L^{\alpha\beta} = \sum_{n=1}^{N} \tilde{L}_n^{\alpha\beta}, \qquad \tilde{\mathbf{L}} = c\sum_{n=1}^{N}\left(\mathbf{p}_n t - \frac{\mathcal{E}_n}{c^2}\mathbf{r}_n\right). \tag{5.46}$$

The crucial property of elements $L^{\alpha\beta}$ is that they are components of the conserving quantity for a closed system. The net angular momentum in the first expression of Equation (5.46) is conserved in any inertial system. It follows then that the $\tilde{\mathbf{L}}$-elements must also be conserved quantities. Indeed, using (5.45) we see that the first term on the right (L' in K′) is conserved; the only way for its counterpart (L in K) on the left to be conserved is for $\tilde{\mathbf{L}}_{\perp}$ to be conserved. The same result can be obtained for the remaining component $\tilde{L}_x$, if we redirect $\mathbf{V}$, say, along the y-direction (Problem 5.7).

We come to the important conclusion that the whole relativistic angular momentum tensor is a conserved quantity. We can therefore write for $\tilde{\mathbf{L}}$

$$\sum_{n=1}^{N}\left(\mathbf{p}_n t - \frac{\mathcal{E}_n}{c^2}\mathbf{r}_n\right) = \text{const.} \tag{5.47}$$

or

$$t\sum_{n=1}^{N}\mathbf{p}_n - c^{-2}\sum_{n=1}^{N}\mathcal{E}_n\mathbf{r}_n = \text{const.} \tag{5.47a}$$

Consider the total energy of the system. For noninteracting particles, we have

$$\mathcal{E} = \sum_{n=1}^{N}\mathcal{E}_n. \tag{5.48}$$

Using this, we can rewrite (5.47) as

$$\frac{\sum_{n=1}^{N}\mathcal{E}_n\mathbf{r}_n}{\mathcal{E}} - \frac{\sum_{n=1}^{N}\mathbf{p}_n}{\mathcal{E}/c^2}t = \text{const.} \tag{5.49}$$

The form of the expression on the left is begging us to introduce new characteristics of the system:

$$\mathbf{R} \equiv \frac{\sum_{n=1}^{N} \mathcal{E}_n \mathbf{r}_n}{\mathcal{E}} \tag{5.50}$$

and

$$\mathbf{V} \equiv c^2 \frac{\sum_{n=1}^{N} \mathbf{p}_n}{\mathcal{E}}. \tag{5.51}$$

Since $\sum_{n=1}^{N} \mathbf{p}_n = \mathbf{p}$ is the net momentum of the system, the last definition can be rewritten as

$$\mathbf{V} = \frac{\mathbf{p}}{\mathcal{E}/c^2} = \frac{\mathbf{p}}{m} = \text{const.} \tag{5.52}$$

(In view of (5.21), $m = \mathcal{E}/c^2$ can be considered as the total relativistic mass of the system.)

Thus, $\mathbf{V}$ can be interpreted as the velocity of motion of the system as a whole. Equation (5.54) in these notations can be written as

$$\mathbf{R} = \mathbf{V}t + \text{const}, \tag{5.53}$$

and the conservation of $\mathbf{L}$ physically means the conservation of velocity of this motion. In other words, whatever happens within an isolated system, there is a certain point in it with position vector $\mathbf{R}(t)$ being the linear function of t, that is, moving uniformly with velocity $\mathbf{V}$. Equation (5.53) is the relativistic generalization of the law of inertia: the center of mass of an isolated system moves in a straight line at a constant speed regardless of any possible processes (e.g., collisions between the particles or particle decay). Equation (5.50) for position vector $\mathbf{R}$ is the relativistic generalization of the concept of center of mass of a system. This equation looks similar to the nonrelativistic definition, but it is essentially different in that instead of the fixed individual masses of the respective constituent particles, we have now their relativistic energies ε_n, which are variables, depending each on the speed of the respective particle. Only in the limit $v_n/c \ll 1$ for each particle, Equation (5.50) reduces to the nonrelativistic expression

$$\mathbf{R} \xrightarrow[v_n \ll c]{} \frac{\sum_{n=1}^{N} m_{0n} r_n}{\sum_{n=1}^{N} m_{0n}}. \tag{5.54}$$

Here m_{0n} is the rest mass of an n-th particle.

The nonrelativistic expression (5.54) holds for any system of particles, whereas (5.50) is defined only for a system of noninteracting particles. In relativistic mechanics, the rest mass of a system of interacting particles is determined not only by their masses and velocities, but also by the energy and momentum of the field they produce.[1)]

1) The short-range forces responsible for particle collisions and decays are usually felt only when the particles are in direct physical contact with one another, that is, at a distance comparable with their sizes. For a system of point particles assumed here, such forces are totally negligible at finite distances.

Thus, in relativity the concept of center of mass and the law of its motion automatically follow from the properties of four-dimensional tensor of angular momentum.

5.5 Force and Motion

The change in our concepts of space and time causes similar changes in our concepts of motion and force. We have seen that velocity of a moving object behaves "paradoxically" when we consider it from different reference frames. In particular, it does not obey the "obvious" law of addition of velocities. We can rephrase this in the language of transformations by stating that the velocity behaves differently with respect to different transformations; so far as we are confined to one reference frame (for instance, we rotate the axes in our three-dimensional space, but remain within the same system K), time t behaves as a scalar quantity and the velocity $\mathbf{v}$ transforms like a regular three-dimensional vector. However, when we perform rotations in space-time involving the time axis (which represents a transition to another reference frame), the velocity does not behave as a four-dimensional vector. This role is taken up by the "representative" – the 4-velocity. We have defined the 4-velocity in Equation (5.1) as $u_i = \mathrm{d}x_i/\mathrm{d}s$, where index "0" corresponds to the time axis and 1, 2, 3 correspond to the three spatial axes, whereas the interval $\mathrm{d}s$ is a scalar under 4-rotations.

The same can be done to describe the dynamical characteristics – momentum, energy, and force. By analogy with the nonrelativistic definition of 3-momentum as a vector with three components $p_\alpha = m_0 v_\alpha = m_0 \mathrm{d}x_\alpha/\mathrm{d}t$, we have defined relativistic 4-momentum as a vector with four components

$$P_i = m_0 u_i = m_0 \frac{\mathrm{d}x_i}{\mathrm{d}s}. \tag{5.55}$$

How do these changes affect the concept of force and accelerated motion?

The basic definition of force is

$$\mathbf{f} = \frac{\mathrm{d}\mathbf{P}}{\mathrm{d}t}. \tag{5.56}$$

Formally, it is similar in some respects to the definition of velocity as $\mathbf{v} = \mathrm{d}\mathbf{r}/\mathrm{d}t$. Like the velocity, force behaves as a three-dimensional vector under restricted Lorentz transformations not involving time (pure spatial rotations). The 3-momentum $\mathbf{P}$ is the product of relativistic mass and velocity:

$$\mathbf{P} = m\mathbf{v} = m_0 \gamma(v)\mathbf{v}. \tag{5.57}$$

Because of the relativistic factor $\gamma(v)$, the relation between the mass, acceleration, and force is more subtle here than in nonrelativistic mechanics. To show it, let us express the force directly in terms of acceleration $\mathbf{a} = \mathrm{d}\mathbf{v}/\mathrm{d}t$. This will require a little algebra.

Putting the expression (5.57) for $\mathbf{P}$ into definition (5.56) and performing differentiation, we obtain

$$\mathbf{f} = m_0 \frac{\mathrm{d}}{\mathrm{d}t}\mathbf{v}\gamma(v) = m_0 \left[\frac{\mathrm{d}\mathbf{v}}{\mathrm{d}t} \cdot \gamma(v) + \mathbf{v}\frac{\mathrm{d}}{\mathrm{d}t}\gamma(v)\right]. \tag{5.58}$$

Now, since $v^2 = \mathbf{v}^2$, the Lorentz factor can be written as $\gamma(v) = (1 - \mathbf{v}^2/c^2)^{-1/2}$. Therefore,

$$\frac{d}{dt}\gamma(v) = \frac{d}{dt}\left(1 - \frac{\mathbf{v}^2}{c^2}\right)^{-1/2} = -\frac{1}{2}\left(1 - \frac{v^2}{c^2}\right)^{-3/2} \cdot \left(-2\frac{\mathbf{v}}{c^2}\frac{d\mathbf{v}}{dt}\right) = \gamma^3(v)\frac{\mathbf{av}}{c^2}. \tag{5.59}$$

Putting this into the previous equation gives

$$\mathbf{f} = m_0\gamma(v)\left[\mathbf{a} + \gamma^2(v)\frac{\mathbf{a}\cdot\mathbf{v}}{c^2}\mathbf{v}\right]. \tag{5.60}$$

Now, let us read this equation. We see that the relation between force and acceleration is not that simple as it is in nonrelativistic mechanics. Its most interesting feature is that the acceleration does not generally point in the direction of the applied force!

We will illustrate relation (5.60) by considering a few special cases.

First, consider slow motions. Then, the second term in the brackets is negligible as compared to the first, and the Lorentz factor is close to 1. The equation then reduces, as it should, to the known nonrelativistic limit $\mathbf{f} = m_0\mathbf{a}$.

Now, let velocity be arbitrary and change only in direction. Geometrically, it means that the moving mass traces out an arch of a circle. The acceleration is accordingly perpendicular to the velocity, and the scalar product $\mathbf{a}\cdot\mathbf{v}$ is zero. Then

$$\mathbf{f} = m_0\gamma(v)\mathbf{a}. \tag{5.60a}$$

Next, consider the case when velocity changes only in magnitude (the force is applied along the velocity). Then $\mathbf{av} = av$, the mass moves in a straight line, and we have

$$f = m_0\gamma(v)\left[a + \gamma^2(v)\frac{v^2}{c^2}a\right] \Rightarrow m_0\gamma^3(v)a. \tag{5.60b}$$

Now, compare the two results. Don't you notice something strange about them?

In nonrelativistic mechanics, we have actually two definitions of mass: as a measure of the amount of matter in a body and as a measure of a body's inertia. The first can be measured, for instance, by weighing the body, and the second can be determined as a ratio of the applied force to the resulting acceleration. Both definitions had gone peacefully together, hand in hand.

However, in relativistic mechanics the situation is different and the two definitions of mass turn out to represent two characteristics with different behavior. The first definition – mass as the amount of matter in a body – applies directly to the body at rest. It tells us what we get if we stop the body, and is represented by a *constant* factor m_0 – rest mass. For instance, if we try to stop a photon in empty space, we get nothing, and we accordingly say that the rest mass of a free photon with definite $\mathbf{P}$ is zero. But we do not actually have to stop a moving object or to catch up with it to measure its rest mass – we can instead measure its energy ε and momentum $\mathbf{P}$ and then find the rest mass as the invariant $m_0 = (\mathcal{E}^2 - P^2c^2)^{1/2}/c^2$. If we apply this procedure to a photon, we will get zero.

The second definition – mass as the amount of body's inertia (call it inertial mass) – has direct and clear physical meaning only in situations when it can be measured as the ratio of the applied force to the resulting acceleration. Generally, such interpretation is not applicable because, according to (5.60), the force and acceleration are not parallel. But it can be applied in special cases like (5.60a) and (5.60b).

We then get two different results and accordingly introduce two different notations for corresponding inertial mass [29]:

$$m_{\perp} = m_0 \gamma(v) \tag{5.61a}$$

and

$$m_{\|} = m_0 \gamma^3(v). \tag{5.61b}$$

The mass $m_{\|}$ turns out to be greater than the mass $m_{\perp}$. Physically, this means that a moving body is more inert in the longitudinal direction than in the transverse direction. When you push it in the direction of its trajectory, the body resists harder than when you push it in the direction perpendicular to its trajectory. The inertial mass is anisotropic!

How can we understand this result? Actually, there is no need for any deeper understanding here, this is how Nature works; the equations we use give the adequate description of motion in excellent agreement with experiment and predict correctly the outcome of other experiments, and this is all one should expect from a theory.

Still, we can offer a somewhat naïve comment.

Relativity forbids any physical body with nonzero rest mass to reach the speed of light. To implement this ban, it has made the inertial mass a variable quantity *increasing* with speed, so that the closer the speed to c, the more vigorous is the body's resistance to a further change of speed. It already resists harder than the rest mass would do when the applied force is perpendicular to the velocity, even though such force cannot change the speed. But when the force pushes *along* the velocity to increase the speed, the resistance is much harder still – as if to make the ban more efficient! It is true that the "ban" works in a strange way – according to (5.61b), the resistance is the same for the force tending to accelerate the body and for the force tending to decelerate the body. Whatever the interpretations, the "longitudinal" inertia increases with speed faster than the "transverse" one!

Equations (5.61) and (5.61b) describe only two special cases. One might need an expression that would determine acceleration in the general case of arbitrary orientation between the velocity and the force. In nonrelativistic mechanics, the corresponding expression $\mathbf{a} = m^{-1}\mathbf{f}$ does not depend on orientation because the mass is a constant. Since this is not so in relativistic mechanics, the derivation of the wanted expression is more difficult. Unfortunately, we cannot factor out the acceleration on the right-hand side of Equation (5.60), because it is entangled with velocity in a scalar product **av**. If we rewrite the equation as

$$\mathbf{a} = \frac{\mathbf{f}}{m_0 \gamma(v)} - \gamma^2(v) \frac{\mathbf{a} \cdot \mathbf{v}}{c^2} \mathbf{v}, \tag{5.62}$$

it would be of little help because **a** is also present in the right-hand side. But we can do the following trick. Multiply (5.62) through by **v** to form a scalar product on both sides.

Since $\mathbf{vv} = v^2$, we have

$$\mathbf{a} \cdot \mathbf{v} = \frac{\mathbf{f} \cdot \mathbf{v}}{m_0 \gamma(v)} - \gamma^2(v) \frac{v^2}{c^2} \mathbf{a} \cdot \mathbf{v}. \tag{5.63}$$

This can be rewritten as

$$\left[1 + \gamma^2(v) \frac{v^2}{c^2}\right] \mathbf{a} \cdot \mathbf{v} = \frac{\mathbf{f} \cdot \mathbf{v}}{m_0 \gamma(v)} \tag{5.64}$$

or

$$\gamma^2(v) \mathbf{a} \cdot \mathbf{v} = \frac{\mathbf{f} \cdot \mathbf{v}}{m_0 \gamma(v)}. \tag{5.65}$$

Put this back into Equation (5.62)! Immediately, we will get the desired expression with **a** on only one side:

$$\mathbf{a} = \frac{1}{m_0 \gamma(v)} \left(\mathbf{f} - \frac{\mathbf{f} \cdot \mathbf{v}}{c^2} \mathbf{v} \right). \tag{5.66}$$

Read this equation. It says that acceleration is determined not only by the force, but also by the velocity of the body, and generally is not parallel to the force.

Exercise 5.1

Consider a point charge q in a homogeneous electric field **E**. A laboratory model of such system can be a charge inside a parallel-plate capacitor with a maintained constant voltage across the plates. The force on the charge will be

$$\mathbf{f} = q\mathbf{E}. \tag{5.67}$$

Suppose first that the charge enters the capacitor through the gap between the plates and is moving parallel to the plates. At this moment, its velocity is perpendicular to the electric field. The field starts to bend the trajectory of the charge and the acceleration of the charge at this moment is, according to (5.66), parallel to the force and perpendicular to the velocity. Its magnitude

$$a_\perp = \frac{f}{m_0 \gamma(v)} = \frac{a_0}{\gamma(v)} \tag{5.68}$$

is determined by the "transverse" mass (5.61a), which is just the relativistic mass (5.14) of the particle moving with the speed v. The quantity $a_0 = f/m_0 = qE/m_0$ here can be called the *proper acceleration* of the charge that is instantly at rest relative to the capacitor (we will see later that in the considered case it is equal to the acceleration of the charge moving within the capacitor along the electric field lines, measured in a comoving reference frame).

Let us now consider another situation, when the charge has been placed between the plates and then released from rest. In this case, the charge starts accelerating along the direction of the electric field and its acquired velocity will always be parallel to the force. Equation (5.66) now gives

$$a = \frac{1}{m_0\gamma(v)} f\left(1 - \frac{v^2}{c^2}\right) = \frac{a_0}{\gamma^3(v)}. \tag{5.69}$$

As is seen from Equation (5.69), the acceleration measured in the rest frame of the capacitor decreases as the inverse cube of the Lorentz factor. It is the acceleration along the velocity, determined by the "longitudinal mass" (5.61b). As the charge speeds up, approaching the light barrier, its acceleration decreases much more rapidly than would do the transverse acceleration at the equal speed under the same force.

How fast does the charge accumulate speed under the circumstances?

Write the previous equation as

$$\frac{dv}{dt} = \frac{a_0}{\gamma^3(v)} \quad \text{or} \quad \frac{d\beta}{(1-\beta^2)^{3/2}} = \frac{a_0}{c} dt, \tag{5.70}$$

where $\beta \equiv v/c$. This is the differential equation, whose direct integration gives

$$\int \frac{d\beta}{(1-\beta^2)^{3/2}} = \frac{a_0}{c} t. \tag{5.71}$$

The remaining integral can be easily taken by changing the variable $\beta = \sin\alpha$. If the motion starts from rest, the result is

$$\frac{\beta}{\sqrt{1-\beta^2}} = \frac{a_0}{c} t. \tag{5.72}$$

Solving this for v gives

$$v = \frac{a_0 t}{\sqrt{1 + (a_0^2/c^2)t^2}}. \tag{5.73}$$

This gives the answer to our question. It shows that while the proper acceleration remains constant in the rest frame of the charge, its acceleration relative to the capacitor decreases with time. At $t \to \infty$, we have $v \to c$.

Denote now as x the distance traveled by the charge. Then, we can take our inquiry a step further writing

$$\frac{dx}{dt} = \frac{a_0 t}{\sqrt{1 + (a_0^2/c^2)t^2}}. \tag{5.74}$$

This is again a differential equation. After separating the variables it is solved even easier than the previous one and yields

$$x = \frac{c^2}{a_0}\left(\sqrt{1 + \frac{a_0^2}{c^2} t^2} - 1\right) \tag{5.75}$$

or

$$x + x_0 = x_0\sqrt{1 + \frac{(ct)^2}{x_0^2}} = \sqrt{x_0^2 + (ct)^2}, \qquad x_0 \equiv \frac{c}{a_0}. \tag{5.76a}$$

Introduce a "shifted" variable $\tilde{x} \equiv x + x_0$. Then, the relation between x and ct will take the form

$$\tilde{x}^2 - (ct)^2 = x_0^2. \tag{5.76b}$$

The corresponding world line is a hyperbola; therefore, such motion is called hyperbolic motion. At sufficiently large t, the world line asymptotically approaches $x \approx ct$, remaining, of course, always less than ct. In terms of the space-time physics, we can say that the world line of the charge asymptotically approaches the photon's world line (Figure 5.2). At very small t, on the other hand, Equation (5.75) reduces to

$$x \approx \frac{1}{2}a_0 t^2. \tag{5.77}$$

This is just what one should expect: within a sufficiently short time interval after the start, the speed of the charged particle remains much less than the speed of light;

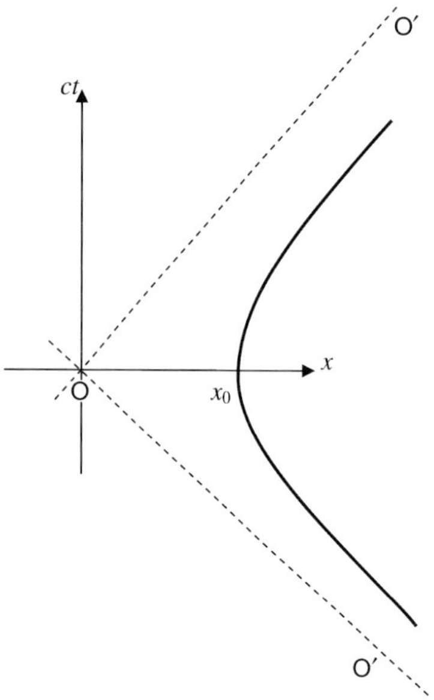

Figure 5.2 The world line of a body in a hyperbolic motion. Physically, it can be an electric charge q in a homogeneous electric field **E** with the initial velocity lined up with **E**. The x-axis is directed along **E**. The graph here corresponds to a charge initially moving in the direction opposite to the electric force $\mathbf{f} = q\mathbf{E}$, slowing down until it reaches the turning point $x = x_0$ and then accelerating in the positive x-direction. Asymptotically, the world line, while remaining timelike, approaches the lightlike lines parallel to OO″ ($t \to \infty$) and OO′ ($t \to \infty$).

accordingly, its mass and thereby its acceleration under the constant force both remain constant. As a result, we obtain a well-known expression for displacement in motion with constant acceleration.

If we eliminate time from Equations (5.73) and (5.75), we obtain the relation between the instantaneous velocity v and distance x:

$$v = c\frac{\sqrt{x(2x_0 + x)}}{x_0 + x} = c\frac{\sqrt{(x_0 + x)^2 - x_0^2}}{x_0 + x} = c\sqrt{1 - \frac{x_0^2}{(x_0 + x)^2}}. \tag{5.78}$$

No matter how large the distance traveled by the charge, its speed remains less than c.

Now, we are in a position to find the relativistic momentum and energy of the moving charge at each instant or each location. For instance, putting expression (5.78) into Equations (5.13) and (5.11), we obtain after some algebra

$$P = m_0 c\sqrt{\frac{a_0}{c^2}x\left(2 + \frac{a_0}{c^2}x\right)}, \qquad \mathcal{E} = m_0 c^2 + qEx. \tag{5.79}$$

Look carefully at the expression for relativistic energy of the particle. Does it make sense to you? It consists of two contributions; as one is from the rest energy, the second one therefore must be the kinetic energy gained in the electric field. The corresponding term qEx is precisely the lost potential energy of the charge that travels a distance x in a homogeneous electric field E. According to the conservation law, all the lost potential energy converts into kinetic energy of the moving charge.

So far we have discussed the properties of force in one reference frame. Our next question is how would the force transform if we switch to another frame of reference? How would the force measurements carried out by different observers relate to each other?

One way to get the answer would be using the analogy with the velocity vector. In the Minkowski space it is represented by 4-velocity $u_i = \mathrm{d}x_i/\mathrm{d}s$. Similarly, we can form a four-dimensional vector (4-force) representing the physical force and use its transformation properties to find the transformation rules for the force **f**.

We define the 4-force in space-time (Minkowski's force) as the derivative of 4-momentum with respect to corresponding interval [29]. Recalling the definition (5.16) of 4-momentum, we have

$$G_i = \frac{\mathrm{d}P_i}{\mathrm{d}s} = m_0 c\frac{\mathrm{d}u_i}{\mathrm{d}s} = m_0 c\frac{\mathrm{d}^2 x_i}{\mathrm{d}s^2} = m_0 c w. \tag{5.80}$$

This is a relativistic generalization of Newton's second law in Lorentz invariant form: the 4-force is the product of the invariant mass m_0 and 4-accelaration w. In view of (5.11) it immediately follows that 4-force is perpendicular to the 4-velocity:

$$Gu \equiv G_0 u_0 - G_\alpha u_\alpha = m_0 c\left(u_0\frac{\mathrm{d}u_0}{\mathrm{d}s} - u_\alpha\frac{\mathrm{d}u_\alpha}{\mathrm{d}s}\right) = \frac{1}{2}m_0 c\frac{\mathrm{d}u^2}{\mathrm{d}s} = 0. \tag{5.81}$$

Let us now express the 4-force G explicitly in terms of the 3-force. Writing $ds = c\,dt/\gamma(v)$ (recall (5.4)) and keeping in mind that $d\mathbf{P}/dt = \mathbf{f}$, we have

$$G_i = \frac{\gamma(v)}{c}\frac{dP_i}{dt} = \frac{\gamma(v)}{c}\left(\frac{dP_0}{dt}, \frac{d\mathbf{P}}{dt}\right) = \frac{\gamma(v)}{c}\left(\frac{dE}{cdt}, \mathbf{f}\right) = \frac{\gamma(v)}{c}\left(\frac{\mathbf{f}\cdot\mathbf{v}}{c}, \mathbf{f}\right). \quad (5.82)$$

Here, we have used the fact that $d\varepsilon/dt = \mathbf{f}\cdot\mathbf{v}$ (the rate of change of the energy of a body is equal to the work done on it by the force $\mathbf{f}$ per unit time. This is a case of a well-known work–energy theorem that holds in both – Newtonian and relativistic mechanics).

Thus, the temporal component of the 4-force turns out to be connected with the work of the 3-force.

Once we have defined G as a 4-vector, we know its transformation properties. We thus can relate the components of the 4-force, measured by two different inertial observers. On the other hand, we can express the components G' of this force in another reference frame in terms of the 3-force $\mathbf{f}'$. This will give us the equations connecting the components of the 3-force as measured in two different frames. The diligent reader can try to do this as an exercise.

But we can obtain the same result more easily by using the definition (5.56) of the 3-force and applying the transformation rules to momentum and time. Consider two frames K and K′ with parallel axes. Suppose the frame K′ slides relative to K along the x-axis with a speed V. Then we have

$$\left.\begin{aligned} &dP_{x'} = \gamma(V)\left(dP_x - \frac{V}{c^2}d\mathcal{E}\right), \quad dP_{y'} = dP_y, \quad dP_{z'} = dP_z \\ &dt' = \gamma(V)\left(dt - \frac{V}{c^2}dx\right) \end{aligned}\right\}. \quad (5.83)$$

Suppose the motion takes place in the plane xy, so that we can consider only x- and y-components of force. Using the above-mentioned relation $d\mathcal{E}/dt = \mathbf{f}\cdot\mathbf{v}$, we find

$$\left.\begin{aligned} &f_{x'} = \frac{dP_{x'}}{dt'} = \frac{dP_x - (V/c^2)d\mathcal{E}}{dt - (V/c^2)dx} = \frac{(dP_x/dt) - (V/c^2)(d\mathcal{E}/dt)}{1 - (Vv_x/c^2)} = \frac{f_x - (V/c^2)\mathbf{f}\cdot\mathbf{v}}{1 - (Vv_x/c^2)} \\ &f_{y'} = \frac{dP_{y'}}{dt'} = \frac{dP_y}{\gamma(V)dt(1 - (Vv_x/c^2))} = \frac{f_y}{\gamma(V)(1 - (Vv_x/c^2))} \end{aligned}\right\}. \quad (5.84)$$

To express the forces in K in terms of the forces in K′, we can just swap primed and unprimed variables and simultaneously change the sign of V, as we have done in obtaining the reversed Lorentz transformation.

$$\left.\begin{aligned} &f_x = \frac{dP_x}{dt} = \frac{dP_{x'} + (V/c^2)d\mathcal{E}}{dt' + (V/c^2)dx'} = \frac{f'_x + (V/c^2)\mathbf{f}'\cdot\mathbf{v}'}{1 + (Vv'_x/c^2)} \\ &f_y = \frac{dP_y}{dt} = \frac{dP_{y'}}{\gamma(V)dt'(1 + (Vv'_x/c^2))} = \frac{f_{y'}}{\gamma(V)(1 + (Vv'_x/c^2))} \end{aligned}\right\}. \quad (5.84a)$$

Consider a few important special cases. Suppose that the transverse component of the applied force does not do work:

$$\mathbf{f}_{\perp} \cdot \mathbf{v}_{\perp} = f_y v_y + f_z v_z = 0. \tag{5.85}$$

This may be the case when either the transverse force or transverse velocity or both are absent, or else they are mutually perpendicular (for instance, when a charged particle is tracing out a helix along the direction of a magnetic field in an inertial system K, and another inertial system K′ is moving along the same direction, taken as the x-direction). In this case, $\mathbf{f} \cdot \mathbf{v} = f_x v_x$, and Equation (5.84) for the longitudinal force reduces to

$$f'_x = f_x. \tag{5.86}$$

The component of force *along* the direction of relative motion of the two systems is in this case unchanged.

If the mass moves together with the reference frame K′, we have $v'_x = 0$, $v_x = V$, and in addition to (5.86), it follows from the second expression of Equation (5.84)

$$f'_y = \gamma(V) f_y. \tag{5.87}$$

If the mass is stationary in K, then $v_x = 0$, $v'_x = -V$, and the second expression of Equation (5.84) reduces to

$$f_y = \gamma(V) f'_y \quad \text{or} \quad f'_y = \frac{f_y}{\gamma(V)}. \tag{5.88}$$

Consider a possible physical situation illustrating this result. If we measure the gravity force acting on a car moving with a speed V on a horizontal track, then the driver of this car would measure a slightly greater force: since the direction of this force is perpendicular to the direction of motion, the force in the driver's reference frame is given by Equation (5.87). The difference is very small because the Lorentz factor is practically indistinguishable from unity in this case. But in the case of relativistic velocities, the difference may be very large. We will see possible dramatic consequences of such a difference in Section 7.5.

The discussed aspect of the Lorentz transformation of relativistic force is intimately connected to another aspect that appeared in the very beginning of this section (Equation (5.60)) but was left unexplained. Now, we will name it explicitly: apart from telling us how the force on an object is measured by observers in different reference frames, the Lorentz transformations tell us that the force on an object in *one and the same* reference frame depends on its velocity. To emphasize this aspect, consider, for instance, the transformation of the transverse force in Equation (5.84) and express v_x on the right in terms of v'_x. Putting expression (3.22) for v_x into the denominator of (5.84) for f_y, we obtain

$$1 - \frac{V v_x}{c^2} = 1 - \frac{V}{c^2} \frac{V + v'_x}{1 + (V v'_x / c^2)} = \frac{c^2 - V^2}{c^2 + V v'_x} = \frac{\gamma^{-2}(V)}{1 + (V v'_x / c^2)}. \tag{5.89}$$

The resulting expression for the transverse force in K′ in terms of the longitudinal component of the body's velocity in the same frame K′ is

$$f'_y = \gamma(V)\left(1 + \frac{Vv'_x}{c^2}\right) f_y. \tag{5.90}$$

Even if the force f_y on a body in K is fixed and does not depend on its velocity v_x, it is velocity-dependent in K′. Equation (5.90) shows that the force f'_y measured by observer in K′ is a linear function of v'_x. At $v'_x \to \mp c$ (an ultrarelativistic particle), the force approaches the limit

$$f'_y \Rightarrow \sqrt{\frac{1 \mp (V/c)}{1 \pm (V/c)}} f_y. \tag{5.91}$$

When $v'_x = -V$ (the body is stationary in K, $v_x = 0$), Equation (5.90) yields the already familiar result (5.88). When $v'_x = 0$ (the body is stationary in K′), we recover the result (5.87).

We will discuss the profound consequences of the velocity dependence of relativistic force in Chapter 7.

Exercise 5.2

Suppose you forget the expression for relativistic mass as a function of velocity, but remember the transformation law for the transverse component of a force, together with the fact that the transverse mass behaves under force as relativistic mass does. Derive the forgotten expression from the information you have.

Solution

Applying the transformation law (5.87) for transverse force to a mass moving with velocity **v** perpendicular to your line of sight gives

$$f_\perp = \frac{f'_\perp}{\gamma(v)}. \tag{5.92}$$

On the other hand, we can write in systems K and K′

$$f_\perp = m_\perp a = m\frac{d^2 r_\perp}{dt^2}, \qquad f'_\perp = m_0 a' = m_0 \frac{d^2 r'_\perp}{dt'^2} = m_0 \gamma^2(v) \frac{d^2 r_\perp}{dt^2}. \tag{5.93}$$

Here we have used the Lorentz invariance of transverse coordinates, the time dilation effect (2.26), and the fact that K′ is the rest frame for the mass m, so that in it $m = m_0$. It follows then from the last two equations that

$$f_\perp = \frac{m}{m_0} \frac{f'_\perp}{\gamma^2(v)}. \tag{5.94}$$

The comparison with (5.90) at $v'_x = 0$ recovers the forgotten result:

$$m = m_0\gamma(v). \tag{5.95}$$

5.6 Mass and Energy

$\mathcal{E} = mc^2$ made the fact plain that energy and matter are one and the same.

James Riordon, *Young Albert E. and the Miracle Year*

Now, we need to discuss in more detail the relation (5.21) between mass and energy. This relation was obtained in Section 5.3 as a result of Equation (5.14), describing how a body's mass depends on its speed. And the latter equation appears to emerge almost out of nothing. Actually, this "nothing" is a lot – it is the realization that the momentum **P** of an object is only a spatial projection of a more general entity – a 4-vector P. The dependence (5.14) follows automatically from this fundamental consequence of the relativity postulates.

Here we will show another way to derive Equation (5.14) and then we will illustrate the resulting new aspects of the concept of mass.

Imagine again, as we have done in Example 3.1, Tom in his rushing spaceship, this time on an exploration mission. He launches two identical surveillance spacecrafts in two opposite directions – one forward and the other backward. Both spacecrafts have been launched with equal speeds v' relative to Tom's ship, so Tom remains exactly in the middle between them as they fly apart, that is, he stays at the center of mass of the whole system.

Alice watches this from her reference frame (inertial system A). In her frame, the spaceship carrying Tom is moving with a speed V. Alice knows that internal forces cannot change the motion of the center of mass of a system, so she expects that Tom's ship will keep on moving in a straight line in the same direction at the same speed V after the launch of the two spacecrafts. Also, since the whole system is symmetrical with respect to the center of mass in Tom's reference frame and Tom remains at this center, Alice expects that in her frame, Tom's ship must remain exactly midway between the two spacecrafts receding from his ship. In other words, if x_+ and x_- are the respective instant distances in A between Tom's ship and the forward- and backward-launched crafts, Alice expects that $x_+ = x_-$.

The first one of these expectations is correct, the second one is wrong. Recall that according to the velocity "addition" theorem, an object launched in the forward direction will move only slightly faster than the spaceship (in Alice's frame) because its speed cannot attain the light barrier; the object launched in the backward direction will be observed by Alice moving significantly slower than the spaceship, or even in the opposite direction (as it definitely does in Tom's frame). Therefore, in Alice's frame, the first object will always be closer to Tom than the second object. The center of mass of the two identical objects is not exactly midway between the objects.

Why is it so? The explanation is pretty obvious: the objects are not exactly identical in Alice's frame! When in motion, they are not the same as when at rest. The relativistic mass of an object must increase with its speed. Thus, because the first object is moving faster in system A than the second object, it has to be more massive, even though their rest masses are equal. In A, the center of mass of the system must therefore be closer to this mass.

By how much is the "symmetry" broken here? This depends on how mass changes with speed. If in system A the distance x_+ is, say, only half the distance x_-, it means that in A, the first object is twice as massive as the second object. If we manage to find these distances, then their quantitative comparison will give us the measure of the difference between the two masses. And knowing their speeds will then give us the information about the relation between the speed of an object and its mass.

Let us calculate the instant distances between Tom and the spacecrafts in Alice's frame. At a moment t of her time, Tom will progress by the distance Vt. The first and the second spacecraft will have the velocities v_+ and v_-, respectively, determined by the velocity "addition" theorem (Equation (3.7)). The distances between Tom and the two spacecrafts will be (Figure 5.3)

$$\begin{aligned} x_+ &= v_+ t - Vt = \left(\frac{v' + V}{1 + (Vv'/c^2)} - V\right)t = \frac{v't}{\gamma^2(v')(1 + (Vv'/c^2))}; \\ x_- &= Vt - v_- t = \left(V - \frac{V - v'}{1 - (Vv'/c^2)}\right)t = \frac{v't}{\gamma^2(v')(1 - (Vv'/c^2))}. \end{aligned} \tag{5.96}$$

(Mind you, each of the expressions $(v_+ - V)$ and $(v_- + V)$ on the left is just the algebraic sum of the velocities of the two different objects – *not* the velocity of one object recorded by two different observers.)

As we had expected, the distances are not the same and the object fired from a moving source (in our case, from Tom's spaceship) in the forward direction remains at all times closer to the source than the object fired in the opposite direction. But our quantitative result tells a lot more than that.

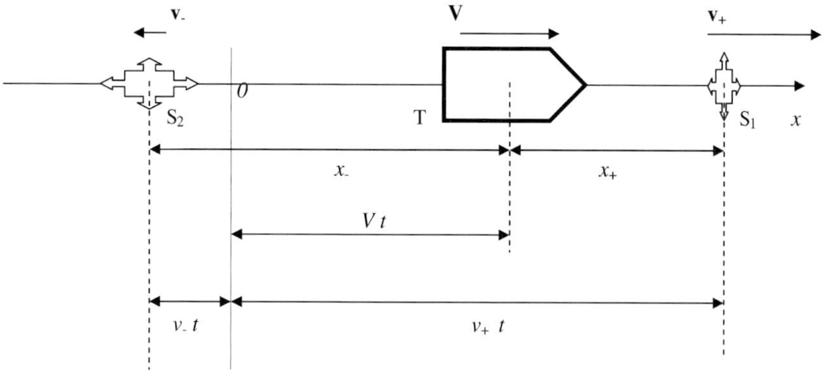

Figure 5.3 Tom's spaceship; S_1 and S_2 – the spacecrafts launched by Tom.

We can now use it for determining the speed dependence of mass (in case we forgot the result (5.14)!).

As we have learned in Section 5.4, the center of mass of a system of N noninteracting masses $m_1, m_2, \ldots, m_i, \ldots, m_N$ moving with velocities $\mathbf{v}_1, \mathbf{v}_2, \ldots, \mathbf{v}_i, \ldots, \mathbf{v}_N$, and instantly positioned at $\mathbf{r}_1, \mathbf{r}_2, \ldots, \mathbf{r}_i, \ldots, \mathbf{r}_N$, is given by

$$\mathbf{R} = \frac{\sum_{i=1}^{N} m_i(v_i)\mathbf{r}_i}{\sum_{i=1}^{N} m_i(v_i)}. \tag{5.97}$$

Apply this to the case when there are only two objects with equal rest masses flying apart with speeds v_+ and v_-. In this case $m_1(v_1) = m(v_+)$, $m_2(v_2) = m(v_-)$. Take the line connecting the masses to be the x-axis. Then, the X-coordinate of the center of mass (occupied by Tom) is obtained from the previous equation as

$$X = Vt = \frac{m(v_+)x_1 + m(v_-)x_2}{m(v_+) + m(v_-)}. \tag{5.98}$$

Since $x_1 = X + x_+$ and $x_2 = X - x_-$, this reduces to

$$m(v_+)x_+ - m(v_-)x_- = 0. \tag{5.99}$$

(You can get the same result if, for each moment of time t, you use an instantaneous (stationary in A!) auxiliary system positioned at the corresponding instant location X.)

Thus,

$$\frac{m(v_+)}{m(v_-)} = \frac{x_-}{x_+}. \tag{5.100}$$

Using (5.96) we get

$$\frac{m(v_+)}{m(v_-)} = \frac{1 + (Vv'/c^2)}{1 - (Vv'/c^2)}. \tag{5.101}$$

At the first glance, this does not look very illuminating; but recall our result (3.14) in Chapter 3! Comparison of (5.101) with this result tells us immediately that the moving masses must relate as

$$\frac{m(v_+)}{m(v_-)} = \frac{\gamma(v_+)}{\gamma(v_-)}. \tag{5.102}$$

In other words, mass must depend on speed as

$$m(v) = m_0\gamma(v), \tag{5.103}$$

which is precisely the result (5.14) obtained in Section 5.3. It is amazing that the introduction of 4-momentum vector in that section produced this result automatically, without any analysis or thought experiments. The realization of the fact that certain characteristics of a system can be written as a 4-vector allows us to see deeper the fabric of reality.

Let us now get back to the more general case (5.97). Suppose that our system is isolated and the individual masses do not interact with each other (except for occasional collisions). Then, their velocities $\mathbf{v}_i = d\mathbf{r}_i/dt$ remain constant between the collisions and only their coordinates $\mathbf{r}_i$ change. Differentiating Equation (5.97) with respect to time, we obtain the velocity of the system's center of mass

$$\frac{d\mathbf{R}}{dt} = \mathbf{V} = \frac{\sum_{i=1}^{N} m_i(v_i)\mathbf{v}_i}{\sum_{i=1}^{N} m_i(v_i)} \tag{5.104}$$

in terms of the individual masses and their velocities. It can be rewritten as

$$\sum m_i(v_i)\mathbf{V} = \sum m_i(v_i)\mathbf{v}_i \tag{5.105}$$

(we have dropped the summation index in the sums). The products on the right are, of course, the individual momenta of the moving constituents of the system. Their sum gives the net momentum of the system. On the other hand, the momentum $\mathbf{P}$ of any system moving with velocity $\mathbf{V}$ is $M\mathbf{V}$, where M is the relativistic mass of the system. Therefore, we can write

$$M\mathbf{V} = \sum m_i(v_i)\mathbf{v}_i = \sum \mathbf{p}_i = \mathbf{P}. \tag{5.106}$$

Comparing the two expressions (5.105) and (5.106), we see that the total mass of the system of noninteracting parts is the sum of their individual relativistic masses

$$M = \sum m_i(v_i). \tag{5.107}$$

The corresponding expressions for the center of mass and its velocity can be written as

$$\mathbf{R} = \frac{\sum_{i=1}^{N} m_i(v_i)\mathbf{r}_i}{M}, \qquad \mathbf{V} = \frac{\sum_{i=1}^{N} m_i(v_i)\mathbf{v}_i}{M}. \tag{5.108}$$

The result (5.107) seems pretty obvious. It looks like the familiar nonrelativistic expression for the net mass of a system. The fundamental difference between the relativistic and nonrelativistic concept of mass becomes clearer if we consider a special case.

Suppose that we are in the rest frame of the system, where $\mathbf{V} = 0$. This means that $\mathbf{P} = 0$, even though the individual particles may be moving. A good example of such a system is a container with gas. The gas as a whole is stationary together with the container, while the molecules constituting the gas are in motion. By definition, the net mass on the left of (5.107) is now the *rest mass* of the system:

$$M \underset{V=0}{=} M_0 = \sum m_i(v_i). \tag{5.109}$$

Since the individual masses may be moving, the rest mass of the system is the sum of the *relativistic* masses of the constituent particles. It is *not* the sum of their rest masses!

In the case of a container with gas, the masses of moving molecules exceed their rest masses, so the rest mass of the gas in the container is *greater* than just the sum of molecular rest masses.

In other words, the rest mass is not an additive characteristic of a system. It is generally not equal the sum of rest masses of system's parts.

How can that be? The law of conservation of mass has been established as far back as eighteenth century. The mass cannot appear out of nothing or disappear without a trace. This law was proved by millions of experiments, and appears to be confirmed daily in our life as well as in the laboratories. According to the nonrelativistic physics, we can heat the matter up to any attainable temperature, compress it up to enormous pressures, combine its components in any possible chemical compositions – this may change the internal energy of the system, but its net mass in all cases will remain the same. However, special relativity tells us that, since the mass and energy are rigidly connected by the direct proportionality equation (5.21), if we heat a container with gas, the gas in the container will become heavier, along with becoming more energetic. Thus, the rest mass of a system can be considered as merely the measure of its internal (rest) energy, and can change together with the latter, if the system is not isolated.

On the intuitive level, we can understand why the hot gas is more massive than the same amount of cold gas if we note that in order to heat the gas we have to pump energy into it. The internal energy of gas increases. But energy is associated with mass and, therefore, gaining energy is gaining mass! Note that, contrary to the non-relativistic view of mass, in the given case this additional mass is not some additional material coming in. It is literally "made up" of the incoming energy divided by c^2. The gas in container becomes more massive not because of the addition of new molecules, but because, on the average, each molecule becomes more massive and it becomes more massive because it is moving faster in the hot gas than in the cold gas.

(The example also illustrates that, when we say "the same amount of gas," we must be very careful to specify what we mean. Here we mean the same sealed container with the same fixed number of molecules. But then we notice that at high enough temperatures, the gas molecules can dissociate into single atoms, so the number of molecules will not remain the same. Since the atoms can be ionized, and even their nuclei can disintegrate, or, contrariwise, merge together at high enough temperatures, we see that the more general formulation "the total number of particles" is not a reliable indicator either. We may be tempted to say "the amount of matter in a sealed container with the walls impenetrable for particles." This does not work either, even though we overtly substituted "gas" with a far broader term "matter," because at high enough temperatures the container itself can disintegrate as well. We conclude that the expression "the same amount of gas" makes perfect sense only within a limited range of physical conditions under which the given gas retains its chemical attributes.)

We can also understand the statement "mass is made up of energy" in the following way. Look again at Equation (5.109) and rewrite it as

$$M \underset{V=0}{=} M_0 = \sum m_i(v_i) = \sum m_{0i}\gamma(v_i). \tag{5.109a}$$

This definition works for arbitrary individual rest masses and arbitrary possible speeds. Imagine, for example, a cloud of particles moving with speeds close to the speed of light. In this case the relativistic mass of each particle by far exceeds its rest mass. Accordingly, the rest mass of the cloud, being the sum of individual *relativistic* masses, is much greater than just the sum of the rest masses of its constituents. Consider an extreme case when our particles are photons. Then, the sum of their rest masses is zero, but the rest mass of the system will not be zero. Thus, the "empty space" within a sealed room with air totally pumped out has a nonzero rest mass, because it is filled with photons of infrared radiation characteristic of the room temperature. We can understand it better if we rewrite (5.109) as

$$M_0 = \sum \frac{\varepsilon_i}{c^2} = \frac{\varepsilon_0}{c^2}, \tag{5.110}$$

where ε_i is the energy of the ith particle. In the given case, the particles are the photons in the container. Each photon has zero rest mass, but, of course, a nonzero energy (or, if you wish, a relativistic mass). Consider the simplest possible case: you have two photons with the same frequency flying apart. Each photon has the zero rest mass. But if you consider the two photons as one system, its rest mass, being the ratio of the net energy to c^2, is not zero. This system of two massless photons is massive!

These examples illustrate two intimately linked properties of the rest mass: first, the rest mass is not an additive characteristic and when you combine a few items together, the rest mass of the resulting system may differ from the sum of the rest masses of its parts (the same automatically refers to the *rest* energy); second, the rest mass of the system can be "made up" of the energy (or, more accurately, of the energies of the constituent parts of the system divided by c^2).

For a system with a nonzero rest mass, there is always a corresponding rest frame where this mass can be measured directly. By definition, the rest frame is a frame where the given system has the zero *net* momentum. Therefore, in the rest frame of the system, for any of its moving parts there has to be a counterpart with the opposite momentum. If you have a bunch of photons moving all in one direction, you cannot find a reference frame satisfying this requirement. Such a system does not have a rest frame. A system without the rest frame cannot have a nonzero rest mass.

Applying this to only one photon, we are tempted to conclude that its rest mass is *always* zero – a photon does not have the rest frame!

This negative statement is true for a free photon moving in a certain direction in empty space. Generally, however, it may be wrong. According to quantum mechanics, there are situations when even a single photon can have a rest frame, and, accordingly, a nonzero rest mass!

Consider a simple optical experiment: a beam of light is incident on a beam splitter (Figure 5.4). This can be considered as a simplified version of the Michelson

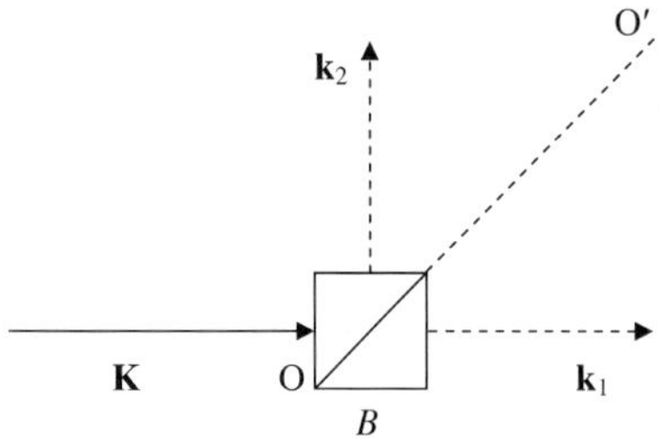

Figure 5.4 An experiment transforming a photon state with a sharply defined momentum **K** into superposition of states with momenta $\mathbf{k}_1$ and $\mathbf{k}_2$. B – beam splitter.

experiment in Section 2.1, which illustrates another important aspect of the relativity theory. The light is partially transmitted by the splitter and partially reflected. As a result, the initial beam is split into two daughter beams. The light intensity in each daughter beam is less than in the initial one and their sum equals the intensity in the initial beam. What happens if we start to dim the incident beam, until it contains only one photon? The conventional wisdom tells us that the photon will be either transmitted or reflected by the splitter. The conventional wisdom is wrong in this case. If we do not make any attempt to watch the photon's progress, the photon will take both paths simultaneously.

Do not think that it will split into two less energetic photons. According to (5.31), the photon's energy is determined by its frequency, which is not changed by the beam splitter. Do you remember the assertion made in Section 5.3 that a photon propagates as a wave? Well, when a wave hits a beam splitter like the one in Figure 5.4, it splits into two waves of the same frequency, regardless of how faint it is. Which means that one and the same indivisible photon associated with the wave, will travel both ways at once! I will stop here and ask those unacquainted with quantum mechanics to just take this for granted and focus on the consequences. Here we have *only one* photon moving simultaneously in two different directions. Therefore, we can find a reference frame, in which the motion of one part of this system will be "balanced" by the motion of its counterpart with the equal and opposite momentum. As is clear from Figure 5.4, this system must move somewhere along the bisector between the two paths provided by the beam splitter. Having boarded this system, we will observe one photon in a superposition between the two opposite directions of motion, with each state contributing its respective energy to the total $\mathcal{E}' = \mathcal{E}_0$. The energy will be nonzero, but the net momentum zero. As a result, a single photon will have a nonzero rest mass, equal to $\mathcal{E}_0/c^2$. The problem is reduced to finding energy $\mathcal{E}_0$ of the superposition in its rest frame K′. I want to leave this problem to the reader (Problem 5.20) and to show here the other way to solve the same problem without looking for the rest frame.

Recall Equation (5.23) in Section 5.3. It tells us not only that the energy and momentum of the system form a 4-vector, but also something no less important: namely, there is no need to board the rest frame of the system to measure its rest mass. We can instead measure its energy $\mathcal{E}$ and the magnitude of momentum $\mathbf{P}$ in a frame where we are and then calculate the rest mass as

$$m_0 = \frac{1}{c^2}\sqrt{\mathcal{E}^2 - P^2c^2} \tag{5.111}$$

(this is equivalent to measuring the magnitude of the 4-momentum by measuring its temporal and spatial projections). In our case, we are in the lab with the beam splitter. The initial state of the photon (before it hits the splitter) is characterized by the frequency ω and the sharply defined wave vector $\mathbf{k}$ (one plane wave) directed toward the splitter. Equations (5.23) and (5.30) in Section 5.3 tell us that the rest mass of the photon in this state is zero. After passing the beam splitter the state of the photon is changed. Although its frequency, and thereby its energy, remains the same, its motion is now a superposition of the two motions with wave vectors $\mathbf{k}_1$ and $\mathbf{k}_2$, respectively, perpendicular to one another (two different plane waves). As we said before, the intensity of each daughter wave is only half the original intensity (assuming the chances of being reflected or transmitted are 50 : 50). Accordingly, the energy carried by either daughter wave is $(1/2)\mathcal{E}$ and the magnitude of each individual momentum is $(1/2)\mathcal{E}/c$. The magnitude of the net momentum is

$$P = \sqrt{P_1^2 + P_2^2} = \frac{1}{\sqrt{2}}\frac{\mathcal{E}}{c}. \tag{5.112}$$

Putting this into (5.111) gives

$$m_0 = \frac{1}{\sqrt{2}}\frac{\mathcal{E}}{c^2} = \frac{1}{\sqrt{2}}\frac{\hbar\omega}{c^2}. \tag{5.113}$$

This is the sought-for solution.

Actually, we can perform a more straightforward version of this thought experiment, producing a photon state with its rest frame directly at rest in the lab, if we just "turn on" the light between the two stationary parallel mirrors shown in Figure 5.5 (we skip the question of *how* the experimenters can "bring in" the light there). The light waves will bounce back and force between the mirrors, but only those with certain wavelengths can survive. The surviving wavelengths (and the corresponding frequencies) satisfy the resonant condition that only the integer number of their half-waves fit into the spacing between the mirrors. Such a system selecting only a discrete set of frequencies of electromagnetic oscillations is called the optical resonator (in case shown in Figure 5.5 this is the Fabri–Perrot resonator [50]). Suppose we have just one such frequency ω. We cannot describe this state by saying that the wave is running first one way and then the other way. As in the previous case with the beam splitter, it is running both ways at once, producing what we call a standing wave. A very appropriate name, considering that the associated electromagnetic energy does not go anywhere, but rather stays bound, only oscillating within the resonator. Now consider again a limiting procedure reducing the light to only one photon between the mirrors. Then the photon is moving simultaneously in the two opposite directions, so that the resonator's rest frame is the *rest frame of this photon*, and the photon's energy divided by c^2 is the photon's rest mass:

$$m_0 = \frac{\hbar\omega}{c^2}. \tag{5.114}$$

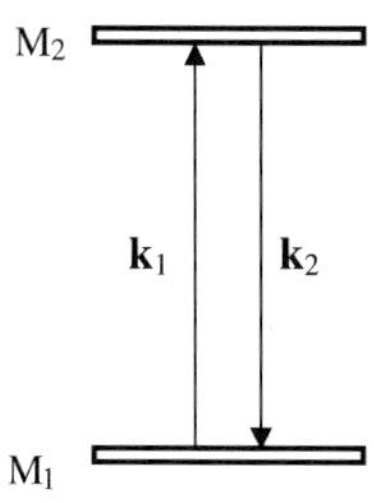

Figure 5.5 An open resonator. M_1 and M_2 – mirrors; $\mathbf{k}_1$ and $\mathbf{k}_2$ – the two possible wave vectors forming an allowed state (eigenstate) of the electromagnetic oscillations in the resonator.

Let us summarize our findings. We have considered one and the same photon in three different possible states: (1) moving in one direction, (2) moving in two mutually perpendicular directions, and (3) moving in two opposite directions. In the first state, the rest mass of the photon is zero, because all the photon's energy goes exclusively into its motion. In the second state, only *part* of the energy goes into translational motion of the photon's center of mass along the bisector OO′ in Figure 5.4; the remaining part of the energy gives rise to the rest mass (5.113) of the photon. In the third state, the center of mass of the photon stands still midway between the mirrors! Nothing is left for translational motion of the state. Accordingly, the rest mass in this case exceeds (5.113) and reaches its maximum (5.114), because all the photon's energy goes into it.

Beware of the hidden pitfall here. I said that one can find the solution for case 2 by switching to the rest frame of the corresponding state. The observer in this frame will see two waves moving in the opposite directions, just as in case 3 (Figure 5.6a). On the other hand, we can start from case 3 and switch to a reference frame moving parallel to the mirrors with appropriate speed so as to see two bouncing waves propagating in the two mutually perpendicular directions (Figure 5.6b). Actually, the latter situation would be a variety of the experiment with the light pulse bouncing first within a stationary cylinder and then in a moving cylinder in Section 2.5, Figure 2.4. The resonator here plays the same role as the cylinder there.

Each such switching to another reference frame demonstrates that the two cases 2 and 3 can be converted to one another by an appropriate Lorentz transformation.

Now, here is the pitfall: it may appear that the rest mass (5.114) found for the photon in the stationary resonator will decrease in a reference frame where the resonator will be moving, because in that frame a certain part of the photon state's energy will go to the translational motion of the center of mass of the state. And vice versa, the rest mass (5.113) found for the photon in the split state produced by a stationary beam splitter may be expected to increase in the reference frame of Figure 5.6a, because in that frame *all* of the photon's state's energy will go to the rest energy.

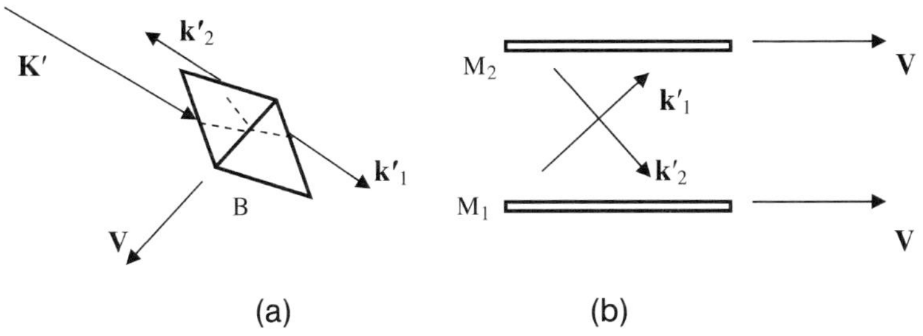

Figure 5.6 (a) The rest frame K′ of the split state (2). The beam splitter is moving in the direction **V**. The photon is in a superposition of the two states characterized by the antiparallel wave vectors $\mathbf{k}_1'$ and $\mathbf{k}_2'$, Lorentz-transformed from the crossed vectors $\mathbf{k}_1$ and $\mathbf{k}_2$ of the lab frame K. (b) The frame K′, moving to the left relative to the rest frame K (the lab) of the state (3). In K′, the resonator is moving to the right with velocity *V*. The photon is in the superposition of the two states characterized by the crossed vectors $\mathbf{k}_1'$ and $\mathbf{k}_2'$, Lorentz-transformed from antiparallel states $\mathbf{k}_1$ and $\mathbf{k}_2$ of the lab frame.

In both situations, our descriptions of the transformed states are correct, but the above conclusions would be wrong. The rest mass is an invariant, it does not depend on RF used for its calculation. It is true that when we switch to a reference frame where the resonator is moving, the rest energy of the trapped photon there becomes only a *fraction* of its total energy and therefore the rest mass is no longer described by Equation (5.114). But it is wrong to think that its numerical value will be less than the value found in the rest frame of the resonator. The formula will be different, but the value will remain the same. It is not the value of the rest energy/mass that decreases because it becomes a fraction, it is the total energy that increases, because the *same* value of the rest energy/mass now becomes a fraction. And it is natural that the energy of an object locked within a resonator (Figure 5.6) or cylinder (Figure 2.4) should increase when it is boosted together with its container. We will see later that the transformation rules for the frequency ω are consistent with this change of energy, so that the fundamental Equation (5.31) holds in any RF.

A similar statement can be made about the state produced by the beam splitter. In all cases, the rest mass of a system is the *invariant characteristic* of this system. It cannot be changed by "boarding another train." Energy (and momentum) of an isolated system can be changed by switching to another train (they are not invariant), but not by evolving the system (they are conserved).

The invariance of the rest mass of a system can be seen immediately from the expression (5.23) for the 4-momentum. Geometrically, the rest mass is (up to the factor c) just the kinematic length of the 4-momentum.

To summarize this part, let us consider a system whose interaction with the environment can be neglected (an isolated system). Some of its physical characteristics may be invariant, some conserved, some additive, and some may be any combination of those adjectives. Note the distinction between an *invariant* quantity (same value in all inertial systems), *conserved* quantity (same value before and after some process) [44], and *additive* quantity (same value as the sum of constituents). Rest mass is invariant and conserved, but not additive; energy is conserved and additive, but not invariant; electric charge is all – additive, invariant, *and* conserved; velocity is neither invariant, *nor* additive, *nor* conserved.

Beware of misleading statements about the rest mass in many otherwise excellent treatments. Namely, they state that the rest mass is invariant (true) but not conserved (wrong!). According to (5.110), the rest mass of a system is the system's rest energy divided by c^2. Nonconservation of the rest mass would mean nonconservation of energy. Obviously, the authors use here the word "nonconserved" instead of the appropriate word "nonadditive."

So far we have restricted ourselves to systems with noninteracting parts. But what if the parts are interacting? Then, the interaction energy also contributes to the net energy and, accordingly, to the net mass of the system; therefore, it must be included into our equations. In the nonrelativistic approximation, the rest mass of the system can be expressed in terms of the energies involved as

$$M_0 = \frac{1}{c^2}\left[\sum_{i=1}^{N} \mathcal{E}_i + \frac{1}{2}\sum_{i\neq}\sum_{k} U_{ik}\right]. \tag{5.115}$$

Here, $\mathcal{E}_i$ is an individual energy of part i and U_{ik} is the interaction energy between the parts i and k. This potential energy is counted with respect to the reference point where the two members of the corresponding pair do not interact, which happens in most cases when they are very far apart. The double sum is taken over all possible pairs. The factor 1/2 takes care of the fact that U_{ik} and U_{ki} are the same thing, so without this factor the interaction energy would be counted twice.

The accurate relativistic expression should include explicitly the energy of the fields associated with the system of interacting particles. According to such general expression, as well as approximation (5.115), the rest mass of the system can be greater or less than the sum of relativistic masses (5.110), depending on the kind of interaction. If the interaction tends to bind the parts together, it contributes a negative term.

As an example, consider the interaction between two neutral atoms in a diatomic molecule such as hydrogen (H_2) or nitrogen (N_2). The potential energy of interaction between the atoms as a function of interatomic distance is plotted in Figure 5.7. It goes to zero when the atoms are far apart, decreases as they get closer, reaches its negative minimum at a certain distance $r = r_0$, and then increases for $r < r_0$. In terms of forces, the atoms attract when $r > r_0$ and repel when $r < r_0$. Classically, the distance r_0 corresponds to the state of stable equilibrium and determines the size of the molecule. In terms of energy and the corresponding mass, the rest mass of the system is equal to the sum of the individual atomic rest masses only when the molecule is

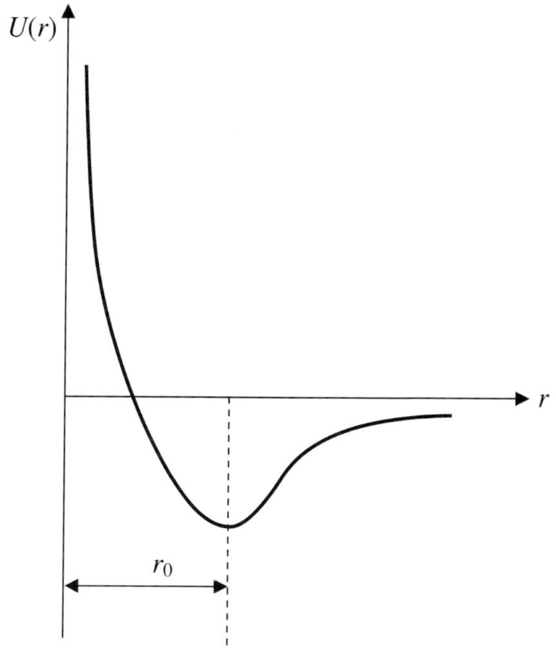

Figure 5.7 The graph shows potential energy versus distance for two atoms. The long-range branch ($r > r_0$) corresponds to attraction force and the short-range one ($r < r_0$) to the repulsion force.

disassembled and the atoms are stationary and far apart. In the vicinity of the equilibrium state ($r \approx r_0$), the negative contribution of the interaction energy is "absorbed" into the net mass, so the rest mass of the molecule is *less* than the sum of the rest masses of individual atoms. The rest mass is minimal at the equilibrium distance $r = r_0$. If a certain agent squeezes the molecule, bringing atoms closer than $r = r_0$, it does work against emerging repulsive forces. This increases the molecule's energy and its rest mass.

For the same reason, a stretched or compressed spring is more massive than the relaxed spring. When you deform an elastic object, the work you do goes into its internal energy, which increases its rest mass. For a spring obeying Hooke's law, the mass increase is measured by

$$\Delta m_0 = \frac{1}{2}\frac{k}{c^2}x^2, \tag{5.116}$$

where k is the spring constant and x is the amount of deformation.

A system of two unlike stationary electrical charges q_1 and q_2 distance r apart has a smaller rest mass than the system of the same charges at rest infinitely far apart; again, the deficiency in mass (mass defect) will be determined by the interaction term (potential energy of interaction divided by c^2):

$$\Delta m = \frac{|q_1 q_2|}{4\pi\varepsilon_0 c^2 r}. \tag{5.117}$$

If the charges are of the same sign, then the rest mass of the system will *exceed* the sum of the rest masses by the same amount.

As in the above case with a molecule, we can describe these results by considering the process of formation of the corresponding system. Suppose that we have two unlike charges at rest and very far apart. Then the rest mass of the system of charges is the sum of their rest masses. If they are allowed to move freely toward each other under mutual attraction, they will accelerate and gain kinetic energy at the cost of the potential energy of the system. If we stop them at a distance r, this kinetic energy will be taken from the system, so its net energy and thereby its rest mass will decrease. Applying the work and energy theorem, it is easy to show that the decrease will be measured by expression (5.117).

On the contrary, if we have the like charges, we will have to do work on them to bring them from infinity to a distance r, therefore, the system of charges separated by a distance r will store more energy and thereby have a greater rest mass than the sum of their individual rest masses. Again, the mass excess will be given by the amount (5.117).

As is known from electrostatics, there can be no stable configuration of charges under electrical forces alone. For two attracting charges making a hydrogen atom – proton and electron – there is no finite separation r between them, at which they could form a stable atom while remaining stationary. They can only do it when in motion. In classical description used before the appearance of quantum mechanics, this was an orbital motion of the electron and the proton around their center of mass, practically equivalent to revolution of the electron around the nucleus. This motion brings in the corresponding kinetic energy, but its contribution is, of course, less than the negative contribution of the interaction energy (otherwise the system would fly apart). As a result, the rest mass of a hydrogen atom is less than the sum of the rest masses of its constituents. A far more accurate quantum-mechanical description of an atom does not change this fundamental result.

The interaction of two electrical charges occurs through the electrical field they produce. This field has energy associated with it. The local energy density η_e associated with the electric field **E** at a given point is proportional to the square of the field:

$$\eta_e = \frac{1}{2}\varepsilon_0 E^2, \qquad (5.118)$$

where ε_0 is the permittivity of free space. If we consider Coulomb's field due to a charge and integrate expression (5.118) over the whole space around the charge, we will obtain the total electrostatic energy of the system. Thus, the field of a charge may be a contributor to its rest mass. The mass of an electron must be, at least partially, "made up" of the energy of its electric field. There were numerous attempts to derive the rest mass of an electron as originating *entirely* from the energy stored in its electrical field. The basic idea looks very simple.

Imagine a stationary electron as a tiny sphere of radius a, with all its charge distributed uniformly over the surface of the sphere. As we know from Gauss's law, the electric field of this system is zero inside the sphere, and outside it is radial, with its magnitude being the same as for a point charge q_e at the center:

$$E = \begin{cases} 0, & r \geq a, \\ \dfrac{q_e}{4\pi\varepsilon_0 r^2}, & r < a. \end{cases} \qquad (5.119)$$

Putting this into (5.118) and integrating over the whole space around the sphere yields the result that the total energy stored in the field is

$$U_e = \frac{e^2}{2a}, \qquad (5.120)$$

where we denoted for simplicity $e^2 \equiv q_e^2/4\pi\varepsilon_0$. All we need now is to adjust the "electron radius" a so as to make the electromagnetic energy (5.120) equal to its measured rest energy

$$m_e c^2 = U_e, \quad \text{so that} \quad m_e = \frac{e^2}{2ac^2}. \qquad (5.121)$$

What we actually know from experiments is the electron's rest mass, so Equation (5.121) can be considered as determining the so-called "classical radius" of the electron. If this description were correct, the electron's mass would have entirely electromagnetic origin and it could be called the electromagnetic mass.

This program has not been totally successful for many reasons and one of them is that an elementary particle like an electron can be described in the framework of the theory of relativity only as a point particle [29, 34]. Since a point has no size, the "electron radius" a must go to zero, and then, if the electron's mass has an entirely electromagnetic nature, it must, according to Equation (5.121), be infinite. This shows that classical electrodynamics alone cannot account for such an observed characteristic of the electron as its mass. Either an electron is not a truly elementary particle or it is not entirely electromagnetic entity; and in any event, in these domains, we have to take into account quantum-mechanical nature of any particle and turn to corresponding generalized descriptions such as the rapidly developing string theory [37]. Whatever new knowledge we can gain in the future, today the idea of deriving the mechanical mass entirely from the field description of an object seems very attractive and is being actively explored [38].

As another example of possible "energy storage" consider magnetic field. The energy density stored in the magnetic field **B** at a given locality is

$$\eta_{\mathrm{m}} = \frac{B^2}{2\mu_0}, \tag{5.122}$$

where μ_0 is the permeability of free space.

The existence of magnetic energy is manifest, for example, in the fact, that even after you disconnect a solenoid from a battery, the current keeps on flowing for a while. It cannot be stopped immediately, just as you cannot stop immediately a loaded truck because of its high inertia. The inertia of charges whose collective directed motion produces the current is only a very minor fraction that cannot account for all the energy (and, accordingly, inertia!) of a circuit with a current. The source of this energy is the magnetic field in the space around the current.

The following example illustrates the situation with *both* kinds of electromagnetic energy present. Consider first a charged capacitor. It is more massive than the same capacitor uncharged. Where does the additional mass come from? When you charge the capacitor, you transfer a certain amount of charges from one plate to another. To do this, you have to do work against the electrical forces that resist separation of charges. This work goes into the energy of the emerging electrical field between the plates. Since any energy has mass associated with it, the mass of the capacitor increases.

Similarly, by turning on the current in a wire, you make the battery do work. A certain (relatively small) fraction of this work goes to increase of kinetic energy associated with the collective motion of the charges producing current, but most of it goes into the magnetic field around the wire. The current-carrying wire is more massive than the same wire without the current – partially because of the mass increase in the charges "marching" along the wire, but mostly because of the additional energy pumped into the associated magnetic field. The longer the wire (if it is straight), the farther in the surrounding space is spread the magnetic field of the current and the greater the amount of energy stored in it. Theoretically, if you have an infinitely long straight wire carrying a current, the rest mass of each unit length of such a wire becomes infinite, because its rest energy goes to infinity (there will be more of it in Chapter 6).

Now, the electric and magnetic energy can convert into one another just as potential and kinetic energy do in mechanics. The conversion of the electric energy into magnetic and back can reiterate periodically, leading to electromagnetic oscillations.

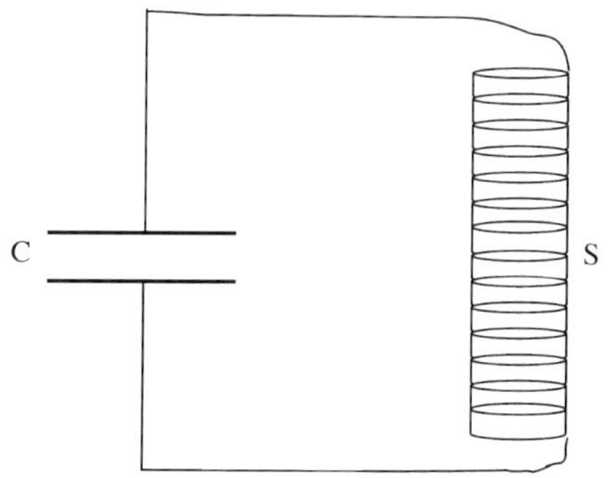

Figure 5.8 An electromagnetic oscillator: C – capacitor; S – solenoid.

The simplest system where such oscillations can occur consists of a capacitor connected to a solenoid (Figure 5.8). Let the capacitor be charged at the initial moment, so there is a strong electric field between its plates and, accordingly, the variable part of the system's energy is concentrated there. Then the charges start flowing away from their respective plates, creating the current in the coil. The electric field, and thereby the electric energy, in the capacitor decreases, but in exchange, there appears current and thereby magnetic energy in the solenoid. When the capacitor becomes totally discharged, the electric energy is zero, but the current and the magnetic energy in the coil reach their maximum. At this moment, all the initial electromagnetic energy of the system is concentrated within the solenoid as the energy of the current and its magnetic field. Accordingly, the rest mass of the coil also reaches its maximum. Because of the magnetic energy associated with this mass (or, if you wish, its inertia), the current keeps on flowing even though the capacitor is now totally discharged. As a result, by the time when the current and its magnetic energy disappear, the capacitor gets recharged and there appears again a strong electric field. The rest mass of the capacitor is now again at its maximum value.

The energy as such does not disappear. It is just playing "hide and seek" with itself, appearing now in this, now in that disguise. The total energy, regardless of its varying forms, remains constant and, with it, so does the total rest mass of the system. But a part of this mass associated with the electromagnetic energy is circulating between the capacitor and the solenoid, so that they alternately become more or less massive. Again, what we perceive here as mass actually represents energy that can be "packed" in different and changing forms.

Consider now another aspect of the problem. We just said that the total energy of the whole electromagnetic oscillator is conserved. This is only approximately true. A more careful look shows that the amplitude of the oscillations gradually decreases with time and eventually the oscillations damp out. This happens even if there are no losses to heat (suppose all the circuit is in a superconductive state). This is because a certain fraction of the electromagnetic energy leaks out in space with each cycle as the electromagnetic radiation. The described oscillator acts as a radiating antenna. The departing radiation carries energy away. This energy and its associated mass propagate in space to the far reaches of the universe. The total energy of the system "antenna + its radiation" (and its total relativistic mass) conserves; the energy of the antenna itself (and its rest mass) decreases.

An atom also can exchange energy with electromagnetic field, playing a role of a microscopic radiating or receiving antenna. In the classical description, an atom with the electron responsible for a given optical transition with emission or absorption of a photon is modeled as a tiny oscillating or rotating dipole. But, according to quantum mechanics, it can either radiate or absorb (depending on initial state of the system) only a certain fixed amount of the electromagnetic energy (a photon!). Each such act changes the energy/rest mass of the atom. Similar changes are associated with chemical reactions, in which the electron configurations of reacting atoms change. But if we use Equation (5.21) to calculate the changes in mass associated with chemical reactions, we will find them extremely small (Problem 5.26). No surprise chemists did not notice these changes and had for a long time believed that the (rest) mass conserves in chemical reactions (Lavoisier's law).

The situation is dramatically different in nuclear transformations for two reasons. First, the mass of an atomic nucleus is more than three orders of magnitude greater than the net mass of all the electrons around it. Accordingly, the overwhelming part of the whole atomic energy is associated with (stored in) the nucleus. Second, the nuclear forces, even though short-ranged, are by far stronger within the range of their action than the electromagnetic forces, and accordingly they can do greater work. As a result, the energy released or absorbed per unit mass in nuclear reactions may be millions of times greater than in any known chemical reactions. Nuclear physics gave the most compelling experimental confirmation of the relativistic mass–energy equivalence. This equivalence, observed as a "mass defect" accompanying enormous energy release in nuclear reactions, was one of the most fundamental experimental discoveries in nuclear physics and physics in general. It gives humanity an enormous power and, as Einstein himself put it, the fate of our civilization critically depends on our wisdom in using this power.

Discussion

So far we have focused more on the aspect that can be called "mass is made up of energy" or "energy is massive." Now, the just mentioned huge energy release that is possible in some nuclear processes, makes us look at it the other way: *mass is energetic*. From this viewpoint, equation $\mathcal{E} = mc^2$, first, states that any (even resting) mass stores energy and, second, determines the maximum possible amount of energy that can be released by a system with the rest mass m_0. If all this energy stored in mass m_0 is released, the initial system as such totally disappears (taking a more cautious stand, we could say, it changes beyond recognition). In the final state we have a quite different system, for example, electromagnetic radiation. Scientists call such a transformation the annihilation (reduction to zero in Latin). One can hear sometimes that in this process mass converts into energy. Such a statement, made by physicists using professional jargon, is by itself harmless when addressed to a professional audience. But we must be careful using such terminology because it can be misleading. If mass and energy are indeed *the same thing* under two different names, then the statement "mass converts into energy" would mean that an entity or an attribute turns into itself, which obviously does not make much sense. If they are not exactly the same thing, then the above statement would be just wrong, because the conversion of mass into energy or vice versa would mean the appearance (or disappearance) of energy – violating the fundamental law of Nature. The true meaning of the equivalence of mass and energy is not in the possibility of changing or converting one into the other. It is in that any change of energy of a system is accompanied by the corresponding change of its mass, and vice versa. Accordingly, an appropriate verbal description of the annihilation would be to say that in this process *all* mass and energy of one state of matter transforms into mass and energy of another state of matter. Consider, for instance, the system of an electron e and its "antiparticle" positron p. We will have more to say about positrons as a relativistic phenomenon later. But for now we just accept the experimentally known fact of their existence and consider possible consequences from the viewpoint of the mass–energy relationship.

A positron is an electron's "antipode" – a particle with exactly the same attributes but opposite sign of electric (and also the so-called "lepton") charges. If both particles are initially at rest and very far apart, then the rest mass of the system (e + p) is the sum of their individual rest masses. Starting from this state, they can "fall" onto one another because of electrical attraction. What is the final outcome? Since their charges are of the opposite signs, the net electric and net lepton charge of the system is zero. Therefore, nothing forbids the system to transform into another state with the zero net charges. In Nature, what is not explicitly forbidden is in principle allowed. The question is whether there exists another state with zero electric and lepton charges. The answer is yes – photons have neither electrical nor lepton charge. And Einstein's mass–energy relationship tells us that the system (e + p) is "energetic" by the mere fact of having a mass, even though in the given case it is only the rest mass! According to this relationship, it can be entirely converted into radiation if no conservation law forbids it. In this process, the mass of the two particles becomes the mass of the radiation and their energy becomes the energy of this radiation. Moreover, since the initial system had a rest mass, it had a rest frame (for instance, our lab). Therefore, according to the theorem about the center of mass (Equation (5.53)) mentioned in the beginning of this section, the outgoing radiation has the same rest frame, and all its mass is the *rest* mass. We have already discussed how this is possible. The initial momentum of the whole system is zero. So must be its final momentum. Since the momentum of a photon in free space is nonzero, the annihilation can occur only with the emergence of at least two photons. In this case, the photons must have the same frequency and move in opposite directions. They will have equal energies and equal but opposite momenta, thus making the net momentum zero. The law of conservation of momentum is satisfied. On the other hand, the net energy of the system after its "annihilation" equals its initial energy $\mathcal{E} = \mathcal{E}_0 = 2m_e c^2$, where m_e is the electron rest mass. Since the net momentum is zero, then applying Equation (5.23) to the final system (the two photons) at $\mathbf{P} = 0$, we see that in this case all electromagnetic energy is, indeed, the rest energy and all its mass is the rest mass. Again, neither of the receding photons has a rest mass of its own, but *the system* of the two has. We have another illustration that the *rest* mass of the system is not equal to the sum of the rest masses of its parts. On the contrary, the total relativistic mass of the system of noninteracting parts, as we have seen, is the sum of the individual relativistic masses.

From this viewpoint, mass can be considered as one "common denominator" of all possible kinds of energy present in a given system (or, more formally, as a numerical characteristic of the net energy). As we have already mentioned, this means that a deeper view of mass would be that it is just a measure of the energy of a system. Formally, we can always choose such a system of units in which the fundamental constant $c = 1$ (we essentially used it in our description of the 4-velocity in Section 5.1). In this case, we would just have $\mathcal{E} = m$. Such a fundamental characteristic of a system as its inertia would be merely a property of bundled energy.

All these examples also show that we can no longer say that the mass is always a measure of the amount of matter in a system. This is obvious from the fact that we can change the amount of energy, and thereby the amount of mass in the system, by just switching to another reference frame. But "the amount of matter," as it is usually understood, is not something that depends on which train you happen to board. Or consider, for instance, two billiard balls of equal rest masses. One of them is stationary and the other one is moving and, therefore, is more massive than the first ball. A head-on collision is possible, in which the first (moving) ball is stopped and the second one "inherits" its momentum and kinetic energy and starts moving with the same velocity as the first ball before the collision. Now the second ball is more massive than the first one. If mass were a measure of the amount of matter, we would have to conclude that in this process a certain amount of matter was transferred from the first ball to the second one. But imagine another inertial observer moving together with the first ball before the collision. For this observer, the first ball was initially at rest and the second ball was moving and therefore was more massive. After the collision, the second ball is stopped and the first one "inherits" its initial momentum and kinetic energy and starts moving with the same velocity as the other ball before the collision. This observer would conclude that the same amount of matter was transferred from the second ball to the first one. But the amount of matter is an absolute characteristic and, therefore, the process of its transfer from system A to system B cannot depend on the choice of reference frame. Consequently, mass cannot be a measure of the amount of matter.

However, one can as forcefully counterargue that the relativistic mass can be a measure of amount of matter in a system. Moreover, one can even use the same argument with the balls, with only two modifications: instead of the mechanical balls, consider two identical particles, say, protons (p) and instead of the elastic collision, consider an inelastic one. An inelastic collision is possible, in which a high-energy proton hits another proton resting in the lab (the target) and produces two more particles: another proton and antiproton [34]. Schematically,

$$\mathrm{p} + \mathrm{p} \rightarrow \mathrm{p} + \mathrm{p} + \mathrm{p} + \mathrm{p}^{-}. \tag{5.123}$$

Here p and p^{-} stand for a proton and an antiproton, respectively. The left-hand side of this equation shows the system before the collision and the right side – the system after collision. Just as in the electron–positron case considered before, the new particles here come in proton–antiproton pair to satisfy the conservation of charges: the electric charge and the so-called baryon charge. And of course, the mass and energy of the new particles appear to come from mass and energy of the bombarding proton. It is difficult to find a more compelling evidence of a potential capacity of relativistic mass to create new matter. And accordingly, relativistic mass *can*, from this viewpoint, be considered as a measure of the amount of matter in the system. To determine the exact amount of matter of a specified kind (in this example – the possible number of new particles with a proton's rest mass), one has to consider the whole process in the center of mass of the two original protons (Problem 5.22). This requires that we specify not only the energy of the bombarding particle, but also the target particle. Only then can we specify the center of mass of the system and the potential outcome of the inelastic collision. As another illustration, I recall a hypothesis that all observable universe might have been originated from the collision of only two ultra-ultra-ultra. . .-relativistic particles! This scenario, crazy as it may seem, does not contradict any known laws of physics and drives the argument to its extreme.

Now, what final conclusion can we draw from the above discussion? As we think of it, we can spot a subtle substitution of terms. In the initial argument, even though we consider formally two colliding particles, we were implicitly trying to associate the initial amount of transferred matter with only one of them, either A or B, depending on reference frame. We came to the conclusion that this would contradict the basic property of matter that it can go either from A to B for all observers or from B to A for all observers. It cannot go from A to B for some observers and from B to A for others. In the counterargument, we focused on another property of matter – its ability to exist in many different forms that can change into one another. In the corresponding example with the inelastic collision, the produced pair of new particles cannot be discussed in terms of their transition from the bombarding particle to the target particle or vice versa. They have emerged from the energy of the *whole system* – the bombarding particle plus the target. Both are absolutely necessary for the process. Without the target, no matter how fragile, the originally stable bombarding particle, no matter how energetic, will not produce anything new. In the example with the colliding protons (Equation 5.123), I said that the mass and energy of the newly born pair $p + p^{-}$ only *appear* to come from the bombarding particle. It will appear to come from the target for the observer moving with the bombarding particle. Actually, it comes from the system as a whole. This is why the above hypothesis of the birth of the universe required at least two particles for start!

Our final conclusion, therefore, can be formulated in the following way: the mass of a *single* particle cannot be considered as a measure of the amount of matter in it, because it can be reduced down to the rest mass by trying to catch up with the particle. In contrast, the mass of a *system* of at least two particles A and B with different velocities can be a measure of the amount of matter in the system, but this amount cannot be associated with only one particle. Even if A is much more "energetic" than B, we have to keep in mind that the property of being more or less energetic is the attribute of the reference frame rather than of the particle itself, and an attempt to catch up with A (change of reference frame) will reverse these attributes. Therefore, neither particle is more important than the other, and the ability of the system to produce new particles does not depend on choice of reference frame. This ability can be evaluated quantitatively most easily by considering the process in the center of mass of the system.

One of the leading motifs in our attempts to get a deeper insight into the fabric of reality is reducing the number of fundamental concepts necessary for a more complete description of the world. The theory of relativity has shown an intriguing possibility to get rid of mass as an independent fundamental property of material objects and demote it to the status of a mere substitute of energy. Already, many physicists avoid the term "relativistic mass" and apply the term "mass" only to the rest mass of a system. As Wilczek has put it [38], one of the goals is to construct a physical theory where all the masses of all known particles could be shown to result only from the corresponding fields. According to such a theory, our world would be the world "... without mass."[2)]

5.7
Mechanics From the Least Action

The concept of 4-interval has provided us with elegant description of the events and processes in space-time. It can do the same in providing us with no less elegant description of mechanics. Its basic advantage is in the appealing symmetry of the resulting expressions, which explicitly indicate their Lorentz-invariance.

In this section, we will illustrate this feature mostly for a free particle and in the next chapter we will generalize this approach to describe particles interacting with electromagnetic field.

We will exploit here the powerful principle of the least action. This principle provides the most general framework for almost all physical theories [29, 39].

According to this principle, for any mechanical system, there is a characteristic S called action, which takes on an extremal value for the actual motion of the system as opposed to any other imaginable motions. If the system consists of only one particle, its action can be represented as the line integral of a certain function Λ (called the Lagrangian) over a path connecting its known starting position at the known initial moment of time and final position at the known final moment of time. In the relativistic description of motion, this converts into the integral over a world line of the particle connecting the two given events M and N (a path integral of the type considered in Section 4.6).

2) Of course, one could as adamantly argue against using the term "energy" in favor of the term "relativistic mass" or just "mass" in all descriptions of the physical phenomena. For example, instead of talking about the "electric field energy," one could talk about the "mass of the electric field." In view of what was said in this section, which term to use would be just a matter of taste. What is actually important is that, as emphasized in the epigraph to this section, relativity has reduced the number of independent concepts here from 2 to 1.

The action is a scalar and, accordingly, the integral must be taken of a scalar (invariant of the Lorentz transformations). The most simple scalar function describing the motion of a free particle between two close points in space-time is the interval between these points. Therefore, the total action S between the two events M and N must be proportional to the line integral along a path connecting these events (the path integral):

$$S = -\kappa \int_M^N ds = -\kappa\, s\{x^j(t)\}. \tag{5.124}$$

This integral depends on chosen path. We want to relate it to the actual motion of the particle, in which it traces out the specific world line between the events (and we know that for a free particle, this is a straight line.) The event M is the passing of the particle by the point $\mathbf{r}_M$ at the moment of time t_M, and the event N is its passing by the point $\mathbf{r}_N$ at a later moment t_N. The proportionality constant must be an invariant characteristic of the particle. We know for certain (Section 4.6) that the integral itself on the right-hand side of (5.124) is maximal along the straight (and timelike) world line. On the other hand, it is convenient to have *minimum* for the action S to obtain correct nonrelativistic limit. This is done by setting the constant κ to be positive and putting the minus sign in our definition (5.124). With this sign, since the line integral is *maximal* along the straight line, S would reach its *minimum* along this line.

Now, using (4.11) we can rewrite (5.124) as the integral over the time interval between the events

$$S = -\int_{t_M}^{t_N} \kappa c \sqrt{1 - \frac{v^2}{c^2}}\, dt. \tag{5.125}$$

Note that unlike the Lorentz factor figuring in a transformation between two reference frames, in the Lorentz factor figuring in (5.125), v is to be considered as the particle's speed in *one and the same* reference frame, and generally before we have established that the actual motion of a free particle is uniform, this speed should be considered as an arbitrary function of time, with only one restriction that for all known particles $|v| \le c$.

In classical mechanics, the coefficient at dt is called the Lagrangian function. The same term is used in relativistic mechanics. Thus,

$$\Lambda\{v(t)\} = -\frac{\kappa c}{\gamma\{v(t)\}}. \tag{5.126}$$

Now, what about the value of κ? We suspect that, as an invariant characteristic of a particle, it should be proportional to m_0. We can prove it in the following way. We know that at $v \to 0$ the classical Lagrangian function for a free particle is

$$\Lambda(v) = \frac{1}{2} m_0 v^2. \tag{5.127}$$

Consider the relativistic expression (5.126) in this limit (we suppress the time dependency of v):

$$\gamma^{-1}(v) \underset{v \ll c}{\rightarrow} 1 - \frac{1}{2}\frac{v^2}{c^2}; \qquad \Lambda(v) \underset{v \ll c}{\rightarrow} -\kappa c + \frac{1}{2}\kappa\frac{v^2}{c^2}. \tag{5.128}$$

But the constant terms in the Lagrangian do not affect the equation of motion and can be discarded. So we can drop the first term in the Taylor expansion (5.128) of $\Lambda(v)$. Then comparing with (5.127) gives

$$\kappa = m_0 c. \tag{5.129}$$

Therefore,

$$S = -m_0 c \int_M^N ds, \qquad \Lambda = -\frac{m_0 c^2}{\gamma(v)}. \tag{5.130}$$

Now that we have a relativistic expression for the action of a free particle, let us see how it works.

In Newtonian mechanics, the momentum can be expressed in terms of the Lagrangian as

$$p = \vec{\nabla}_v \Lambda = \frac{\partial \Lambda}{\partial \mathbf{v}}, \quad \text{so that} \quad p_\alpha = \frac{\partial \Lambda}{\partial v_\alpha}. \tag{5.131}$$

(To remind you, here the Greek index α ranges from 1 to 3, standing for x, y, z, respectively.)

Applying this to the relativistic Lagrangian (5.130) gives us already familiar result:

$$\mathbf{P} = m_0 \gamma(v)\mathbf{v}. \tag{5.132}$$

The energy of the system is defined as

$$\mathcal{E} = \mathbf{P} \cdot \mathbf{v} - \Lambda. \tag{5.133}$$

Again, applying this to (5.130), we obtain

$$\mathcal{E} = m_0\gamma(v)\mathbf{v}\cdot\mathbf{v} + \frac{m_0 c^2}{\gamma(v)} = m_0 c^2 \gamma(v)\left[\frac{v^2}{c^2} + \frac{1}{\gamma^2(v)}\right] = m_0\gamma(v)c^2 = mc^2, \tag{5.134}$$

also a familiar result.

Combining this with (5.132) recovers Equation (5.29):

$$\mathbf{P} = \frac{\mathcal{E}}{c^2}\mathbf{v} \tag{5.135}$$

For a photon, or graviton, or any massless particle with $m_0 = 0$ and accordingly $v = c$, this reduces to

$$\mathbf{P} = \frac{\mathcal{E}}{c}\hat{\mathbf{n}} \tag{5.136}$$

($\hat{\mathbf{n}}$ is the unit vector in the direction of motion of the particle).

Thus, the new formulation describes all the results obtained in Section 5.3. In addition, it allows us to represent all these results explicitly in the Lorentz-invariant form, as the conditions imposed onto corresponding 4-vectors by the least action principle:

$$\delta\int_M^N ds = 0. \tag{5.137}$$

According to our result (4.88) in Section 4.7, we can write

$$ds = \sqrt{dx_j dx^j} \tag{5.138}$$

and

$$\delta\int_M^N ds = \int_M^N \delta\, ds = \int_M^N \frac{dx_j\, \delta dx^j}{ds}. \tag{5.139}$$

But $dx_j/ds = u_j$. Using this and multiplying by $(-m_0 c)$ gives the variation of the action S (5.130) in the form

$$\delta S = -m_0 c\int_M^N u_j\, \delta dx^j. \tag{5.140}$$

Now, we can use the fact that for any continuous function $f(q)$ the operations $d\,\delta f(q)$ and $\delta df(q)$ are commutative. Indeed, writing the differential of f as $df = f' - f$, where the value f' is infinitesimally close to f, we have

$$\delta df = \delta(f' - f) = \delta f' - \delta f.$$

But the last expression just defines the differential of variation:

$$\delta f' - \delta f \equiv d\delta f,$$

so that

$$\delta df = d\delta f.$$

Applying this to $f(q) = x(s)$, we can write the integral (5.140) as

$$\delta S = -m_0 c\int_M^N u_j \delta dx^j = -m_0 c\int_M^N u_j d\delta x^j. \tag{5.140a}$$

The last integral can be taken by parts:

$$\delta S = -m_0 c u_j \delta x^j|_M^N + m_0 c\int_M^N \delta x^j \frac{du_j}{ds} ds. \tag{5.141}$$

We can extract two kinds of information from this expression. First, we can obtain the equation of motion. To this end, we consider a bundle of mathematically possible trajectories connecting the fixed points (events in space-time) M and N (Figure 4.19). At these points, the variations are zero:

$$\delta x^j|_M = \delta x^j|_N = 0. \tag{5.142}$$

We are left with

$$\delta S = m_0 c \int_M^N \delta x^j \frac{du_j}{ds} ds. \tag{5.143}$$

The variation δS must go to zero for trajectories close to the actual trajectory. This can be the case only if the actual trajectory satisfies the equation

$$\frac{du^j}{ds} = 0, \quad \text{that is,} \quad u^j = \text{const.} \tag{5.144}$$

If we express the components of 4-velocity in terms of the 3-velocity **v** according to (5.5b), we will obtain the known equation of motion of a free particle, **v** = const. But in (5.144) it appears as a 4-vector equation in the 4-velocity space.

Second, we can find variation δS as a function of x_j. To illustrate this case, we can go again to Figure 4.19, but now we will keep fixed only the point M, that is, require that only the first term in Equation (5.142) be zero; in addition, we consider out of all possibilities only the *actual* trajectory, for which the second term (the integral) in (5.141) is zero (Figure 5.9). The remaining end point N is now neither fixed nor totally free; it is allowed to slide along the *actual* trajectory either toward, or away from M, so

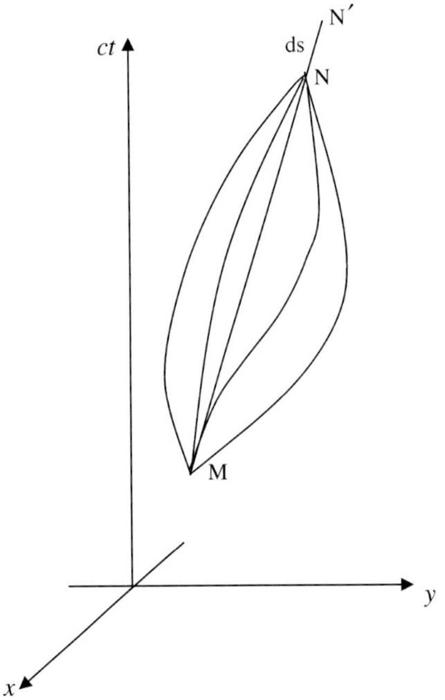

Figure 5.9 A bundle of the world lines connecting two events M and N in space-time. In the absence of external forces, the straight world line corresponds to the *actual* path of a particle. Here, the action S is considered only as a function of point on this path. Once the path is fixed, such point can be defined by its distance (kinematic length) s from the point M along this path, which is, in turn, a function of coordinates x^j. Thus, $S = S(s) = S(x^j)$.

that corresponding variations δx_j become actually the projections of the small segment ds of the actual trajectory onto the respective axes. Analytically, this gives us the equation

$$\delta S = -m_0 c u_j \delta x^j. \tag{5.145}$$

In this equation, the scalar δS is represented as a dot product of two 4-vectors: covariant vector $m_0 c u_j$ and corresponding contravariant 4-displacement δx^j. In vector $m_0 c u_j$, we recognize the familiar 4-momentum P. In view of the above remark about the physical meaning of the variations δx^j, we can now determine the covariant components of P from (5.145) as

$$P_j = -\frac{\partial S}{\partial x^j}. \tag{5.146}$$

In classical mechanics, the three partial derivatives of action with respect to the spatial coordinates are, indeed, the corresponding components of momentum **P**:

$$P_\alpha = -\frac{\partial S}{\partial x^\alpha}, \quad (\alpha = 1, 2, 3), \quad \text{or} \quad \mathbf{P} = -\vec{\nabla} S, \tag{5.147}$$

and the negative time derivative is the particle's energy:

$$\mathcal{E} = -\frac{\partial S}{\partial t}. \tag{5.148}$$

The new element of the fundamental importance brought in here by relativity is that now momentum and energy together form a 4-vector in Minkowski's space-time. In view of the comments (Equations (4.82) – (4.85)) in Section 4.7, the co- and contravariant components of this vector are related to the physical energy and momentum as

$$P_j = m_0 c u_j = \left(\frac{\mathcal{E}}{c}, -\mathbf{P}\right); \qquad P^j = m_0 c u^j = \left(\frac{\mathcal{E}}{c}, \mathbf{P}\right). \tag{5.149}$$

From these expressions, using Equation (5.2a) $u_j u^j = 1$, we obtain the equation for the 4-momentum of a free particle in the covariant form

$$P_j P^j = m_0^2 c^2. \tag{5.150}$$

This is just another form of the familiar result (5.23), but in this form it is explicitly emphasized that energy and momentum of a body are not separate characteristics but the components of a more general entity – a four-dimensional momentum.

Combining the last equation with (5.147) yields the relativistic generalization of the Hamilton–Jacoby classical equation:

$$\frac{\partial S}{\partial x^j}\frac{\partial S}{\partial x_j} = m_0^2 c^2 \tag{5.151}$$

or, in the "unfolded" form,

$$\frac{1}{c^2}\left(\frac{\partial S}{\partial t}\right)^2 - \left(\frac{\partial S}{\partial x}\right)^2 - \left(\frac{\partial S}{\partial y}\right)^2 - \left(\frac{\partial S}{\partial z}\right)^2 = m_0^2 c^2. \tag{5.152}$$

This form of equation is evidently Lorentz-invariant.

If a particle is in an external static field, it has also a potential energy U, so that the total energy is

$$\mathcal{E} = \sqrt{m_0^2 c^4 + p^2 c^2} + U \tag{5.153}$$

or

$$(\mathcal{E} - U)^2 - p^2 c^2 = m_0^2 c^4. \tag{5.153a}$$

Using again (5.147), (5.148), and (5.150), we obtain the relativistic Hamilton–Jacoby equation for this more general case:

$$\frac{1}{c^2}\left(\frac{\partial S}{\partial t} + U\right)^2 - \left(\frac{\partial S}{\partial x}\right)^2 - \left(\frac{\partial S}{\partial y}\right)^2 - \left(\frac{\partial S}{\partial z}\right)^2 = m_0^2 c^2. \tag{5.154}$$

5.8
Relativistic Motion in Coulomb's Field

The material of the previous section is pretty general and therefore looks rather abstract. Here, we will illustrate the general concepts by considering a special case – the motion of a charged particle in electric (Coulomb) field of a stationary point source. Since we want to know what new results relativity brings in as compared to the classical results, it is worthwhile to refresh our memory with a brief review of the classical results. Let us first write down the nonrelativistic expressions for the energy and angular momentum of a particle in a central field [39]. According to the definition of the angular momentum

$$\mathbf{L} = \mathbf{r} \times \mathbf{p}, \tag{5.155}$$

the vectors **L** and **r** are mutually perpendicular, and the conservation of **L** means that during the particle's motion its position-vector always remains in one fixed plane perpendicular to **L**.

Using the polar coordinates in this plane with the origin at the center of the field, we can write for the magnitude of **L**

$$L = r p_\varphi = m r^2 \dot{\varphi} = \text{const}, \tag{5.156}$$

or

$$\dot{\varphi} = \frac{d\varphi}{dt} = \frac{L}{mr^2}. \tag{5.156a}$$

Now, putting (5.156) into the expression for the total energy of the particle, we obtain

$$\mathcal{E} = \frac{1}{2} m(\dot{r}^2 + r^2 \dot{\varphi}^2) + U(r) = \frac{m\dot{r}^2}{2} + \frac{L^2}{2mr^2} + U(r). \tag{5.157}$$

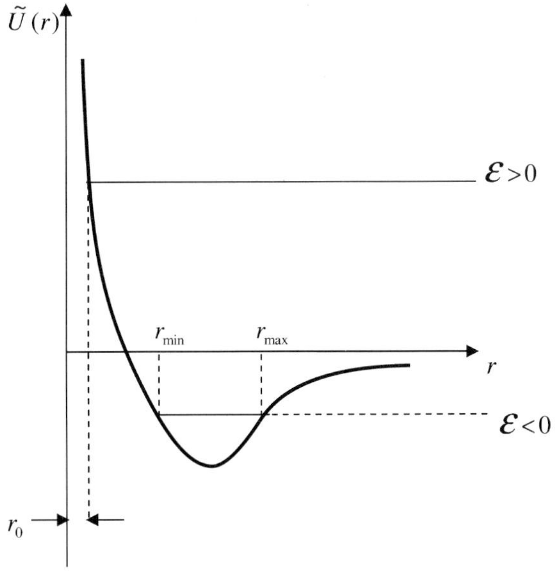

Figure 5.10 The graph of the effective potential energy (5.158) (for the motion in Coulomb's field), which includes the kinetic energy of orbital motion.

Here, the first term on the right is the kinetic energy of the radial motion of a particle and the second term is the kinetic energy of its angular (orbital) motion expressed in terms of its angular momentum. Since this term is an explicit function of r, it is convenient to combine it with the potential energy $U(r)$ and call this combination the *effective potential energy* (Figure 5.10):

$$\tilde{U}(r) \equiv U(r) + \frac{L}{2mr^2}, \tag{5.158}$$

so that

$$\mathcal{E} = \frac{1}{2} m\dot{r}^2 + \tilde{U}(r). \tag{5.159}$$

We find from this equation the radial component of the particle's velocity

$$\dot{r} \equiv \frac{\mathrm{d}r}{\mathrm{d}t} = \sqrt{\frac{2}{m}(\mathcal{E} - \tilde{U})}. \tag{5.160}$$

This expression shows that we can formally consider the radial component of motion in the orbital plane as one-dimensional motion in the field with the effective potential energy given by (5.158). Now, if we combine Equations (5.156a) and (5.160), we obtain

$$\frac{\mathrm{d}\varphi}{\mathrm{d}r} = \frac{L/r^2}{\sqrt{2m(\mathcal{E} - \tilde{U}(r))}}. \tag{5.161}$$

This differential equation can be recast in the integral form to give the expression for the particle's trajectory $\varphi(r)$:

$$\varphi(r) = \int_{r_0}^{r} \frac{L/r^2}{\sqrt{2m(\mathcal{E} - \tilde{U}(r))}} dr. \tag{5.162}$$

Here, r_0 is the particle's radial distance from the origin along the direction from which we start plotting φ.

This is enough to grasp the most essential features of particle's motion. Let us apply these results to the motion in the Coulomb's or gravitational field. In this case,

$$U(r) = \pm \frac{\alpha}{r}. \tag{5.163}$$

Here,

$$\alpha = \frac{|Qq|}{4\pi\varepsilon_0} \tag{5.164}$$

for a charge q in the field of the charge Q and the "–" sign is taken for the Coulomb interaction between the unlike charges. The same equation (5.163) with $\alpha = GMm$ and the minus sign describes the gravitational interaction of masses M and m in the nonrelativistic limit.

Let us first consider the radial motion. Its boundaries are determined by the equation

$$\mathcal{E} = \tilde{U}(r). \tag{5.165}$$

As seen from (5.160), its roots define the so-called turning points, at which the radial velocity becomes zero. This does not mean, however, that the particle comes to an instant stop because it keeps moving along the local $\hat{\boldsymbol{\varphi}}$-direction.

If Equation (5.165) has only one root, then the particle's motion is unbounded (Figure 5.10). This corresponds to the positive total energy. In astronomy, the corresponding motion is the motion of an object approaching a star (which could be our Sun) from infinity, bending around it, and receding to infinity again. In the particle physics, this corresponds to an electron scattering by an atomic nucleus (Rutherford's famous experiment).

If the equation has two different roots, r_{min} and r_{max}, then the particle's motion is restricted to the region

$$r_{min} \leq r \leq_{max}. \tag{5.166}$$

As seen from the graph of the effective potential energy, this can only be the case for the negative total mechanical energy, which is possible only in the field of attraction. The corresponding state is the bound state (e.g., the motion of a planet or periodic comet orbiting the Sun; in the atomic (nonquantum) physics, the motion of an electron in a hydrogen atom).

Is it possible for a particle in the field (5.163) of attraction to fall onto the center? The answer given by the classical physics (neglecting the radiation loss of an accelerated charge or mass) is *no* (with the obvious exception of the zero angular momentum when the particle's motion is strictly radial.) This negative result is clear from Figure 5.10 and the analysis of the effective potential energy: as $r \to 0$, the positive "centrifugal" (orbital kinetic) energy exceeds in magnitude the potential energy of attraction.

Another important property refers to the orbital (or angular) component of the motion of a bound particle. What is the angle swept out by the particle's position-vector as the particle makes one complete cycle in its radial motion? Such a cycle is completed if, for instance, the particle is launched in the direction perpendicular to its position vector, to move farther away from the center, and then returns to the original position. The answer is given by integrating (5.162) between $r_{\min}$ and $r_{\max}$ and doubling the result:

$$\Delta\varphi = 2\int_{r_{\min}}^{r_{\max}} \frac{L/r^2}{\sqrt{2m(\mathcal{E} - \tilde{U}(r))}}\,dr. \tag{5.167}$$

Since $r_{\min}$ and $r_{\max}$ are the roots of Equation (5.165), it may be convenient, to avoid the appearance of the fictitiously diverging integrals, to rewrite (5.164) as

$$\Delta\varphi = -2\frac{\partial S(r_{\min}, r_{\max})}{\partial L} = -2\frac{\partial}{\partial L}\int_{r_{\min}}^{r_{\max}} \sqrt{2m(\mathcal{E} - \tilde{U}(r))}\,dr, \tag{5.168}$$

where S is the classical action between perihelion and aphelion of the orbit.

If you take this integral with the *attractive* potential (5.163), the result will be exactly 2π. Geometrically, this means that the trajectory of the particle is a closed curve: as the particle returns to its initial position in its radial motion, the position-vector sweeps out the full angle 2π, and the particle returns to the initial position in its angular motion as well. In fact, we know that the trajectory of a bound particle in the Coulomb's or Newton's field is an ellipse.

It turns out that this is, in nonrelativistic domain, a unique characteristic specific for only two kinds of the potential field: $U(r) = \alpha/r$ with $\alpha < 0$ and $U(r) = kr^2$ with $k > 0$ (spatial oscillator) [39].

Now, we are in the position to compare this with relativistic predictions.

Relativistic motion in the *gravitational* field is accurately described by the general theory of relativity. One of its predictions was the precession of the elliptic planetary orbits, which has explained well-known precession of the orbit of Mercury. Another startling prediction was the existence of the black holes – probably one of the most exotic states known to man. An object passing by a black hole with a sufficiently small impact parameter will spiral toward the center and eventually get swallowed up by it. On the other hand, as is clear from the existence of stable orbits in the nonrelativistic mechanics, stable motions around the center (black hole or not) are also possible under certain conditions. However, the corresponding orbits are not closed. When a satellite returns to its, say, perihelion, its angular position is different from the previous one. The third interesting prediction is for the unbounded motion: there is a region of the impact parameters $b_c < b < b_m$ for which the passing particle undergoes the so-called spiral scattering; in contrast to regular scattering, when a particle just passes by the center and gets deflected in the scattering field, in the case of the spiral scattering the particle approaching the center starts orbiting around it along the trajectory with decreasing radius, thus spiraling toward the center. This spiraling, however, is not followed by the falling onto the center. After a certain amount of turns depending on b and p, the distance to the center reaches a certain minimal value also determined by b and p, the spiral starts unwinding and the particle recedes from the center. The winding and unwinding branches are symmetric with respect to the perihelion. The last wound of the unwinding spiral gradually changes to an asymptotic trajectory symmetric with the incident one [40–43].

I want to emphasize that all these effects have been predicted by the *general* theory of relativity. One of them – the motion of perihelion of Mercury – had been known long before the appearance of relativity theory, but it could not have found a compelling explanation in Newtonian mechanics.

In view of these results, it is interesting to ask, what does *special* relativity say about motions in the Coulomb's field of attraction?

We can again use the formalism of the relativistic Hamilton–Jacoby equation

$$\left(\frac{\partial S}{dt}+\frac{\alpha}{r}\right)^2-(\vec{\nabla}S)^2=m_0^2c^4 \tag{5.169}$$

and use the polar coordinates in the plane of motion with the origin at the source of the Coulomb's field. The gradient of S in these coordinates is

$$\vec{\nabla}S=\frac{\partial S}{\partial r}\hat{\mathbf{r}}+\frac{\partial S}{r\partial\varphi}\hat{\varphi}, \tag{5.170}$$

where $\hat{\mathbf{r}}$ and $\hat{\varphi}$ are the corresponding unit vectors, so the equation takes the form

$$\left(\frac{\partial S}{dt}+\frac{\alpha}{r}\right)^2-\left(\frac{\partial S}{\partial r}\right)^2-\frac{1}{r^2}\left(\frac{\partial S}{\partial\varphi}\right)^2=m_0^2c^4. \tag{5.171}$$

We can look for a solution in the form [29,39]

$$S=-\mathcal{E}t+L\varphi+S_r(r), \tag{5.172}$$

where $\mathcal{E}$ and L are, respectively, the total energy and the magnitude of the angular momentum of the particle and $S_r(r)$ is the contribution to the net action from the radial motion:

$$S_r(r)=\int p_r(r)\mathrm{d}r. \tag{5.173}$$

As we found in the previous section (and as is also seen from (5.113)),

$$\frac{\partial S}{\partial t}=-\mathcal{E},\qquad \frac{\partial S}{\partial\varphi}=L,\qquad \frac{\partial S}{\partial r}=p_r, \tag{5.174}$$

where p_r is the radial component of momentum. Therefore, the relativistic Hamilton–Jacoby equation is equivalent to the expression of the relativistic energy and momentum for a particle in a potential field:

$$\frac{1}{c^2}(\mathcal{E}-U(r))^2-p_r^2-\frac{L^2}{r^2}=m_0^2c^2. \tag{5.175}$$

It follows

$$p_r = \pm\sqrt{\frac{1}{c^2}(\mathcal{E} - U(r))^2 - \frac{L^2}{r^2} - m_0^2c^2}. \qquad (5.176)$$

By analogy with the classical formula (5.160), we can recast this in the form

$$p_r(r) = \frac{1}{c}\sqrt{\tilde{\mathcal{E}} - \tilde{U}(r)}, \qquad (5.177)$$

where

$$\tilde{\mathcal{E}} \equiv \mathcal{E}^2 - m_0^2c^4, \qquad \tilde{U}(r) \equiv \frac{(Lc)^2}{r^2} + U(2\mathcal{E} - U). \qquad (5.178)$$

In contrast to the classical mechanics, the relativistic effective potential function $\tilde{U}(r)$ depends not only on angular momentum, but also on energy of the system.

Let us apply the obtained results to the Coulomb's field of attraction (Equation (5.162) with the minus sign). In this case,

$$\tilde{U}(r) = \frac{L^2c^2 - \alpha^2}{r^2} - 2\frac{\alpha\,\mathcal{E}}{r}. \qquad (5.179)$$

The graph of the effective potential function $\tilde{U}(r)$ for this case is shown in Figure 5.11. We can easily see from both – the graph and Equation (5.179), that there are three distinct cases:

$$\left.\begin{array}{ll} L > L_c & \text{(a)} \\ L < L_c & \text{(b)} \quad L_c \equiv \dfrac{\alpha}{c} \\ L = L_c & \text{(c)} \end{array}\right\}. \qquad (5.180)$$

Case (a) can, in turn, be separated into two different cases depending on the total energy of the system. If the total energy exceeds or equals the rest energy ($\tilde{\mathcal{E}} \geq 0\ (\mathcal{E} \geq m_0c^2)$), then the system is unbound, that is, the motion is infinite; in astronomy, this corresponds to a particle approaching the center of attraction from infinity, reaching the point of the closest approach (the turning point), and then receding to infinity again after being deflected through a certain angle; in electrostatics, this corresponds to the particle scattering.

If the total energy is less than the rest energy ($\tilde{\mathcal{E}} < 0\ (\mathcal{E} < m_0c^2)$), then the system is bound and the motion takes place within a finite area of the orbital plane. In astronomy, this corresponds, for example, to a planetary system; in electrostatics, to an electron permanently orbiting around the nucleus.

In this respect, case (a) does not differ much from the nonrelativistic analogue. In cases (b) and (c), however, the relativistic predictions are totally different from the classical ones. Both – Equation (5.180) and Figure 5.11b – show something totally new: the particle's motion is not bounded from "below": there is nothing to stop it from getting infinitesimally close to the center. Regardless of the particle's energy, it spirals toward the center and "falls" onto it within a finite time interval after being released from a certain initial position satisfying conditions (b) or (c).

We thus come to a fundamental conclusion that the relativistic motion in the Coulomb's field is critically determined by the interaction constant α: this constant

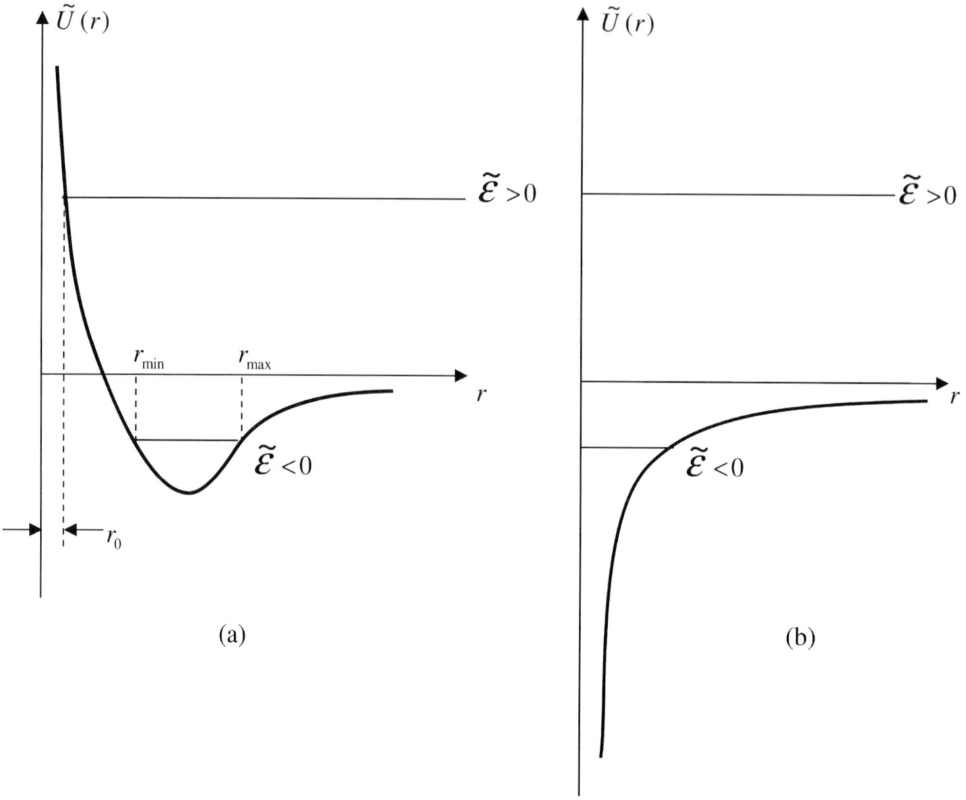

Figure 5.11 The graph of the *relativistic* effective potential energy (5.178) for the case of the Coulomb's field of attraction. (a) The particle's angular momentum exceeds the critical value $L_c = \alpha/c$. In this case, the particle can have stable states. These states are unbound with one turning point r_0 when ($\hat{\mathcal{E}} > 0$), that is, $\mathcal{E} > m_0c^2$, or bound when $\hat{\mathcal{E}} < 0$, that is, $\mathcal{E} < m_0c^2$, with two turning points r_{min} (perihelion) and r_{max} (aphelion). Regardless of the value of $\hat{\mathcal{E}}$, the center of attraction (the origin) is inaccessible for the particle. (b) The particle's angular momentum is equal to or less than the critical value. Then the particle, regardless of its energy, is bound to pass through the center of attraction.

introduces the crucial value of the angular momentum $L_c = \alpha/c$ such that all particles with $L < L_c$ get "sucked into" the center regardless of their energy. If the energy exceeds the rest energy, the corresponding particle can be thought of as coming from infinity and being "absorbed" by the center. If the total energy of the particle is less than its rest energy, the particle's orbital motion is unstable with respect to the falling onto the center.

The process of scattering, which corresponds to the first of these cases, is described quantitatively by the so-called cross section of a certain outcome [29, 39]. Geometrically, it can be visualized as an area in the plane perpendicular to the monoenergetic flux of the incident particles, such that all the particles of the incident beam, which happen to cross this area, are bound for this outcome. We are now interested in the outcome "absorption." The probability of this outcome for a randomly chosen particle in the incident beam is equal to the probability that the particle's angular momentum turns out to be less than L_c. As we mentioned before, for a fixed energy, the particle's angular momentum is determined by the impact parameter b. The impact parameter is the minimal distance between the incident particle and the center in the absence of the interaction between them. In the presence of the interaction, it is the distance between the particle and the symmetry axis of the incident beam when the particle is far away from the center. The particle's angular momentum can be expressed in terms of the impact parameter as $L = bp$, where p is the particle's initial momentum. Therefore, applying Equation (5.176) to this case gives

$$b = \frac{L}{p} = \frac{Lc}{\sqrt{\mathcal{E}^2 - m_0^2 c^4}}, \qquad b_c = \frac{L_c}{p} = \frac{\alpha}{\sqrt{\mathcal{E}^2 - m_0^2 c^4}}, \qquad \mathcal{E} \geq m_0 c^2. \tag{5.181}$$

All the particles with the impact parameters $b < b_c$ are doomed to absorption by the center. They can be visualized as initially moving within a fictitious "tube" of radius b_c (Figure 5.12). Its cross-sectional area is $\sigma = \pi b_c^2$, or, in terms of characteristics of the system:

$$\sigma_c = \frac{\pi \alpha^2}{\mathcal{E}^2 - m_0^2 c^4}, \qquad \mathcal{E} \geq m_0 c^2. \tag{5.182}$$

In the special case $\mathcal{E} = m_0 c^2$, we have $\sigma_c = \infty$. Does it make sense to you? It does if you recall that this case would, classically, correspond to the motion along a parabolic trajectory and such a trajectory asymptotically does not remain at finite distance from the symmetry axis, that is, we cannot attribute to it any finite impact parameter.

Note that b and σ_c are defined only for $\mathcal{E} > m_0 c^2$ (unbounded motion). At $\mathcal{E} > m_0 c^2$, the impact parameter in (5.176) becomes imaginary – this characteristic loses its meaning since the motion of a bound particle cannot be described as scattering.

Let us consider such a situation in more detail (case (a)). It corresponds to a bound state with an orbit of a finite size. As we have mentioned before, in the attractive field (5.163) the classical motion along such an orbit is periodic and the orbit is an ellipse. Is this true in relativity? The answer is no. An interesting thing is that we can predict

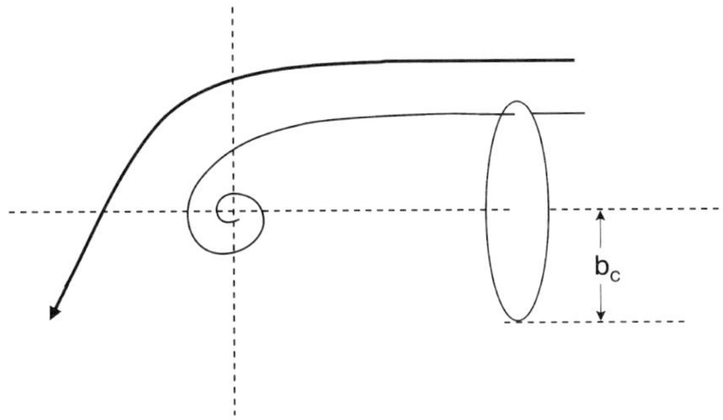

Figure 5.12 Scattering of parallel beam of particles incident on the Coulomb's center. The particles with an impact parameter $b > b_c$ undergo regular scattering (their trajectories are just deflected by the field of Coulomb's center). The particles with $b < b_c$ (that is, initially approaching the center within a fictitious "tube" of radius b_c) wound up bumping into the origin. We can consider them as being absorbed by the scatterer and accordingly call the tube's cross-sectional area $\sigma_c = \pi b_c^2$ the cross section of absorption.

this answer qualitatively without solving any equations. To do this, just recall the weird properties of the mass studied in Section 5.5. The mass of a moving object depends on its speed and since speed changes in the external field, the mass becomes position-dependent. In addition, it responds differently to a force acting along the trajectory and to the force of the same magnitude acting perpendicular to the trajectory. We describe this difference by introducing terms "transverse mass" (which is numerically equal to the relativistic mass) and "longitudinal mass" (Equations (5.61a) and (5.61b)). This is a purely relativistic phenomenon and it makes all the difference. Namely, we can consider the periodicity (or synchrony) of motion along the coordinates r and φ in the given field as the result of the fine-tuning of the corresponding accelerations. The actual force in a central field is always in radial direction. For any point on the orbit, we can break up this force into longitudinal component (parallel to the instant velocity) and transverse component (perpendicular to the instant velocity). In a nonrelativistic motion, this results in the identical breaking up for the components of acceleration. The acceleration determines the curvature of the orbit. Therefore, the fact that the orbit in the Coulomb's field is closed is, in fact, the result of a fine-tuning between the longitudinal and transverse components of acceleration in this field. Since in relativistic motion, the ratio of these components is no longer equal to the ratio of the corresponding components of the force, this tuning is destroyed; we can therefore expect that the trajectory may lose its property of being elliptical, or even of being closed.

This is precisely what happens.

To calculate the exact trajectory, we write expression (5.172) with $p_r(r)$ from Equation (5.177). The trajectory is defined by the equation $\partial S/\partial L = \text{const}$. Taking the derivative gives

$$\varphi = \int\limits_{r_{\min}}^{r_{\max}} \frac{Lc/r^2}{\sqrt{\tilde{\mathcal{E}} - \tilde{U}(r)}} dr. \tag{5.183}$$

This looks similar to the nonrelativistic expression (5.167), with the important distinction that $\tilde{\mathcal{E}}$ and $\tilde{U}(r)$ are determined by (5.178). For the Coulomb's field, the integral can be taken and the resulting expression $r(\varphi)$ for $L < L_c$ and $\mathcal{E} \leq m_0 c^2$ confirms our prediction: at the change of the angle φ by 2π, the radial distance from the center does not return to its original value. And vice versa, if we consider the inverse function $\varphi(r)$, it turns out to be multivalued, and the difference between the two values corresponding to the same r is not a multiple integer of 2π. Physically, this means that the orbit is not a closed curve. We can still describe such a motion as motion along an ellipse, but the ellipse itself is turning in its plane instead of being fixed. We say that the orbit is precessing (Figure 5.13).

The precession of the planetary orbits discovered in the general relativity is already contained in the special relativity! In a way, this effect can be considered as a direct consequence of speed dependence and "anisotropy" of mass discussed in Section 5.5. The fact that these properties enabled us to predict the precession of orbits in Coulomb's field before doing any calculations demonstrates relevance and usefulness of the concept of the relativistic mass.

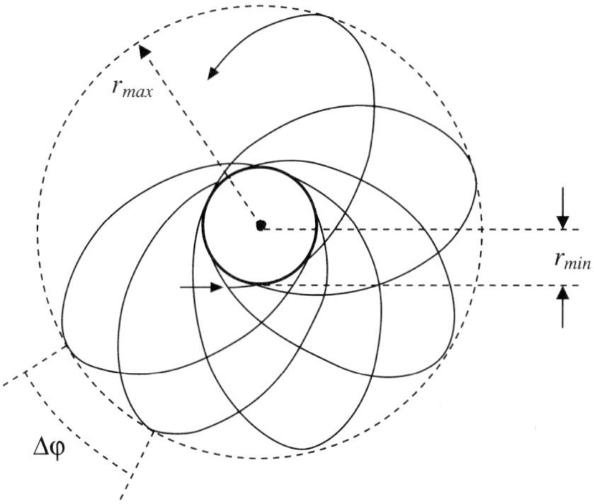

Figure 5.13 Relativistic orbit of a bound charge with $L > L_c$ in Coulomb's field.

Problems

5.1 Find the components of 4-velocity of a regular jetliner moving north–west at a speed 760 miles/h (identify the x-axis with south direction, y-axis with east direction and a local vertical with the z-direction).

5.2 An electron has an instant speed 253 000 km/s in the direction making 30° with the x-axis in the x–z plane. Find
(a) all three components of its velocity v;
(b) all four components of its 4-velocity u.

5.3 Find all components of 4-velocity u
(a) for a stationary object;
(b) for a photon moving in free space along the y-direction;
(c) compare their norms.

5.4 Using Equation (5.5), find all components of 4-velocity of a hypothetical superluminal particle moving along the y-direction with a speed $v = 3.7c$. Interpret your result.

5.5 Find the components of 4-velocity for a particle moving along the y-direction with an infinite speed.
(*Hint*: Write down the expression (5.5) for a finite $v > c$ and then use the limiting procedure $v \to \infty$.)

5.6 A particle with a rest mass m_0 is moving in a box, where it bounces up and down from the bottom and the top, respectively. The speed of its motion is v_z.
(a) Find the relativistic mass of the particle.
(b) Consider an observer moving at a speed v_x along the x-direction and find the particle's relativistic mass in the rest frame of this observer.

5.7 Show that vector $\tilde{\mathbf{L}}$ defined by Equation (5.41) is a conserved quantity.

5.8 An electron enters the gap between the plates of a parallel-plate capacitor through the small hole in one of the plates. At this moment, it is moving along the electric field $\mathbf{E}$ with velocity $\mathbf{v}_0$. Find its velocity $\mathbf{v}(t)$ at later moments as a function of t.

5.9 (a) Find the asymptotic line for the world line of a charge in a homogeneous electric field.
(b) As seen from Figure 5.2, no photon shot from the origin after a certain critical moment $t_c > 0$ will ever catch up with the charge q. For given q, m_0, $\mathbf{E}$, find t_c.

5.10 At a certain moment of time, the charge q of the rest mass m_0 is moving at an angle 45° to the electric field $\mathbf{E}$ directed along the x-axis in the xy-plane. Find the x- and y-components of the charge's acceleration.

5.11 (a) Derive Equation (5.27), $v = d\mathcal{E}/dp$.
(b) Derive the more general equation $\mathbf{v} = \vec{\nabla}_p \mathcal{E}$, where the del–operator $\vec{\nabla}_p$ denotes the gradient with respect to the variables p_α, that is, in the momentum space.

5.12 Show that the equation $d\mathcal{E}/dt = \mathbf{f} \cdot \mathbf{v}$ holds in relativity.

5.13 Derive the transformation rules for 3-force $\mathbf{f} = d\mathbf{p}/dt$ from transformation rules for Minkowski's 4-force G_j.

5.14 A rocket with the rest mass of 120 ton is moving past the Earth at $V = 200\,000$ km/s.
(a) Find the total relativistic energy and momentum of the rocket.
(b) Find the total energy and momentum of the rocket in its rest frame.

5.15 Using the relativistic relation between energy $\mathcal{E}$ and momentum P, express $\mathcal{E}$ in terms of P and the rest mass m_0. Expand this expression into a Taylor series in P and identify in this expansion the terms corresponding to the rest energy and classical kinetic energy.

5.16 The energy of an ultrarelativistic particle is measured to be $\varepsilon = 10^{-4}$ J and its momentum is 7.2×10^{-14} kg m/s. Find the rest mass of the particle.

5.17 You are sitting on a bridge and observing a relativistic vehicle moving down the bridge at a speed 250 000 km/s. The gravitational force on the vehicle (measured by you) is 1.5×10^4 N.
(a) What is the gravitational force on the vehicle as measured by its driver?
(b) The gravitational force on your car, which is parked on the side of the bridge, is 3×10^3 N. What is this force as measured by the driver of the vehicle?

5.18 An electron has been accelerated so that its kinetic energy is nine times its rest energy [44].
(a) What is its speed?
(b) What is its momentum? (Electron's mass is 9.1×10^{-31} kg.)

5.19 Two electrons approach each other head-on at $0.5c$ relative to the reference frame K [44].

(a) What is the total kinetic energy of the electrons in K?

(b) What is the total kinetic energy of the system in the rest frame of one of the electrons?

5.20 Find the rest mass of a photon in the superposition of the two split states shown in Figure 5.4, by switching to the *rest frame* of this state.

5.21 In Tom's exploration mission, he launches two identical surveillance spacecrafts with equal speeds v' relative to his spaceship, one in the forward direction and the other in the backward direction. Since Tom's spaceship keeps on moving with the same velocity $\mathbf{V}$ in Alice's reference frame, its momentum remains constant. Therefore, in the Alice's frame, the momenta of the spacecrafts after the launch satisfy the condition $\mathbf{p}_+ = -\mathbf{p}_-$. On the other hand, $\mathbf{p}_+ = m(v_+)\,\mathbf{v}_+$ and $\mathbf{p}_- = m(v_-)v_-$, so that

$$\frac{p_+}{p_-} = \frac{m(v_+)\,v_+}{m(v_-)\,v_-} = 1,$$

and

$$\frac{m(v_+)}{m(v_-)} = \frac{v_-}{v_+}$$

Since $v_- < v_+$, we conclude that a fast moving object is *less* massive than the slowly moving object of the same rest mass. Find and explain the source of this nonsense.

5.22 Consider a system of two stationary noninteracting masses m_{01} and m_{02}.

(a) What is the rest mass M_0 of this system?

(b) What happens with M_0 if the mass m_{01} gets a boost with velocity $\mathbf{v}_1$ while the mass m_{02} remains at rest?

(c) What happens with M_0 if the mass m_{02} gets boosted to velocity $\mathbf{v}_2$, while the mass m_{01} remains at rest?

(d) What happens with M_0 if both masses are boosted to the velocities $\mathbf{v}_1$ and $\mathbf{v}_2$ respectively?

(e) From your general result in (d), find what happens with M_0 in the special case $\mathbf{v}_1 = \mathbf{v}_2$?

(f) Show that the answer to question (e) is consistent with the invariance of the rest mass.

5.23 A parallel-plate capacitor with the plate area A and the separation d between the plates is charged to a potential difference Φ across the plates.

(a) By how much has the rest mass of the capacitor increased?

(b) How will the answer change for a spherical capacitor with radii a and b of the shells?

(c) Suppose that the outer spherical shell in (b) has been removed and the remaining inner shell is so thin that its rest mass is negligible. This shell is charged again to the same potential Φ as before (now relative to infinity).

What is the rest mass of the shell after being charged? Express your answer, first, in terms of potential Φ and then in terms of the corresponding charge Q on the shell.

5.24 Two identical solenoids of length L and radius a, with N turns/m, are balanced on the balance scale. Now a current I is being passed through one of the solenoids. By how much will this solenoid become heavier than the one without current?

5.25 Two coaxial infinitely long cylindrical shells have radii a and b ($a < b$). The outer shell is made of a superconducting material (the magnetic field **B** does not penetrate into a superconductor). The rest mass of the unit length of the inner shell without current is λ_0. When a current I flows along the shell, its rest mass increases.

(a) Find the ratio $\lambda_0(I)/\lambda_0$ and plot it as a function of b. What happens if the radius of the outer shell goes to infinity?

(b) Plot the above ratio as a function of a. What happens if a approaches zero?

5.26 (a) The "ionization potential" (the minimal energy necessary for ionization) of a hydrogen atom is 13.6 eV (1 eV is the energy acquired by an electron passing across the potential difference of 1 V). Find the mass defect of the hydrogen atom.

(b) Find the mass difference corresponding to an optical transition in a He–Ne laser accompanied by emission of the red light with the wavelength $\lambda = 5663$ Å (1 Å $= 10^{-10}$ m).

(c) In a certain chemical reaction 103 cal/kg of heat is released. Find the corresponding mass defect per each kilogram of the reacting chemicals.

5.27 A thin ring of radius R has a rest mass m_0. The ring is brought to a rapid rotation with an angular velocity ω about the symmetry axis perpendicular to its plane, so that the center of mass of the ring remains stationary. Find the rest mass of the ring in this state.

5.28 You have a disk, a spherical shell, and a solid sphere – all three of the same radius R and rest mass m_0 in the stationary state. Now each of the objects is brought to a state of rotation about axis of the azimuthal symmetry. Each object is rotating as a rigid body with an angular velocity Ω. Find the rest mass of (a) disk; (b) shell; and (c) sphere in this state.

5.29 In the previous problem, find the rotational inertia and angular momentum of all three bodies.

5.30 Suppose you have an EM oscillator shown in Figure 5.9 with the distance D between the centers of the capacitor C and the inductance L. The capacitor is charged to the voltage V.

As we know, the electromagnetic part of the oscillator's mass owing to the initial voltage will flow periodically between the capacitor and the inductance. Describe quantitatively the corresponding motion of the "purely mechanical" part of the oscillator. Find the amplitude and the frequency of this motion.

5.31 In a deeply inelastic collision, a high-energy proton colliding with another proton within a target can produce a pair proton–antiproton.
(a) What minimal kinetic energy of the bombarding proton is required for this to happen?
(b) How will the answer change if the target is an electron?
(c) If the bombarding particle is an electron?
(d) If both the target and the bombarding particle are electrons?

5.32 As you know, a free mu-meson is unstable and decays into an electron and neutrino, with an average lifetime of only 10^{-6} s. Suppose you have accelerated the initial muon so that its relativistic mass is 40 times greater than the rest mass of a proton.
(a) What will the muon's average lifetime be in this state?
(b) Will you expect the creation of new particles (e.g., protons and antiprotons) in the process of its decay? If yes, how many proton–antiproton pairs do you expect to observe? If no, explain why.

5.33 Can a free electron accelerated to a certain energy emit or absorb photons? Whatever your answer is, prove it.

5.34 Prove the "variation theorem": $\delta \mathrm{d}f(q) = \mathrm{d}\delta f(q)$.

5.35 (a) Find the radial part of relativistic action (5.172) for a charged particle in Coulomb's field by integrating the radial component p_r of the particle's momentum.
(b) Find the particle's trajectory from the equation $(\partial S/\partial L) = \text{const}$ or by taking the integral (5.183).
(c) For the case $\mathcal{E} < m_0 c^2$, $L > L_c$, show that the particle's orbit is generally not closed.

5.36 Find the change in the polar angle corresponding to the complete cycle in the radial motion of a particle in the field (5.163). Consider the integral (5.167) between the perihelion and aphelion of the orbit, taking into account that these points are the roots of Equation (5.165) for this field.
(*Hint*: Consider (5.167) as the proper integral between $r_{\min} + \delta$ and $r_{\max} - \delta$ and then let $\delta \to 0$.)

5.37 Find the scattering angle for a charge passing through a Coulomb's field of a nucleus.
(*Hint*: The scattering angle $\chi = \pi - 2\varphi_0$, where $2\varphi_0$ is the angle between the two asymptotes of the charge's trajectory.)

5.38 Find the "fall time" for a charge with $L < L_c$ released from a distance r_m in the Coulomb's attraction field.
(*Hint*: The dependence $r(t)$ is determined by the equation $\partial S/\partial E = \text{const}$.)

6
Relativity at Work II: Electromagnetism and Optics

Nothing will come out of nothing.
Shakespeare, *King Lear*

6.1
Electric Field and Something Else (The Origin of Magnetic Field)

Historically, special relativity (SR) has emerged from the attempts to reconcile the electromagnetic (EM) theory with numerous experimental data and thought experiments involving moving media [32]. In this chapter we show that once we accept Coulomb's law as an axiom, all the rest of electromagnetism can be deduced from relativistic postulates.

The classical theory of electricity is based on the concept of electric charge and electric field. These two are related by Gauss's law stating that the electric field flux through any closed surface is equal to q/ε_0, where q is the enclosed net charge and ε_0 is permittivity of free space. As to magnetism, originally it had been thought to originate from the two kinds of opposite magnetic charges (or "poles") – northern and southern magnetic poles. The search for magnetic poles (or monopoles [44, 45]) has been for decades an ongoing scientific endeavor. However, no magnetic charges have been found so far. Therefore, the magnetism at first appears to emerge out of nowhere: in contrast to electricity, there is apparently no such thing as the magnetic charges, and yet there are magnetic forces, and their carrier – the magnetic field.

The experiments show that the ultimate source of magnetic field is also electric charge, only when in a state of *motion*.

But the state of motion is *not* an *intrinsic* characteristic of a charge – it depends on a reference frame (RF). Suppose we have a stationary electric charge. It produces *only* an electric field around itself. Then we can produce a magnetic field as well, just by switching to another reference frame where the charge is moving. Since motion is relative, this means that, at least until we discover magnetic monopoles, all magnetic fields can be considered as a *purely relativistic effect*.

We should not take it too simplistically, though. It is not true that in the rest frame of a charged body (where the body's momentum $\mathbf{P} = 0$) we can *always* observe *only* electrical field. A spinning charged body produces the magnetic field in its rest frame. The only frame free of this field is the one rotating together with the charged body.

Special Relativity and How it Works. Moses Fayngold

ISBN: 978-3-527-40607-4

The body is stationary in this frame (the body's linear momentum **P** and angular momentum **L** are *both* zero), and accordingly, its field is purely electrostatic, but this frame is not inertial! Moreover, as we will find later, in the weird world of quantum mechanics, there can be found no reference frames, inertial or not, that would be free of a magnetic field due to a single electron. Even in classical world, a current-carrying wire produces magnetic field that is also observed in *all* reference frames. These examples illustrate the subtlety of the phenomenon we are going to discuss. What is most important here, however, is the statement that at least in classical domain, we can always find a certain motion of electrical charges as a source of the corresponding magnetic field.

If this is true, then magnetic phenomena can be deduced from electric phenomena merely as a result of Lorentz transformations. Let us see how it works.

We will consider a few simple situations, actually, a few textbook problems, but entirely from the viewpoint of the theory of relativity.

Situation 1: Travels Within a Capacitor

Suppose we have a point charge q in a parallel-plate capacitor (Figure 6.1a). It experiences a force

$$\mathbf{f} = q\mathbf{E}, \tag{6.1}$$

where **E** is the electric field between the plates. Coulomb's force on a charge does not depend on its state of motion, therefore the force (6.1) does *not* change if the charge is moving.

Suppose the charge does start moving along the electric field lines, that is, in the direction perpendicular to the capacitor's plates (Figure 6.1b). In the rest frame of the capacitor, the force on the charge is determined, as before, by Equation (6.1).

Let us now introduce an observer sitting on the charge. For this observer (call him Tom), the charge remains stationary, and instead he sees the capacitor moving in such a way that one and the same electric field line slides through the fixed charge (Figure 6.1c). The transformation law (5.86) for longitudinal force predicts that Tom will in this situation measure the same force as we do. Since the electric force $\mathbf{f} = q\mathbf{E}$, and the charge is invariant, this immediately translates into the

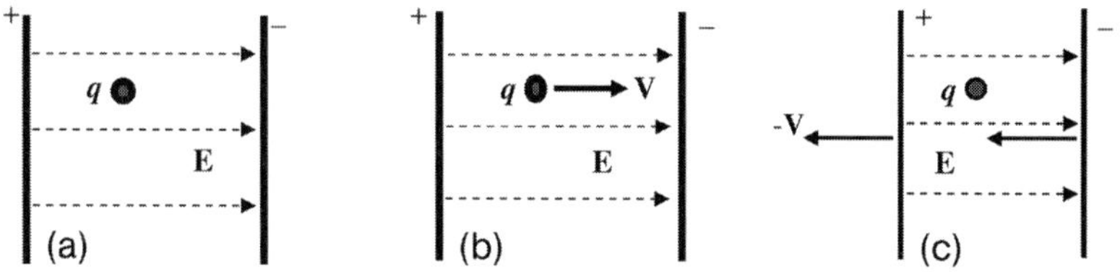

Figure 6.1 (a) Point charge q at rest inside of a charged parallel-plate capacitor; (b) the same charge moving along the electric field lines; (c) the charge is stationary, but the capacitor is now moving in the opposite direction. In all three cases the electric force on the charge is the same: in cases (a) and (b) – because the electric force does not depend on the state of motion of a test charge; in case (c) – because the electric field between the plates remains the same as in (a) and (b).

statement that the longitudinal component of the electric field **E** is the same in both reference frames:

$$E_{||} = E'_{||}. \tag{6.2}$$

This conclusion is consistent with electrostatics. The electric field **E** within the capacitor is produced by the surface charges on the capacitor's plates and it is homogeneous. From Tom's perspective, these charges do not change, nor do the plates (their area is not affected by motion in the given arrangement). Therefore, the motion of the plates does not change the field between them.

Strictly speaking, in Tom's RF the field is due to the *moving* charges, which takes us beyond electrostatics. As we will see later, the field at a given moment of time originates from positions of the corresponding charges at the earlier moments (retardation effect). But if we know this field in a RF where these charges are stationary, we can determine it in another RF where these charges are moving by using the Lorentz transformations instead of its direct calculation "by brute force" in the new frame. The Lorentz transformations take care of all the effects of this motion automatically. In the given case we conclude from (6.1) that the longitudinal (parallel to the relative velocity) component of the electric field is not affected by motion of the observer.

Now let us change the mutual orientation of **E** and **V**, say, by turning the capacitor through 90° so that the charge q, which had, in the first experiment, been moving *along* the electric field lines, will now move *perpendicular* to the electric field lines (Figure 6.2a, with the capacitor turned). The same laws of electrostatics tell us again that we in the lab should not expect any change in the acting force on the charge. And the lab experiment confirms this expectation.

But this is not so for Tom this time! Relative to Tom, the charge q is stationary, but the capacitor is moving to the left with velocity **V** (Figure 5.2b). And, according to Tom's measurements, the electric force will increase. Tom will measure the force on the charge to be greater by a factor $\gamma(V)$ than the force measured by us. What is the origin of this difference?

Note that this time the force is perpendicular to the instant direction of motion. The transformation rules (5.84) tell us that the transverse component of a force on an

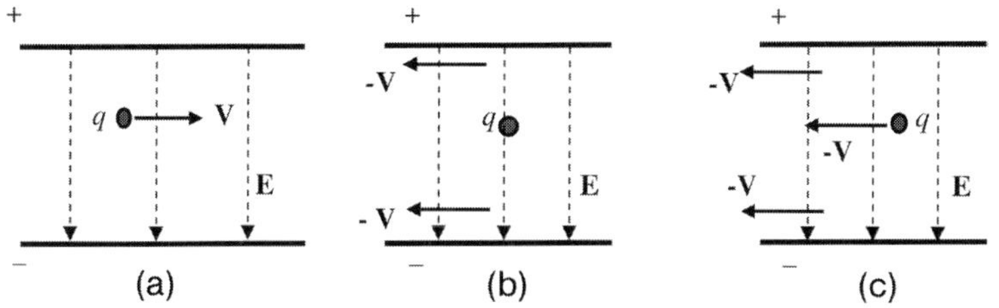

Figure 6.2 Moving across the electric field. (a) Charge q is moving in K (the rest frame of the capacitor); (b) the same process observed in K′ (comoving with the charge); (c) charge q remains stationary in K and is observed from K′.

object in its rest frame is $\gamma(V)$ times greater than in the frame where the object is moving:

$$f'_\perp = \gamma(V) f_\perp > f_\perp. \tag{6.3}$$

Accordingly, the transverse component of the electric field $\mathbf{E}'$ in the moving reference frame K′ increases by the same factor:

$$E'_\perp = \gamma(V) E_\perp > E_\perp, \tag{6.4}$$

where unprimed $\mathbf{E}$ is the field in the lab frame (the rest frame K of the capacitor).

But this is a formal explanation. What real physical phenomenon is responsible for this? Taking a closer look, we notice that the surface charge density σ on the capacitor's plates, measured by Tom, must increase by the factor of $\gamma(V)$ due to Lorentz contraction of the plates along the direction of their motion: the same amount of the plate's charge is now distributed over the plates with their length shortened. The increased surface charge density produces proportionally greater electric field between the plates. Thus, the relativistic length contraction makes the electric field component $E_\perp$ transform as (6.4), *which is consistent with the transformation rule* (6.3) *for the corresponding force*. In other words, the electrostatics together with the relativistic length contraction (due to the motion of the capacitor) produces *the same result* as the relativistic transformation (5.86) alone without any reference to length contraction. (Note that in this case, if we wish to appeal to electrostatics, we do not have to bother about the retardation effects because the motion *along* the plates preserves the uniformity of the charge distribution.)

However, this honeymoon of relativity with electrostatics lasts only when the test charge q is stationary in K′ and breaks the moment the charge q starts *moving* in K′. At this moment, the prediction of the electrostatics (even with the account taken of the length contraction effect) departs from the relativistic prediction (5.84). As an example, consider Tom moving between the capacitor's plates parallel to them while the charge q remains stationary inside the capacitor. Then this charge is *moving* relative to Tom with the speed $v'_x = -V$ (Figure 6.2c). Now, it so happened that Tom had read about relativity *before* having learnt anything essential about magnetism; so, by the time of this experiment he had known the electrostatics and basics of relativity, but almost nothing about magnetism. According to electrostatics, Tom expects the result (6.3) to hold. The electric field remains the same as in case shown in Figure 6.2b, and the electrostatics, being insensitive to the motion of a test charge, predicts that the force acting on it must also remain the same.

In contrast, the motion-sensitive relativity predicts that the force on a moving object may differ from the force on the stationary object in the same surroundings. Indeed, the transformation rules (5.84) demand that the transverse force on the *moving* charge must be

$$f'_\perp = \frac{f_\perp}{\gamma(V)} < f_\perp. \tag{6.5}$$

Thus, *electrostatics* says that the force on the charge q in K must be *greater* than it is in K, while the *relativistic prediction* (6.5) is that now it must be *less* than it is in K. This is a logical contradiction since it gives two different results for a force on the same object, calculated by two different methods. If we are to believe in special theory of relativity (STR), then the only conclusion we can make is that electrostatics *alone* is not consistent with relativity. Where does the inconsistency lie? The relativistic equation (5.84), and thereby (6.5), refers to the *net* force, whereas (6.3) refers only to the *electrostatic* force. There is only electric field in K. Since (6.5) that we believe is correct differs from (6.3), we conclude that the electrostatic force (6.3) is *not the only actor* in the play in system K′. *Relativity demands* that there must be an additional, nonelectrostatic, force on q. Overlooking this additional contribution to the net force in K′ leads to contradiction.

Thus, the only way for Tom to account for the discrepancy between the *increased* electric field and accordingly, electric force (6.3), and *decreased* net force (6.5) on the moving charge is to admit the appearance of some additional force in his reference frame. No such force is there in K, where the plates are stationary, but it is there in K′, where they are moving. Even so, the new force acts only when the test charge q also is *moving relative to Tom*. Therefore, the new (nonelectrical) force must arise from the *moving* electrical charges and can be detected only by a *moving* test charge. We came across a new kind of force depending on velocities of both – its source *and* its detector. A remarkable but foreseeable symmetry! Relativity cannot discriminate between motion of a source and motion of a detector.

Having recognized this, we can now employ the same relativity for *calculating* the nonelectrostatic force necessary for the net result to be compatible with relativistic prediction (6.5). The additional force must be equal to the difference between the *net* force (6.5) and the purely electrical contribution (6.3):

$$\Delta f'_{\perp} = f_{\perp}\left[\gamma(V) - \gamma^{-1}(V)\right] = f_{\perp}\gamma(V)\left[1 - \frac{1}{\gamma^2(V)}\right] = \frac{V^2}{c^2}\gamma(V) f_{\perp}. \tag{6.6}$$

The direction of the additional force must be *opposite* to the electric field to compensate for the increase of this field in the moving capacitor and to insure nevertheless the decrease of the net force.

We have arrived at the crucial moment: *The Emergence of Magnetic Field.*

The factor $\gamma(V) f_{\perp}$ in (6.6) is, according to (6.3) and (6.4), equal to the electric force in K′. Therefore,

$$\Delta f'_{\perp} = \frac{V^2}{c^2} q E'_{\perp}. \tag{6.7}$$

Thus, a nonelectric force, emerging in K′, must stand in proportion to its counterpart – the electric force (6.3) in this reference frame:

$$f'_{\text{Non-El}} = \Delta f'_{\perp} = \frac{V^2}{c^2} f'_{\text{El}} = qV\frac{V}{c^2} E'_{\perp}. \tag{6.8}$$

In the spirit of electrostatics, which is a field theory, we say that there must be a corresponding *nonelectric* field responsible for this force. Let us call this new field the *magnetic field* and denote it as B' (we have primed it because it emerges (in our notations) in the primed reference frame K′). Accordingly, we will call the corresponding nonelectric force (6.8) the *magnetic force*. By analogy with electrostatics, where we define the electric field so that its product with the (test) charge gives the electric force, we can now define the field B' so that its product with $q\mathbf{V}$ gives the magnetic force. The inclusion of **V** here is necessary because, as we have already emphasized, in this case it is a *moving* charge that does the testing. The condition of its motion is absolutely essential, since a stationary charge can detect only the electric field. The charge that is moving faster is more sensitive to the magnetic field than the same charge moving slower. Therefore, it is the *product* $q\mathbf{V}$ of the charge and its velocity that we must take as the testing device to measure the new field.

Now notice that the magnetic force, as any other force, is a vector, and the *directions* of $\mathbf{f}'_{\text{Non-El}}$ and **V** are mutually perpendicular. This condition can be satisfied only if the new field is a *vector* field, that is, $B' \to \mathbf{B}'$, and $\mathbf{B}'$ is in the direction perpendicular to both – the velocity **V** and the force. Then we must rewrite the above equation in the form of the cross product:

$$\mathbf{f}'_{\text{magn}} = q\mathbf{V} \times \mathbf{B}', \tag{6.9}$$

where we have defined

$$\mathbf{B}' = -\frac{\mathbf{V}}{c^2} \times \mathbf{E}'_{\perp}, \tag{6.10}$$

or, according to (6.3):

$$\mathbf{B}' = -\gamma(V)\frac{\mathbf{V}}{c^2} \times \mathbf{E}_{\perp}. \tag{6.11}$$

Combining the result (6.9) for the magnetic force with the original expression (5.1) for the electric force we arrive at correct and general expression for the *net* electromagnetic force on a charge:

$$\mathbf{f}_{\text{net}} = q(\mathbf{E} + \mathbf{V} \times \mathbf{B}). \tag{6.12}$$

We have dropped the primes here to emphasize that this is a general statement, true in any inertial reference frame, where there is an electric and magnetic field and a moving charge.

The "magnetic contribution" (6.9) derived by Tom is known as the Lorentz force on a charge, moving in a magnetic field. Historically, it was obtained from the experiments, but as we see, it is also a logical consequence from the relativistic postulates, and can be derived as such. The general equation (6.12) for the net force is also referred to in some textbooks as Lorentz force law, even though the first (purely electric) term had been known long before Lorentz. When applied to the system K′ considered above, the general expression (6.12) recovers result (6.5) for the relativistic force in the above-considered situation in Figure 6.2c. This restores the harmony in our picture of the world, but on a higher level – relativity and electromagnetism, rather than relativity and electrostatics only.

As a by-product, Tom obtained Equation (6.11), which expresses the magnetic field in K′ in terms of the electric field in K. Together with (6.2), it is an embryo of the Lorentz transformations for the electromagnetic field.

Situation 2: Sliding Down a Wire

The arrangement in this situation is more convenient for deriving the second fundamental law of magnetostatics – Ampere's law.

Suppose we have a uniformly charged wire with linear charge density (charge per unit length) λ (Figure 6.3a). We want to find the electric force f on a point charge q at a distance r from the wire. We know from electrostatics [44] that the electric field due to an infinite straight wire is pointing away from or to the wire depending on the sign of its charge and its magnitude at a distance r from the wire is

$$E_{\perp} = \frac{\lambda}{2\pi\varepsilon_0 r}. \tag{6.13}$$

The force on q is

$$f_{\perp} = qE_{\perp} = q\frac{\lambda}{2\pi\varepsilon_0 r}. \tag{6.14}$$

The labels "⊥" here indicate that the field and corresponding force are perpendicular to the wire.

Denote the rest frame of this system as K. Now consider the same system from the viewpoint of Tom in a reference frame K′ moving along the wire with a speed V. The wire is moving with respect to Tom with the speed $-V$ (Figure 6.3b). What will be the force on q as measured by Tom in K′?

As in the previous situation, the transformation rules (5.86) yield different answers depending on state of motion of the charge. For instance, if the charge is moving

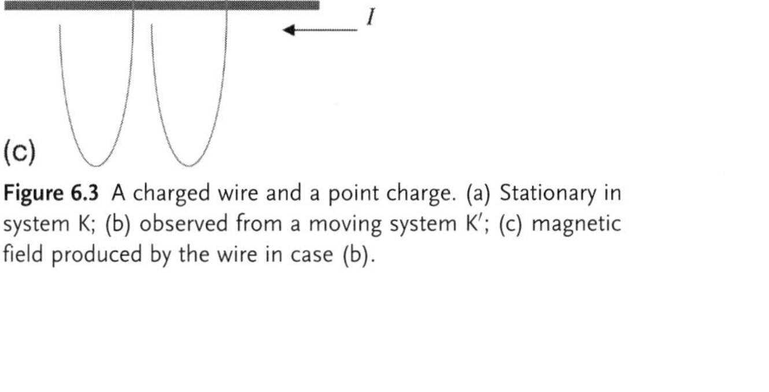

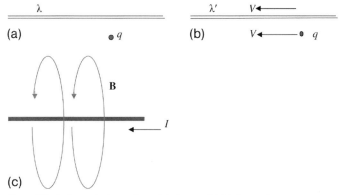

Figure 6.3 A charged wire and a point charge. (a) Stationary in system K; (b) observed from a moving system K′; (c) magnetic field produced by the wire in case (b).

together with Tom, the transformation rules predict the increase of the force (6.14) by a factor of $\gamma(V)$ and we recover Equations (6.3) and (6.4). The only difference is that now $f_\perp$ and $E_\perp$ can have any direction $\hat{\mathbf{s}}$ in a plane perpendicular to the wire; they are vectors in two-dimensional space, and accordingly we can write the corresponding equations in the vector form:

$$\mathbf{E}'_\perp = \gamma(V)\mathbf{E}_\perp, \tag{6.15}$$

$$\mathbf{f}'_\perp = q\mathbf{E}'_\perp = \gamma(V)\mathbf{f}_\perp. \tag{6.16}$$

Again, this is consistent with the change in the linear charge density due to the length contraction:

$$\lambda' = \frac{q'_{l'}}{l'} = \frac{q_l}{l'} = \gamma(V)\lambda. \tag{6.17}$$

If, however, the charge remains stationary in K, the transformation rules give the result

$$\mathbf{f}'_\perp = \gamma^{-1}(V)\mathbf{f}_\perp, \tag{6.18}$$

which appears to flatly contradict (6.16). The contradiction would remain an unsolved puzzle for Tom if he would try to find purely electrostatic explanation to it. But Tom has already learned something from the previous situation. Acting as before, he calculates the difference between (6.18) and (6.16) and attributes it to the magnetic force on the moving charge in his reference frame. The corresponding magnetic field is related to the electric field of the wire by the same Equations (6.10) and (6.11). As in Situation 1 with moving plates, its physical origin is the motion of the wire producing the current

$$I' = \frac{q'_l}{t'} = \frac{\lambda' l'}{t'} = \lambda' V. \tag{6.19}$$

(Here, q'_l is the amount of charge in a length element l' and t' is the time in K′ it takes the element to shift by its own length.)

But we can get more than just recover the same results as in Situation 1. Using (6.13) and (6.15), (6.16), and (6.19), we obtain

$$B' = \frac{V}{c^2}\frac{\lambda'}{2\pi\varepsilon_0 r} = \frac{1}{c^2}\frac{I'}{2\pi\varepsilon_0 r}. \tag{6.20}$$

Suppose, we introduce a new constant

$$\mu_0 \equiv \frac{1}{\varepsilon_0 c^2} \tag{6.21}$$

(it is called magnetic permeability of free space). We then immediately recognize in (6.20) the known expression for the magnetic field around a straight wire carrying current I', a distance r away from the wire:

$$B' = \frac{\mu_0}{2\pi}\frac{I'}{r}. \tag{6.22}$$

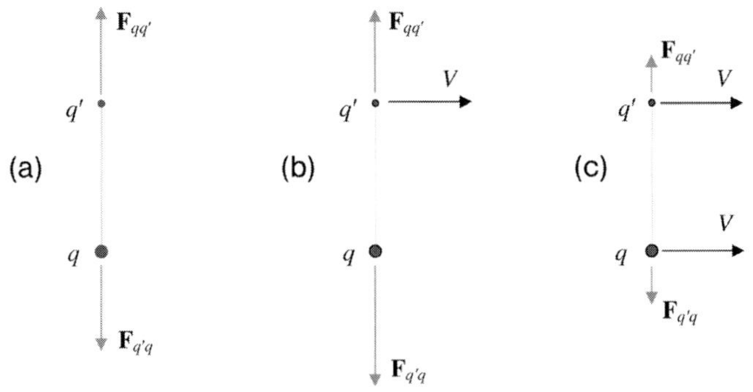

Figure 6.4 Two point charges and the net forces on them. (a) Both charges are stationary in K; (b) q' is moving in K; (c) both charges are moving together.

In vector notation

$$\mathbf{B}' = \frac{\mu_0}{2\pi}\frac{I'}{r}\hat{\boldsymbol{\varphi}}, \tag{6.23}$$

where $\hat{\boldsymbol{\varphi}}$ is the unit vector along the azimuthal coordinate line (in order for the magnetic force to point to the current in wire A, the corresponding magnetic field must be in the direction perpendicular to both – the wire and the position vector from the wire to the observation point). Figure 6.4 depicts the magnetic field lines around the current-carrying wire.

Alternatively, rewriting (6.22) as

$$\mu_0 I' = B' 2\pi r = \oint B'_s \mathrm{d}l, \tag{6.24}$$

we recognize the familiar form of the *Ampere's law* for a current-carrying wire (Figure 6.3c).

Situation 3: Playing with Point Charges

Consider now the system of two *point* charges q and q', a distance r apart (Figure 6.4). We will play three simple games with them: first, we will set both charges stationary in K; second, we will set one of the charges moving; and third, we will set both charges moving with one common velocity.

(a) In the first case, Coulomb's force on the charges is

$$\mathbf{f}_q = \frac{qq'}{4\pi\varepsilon_0 r^2}\hat{\mathbf{r}} = -\mathbf{f}_{q'}, \tag{6.25}$$

where $\hat{\mathbf{r}}$ is the unit vector drawn from q' to q.

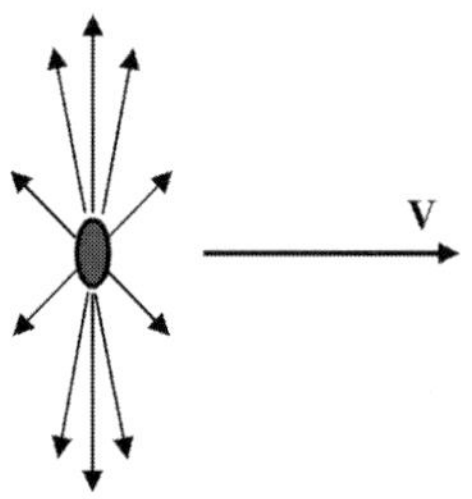

Figure 6.5 The electric field of a uniformly moving charge $q > 0$. The field remains radial, but no longer spherically symmetrical. Its symmetry is "downgraded" from spherical to axial (about the direction of motion). Only a few electric field vectors in the plane containing the velocity vector **V** are shown. For a negative charge the field would be converging.

(b) In the second case, suppose the charge q' moves to the right with a speed V. The electric force in K on this charge will remain the same. The force on q, on the contrary, will change, because q is now in the electric field of a *moving* source.

We can find this force in the following way. Let us contact an observer in system K′ moving together with q'. In his system, q' is stationary and q is moving to the left with the speed V. The situation in K′ is symmetric to the situation in K. Since all inertial reference frames are equivalent, the observer in K′ concludes that, in his frame, the force f'_q on q must have the same magnitude as the force on q' in K, that is, it must also be described by Coulomb's law (6.25). Therefore, the sought-for force on q *measured in K* is given by

$$f_q = \frac{f'_q}{\gamma(V)(1+(Vv'_x/c^2))}\underset{v'_x \to -V}{\Rightarrow} \gamma(V) f'_q = \gamma(V)\frac{qq'}{4\pi\varepsilon_0 r^2} \tag{6.26}$$

(v'_x being the longitudinal component of velocity of charge q in K′, in our case equals $-V$).

Thus, when q' is moving, while q remains in place, the interaction forces on the two charges measured in one system K are not equal in magnitude: Newton's third law for particles does not hold! The force on stationary charge q is greater than the force on moving charge q' (Figure 6.4b).

In the spirit of the field theory, we interpret the force on a point charge at a given location as the effect of the field due to other charges. Therefore, we attribute the increase of the force on q to the increase of the electric field produced by a moving charge q'. This local field is perpendicular to the direction of motion, so we will write it as

$$\mathbf{E}_\perp = \mathbf{E}'_\perp \gamma(V) = \gamma(V)\frac{q'}{4\pi\varepsilon_0 r^2}\hat{\mathbf{r}}. \tag{6.27}$$

When we "map" an electric field graphically by drawing the electric field lines, the lines are packed closer together where the field is stronger. More precisely, the number of lines crossing the unit area perpendicular to their local direction stands in proportion to the local field strength. Therefore, if we consider a spherical surface centered at the instantaneous position of a moving charge q', we will see the electric field lines packed $\gamma(V)$ times closer together on the equator than they are in case of stationary charge. On the other hand, the depicted lines do not disappear or pop up into existence out of nothing just because the charge started moving. As we have emphasized in the beginning of this section, the total "number" of lines is an invariant determined entirely by the amount of charge in the source (Gauss's law). The total number of lines for a moving charge is the same as for the stationary charge and the increase in their concentration on the equator must come at the expense of their concentration in the polar regions (ahead and behind of moving charge, Figure 6.5). We conclude that the field $E_{||}$ at points N and S ("North" and "South" poles in Figure 6.5) must be weaker for the moving charge than for the same stationary charge.

This seems to contradict our conclusion in (6.2) that $E'_{||} = E_{||}$! But it does not. The conclusion illustrated in Figure 6.5 refers to two forces (one transverse and one longitudinal) applied to *two different test charges* in *one* reference frame. Equation (6.2) refers to *one* longitudinal force on the same test charge observed from *two different reference frames.*

We can illustrate the whole thing in the following way. Place two K′ observers – Nancy at the Northern pole and Sam at the Southern pole of a sphere of radius r' centered at the charge q' (Figure 6.6). They measure their respective field magnitudes to be

$$E'_{||} = \frac{q'}{4\pi\varepsilon_0 r'^2}. \tag{6.28}$$

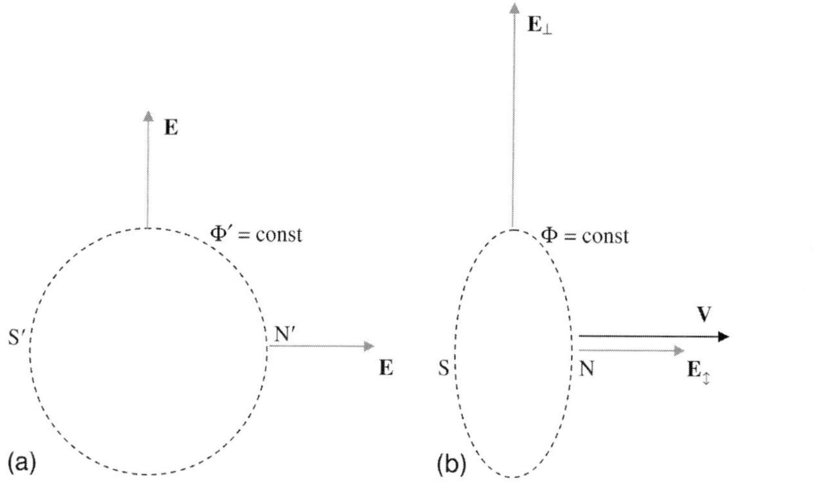

Figure 6.6 (a) An equipotential surface for a point charge q' in its fest frame K′. Both the potential and the field magnitude are constant over the surface; (b) the same surface observed in the frame K, where the charge is moving. The field is flattened along the direction of motion by a factor of $\gamma(V)$. The field at "polar" points N and S is weaker than the field at the equatorial points of the surface (points in the plane perpendicular to the line of motion), even though these points are closer to the charge than the surface's equator, due to the length contraction. The electric field of a moving charge is not described by Coulomb's law (6.25).

In reference frame K, according to (6.2), the measurements must give the same result. But this does not contradict our conclusion about the decrease of the electric field in the polar regions because in K, both Nancy and Sam are closer to q' due to the Lorentz contraction. In K, their distance from the center, being along the direction of motion, is $r'/\gamma(V)$, so that the moving sphere with N and S drawn closer together is an oblated spheroid. No surprise that the corresponding local field $E'_{||}$, being measured in K′ at r' meters from q', is quite "normal" for Nancy and Sam, but is abnormally weak for us, who find these points closer to q' by a factor of $\gamma(V)$ and still find the field equal to $E'_{||}$, rather than $\gamma^2(V)E_{||}$ dictated by the electrostatics (Figure 6.6b). In other words, the field of the moving charge indeed turns out to be "diluted" in longitudinal directions and "concentrated" in transverse directions. Quantitatively, consider the locus of points with the same electric potential around the moving charge. In frame K′, it is a sphere centered at the charge. In frame K, it turns out to be an ellipsoid of revolution flattened along the direction of motion, with the longitudinal axis decreased by a factor of $\gamma(V)$. This effect can be considered as a specific manifestation of the Lorentz contraction for the electric field.

Some readers may still be not satisfied with this. Good for you! Indeed, if the equipotentials are flattened in the described way, then two different but close equipotentials Φ'_1 and Φ'_2 associated with q' will be the concentric spheres centered at q' in K′, and flattened ellipsoids Φ_1 and Φ_2 around q' in K (Figure 6.7). Now recall that the electric field is the (negative) gradient of the potential. Therefore, its magnitude is inversely proportional to the distance between the two close equipotentials. Since this distance is constant in K′, we have there the same magnitude of the electric field at the same distances from q' – the familiar Coulomb's law describing spherically symmetrical field around a point charge. By the same token, in K the electric field **E** on the corresponding equipotential must have a *greater* magnitude

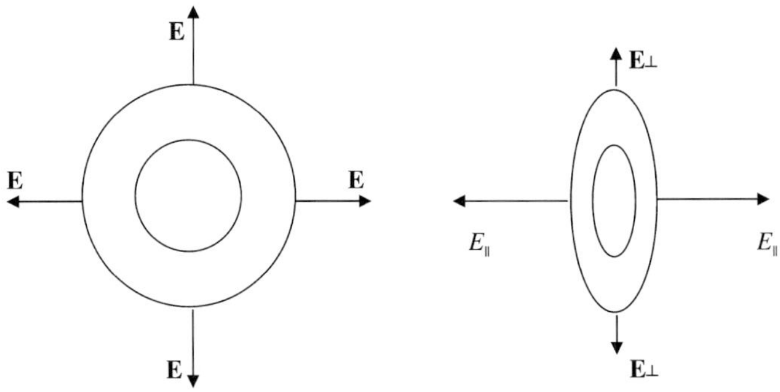

Figure 6.7 (a) Two equipotentials around a stationary point charge. The corresponding electric field is spherically symmetrical; (b) the same equipotentials observed in another reference frame, where the charge is moving. The part of electric field corresponding to the potential Φ is stronger on the symmetry axis and weaker on the equator. The *total* electric field of a moving charge is weaker on the symmetry axis and stronger on the equator.

along the direction of motion and *smaller* magnitude on the "equator," which is quite the opposite to our previous conclusion.

If you happen to know both electricity and magnetism, you will eventually resolve this contradiction by recalling that there is such thing as the magnetic vector potential. If you do not, you will come to know it now. Based on our previous thought experiments, we again conclude that since the electrostatic alone contradicts the relativistic prediction (6.27) and (6.28), it is not the whole story. Just as relativity demands the emergence of a magnetic field around the moving charge, it demands that this field must also contribute somehow to the electric field of this charge. In other words, the electric field is generally not determined by the electric potential function alone. There must exist also the magnetic vector potential associated with the magnetic field and this potential determines magnetic field and a certain fraction of the electric field. If we take account of this additional term, the resulting electric field will be equal to (6.27) and (6.28).

Let us stop at this, to return to the problem later on the quantitative level, and meanwhile consider the next stage of our thought experiment.

(c) Let now *both* charges q and q' move together in K. What is the force on q in this case? It is given by Equation (5.84a) for a transverse force on an object in K in terms of the force on it as measured in K′. If K′ is a comoving system, then both charges in it are stationary, so that $v'_x = 0$, and the force on q is given by Coulomb's law. Putting this information into (5.84a) yields

$$f_q = \gamma^{-1}(V)\frac{qq'}{4\pi\varepsilon_0 r^2}. \tag{6.29}$$

Very interesting! When the charge q is stationary, the passing charge q' exerts on q a force $\gamma(V)$ times *greater* than corresponding Coulomb's force for the same distance (case (b), Equation (6.26)). But when q is also moving with the same velocity, the force on q becomes $\gamma(V)$ times *weaker* than the standard Coulomb's force. The force on q now turns out to be depending not only on its surroundings producing corresponding local field, but also on the state of motion of q itself. We already know that this is specific for a charge in a magnetic field. This magnetic field is due, of course, to the *moving* charge q'; it is not felt by q so far as q remains stationary. But as it starts moving, there appears a nonzero Lorentz force on it. We conclude that the discrepancy between cases (6.26) and (6.29) must be attributed to this force. We can now find this additional force to determine the corresponding magnetic field **B**.

Let both charges be positive. Then, because the force (6.29) is less than the force (6.26), the decrease must be due to magnetic attraction. Quantitatively,

$$f_{\text{Lor}} = \left[\gamma(V) - \gamma^{-1}(V)\right]\frac{qq'}{4\pi\varepsilon_0 r^2} = \gamma(V)\frac{V^2}{c^2}\frac{qq'}{4\pi\varepsilon_0 r^2}. \tag{6.30}$$

Comparing this with already known expression (6.9) for the Lorentz force law (rewritten for the system K) and with (6.27) for the electric field of moving charge q' gives

$$\mathbf{B} = \gamma(V)\frac{q'}{4\pi\varepsilon_0 r^2}\hat{\mathbf{n}} \times \frac{\mathbf{V}}{c^2} = \frac{\mathbf{V}}{c^2} \times \mathbf{E} = \gamma(V)\frac{\mathbf{V}}{c^2} \times \mathbf{E}'. \tag{6.31}$$

Finally, since $\varepsilon_0 c^2 = 1/\mu_0$, we can rewrite (6.31) as

$$\mathbf{B} = \gamma(V)\frac{\mu_0}{4\pi}\frac{q'\mathbf{V} \times \hat{\mathbf{r}}}{r^2}. \tag{6.32}$$

In the limit of slow motion, $\gamma(V) \to 1$ and Equation (6.32) reduces to the familiar form of the Biot–Savart law for a magnetic field due to a moving point charge.[1] The exact general expression for the magnetic field of an arbitrary moving point charge is given in Problem 6.1 at the end of this chapter.

Thus, as we have already suspected in the concluding comments to the previous case (b), not only a current, but also a single moving point charge produces magnetic field. And again, since any charge can be made "moving" by just switching to another reference frame, the magnetic field can be called into existence by changing reference frame. Relativity decrees the necessity of magnetic field to maintain the consistency of transformational properties of force.

Let us summarize this part: in Situation 1 we have found, among other things, the electric and magnetic fields in system K′ and expressed them in terms of **E** in K

$$\mathbf{E}'_\perp = \gamma(V)\mathbf{E}_\perp, \qquad \mathbf{B}'_\perp = -\gamma(V)\frac{\mathbf{V}}{c^2} \times \mathbf{E}_\perp \tag{6.33}$$

(first equation here is (6.4) in vector form), and also we found the expression for the magnetic force on a moving charge (Lorentz force law). In Situation 2, we derived the Ampere's law. In Situation 3, we have derived Biot and Savart's law.

We have found that all three basic laws of magnetostatics – the Lorentz force law, the Ampere's law, and the Biot–Savart law – are merely the consequences of the relativistic postulates.

The situations we have considered in this section are rather special: even though the electric field in all three situations is observed in both systems K and K′, the magnetic field appears only in one of them as a result of the Lorentz transformation for *electric* force. In this respect, the magnetic field is not an independent entity, but rather a transformed aspect of the electric field. This role of the original electric field **E** as a potential "source" of the magnetic field is manifest in any reference frame moving in a direction perpendicular to **E**, or, more generally, having a component of **V** perpendicular to **E**. Accordingly, we can sometimes literally "create" magnetic field "out of nothing" by switching to another reference frame, for instance, just by boarding a train. Of course, this "nothing" is not the exact nothingness – any electric

1) The Biot–Savart law is not actually an exact law of physics. It is a useful description of the magnetic field produced by an element of a continuous system of charges in stationary collective motion [44, 45].

field ultimately requires the (current or past) presence of the electric charges. Boarding the train, we only change the state of motion of these charges. This shows that the magnetic field is a purely relativistic phenomenon.

6.2 Magnetic Field and Something Else ...

The conclusion we made in the previous section may produce an impression that only electric field is a primary entity existing in its own right, whereas the magnetic field is something secondary.

Here we will as forcefully argue that, quite the contrary, in some situations the electric field can be considered as a secondary relativistic effect – as a disguised manifestation of magnetic force to an observer in another reference frame. In the next section, we will discuss both views as aspects of a bigger picture.

Consider again the situation with only one kind of field in system K, but this time it will be the magnetic field. Let this field be uniform. Such a field can be produced inside a long straight tightly wound solenoid, carrying a steady current I (a straight solenoid flown around by a current plays the same role in producing a uniform magnetic field as a charged parallel-plate capacitor does in producing the uniform electric field between the plates). The magnetic field $\mathbf{B}$ points parallel to the symmetry axis of the solenoid (Figure 6.8a).

Let us first find the strength of this field. We know the current I and the number of turns of the wire per unit length N of the solenoid. If there is no magnetic field outside the solenoid, then applying the Ampere's law (6.8) to a narrow rectangular loop stranded across the wall of the solenoid, we find (Problem 6.6)

$$\mathbf{B} = \mu_0 N I \hat{\mathbf{z}}, \tag{6.34}$$

where the direction of the unit vector $\hat{\mathbf{z}}$ is related to the sense of current around the solenoid by the right-hand rule.

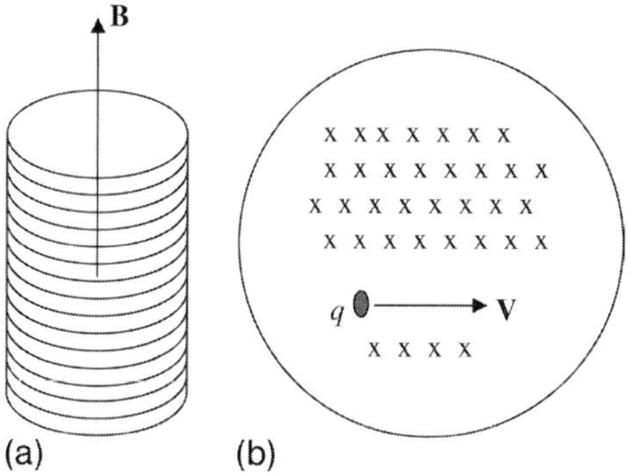

Figure 6.8 (a) A tightly wound solenoid with a homogeneous magnetic field inside; (b) a charge q moving across the magnetic field within the solenoid. The solenoid is now laid horizontal and its magnetic field points into the page.

Now we can obtain the transformation rule for the z-component of the magnetic field. Consider another inertial RF K′ moving along the direction of the field with the speed V. This motion does not break the axial symmetry of the system. Under these conditions, there can be no transverse magnetic field in the frame K′. Indeed, a component of the field perpendicular to the symmetry axis of the solenoid would break the axial symmetry, in contradiction with the experimental conditions. Thus, in this case, $B'_{\perp} = B_{\perp} = 0$. As to B'_z, it can be determined by the same Ampere's law, which holds in K′ as well as in K. Applying (6.34) in K′, we have

$$B'_z = \mu_0 N' I'. \tag{6.35}$$

But, due to the relativistic length contraction, the same amount of turns will fit into shorter length of the solenoid in system K′; therefore, $N' = \gamma(V)N$. On the other hand, due to time dilation, the current measured in K′ is more "sluggish" – it is less than in K by the same factor. The two relativistic effects work in this case in the opposite directions, leaving the z-component of the field invariant:

$$B'_z = B_z. \tag{6.36}$$

The longitudinal (parallel to the relative velocity of the two inertial frames) component of the magnetic field does not change under the Lorentz transformation. This conclusion is similar to the result (6.5) for the electric field.

Our next question is what happens if a point charge q is shot with a velocity V along the direction of **B**. The answer is that the charge will keep on sliding down this direction with the same speed and there is no electric field to act on it; as to the magnetic field, we know from Equations (6.9) and (6.12) that it exerts no force on the charge moving parallel to the field. An inertial observer in reference frame K′ comoving with the charge will see the charge staying in place, while the solenoid is moving relative to the observer in the opposite direction. He concludes that the net force on the charge is zero, as it is in the rest frame of the solenoid. The same, of course, follows from the transformation rules (5.84).

But the net force may be the sum of both – magnetic and electric forces. In our case the magnetic force is zero, because the charge is stationary in system K′. Since the net force is also zero, it follows immediately that there is no electric field in K′. The motion along the direction of magnetic field does not cause the emergence of an electric field.

Suppose now that we have launched a point charge with the initial velocity **V** perpendicular to the field **B** (Figure 6.8b). We know that there must appear the magnetic Lorentz force on the charge:

$$\mathbf{f}_{\perp} = \mathbf{f}_{\mathrm{Lor}} = q\mathbf{V} \times \mathbf{B}. \tag{6.37}$$

The label "⊥" here indicates that the force is perpendicular to the direction of motion of the charge. As we have realized in the previous section, the existence of such a force is required by relativity.

Consider now the same charge from another inertial reference frame K′, where it is (at a given moment) stationary. Applying Equation (5.84) with $v_x = V$ to this case gives the expression for the transverse force on the charge in K′:

$$f'_{\perp} = \frac{f_{\perp}}{\gamma(V)(1-(V^2/c^2))} = \gamma(V) \quad f_{\perp} > f_{\perp}. \tag{6.38}$$

Using (6.37) gives

$$\mathbf{f}'_{\perp} = \gamma(V) q \mathbf{V} \times \mathbf{B}. \tag{6.39}$$

Now, suppose that we know something about magnetism and almost nothing about electricity. How then are we going to interpret the force (6.38) and (6.39) in system K′? Obviously, it is *not* a magnetic force for a K′ observer moving together with the charge, since the charge q is at rest in K′. The only plausible conclusion will be that, since there is a nonmagnetic force on the charge, there has to be a nonmagnetic field responsible for this force. We can then call this new field the electric field and say that it appears whenever an observer is moving relative to a system with existing magnetic field in a direction perpendicular to the field. Since the charge is now stationary in K′, the most natural measure of the electric field will be the force per unit charge:

$$\mathbf{E}'_{\perp} \equiv \frac{\mathbf{f}'_{\perp}}{q} = \gamma(V)\mathbf{V} \times \mathbf{B}. \tag{6.40}$$

The electric field appears here merely as a secondary entity – a purely relativistic effect. It would not exist without magnetic field **B** in K. On the other hand, there has to be a physical source in K′ responsible for the observed electric field there. We will discuss the nature of this source in the next section.

Knowing already some relativity, the reader may suspect that the magnetic field measured in K must also be observed as such in K′ without disguise. The special case considered here, however, does not provide us with an opportunity to find *all* components of the magnetic field in K′ because this field can be detected by a *moving* charge, whereas, by our choice of K′, the charge q is now stationary in K′.

In order to find the magnetic field $\mathbf{B}'_{\perp}$ in K′, consider a more general case when the velocity of the charge is parallel but not equal to the relative velocity between the systems K and K′: $v_x \neq V$. Then the force on q in K′ is

$$\mathbf{f}'_{\perp} = \frac{\mathbf{f}_{\perp}}{\gamma(V)(1-(Vv_x/c^2))} = \frac{q\mathbf{v} \times \mathbf{B}}{\gamma(V)(1-(Vv_x/c^2))}. \tag{6.41}$$

Since the charge is now moving in K′, the force on it given by (6.41) must be a combination of both – electric and magnetic forces

$$\mathbf{f}'_{\perp} = \mathbf{f}'_{\mathrm{El}} + \mathbf{f}'_{\mathrm{Mag}} = q\mathbf{E}' + q\mathbf{v}' \times \mathbf{B}' \tag{6.42}$$

(the *magnetic* contribution must stand in proportion to the speed v' of the charge in the *reference frame* K′ (Equation (6.37))). We already know that $\mathbf{E}'$ is given by Equation (6.40), so the electric force on q is

$$\mathbf{f}'_{\mathrm{Electr}} = q\mathbf{E}' = q\mathbf{E}'_{\perp} = q\gamma(V)\mathbf{V} \times \mathbf{B}. \tag{6.43}$$

Then the magnetic force in K′ can be found as the difference between the net force (6.41) and the electric force (6.43):

$$\mathbf{f}'(\text{Magn}) = \mathbf{f}' - \mathbf{f}'(\text{El}) = \frac{q\mathbf{v} \times \mathbf{B}}{\gamma(V)(1 - (Vv/c^2))} - q\gamma(V)\mathbf{V} \times \mathbf{B}$$

$$= q\left[\frac{\mathbf{v}}{\gamma(V)(1 - (Vv/c^2))} - \gamma(V)\mathbf{V}\right] \times \mathbf{B}.$$

Simple algebra within the brackets leads to the result

$$\mathbf{f}'(\text{Magn}) = q\gamma(V)\frac{\mathbf{v} - \mathbf{V}}{1 - (Vv/c^2)} \times \mathbf{B}. \tag{6.44}$$

Looking at the vector factor multiplying $\mathbf{B}$, we recognize that it is just the relative velocity $\mathbf{v}'$ of the charge in K′. Thus,

$$\mathbf{f}'(\text{Magn}) = q\gamma(V)\mathbf{v}' \times \mathbf{B}. \tag{6.45}$$

Comparing this with (6.42) gives

$$\mathbf{B}'_{\perp} = \gamma(V)\mathbf{B}. \tag{6.46}$$

Thus, with only the electric field in K, we have in K′

$$\mathbf{E}'_{\perp}(\mathbf{E}) = \gamma(V)\mathbf{E}_{\perp}, \qquad \mathbf{B}'_{\perp}(\mathbf{E}) = \gamma(V)\frac{\mathbf{V}}{c^2} \times \mathbf{E}_{\perp}. \tag{6.47}$$

With only the magnetic field in K, we have in K′

$$\mathbf{E}'_{\perp}(\mathbf{B}) = \gamma(V)\mathbf{V} \times \mathbf{B}_{\perp}, \qquad \mathbf{B}'_{\perp}(\mathbf{B}) = \gamma(V)\mathbf{B}_{\perp}. \tag{6.48}$$

When both fields are present, we have

$$\mathbf{E}'_{\perp} = \gamma(V)(\mathbf{E}_{\perp} + \mathbf{V} \times \mathbf{B}_{\perp}), \qquad \mathbf{B}'_{\perp} = \gamma(V)\left(\mathbf{B}_{\perp} + \frac{\mathbf{V}}{c^2} \times \mathbf{E}_{\perp}\right). \tag{6.49}$$

We have derived Lorentz transformations for the electric and magnetic field. Adding here

$$\mathbf{E}'_{||} = \mathbf{E}_{||}, \qquad \mathbf{B}'_{||} = \mathbf{B}_{||}, \tag{6.50}$$

we complete the transformation rules for all components of the electromagnetic field. The ultimate source of these rules is the relativistic nature of space and time.

Let us summarize the big lesson we have learnt here.

If we know *all* the forces on an object in reference frame K, we can find the net force on it in this frame without looking out to other frames. But if we do not, we can still determine the correct net force by performing the Lorentz transformation from another reference frame where the net force on the object is known. Not only will the Lorentz transformation take care of the correct result, but also tell us if something is missing in our compendium of known forces in the original reference frame. It was in this way that Tom in Section 6.1 has derived the basic laws of magnetism from relativity and Coulomb's law only. Historically, these laws have been discovered as a generalization of the experimental facts, but they can be as well deduced without any experiments, as a logical consequence from the relativity postulates.

6.3 Is Electric Charge Invariant? (The Subtleties of "Neutrality")

Here we will consider some interesting implications of the relativity of fields, which concern the invariance of an electric charge. If the electric charges are the ultimate source of both electric and magnetic fields, then the first of Equations (6.49) appears to contradict it. Consider an infinite, straight, current-carrying wire, whose lattice is stationary in K (Figure 6.9a). Suppose that the wire is *electrically neutral* in K. Then in frame K there is only magnetic field **B** around the wire

$$\mathbf{B} = \frac{\mu_0}{2\pi} \frac{I}{s} \hat{\varphi}, \tag{6.51}$$

and there is no electric field. But what about another reference frame K′ moving, say, up the wire? Denote the velocity of this motion as $\mathbf{V} = V\hat{\mathbf{z}}$, where $\hat{\mathbf{z}}$ is a unit vector in the direction of the wire. According to (6.48), in system K′, apart from magnetic field

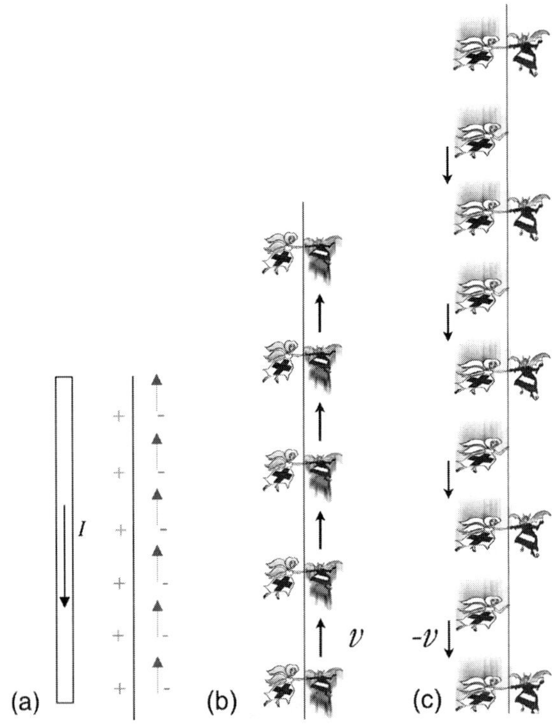

Figure 6.9 (a) A length element of a neutral current-carrying wire; (b) a handshaking ritual establishing (or verifying) the neutrality of the wire in the rest frame of its crystalline lattice; (c) the same procedure establishing (or verifying) the nonzero net electric charge on the wire in the rest frame of its electrons.

$\mathbf{B}' = \gamma(V)\mathbf{B}$, will appear the electric field pointing, for a given arrangement, toward the wire:

$$\mathbf{E}' = \gamma(V)\mathbf{V} \times \mathbf{B}. \tag{6.52}$$

It is easy to understand the origin of **B** in K, in terms of the electric charges as the ultimate source. Indeed, even though the net charge density is zero in a neutral wire, one component of the net charge is in a state of collective motion and since any moving charge produces a magnetic field, the origin of this field is clear. But what produces the *electric* field (6.52) observed in K′? As we have emphasized in the beginning of this chapter, for any electric field there always has to be an electrical charge as its ultimate source. How does it sit together with the condition that the wire is electrically neutral?

There can be only one answer to this question. The wire is neutral in K and accordingly there is no electric field there. If in K′ there is a long-range electric field pointing away from (or toward the) wire, then the wire must be electrically charged in this system. The field (6.52) can be written in a way showing explicitly its dependence on charge density. Indeed, combining (6.52) with (6.51) gives

$$\mathbf{E}' = -\gamma(V)V\frac{\mu_0}{2\pi}\frac{I}{s}\hat{\mathbf{s}}, \tag{6.53}$$

where $\hat{\mathbf{s}} = \hat{\boldsymbol{\varphi}} \times \hat{\mathbf{z}}$ is the unit vector pointing radially away from the wire. Further, comparing this with the electrostatic expression (6.13), we find the linear charge density of the wire in K′:

$$\lambda' = -\gamma(V)\frac{V}{c^2}I. \tag{6.54}$$

Equation (6.54) says that each length element of the wire is indeed electrically charged and the linear charge density is in proportion to the current in K. The net charge of the wire is zero in K and nonzero in K′. The charge is not invariant! Moreover, since the wire is infinitely long, its *net* charge is more than just nonzero – it is infinite in K′! And if, in addition, $V \to c$, then even the charge of any finite length element of the wire tends to infinity! The charge invariance – a sacred dogma of current theoretical physics – appears to be violently crushed in relativity. Or rather, since we so strongly believe in charge invariance, this contradiction appears to be a death sentence to relativity itself. What are we going to choose?

The answer is – neither. The dilemma we have come to originates from the incorrect boundary conditions and from fuzziness of our verbal statements, not from any logical inconsistency within the framework of relativity.

Consider first the boundary conditions. The incorrect element in the conditions is the assumption of the *infinitely long* straight wire. This is an obvious idealization. There is no such thing in reality. As any other idealization (e.g., a point mass or charge, an absolutely rigid body, etc.), it must be treated with caution. It is a quite legitimate approximation for calculating, say, the electric or magnetic field around the wire. This approximation works fine if we remain so close to the wire that we do not see its end points. For distances s sufficiently smaller than the actual length l of the wire, we can say that its end points are infinitely far away. But if we forget about this limitation and extend (6.53) that holds only for $s \ll l$ to arbitrary s, then already an attempt to calculate the magnetic or electric *energy* (and thereby the rest mass) per unit length, as we know from Problem 5.25, gives the infinite result. Let alone the *net* energy and the rest mass of the whole system.

We do not get very much excited about the infinite energy resulting from the assumption of the infinite length. This result follows from the fact that an infinitely long charged or current-carrying wire *already has* an infinite charge or infinite kinetic energy of marching electrons, to begin with. No surprise that the infinity in the initial conditions may sooner or later pop up in a final result.

The origin of the infinite electric charge of the whole wire as observed in K′ is more subtle than the origin of the infinite energy of its electric and magnetic field. However, it can also be attributed to the infinite length of the wire. We can understand it on the qualitative level in the following way.

Imagine that we use an exotic ritual of handshaking for counting the charges. Let us model our wire as a geometrical straight line, with two periodic arrays of equal-magnitude charges arranged along the line. The plus charges are all stationary, while the minus charges are all moving along the line with the same speed. The distance between any two neighboring moving charges is adjusted so that at any moment it is equal to the distance between two neighboring stationary charges (Figure 6.9b). Therefore, whenever a moving "−" passes by a stationary "+", the next nearest "−" down the line passes by the next stationary "+".

We want to make sure that the whole system is electrically neutral. This requires an equal number of charges of both signs. But what do "equal numbers" mean, when the number of charges of either sign is infinite? Mathematicians have developed a recipe to deal with this kind of problem [46,47]. It is widely used in the sets theory to compare the infinite sets and is known as a principle of "one-to-one correspondence": if we can find such a correspondence between the respective elements of the two sets S_1 and S_2, that for any element of S_1 there is only one element of S_2 and vice versa, then the two sets are said to have equal power. Our two sets of plus and minus charges is the simplest case of discrete infinite sets. The infinite discrete sets are all of the same power. They are called denumerable (countable) because they can be numbered ("counted") by establishing one-to-one correspondence between the elements of the set and the elements 1, 2, 3, ... of the infinite set of integers.

The handshake procedure is a physical embodiment of the one-to-one correspondence applied to our two sets of charges. Let us require that each minus exchanges a handshake with corresponding plus it is passing by. This requirement, together with the initial arrangement regarding the distances, insures that *all* minuses exchange each a handshake with a respective plus, and vice versa, at one and the same moment of time in the stationary frame K. As a result, we know that the net charge of the whole infinite system of charges is exactly zero. We have effectively made sure that the number of pluses in our infinite system is exactly equal to the number of minuses without actually counting them!

Now, consider the same system from the viewpoint of an observer moving, say, together with the minus charges. Due to relativity of simultaneity, for this observer the handshakes do not occur all at the same time, and the distance between two neighboring minuses is not the same as between two neighboring pluses; the former is greater than the latter. This alone already makes the situation, at best, ambiguous. A mathematician, if he so desires, can still establish one-to-one correspondence between both sets; but for a physicist concerned with what is happening within a finite span of length, in the first place, it is clear that the "linear density" (number of items per unit length) in K′ is greater for the plus charges and less for the minus charges. Indeed, if we denote the distance in K between the two neighboring pluses as d, then in K′ the distance between them will be *less* than d by the corresponding Lorentz factor $\gamma(V)$, whereas the distance between the two neighboring minuses will be *greater* than d by the same factor. Our line turns out to be "strewn" denser with pluses than with minuses (Figure 6.9c). And, since both arrays are uniform, this converts into a statement that the line is positively charged, and the net charge of the whole infinite system is positive and infinite. Obviously, the infinite value of the *net* charge originates from the infinite length of the wire, or, which is the same, from the infinite number of charges of either sign.

This conclusion about the nonzero and even infinite net charge sounds as total nonsense in view of the fact that the handshaking procedure has established beyond any doubt, albeit without actual counting, the exact equality between the numbers of charges of both signs. Clearly, switching to another reference frame cannot change the numbers of the observed items – recall the example with the passengers inside a car in Chapter 1!

The difference between the current situation and that in Chapter 1 is, again, that now the number of items is infinite. And changing the way you compare two "equal" infinite sets may produce miracles. Here is another example.

Imagine a version of a situation described in a story by one of the greatest science fiction writers, Stanislaw Lem [48]. You represent our Milky Way galaxy in a Universal Congress of cosmologists. Each galaxy has sent its representative to the Congress; since the number of galaxies in the universe is assumed to be infinite, so is the number of participants. In order to accommodate all the guests, a Universal Cosmo Hotel had been built, with the infinite number of rooms. Each guest was assigned a separate room and after everybody had signed in, all the rooms of the hotel were occupied; there was not a single room left free and not a single guest without a room. It is difficult to imagine a better example of one-to-one correspondence. The number of guests in this example is exactly the same as the number of rooms.

But here comes a crunch. Because of the extreme complexity of the problems discussed on the Congress (the cosmological expansion, dark energy, the concept of multiuniverse, the possibility of quantum tunneling between two different universes, etc.), the proceedings ran out of schedule; a long planned Universal Congress of mind-readers is to start tomorrow and the hotel is still all occupied with the cosmologists. Each room can accommodate only one guest, but tomorrow the hotel has to accommodate an additional infinite number of guests – one from each galaxy.

After a short brainstorming, the management has found a simple solution. You are asked to relocate from room 1 to room 2; your former next-door neighbor is moved from room 2 to room 4; the resident of room 3 moved to room 6, and so on, so that a resident of room No. n is reassigned to the room No. $2n$. As a result, each cosmologist is now assigned a room with an even number, and not a single one of them is evicted into the interstellar space, since the amount of the even numbers is infinite. At the same time, all the rooms with odd numbers are made vacant and became available to accommodate an infinite number of the new guests, since there are infinite amount of the odd numbers as well. Thus, all the cosmologists, whose number was exactly equal to the number of all rooms, have now occupied only half of this number, and each in one separate room, as before. In mathematical language, the new arrangement establishes a one-to-one correspondence between the set of cosmologists and set of even numbers; the remaining infinite set of the odd numbers can be brought to a one-to-one correspondence with the infinite set of arriving mind-readers.

Now, to bring it closer to our situation with the infinite wire with plus and minus charges imagine that the door of each room has a negative unit charge and each arriving guest, when signing in, is given a tab with a positive unit charge, with the instruction to stick it onto the door of the room assigned to the guest. The moment the positive tab gets stuck to the door, the corresponding room is made electrically neutral, which results in turning down the small light with this room's number on the front desk. In this way, the administration of the hotel at any moment knows whether a room is occupied or vacant. When all of the rooms are occupied, there are no lights on the switchboard and the management knows that the financial plan is doing OK without any actual counting the rooms.

Now, consider and compare the two moments of time – one right before and the other right after the rearrangement of the cosmologists described above. Before rearrangement, all the rooms of the hotel were occupied; the switchboard on the front desk was all dark, indicating that the net electric charge of the whole hotel was zero: each room had a positive tab plastered on its negative door and was thus electrically neutral. This is just another version of the handshaking ritual between plus and minus charges in an infinitely long wire.

The situation changes dramatically right *after* the rearranging (but before the arrival of the mind-reading crowd). Assume that in the process of relocation each guest takes the tab off the door of his/her initial room and puts it on the door of the new room. Then after the rearrangement only rooms with even numbers have the positive tabs on their doors and are neutralized. The rooms with the odd numbers now have their negative doors active. The total numbers of the positive tabs and negative doors remains the same as before – none of them have been removed from the system or added to it; and yet the net charge of the hotel is now infinitely negative and the front desk is brightly illuminated by an infinite number of lights to indicate this fact!

This example shows that the net electric charge does not conserve even in one reference frame. The origin of this "misbehavior" of charges is, of course, the actual infinity of their numbers. Similarly, the violation of the electric charge *invariance* for the infinitely long wire has more to do with the *infinite* number of charges in it than with Relativity as such. In cases like this, the fact that the amounts of charges of either sign are "made" equal and thereby the net charge is zero in system K does not guarantee that it will remain zero in another frame, with different procedure for establishing one-to-one correspondence. Just as "nothing will come out of nothing," so *everything* can come out of infinity. The subtraction of two *actual* infinities is not a legitimate operation, since it does not produce a uniquely defined result.

Thus, we can shrug off the invariance violation for a system including an infinite wire. But this does not seem to let us out of the woods yet. Equation (6.54) describes a similar violation for a wire of *finite* length! Once the net linear charge density is nonzero in K′, so will be the net charge of any finite stretch of the wire, while it is zero in K for this stretch. This kind of violation comes from the above-mentioned ambiguity of human language, because we attempt to consider a stretch of wire as a separate system in its own right, whereas in the described situation it is only a part of the whole wire (see discussion below).

Let us describe the situation quantitatively, for an *arbitrary* relative speed V between the two reference frames.

Suppose the conducting wire is metallic, with the positive ions constituting the crystalline lattice with the proper volume charge density ρ_0. Corresponding linear charge density is

$$\lambda_+ = \lambda_+^0 = \rho_0\sigma \equiv \lambda_0, \tag{6.55}$$

where σ is the cross-sectional area of the wire. The observed current in system K is produced by collective motion of the free electrons with the drift velocity v_d. Since the wire is electrically neutral in the lab (system K), the magnitude of the linear charge density of the marching electrons must be also equal to (6.55). But as you start sliding down the wire, the fragile balance of the plus and minus-charges breaks because, just as in special case with $V = v_d$, the two charge densities in the new RF will enter the equation with different coefficients.

Let us express the phenomenon in terms of the proper linear charge density λ_0 of the ion lattice. The magnitude of the linear charge density of the marching electron cloud in system K is the same as that of the ions:

$$|\lambda_-| = \lambda_0. \tag{6.56}$$

Due to Lorentz contraction accompanying this motion, this charge density is greater than corresponding *proper* charge density of the electron cloud by the factor of $\gamma(v_d)$. Therefore,

$$\lambda_-^0 = -\frac{\lambda_0}{\gamma(v_d)}. \tag{6.57}$$

In system K′ we have

$$\lambda'_+ = \lambda_0\gamma(V), \quad \lambda'_- = \lambda_-^0\,\gamma(v') \tag{6.58}$$

where v' is the relative velocity of the electron drift in system K′:

$$v' = \frac{V \pm v_d}{1 \pm \frac{V v_d}{c^2}}$$

The "–" sign here corresponds to $V\uparrow\uparrow v_d$, and the "+" sign corresponds to $V\uparrow\downarrow v_d$.

Using (6.57) and the known property of the Lorentz factor $\gamma(v')$ found in Section 3.1 (Equation (3.12)), we can rewrite (6.58) in the form

$$\lambda'_{+} = \lambda_0 \gamma(V), \quad \lambda'_{-} = \lambda^0_{-} \gamma(v') = -\lambda_0 \gamma(V)\left(1 \pm \frac{V v_d}{c^2}\right). \tag{6.59}$$

The net charge density in K′ is

$$\lambda' = \lambda'_{+} + \lambda'_{-} = \pm \gamma(V) \frac{V v_d}{c^2} \lambda_0. \tag{6.60}$$

Not only is it not a zero, but also its nonzero value turns out to be just right for the resulting electric field to be consistent with the known result of the Lorentz transformation. Indeed,

$$E' = \frac{\lambda'}{2\pi\varepsilon_0 s} = \frac{\lambda_0 V v_d \gamma(V)}{2\pi\varepsilon_0 c^2 s}. \tag{6.61}$$

But $\lambda_0 v_d = I$ is the current in system K and $c^{-2} = \mu_0 \varepsilon_0$. Therefore,

$$\mathbf{E}' = \frac{\mu_0}{2\pi} \gamma(V) \mathbf{V} \times \frac{I}{s} \hat{\varphi} = \gamma(V) \mathbf{V} \times \mathbf{B}. \tag{6.62}$$

This is the Lorentz transformation (6.52) for the case with only the magnetic field in system K. Our description of the phenomenon seems to be self-consistent.

This is all very well, but how about the invariance of the electric charge? Now this invariance is violated for any *finite* peace of wire.

The answer to this puzzle is that a piece of wire with a steady current in it is as illegal an object as an infinite wire, although for another reason: you cannot get a steady march of the electrons in a wire with two open ends.[2)] Which means that such a wire is not an isolated system of charges. When we say that the electric charge is an invariant characteristic of a system, we mean the charge in an isolated system. If the system is not isolated, the statement of the charge invariance may not apply. No surprise that it does not hold for a wire with the two end points.

The only way to maintain a steady drift of the charge carriers is to bend the wire into a closed loop. Then you can either maintain the steady current in the loop by including the battery, or by creating the wire in a superconducting state. In either case, we obtain the system, which can be considered as closed or isolated system. Suppose again that such a system is electrically neutral in its rest frame. Then, due to the charge invariance, it must remain neutral in any other reference frame. If the theory of relativity does give a self-consistent description of the world, it must meet this requirement.

Let us check it. Suppose we have a rectangular superconducting loop with the sides a and b (Figure 6.10). Denote, as before, the proper linear charge density of the lattice

2) The "illegal" character of a segment **dl** with steady current I in it (the so-called current element $\delta\mathbf{I} = I\mathbf{dl}$) is also manifest in the fact that the magnetic interaction between two such elements does not generally obey Newton's third law [45]. As was mentioned in Section 6.1, in such cases rigorous calculation of forces must also include the momentum of the EM field around the interacting objects.

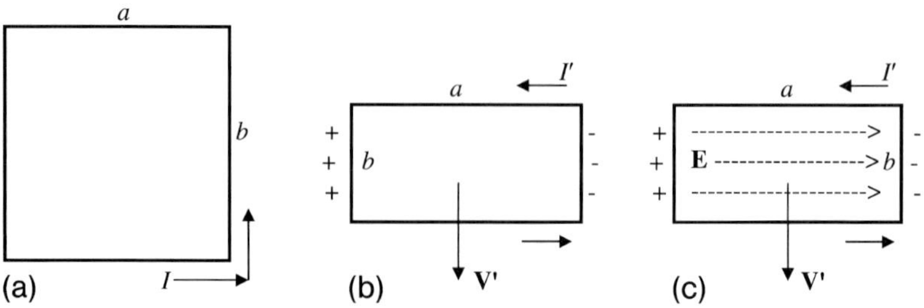

Figure 6.10 (a) A current loop in its rest frame K; (b) the same loop observed in another reference frame moving up relative to K; (c) the origin of the electric field: it emerges as a result of redistribution of charges observed in the moving current loop.

as λ_0. The electrons are drifting clockwise along the loop with speed $v_{\rm d}$. Their proper charge density λ_0^- is related to λ_0 by Equation (6.57).

Let us now calculate the charges on the loop's sides. The positive charge is

$$Q_+ = 2(a+b)\lambda_0. \tag{6.63}$$

The negative charge

$$Q_- = 2(a+b)\lambda_- = -2(a+b)\lambda_0 = -Q_+, \tag{6.64}$$

so that the net charge is, indeed, zero in K.

Consider now a system K′ moving up with a speed V. In this system

$$a' = a, \qquad b' = \frac{b}{\gamma(V)}. \tag{6.65}$$

Next we want to determine the primed linear charge densities in this system. We denote as $\lambda'_+(a)$ and $\lambda'_+(b)$ the linear charge densities of the lattice in sides a and b, respectively. From the given conditions it follows immediately

$$\lambda'_+(a) = \lambda_0, \qquad \lambda'_-(b) = \lambda_0\gamma(V). \tag{6.66}$$

Therefore,

$$Q'_+(a) = 2\lambda_0 a, \qquad Q'_+(b) = 2b'\lambda'_+(b) = 2\frac{b}{\gamma(V)}\lambda_0\gamma(V) = 2b_0\lambda_0. \tag{6.67}$$

The net positive charge in K′ is

$$Q'_+ = Q'_+(a) + Q'_+(b) = 2(a+b)\lambda_0 = Q_+. \tag{6.68}$$

The net charge of the whole lattice of the loop is invariant.

Determination of the linear densities of the marching electrons in K′ is a little more involved. Obviously, since $a' = a$, the densities of the negative charge carriers in K′ are the same as in K in both top ($a_{\rm T}$) and bottom ($a_{\rm B}$) sides of the loop:

$$\lambda'_-(a_{\rm B}) = \lambda'_-(a_{\rm T}) = -\lambda_+(a) = -\lambda_0. \tag{6.69}$$

As to the vertical sides, we now have to distinguish between the negative charge densities in the left side b (b_L) and right side b (b_R): in the rest frame of the loop, the electrons in b_L are moving up (in the same direction as the frame K′), and in b_R they are moving down (in the direction opposite to that of K′). Accordingly, their vertical speeds relative to K′ are different:

$$v'_L \equiv v'(b_L) = \frac{V - v_d}{1 - (Vv_d/c^2)}, \qquad v_R \equiv v'(b_R) = \frac{V + v_d}{1 + (Vv_d/c^2)}. \tag{6.70}$$

The corresponding densities are

$$\lambda'_-(b_L) = \lambda_0^- \gamma(v'_L) = -\frac{\lambda_0}{\gamma(v_d)}\gamma(v'_L), \quad \lambda'_-(b_R) = \lambda_0^- \gamma(v'_R) = -\frac{\lambda_0}{\gamma(v_d)}\gamma(v'_R). \tag{6.71}$$

Using (6.70) and known property (3.17) of the Lorentz factor, we obtain

$$\lambda'_-(b_L) = -\lambda_0\gamma(V)\left(1 - \frac{Vv_d}{c^2}\right), \quad \lambda'_-(b_R) = -\lambda_0\gamma(V)\left(1 + \frac{Vv_d}{c^2}\right). \tag{6.72}$$

Comparing with (6.67) shows that the total charge of all marching electrons in vertical sides is not invariant with respect to transformation from K to K′. Its magnitude decreases in b_L and increases in b_R. As a result, these sides are *not* neutral in K′. Their respective net charges are

$$\left.\begin{aligned} Q'(L) &= Q'_+(b_L) + Q'_-(b_L) = b\,\lambda_0\frac{Vv_d}{c^2} = Q_+(b)\frac{Vv_d}{c^2} \\ Q'(R) &= Q'_+(b_R) + Q'_-(b_R) = -b\,\lambda_0\frac{Vv_d}{c^2} = -Q_+(b)\frac{Vv_d}{c^2} \end{aligned}\right\}. \tag{6.73}$$

For the arrangement shown in Figure 6.10, the left side is positively charged and the right side has a negative net charge of the same magnitude. Accordingly, in K′ the loop produces not only magnetic field, as in K, but also the electric field. This field is the sum of the fields due to the charged individual sides b_L and b_R (Figure 6.10c). Between the sides, the general "flow" of the field lines is from left to right.

In contrast to the case with the infinite wire, the "violation" of the charge invariance here is a real physical effect, but I nevertheless take it into quotation mark because it is not really a violation of anything. As was emphasized before, a side of the loop is not by itself a closed system. Only the loop as a whole has this status. And its *net* charge is zero in K′ as it is in K:

$$Q' = 2Q'(a) + Q'(L) + Q'(R) = 0. \tag{6.74}$$

As we have already stressed in Section 5.6, the electric charge of any finite closed system is both – conserved and invariant characteristic of the system. But relativistic invariance of the charge is a more subtle property than it was thought in prerelativistic physics. According to relativity, different observers will measure the same *net* charge of a closed system, but they may see *quite different distributions* of this charge between parts of the system. As a consequence of relativity of time, the *charge distribution* is also a relative characteristic. Boarding another train, you may measure a different charge distribution in the same system you had observed from the platform. It is this redistribution of charges associated with change of reference frame, which is ultimately responsible for the emergence of the electric field when we move across a region with magnetic field.

This resolves the puzzle with the charge invariance "violation." As a by-product, we have found the answer to the question about the origin of the electric field (Section 6.2) seen by an observer drifting across the region with a magnetic field. When we move within a solenoid in a direction perpendicular to its symmetry axis as in Figure 6.7b, the opposite sides of each turn (one side above and the other below us) have opposite charges on them. The charge distribution on the sides is different from the case considered here because the turns are not rectangular. But the outcome is the same – we see the electric field perpendicular to both – the direction of motion and the magnetic field.

The discussed effect lies at the core of another spectacular prediction of the theory of relativity – that a moving magnetic dipole $\mathcal{M}$ develops an electric dipole moment [49]

$$\mathbf{p} = \frac{\mathbf{v} \times \mathcal{M}}{c}. \tag{6.75}$$

Indeed, the appearance of the two opposite charges on the vertical sides of the loop in Figure 6.10 means, among other things, that the loop acquires a nonzero electrical dipole moment perpendicular to both – the direction of motion and the magnetic moment $\mathcal{M}$ of the loop. As to the magnitude, for the simple model considered here it can be easily shown to be exactly equal to (6.75) (Problem 6.13).

6.4 Maxwell's Equations

We are now in a position to formulate the complete set of equations describing behavior of the EM field, that is, relating the electric and magnetic field to one another and to their sources – charges and currents.

Let us start first with Coulomb's law:

$$\mathbf{E}(\mathbf{r}) \equiv \frac{\mathbf{f}_q}{q} = \frac{q'}{4\pi\varepsilon_0 r^2}\hat{\mathbf{r}}. \tag{6.76}$$

It is assumed here that the origin of the coordinate system is set at the charge q' producing the field.

There is no reason, apart from our convenience, in doing this, which becomes evident if we have more than one sources of the field. In a more general case, when the source is at an arbitrary position $\mathbf{r}'$, we can write

$$\mathbf{E}(\mathbf{r}) \equiv \frac{\mathbf{f}_q}{q} = \frac{q'}{4\pi\varepsilon_0 \mathscr{r}^2}\hat{\mathscr{r}}. \tag{6.77}$$

Here $\mathscr{r} \equiv |\mathbf{r} - \mathbf{r}|$ is the separation distance between the source and the field points, and

$$\hat{\mathscr{r}} \equiv \frac{\mathscr{r}}{\mathbf{r}} \tag{6.78}$$

is the unit separation vector, directed from the source to the field point. Relation (6.77) can be recast in the more general form – the Gauss's law [44], which holds for both – discrete and continuous charge distribution:

$$\Psi_e \equiv \oiint_A \mathbf{E} \cdot \mathbf{da} = \frac{Q}{\varepsilon_0}. \tag{6.79}$$

Here Ψ_e is the electric flux through a closed surface containing the given charge distribution and Q is the net charge of this distribution. An important notice: the electric field $\mathbf{E}$ evaluated at the surface points on A is the net field produced by both the charges inside the surface and all the charges outside; the field of the latter does not contribute to the net flux. Now, if we express the net charge as the integral of the charge density distribution $\rho(\mathbf{r}')$ over the volume V enclosed by the surface A

$$Q = \iiint_V \rho(\mathbf{r}')\mathrm{d}\mathbf{r}', \tag{6.80}$$

and use the most general definition of the divergence

$$\operatorname{div} \mathbf{E} \equiv \vec{\nabla} \cdot \mathbf{E} \equiv \lim_{\Delta V \to 0} \frac{\oiint_A \mathbf{E} \cdot \mathbf{da}}{\Delta V}, \tag{6.81}$$

we can write the Gauss's law in the differential form

$$\vec{\nabla} \cdot \mathbf{E} = \frac{\rho}{\varepsilon_0}. \tag{6.82}$$

This forms the first of Maxwell's equations.

Let us now turn to the connection between the magnetic field and its sources. The magnetic charges, if they existed, would produce, like electric charges, the radial magnetic field connected with its source by the law similar to (6.82), namely

$$\vec{\nabla} \cdot \mathbf{B} = \mu_0 \tilde{\rho}. \tag{6.83}$$

Here we have denoted with "tilde" the hypothetic magnetic charge density distribution. However, such magnetic charges (so-called monopoles) up to now have not been discovered. Therefore, we can set in Equation (6.83) the magnetic charge density to be identically zero. In other words, magnetic field has the zero divergence:

$$\vec{\nabla} \cdot \mathbf{B} = 0. \tag{6.84}$$

This is the second of Maxwell's equations. It is essentially the statement that there are no monopoles (more cautiously, all known magnetic field configurations are not produced by monopoles), which turns our attention to the second (actually existing!) source of magnetic field: electric current. As we found in Section 6.1, the connection between the two, required by relativity in a stationary state, is expressed by Ampere's law (6.24). Introducing the current density vector

$$\mathbf{j} \equiv \lim_{\Delta a \to 0} \frac{\Delta I}{\Delta a} \hat{\mathbf{n}}, \tag{6.85}$$

where ΔI is the amount of current through area element Δa perpendicular to the direction $\hat{\mathbf{n}}$ of motion of the corresponding local charge, we have

$$\mu_0 \iint_A \mathbf{j} \cdot \mathbf{da} = \oint_L \mathbf{B} \cdot \mathbf{dl}. \tag{6.86}$$

Apart from being a law of Nature, this is one step short of being a good physical illustration of Stokes theorem: a surface integral of the curl of a vector field **B** over an area A is equal to the line integral of **B** over the loop L enclosing this area (let us call it the circulation of **B**). To make this step, we can just write down the definition of the curl

$$\text{curl}\,\mathbf{B} \equiv \vec{\nabla} \times \mathbf{B} \equiv \lim_{\Delta a \to 0} \frac{\oint_L \mathbf{B} \cdot \mathbf{dl}}{\Delta a} \hat{\mathbf{n}}, \tag{6.87}$$

where Δa is the small area enclosed by the loop L and $\hat{\mathbf{n}}$ is a unit vector perpendicular to the area Δa and connected with the direction of integration along the loop by the right-hand rule.Then it follows immediately

$$\vec{\nabla} \times \mathbf{B} = \mu_0 \mathbf{j}. \tag{6.88}$$

This is the differential form of the Ampere's law.

Expression (6.88), however, does not form a complete Maxwell's equation. Look carefully at its original integral form. There is an infinite number of different surfaces $A, A', A'', \ldots$, enclosed by the same loop L and according to the Stokes theorem any one of them is equally legitimate. Applied to our situation, it says that it does not matter what surface to consider. And indeed, if we have an uninterrupted steady current, the Ampere's law works fine. The same current flows through any surface enclosed by the loop surrounding the current (Figure 6.11). As we will see later, this is a manifestation of the charge conservation for a stationary case. But what if the current is not steady or is interrupted?

Suppose you are charging the capacitor as in Figure 6.12. Even if the current is maintained steady during the process of charging, it is interrupted at the capacitor's

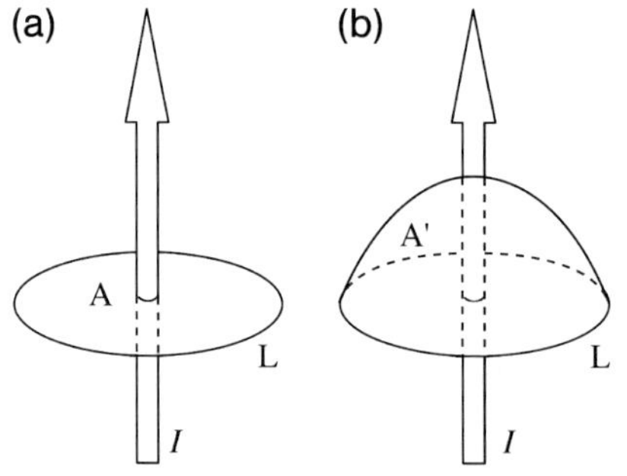

Figure 6.11

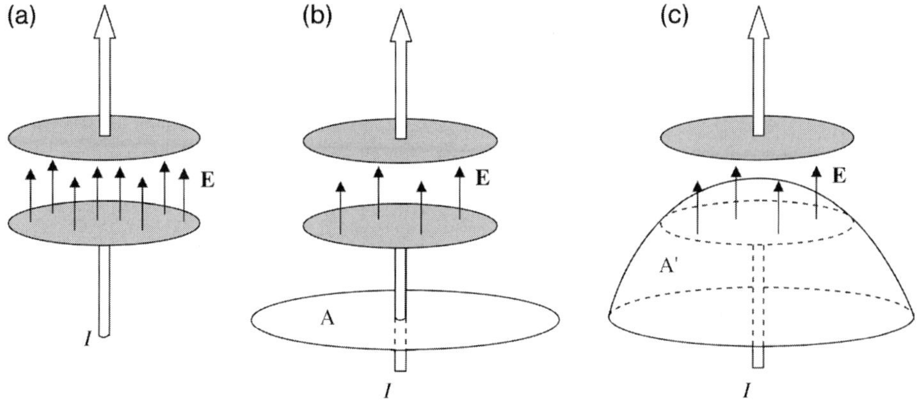

Figure 6.12 Charging a capacitor.

plate. If you choose the surface A so that it goes across the current wire, the Ampere's law gives the correct magnetic field around the wire; but if you choose another (but enclosed by the same loop L) surface A′ that straddles between the plates, then there is no current through it and therefore the circulation in (6.86) should be zero. For the axially symmetrical case considered here this would mean that magnetic field is also zero. Thus, the Ampere's law applied to this situation says that there is magnetic field around the wire, if you consider the surface A, and that there is no field there if you consider the surface A'. Obviously, the Ampere's formulation is not complete. There must be something between the plates that crosses the surface A' instead of current; something takes over where the current fails. What is it?

Consider what happens with the capacitor plate in this process. The plate intercepts the current and accommodates the arriving charge, which is spreading over its surface. As a result, there appears the surface charge density that produces the changing electric field between the plates according to the equation[3])

$$\mathbf{E} = \frac{\sigma}{\varepsilon_0}\hat{\mathbf{n}}. \tag{6.89}$$

The electric field flux in this situation is (A is now the area of the plate)

$$\Psi_e = EA = \frac{\sigma A}{\varepsilon_0} = \frac{q}{\varepsilon_0}, \tag{6.90}$$

and its time derivative

$$I \equiv \varepsilon_0 \frac{d\Psi_e}{dt} = \frac{dq}{dt} = \varepsilon_0 \frac{\partial E}{\partial t} A \equiv I_D. \tag{6.91}$$

3) To simplify the discussion, we assume here the change slow enough so that the surface charge density has time to spread uniformly over the plates and accordingly be treated as the function of time only, but not position; in this approximation, the electric field between the plates can be considered as uniform and described by (6.89). In the differential form, the approximation becomes exact.

The first equation here describes correctly the actual current I in the wire. The second equation describes what is going on between the plates, and shows that, even though there are no moving charges there, the *rate of change of the electric flux is numerically equal to the current in the wire.* In a way, this rate of change serves here as a representative I_D of the absent current. This property is reflected in its name – the displacement current. Both – the concept and its name were introduced by Maxwell. It follows that the corresponding density

$$\mathbf{j}_D \equiv \varepsilon_0 \frac{\partial \mathbf{E}}{\partial t} \tag{6.92}$$

plays the role of the real current density in a place where the actual current is stopped (or, more generally, is different from its value in adjacent place along the current flow). Based on these observations, Maxwell generalized the Ampere's law by postulating that the complete expression for the circulation of **B** must consist of two contributions: from the actual current and from the displacement current

$$\oint_L \mathbf{B} \cdot \mathrm{d}\mathbf{l} = I + I_D = I + \varepsilon_0 \frac{\mathrm{d}\Psi_e}{\mathrm{d}t}, \tag{6.93}$$

or, in the differential form,

$$\vec{\nabla} \times \mathbf{B} = \mu_0 \left(\mathbf{j} + \varepsilon_0 \frac{\partial \mathbf{E}}{\partial t} \right). \tag{6.94}$$

The quantity $\mathbf{D} = \varepsilon_0 \mathbf{E}$ is accordingly called the electrical displacement and we write in (6.94) the partial derivative to indicate that once we go local, we evaluate the rate of change of **E** at a *fixed position* in space.

Equation (6.94) is the Maxwell's third equation.

Now we are prepared for derivation of the fourth and last of Maxwell's equations.

Actually, we can predict its general form right away from the symmetry considerations.

In (6.94), the curl of **B** is determined by the electric current density and the rate of change of the electric field at a given point. Similarly, the curl of **E** must be determined by the *magnetic* current density $\tilde{\mathbf{j}}$ and the rate of change of the magnetic field at this point:

$$\vec{\nabla} \times \mathbf{E} = \mu_0 \tilde{j} + \kappa \frac{\partial \mathbf{B}}{\partial t}. \tag{6.95}$$

But since magnetic currents do not exist because of the absence of magnetic charges, the equation reduces to

$$\vec{\nabla} \times \mathbf{E} = \kappa \frac{\partial \mathbf{B}}{\partial t}. \tag{6.96}$$

The only remaining unknown here is a proportionality constant κ. To determine it, let us consider the following situation: a rectangular conducting loop L with the area $A = ab$ is moving to the right with a speed V across a region with uniform magnetic field **B** pointing away from us into the page (Figure 6.13). Suppose that Alice is

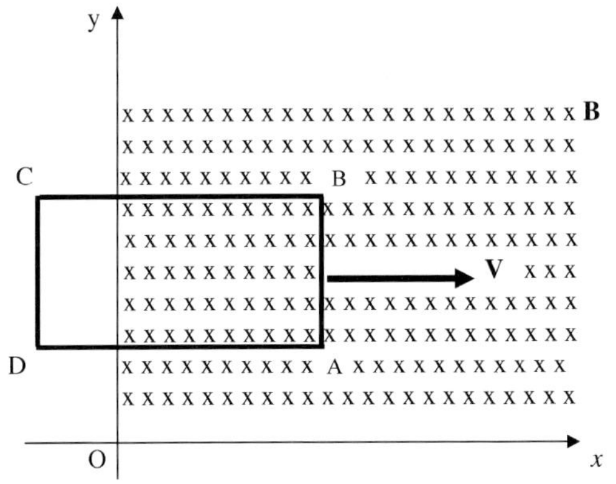

Figure 6.13 Alice and Tom's experiment.

recording the results. When the leading side $AB = a$ enters the field region, there appears the Lorentz force on each charge in it. Using the right-hand rule, we find the direction of the force up for the positive charges and down for the negative charges. In both cases this tends to set up the current in the side AB from A to B. Until the loop is entirely inside the field region, this force does not act in the trailing side CD. As a result, the net line integral of this force (per unit charge) over L is nonzero and equal to

$$\mathcal{E} = \oint_L \frac{\mathbf{f}_L}{q} \cdot \mathrm{d}\mathbf{l} = \oint_L (\mathbf{V} \times \mathbf{B})\mathrm{d}\mathbf{l} = aVB. \tag{6.97}$$

Here, we used the fact that the vectors **V** and **B** in the integrand are perpendicular to one another and both are constants and the horizontal sides BC and AD of the loop do not contribute to the result. Now, since $V = \mathrm{d}x/\mathrm{d}t$, where x is an instant position of the side AB, the expression reduces to

$$\mathcal{E} = aVB = a\frac{\mathrm{d}x}{\mathrm{d}t}B = B\frac{\mathrm{d}(ax)}{\mathrm{d}t} = \frac{\mathrm{d}BA(x)}{\mathrm{d}t} = -\frac{\mathrm{d}\Psi_{\mathrm{m}}}{\mathrm{d}t}. \tag{6.98}$$

Here $A(x)$ is the instant area within A, filled with the magnetic field, so that Ψ_{m} is the instant magnetic flux through the loop. The net result (the force per charge circulation around the loop) is frequently called the electromotive force (EMF). It turns out to be equal to the negative rate of change of this flux. The minus sign here determines the relationship between the directions of the change of flux and the EMF induced by this change. We will formulate this relationship more rigorously in the next section.

At this point we notice Tom riding on the loop and ask him of his account of the same process. It turns out pretty close to that of Alice's, with the main distinction that what is the Lorentz force for Alice is the electric force for Tom. We go back to Section 6.2 and indeed see Equation (6.40) relating the magnetic field **B** in Alice's frame of reference K to electric field $\mathbf{E}'$ in Tom's frame K′. Thus, Tom observes, apart from $\mathbf{B}'$, the up directed electric field arranged in the plane of the loop. The effect of this field forcing the charges to move around the loop is, as in case of the Lorentz force in Alice's frame, also described by the circulation of the field:

$$\mathcal{E}' \equiv \oint\limits_{L'} \mathbf{E}' \cdot \mathrm{d}\mathbf{l}'. \tag{6.99}$$

In our case, when $\mathbf{E}'$ is described by (6.40), the integrand reduces to

$$\mathbf{E}' \cdot \mathrm{d}\mathbf{l}' = \gamma(V)(\mathbf{V} \times \mathbf{B}) \cdot \mathrm{d}\mathbf{l} = \gamma(V) V B \,\mathrm{d}l. \tag{6.100}$$

In Tom's RF the speed $V = \mathrm{d}x'/\mathrm{d}t'$, where x' is the instant position of the leading edge of the magnetic field region moving to the left in his frame. Also, in his frame, as in Alice's one, the integration reduces to the line integral over the side $\mathrm{AB} = a$. Therefore,

$$\mathcal{E}' = \frac{\mathrm{d}x'}{\mathrm{d}t'} \gamma(V) B l. \tag{6.101}$$

But, according to (6.48), $\gamma(V)B = B'$ (we drop the subscript "$\perp$" here) and $x'a = A'(x')$. Therefore,

$$\mathcal{E}' = \frac{\mathrm{d}}{\mathrm{d}t'}(B' A'(x')) = -\frac{\mathrm{d}\Psi'_{\mathrm{m}}}{\mathrm{d}t'}. \tag{6.102}$$

Relationship (6.102) is known as a special case of Faraday's law. Together with (6.98) it can be formulated as the universal flux rule: any change of the magnetic flux through a loop induces the EMF given by (6.102) [44]. The differential equivalent of (6.102) is (dropping primes)

$$\vec{\nabla} \times \mathbf{E} = -\frac{\partial \mathbf{B}}{\partial t}. \tag{6.103}$$

Comparison with (6.96) shows that the constant $\kappa = -1$. Equation (6.103) is the last of Maxwell's equations. We can now write down all of them together

$$\vec{\nabla} \cdot \mathbf{E} = \frac{\rho}{\varepsilon_0}, \quad \vec{\nabla} \times \mathbf{E} = -\frac{\partial \mathbf{B}}{\partial t}, \quad \nabla \cdot \mathbf{B} = 0, \quad \vec{\nabla} \times \mathbf{B} = \mu_0 \left(\mathbf{j} + \varepsilon_0 \frac{\partial \mathbf{E}}{\partial t} \right). \tag{6.104}$$

Discussion

We have already mentioned the asymmetry between $\vec{\nabla} \cdot \mathbf{E}$, on the one hand, and $\vec{\nabla} \cdot \mathbf{B}$, on the other. Physically this means the existence of the electric charges ρ and the absence of the magnetic charges $\tilde{\rho}$ (monopoles). Accordingly, we have no magnetic current density $\tilde{\mathbf{j}}$ in Equation (6.103). Their absence looks like the gaping holes in the equations.

The sign difference in the time derivatives on the right in the two equations may also appear surprising. Why, indeed, should these two derivatives come with the opposite signs?

Below we will show that the minus sign in Equations (6.102) and (6.103) has a deep physical meaning.

6.5
Lenz's Law

To see the implications of the sign difference in the two time derivatives in Maxwell's equations, let us consider the Alice–Tom joint experiment in the previous section in more details.

First we analyze the connection between the EMF and the change of magnetic flux. Equation (6.102) derived by Tom for the special case turns out to be much more general. In its final form it does not contain any information about what causes the change of flux. All it tells us is that the EMF around a loop is equal in magnitude to the rate of change of the flux through the loop. It says nothing about what causes this change. It may be the arrangement used in the Alice–Tom experiment; it may be the change in time of magnetic field itself, straddling through a loop fixed in space; it may be the loop's rotation in a constant magnetic field – a prototype of an electric motor or a unit in an electric power generator; it may be deformation of the loop, which changes its interception area; it may be a combination of some or all of the above; similarly, in expressions (6.97)–(6.99) for EMF we may have just a line integral around a closed path without any physical loop at all (Figure 6.14). The main thing is change of flux, regardless of its cause, through a closed path, regardless of its shape. Therefore, the conclusions we will obtain will have a very general nature.

Let us first determine the flux and its change in the Alice–Tom experiment as functions of time. Take the moment when the side AB enters the magnetic field region as the zero moment in K. Then the flux is zero for all $t \leq 0$ and increases linearly with time between $t = 0$ and $t_a = a/V$ (the time it takes the whole loop to enter the field region). After this moment the flux remains constant until the side AB reaches the opposite end of the field region. This moment depends on size of the region (we assume it to be greater than the size of the loop). After it, the flux decreases linearly until it reaches zero when the trailing side CD exits the

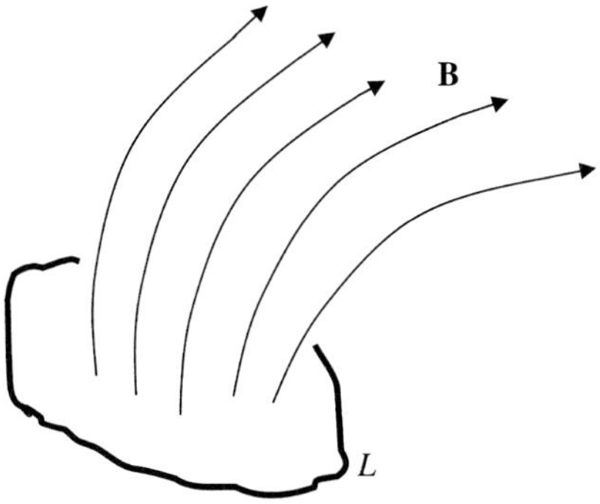

Figure 6.14 An arbitrary closed curve L. If there is a changing magnetic flux $\Psi_m(t)$ through L, there is an electromotive force E_m around the loop L.

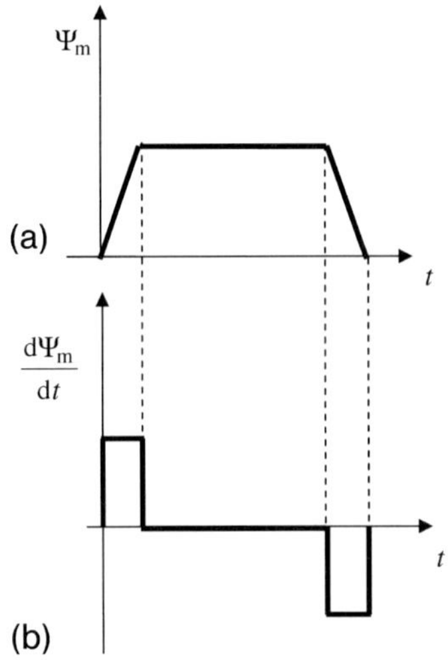

Figure 6.15 The magnetic flux through the moving loop (a), and its time derivative (b) in the Alice–Tom experiment.

region. The plots of the time dependence $\Psi_m(t)$ and $d\Psi_m(t)/dt$ are shown in Figure 6.15.

According to (6.98), when Ψ is constant, there is no EMF. Physically, it is because the directions of the EMF in both sides of the loop – AB and CD – coincide, tending to drive the current up in both sides. These two tendencies cancel each other out leading to the zero net EMF in the whole loop.

Now let us consider the origin of the minus sign in the Faraday's law. According to this sign, the EMF under the above conditions is driving the current in the loop, as we noted before, in the counterclockwise direction. But the current in the loop produces the magnetic field and thereby the magnetic flux of its own. And, applying the right-hand rule, you can see that its direction (more accurately, the direction of the corresponding magnetic field) is *opposite* to the original field. The secondary flux tends to decrease the growth of the primary flux. The net flux is increasing slower than it would do without the secondary one.

Now, what happens during the last stage when the loop is exiting the field region? In this case, the magnetic flux through the loop is *decreasing* and, accordingly, the current in the loop starts flowing in the clockwise direction. The physical reason for this is that the leading edge AB is now outside the field region and there is no motional EMF generated in it. In the trailing edge CD the induced current is, as before, in the up direction, but without the counterbalance from the side AB, it leads to the *clockwise* current along the loop. But this reverse of the current accordingly *reverses* the secondary field produced by it. The secondary field is now pointing in *the same* direction as the primary one, as if trying to keep the corresponding flux from reduction. In the upshot, the produced EMF is always in such a direction as to decrease the change of flux. This rule, determining the direction of the induced EMF, is known as the Lenz's law. As Griffiths put it in his book on electromagnetism [44]: *Nature Abhors Change in Flux.*

In materials in a superconducting state this response is so strong that the net flux through a superconducting loop remains constant [34]. Thus, the first aspect of the minus sign in (6.98) is the tendency to conserve the magnetic flux.

The second aspect may be even more interesting. To see it, let us reverse the argument and instead of questioning the validity of the minus sign in (6.103), question the validity of the plus sign in the last of Equations (6.104)? Maybe the symmetry in the Maxwell equations would be more complete if we, instead of plus, put the minus sign there as well? Or, putting it in another way: just as we have found that the minus sign in (6.103) is justified, what if anything justifies the plus sign in the last of Equation (6.104)?

Let us play a little with this equation. Take divergence of it. We know from the vector calculus that the divergence of any curl is zero. So we will get the equation with the zero on the left:

$$0 = \vec{\nabla} \cdot \mathbf{j} + \varepsilon_0 \vec{\nabla} \cdot \frac{\partial \mathbf{E}}{\partial t}. \tag{6.105}$$

Interchanging ordering of the differential operations in the last term and combining with (6.82), we obtain

$$\vec{\nabla} \cdot \mathbf{j} + \frac{\partial \rho}{\partial t} = 0, \tag{6.106}$$

or

$$\vec{\nabla} \cdot \mathbf{j} = -\frac{\partial \rho}{\partial t}. \tag{6.107}$$

This equation has an obvious physical interpretation: any divergence of the current density at a given point is accompanied by the change of the charge density at this point. Suppose, for instance, that $\vec{\nabla} \cdot \mathbf{j}$ is positive. Physically this means the exodus of the positive charges from (or the convergence of the negative charges to) the point in question. In either case, this must be accompanied by the decrease in the charge density at this point. But the time derivative of a decreasing quantity is negative, whereas Equation (6.107) demands that the right-hand side be positive. This condition is taken care of by the minus sign before the derivative in (6.107). Thus, this sign, or, which is the same, the plus sign in (6.106) and (6.92) reflects the conservation of the electric charge. The plus in (6.92) is as justified and as important as the minus in (6.102).

We may appreciate this result even more if we notice that the current density and the volume charge density in Equation (6.106) together form a 4-vector:

$$(c\rho, \mathbf{j}) = j. \tag{6.108}$$

Indeed, if we take the definition of volume charge density $dq = \rho\, dV$ and multiply it by dx^l

$$dq\, dx^l = \rho\, dV dx^l = \rho\, dV dt \frac{dx^l}{dt}, \tag{6.109}$$

we will have a 4-vector on the left (since the four components dx^l, $l = 0,1,2,3$ form a 4-vector, whereas dq is a scalar). Therefore the expression on the right has to be also a 4-vector. But $Vdt = d\Omega/c$ (where $d\Omega$ is a four-dimensional volume element) is a scalar. Therefore, $\rho(dx^l/dt) = (c\rho, \mathbf{j})$ is a 4-vector.

Using this, Equation (6.107) can be recast in the following elegant form

$$\frac{\partial(c\rho)}{\partial x^0} + \frac{\partial j^\alpha}{\partial x^\alpha} = 0 \tag{6.110}$$

or

$$\frac{\partial j^l}{\partial x^l} = 0, \tag{6.111}$$

or, using the four-dimensional differential del operator introduced in Section 4.8,

$$\Box \cdot j = 0. \tag{6.112}$$

We conclude that the sign difference between the Faraday's law and the corresponding part of the Maxwell's equation (6.104) (let us call it the *displacement current law*) represents the true symmetry between the electric and magnetic aspects of the electromagnetic field. This conclusion is supported by an interesting observation: if the monopoles do exist (and are discovered someday), the monopole charge must, in contrast to the electric charge, be a pseudoscalar, rather than a scalar quantity (Problem 6.10).

The final (and the most important for our subject) aspect of this sign difference (or of the actual symmetry represented by it) refers to the mere existence of light. It will be discussed later in the "optical" part of this chapter.

6.6
4-Potential, Gauge Transformations, and Gauge Invariance

In this section we will make a step to a four-dimensional formulation of electrodynamics showing explicitly the Lorentz invariance of its equations.

We start with the set of the Maxwell's equations and see that in this system the electric and magnetic field are coupled; in other words, this is a system of coupled equations for two vector functions **E** and **B**. Our first attempt will be to decouple them to solve separately for **E** or **B**.

We first take a curl of Equation (6.102):

$$\vec{\nabla} \times (\vec{\nabla} \times \mathbf{E}) = -\frac{\partial}{\partial t} \vec{\nabla} \times \mathbf{B}. \tag{6.113}$$

Combining this with (6.103) gives

$$\vec{\nabla} \times (\vec{\nabla} \times \mathbf{E}) = -\frac{\partial}{\partial t}\left(\mu_0 \mathbf{j} + \mu_0 \varepsilon_0 \frac{\partial \mathbf{E}}{\partial t}\right). \tag{6.114}$$

Now we have the equation for the electric field only, with **B** eliminated. We can rewrite it in a more convenient form applying the known identity for double curl from the vector calculus:

$$\vec{\nabla} \times (\vec{\nabla} \times \mathbf{E}) \equiv \vec{\nabla}(\vec{\nabla} \cdot \mathbf{E}) - \nabla^2 \mathbf{E}. \tag{6.115}$$

Using this and rearranging gives

$$\nabla^2 \mathbf{E} - \mu_0 \varepsilon_0 \frac{\partial^2 \mathbf{E}}{\partial t^2} = \vec{\nabla} \frac{\rho}{\varepsilon_0} + \mu_0 \frac{\partial j}{\partial t}, \tag{6.116}$$

or, recalling that according to (6.21), $\mu_0 \varepsilon_0 = c^{-2}$:

$$\nabla^2 \mathbf{E} - \frac{1}{c^2} \frac{\partial^2 \mathbf{E}}{\partial t^2} = -\Box^2 \mathbf{E} = \vec{\nabla} \frac{\rho}{\varepsilon_0} + \mu_0 \frac{\partial \mathbf{j}}{\partial t}. \tag{6.117}$$

The similar equation can be obtained for **B**.

The left-hand side in (6.117) is the familiar D'Alambertian operator applied to **E**. The emergence of the D'Alambertian on the left, as opposed to the Laplacian operator, is manifestation of relativity; it shows that the effect of any change of the charges and currents on the right spreads over surrounding space with the speed of light and accordingly affects the field at later moments. The expression on the right-hand side, however, is still pretty messy. Only in case of free space without charges and currents the solution is simple and we will discuss this in later sections devoted to light.

Here we are interested in the general picture. A general solution of this equation must contain an explicit expression of the field in terms of the distribution and motion of its sources. It shows that the field depends not only on spatial distribution of charges and currents (i.e., charge velocities!), but also on their accelerations, because it contains the time derivative of the current density. This is precisely what we noted in Section 6.1. But we also noted there that the description in terms of potentials would be simpler because the latter do not depend on charge accelerations.

So let us see if we can recast the Maxwell equations in terms of the potentials. Let us start with Equation (6.84). We know that the divergence of any curl is identically zero. Therefore, we can always satisfy this equation by introducing a new vector field **A** such that

$$\mathbf{B} = \vec{\nabla} \times \mathbf{A}. \tag{6.118}$$

The vector **A** is already familiar to us as the magnetic vector potential.

Now turn to the electric field. In a static situation, it is conveniently expressed as the negative gradient of the electrical potential Φ:

$$\mathbf{E} = -\vec{\nabla} \Phi. \tag{6.119}$$

But the scalar potential function Φ alone is not enough in the dynamical situation. We can see it immediately by taking curl of **E**: in case of (6.119) we will get zero – a known result in static situation. But the Faraday's law shows that curl of **E** may be nonzero as well. So to obtain the general expression for the electric field in terms of potentials, let us reverse our argument and start with the differential form of Faraday's law. Putting there the expression (6.118) yields

$$\vec{\nabla} \times \mathbf{E} = -\frac{\partial}{\partial t}\vec{\nabla} \times \mathbf{A} = \vec{\nabla} \times \left(-\frac{\partial \mathbf{A}}{\partial t}\right). \tag{6.120}$$

We could infer from here that the part of **E** produced by changing magnetic field is the negative time derivative of the vector potential. Then, adding to it another part from (6.119), which describes contribution from the electric charges, we will obtain the desired general expression.

More formally, we can do it by noticing that we can add inside of parentheses in (6.120) the gradient of an arbitrary potential function, since the curl of any gradient is identically zero. Then consulting with (6.119) we can identify this function with Φ. The final result for both fields is

$$\mathbf{E} = -\vec{\nabla}\Phi - \frac{\partial \mathbf{A}}{\partial t}, \qquad \mathbf{B} = \vec{\nabla} \times \mathbf{A}. \tag{6.121}$$

Look carefully at the first of these equations. Notice that Φ/c and the components of **A** have the same dimensionality. It looks like the quantity Φ/c and **A** together form a 4-vector. If this is true, then the electric potential function Φ and magnetic vector potential **A** are parts of a more general entity – 4-potential:

$$\mathsf{A} = \left(\frac{\Phi}{c}, \mathbf{A}\right). \tag{6.122}$$

Before finding a compelling and at the same time simple argument for this, note that this entity is more subtle than, say, 4-displacement, 4-velocity, or 4-momentum. As a characteristic of a physical system, it is not uniquely defined. We know already from electrostatics (i.e., from Equation (6.119)) that the electric potential function is only defined up to an additive constant. This reflects freedom of choice of the reference point to which we ascribe the zero potential. I can define the potential energy of the book on my table with respect to the tabletop, or to the floor of my room, or with respect to the sea level. It is not potential itself – the potential difference between two points in space is what matters.

It turns out that in our more general picture there exist an infinite number of different 4-potentials, all corresponding to the same observable field **E** and **B**. Indeed, take a closer look at (6.118). What happens if we switch from **A** to **A**′ such that

$$\mathbf{A} \to \mathbf{A}' = \mathbf{A} + \vec{\nabla}\chi, \tag{6.123}$$

where $\chi = \chi(\mathbf{r}, t)$ is an arbitrary scalar function? Well, since curl of any gradient is identically zero, the new magnetic vector potential **A**′ will produce the same magnetic field as before. But since we are beyond the domain of the electrostatics, the electric field is defined by (6.121) instead of (6.119) and may be affected by the transformation (6.123). Indeed, we would have

$$\mathbf{E}' = -\vec{\nabla}\Phi - \frac{\partial A'}{\partial t} = \mathbf{E} - \vec{\nabla}\frac{\partial \chi}{\partial t} \neq \mathbf{E}. \tag{6.124}$$

But we immediately notice that if, simultaneously with (6.123), we transform Φ according to

$$\Phi \to \Phi' = \Phi - \frac{\partial \chi}{\partial t}, \tag{6.125}$$

then $\mathbf{E}' = \mathbf{E}$ and the electric field will remain the same as before. It is important to note that both transformations – (6.123) and (6.125) – are to be performed concurrently (all the more so that $\mathbf{A}$ and Φ are parts of the same 4-vector!), so let us write them down in one equation:

$$\mathbf{A} \to \mathbf{A}' = \mathbf{A} + \vec{\nabla}\chi, \qquad \Phi \to \Phi' = \Phi - \frac{\partial \chi}{\partial t}. \tag{6.126}$$

Now, if we recall that Φ/c forms the zeroth component A_0 of the 4-potential A, we can rewrite this in a more elegant form,

$$A \to A' = \left(A^j + \frac{\partial \chi}{\partial x^j}\right)\hat{\mathbf{x}}^j = A + \Box\chi, \tag{6.127}$$

where we have used our notation (4.118) for the 4-gradient.

Now, the 4-gradient $\Box\chi$ is a 4-vector. Within the conventional formalism, a vector can be added only to a vector.[4)] Therefore, the above equation can be self-consistent only if A is a 4-vector. As an illustration, consider a special case when the original potentials $\Phi = 0$ and $\mathbf{A} = 0$. Then Equation (6.127) produces a set of the transformed components A' forming a 4-vector. This justifies the above assumption about the 4-vector nature of the set $(\Phi, \vec{A})$.

Transformation (6.127) is called the gauge transformation. Since $\mathbf{E}$ and $\mathbf{B}$ are invariant under these transformations, so are Maxwell's equations. This kind of invariance is called the gauge invariance. Note the difference between the Lorentz invariance and the gauge invariance. The former is the invariance with respect to coordinate transformations. The latter is the invariance with respect to the gauge transformations. This invariance gives us a certain freedom in choice of the optimal 4-potential leading to most simple form of equations expressed in terms of A.

We have come to a point where we are properly equipped to derive these equations.

Step 1 – express the field in terms of the potentials using (6.121) and put it into the Maxwell's equations (6.104). Then the two equations will be satisfied automatically. The remaining two read

$$-\nabla^2\Phi - \frac{\partial}{\partial t}\vec{\nabla}\cdot\mathbf{A} = \frac{\rho}{\varepsilon_0} \tag{6.128}$$

and

$$\vec{\nabla}\times(\vec{\nabla}\times\mathbf{A}) = \mu_0\mathbf{j} - \mu_0\varepsilon_0\frac{\partial}{\partial t}\left(\vec{\nabla}\Phi + \frac{\partial \mathbf{A}}{\partial t}\right) \tag{6.129}$$

4) There is a new, very powerful, and elegant formalism based on Clifford algebra and developed by D. Hestenes [50], which can operate with generalized quantities being a combination of a vector and a scalar (much like a complex number is a combination of a real and an imaginary number). This formalism allows to present physics in a more simple and straightforward way. If applied to the case in question, it also shows that the set $(\Phi, \mathbf{A})$ forms a pure 4-vector.

or, using again the identity (6.115) and rearranging,

$$\nabla^2\Phi + \frac{\partial}{\partial t}\vec{\nabla}\cdot\mathbf{A} = -\frac{\rho}{\varepsilon_0}, \tag{6.130}$$

$$\nabla^2\mathbf{A} - \frac{1}{c^2}\frac{\partial^2\mathbf{A}}{\partial t^2} - \vec{\nabla}\left(\vec{\nabla}\cdot\mathbf{A} + \frac{1}{c^2}\frac{\partial\Phi}{\partial t}\right) = -\mu_0\mathbf{j}. \tag{6.131}$$

It does not look too simple and therefore we make step 2: exploit the fact that the 4-potential is not single-valued and choose the functions **A** and **Φ** so that they satisfy the requirement:

$$\left(\vec{\nabla}\cdot\mathbf{A} + \frac{1}{c^2}\frac{\partial\Phi}{\partial t}\right) \equiv \Box\cdot A = 0. \tag{6.132}$$

(i.e., we require the 4-divergence of A to be zero.)

Immediately, the previous two equations simplify to

$$\left.\begin{aligned} \nabla^2\Phi - \frac{1}{c^2}\frac{\partial^2\Phi}{\partial t^2} &= -\frac{\rho}{\varepsilon_0} \\ \nabla^2\mathbf{A} - \frac{1}{c^2}\frac{\partial^2\mathbf{A}}{\partial t^2} &= -\mu_0\mathbf{j} \end{aligned}\right\}. \tag{6.133}$$

They form a system of the four inhomogeneous D'Alambert equations explicitly showing that the field is generated by the sources – charges and currents.

Now, recall from Section 6.5 that (ρc, **j**) form a 4-vector j of current density. This allows us to rewrite (6.133) in a simple and elegant form

$$\Box^2 A = -\mu_0 j. \tag{6.134}$$

This simple equation (actually, four equations) contains all electrodynamics. And we can appreciate it even better if we notice that it is explicitly Lorentz-invariant.

It remains to prove that the condition (6.132), leading to this beautiful result, can be satisfied.

Suppose that initially it is not satisfied, so that

$$\vec{\nabla}\cdot\mathbf{A} + \frac{1}{c^2}\frac{\partial\Phi}{\partial t} \equiv \Box\cdot A = \lambda(\mathbf{r},t) \neq 0, \tag{6.135}$$

where $\lambda(\mathbf{r}, t)$ is a scalar function of position in space-time. Now switch from A to A' using gauge transformation (6.127). Then our initial expression (6.135) will take the form

$$\vec{\nabla}\cdot\mathbf{A}' + \frac{1}{c^2}\frac{\partial\Phi'}{\partial t} = \vec{\nabla}\cdot(\mathbf{A} + \vec{\nabla}\chi) + \frac{1}{c^2}\frac{\partial}{\partial t}\left(\Phi - \frac{\partial\chi}{\partial t}\right)$$

$$= \vec{\nabla}\cdot\mathbf{A} + \frac{1}{c^2}\frac{\partial\Phi}{\partial t} + \nabla^2\chi - \frac{1}{c^2}\frac{\partial^2\chi}{\partial t^2} = \lambda(\mathbf{r},t) + \Box^2\chi \tag{6.136}$$

or, in four-dimensional notations,

$$\Box\cdot A' = \lambda(\mathbf{r},t) + \Box^2\chi(\mathbf{r},t). \tag{6.137}$$

Since χ is an arbitrary function of $\mathbf{r}, t$, we can always choose it so that $\lambda(\mathbf{r}, t) + \Box^2\chi(\mathbf{r}, t) = 0$,

or

$$\Box^2\chi(\mathbf{r}, t) = -\lambda(\mathbf{r}, t). \tag{6.138}$$

In other words, it suffices to use the gauge transformation with a function χ being a solution of the inhomogeneous D'Alambert equation (6.138). This automatically leads to the condition that the 4-divergence $\Box \cdot A'$ is zero. This condition is called Lorentz condition, or Lorentz gauge. Then the transformed 4-potential A', while describing the same physical field, will (upon dropping the primes) satisfy Equation (6.134).

A beautiful thing about the Lorentz gauge is its Lorentz invariance. It is evident from the structure of the Lorentz condition: it is condition on the value of 4-divergence $\Box \cdot A'$, which is a scalar in space-time. Therefore, if it is satisfied in one reference frame, it will be automatically satisfied in all other reference frames.

6.7 Electrodynamics and The Principle of the Least Action

Now we can exploit to a further extent the ideas used in Section 5.7 for invariant formulation of mechanics. Here we will extend the corresponding approach to obtain a similar formulation of electrodynamics, which will also incorporate the results already obtained in the previous sections.

All the electrodynamics is about particle–field interaction. Therefore, first, there must be characteristics of a particle – its rest mass and charge; the characteristic of its motion – its world line; and the characteristic describing the EM field – its 4-potential A.

Before formulating the theory using these characteristics, let us recall its original formulation using Maxwell's equations, which relate electric and magnetic fields $\mathbf{E}$ and $\mathbf{B}$ to their sources – charges and currents:

$$\vec{\nabla} \cdot \mathbf{E} = \frac{\rho}{\varepsilon_0}, \qquad \vec{\nabla} \times \mathbf{E} = -\frac{\partial \mathbf{B}}{\partial t}, \tag{6.139}$$

$$\vec{\nabla} \cdot \mathbf{B} = 0, \qquad \vec{\nabla} \times \mathbf{B} = \mu_0 \mathbf{j} + \mu_0 \varepsilon_0 \frac{\partial \mathbf{E}}{\partial t}. \tag{6.140}$$

As in Section 5.7, our starting point will be the principle of the least action. But here we have to extend it to particles interacting with the field.

After including interaction, the generalized action must, of course, remain a scalar. Since it describes motion of a particle in the field, this scalar must be proportional to both – motion characteristics of the particle $\mathrm{d}s$ and field characteristic A.

The simplest possible invariant proportional to product of $\mathrm{d}s = (\mathrm{d}x^0, \mathrm{d}\mathbf{r})$ and $A = (A_0, \mathbf{A})$ is their dot (inner) product:

$$A\,\mathrm{d}s = A_j\,\mathrm{d}x^j = \Phi c\,\mathrm{d}t - \mathbf{A} \cdot \mathrm{d}\mathbf{r}. \tag{6.141}$$

The expression for the action of the system charge-field can be written as

$$S_{\mathrm{EM}} = -\frac{q}{c}\int_{\mathrm{M}}^{\mathrm{N}} A_j \, \mathrm{d}x^j = \frac{q}{c}\int_{\mathbf{r}_{\mathrm{M}}}^{\mathbf{r}_{\mathrm{N}}} \mathbf{A}\cdot \mathrm{d}\mathbf{r} - q\int_{t_{\mathrm{M}}}^{t_{\mathrm{N}}} \Phi \, \mathrm{d}t. \tag{6.142}$$

The components of the 4-potential A in Equation (6.142) are to be evaluated at the points of a world line s of the charge q.

Next we need to add the purely mechanical action of the charged particle itself (Section 5.7). We then obtain the full expression for the action of a charge in an EM field:

$$S = S_{\mathrm{Mech}} + S_{\mathrm{EM}} = -\int_{\mathrm{M}}^{\mathrm{N}} m_0 c \, \mathrm{d}s + \frac{q}{c}\int_{r_{\mathrm{M}}}^{r_{\mathrm{N}}} \mathbf{A}\cdot \mathrm{d}\mathbf{r} - q\int_{t_{\mathrm{M}}}^{t_{\mathrm{N}}} \Phi \, \mathrm{d}t. \tag{6.143}$$

Finally, by writing $\mathrm{d}\mathbf{r} = \mathbf{v}\mathrm{d}t$ and $\mathrm{d}s = c\mathrm{d}t/\gamma(v)$, we can collect all three terms on the right in one integral over time:

$$S = -\int_{t_{\mathrm{M}}}^{t_{\mathrm{N}}} \left(\frac{m_0 c^2}{\gamma(v)} + q\Phi - \frac{q}{c}\mathbf{A}\cdot\mathbf{v} \right) \mathrm{d}t. \tag{6.144}$$

The integrand in this expression is the Lagrange's function:

$$\Lambda = -\frac{m_0 c}{\gamma(v)} - q\Phi + \frac{q}{c}\mathbf{A}\cdot\mathbf{v}. \tag{6.145}$$

This expression differs from the purely mechanical Lagrangian (5.130) by an additional term

$$\Lambda_{\mathrm{EM}} = -q\Phi + \frac{q}{c}\mathbf{A}\cdot\mathbf{v} \tag{6.146}$$

describing particle's interaction with the field.

The vector derivative of Λ with respect to the components of the velocity $\mathbf{v}$ is the generalized momentum of the particle [29,34]. Denote it as $\mathbf{\Pi}$:

$$\mathbf{\Pi} = \frac{\partial \Lambda}{\partial v} = \vec{\nabla}_v \Lambda. \tag{6.147}$$

Taking the "velocity gradient" of the reciprocal Lorentz factor in (6.145) gives

$$\vec{\nabla}_v \gamma^{-1}(v) = \vec{\nabla}_v \sqrt{1 - \frac{v^2}{c^2}} = \gamma(v)\frac{\mathbf{v}}{c^2}. \tag{6.148}$$

Therefore,

$$\mathbf{\Pi} = m_0 \gamma(v)\mathbf{v} + \frac{q}{c}\mathbf{A} = \mathbf{P} + \frac{q}{c}\mathbf{A}. \tag{6.149}$$

We see that the generalized momentum consists of two contributions: familiar mechanical momentum $\mathbf{P} = m\mathbf{v}$ and the additional term $(q/c)\mathbf{A}$. The energy of the particle (the Hamilton function) is, just as in case (5.133),

$$H = \mathbf{v}\cdot \frac{\partial \Lambda}{\partial v} - \Lambda = \mathbf{v}\cdot \mathbf{\Pi} - \Lambda. \tag{6.150}$$

Putting here the previous expressions for Λ and Π gives

$$H = m_0\gamma(v)c^2 + q\Phi = mc^2 + q\Phi = \mathcal{E} + q\Phi. \tag{6.151}$$

This is precisely what one would expect: the energy of the particle in the field consists of two contributions: its relativistic energy mc^2 and the interaction energy in the electrical "scalar" potential field Φ. There is no contribution due to interaction with the magnetic field, since the latter does not do work on the charge.

The famous relation (5.153) between energy and momentum can now be expressed in terms of the Hamilton function, generalized momentum, and the 4-potential, if we put into it Equations (6.149) and (6.151):

$$(H - q\Phi)^2 - c^2\left(\mathbf{\Pi} - \frac{q}{c}\mathbf{A}\right)^2 = m_0^2c^4. \tag{6.152}$$

Solving this for H gives the relativistic expression for the Hamilton function in the EM field in terms of generalized momentum:

$$H = \sqrt{m_0^2c^4 + c^2\left(\mathbf{\Pi} - \frac{q}{c}\mathbf{A}\right)^2} + q\Phi. \tag{6.153}$$

Finally, we can write the Hamilton–Jacoby equation for a particle in the field. It is obtained by the replacement

$$\mathbf{\Pi} = \vec{\nabla} S, \qquad H = -\frac{\partial S}{\partial t}. \tag{6.154}$$

Then we obtain from (6.152)

$$\left(\frac{\partial S}{\partial t} + q\Phi\right)^2 - c^2\left(\vec{\nabla} S - \frac{q}{c}\mathbf{A}\right)^2 = m_0^2c^4. \tag{6.155}$$

Now we are in a position to identify the electric and magnetic fields in terms of 4-vector A. This will require some more mathematics.

Suppose we have a charge so small that its own field is negligible with respect to the external field. As in Section 5.7, we can obtain the equation of motion of such a charge in the given external electromagnetic field from the principle of the least action. The difference from Equations (5.140)–(5.144) used in Section 5.7 is that now the action must include both – mechanical and field contributions, that is, have the form (6.144). Acting as in Section 5.7, we obtain the Lagrange equation of motion:

$$\frac{\mathrm{d}}{\mathrm{d}t}\frac{\partial \Lambda}{\partial v} = \frac{\partial \Lambda}{\partial r}, \qquad \text{or} \qquad \frac{\mathrm{d}}{\mathrm{d}t}\vec{\nabla}_v\Lambda = \vec{\nabla}\Lambda. \tag{6.156}$$

But according to (6.145) and (6.147),

$$\frac{\partial \Lambda}{\partial v} = \mathbf{\Pi}, \qquad \text{and} \qquad \vec{\nabla}\Lambda = \frac{q}{c}\vec{\nabla}(\mathbf{A}\cdot\mathbf{v}) - q\vec{\nabla}\Phi. \tag{6.157}$$

Now, recall the known equation of vector calculus for the gradient of the dot product of two vectors:

$$\vec{\nabla}(\mathbf{A}\cdot\mathbf{v}) = (\mathbf{A}\cdot\vec{\nabla})\mathbf{v} + (\mathbf{v}\cdot\vec{\nabla})\mathbf{A} + \mathbf{A}\times(\vec{\nabla}\times\mathbf{v}) + \mathbf{v}\times(\vec{\nabla}\times\mathbf{A}). \tag{6.158}$$

Since variable vectors **r** and **v** are mutually independent, this reduces to

$$\vec{\nabla}(\mathbf{A}\cdot\mathbf{v}) = (\mathbf{v}\cdot\vec{\nabla})\mathbf{A} + \mathbf{v}\times(\vec{\nabla}\times\mathbf{A}). \tag{6.159}$$

Thus,

$$\vec{\nabla}\Lambda = \frac{q}{c}(\mathbf{v}\cdot\vec{\nabla})\mathbf{A} + \frac{q}{c}\mathbf{v}\times(\vec{\nabla}\times\mathbf{A}) - q\vec{\nabla}\Phi, \tag{6.160}$$

and the Lagrange equation takes the form

$$\frac{\mathrm{d}}{\mathrm{d}t}\left(\mathbf{P} + \frac{q}{c}\mathbf{A}\right) = \frac{q}{c}(\mathbf{v}\cdot\vec{\nabla})\mathbf{A} + \frac{q}{c}\mathbf{v}\times(\vec{\nabla}\times\mathbf{A}) - q\vec{\nabla}\Phi. \tag{6.161}$$

But the full time derivative

$$\frac{\mathrm{d}\mathbf{A}}{\mathrm{d}t} = \frac{\partial\mathbf{A}}{\partial t} + (\mathbf{v}\cdot\vec{\nabla})\mathbf{A} \tag{6.162}$$

and we finally obtain

$$\frac{\mathrm{d}\mathbf{P}}{\mathrm{d}t} = -\frac{q}{c}\frac{\partial\mathbf{A}}{\partial t} - q\vec{\nabla}\Phi + \frac{q}{c}\mathbf{v}\times(\vec{\nabla}\times\mathbf{A}). \tag{6.163}$$

Now let us read this equation. The time derivative of momentum on the left is the net force on the particle. We know from the thought experiments carried out together with Tom in Section 6.1 that the net force on the charged particle consists of two contributions: one due to the electric field and the other due to the magnetic field (Lorentz force law, Equation (6.12)). Comparing with that equation recovers in a more formal way already familiar relations (6.121) between the potentials and fields. Using these relations, we bring (6.163) to the form

$$\frac{\mathrm{d}\mathbf{P}}{\mathrm{d}t} = q\mathbf{E} + q(\mathbf{v}\times\mathbf{B}). \tag{6.164}$$

This is nothing else but the equation of motion of a charged particle under the Lorentz force (6.12).

Now, using $\mathcal{E} = \sqrt{m_0^2c^4 + p^2c^2}$ from (5.24), it is easy to get

$$\frac{\mathrm{d}\mathcal{E}}{\mathrm{d}t} = \mathbf{v}\cdot\frac{\mathrm{d}\mathbf{p}}{\mathrm{d}t}. \tag{6.165}$$

Putting here $\mathrm{d}\mathbf{p}/\mathrm{d}t$ from (6.164), and noting that $(\mathbf{v}\times\mathbf{B})\cdot\mathbf{v} = 0$, we obtain

$$\frac{\mathrm{d}\mathcal{E}}{\mathrm{d}t} = q(\mathbf{E}\cdot\mathbf{v}). \tag{6.166}$$

This is essentially work–energy theorem: change in the relativistic energy of a charged particle is equal to work of the electric force on the particle.

Thus, in a bigger picture that we have obtained, the already familiar results follow from the more general principle of the least relativistic action. Note that, as we had found earlier, the electric field is determined by both – electric *and* magnetic potentials. Looking back at Figure 6.7, we see that it does not show the contribution from the magnetic field, so no surprise it contradicts the correct picture shown in Figure 6.6. Now, that we have the exact equation for both contributions into **E**, we can see how this works.

We start with writing the potentials in K′:

$$\Phi' = \frac{q'}{4\pi\varepsilon_0 r'}, \qquad \mathbf{A}' = 0. \tag{6.167}$$

In K, we get the corresponding potentials automatically, applying general transformation rules for 4-vectors:

$$\begin{aligned} \Phi &= \gamma(V)\left(\Phi' + \frac{V}{c}A'_x\right) = \gamma(V)\Phi', \\ A_x &= \gamma(V)\left(A'_x + \frac{V}{c}\Phi'\right) = \frac{V}{c}\gamma(V)\Phi', \\ A_y &= A_z = 0. \end{aligned} \tag{6.168}$$

Putting (6.167) into this equation gives the expression for Φ:

$$\Phi(\mathbf{r},t) = \gamma(V)\frac{q'}{4\pi\varepsilon_0\sqrt{x'^2 + r'^2_\perp}}, \quad \text{where} \quad r'^2_\perp = y'^2 + z'^2. \tag{6.169}$$

The expression for **A** will be different only in the coefficient, so we can restrict calculation to Φ only and then use (6.168).

We need to find Φ as a function of coordinates of *system K*. Therefore we have to express the primed coordinates in the above equation in terms of **r**, *t*. Using the inverse Lorentz transformation, we obtain

$$\Phi(r,t) = \gamma(V)\frac{q'}{4\pi\varepsilon_0\sqrt{\gamma^2(V)(x - Vt)^2 + r^2_\perp}}, \qquad \mathbf{A} = \left(\frac{V}{c}\Phi, 0, 0\right). \tag{6.170}$$

Now we have to take differential operations

$$\begin{aligned} -\vec{\nabla}\Phi &= \gamma(V)\frac{q'}{4\pi\varepsilon_0}\frac{\gamma^2(V)(x - Vt)\hat{\mathbf{x}} + \mathbf{r}_\perp}{\left[\gamma^2(V)(x - Vt)^2 + r^2_\perp\right]^{3/2}}, \\ \frac{\partial \mathbf{A}}{\partial t} &= \frac{V^2}{c^2}\gamma(V)\frac{q'}{4\pi\varepsilon_0}\frac{\gamma^2(V)(x - Vt)\hat{\mathbf{x}}}{\left[\gamma^2(V)(x - Vt)^2 + r^2_\perp\right]^{3/2}}. \end{aligned} \tag{6.171}$$

Note that the contribution from the magnetic vector potential has only the *x*-component and is opposite in sign to the contribution from Φ. This takes the resulting electric field from that depicted in Figure 6.7 to that shown in Figure 6.5. The resulting analytical expression is

$$\mathbf{E} = \gamma(V)\frac{q'}{4\pi\varepsilon_0}\frac{(x - Vt)\hat{\mathbf{x}} + \mathbf{r}_\perp}{\left[\gamma^2(V)(x - Vt)^2 + r^2_\perp\right]^{3/2}}. \tag{6.172}$$

Similarly, taking the curl of **A** gives the correct expression for the magnetic field

$$\mathbf{B} = \mathbf{B}_\perp = \frac{V}{c^2}(\hat{\mathbf{x}} \times \mathbf{E}). \tag{6.173}$$

The magnetic field of the charge is everywhere perpendicular to the direction of its velocity and to the position vector of the field point.

Of course, we could obtain the same results more easily by performing the Lorentz transformation directly from the known field in K′ to K. But it is not always that we know the field in K′, to begin with. And what if the charge does not move uniformly? In this case, the system comoving with the charge will be different at different moments of time. Accordingly, the velocity **V** in Equations (6.170) and (6.171) will be also a variable depending on *t*. As a result, in the time derivative of **A** there will appear the term $\dot{\mathbf{V}}$, that is, the acceleration of the charge. Thus, the electric field of a charge generally depends on its instantaneous acceleration, whereas the potential depends only on the instantaneous velocity. The field is no longer described by Coulomb's law. So we generally do not know the initial field in K′ to begin with. The same is true for the potential, but the latter is much easier to find. All four components of the potential are the solutions of the D'Alambert equation (6.134) with given charge and current distribution. The solution of these equations is described in detail in many courses on EM, see, for example, [29,44,45].

In the remaining part of this section, we will make some more comment on the gauge invariance.

If we know Φ and **A**, we can determine the physical fields **E** and **B** that determine the experimentally measurable forces on a particle. The opposite, however, is not true. Knowing the fields **E** and **B** does not determine uniquely the 4-potential *A*. There is an infinite set of different 4-potentials $A, A', A'', \ldots$, all producing the same physical EM field **E**, **B** via operation (6.121). We can show it in the following way. Suppose that in Equation (6.141) for the electromagnetic part of the action we change the 4-vector *A* to

$$A_j' = A_j - \frac{\partial \chi}{\partial x^j}. \qquad (6.174)$$

Here $\chi(x^j)$ is an arbitrary scalar function of time and position in Minkowski's space-time and the added term is the 4-gradient of this function. The line integral of any gradient depends only on the end points of the integration line. The corresponding new action will be

$$S \Rightarrow S' = \int_{\mathrm{M}}^{\mathrm{N}} \left(-m_0 c\mathrm{d}s - \frac{q}{c} A_j \mathrm{d}x^j + \frac{q}{c}\mathrm{d}\chi \right) = S + \frac{q}{c}[\chi(\mathrm{N}) - \chi(\mathrm{M})]. \qquad (6.175)$$

It is different from *S* by a function depending only on points M and N and independent of the shape of the path connecting these points. Since this function is constant with respect to variation of the path between the fixed end points M and N, its presence in the expression (6.175) does not affect the equations of motion. We see that two 4-vectors *A* and *A′* differing by the 4-gradient of an arbitrary scalar function in space-time correspond to the same physical electromagnetic field. You can check it directly by writing *A* as $(\Phi, \mathbf{A})$ and x^j as $(ct, \mathbf{r})$ so that transformation (6.174) takes the form familiar from Section 6.6:

$$\Phi' = \Phi - \frac{1}{c}\frac{\partial \chi}{\partial t}, \qquad \mathbf{A}' = \mathbf{A} + \vec{\nabla}\chi. \qquad (6.175\text{a})$$

If you now put the new potentials Φ', $\mathbf{A}'$ instead of Φ, $\mathbf{A}$ into Equation (6.121), you will get the same field $\mathbf{E}' = \mathbf{E}$, $\mathbf{B}' = \mathbf{B}$.

Only those characteristics of the EM field, which are invariant with respect to transformation (6.174) have the physical meaning (i.e., can result from a clearly specified and executable experimental procedure). This property is called the gauge invariance. It is a consequence of the fact that the EM field is uniquely described by a 4-vector A satisfying the D'Alambert equation.

6.8 Electrodynamics in Tensor Notations

We start this section by taking a closer look at the relations (6.121). Let us write them down in components, using x^0, x^1, x^2, x^3 for ct, x, y, z, respectively:

$$E_1 = \frac{\partial A_0}{\partial x^1} - \frac{\partial A_1}{\partial x^0}, \quad E_2 = \frac{\partial A_0}{\partial x^2} - \frac{\partial A_2}{\partial x^0}, \quad E_3 = \frac{\partial A_0}{\partial x^3} - \frac{\partial A_3}{\partial x^0}$$

and similar for $\mathbf{B}$:

$$B_1 = \frac{\partial A_3}{\partial x^2} - \frac{\partial A_2}{\partial x^3}, \quad B_2 = \frac{\partial A_1}{\partial x^3} - \frac{\partial A_3}{\partial x^1}, \quad B_3 = \frac{\partial A_2}{\partial x^1} - \frac{\partial A_1}{\partial x^2}. \tag{6.176}$$

The apparent contradiction with (6.121) (the sign difference between the zeroth component of A there and here) is explained by noticing that here we use the *covariant* components of the 4-vector A: $A_l = (\Phi/c, -\mathbf{A})$. Already here the attentive reader may see an emerging symmetry between $\mathbf{E}$ and $\mathbf{B}$, which is not immediately seen in the original 3D equation (5.104). To reveal this symmetry in full, note that all expressions on the right in (5.155) are the derivatives of the covariant components of A with respect to the contravariant components of a 4-displacement s. We know from Section 4.7 that such derivatives form the components of a covariant tensor of the second rank. In our case, the derivatives are combined so as to make the tensor antisymmetric. We can now introduce explicitly this tensor and the corresponding notation for it:

$$F_{kl} = \frac{\partial A_l}{\partial x^k} - \frac{\partial A_k}{\partial x^l}. \tag{6.177}$$

This is an electromagnetic field tensor. By definition, all its diagonal elements are zero. The physical meaning of each component of the tensor follows directly from its definition and relations (6.176). We can write the result as a matrix in which index k numbers the rows and l numbers the columns:

$$F_{kl} = \begin{pmatrix} 0 & E_x & E_y & E_z \\ -E_x & 0 & -B_z & B_y \\ -E_y & B_z & 0 & -B_x \\ -E_z & B_y & B_x & 0 \end{pmatrix}. \tag{6.178}$$

Just as time and position $(ct, \mathbf{r})$, or energy and momentum $(\varepsilon/c, \mathbf{P})$, or charge and current density $(\rho c, \mathbf{j})$, or the electric and magnetic potential $(\Phi/c, \mathbf{A})$ form their respective 4-vectors s, P, j, A, the electric and magnetic field vectors are merely the components of a single and more general entity – the electromagnetic field tensor:

$$(\mathbf{E}, \mathbf{B}) = F_{kl}. \tag{6.179}$$

Similarly, we can define the contravariant tensor $F^{kl} = (-\mathbf{E}, \mathbf{B})$ and determine the corresponding matrix form (Problem 6.17).

Let us express the electric and magnetic field in Maxwell's equations in terms of the F-tensor. Using (6.178), we obtain after some rearranging the tensor equations for F in a very elegant form:

$$e^{iklm} \frac{\partial F_{lm}}{\partial x_k} = 0, \qquad \frac{\partial F^{ik}}{\partial x^k} = -j^i. \tag{6.180}$$

Here e^{iklm} is completely antisymmetric 4-rank unit tensor. The reader can also check by direct inspection that Equations (6.180) are equivalent to the first and the second pair of Maxwell's equations. But the form (6.180), apart from being more elegant, explicitly shows that all electrodynamics is described by Lorentz-invariant tensor equations in space-time.

It may be instructive to obtain some invariant characteristics of the EM field and equations for a particle in this field from the least-action principle. Actually, this will be the extension of the previous section made in 4-tensor notations. Using the expression (6.144), we can write the principle of the least action in these notations written as

$$\delta S = -\delta \int_{\mathrm{M}}^{\mathrm{N}} \left(m_0 c \mathrm{d}s + \frac{q}{c} A_l \mathrm{d}x^l \right) = 0. \tag{6.181}$$

Since $\mathrm{d}s = \sqrt{\mathrm{d}x_l \mathrm{d}x^l}$, the variation is

$$\delta S = -\int_{\mathrm{M}}^{\mathrm{N}} m_0 c \mathrm{d}x_l \frac{\mathrm{d}\delta x^l}{\mathrm{d}s} + \frac{q}{c} A_l \mathrm{d}\delta x^l + \frac{q}{c} \delta A_l \mathrm{d}x^l = 0. \tag{6.182}$$

The first term in the integrand can be written as $m_0 c u_l \mathrm{d}\delta x^l$, where u_l are the covariant components of 4-velocity. Then the first two terms here can be integrated by parts:

$$-\left(m_0 c u_l + \frac{q}{c} A_l \right) \delta x^l \Big|_{\mathrm{M}}^{\mathrm{N}} + \int_{\mathrm{M}}^{\mathrm{N}} \left(m_0 c \mathrm{d}u_l \delta x^l + \frac{q}{c} \delta x^l \mathrm{d}A_l - \frac{q}{c} \delta A_l \mathrm{d}x^l \right) = 0. \tag{6.183}$$

The first term here is zero since the variations at points M and N are zero. Next, we write

$$\delta A_l = \frac{\partial A_l}{\partial x^k} \delta x^k, \qquad \mathrm{d}A_l = \frac{\partial A_l}{\partial x^k} \mathrm{d}x^k, \tag{6.184}$$

so that

$$\int_M^N \left(m_0 c \mathrm{d}u_l \delta x^l + \frac{q}{c} \frac{\partial A_l}{\partial x^k} \delta x^l \mathrm{d}x^k - \frac{q}{c} \frac{\partial A_l}{\partial x^k} \mathrm{d}x^l \delta x^k \right) = 0. \tag{6.185}$$

Writing $\mathrm{d}u_l = (\mathrm{d}u_l/\mathrm{d}s)\mathrm{d}s$, $\mathrm{d}x^l = u^l \mathrm{d}s$, and interchanging the dumb indexes l and k in the third term, we can bring this to the form

$$\int_M^N \left[m_0 c \frac{\mathrm{d}u_l}{\mathrm{d}s} - \frac{q}{c} \left(\frac{\partial A_k}{\partial x^l} - \frac{\partial A_l}{\partial x^k} \right) u^k \right] \delta x^l \mathrm{d}s = 0. \tag{6.186}$$

Condition (6.186) must hold for arbitrary variations δx^l. This can only be the case if the integrand is identically zero:

$$m_0 c \frac{\mathrm{d}u_l}{\mathrm{d}s} - \frac{q}{c} \left(\frac{\partial A_k}{\partial x^l} - \frac{\partial A_l}{\partial x^k} \right) u^k = 0. \tag{6.187}$$

Here we recognize in parentheses our newly introduced entity – the electromagnetic field tensor.

Using it, we can rewrite the equation in the form

$$m_0 c \frac{\mathrm{d}u_k}{\mathrm{d}s} = \frac{q}{c} F_{kl} u^l. \tag{6.188}$$

On the left-hand side, we see the product of the particle's rest mass and 4-acceleration, or, which is the same, the derivative of the 4-momentum with respect to s (Minkowski's 4-force). Therefore, the expression on the right must also be the force. Indeed, substituting the components F_{kl} with their expressions (6.178) in terms of the electric and magnetic fields, we will see that it is 4-vector of the Lorentz force. Thus, Equation (6.188) is nothing else but the equation of motion of a charged particle in the EM field in the 4D form. And getting back to 3D notations, we see that three spatial components of (6.188) ($k = 1,2,3$) are identical to the vector equations of motion and the temporal component ($k = 0$) is essentially the work–energy theorem known from mechanics.

Now, what happens if we consider in the variation δS only the actual trajectories? In this case the action S becomes the *actualized* least action, that is, the actual action of a system considered only as a function of the upper limit of the particle's actual trajectory. The second term in (6.185) becomes zero, and the first term, in which the upper limit is now considered as a variable, gives the differential of action as a function of coordinates. In other words, the variation of S becomes the differential of S and variations of the coordinates become the differentials of coordinates:

$$\delta S \rightarrow \mathrm{d}S, \qquad \delta x \rightarrow \mathrm{d}x, \tag{6.189}$$

and Equation (6.183) transforms into

$$\mathrm{d}S = -(m_0 c u_l + q A_l) \mathrm{d}x^l. \tag{6.190}$$

It follows

$$-\frac{\partial S}{\partial x^l} = m_0 c u_l + q A_l = p_l + q A_l. \tag{6.191}$$

But the 4-vector on the left is the 4-vector of the generalized 4-momentum $\tilde{P}$ of a particle (the momentum composed of the regular 4-momentum p and the 4-potential A). Putting into (6.191) the corresponding expressions and switching to the contravariant components, we obtain

$$\tilde{P}^l = \left(\left(\frac{\mathcal{E}_r}{c} + \frac{q}{c}\Phi\right),\quad \mathbf{p} + \frac{q}{c}\mathbf{A}\right), \tag{6.192}$$

where $\mathcal{E}_r$ is the relativistic energy of the particle. Using (6.149) and (6.151), we can write this as

$$\tilde{P}^l = \left(\frac{\mathcal{E}}{c},\ \tilde{\mathbf{P}}\right), \tag{6.193}$$

where $\mathcal{E}$ is already the total energy of the particle, including its interaction energy with the field.

As one would expect, the temporal component of $\tilde{P}$ is the total energy of the particle and the spatial part is its generalized 3-momentum.

You remember, we stated in Chapter 1 that an important aspect of the relativity theory is revealing the absolute (invariant) characteristics of physical systems. One of the advantages of tensor notations is that they allow us to easily find such characteristics (relativistic invariants). Thus, for a quantity represented by a 4-vector Q^l (recall that a vector is a first rank tensor), the corresponding relativistic invariant is the dot product $\mathrm{I} = Q_l Q^l = Q_0^2 - \mathbf{Q}^2$. Applying this to, say, 4-momentum, we obtain the rest mass of a particle as its relativistic invariant.

Here we can extend this to finding the *field* invariants using the field's representation by the tensor F_{kl}. We can compose two invariant quantities from the components of this tensor

$$I_1 = F_{kl}F^{kl}, \qquad I_2 = e_{iklm}F^{ik}F^{lm}. \tag{6.194}$$

Here e_{iklm} is the perfectly antisymmetric unit tensor (see Section 4.7). Note that only the first of these characteristics forms a true scalar, whereas the second one is a pseudoscalar. Switching to less elegant but more direct three-dimensional form by using (6.178), we obtain the two invariants in terms of the field vectors:

$$I_1 = E^2 - c^2B^2, \qquad I_2 = \mathbf{E}\cdot\mathbf{B}. \tag{6.195}$$

The pseudoscalar nature of the second invariant is immediately seen, since the magnetic field is a pseudovector. Also, if $E = cB$ in one reference frame, this holds in all other reference frames; and if the electric and magnetic fields are mutually perpendicular at a certain point, they remain so in all reference frames.

All these results follow from the least-action principle (6.181) and (6.182) for a particle in the electromagnetic field. This action contains a purely mechanical part S_m describing a free particle and the interaction part S_{mf}, describing the particle's interaction with the field. But the full action must also include the third part S_f, representing the field itself. Equation (6.195) suggests a simple way to find this part, thus completing the mathematical description of the electromagnetic phenomena.

Namely, since the action is a scalar, it must be the 4-integral of a scalar function (the Lagrangian density). The quantity I_1 is such a scalar! Also, we know from mechanics that the Lagrange function is the *difference* between the kinetic and potential energy of the system. By analogy, we can assume that the Lagrange function of the electromagnetic field considered as a system must be also such difference. The quantity I_1 is, indeed, the difference of the two terms.

Thus, we can write

$$S_f = \int F_{kl} F^{kl} d\Omega = \int (E^2 - c^2 B^2) d\Omega, \tag{6.196}$$

where $d\Omega = dV dt$ is the volume element of space-time. This, in turn, suggests that we can interpret the two terms in the integrand as the energy densities due to the electric and magnetic parts of the electromagnetic field. The *total* energy density of the EM field must be the *sum* of these terms. Indeed, as we know from the electromagnetic theory, the total energy density is

$$\eta_{EM} = \frac{1}{2}\left(\varepsilon_0 E^2 + \frac{1}{\mu_0} B^2\right) = \frac{1}{2}\varepsilon_0 (\mathbf{E}^2 + c^2 \mathbf{B}^2). \tag{6.197}$$

If we take the time derivative of $\eta_{EM}$and use Maxwell's equation in the vector form (6.139) and (6.140), we will obtain

$$\frac{\partial \eta_{EM}}{\partial t} = \varepsilon_0 \left(\mathbf{E}\frac{\partial E}{\partial t} + c^2 \mathbf{B}\frac{\partial \mathbf{B}}{\partial t}\right) = -\mathbf{j}\cdot\mathbf{E} + \mathbf{E}\cdot(\vec{\nabla}\times\mathbf{B}) - \mathbf{B}\cdot(\vec{\nabla}\times\mathbf{E}). \tag{6.198}$$

But, from the vector calculus (see, e.g., [44,45]), we get

$$\mathbf{E}\cdot(\vec{\nabla}\times\mathbf{B}) - \mathbf{B}\cdot(\vec{\nabla}\times\mathbf{E}) = -\vec{\nabla}\cdot(\mathbf{E}\times\mathbf{B}). \tag{6.199}$$

Therefore,

$$\frac{\partial \eta_{EM}}{\partial t} = -\mathbf{j}\cdot\mathbf{E} - \vec{\nabla}\cdot\mathbf{S}, \tag{6.200}$$

where

$$\mathbf{S} = \frac{1}{\mu_0}(\mathbf{E}\times\mathbf{B}). \tag{6.201}$$

Equation (6.200) has clear physical interpretation. The left side is the rate of change (per unit volume) of the field energy; the term $\mathbf{j}\cdot\mathbf{E}$ is work per unit volume per unit time done by the field on the charge (power density),

$$\mathbf{j}\cdot\mathbf{E} = \frac{\partial \mathcal{E}_r}{\partial t}, \tag{6.202}$$

that is, the rate of change of the relativistic energy of the particles producing current. Therefore, the vector $\mathbf{S}$ is the flux density of the EM energy. This vector is called the Poynting vector.

Thus, Equation (6.200) expresses the conservation of the total energy of the whole system particles + field.

6.9
Running Waves

Let there be light . . .

Genesis

Here we consider a special but very important case of EM field far from its sources. This means that the corresponding region of space-time does not include charges or currents. It is in this domain where we can discuss and fully appreciate the second aspect of the minus sign in Faraday's law. The general equations (6.104) for the field or (6.133) and (6.134) for the 4-potential reduce to homogeneous equations

$$\nabla^2 \mathbf{E} - \frac{1}{c^2}\frac{\partial^2 \mathbf{E}}{\partial t^2} = 0, \tag{6.203}$$

and similarly for vector **B** or the 4-vector A. This is an example of a *wave equation* in Cartesian coordinates,

$$\nabla^2 \Psi - \frac{1}{u^2}\frac{\partial^2 \Psi}{\partial t^2} = 0, \tag{6.204}$$

where Ψ stands for an arbitrary physical quantity characterizing a given process. It may be, for example, local (and instant) air pressure in an acoustic wave or a local displacement from the equilibrium position in a wave on a string; u is the speed of the corresponding wave, which will be confirmed when we find a solution of the equation.

Tracking down the origin of this equation, we find it in the sign difference between the time derivatives of the electric and magnetic field, that is, in the minus sign in the Faraday's law. Without this minus, the resulting equation for the field in vacuum would be

$$\nabla^2 \mathbf{E} + \frac{1}{c^2}\frac{\partial^2 \mathbf{E}}{\partial t^2} = 0. \tag{6.205}$$

Although looking similarly to the D'Alambert equation, Equation (6.205) is fundamentally different from it and its solutions would describe anything but waves. Since light is electromagnetic waves, we see that, mathematically, the minus sign in the Faraday's law is, apart from its other important function considered before, also ensuring the existence of light.

Now let us go back to the *wave* Equation (6.204) and find its general solutions. To this end, consider the most simple case when the wave function Ψ depends only on one spatial coordinate, say, x (and, of course, on time t). The corresponding wave is called a plane wave, since its wave front (or the surface of a constant perturbation) is a plane perpendicular to the propagation direction x. Equation (6.204) simplifies to

$$\frac{\partial^2 \Psi}{\partial x^2} - \frac{1}{u^2}\frac{\partial^2 \Psi}{\partial t^2} = 0. \tag{6.206}$$

To find a solution, we first rewrite (6.206) as

$$\left(\frac{\partial^2}{\partial x^2} - \frac{1}{u^2}\frac{\partial^2}{\partial t^2}\right)\Psi = 0, \tag{6.207}$$

that is, consider the left-hand side as the result of action of the reduced D'Alambertian operator on function $\Psi(x, t)$.

One of the advantages of the concept of operator is that in many situations it can be treated as a regular expression (e.g., function of variables). In our case, we can formally represent our operator in the form

$$\left(\frac{\partial}{\partial x} + \frac{1}{u}\frac{\partial}{\partial t}\right)\left(\frac{\partial}{\partial x} - \frac{1}{u}\frac{\partial}{\partial t}\right)\Psi = 0. \tag{6.208}$$

The structure of this equation suggests introducing the new variables:

$$\eta = x + ut, \qquad \xi = x - ut. \tag{6.209}$$

In these variables, the equation takes the form

$$\frac{\partial^2 \Psi}{\partial \eta \, \partial \xi} = 0, \tag{6.210}$$

which can be easily solved. Indeed, it follows immediately from (6.210) that

$$\frac{\partial \Psi}{\partial \xi} = F(\xi), \tag{6.211}$$

where $F(\xi)$ is an arbitrary function depending only on ξ (and thereby constant with respect to η). Integrating (6.211) over ξ yields

$$\Psi(\eta, \xi) = \int F(\xi) d\xi + C(\eta). \tag{6.212}$$

The second term on the right is a constant with respect to ξ (which does not preclude it to be an arbitrary function of η). Denoting the integral as $\Psi_1(\xi)$, and $C(\eta)$ as $\Psi_2(\eta)$, and returning to the original variables, we obtain

$$\Psi(x, t) = \Psi_1(x - ut) + \Psi_2(x + ut). \tag{6.213}$$

The general solution of the wave equation is the sum of two contributions. The first contribution describes a wave of an arbitrary shape running in the positive x-direction; the second one represents a wave traveling in the opposite direction. Makes perfect sense – both directions are equally possible and, accordingly, both are represented in the general solution. Each wave in (6.213) depends on its respective combination of variables $(x - ut)$ or $(x + ut)$. If variables x and t change in such a way that these combinations remain constant, then the respective perturbation Ψ also remains constant. Mathematically: if $x - ut = x_1 = \text{const}_1$, that is, $x = x_1 + ut$, then the corresponding perturbation Ψ_1 is running along x with the speed u. The same is true for the second wave traveling in the opposite direction. Thus u is, indeed, the propagation speed of the wave in a given system.

I want to emphasize an important point here. We know that any shape can be considered as (or represented by) a superposition of harmonic waves of different frequencies. Each such harmonic can be visualized as a graph of an uninterrupted sine or cosine function of x, running along the x-direction. The resulting shape will preserve its initial form $\Psi_{1,2}(x, 0)$ for all later moments $t > 0$, if the constituent harmonics all run at the same speed. In other words, the above simple theory works only if the propagation speed does not depend on frequency. This is a very strict requirement, which usually does not hold. But it holds for light in vacuum. In this case any optical perturbation of any shape propagates through space as a "rigid" object, that is, retaining its original shape.

Since the monochromatic waves can be considered as the simple building blocks of any function, it is natural to consider the properties of monochromatic electromagnetic waves. In other words, we now want to consider the basic properties of the monochromatic solutions of Equation (6.205). This can be done in two steps. First, we can simply consider the special case when the functions Ψ_1 and Ψ_2 of the general solution (6.213) are monochromatic, and represent them in the exponential form

$$\Psi_{1,2}(x \pm ut) = \Psi_{1,2} e^{ik(x \pm ut)} = \Psi_{1,2} e^{i(kx \pm \omega t)} \tag{6.214}$$

(here we use $u = \omega/k$). Then we generalize this onto the three-dimensions of space ($x \to \mathbf{r}$, $k = k_x \to \mathbf{k}$) and replace a scalar potential with the, say, electric field vector $\mathbf{E}$, since both satisfy the same D'Alambert's equation:

$$\mathbf{E}_{1,2}(\mathbf{r} \pm \mathbf{u}t) = \mathbf{E}_{1,2} e^{i(k \cdot r \pm \omega t)}. \tag{6.215}$$

Here we in addition require that $u = c$ for all frequencies. This corresponds to the wave Equation (6.205). Putting there one of our harmonic solutions (6.215), we get

$$k^2 = k_x^2 + k_y^2 + k_z^2 = \frac{\omega^2}{c^2}. \tag{6.216}$$

This is known as the special case of the dispersion relation – for light in vacuum.

This is all that we can infer from the D'Alambert equation. We have lost part of the information when we decoupled the Maxwell's equations. To recover all information stored there, let us go back to these equations and see what they tell us in monochromatic case. We use the known (easily proved) properties, that if $\mathbf{E} = \mathbf{E}_0 \exp(\mathbf{kr} \pm \omega t)$, then

$$\vec{\nabla} \cdot \mathbf{E} = i\mathbf{k} \cdot \mathbf{E}, \qquad \vec{\nabla} \times \mathbf{E} = i\mathbf{k} \times \mathbf{E} \tag{6.217}$$

and similar for $\mathbf{B}$. Then the Maxwell's equations for field in vacuum tell us that

$$\mathbf{k} \cdot \mathbf{E} = 0, \qquad \mathbf{k} \cdot \mathbf{B} = 0; \qquad \mathbf{k} \times \mathbf{E} = -\omega \mathbf{B}, \qquad \mathbf{k} \times \mathbf{B} = \frac{1}{c^2} \omega \mathbf{E}. \tag{6.218}$$

It follows immediately that $E = cB$ and the vectors $\mathbf{k}$, $\mathbf{E}$, $\mathbf{B}$ form the right triplet of the mutually orthogonal vectors (Figure 6.16), and we also recover Equation (6.216).

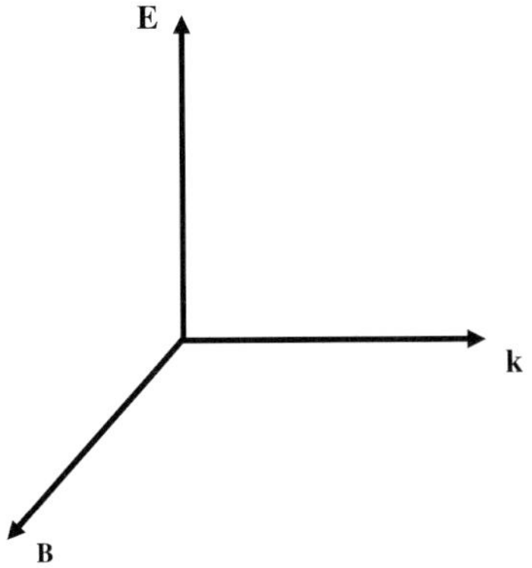

Figure 6.16

6.10 The Relativistic Doppler Effect I

We found the equations describing the running EM waves (light) in the form of the plane waves. One of the basic characteristics of a wave is its frequency and also its propagation vector **k**. How do these characteristics change under the Lorentz transformations?

One of the ways to the answer is to note that both characteristics together form a 4-vector. Indeed, the expression (phase)

$$\Phi \equiv \omega t - \mathbf{k} \cdot \mathbf{r} \tag{6.219}$$

in Equations (6.214) and (6.215) of the previous section is a scalar function of a point in space-time. The point itself specifies a 4-position-vector (4-displacement) $s = (ct, \mathbf{r})$. This shows that combination

$$k \equiv \left(\frac{\omega}{c}, \mathbf{k}\right) \tag{6.220}$$

has to be a 4-vector as well, with components[5)]

$$k^0 = \frac{\omega}{c}, \qquad k^1 = k_x, \qquad k^2 = k_y, \qquad k^3 = k_z. \tag{6.221}$$

The scalar phase function Φ is the dot product of s and k:

$$\Phi = ks = k_j s^j = \omega t - \mathbf{k} \cdot \mathbf{r}. \tag{6.222}$$

Once we know that k is a 4-vector, we can immediately write down the transformation law for it:

$$\omega = \gamma(V)(\omega' + Vk'_x), \quad k_x = \gamma(V)\left(k'_x + \frac{V}{c^2}\omega'\right), \quad k_y = k'_y, \quad k_z = k'_z. \tag{6.223}$$

5) According to (6.216), this vector has the zero norm. In other words, it is the null vector.

Denote as θ the angle between the directions **k** and **V** and θ' the angle between **k**$'$ and **V**. Then we have

$$k_x = k\cos\theta = \frac{\omega}{c}\cos\theta, \qquad k'_x = k'\cos\theta' = \frac{\omega'}{c}\cos\theta' \tag{6.224}$$

and the transformation (6.223) can be written as

$$\omega = \omega'\gamma(V)\left(1 + \frac{V}{c}\cos\theta'\right). \tag{6.225}$$

This answers the posed question. The answer depends on angle between the radiation direction and relative velocity **V**.

Suppose that the primed system K$'$ is the rest frame of the source of light and K is the rest frame of the detector of this light. Then ω' is the proper frequency ω_0 of the source and **V** is the source's velocity in K. In practical applications, it is more convenient to express the transformed frequency ω observed in K (i.e., the frequency recorded by the detector) in terms of the angle θ rather than θ'. To this end, we use the second Equation (6.223) in the form

$$\omega\cos\theta = \gamma(V)\,\omega'\left(\cos\theta' + \frac{V}{c}\right). \tag{6.226}$$

Dividing the last two equations one by another gives

$$\cos\theta = \frac{\cos\theta' + (V/c)}{1 + (V/c)\cos\theta'}, \qquad \text{or} \qquad \cos\theta' = \frac{\cos\theta - (V/c)}{1 - (V/c)\cos\theta}. \tag{6.227}$$

The reader can recognize in these expressions the familiar result (3.16) obtained from transformation of velocities.

Finally, putting the last expression into (6.225) and redenoting $\omega' \to \omega_0$ (commonly used notation for the proper frequency of the source), we obtain after simple manipulations

$$\omega = \frac{\omega_0}{\gamma(V)(1 - (V/c)\cos\theta)}. \tag{6.228}$$

This is a general expression describing the Doppler effect (the frequency shift due to relative motion of the source and detector) as a function of V and θ.

6.11
The Origin of Light

We said in the beginning of this chapter that the electrical charges and their motion (currents) are the ultimate source of any electromagnetic field. Therefore even the field of a running electromagnetic wave, apparently existing on its own, must have its source. We want now to trace down and explore the nature of this source. Since the electromagnetic waves constitute light, we can pose the question: what is the origin of light?

It cannot be just a stationary charge – this is only the source of the radial Coulomb's field attached to it. Nor is it even a uniformly moving charge: we know that the latter produces only the radial electric field (flattened in the longitudinal direction) and the circumferential magnetic field. Both fields appear to be "rigidly" attached to the charge (i.e., they move together with it) and fall off in magnitude inversely proportional to the square of the distance from it.

Thus, the only way the light can be produced is by *accelerating* charge.

Let us look into the mechanism of this process, first on a qualitative level [3,51]. We will utilize the concept of the field lines and use the two known facts:

(1) The field lines are continuous unless interrupted by a point charge.
(2) Any field perturbation propagates from its source through space with the invariant speed c.

Now consider a point charge which remains stationary until the moment $t=0$. Up to this moment there is only Coulomb's field around the charge. At the moment $t=0$, the charge is "kicked" by an instant force lasting for a negligibly short time interval $\Delta t \to 0$ (say, a collision with another particle) accelerating it to a certain velocity v along the x-direction. In order for this force to do so, it must have the finite impulse, that is, it must be very large at the time of its action:

$$f = \frac{\Delta P}{\Delta t} \tag{6.229}$$

(here ΔP is the corresponding momentum transferred to the charged particle). Very soon after the zero moment the particle is free again and moves with a constant speed v acquired in the collision. Accordingly, it is now surrounded by another field – that of a *moving* charge (Figure 6.18). Initially this new field is formed only in the immediate vicinity of the moving charge, but as time goes on, it spreads outward from the origin with the speed c. So at a moment $t > 0$, the whole space consists of two different fields: outside the sphere of radius $r = ct$ we have the original Coulomb's field of the stationary charge, centered around the origin, and within this sphere the field of the moving charge centered around its instant position $x = vt$. The second part of this sentence means that the electric field lines of the corresponding field remain radial, but they are diverging from (or converging at) the actual instant position of the charge. Also, they are concentrated toward the equatorial plane perpendicular to the x-direction and moving together with the charge. In addition, the sphere is filled with the circumferential magnetic field around the x-axis. Note that, *from the viewpoint of electrostatics alone*, the exterior Coulomb's field is now the "wrong" field, since its field lines remain radial with respect to the origin, where there is no charge anymore. But the far field does not know this, since the information about the collision has not yet reached it by the moment t, so it is the right field from the viewpoint of electrodynamics. The exterior field is separated from the interior one by a spherical surface (more accurately, a spherical layer of negligibly small average thickness $\delta r = c\Delta t$) expanding with the speed of light.

What happens on the surface itself? The answer comes from the requirement that the field lines be continuous. For the magnetic field, this requirement is satisfied automatically, since there is no magnetic field outside the sphere and the one within

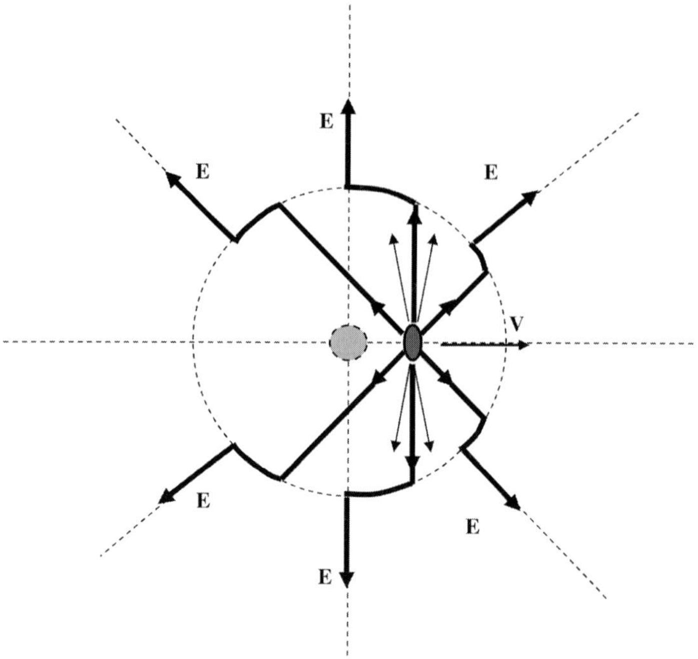

Figure 6.17 A simplified diagram showing formation of *kinks* in the electric field lines due to acceleration of the source. Selected lines are highlighted in bold. Shown in the center is the initial state of the charge (stationary, $t = 0$). At this moment the charge is subject to an infinite force, which instantly accelerates it to a speed V. At $t > 0$, the charge is moving uniformly at this speed. The instant change of motion accordingly changes the field. The spherical boundary centered at the origin and enclosing the new field corresponding to the moving charge expands with the invariant speed c. Physically, this expanding surface is filled with kinks, some of which are shown here.

is already continuous.[6] For the electric field, the continuity is maintained by the emergence of kinks connecting the ends of the respective field lines that otherwise would be disrupted (Figure 6.17). For example, consider the interior electric field line perpendicular to the x-axis. Clearly, it is a part of the original Coulomb's field line also perpendicular to the x-axis. By a moment t, this part is shifted along x by a distance ct. But, due to the continuity, the shifted segment is connected to the "wrong" part by the arch of the circle in Figure 6.17. We thus observe the appearance of the system of kinks filling out all the surface of the sphere. From Figure 6.17 we can infer the following properties of the kinks:

6) Do not confuse the continuity of the magnetic field lines with the continuity of the magnetic field itself. The latter may be discontinuous here: in the limit $\Delta t \to 0$, it is zero outside the sphere and nonzero on (and within) the sphere. But since the magnetic field is circumferential, while the discontinuity is observed along the radial directions, this discontinuity does not affect the continuity of the magnetic field lines.

1. Each kink is perpendicular to the corresponding local direction of propagation of the sphere.
2. The kinks as observed from the origin are maximal in the directions perpendicular to the x-direction, that is, perpendicular to the instant acceleration of the charge. They have zero size (there are no kinks) *along* this direction. Generally, the length of each kink is proportional to $\sin\theta$, where θ is the angle between x and the corresponding local radial direction.
3. All the kinks are moving away from the origin with the speed of light.

Now we can draw some physical conclusions from this picture. The kinks in the electric field lines can have only one interpretation: they represent the new kind of the electric field – the field perpendicular to the respective radial directions and thereby to the propagation directions of perturbation (the transverse field). For a fixed distance r from the origin, the relative lengths of the kinks represent relative magnitudes of the corresponding transverse field. Therefore, the electric field strength of the perturbation moving in direction θ is proportional to $\sin\theta$. The same is true for a fixed radial direction: the lengths of two kinks at r_1 and r_2 relate as r_1/r_2; but in this case, we must rescale this ratio according to the radial dependence of the original Coulomb's field. Denoting the transverse and radial components of the field as $E_\perp(r)$ and $E_r(r)$, respectively, we will have

$$\frac{E_\perp(r_1)}{E_\perp(r_2)} = \frac{r_1}{r_2}\frac{E_r(r_1)}{E_r(r_2)} = \frac{r_1}{r_2}\left(\frac{r_2}{r_1}\right)^2 = \frac{r_2}{r_1}. \tag{6.230}$$

This means that the transverse field falls off as the inverse distance from the origin: $E_\perp(r) \sim r^{-1}$.

At each point on the sphere, there is also the magnetic field perpendicular to both, $\mathbf{r}$ and $\mathbf{E}_\perp(r, \theta)$. And, as we know from the picture of the field of a moving charge (Section 6.2), this magnetic field is at each point also proportional to $E_r(r)$ and thereby to $E_\perp(r, \theta)$. Both these properties are consistent with the properties (6.218) of the running waves found in Section 6.9.

Summarizing this part, we can write

$$\mathbf{B}_\perp(r,\theta) \perp \mathbf{E}_\perp(r,\theta), \qquad B_\perp(r,\theta) \sim E_\perp(r,\theta) \sim \frac{\sin\theta}{r}. \tag{6.231}$$

In contrast to the radial component of the field, which remains attached to (moves together with) the charge within and is stationary outside the sphere, the transverse component is *moving away* from the charge. Therefore it represents the field detached from its source and starting independent existence. Since it falls off with the distance slower than the corresponding Coulomb's field, it becomes dominating and then practically the only field observable sufficiently far from the charge. It is radiation field – it is light!

Furthermore, applying the expression (6.199) with E and B from (6.231), we see that the energy flux density associated with the radiation field is inversely proportional to the square of the distance from the charge. Therefore the net flux of the electromagnetic energy through a surface surrounding the charge does not depend on the size of the surface, whose area is proportional to r^2. The amount of energy radiated by the source into free space during a time interval Δt neither decreases nor increases as it propagates farther and farther away – it conserves.

All the listed properties inferred from our qualitative description are known to be the properties of light radiated by an accelerated charge. Which means, among other things, that the minus sign in (6.103), leading to existence of light, had been there by the very first moment of creation – the moment of the Big Bang!

6.12
Aberration of Light – 2

Let us meet again with our inertial observers – Mr. O'Bryan and the engineer from Section 4.4.

We want Mr. O'Bryan to replay his mission 2 (boarding system K'_x), while the engineer remains in system K. Now they are both focused on the motion of the same element of the water surface in the elevator's fish tank (keep in mind that the water surface in that experiment represented any other horizontal surface in the lift).

As the elevator rises, the surface element moves vertically as seen by Mr. O'Bryan and sideways as seen by the engineer (Figure 6.18). The engineer denotes the velocity of the surface element as **v**; he measures $v_x$for its horizontal component and $v_y = v'_y/\gamma(v_x)$ for the vertical component. So the velocity vector makes an angle θ with the vertical, where

$$\tan\theta = \frac{u}{v} = \frac{u}{v'}\gamma(u). \qquad (6.232)$$

Comparison with (4.51) shows that $\theta \neq \alpha_x$, that is, the direction of **v** is not generally perpendicular to the surface element. Well, we have already become psychologically prepared to various manifestations of relativity and here is just another one: the water surface moves perpendicular to itself as seen by Mr. O'Bryan, and not perpendicular to itself as observed by the engineer.

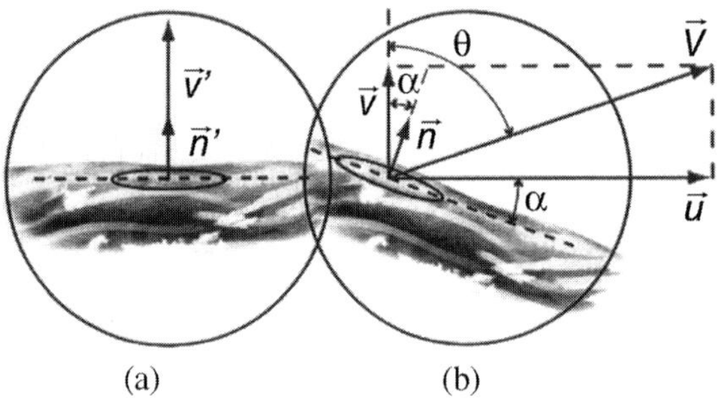

Figure 6.18 Relative orientation of the moving surface element and its velocity (a) in the ship's reference frame and (b) in the shore's reference frame.

Is there at least a small island of absoluteness in this ocean of relativity? Does there exist such a speed of elevator's rise at which the water surface would move perpendicular to itself in all inertial reference frames? And if it does, what is its value?

Let us introduce a unit vector **n** perpendicular to the water surface in the engineer's reference frame. By definition, vector **n** makes an angle α_x with the vertical. So the mathematical condition for our question and the sought-for value of v is $\mathbf{v}||\mathbf{n}$, that is, $\tan\alpha_x = \tan\theta$, or

$$\frac{v_x}{v_y} = \gamma^2(v_x)\frac{v_x v_y}{c^2}. \tag{6.233}$$

Since $v'_y = \gamma(v_x)v_y$, the last equation is satisfied for $v'_y = c$. The elevator in Mr. O'Bryan's RF should rise (or sink) with the speed of light.

Set aside the question if the elevator can do such a feat; let us just focus on the result itself. The value of c is an invariant of Lorentz transformations – it is absolute. We therefore suspect that in this case the magnitude of **v** (the surface element's speed as measured by the engineer) must also remain equal to c. Let us check it. For $v'_y = c$, the transverse component of **v** measured by the engineer will be $v_y = v'_y\gamma^{-1}(v_x) = c\gamma^{-1}(v_x)$. The magnitude of **v** is therefore

$$v = \sqrt{v_x^2 + v_y^2} = \sqrt{v_x^2 + c^2\gamma^{-2}(v_x)} = c. \tag{6.234}$$

A remarkable result! *If* the elevator could rise (or sink) with the speed of light in vacuum, then any shore-based observer would also see the water surface in its fish tank moving with the same speed in the direction perpendicular to its surface. The surface's property of moving perpendicular to itself becomes absolute when it acquires the absolute speed c. But since the surface is now tilted, the *direction* of its motion must also be tilted by the same angle with respect to vertical direction.

Now, if the reader compares the above conclusion to that illustrated in Figure 2.4 in Chapter 2 from quite a different viewpoint, he will find them identical. No wonder! Once we have admitted the possibility for the water surface to move with the speed of light, we have naturally arrived at the result shown in Figure 6.18 for a light wave.

The water cannot, of course, move with the speed c, but it can, in principle, move with the speed arbitrarily close to c. That would definitely be an utterly exotic situation. But, as we have emphasized before, the equations we used describe the same effect for any moving surface, which is horizontal in K'_x. We can therefore, first, substitute the water surface with a wave front of any nature. Instead of rising elevator we might just consider a light wave propagating up in a transparent medium stationary in frame K'_x. We may consider such a frame moving much faster than Mr. O'Bryan's ship. In the last two cases, the values of α and θ may be quite measurable. Consider, for example, the light wave rising from a bulb in the water tank of a spaceship from a Star Trek serial. The spaceship moves horizontally with the speed 300 km/s (which is quite real for an advanced technology and very slow by a science fiction standard). The speed of light in water is about 1.33 times less than c. For these conditions, Equations (4.51) and (6.232) give $\alpha = 7.5\times10^{-4}$ and $\theta = 1.33\times10^{-3}$, respectively. The difference in values for α and θ shows that vector **v** does not make a right angle with the wave surface. This result illustrates our previous conclusion: the surface element or the element of the wave front moving perpendicular to itself in one reference frame does not generally move perpendicular to itself in another inertial reference frame. But for one exception: when an element moves with the invariant speed *c*.

We can summarize these results in the following way. Generally, the wave front in a moving medium is not necessarily perpendicular to the wave velocity **v**. For instance, the light waves diverging from a bulb in a water tank inside a spaceship with Mr. O'Bryan aboard would be spherical for him and nonspherical for us, and this nonsphericity could be measured. Mr. O'Bryan interprets what he sees as yet another manifestation of the principle of relativity: he cannot tell whether his ship is at rest or on the move by observing waves in ship's water tank. We, together with the engineer, would interpret our observations as a kinematic effect: we see the very same waves diverging nonspherically because the medium supporting the waves (water in the tank) is itself moving with the ship. Naturally, this motion singles out a special direction among all the others – the direction of the ship's velocity, that of $u = v_x$. As a result, some physical properties of such moving medium (e.g., its ability to transmit waves) depend on the angle between **u** and the wave vector **k**. We call such a medium nonisotropic (or anisotropic). Our conclusion is that any isotropic medium, when moving, is equivalent to a fictitious stationary anisotropic medium. In this respect, it can resemble certain type of crystals, whose physical properties are different for different directions. Or else it resembles a fluid at rest in an external electric or magnetic field.

But if a disturbance in a fluid can propagate, as light in vacuum, with the speed *c*, this propagation, according to our equations, will not be affected by translational motion of the fluid. Whether this fluid moves or not becomes immaterial, since the motion of this medium will not be revealed in observations. Why then talk about the medium at all? According to all experiments, a disturbance propagating with $v' = c$ can in all respects be considered as propagating just in vacuum. Stop talking of the medium! It was just what happened to ether about 100 years ago. Since the year 1905 people almost never mention ether. They refer to vacuum when talking about light propagation in the intergalactic space or in a laboratory container with all air pumped out. It is true that, as has been found out later, vacuum is not an absolute emptiness and all the particles, photons included, can be described as excited states of vacuum. But this is a different story. And no matter the revealed complexity of vacuum, when it comes to light's motion, its description does not require any carrier necessary to support the light waves. In this respect (as we can see things today), light exists and propagates in its own right!

Well, what happens if we apply our equations to light in vacuum? Consider Mr. O'Bryan this time on a space mission trying to signal with a laser pulse from his spaceship to the engineer. The pulse is fired in the direction perpendicular to the ship's motion (Figure 6.19). But the engineer sees the pulse propagating in slightly different direction that makes an angle θ with the perpendicular line. Accordingly, he observes the wave front of the pulse tilted through the same angle (since $\alpha = \theta$ in this case). He finds the angle by putting in equations $v_x = u$, $v'_y = c$:

$$\tan\alpha = \tan\theta = \frac{u}{c}\gamma(u). \tag{6.235}$$

He has, accordingly, to tilt his detector to achieve better acceptance of the pulse.

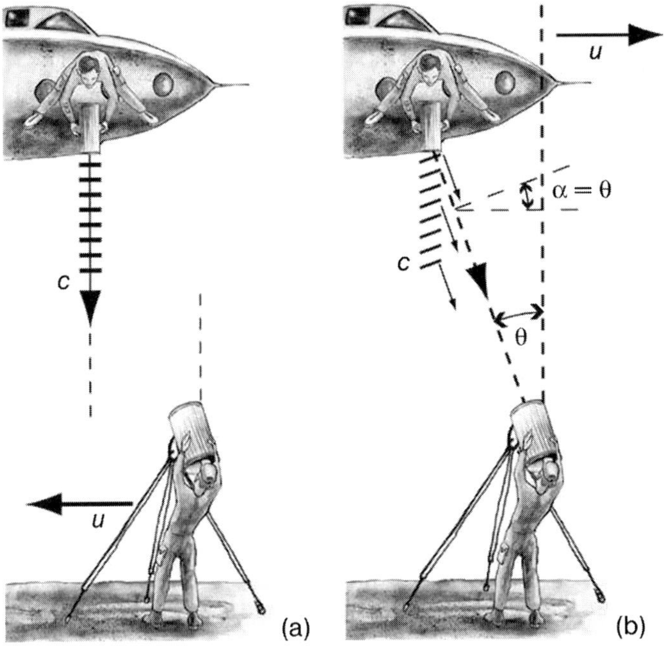

Figure 6.19 Laser signaling from the spaceship to the engineer (a) observed by Mr. O'Bryen and (b) observed by the engineer. In the latter case both – the propagation vector and the wave front of the light waves are tilted by the same angle θ. This is aberration of light, which, in the frame of its source, was emitted along the vertical direction.

Now, substitute the spaceship with a distant star moving in the direction transverse to the line connecting it with the engineer (or an astronomer, for that matter). Then we will get the same result for the light from that star! The astronomer will have to tilt his telescope to achieve better image of the star (or the star's image will be shifted slightly from the center of the vision field). There is a subtle point here, though: in case of light the astronomer cannot tell *relative to what* he has to tilt his telescope, since the actual line of sight to the star is his only reference direction. The effect can only be observed if the astronomer's reference frame changes the direction of its motion. With the position of the telescope rigidly fixed in this reference frame, the change in its motion will cause the star's image slightly shift with time. This shift can be found by comparing the star's photographs made at different times. But precisely such a situation is realized on Earth, which changes the direction of its translational motion due to its orbiting around the Sun. Such motion must be manifest in periodic (with the period of 1 year), circular, or elliptic motion of a star's image in the photographic films obtained with a fixed telescope. This phenomenon was discussed in Section 3.4 from a quite different viewpoint – as the result of relativistic addition rule for velocities. As mentioned there, it was long ago noticed by astronomers and known as *aberration of light* [28,29]. But what is less known is this: while the deviation of the light's velocity vector from the perpendicular line can be easily explained already in terms of nonrelativistic addition of velocities, the corresponding tilt of the wave front, which remains perpendicular to the velocity vector, is *purely relativistic effect*. It is because of this effect that we found the electric and magnetic field vectors in a light wave in vacuum to be always perpendicular to the wave vector **k**, according to Equations (6.218) in Section 6.9. Actually it is another manifestation *of relativity of time*. This is how relativity works in determining basic laws of optics.

6.13
Why Do Droplets Sparkle? Or Can We Trap the Light in a Droplet?

Have you ever been fascinated by pure and intense sparkling of dewdrops early in summer morning or drops after a brief summer shower? What causes this spectacular effect?

We understand that what we see is the light coming from a droplet. Physicists can name three distinct processes leading to it: light scattering, reflection, and emission. For an emission, there has to be an autonomous source of light inside a drop. Light directly reflected from the drop's surface is diverging from it as a spherical wave, whose intensity falls off with distance r as $(a/r)^2$, where a is the radius of the droplet. For small droplets this effect is rather small. Light scattering may produce sometimes a more intense beam. It is a very rich and complex process responsible, among other things, for rainbow and optical glory effect [52]. Its complete description necessary for full understanding of all details is a notoriously difficult problem [53].

But there are some aspects in these phenomena that can be understood within simple undergraduate level physics. They will be sufficient for us to answer our initial question.

To this end, we can refer to Alice, Tom, and Peter from book [1]. We want to continue their discussion in Section 6.9 of that book, concerning the behavior of light incident on the interface between two mediums with different optical characteristics. Those who read their discussions may remember that each of them has outlined a different approach to derivation of Snell's law. All three approaches were equally appealing.

Here we outline yet another approach, which is better suited for our specific purpose. This approach is based on the fact that light is a special case of an EM field. According to EM theory, running EM waves constituting light must be continuous of the interface between two transparent mediums. Let these mediums be water and air, or glass and air. Since the speed of light in air is practically the same as in vacuum, we attribute to it the index of refraction $n = 1$.

Suppose now that light of a fixed frequency is incident from glass onto glass–air interface (Figure 6.20). Due to above-mentioned continuity of the field, the succession of reiterating crests and troughs running in glass with speed $u = c/n$ is matched to similar succession in air running with the speed c. We see that the only possible way for the two different successions to match at each point on the interface and at each moment of time is for the transmitted succession to change its direction in a certain way. This change is easy to find. The wavelength is a distance traveled by a wave with phase velocity u in one period T. Applying this to our case (glass and air), we have

$$\lambda_A = cT, \qquad \lambda_G = uT = \frac{c}{n}T. \tag{6.236}$$

On the other hand, as is evident from Figure 6.20,

$$\lambda_A = A_1A_2\cos\alpha_A, \qquad \lambda_G = A_1A_2\cos\alpha_G, \tag{6.237}$$

where α is an angle between a propagation direction and the interface. Combining these two equations immediately yields

$$\frac{\lambda_A}{\lambda_G} = n = \frac{\cos\alpha_A}{\cos\alpha_G}.$$

The angles α_G and α_A, respectively, are complementary to the angle θ_G of incidence and θ_A of refraction for the corresponding waves. Therefore,

$$\frac{\sin\theta_G}{\sin\theta_A} = \frac{1}{n}. \tag{6.238}$$

This is Snell's law for refraction on the glass–air interface. If we know n and the incidence angle θ_G, we can predict the angle θ_A. The law (6.238) tells us that the propagation vector k deflects farther away from the surface normal $\hat{\mathbf{n}}$ when light passes from optically denser medium to a medium with lower optical density.

We will be interested here in a special case when the incidence angle from glass is such that the angle α_A becomes equal to 90°. In this case, the transmitted light in air is running along the interface. The electromagnetic energy carried by the transmitted wave is not departing from the interface, but rather flows along it. On the contrary, since there remains a nonzero normal component of the propagation vector in glass and, accordingly, a certain amount of energy is delivered by the incident wave each second on each area element of the interface. Since there is no passage trough anymore, *all* the incident energy, rather than its fraction, is being thrown back into glass. The intensity of the reflected wave is equal to the intensity of the incident wave. The corresponding incidence angle, at which this condition is attained, is called the critical angle. Denote it as θ_c. Its value for medium air is obtained from (6.238) at $\theta_A = 90°$:

$$\sin\theta_c = \frac{1}{n}. \tag{6.239}$$

What happens if $\theta_G > \theta_c$? It seems from Figure 6.20 that there is no way to match the incident wave with any wave in air. To see it with maximal clarity, consider an extreme case, when the wave in glass is incident at an angle very close to 90°. In order for it to be matched by the wave emerging into air, the latter must crawl along the interface and have its wavelength and speed, respectively, equal to the wavelength and speed of the wave in glass. In other words, the light in air must in this case have its wavelength less by a factor of n than its natural value $\lambda_A = c/2\pi\omega$, and move slower than light in free space by the same factor. It may be tempting to assume that in this case there forms no wave in air at all. But this would not satisfy the continuity conditions either. The total

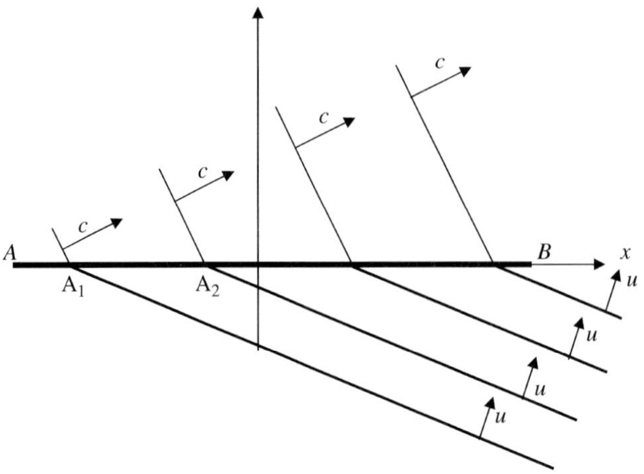

Figure 6.20 The behavior of light on the glass–air interface (AB). A plane wave incident from glass emerges into air propagating in different direction governed by Snell's law. The change of direction is determined by the requirement that the successions of waves in both mediums should match on the interface. The propagation speed in glass is $u = c/n$.

reflection of the incident energy into glass does not necessarily mean that there is no penetration of the incident *field* into air. It does penetrate and does move slower than "normal" light. Well, in this case, what is the difference between such light and the "normal" light wave?

To see the difference, denote x as the direction of a straight line in the interface, parallel to the plane of incidence, and z as the normal direction to the interface. Now, let us write the expression for the propagation vectors of incident and the transmitted waves in unit vector notation:

$$\mathbf{k}_{\mathrm{A}} = k_x \hat{\mathbf{x}} + k_z \hat{\mathbf{z}}, \qquad k_{\mathrm{A}}^2 = k_x^2 + k_z^2 = \frac{\omega^2}{c^2}, \tag{6.240}$$

$$\mathbf{k}_{\mathrm{G}} = k_{\mathrm{G}x} \hat{\mathbf{x}} + k_{\mathrm{G}z} \hat{\mathbf{z}}, \qquad k_{\mathrm{G}}^2 = k_{\mathrm{G}x}^2 + k_{\mathrm{G}z}^2 = \frac{\omega^2}{c^2} n^2. \tag{6.241}$$

The matching condition on the interface requires that the x-components of both vectors be equal:

$$k_x = k_{\mathrm{G}x}. \tag{6.242}$$

Using this and eliminating k_x from the system (6.240) and (6.241) yields

$$k_z = \sqrt{k_{\mathrm{G}z}^2 - (n^2 - 1)\frac{\omega^2}{c^2}}. \tag{6.243}$$

Putting here $k_{\mathrm{G}z} = k_{\mathrm{G}} \cos\theta_{\mathrm{G}} = (\omega/c)\cos\theta_{\mathrm{G}}$, we obtain the expression for k_z in terms of the index of refraction and the angle of incidence

$$k_z = \frac{\omega}{c}\sqrt{1 - n^2 \sin^2\theta_{\mathrm{G}}}. \tag{6.244}$$

Now let us read it. It says that when the angle of incidence exceeds the critical angle, k_z becomes imaginary, $k_z \equiv i\kappa$, with positive κ. How should we interpret this result?

Consider the expression for the transmitted wave:

$$\mathbf{E} = \mathbf{E}_0 \mathrm{e}^{i(k \cdot r - \omega t)} = \mathbf{E}_0 \mathrm{e}^{i(k_x x + k_z z - \omega t)}. \tag{6.245}$$

In the domain of total internal reflection, it takes the form

$$\mathbf{E} = \mathbf{E}_0 \mathrm{e}^{-\kappa z} \mathrm{e}^{i(k_x x - \omega t)}. \tag{6.246}$$

This expression has a very simple physical meaning. It describes a wave running along the interface, with the amplitude exponentially decreasing as a function of z. The energy flux (the Poynting vector, Section 5.8) is most intense at the interface and drops very fast as z increases. A wave with such behavior is called the evanescent wave. What is most interesting is the speed of the wave given by the ratio

$$u = \frac{\omega}{k_x}. \tag{6.247}$$

In view of (6.242) and (6.241), this is equal to

$$u = \frac{c}{n \sin \theta_G} = c \frac{\sin \theta_c}{\sin \theta_G}. \tag{6.248}$$

Since at the total internal reflection the angle of incidence is greater than the critical angle (6.239), the speed is less than c. And note that this is a light wave in vacuum. It turns out that it is possible to force the light in vacuum to move slower than light (i.e., slower than c).

But it can do it only by converting from a plane wave whose amplitude is the same everywhere into an evanescent wave.

This result shows that it is not generally correct to identify the speed of light in vacuum with the invariant speed c.

Imagine now that we have a slab of glass thick enough to contain many wavelengths for a given frequency. Then we can have the same effect of the total internal reflection on both its faces, so the light will, under the described conditions, form a system of the crossed waves within the slab and two symmetric evanescent waves outside the slab (Figure 6.21). The whole system is moving along the slab slower than c, with most of its energy trapped within the slab. This is the simplest model of a dielectric waveguide. The actual optical waveguides used in fiber optics are much more complicated (thin coated optical fibers), but the principle used is the same as just described here.

We can also describe the discussed phenomenon by saying that a transparent dielectric material attracts light considered as a flux of particles. In the language of potential function, such a material can be described as a region with negative potential. We will see later (Section 11.6 and Appendix F) that this approach may be used for description of such processes as superluminal quantum tunneling.

Now we can discuss the crucial part of this section. Consider a transparent spherical bead. It may be a glass or plastic sphere or a water droplet. From what we have just learnt, we might expect that such a sphere can trap the light. The origin of

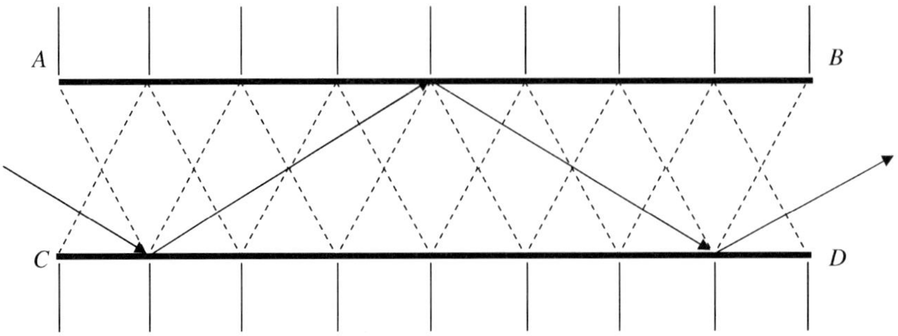

Figure 6.21 A simple model of an optical waveguide – a slab of dielectric material can trap light if the angle of incidence on two parallel horizontal faces of the slab exceeds the critical angle; the trapping works only for direction perpendicular to the slab; the light remains free to travel along the slab. Outside the slab (on both sides – AB and CD) – there is an accompanying evanescent wave traveling along the slab with subliminal speed. A possible optical signal carried by the wave in the slab can also propagate only with a subliminal speed.

the trapped light can be different. It can be produced inside by some photochemical reaction; or it can be trapped from the outside. The latter mechanism, however, must include a few stages, since a direct one-staged trapping from the outside is impossible. Indeed, the electrodynamics is, as we know, invariant under time reverse: if a certain process is possible, then its time-reversed counterpart is equally possible. Therefore, if a certain kind of light can enter the given material, it can as well leak out of it. This is similar to a known behavior of a mechanical system. For instance, an electron from an accelerator, which approaches a nucleus, cannot be trapped by it despite the attracting potential. A chunk of ice entering the solar system from far reaches of the universe cannot be trapped by the Sun and become a member of the solar system. In both cases, trapping requires an additional process leading to the loss of a sufficient fraction of the initial energy of an invading object. Similarly, in order for the incoming light to be trapped, there has to be an additional process appropriately changing the state of the entering light.

However, there is no need to discuss what kind of process that might be, since we are going to show that light cannot, in principle, be trapped by a dielectric sphere. This is equivalent to statement that light cannot be brought to a stable bound state unless it is absorbed.

So let us assume the opposite and consider the case when the light from the very beginning was produced within such a sphere, say, as a result of a certain optical transition in an ingrained source S (Figure 6.22). Let the resulting photons be emitted at a small gliding angle to the surface of the sphere. As in the case of a straight interface, we expect the total internal reflection. Let, in addition, the angle is such that the trajectory of the classical ray representing the wave forms a closed polygon. Finally, if the wavelength is such that the perimeter is multiple integer of λ, then the state forms an optical resonance. We will see later that this is similar to the Bohr–Sommerfeld condition for a wave associated with an electron, to form a stationary state in an atom. By this analogy, a dielectric sphere with a photon (or EM wave) trapped by the total internal reflection can be called the photonic atom [54].

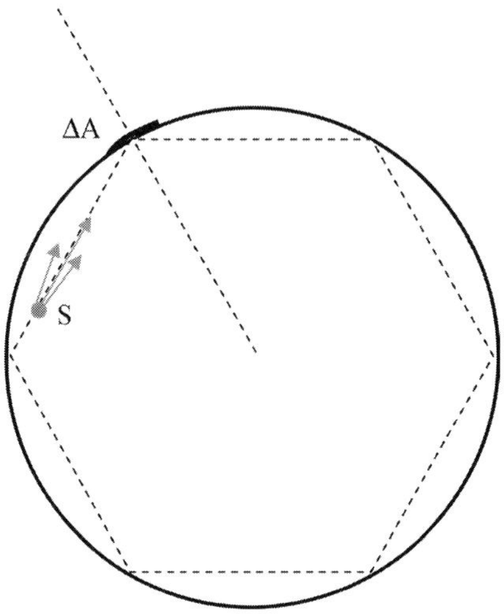

Figure 6.22 A photon emitted from a source within a dielectric sphere and incident from the inside on a small area ΔA of the spherical surface at an angle $\theta > \theta_c$. Under certain conditions, such a photon might be expected to remain trapped within the sphere, but it does not.

This analogy, however, lasts a very short time, because, in contrast to a case with flat interface considered above, now the reflection will not be total. The radiation will be leaking out. This seems strange, because the mechanism of total internal reflection seems to follow from very general principles – it is an exact solution of Maxwell's equations.

The explanation of this difference lies in the curvature of the interface. The same Maxwell's equations that predict the possibility of trapping of light between two parallel flat interfaces filled with a dielectric predict the impossibility of the perfect trapping in the same dielectric with a curved boundary. What is the reason for this?

We can understand it on qualitative level in the following way. The perfect trapping (total internal reflection back into the dielectric material) is accompanied by formation of an evanescent wave running along the interface. If the interface is flat, there arises no problem – the wave can run down it with constant phase velocity $u < c$. However, if the interface is curved, the wave is forced to accordingly turn its direction. The light can do it only up to a certain radial distance $r = r_c$ from the center of curvature, at which its linear phase velocity reaches the value $u = c$. Once this barrier is reached, the farther curving becomes impossible, since the phase velocity of the wave in the free space would become superluminal. At a distance $r > r_c$, the surface of constant phase stops rotating around the center and keeps on moving in straight line with phase velocity c at all points on the surface. In other words, beyond the critical distance r_c the light stops following the curved boundary and propagates along the straight lines. This means that at this point the light detaches from curved trajectories bending around the interface and carries its energy away from the sphere (Figure 6.23). This light travels freely through space and we can see the sparkling droplet.

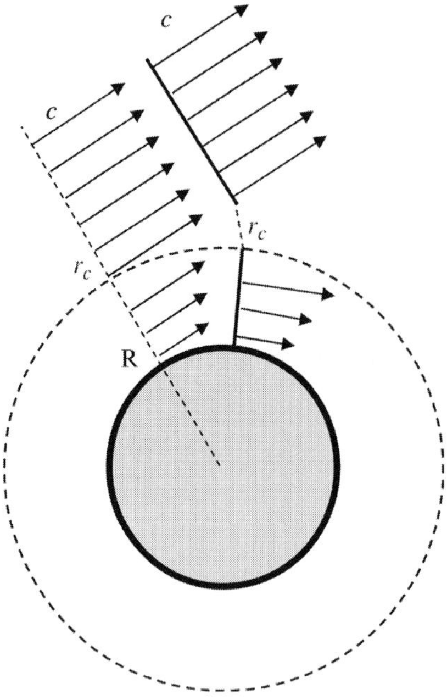

Figure 6.23

We can evaluate the distance r_c at which the wave brakes in two parts from the expression for the wave function. This time we write the expression in spherical coordinates $(x, y, z) \to (r, \theta, \varphi)$ and take into account that the azimuthal and temporal variables can be separated from the polar and radial variables θ and r:

$$E(r, \theta, \varphi, t) = E(r, \theta)\Phi(\varphi, t), \tag{6.249}$$

where

$$\Phi(\varphi, t) = \mathrm{e}^{i(m\varphi - \omega t)} = \mathrm{e}^{im(\varphi - (\omega/m)t)}. \tag{6.250}$$

Here m is number of waves fitting into the circumference. This number has to be integer in order to satisfy the continuity requirement for the wave.

For a radial distance r, the angle φ can be expressed in terms of the length l of the corresponding arc, $\varphi = l/r$. Putting this into (6.240), we obtain

$$\Phi(\varphi, t) \to \Phi(l, r, t) = \mathrm{e}^{i(m/r)(l - u_m t)}, \tag{6.251}$$

where

$$u_m \equiv \frac{\omega}{m} r \tag{6.252}$$

is the linear phase velocity of the wave for a given radial distance. We see that for the same evanescent wave orbiting the center, its phase velocity linearly increases with distance of the corresponding point on the wave surface. The latter can, in this case, be modeled as a door rotating on its hinge. Many of you have passed through such doors in some buildings or in a subway station. But such a door cannot be arbitrarily large [1,29]. If you want to make it arbitrarily large *and* maintain it flat and rotating, then at a certain distance from the hinge it will break into two flat boards. One will remain hinged and rotating, while the other instead of rotational motion will be in translational motion (Figure 6.23). For an ideal (infinitely thin) door representing a phase surface, the breakup will take place at a point where its linear velocity u_m reaches the value of c. Putting this value into (6.252), we obtain

$$r_c = m\frac{c}{\omega}. \tag{6.253}$$

Thus, the formed "translational" part of the broken door (i.e., now a regular plane wave) will move, as it should, with the speed c in a free space.[1)] It can travel sufficiently far away from the drop and get into your eyes if you happen to be in an appropriate position. And you will see a radiant sparkle!

This effect arises because relativity, which somehow allows light in vacuum to move under some circumstances slower than invariant speed, under no circumstances allows light in vacuum to move faster than the invariant speed. As a result, a droplet, even if it could be made totally nonabsorbing, is principally unable to hold any radiation born in it, let alone the radiation that happens to have come from the outside world.

Problems

6.1 Suppose you measure the field due to a moving point charge. The field at an observation point $\mathbf{r}$ at a moment t is determined by the instant position $\mathbf{r}'$ and state of motion of the charge at a previous ("retarded") moment t', for which the travel time of an electromagnetic signal from $\mathbf{r}' = \mathbf{r}_0(t')$ to $\mathbf{r}$ is equal to $t - t'$:

$$R' = |\mathbf{r} - \mathbf{r}'| = c(t - t'). \tag{6.254}$$

The general expression for the electric field due to a uniformly moving electric charge q is

$$\mathbf{E}(\mathbf{r}, t) = \frac{q}{4\pi\varepsilon_0}\gamma^{-2}(v')\frac{\mathbf{R}' - (\mathbf{v}'/c)R'}{(R' - (\mathbf{v}'/c)\mathbf{R}')^3}, \qquad \mathbf{B}(\mathbf{r}, t) = \frac{1}{c}\hat{\mathbf{R}}' \times \mathbf{E}(\mathbf{r}, t), \tag{6.255}$$

where the primes in the right-hand side denote the corresponding variables taken at the retarded moment of time t' and capped vectors indicate the corresponding unit vectors.

(a) Show that Equation (6.254) has only one root.
(b) Express the fields in terms of *current* position $\mathbf{r}_0(t)$ of the charge.

1) The word "translational" here is not exactly accurate, because the corresponding wave is not exactly a plane wave. Very fast it becomes spherical – after all, it has originated from the evanescent wave "attached" to a small droplet.

(c) Show that in a special case when the direction to current position from the observation point makes the right angle with the velocity of the charge, these expressions reduce to Equations (6.27) and (6.32) in the text.

6.2 The general expression for the electric field due to an arbitrary moving charge q is

$$\mathbf{E}(\mathbf{r},t)=\frac{q}{4\pi\varepsilon_0}\left(1-\mathbf{v}'\cdot\frac{\hat{\mathbf{R}}'}{c}\right)^{-3}\left[\gamma^{-2}(v')\frac{\hat{\mathbf{R}}'-(\mathbf{v}'/c)}{R'^2}+\frac{\hat{\mathbf{R}}'\times\left[(\hat{\mathbf{R}}'-(\mathbf{v}'/c))\times\mathbf{a}'\right]}{c^2 R'}\right],$$

$$\mathbf{B}(r,t)=c^{-1}(\hat{\mathbf{R}}'\times\mathbf{E}(\mathbf{r},t)), \tag{6.256}$$

where $\mathbf{a}'$ is the acceleration of the charge at the retarded moment; the rest notations are the same as in the previous problem.

(a) Apply this equation to the case of a uniform circular motion (radius of the circle is R) to find the electric field at the center of the circle.
(b) Using result (a), find the expression for the magnetic field at the center.
(c) Show that this expression is consistent with Biot and Savart law.

6.3 Generalize the results of the previous problem to find the (a) electric and (b) magnetic field of the same charge on the symmetry axis passing through the center of the circle, a distance z away from the plane of the circle.

6.4 Using Equation (6.256) of Problem 6.2 for the electric field of an arbitrary moving charge,

(a) Find the electric field flux through an arbitrary closed surface containing the charge. Show that the flux is the same as for the stationary charge.
(b) Show the same by transforming to the instantaneous rest frame of the charge.

6.5 Find the electric and magnetic fields in an instantaneous rest frame of an arbitrarily moving charge

(a) for an instant when a charge that had been stationary is suddenly "hit" by an external force;
(b) for an instant when a charge moving within a parallel-plate capacitor toward the plate with the charge of the same sign is brought to a stop before turning back (i.e., a charge at a turning point of its motion);
(c) is the condition "... in an instantaneous rest frame" accurately defined in case (a)?

6.6 Find the magnetic field inside a solenoid with N turns per unit length, flown around by a current I.

6.7 In the previous problem, find the magnetic field if the solenoid is moving along its symmetry axis with a speed v.

6.8 Suppose you are inside the solenoid described in Problem 6.6, but now you are moving perpendicular to its symmetry axis with velocity $\mathbf{v}$.

(a) Find the magnetic field recorded by your devices.
(b) What, if any, electric field will you observe?

6.9 A particle of charge q and rest mass m_0 is moving in an external electromagnetic field. Express the particle's acceleration **a** in terms of local electric field **E** and magnetic field **B**.

6.10 (a) Is a quantity **B** describing magnetic field a true vector or a pseudovector?
(b) Suppose a quantity q_m describes the amount of charge of a monopole. Is this quantity a scalar or a pseudoscalar?
Explain your answers in (a) and (b).

6.11 "Magnetic field does no work."
Although physics is based on anything but dogmas, the above statement sounds as a dogma due to definition of the Lorentz force.
(a) If magnetic monopoles are discovered someday, will this dogma remain true?
(b) Suppose there exists a parallel universe with all electric charges replaced by monopoles. Will there be an equivalent dogma in such a universe? Explain your answers.
(c) Write the Maxwell's equations for such universe.

6.12 A charged horizontal plate with a charge density σ starts moving along the x-direction with a speed v relative to the lab.
(a) Find the electric and magnetic field of the moving plate.
(b) In addition, you now also start moving in the same direction with a speed u relative to the lab. Find the electric and magnetic field of the plate observed by you.
(c) The condition (b) is changed so that you are now moving only along the y-direction with a speed w relative to the lab, while the plate keeps on moving as before. Find the electric and magnetic field observed by you in this case.

6.13 Prove Equation (6.75) using the simple model of a rectangular loop flown around by a current I.

6.14 Consider an equipotential surface $\Phi = \text{constant}$ for a stationary point charge. This is also the surface of constant magnitude of the electric field around the charge.
(a) Describe the shape of this surface in a RF moving with velocity V relative to the charge.
(b) Will the magnitudes of the electric field at different points of this surface remain equal? Explain your answer.

6.15 Prove that electric and magnetic field due to a given system of charges and currents do not change under gauge transformations.

6.16 A relativistic spaceship from a Star Trek serial is moving "horizontally" past the Earth at 99% of the speed of light and the water rises in the ship's swimming pool at 5 m/s.

(a) Find the tilt of the water surface with respect to the horizontal, as observed from Earth.

(b) Reverse the conditions: let now the spaceship "crawl" at 5 m/s and the light wave move up in the ship's atmosphere at 99% of the speed of light in vacuum. Find now the tilt of the propagating wave front with respect to the horizontal.

(c) Let the spaceship and the light wave inside move (in mutually perpendicular directions, as before) each at 99% of the speed of light in vacuum. Find the tilt of the wave front in this case.

6.17 Express the contravariant EM field tensor F^{kl} in terms of the field components of **E** and **B**.

6.18 Two charges q_1 and q_2 are distance d apart. The charges simultaneously start moving at a common speed V along their separation line.

(a) Write Coulomb's law for the interaction force between the stationary charges.

(b) Find the interaction force between the charges after they have started moving.

(c) Compare and explain the results.

7
Relativistic Paradoxes

How wonderful that we have met with paradox. Now we have some hope of making progress.

Niels Bohr

7.1
Seeing and Observing (on the Appearance of Fast-Moving Objects)

Experience is the name everyone gives to their mistakes.

Oscar Wilde

The relativistic effects following from Einstein's relativity principle have caused long-standing controversies in scientific literature. Critics thought that these effects lead to paradoxes.

A paradox is an argument that starts with apparently acceptable assumptions and leads by apparently valid deductions to an apparent contradiction [54]. Most of the paradoxes in relativity arise from errors in our reasoning or our understanding of the theory. Some of them are based on confusion between different concepts or operations we use to study the relativistic effects. But none of them so far is found to show any flaw in the theory itself. Therefore, we do not improve the theory of relativity resolving such paradoxes, but we achieve its better understanding.

In this section, we will discuss one of the common sources of relativistic paradox – the confusion between seeing and observing. By seeing we usually mean not only visual images obtained by our eyes but, more generally, any perception of reality given directly by our senses. In some cases and to a certain (very limited) degree, seeing may constitute an observation. But generally, we must be very careful distinguishing one from the other – especially when we, in our loose jargon, use the word "to see" while actually we mean "to observe."

Let us start with simple examples illustrating the cases when "seeing" is something quite different from "observing." Here is an excerpt from a letter denying the reality of the Lorentz contraction because we do not see it in the moving bodies, especially, the celestial objects:

Special Relativity and How it Works. Moses Fayngold

ISBN: 978-3-527-40607-4

> ... our eyes ... never observe in other celestial bodies the contraction which ought to be caused by velocity. In this case we would have to assume that our observations are not correct, and that our brain will compensate the visual physical effects of velocity. We would then lose the faculty of observation, which would also be equivalent to the end of natural science.

This letter illustrates a typical confusion between "seeing" and "observing." It is not our observations that are wrong; it is our direct interpretations of what we see that is frequently wrong. In essence, the above quotation is akin to stating that natural science ends when we have to stop believing our senses. Well, if we look at the history of natural science, we will see that it is often when we do stop believing them that the true science *begins*. It is a commonplace that our senses often mislead us. Our senses witness that the Earth is absolutely motionless, which is wrong. They tell us that the Earth is flat, which is wrong. Our eyes tell us that the Sun and the Moon are of the same size and both are much smaller than our bedroom, which is wrong. They tell us that the stars are all at the same (although undefined) distance from us, which is wrong. They also tell us that all the stars are much smaller than the Moon, which is wrong. Our senses witness that matter, of which all things are made, is continuous, which is wrong.

It was only when people went beyond these direct evidences of senses that they gained insight into the nature of things. Scientific observations often show that what we see is something quite different from the actual reality. The quoted citation about celestial bodies only shows that we sometimes need to question direct evidence that our eyes afford us. We may need to complement this evidence with additional sources of information.

One of such sources is the notion that light takes time to travel the distance between the points of its emission and absorption. Here, we consider a few simple situations illustrating the difference between what we see and what we really observe.

Consider a very thin rod of length l along our line of sight, but moving perpendicular to it with a speed v (Figure 7.1). We want to make a photograph (an instant snapshot) of the rod in motion. Suppose the photograph is taken when the rod is exactly along our line of sight. To make sure that the snapshot was really taken at a certain moment, say, at 8:45:10.3 a.m. of our local time, the exposure time must be negligibly small and the illumination (or the luminescence of the rod itself) must be sufficiently bright. Let me tell you the outcome right upstart: the rod on the photograph will appear tilted to the line of sight. We expect the photographic image to be just a dot (the edge of the rod closest to us, or, more rigorously, the normal projection of the rod onto the film along the instant line of sight, Figure 7.1a). We will actually see the image of the whole rod, with both of its end points – the one that is father away will be trailing on the image and the one closer to us leading (Figure 7.1b). Our conclusion based only on this evidence of senses would be that the rod is tilted backward. This conclusion would be wrong. We see only an optical effect that masks actual spatial orientation of the rod. This orientation was assumed to be vertical in Figure 7.1 and remains vertical – it cannot change only because we are taking a

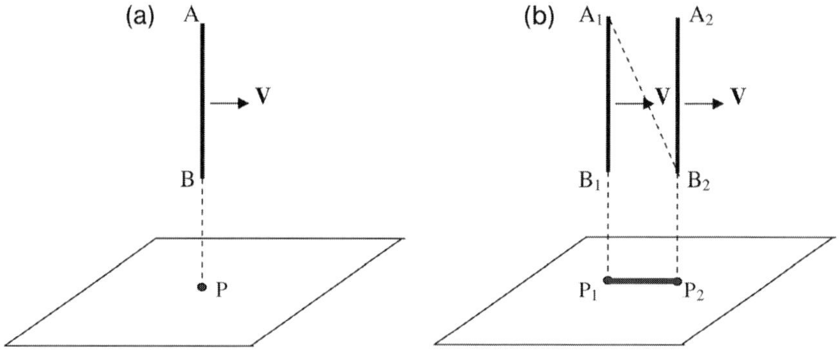

Figure 7.1 (a) Actual instant position of a moving rod at the moment of snapshot and its naively expected image (its geometrical projection P onto the film). (b) Actual image of the rod. It is its optical projection P_1P_2 made by the photons emitted or scattered from the rod. Since their speed is finite, the *optical* image on the photograph corresponds to the geometrical image that would be produced by a *tilted rod* A_1B_2. The exposure is assumed very short. (Actual process involving an optical system with lenses and other elements is more complicated, but the essence is the same.)

snapshot. What we see is an *apparent* orientation caused by optical retardation because of the finiteness of the speed of light. It is not even a relativistic effect. It was known long before the theory of relativity. But it becomes noticeable and even prominent at a speed of the rod comparable to the speed of light, that is, in the relativistic domain.

It is an instant image made by the photons arrived from the different points of the rod. Even though all these photons have arrived at the camera simultaneously, in order for them to do so, the photons from the farther parts of the rod had to depart from those respective parts earlier than the photons from the closer parts. Thus, each photon has departed from its respective point at a different moment of time. But, due to the rod's motion, it was in different positions at different moments. Accordingly, the arrived photons bring in the information about these different positions by landing at different parts of the photographic film. The photons from the farther edge of the rod were emitted earlier, when the rod was in position 1 and, accordingly, they enter the camera from the corresponding direction. The photons from the close edge of the rod were emitted later, when the rod was at position 2. Accordingly, they enter the camera from this slightly different direction. The photons from the intermediate points of the rod enter the camera from the corresponding intermediate directions. When we develop the film, we see the rod on the photograph as if it were, at least partially, oriented along the direction of its motion.

The next thing to stress is that this effect has nothing to do with known fuzziness of some photographs of moving objects. We can sometimes see such fuzziness, for example, in a photograph of a dancer or skater during a movement. This fuzziness is due to the significant exposure (the time the camera is open). Let us look briefly into this effect. If we first take a snapshot of a stationary rod in the above specified position (edge on, strictly along the line of sight), the image will be a dot no matter how long

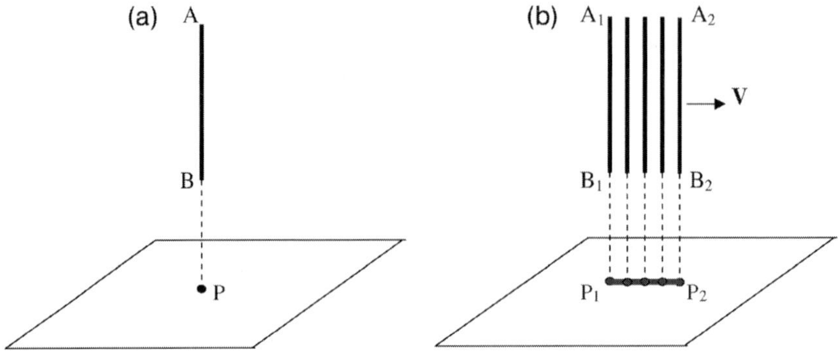

Figure 7.2 The image of a stationary rod (a) and moving rod (b) under changed conditions. The speed of the rod is now so small that the optical retardation is negligible. Instead, the exposure is long, so the rod travels many times its thickness during this time. Accordingly, its image is extended along the line of motion. Hence, the difference in the interpretation of the image. In Figure 7.1 the image shows all parts of the rod between its edges. In contrast, the image here is a succession of reiterated images of the same edge B at different positions of the rod. In both cases, the resulting output (the photographic image) is misleading and its passive seeing falls short of scientific observation. However, the correct interpretation based on the information about the imaging conditions promotes "seeing" to the status of "observing."

the exposure is (Figure 7.2a). If we take a snapshot of the moving rod, the image will be fuzzy (Figure 7.2b). This is merely a succession of the instant images (or instant snapshots) of the same bottom edge of the moving rod, which passes through a range of positions during the exposure. It extends if the exposure increases and shrinks down to zero if the exposure time goes to zero.

Consider another situation: the moving rod at the moment of the snapshot was oriented perpendicular to both the line of sight and the direction of its motion (Figure 7.3). Then the photons from the edges of the rod entering the camera simultaneously with those from the middle were emitted earlier since they had to travel a greater distance. Accordingly, the rod was in a slightly retarded position at that earlier moment, so the respective photons landed on the film from slightly different directions than the photons from the middle of the rod. The same happens with the photons from the intermediate parts of the rod. As a result, the photographic image will be that of a bent rod, whereas the real rod is straight.

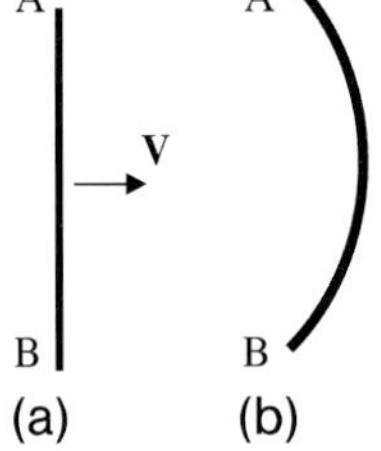

Figure 7.3 (a) Actual shape of a moving rod oriented perpendicular to its velocity and observer's line of sight (i.e., the line of sight is now perpendicular to the page). (b) The photographic image of this rod, made under very short exposure. Owing to optical retardation, the light from the edges entering the optical system together with light from the middle was emitted earlier then the light from the middle. For a sufficiently fast-moving rod, the resulting image is a curved line, even though the rod itself is straight.

In both examples, the actual shape is distorted due to the difference in traveling time from different parts of the object. The result for a *moving* object is that its different parts are recorded from its different positions. This is an optical effect that can be called the *optical retardation*. It has nothing to do with the relativistic length contraction. The latter is the *real contraction* of the object itself, recorded (or observed) in a laboratory where this object is moving. It is a manifestation of relativity of space and time, or, which is the same, of the space-time geometry.

In the case when we use optical devices for observations, the conclusion about the relativistic length contraction is made *after all corrections for change in visual appearance of the body are taken into account*.

The reader can find many other interesting effects associated with imaging of moving bodies and corresponding references in a series of publications by Boris Bolotowsky [55,56].

One might get an impression from the above examples that the described distortions of the actual shape of an object in the process of image formation occur only when the imaged object is in motion with respect to the detecting device. Actually, the deceptive images can be obtained from a stationary object as well, if it is illuminated by a very short light impulse. Consider, for instance, an attempt to photograph a wall right in front of us, but in complete darkness. In this case, we illuminate a wall using a flash. Suppose that both the flash and the exposure are very short. In this case, the resulting image will depend on time discrepancy between the flash and the exposure, as well as on the distance between the camera and the wall. Depending on these parameters, we can get one of the photographs shown in Figure 7.4. If the wall is ideally reflecting, we will get a photograph either uniformly dark, or with a bright spot in the middle.

On the other hand, in many cases even an accurate observation, while reflecting an objective experience of an observer or objective aspect of an event, is actually an *appearance* rather than actual effect. Such an observation is the observation of something apparent.

In physics "apparent" is close to "illusory," it is an observation that is not confirmed by other observations of the same phenomenon, even in the same reference frame (RF). Here are a few examples.

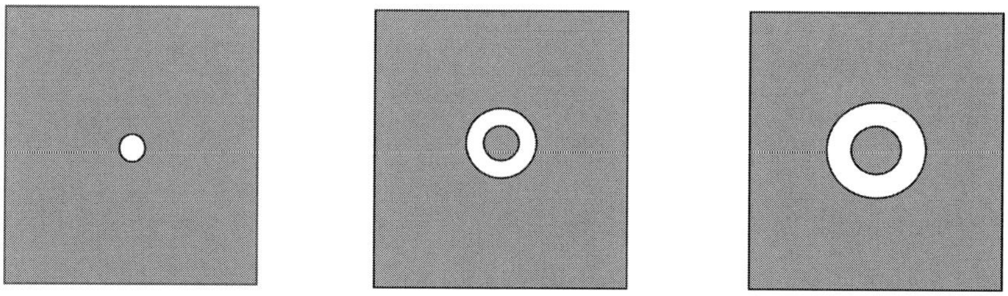

Figure 7.4 Instant snapshots of a wall with diffusively reflective surface illuminated by a very short pulse.

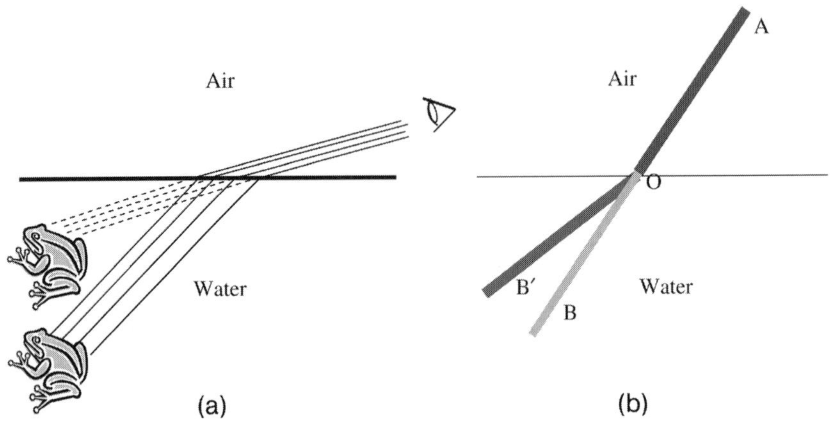

Figure 7.5 (a) An object under water seems to an observer watching from air to be closer to the water surface than it really is. (b) A stick partially immersed in water seems broken at the surface (AO + OB′), with the immersed part closer to the surface. The stick itself (AOB) remains, of course, straight.

I watch a frog in a pond and I see it at a certain position closer to the water surface than it actually is (Figure 7.5). The observed position of the frog is in this case only an apparent position. How do I know? First, by studying the refraction of light at the interfaces, and second, from direct experiment. I can locate the true position of the frog by touch. If I try to do it with my spike, following the deceptive information from my vision only, I will miss. This missing will show me that the observed optical image of the frog (as well as the underwater part of the spike) is only apparent, that is, it does not give me the true locations of these bodies. Some ancient tribes hunting fish in shallow waters must know this difference between the real and the apparent.

I see myself in the mirror and I know that even though the act of seeing is real, the location of what I see is only the appearance. All other observers of my mirror image, *even those moving relative to the mirror,* will measure only my *apparent* position by observing my reflection in the mirror. In this case, the position observed in the described experiment is not real, even though all observers obtain it, stationary or moving alike. To measure my real location, all participants of this experiment must check their observations using also some other tools that would reveal the presence of the mirror and then make corresponding corrections for the reflection.

When I use the word "observing" or "observation," I mean a scientific experiment performed in such a way as to reveal all possible distorting factors and make corrections for them. The result of such observation is the conclusions made after all possible sources of error are identified and the corresponding corrections are made.

In the next section, we consider possible experimental procedures for measuring the length contraction, including situations when the studied object is three dimensional.

7.2 Can We See the Relativistic Length Contraction?

As a natural extension of the previous section, we will discuss here the difference between the actual and the visual aspects in the Lorentz contraction effect. In particular, we will focus on the question what would constitute actual observation or measurement of this effect and whether it is possible to see it in optical experiments.

There are many observable effects depending on length contraction, for instance, in high-energy physics. For example, in a well-known paper [58], the amount of the secondary particles born in a high-energy collision of two nuclear particles was estimated in the rest frame of the center of mass, based on the fact that the colliding particles had each in this frame a shape of the disk caused by Lorentz contraction. Accordingly, the secondary particles born in the collision must emerge from the disk-shaped region of space affecting their resulting angular distribution. The observations confirmed the theoretical models of collision based on this effect. Another example is E. Fermi's famous paper [59] (as far back as 1924!), in which atomic excitation or ionization by a high-energy charged particle had been considered as caused by the particle's Lorentz-contracted field. The corresponding predictions have also been confirmed in numerous experiments.

In principle, the size of a rod moving along its length can be directly measured by marking the instant positions of its edges simultaneously in the reference frame where the rod is moving and then measuring the distance between the marks, but such a procedure is very difficult to carry out.

One can suggest various schemes for such measurements using photography. First, taking a snapshot of a linear object. Imagine a very thin rod moving along its length in the direction perpendicular to the instant line of sight (Figure 7.6). Make the rod short enough so that you could neglect the time difference in light traveling to the camera from the center and the edges of the rod, respectively. Take a snapshot. For a sufficiently short rod, an appropriate snapshot with short exposure and bright illumination may suffice.

If the rod is long, then the travel time from its center to the camera will be noticeably different from that of the edges and it would take additional information and computational algorithms to recover the actual size of the rod. In this case, it would be better to find another direct experimental procedure, or at least, a thought experiment. For instance, the rod can be illuminated by a flat narrow light pulse with a broad front moving perpendicular to the direction of the motion of the rod (Figure 7.7). Then the length of the shadow cast by the rod onto the screen parallel to it will give the length of the rod, if the frequency range of the pulse is such that the diffraction on the edges of the rod is negligible. The shadow can be measured by making the screen light-sensitive and measuring its unexposed part.

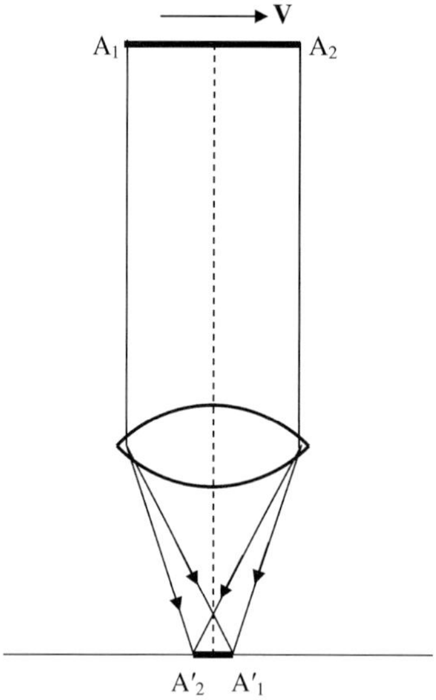

Figure 7.6 A simplified diagram for optical imaging of a rod moving *along* its length. Such imaging can be used to measure Lorentz contraction for *sufficiently short rods*, since in this case the retardation effects are small.

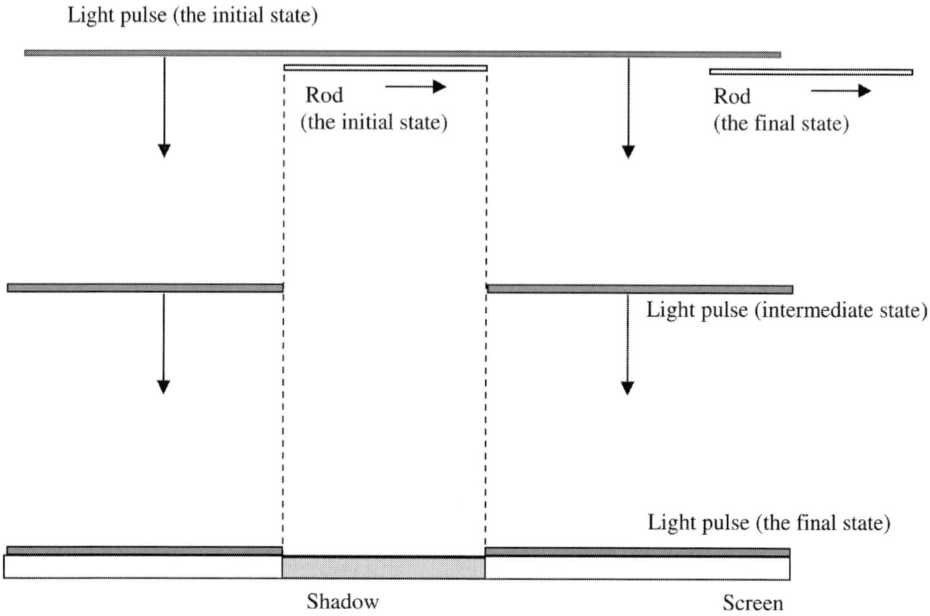

Figure 7.7 A possible experimental scheme for measuring the length of a moving rod. A part of a very narrow laser pulse incident from above is blocked by the rod. The remaining part of the pulse has a gap whose size is equal to the size of the rod in the given reference frame. When the pulse illuminates the screen, there remains a shadow whose size is the measured length of the moving rod.

Repeat such experiment for different speeds of the rod. Then compare the results with the snapshot of the same rod at the corresponding position at rest. You will see that for each value of the rod's velocity v its length l is decreased relative to its proper (rest) length l_0 according to

$$l = \frac{l_0}{\gamma(v)}. \tag{7.1}$$

This is a pure length contraction, since the experiment has been performed under conditions excluding any possible influence of the optical retardation.

In the case of a three-dimensional object, you cannot take pictures without interference with the shape change. In this case, a possible way to get rid of this "shape effect" in physical experiment is to mark simultaneously (in the reference frame where the measurements are made) the instant positions of the front and the rear points of the moving object and then measure quietly the distance between the markings. You will get the same result (7.1). This is again the pure length contraction, since the distorting effect of the optical retardation did not, under given conditions, have a chance to manifest itself. As noted above, the real performance of such an experiment is associated with great difficulties and it has never been done in the described form. It is just a thought experiment. In real experiment, one might take pictures, but then one should take into account the retardation effect on the appearance of a three-dimensional body.

As mentioned in the previous section, a change in the appearance, on the one hand, and the relativistic length contraction, on the other, are *different* phenomena, having nothing in common. Strictly speaking, the former is *not* even a relativistic effect – it originates from the finiteness of the speed of light, *not* from its invariance. An acoustic image of a jetliner in flight, produced by the emitted sound waves, would also be a distorted version of its actual shape, as is an optical image of a fast-moving body. Ironically, the shape distortion of a moving body due to the optical retardation effect tends to make the object look *longer*, not shorter, in the longitudinal direction. Therefore, the optical retardation may totally mask the actual length contraction. Moreover, if the moving object has a sufficient spatial depth, the optical retardation will dominate over length contraction and the photographic image of such an object will suggest that the object is *longer* than its proper length along the direction of motion.

What really happens can be readily seen by tracing, as we have done in the previous section, all rays arriving simultaneously at the camera from different points on the surface of the moving body. In doing this, we have to take into account that while the emitted photons are moving along the corresponding ray directions, so is the body along its own path. Then we will see that, owing to the advance of the body's front, some of the rays from the farther part of the front will be obstructed by the parts that are closer to us. The more remote parts will therefore be hidden from view, as if the front side were turned away from the observer. By contrast, the rear side, which is retrieving from the paths of the emitted rays due to the body's motion, will appear to turn toward the observer. The overall result is that if there were no length contraction, the observer would see the object distorted and stretched out, rather than flattened, in the direction of its motion. With both effects at work, the observed length will be the result of their superimposing.

As an example, let us consider a solid cube of a side a moving with velocity **v** parallel to one of the sides and observed by Peter in the direction instantly perpendicular to **v** (Figure 7.8a). As we have found above, the side of the cube facing Peter (let it be the bottom, i.e., Peter sees the cube passing above him) under given conditions is observed Lorentz-contracted, not obscured by retardation effect, so its length along the direction of motion will be $a' = a/\gamma(v)$. On the other hand, the rear side will appear tilted, since the light from its farther edge (entering the camera simultaneously with light from the bottom edge) left the edge when it was at an earlier position E′ H′. (This tilt due to the optical retardation has nothing to do with the real tilt of the rising water surface or a chandelier in a moving elevator observed by the engineer in Section 4.4.) And the front edge will not be seen on the snapshot made at this moment, since the light coming from it onto the camera will be blocked by the body of the cube. As a result, the cube on the photograph will appear as if it were skewed under a sheer force applied to its bottom and top faces (Figure 7.8b).

We can easily describe this effect quantitatively, assuming that the distance z to the cube is much greater than its size, that is, $z \gg a$. In this approximation, all the light rays from different points of the moving cube to the camera can be considered parallel. Then the ray from the top rear edge of the cube will enter the camera

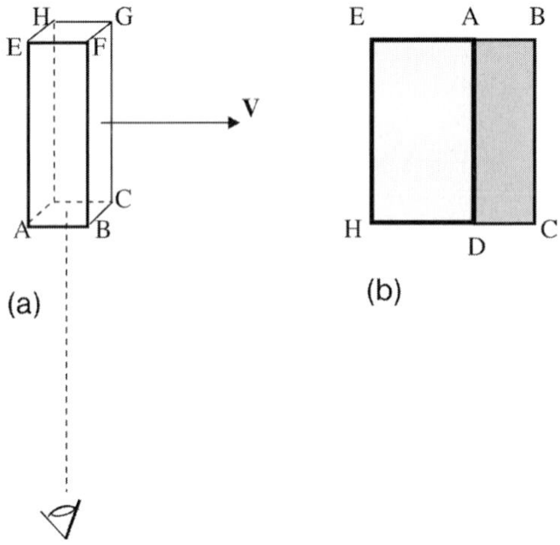

Figure 7.8 Optical imaging of a moving cube. (a) The actual position and shape of the moving cube at the moment of a snapshot. The cube is photographed from beneath (actual distance from the observer to the cube is assumed to be much greater than its size). (b) The image of the bottom of the cube on the photograph. ABCD is the Lorentz-contracted face of the cube; ADEH is the image of the rear face of the cube formed through the same mechanism as in Figure 7.1: due to optical retardation, the face is seen as tilted with respect to the line of sight. Such an image could also be obtained from a stationary object produced by flattening and skewing the initial cube.

together with the ray from the bottom rear edge, if it was emitted earlier by the time interval

$$\Delta t_r = \frac{a}{c}. \tag{7.2}$$

During this time interval, the cube will travel the distance

$$\Delta x = v\Delta t_r = a\beta, \qquad \beta \equiv \frac{v}{c}. \tag{7.3}$$

The net size of the cube on the photograph (along the direction of motion) will be

$$\tilde{a}' = a' + \Delta x_r = (\gamma^{-1} + \beta)a. \tag{7.4}$$

It is greater than just Lorentz-contracted length a/γ. We see that the light retardation effect is, indeed, masking the relativistic length contraction effect because it tends to *extend* the image along the direction of motion. In particular, if the cube is moving sufficiently fast so that $\gamma^{-1} \to 0$, $\beta \to 1$, we will have $\tilde{a}' \to a$, so that no contraction will be seen on the photograph.

Now, let us do some exercises. Suppose that the bottom and the rear of the cube are of different colors or shades. Then, the picture you will see on the photograph looks like that in Figure 7.9, with distinct faces clearly seen. If you are evaluating such a photograph, knowing that the cube was not subject to any forces that might have skewed it, you might wonder whether it is possible to interpret this as the result of rotation of the cube? Indeed, suppose that the cube has just been rotated through an angle φ about an axis perpendicular to the direction of its motion and the line of sight. Then, the rectangle ABCD on the photograph would be the projection of the tilted

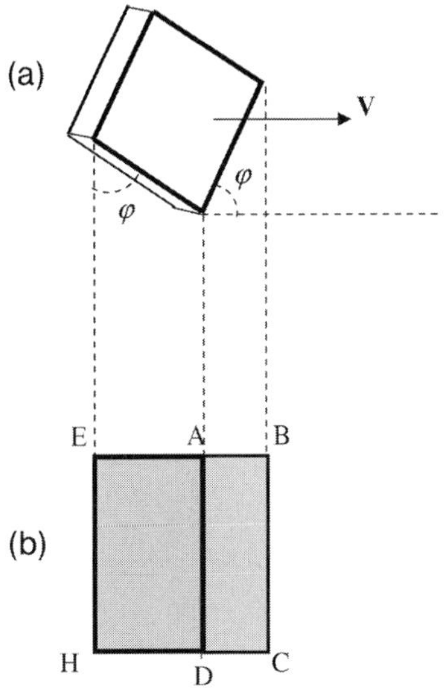

Figure 7.9 Optical imaging of a moving cube.

bottom onto the horizontal plane and the rectangle ADHE could be the projection of the tilted rear face onto this plane (Figure 7.9b). Then you could write

$$a' = a\cos\varphi, \qquad \Delta x_r = a\sin\varphi, \tag{7.5}$$

so that

$$\tilde{a} = a' + \Delta x_r = (\cos\varphi + \sin\varphi)a. \tag{7.6}$$

Comparing with (7.4), we see that

$$(\gamma^{-1})^2 + \beta^2 \equiv 1, \tag{7.7}$$

so that the coefficients γ^{-1} and β can, indeed, be interpreted as the cosine and sine function of a certain angle φ related to the speed v by $\sin\varphi = \beta$.

Thus, the optical retardation of light scattered from a continuously illuminated moving object, having spatial depth along the line of sight of the observer, first, "opposes" the Lorentz contraction in the photographs of this object and second, makes the image seen as if the object were rotated (the Terrell effect [60]). The apparent rotation is due to the fact that different points on the image correspond to different positions of the object at which it would be seen in different projection onto the picture plane.

Figure 7.10 shows the artist's visualization of this effect, by Roland Wengenmayr. It depicts the hypothetical snapshots of Einstein passing by on a bike, made at the moment of closest approach (when the line of sight is perpendicular to velocity). As seen from the images, due to the Terrell effect we can see Einstein's back even though actually we face only his side.

Figure 7.10 Computer simulation of hypothetical short exposure snapshots of Einstein passing by on the bike. At sufficiently high speed, the Terrell effect produces the image showing Einstein's back. This tends to mask the Lorentz contraction (courtesy R. Wengenmayr).

In the case of a spherical body, the considered "opposing" effects completely balance each other on the photographic film for any speed. In this special case, the Terrell effect results in an image of a moving sphere, which is totally undistorted.

From the level we have reached in our study of the phenomenon, this is no surprise, considering that rotation of a sphere through any angle does not change its projection. Therefore, one would see a perfect sphere no matter what the velocity. There is no contradiction with anything here, since we are just having in this example a combined result of two different and opposing effects. Obviously, when we sometimes write in our loose jargon about "seeing" a moving body Lorentz-contracted, we mean only one out of the two considered effects – the length contraction only.

Now, suppose we receive a letter with a statement:

> The so-called relativistic length contraction is a myth, which is not supported by a scientific evidence. We never observe in other celestial bodies the contraction which ought to be caused by their motion.

What should we answer to the author?

We could emphasize two points.

First, even without any reference to the optical retardation, the absence of observed contraction in celestial bodies only proves that the theory of relativity is correct in its statement that for objects moving much slower than light, the length contraction must be negligible. The typical relative velocities of celestial bodies are of the order of magnitude of 30 km/s (the orbital velocity of Earth relative to the Sun). If you plug this number into Equation (7.1), you will come up with the result that the relative change in the size of the celestial body because of the length contraction must be about 5×10^{-9}, which is hopelessly beyond today's detection limits for astronomical objects. Second, even if the changes were greater, they would, for the spherical bodies, be masked by the appearance-changing mechanism (optical retardation) as described above. This can be formulated as a limited "shape invariance" in optical observations (it is limited because it holds for spheres only).

Thus, the whole phenomenon, brought up by the author of the letter to testify against the theory of relativity, turns out to testify in its favor. For, without the relativistic length contraction, out of the two opposing effects considered above, only one mechanism – that of the retardation – would be left, with the result that all moving spherical bodies would acquire the appearance of ellipsoids *elongated* along the directions of their velocities. For very fast bodies, we would see them shaped as cucumbers or sausages, rather than spheres. With this theoretical prediction of *classical* physics, we would stand aghast under the night sky trying to explain why the celestial bodies still seem to remain spherical. We would have to invent the theory of relativity to account for this effect. It is only because of the sluggishness of planets and stars that classical physics did not crash in the 1805 or even earlier, when the speed of light was found to be finite and measured with sufficient accuracy.

Thus, our final answer to the question if it is possible to see the length contraction, may be as follows:

> Yes! In the case of the fast-moving spheres *we can see the length contraction by not seeing it*! For, if there were no contraction, the

light retardation would cause the visual shape of any such sphere to change into the corresponding ellipsoid of revolution about its respective velocity. The absence of this effect can only be explained by the actual contraction of the moving sphere along the same direction and in the same proportion as the optically caused image extension.

7.3
The Lorentz Contraction Paradox

Consider two equal sticks A and B with a proper length l_0 each. Both sticks are oriented parallel to one another, but belong to two different reference frames K and K′, respectively. If both sticks are oriented perpendicular to the direction of relative motion of the two systems, their lengths at a certain moment coincide (Figure 7.11a). If the sticks are oriented *along* this direction, their lengths turn out to be different. In K, the stick B is shorter than A by a factor of $\gamma(V)$; in K′, A is shorter than B by the same factor. We are talking now about the comparison of *the same* pair of sticks. And yet the two observers (Alice and Tom) doing the measurements get contradictory results.

To clarify this apparent contradiction, we need to scrutinize the measurement procedure used in the observations. Consider this procedure from the viewpoint of K. Let at the zero moment of the K-time the trailing end of stick B be coincident with the left end of A at the origin (Figure 7.11b). Alice in K measures the instant position of the leading end to be at a distance

$$l = \frac{l_0}{\gamma(V)} \tag{7.8}$$

from the origin. Tom's stick B, which is in motion, is obviously shorter than A. How is this consistent with Tom's result that B must be longer than A? Tom argues that his stick B must be longer because A is Lorentz-contacted in his frame K′ by the same rule (7.8). If both observers' measurements are being performed under identical conditions, then Alice should also see Tom's stick being longer than hers, rather than being shorter. Since the length of Alice's stick in K is l_0, then she must observe Tom's stick to have the length

$$L = l_0\gamma(V). \tag{7.9}$$

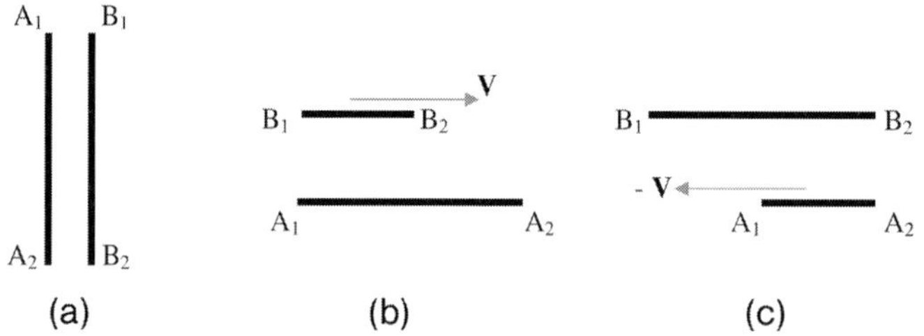

Figure 7.11

This is *not* what she actually measures. What then is the basis for Tom's claim and can we, together with Alice, visualize this basis?

In a nutshell, to measure the length of Tom's stick in K, its end points must be recorded in K simultaneously by *Alice's* time; similarly, the end points of Alice's stick must be marked simultaneously in K′ by *Tom's* time. But simultaneity is a relative characteristic! It is not the same for different observers. Hence, their results are not the same. There is no contradiction in it since these different results refer to different RFs.

Now, this is just a general statement. To get a better understanding, let us consider what happens with time in this measurement.

Let B_1 and B_2 be, respectively, the left and right ends of the stick B. Suppose that when the left ends of both sticks coincide, both local clocks read the moment $t_{A_1} = t'_{B_1} = 0$. Owing to the relativity of simultaneity, the local time in K′ at the instantaneous position of B_2 at $t = 0$ in K will be different from zero. According to our result in (2.56), a leading moving clock distance l ahead of the trailing one must read the earlier time than corresponding trailing clock. If the clock B_1 now reads zero, then the clock B_2 must read

$$t'_{B_2} = -\gamma(V)\frac{V}{c^2}l = -\frac{V}{c^2}l_0. \tag{7.10}$$

Alice can in principle record the readings of Tom's both clocks B_1 and B_2. From the difference in readings of these clocks at one and the same moment of *her* time, she could accordingly conclude that Tom's clocks had not been synchronized. But knowing already something about relativity, she merely attributes this discrepancy to the relativity of time. Moreover, it occurs to her that these two events do not constitute the length measurement in *Tom's* reference frame, since they happen at different moments of his time. She attributes *this* fact to relativity of space. "In order to do a length measurement, Tom must mark the instantaneous positions of the edges of *my* stick at the same moment of time in *his* space," she thinks. If the left ends of both sticks coincide at the zero moment, Tom must mark the right end A_2 also at the zero moment of his time. This means that his clock B_2 must also read the zero time at the moment of marking. Naturally, Alice asks a question where this clock will be at this later moment, when it will read the zero time. She makes a simple calculation. Tom's time between the moments $t' = t'_{B_2}$ and $t' = 0$ is given by Equation (7.10). According to the time dilation effect, this corresponds to a longer time interval Δt in the frame K:

$$\Delta t = \gamma(V)|t'_{B_2}| = \gamma(V)\frac{V}{c^2}l_0. \tag{7.11}$$

Since the clock B_2 is moving with the speed V, it will have traveled the distance

$$x = V\Delta t = \gamma(V)\frac{V^2}{c^2}l_0. \tag{7.12}$$

Therefore, the distance in K between the instant positions of the end points of stick B, which would be simultaneous in K′, is given by

$$L = l + x = \frac{l_0}{\gamma(V)} + \gamma(V)\frac{V^2}{c^2}l_0 = l_0\gamma(V). \tag{7.13}$$

It is precisely $\gamma(V)$ times greater than the length of Alice's stick, in total agreement with Tom's claim!

This result does not mean that L is the proper length of Tom's stick – this would contradict the initial conditions, according to which both sticks have the same proper length l_0. Despite the fact that L is the distance between the end points of B taken at the same moment of Tom's time, it cannot be the proper length of Tom's stick, because it is *measured by Alice in K, not by Tom in K'*. Moreover, since B is stationary in Tom's frame, the requirement that the end points of B be marked at the same moment of his time is not even essential for Tom. On the other hand, the described procedure for obtaining L does not constitute the length measurement of B in Alice's reference frame, because in this procedure, the end points are not taken at the same moment of *her* time. What is true for both observers is that the ratio of L and l_0 is $\gamma(V)$. In Tom's reference frame, the length L and length l_0 are both Lorentz-contracted by a factor of $\gamma(V)$, so that L becomes L_0 and l_0 becomes $l_0/\gamma(V)$, and we recover Tom's original statement that in his frame the stick B is $\gamma(V)$ times longer than stick A.

We may also look at it from slightly different perspective. Namely, ask where is now (at the zero time in K) the local primed clock on Tom's stick (Figure 7.11) that will show the zero time when passing by the end point A_2?

Denote the unknown position of this clock x'. This is its distance in K' from the left end of Tom's stick, or, which is the same, its instantaneous distance from the origin and the left end of Alice's stick. According to (2.56), the current reading of this clock is

$$t'(x') = -\frac{V}{c^2}x'. \tag{7.14}$$

The current distance between this clock and the origin in K is $x = x'/\gamma(V)$, its current distance from A_2 is $l_0 - x'/\gamma(V)$. The time in K to travel this distance is

$$t = \frac{l_0 - (x'/\gamma(V))}{V}. \tag{7.15}$$

According to time dilation effect, this time is related to the proper time t' of the primed clock by

$$t = \gamma(V)|t'| \tag{7.16}$$

and therefore

$$\gamma(V)\frac{V}{c^2}x' = \frac{l_0 - (x'/\gamma(V))}{V}. \tag{7.17}$$

Solving for x', we find

$$x' = \frac{l_0}{\gamma(V)} \quad \text{or} \quad x = \frac{l}{\gamma(V)}. \tag{7.18}$$

This resolves the paradox.

7.4 Predicaments of Relativistic Train

In this section, we shall discuss what had at first emerged as an apparently unsolvable paradox. The paradox is closely linked to the relativity of length, which we have considered before. We will hereafter consider the paradox as a dispute between the two opposing sides.

... The general public was alerted to the implications of the length contraction paradox after a superpower on planet Rulia had come forth with an ambitious Project RT (Relativistic Train). The problem first popped up with the question: what would happen when such a train has to cross a canyon or river. Here, I can give a very brief and simplified description of the problem, retaining only the most essential details. Imagine a train that has to pass a deep canyon. The train just fits across the canyon, so that its proper length L_t is equal to the proper length L_b of the bridge (Figure 7.12):

$$L_t = L_b = L_0. \tag{7.19}$$

Here, L_0 stands for the common proper length of the bridge and the train. Originally, the bridge had been designed to sustain the train's weight $W_0 = m_0 g$, where m_0 is the rest mass of the train and g is the acceleration because of gravity on Rulia. The preliminary tests at low speeds were quite successful. The train had smoothly passed the bridge.

Now, imagine this train moving with a speed close to c. Then it could fly over the canyon with even no bridge at all. The crossing time would be so small that the train would have practically no time to fall down by a tiniest increment. Let us, for example, estimate the distance the train on Earth would fall while flying across a canyon 1 km wide at a speed $V = 200\,000$ km/s, that is, two-thirds the speed of light. The time it takes the front of the train to cross the canyon is $t = L_0/V = 5 \times 10^{-6}$ s. Denoting the vertical direction as y and entering the data into the equation $y = (1/2)g_E t^2$ with $g_E = 9.8\,\mathrm{m/s^2}$, we find $y = 1.23 \times 10^{-10}$ m, which is about the size of an atom. It is smaller than the finest irregularities of the rails' surface and definitely less than the

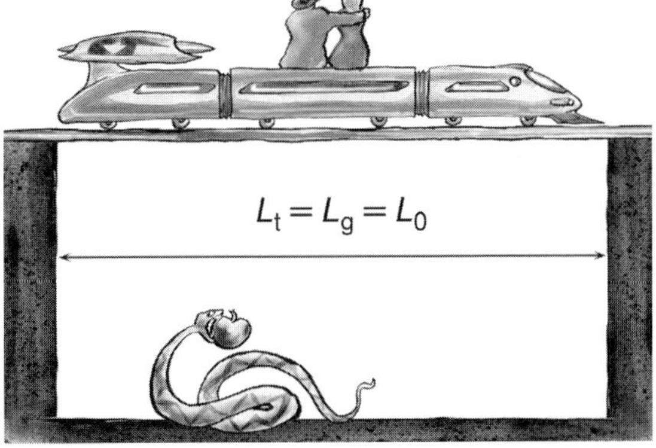

Figure 7.12

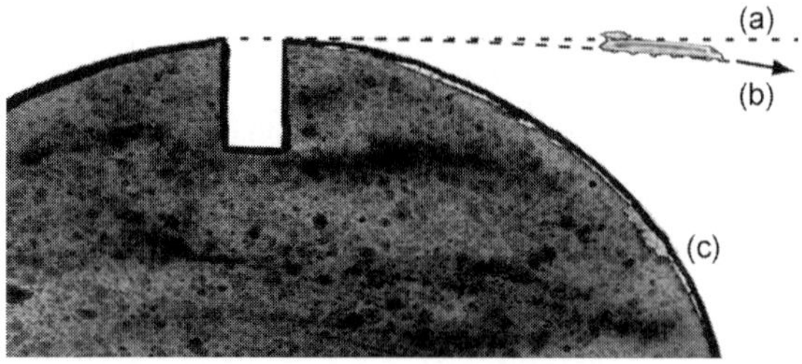

Figure 7.13 An escape trajectory of a relativistic object. The trajectory's curvature is highly exaggerated: (a) the tangent to the planet's surface; (b) the escape trajectory; (c) one of the planet's meridians.

distance the Earth's spherical surface curves away from the flat plane over a horizontal shift of 1 km. In other words, the curvature of the train's path would be less than the curvature of the Earth's surface. Such a train would, with even no bridge in place, fly off the Earth rather than go down, since its speed would by far exceed the escape speed for Earth (which is just 11.2 km/s) (Figure 7.13). Apart from these, there is a finite *time factor* associated with the breakup of a bridge or any other system under an excessive force. A certain time is needed for a bridge to disintegrate and let the train down; this is the time necessary for a given force to do corresponding work. And this time for any real bridge is considerably greater than 5×10^{-6} s needed for the train to cross the bridge in our example.

But we are now concerned with the conceptual aspect of the problem rather than with technical ones. Therefore, to avoid otherwise important technical details complicating the problem, we will represent a real system by its idealized model. We will make the following assumptions: both the train and the deck of the bridge are infinitely thin, so that in the case of a crash it would take no time for the train to fall through the deck; the deck itself is ideally straight. In other words, Rulia is so huge that its surface is curved *less* than would be the trajectory of relativistic train over the canyon even without any bridge. We may therefore not bother about the escape preventing the crash, since there would be no escape for the train under these conditions. Astrophysicists would say that such a planet just falls short of becoming a huge black hole. For us, however, the assumption about the Rulia's size just means that to a high accuracy we can consider the corresponding area of the planet's surface as flat and the gravitational field within this area homogeneous. Accordingly, the train's trajectory in case of a crash would be that of a projectile – a well-known parabolic path from the introductory college physics. And we also suppose that the bridge is made of a highly idealized material that responds instantaneously to an applied force. It breaks instantaneously (in its own reference frame) under the whole train when the load reaches a certain critical limit. A model of such a bridge might be the one consisting of a number of sections, each suspended from supporting cable, and only loosely connected to the neighboring sections (Figure 7.14). Then, the

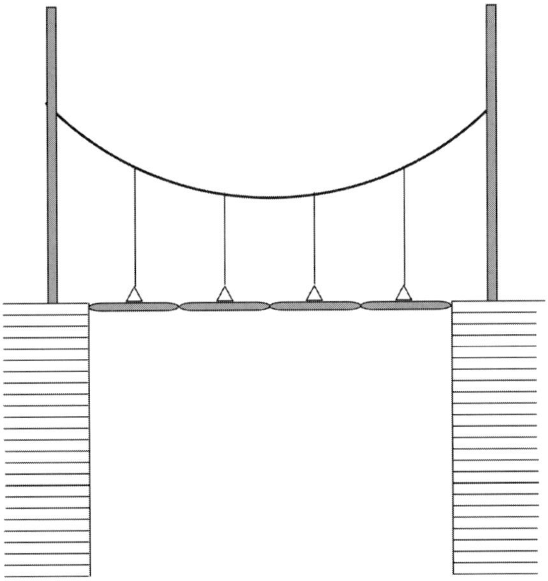

Figure 7.14

section will break down when its load reaches the breaking point of its respective cable.

We assume that the breaking point of the whole bridge is the sum of breaking points of all the cables holding it and this sum exceeds the weight of the stationary train almost by a factor of $\gamma^2(V)$. This was considered more than enough for the bridge to sustain the weight of the *moving* train, which was expected to exceed its proper weight W_0 only by a factor of $\gamma(V)$.

To further simplify the following treatment, we suppose that the Lorentz-contracted train just fits into one section. In other words, the train's velocity will be such that the corresponding Lorentz factor is an integer equal to the number of sections.

We also assume that all technical problems associated with design and launch of relativistic train have been successively solved.

Now, after all these assumptions, the question is: how would the motion with relativistic velocity affect the train and the bridge? Will the bridge sustain the train or does it need to be reinforced? Will the train cross the canyon safely?

An international team of experts from Earth had been invited to address the problem. Among the team members there was an engineer who had once worked at Superconducting Super-Collider (SSC) and then carried out some interesting experiments with Mr. O'Bryan described in Section 4.4. The engineer's opinion was unequivocal: the bridge under given conditions would not sustain the load and the train would crash. The underlying reasoning is simple and straightforward.

"At a speed V close to c," the engineer said, "the train will undergo the longitudinal length contraction, so its length will be

$$l_t = \gamma^{-1}(V) L_0 \ll L_0, \tag{7.20}$$

and its weight will accordingly increase owing to relativistic increase of the mass:

$$W = mg = m_0\gamma(V)g = W_0\gamma(V) \gg W_0. \tag{7.21}$$

The bridge would hold this weight easily if the weight were distributed uniformly *over its whole length*. However, due to contraction, all this weight will fall only on one section of the bridge, constituting only $1/\gamma(V)$ fraction of its whole length. Accordingly, this section will be able to hold only the corresponding fraction of what the whole bridge can hold. The whole bridge can hold a little less than $\gamma^2(V)\,W_0$, so one section can hold a little less than $\gamma(V)\,W_0$. But the actual weight of the train will be exactly $\gamma(V)\,W_0$, that is, it will go beyond the strength limit of the cable supporting the section. Therefore, the cable will snap and the section must collapse. The train will crash, smashing against the opposite wall of the abyss."

This conclusion is illustrated in Figure 7.15, in which, of course, the distance the train would fall is highly exaggerated, to emphasize the result.

The engineer has encountered a formidable opposition from the chief expert, who turned out to be one of the staunchest proponents of the project. The chief expert was a very important man, so important that nobody had ever either heard or dared to ask about his real name. Everybody had respectfully referred to him as Mr. Ex, emphasizing his indisputable expertise in the field of relativistic engineering.

"Ladies and Gentlemen," Mr. Ex said amidst the awed hush. "If you want to see a real picture of the phenomenon, look at it from the viewpoint of the train's passenger.

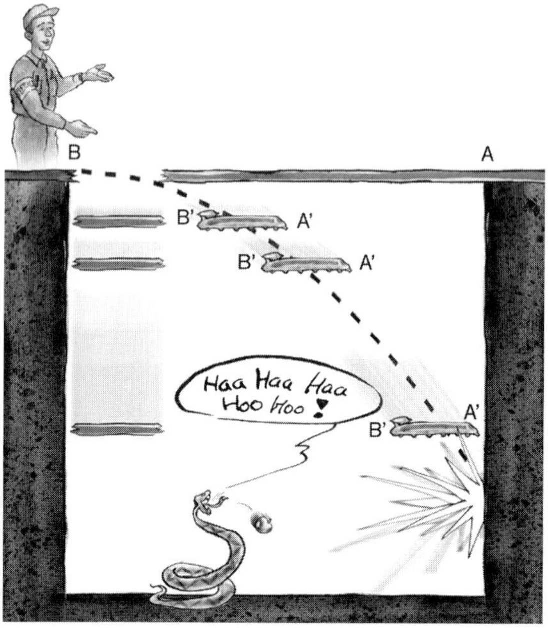

Figure 7.15 The crash of relativistic train as expected to be observed by the engineer. The contracted and overweight train collapses the underlying part of the bridge. The train slips down through the formed gap and traces out a parabola (whose curvature is highly exaggerated).

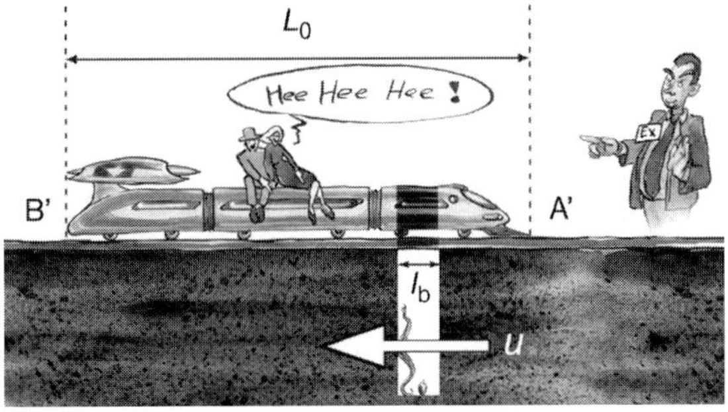

Figure 7.16 The canyon with the bridge as observed by a train's passenger (in Mr. Ex's presentation).

You then will see something different from the scenario conjured up by Mr. Fletcher (this was the engineer's name). The canyon with the bridge passes rushing by your train. The width of the canyon and the length of the bridge undergo relativistic length contraction:

$$l_b = \gamma^{-1}(V) L_0 \ll L_0. \tag{7.22}$$

You can see from this equation that they are much shorter than the train."

Mr. Ex has illustrated this result by Figure 7.16 showing a ridiculously narrow crevice under the train.

"The train's weight," he went on after a significant pause, "will increase from its rest value W_0 up to the value of $W = \gamma(V) W_0$, as stated by Mr. Fletcher, but for a different reason. Since the planet is now moving relative to the train, the planet's mass M_0 undergoes relativistic increase:

$$M = \gamma(V) M_0. \tag{7.23}$$

This will cause the increase of both the planet's gravitational pull and the corresponding acceleration due to gravity

$$g' = G\frac{M}{R_0^2} = G\frac{M_0}{R_0^2}\gamma(V) = g\gamma(V), \tag{7.24}$$

where R_0 is the planet's radius. The resulting weight of the train will be

$$W' = m_0 g' = W_0 \gamma(V) = W. \tag{7.25}$$

To Mr. Fletcher's credit, this particular piece of information about train's weight in his report turned out to be correct. But, Ladies and Gentlemen," Mr. Ex concluded triumphantly after another significant pause, "the point is that only a small fraction of this weight will fall atop the bridge, the rest being supported by the ground. As you can see from Figure 7.16, this fraction is the same as the shaded fraction of the train's proper length L_0 that fits into the contracted bridge:

$$\frac{l_b}{L_0} = \gamma^{-1}(V). \tag{7.26}$$

The resulting load on the bridge will be only $\gamma^{-1}(V)W = W_0$. It is by far less than the breaking point of the bridge. Mister Fletcher also brought up an argument that, in his view, the train's weight, due to the Lorentz contraction, will be concentrated only on one section of the bridge. But even so, the load would be almost by a factor of $\gamma(V)$ less than the breaking point of one section which is almost $\gamma(V)W_0$. Moreover, if you look again at Figure 7.16, showing *the passenger's* perspective, you will see that one section of the bridge needs to support much smaller fraction of the train's weight than the whole bridge. If the bridge needs only to support $1/\gamma(V)$ of the whole train, one section indicated by Mr. Fletcher needs to support only $1/\gamma^2(V)$ of the whole train. The corresponding load on the section will be equal to $W_0/\gamma(V)$. Again, the actual load on the section will be $\gamma^2(V)$ times less than its breaking point $\gamma(V)W_0$. You see that the safety of the train is ensured by a huge margin."

Mr. Ex made a third pause and fired his last victorious shot:

"We come *unavoidably* (he stressed the word) to a conclusion that the train will pass safely across the canyon – for two reasons. Geometrically, the train cannot go down because it just does not fit into the canyon's width; physically, it cannot go down because only a fraction of its full weight bears upon the bridge and the fraction by far less than the breaking point of the bridge. Geometry and physics, the two most general and established sciences about Nature, both give the same answer."

The final glorious scene followed after the engineer had been given word for a reply.

"Sir," said the engineer, "as we all know, relativity grants equal rights to all inertial observers. Therefore, a ground-based observer deserves the same respect as the passenger of the train, and his conclusion must be considered with equal attention. And if you do not see any error in my reasoning, then ..."

"Young man," Mr. Ex snapped, "it is not my job to look for errors in your reasoning. You better find one in mine."

Mr. Ex probably tried to imitate the famous Russian physicist Lev Landau who had allegedly been the author of the above aphorism. And, although the engineer had suspected that Mr. Ex was not a Landau, he could not, at the height and heat of the moment, spot any error in his opponent's argument. After a moment's silence, the audience burst into applause, and the project had been accepted. To make his victory more impressive, Mr. Ex even volunteered to board the experimental train (which was originally planned to be operated under remote control) and personally carry out all the measurements during the first relativistic test. It was decided to design a special cockpit for Mr. Ex. Under insistence of the Safety Board, the cockpit was to be installed at the rear of the train and equipped with an ultrafast catapulting system.

The night after the meeting, the engineer found what he believed to be an inconsistency in Mr. Ex's treatment of the problem. He wrote down the detailed description of his solution and mailed it to Mr. Ex together with the letter of his resignation from the project.

The essential parts of the engineer's solution ran as follows.

In the problem discussed, we have a logical paradox that goes beyond simple relativity of length. The latter paradox is easily resolved by saying that statement "The train is shorter than the bridge" is correct and the statement "The bridge is shorter than the train" is also correct. If you ask how the two mutually exclusive statements may be both correct, the answer is that they are not mutually exclusive, because the physical property "length" is not absolute. It depends on the reference frame used when measuring the length of an object. The above two statements correspond to the two different reference frames. The first is associated with the bridge and the second one with the train. Therefore, there is no paradox here, and both observers may be right, with the caution that each speaks for his or her own reference frame.

But when we speak about what *happens* to the train, we come to a statement of a different nature. It is a statement about an *occurrence*, not about a physical property that can be relative. An occurrence is something absolute, something that can be recorded by *any* observer. The reader can recall the examples with the number of passengers in a car or chemical composition of the bullet in Chapter 1. Those were the examples of an absolute physical fact. Likewise, the train's crash (or its absence) is an absolute physical fact. If the train crashes, the event can be observed from any reference frame. If not, then nobody will observe any crash, irrespective of the reference frame. The train either crashes or does not, without any reference to a system of reference. Therefore, if one observer predicts that a certain condition will cause the train's crash, while the other one holds that this condition ensures safety, someone must be wrong. Who is?

In his report, the engineer indicates three errors in Mr. Ex's argument. The first error was the statement that the gravity force on the train was the same in the train's reference frame as in the engineer's one (Equations (7.24) and (7.25)). That was wrong because it did not take into account the length contraction of the planet itself and its field. Mr. Ex wrote Equation (7.24) (the Newtonian expression for g'), which only holds for a static spherically symmetrical source of gravity. But because of the planet's motion, it is not a static source in the train's reference frame. Apart from the Lorentz contraction of the planet, one has to take account of the retardation effect similar to those discussed in Sections 6.9, 6.10, 7.1, and 7.2 for the electromagnetic phenomena. According to relativity, all known interactions, regardless of their physical nature, have common propagation speed c. Therefore, as we have already learnt, a force exerted on an object by a system of *moving* masses at a given instant is not determined by the positions of the masses at the same instant. A proper account should be taken of the *retardation time* needed for a gravitational perturbation from each moving mass to reach the object. If we want to get a quantitative result, we have to write down the relativistic expressions for all retarded forces owing to all small masses constituting the moving planet and sum them up, that is, perform the integration. In the case of a point mass, we would need to find the corresponding *retarded* position of this mass.

But in all cases we can get the result much more easily by just applying the general transformation rules for force. As we found out in Chapter 5, these rules automatically take account of all possible interactions and their retardation.

As the train's passenger, one will be interested in the component of the gravitation force transverse to the direction of train's motion. According to (5.19), this component is expressed in terms of the corresponding component of the same force observed from Rulia as

$$F'_g = \gamma(V) F_g. \tag{7.27}$$

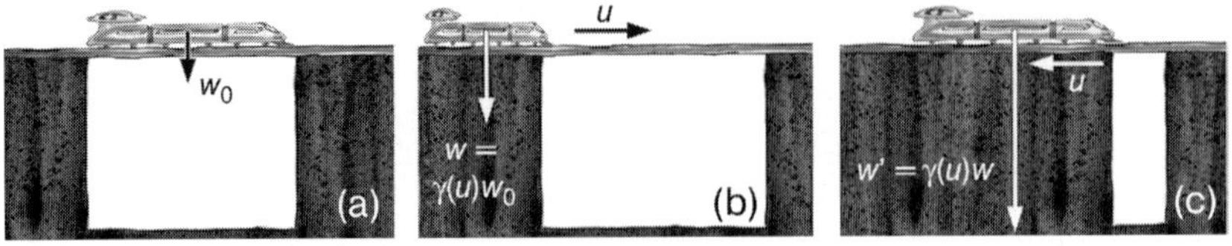

Figure 7.17 The gravity force (a) on the stationary train; (b) on the moving train ($\nu \neq 0$) in the engineer's RF; (c) on the train in Mr. Ex's RF ($\nu \neq 0$) Rulia is moving.

In other words, Equations (7.24) and (7.25) written by Mr. Ex are wrong. The transverse force is *not* the same in the two different inertial frames. If the force on the train is $F_g = W = \gamma(V) W_0$ in the engineer's reference frame, it would be

$$F'_g = W' = \gamma(V)W = \gamma^2(V)W_0 \gg W \tag{7.28}$$

in Mr. Ex's reference frame (Figure 7.17).

There is a deep irony in the fact that just when Mr. Ex had so condescendingly acknowledged correctness of the engineer's conclusion about the gravity force being equal to $\gamma(V)W_0$, *he was himself wrong* in applying this conclusion to *his* reference frame. As we can see from the transformation rules (5.19), the gravity force on the train in Mr. Ex's reference frame is much greater than in the engineer's. Therefore, even a small fraction of the train that fits into the bridge will, contrary to Mr. Ex's expectations, cause a greater burden than the bridge can hold. This was the first of Mr. Ex's blunders.

His second error was in his unspoken assumption that the breaking point of the bridge (or one of its sections) was *the same* in both reference frames – his and the engineer's. Equations (5.19) tell us that a force on an object, exerted perpendicular to the direction of relative motion of the two reference frames, is maximal in the rest frame of the object. In the case of the bridge, its rest frame is the planet Rulia. All observers moving relative to Rulia in the same direction as the train will measure a *smaller* force on the bridge than residents of the planet. If the tests on Rulia have shown that the breaking point for the bridge was F_c, the passenger of the train would measure for this force the value $F'_c = F_c/\gamma(V)$. According to our initial assumption, F_c was equal to $\gamma^2(V)W_0$. Therefore,

$$F_c = \gamma^2(V)W_0, \qquad F'_c = \frac{F_c}{\gamma(V)} = \gamma(V)W_0 \tag{7.29}$$

for the whole bridge and

$$f_c = \gamma(V)W_0, \qquad f'_c = \frac{f_c}{\gamma(V)} = W_0 \tag{7.30}$$

for one section. Now, recall that the weight of the train in Mr. Ex's frame is $\gamma^2(V)$ W_0, its fraction falling on the bridge is $\gamma(V)W_0$, and its fraction falling onto one section of the bridge is just W_0. But even this small fraction turns out, according to the initial conditions discussed above, to be slightly exceeding the breaking point in this reference frame. Thus, analyzing the situation in Mr. Ex's reference frame, we arrive at precisely the same result that the engineer obtained analyzing the situation from *his* perspective – that the weight falling on one section of the bridge slightly exceeds its breaking point. The specific physical characteristics of the system are different for different observers, but the outcome of the process is the same.

This is where the transformation rules (5.19) for the force manifest themselves in all their glory: the train is moving relative to Rulia, therefore the gravity force on the train is $\gamma(V)$ times *greater* in its rest frame than it is in the rest frame of Rulia; the bridge is stationary relative to Rulia, therefore the breaking force for the bridge is $\gamma(V)$ times *smaller* in the train's rest frame than in the rest frame of Rulia. It is easier to break a moving object than a stationary one!

The third error of Mr. Ex was in his "geometrical" treatment of the problem. According to Mr. Ex, the fact that the train in his reference frame will not fit into the narrow crevice must prevent the train from crashing. But it does not. The train can crash piecemeal, bending and going down the crevice in small increments, one at a time (Figure 7.18). From a passenger's perspective, the fall would be accompanied by a continuous deformation. First the front end goes down to get smashed against the opposite wall of the crevice; this gives room for the following parts to do the same; and this goes on until the whole train is swallowed up by the abyss. Instantaneous snapshots of the process show that the falling train consists nearly all the time of two different parts: one still horizontal and one dangling above the crevice. The first one is straight and the second one is curved down. The closer to the front, the steeper the slope of the curve. After the front touches the opposite wall of the crevice, there emerges also a third part, common for all observers, the one smashed against the wall, but we do not consider it.

Since the moment the train's front dives down, we can no longer speak about "the train's reference frame," because the train is no longer one rigid body. It is more like a fluent combination of rods and hooks, continuously changing from more "roddish" to more "hookish." We shall under these circumstances refer to Mr. Ex's reference frame, rather than the train's one, since Mr. Ex, presumably located at the train's rear, would keep moving in a straight line nearly all the time till the very end (Figure 7.18).

Now, pit this picture against the one observed by the engineer: the train going down all at once, in one piece, keeping the horizontal position all the time. One and the same thing is a totally straight rod for some observers and continuously changing combination of straight and curved lines for others. How can it be? Well, it is just one of those weird pictures of the world which, for all their strangeness, turn out to be correct and reasonably explainable. The shape of an object is a relative physical property. What we call shape is just a three-dimensional cross section, cut trough an absolute four-dimensional structure in the Minkowski's world. Such sections are different for different "cuts" just as are two-dimensional sections cut through a three-dimensional object (Figure 7.19). The way the object looks *now* is not the same for

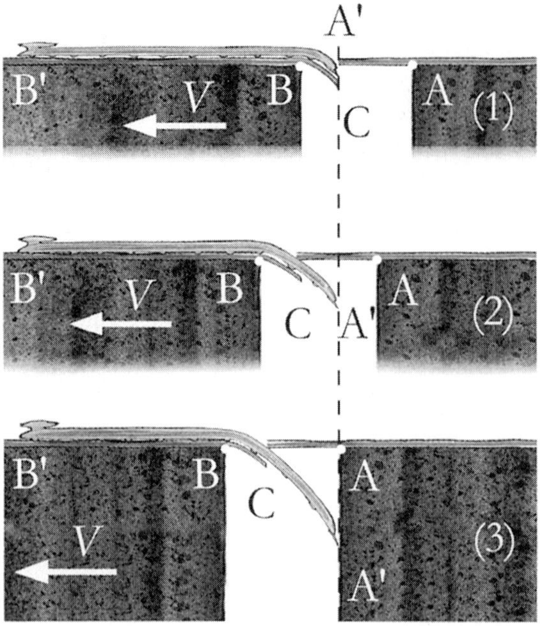

Figure 7.18 The train crash to be observed by Mr. Ex (engineer's presentation). Three successive moments of crash are depicted here as (1), (2), and (3). The train is slipping down through the gap in the moving bridge. C is the broken fraction of the bridge.

everyone because *now* is not one instant for everyone. In our case, the situation can be made clear if we represent the train by a moving segment and draw its *worldsheet* in space-time (Figure 7.20). It is clearly seen from Figure 7.20 that what Mr. Ex would observe *now* are different moments in the train's history observed by the engineer,

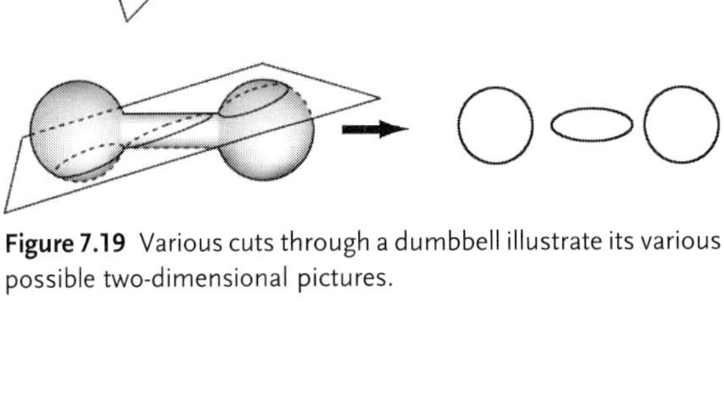

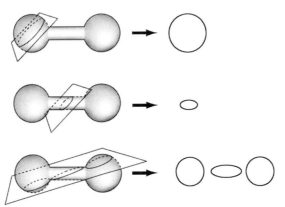

Figure 7.19 Various cuts through a dumbbell illustrate its various possible two-dimensional pictures.

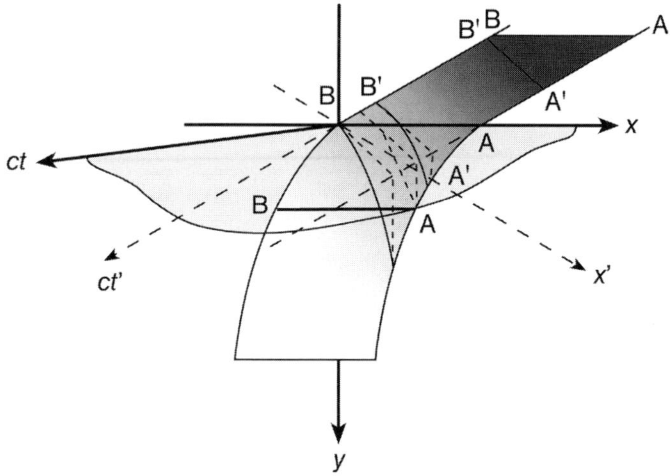

Figure 7.20 The "worldsheet" of the train. Its instantaneous cuts in the engineer's reference frame (K) and that of Mr. Ex's (K′) give the shapes of the train seen by the two observers. All cuts in K are straight segments parallel to the *x*-axis (three successive lines ab – before, at, and after the moment of gap formation in the bridge.) The cuts in K′ are intersections with vertical planes parallel to *x*′-axis. Their shapes depend on time *t*′. The first line A′B′ (when looking in the direction of increasing *ct*′) is a straight segment; it represents the train's shape before the train reaches the bridge. The second line is composed of the straight and curved parts and gives the shape of the train soon after the bridge had started to collapse (the collapse does not happen at one instant in K′). The third cut is a curved line; it represents the train's shape when its rear passes the edge B.

and vice versa. The two events A′ and B′ observed simultaneously by Mr. Ex occur at different moments in the engineer's reference frame: the event A′ at the front happens *after* the event B′ at the rear. This means that the front was farther down the crevice than the rear when both events were observed simultaneously by Mr. Ex. In other words, Mr. Ex observes the rear of the train at different horizontal levels. Therefore, the line going from A′ to B′, that is, the form of the train, turns out to be tilted in his reference frame. The situation is in this respect similar to that with a rising surface in Section 4.4. The difference is only in the initial conditions: first, the "water" (the train) is sinking now and sinking with the acceleration; second, the train remains horizontal in the engineer's reference frame, rather than in Mr. Ex's (we also have Mr. Ex instead of Mr. O'Bryan, but this is irrelevant to physics). Now, if the point A′ is "captured" *after* point B′ in the engineer's reference frame, it is sinking faster. We know that the tilt is proportional to the velocity v of sinking. Therefore, the tilt at A′ is steeper than that at B′. Thus, we get, at least on the qualitative level, an account of the observed features of the train's crash in Mr. Ex's reference frame.

The engineer's letter also contained a more detailed quantitative description of the transition process between the straight and the "hooked" parts of the train. I bring it along here for the need of more sophisticated readers. Let us place the origins of the two discussed reference frames at points B and B′ respectively and start counting time from the moment when the origins of both systems coincide. Then the train's vertical displacement in the ground-based reference frame K is

$$y = \begin{cases} 0, & t<0, \\ -\dfrac{1}{2}gt^2, & t>0. \end{cases} \tag{7.31}$$

In the passenger's reference frame S′, the same displacement can be obtained by just applying Lorentz transformation to y and t (note that transverse displacement is the same in both systems):

$$y' = y = 0 \quad \text{for} \quad t' + \frac{V}{c^2}x' < 0, \tag{7.32a}$$

$$y' = y = -\frac{1}{2}g\gamma^2(V)\left(t' + \frac{Vx'}{c^2}\right)^2 \quad \text{for} \quad t' + \frac{V}{c^2}x' > 0. \tag{7.32b}$$

As we see, an observer comoving with the train's rear would measure the acceleration of the falling part of the train as $g' = \gamma^2(V)g$. The instantaneous velocity of the train's fall at a point x' is

$$v'(x', t') = 0 \quad \text{for} \quad t' + \frac{V}{c^2}x' < 0, \tag{7.33a}$$

$$v'(x', t') = -\frac{1}{2}g'\left(t' + \frac{V}{c^2}x'\right) \quad \text{for} \quad t' + \frac{V}{c^2}x' > 0. \tag{7.33b}$$

One might argue that the expressions (7.33) cannot be true because they give $v'(x', 0) = -(V/c^2)x' \neq 0$ for the initial velocity of the fall at the zero moment $t' = 0$, whereas the fall in a given case begins from rest ($v' = 0$). But the zero moment $t' = 0$ is *not* the initial moment of the fall for a point x' on the train. Recall that the zero moment has been determined as the moment when the *rear* of the train B′ coincides with the edge B of the bridge. By this time all the points x' to the front (i.e., satisfying the condition (7.32b)) have already fallen down a certain distance and acquired nonzero vertical velocity. This velocity is given by the second term in (7.33b).

The instantaneous shape of the train in the system S′ is automatically described by (7.32) as consisting of two parts. One is straight and another one is curved down with its tilt increasing for greater x'. The expression for a local tilt at a moment t' is obtained from (7.32) by taking the derivative

$$S'(x', t') \equiv \frac{dy'}{dx'} = 0 \quad \text{for} \quad t' + \frac{V}{c^2}x' < 0, \tag{7.34a}$$

$$S'(x', t') \equiv \frac{dy'}{dx'} = -g'\frac{V}{c^2}\left(t' + \frac{V}{c^2}x'\right) \quad \text{for} \quad t' + \frac{V}{c^2}x' > 0. \tag{7.34b}$$

Using (7.33b), the last equation can be rewritten as

$$S'(x', t') = \tan\alpha = 0 \quad \text{for} \quad t' + \frac{V}{c^2}x' < 0, \tag{7.35a}$$

$$S'(x', t') = \tan\alpha = -\frac{Vv'}{c^2} \quad \text{for} \quad t' + \frac{V}{c^2}x' > 0 \tag{7.35b}$$

in direct agreement with the engineer's result for the tilt of moving water surface in Section 4.4.

The transition point x'_V between the "roddish" and "hookish" parts of the train is determined by condition $t' + (V/c^2)x'_V = 0$, that is,

$$x'_V = -\frac{c^2}{V}t'. \tag{7.36}$$

For each moment $t' < 0$, all x' to the left of x'_V belong to the straight segment; all x' to the right of this point belong to the "hooked" part. The succession of different snapshots of the process at different moments in both systems is shown in Figures 7.15 and 7.18. While the train falls as one horizontal piece in Figure 7.15, it slides gradually down in Figure 7.18 through the narrow slot in the bridge. In both reference frames, the slot is the $\gamma^{-1}(V)$ fraction of the whole bridge. This picture completes the engineer's description of the process as expected to be observed from the train's cockpit.

... A few days later the engineer received an official note of acceptance of his resignation. He was surprised to find an attachment written personally by Mr. Ex. The letter started as follows:

"Sir,

since you have made an attempt to find an error in my calculations, I am now returning the favor by indicating one in yours."

"Hm," grinned the engineer, "so we are promoted from a young man to Sir along with our resignation. What a twist of a career!"

But what he read next, withered his grin.

"Look at Figure 7.21. It is drawn following your own description of the train's fate as observed from its rear. Clearly, the separation point between the two parts of the train coincides with the point B (the left edge of the bridge). It is at this edge

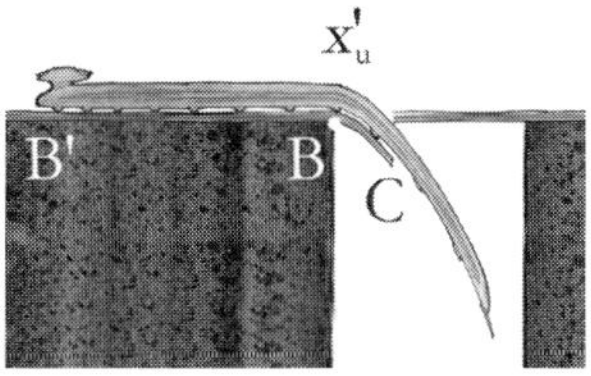

Figure 7.21 An instant in the process of train crash depicted by Mr. Ex in his response to the engineer's letter. The separation point x'_u between the straight (horizontal) and curved part of the train appears to coincide with edge B of the bridge. They appear to slide *together* to the rear of the train in Mr. Ex's reference frame.

where the train starts bending down in your scenario. As time goes by, the point x'_V and the edge B both slide together toward the rear of the train. Your Equation (7.36) states that they both move down the track with a speed

$$u = \frac{x'_V}{t'} = \frac{c^2}{V}. \tag{7.37}$$

This result of yours is wrong for two reasons. First, the speed u is greater than the speed of light! It thus follows from your treatment that the edge B moves faster than light, which is in flat contradiction with the theory of relativity. Second, this result is also in ridiculous contradiction with the initial conditions of the problem, according to which the speed of the edge is less than c and equals V. And, finally, your beautiful pictures in Figures 7.18 and 7.20 of the bent train may well represent a stick of putty, but not a real train. Our train is rigid, it presents stiff resistance against any deforming force. Therefore, your pictures contradict reality. This shows that all the rest of your reasoning had been wrong and you could have made a better use of your free time by visiting a nightclub. I wish you all success in whatever career you choose."

All the rest of the day the engineer was deep in thought. Mr. Ex's argument about Equation (7.36) was irrefutable. The engineer must have made a grave error. And his pictures in Figures 7.18 and 7.20 indeed seem incompatible with train's rigidity. Where could he have gone wrong? Up to this moment everything seemed to fall so neatly in place. But after this moment ...

Instead of visiting a nightclub, the engineer spent the night over his papers. The next morning he mailed his second letter to Mr. Ex. It is not known for sure whether Mr. Ex had enough time to give full consideration to this letter ...

A few months of preparation had passed with much fuss and fanfare. But on the evening of the day of test the mass media reported about miserable failure of the Project RT. The train had crashed into the canyon in the very first test. The only system that proved to be efficient was the catapult that launched the cockpit with its passenger into space and then delivered it back to Earth. Mr. Ex was found to be safe and sound in all respects except that he was unable to speak for a considerable length of time during which an investigation had been carried out. The investigation found, among other things, the engineer's letters with the full account of the problem ...

The engineer's account, together with the comments of other experts, was published in the "Final Report" of the investigating committee. The following gives a simplified description of the engineer's last letter to Mr. Ex. The letter starts with the acknowledgment that Mr. Ex's statement about the superluminal velocity of the separation point x'_V (let us call it point D) was irrefutable. It follows directly from Lorentz transformations. The point D does move faster than light. But this does not in any way undermine the validity of the engineer's results, because point D does *not* coincide with the edge B, except for the very last moment when they both merge at the rear of the train. Let us take a close look at Figure 7.22. It corresponds to train's velocity $V = 0.968c$, for which $\gamma(V) = 4$. Accordingly, the train gets contracted down to 1/4 of its proper length (and of the bridge's length) in the engineer's reference frame. The equal fraction of the bridge breaks under the train and falls down into the canyon. In Mr. Ex's reference frame it is *the bridge* that shrinks down to 1/4 of its proper length; therefore, the gap in the bridge, being only 1/4 of the whole bridge, is just

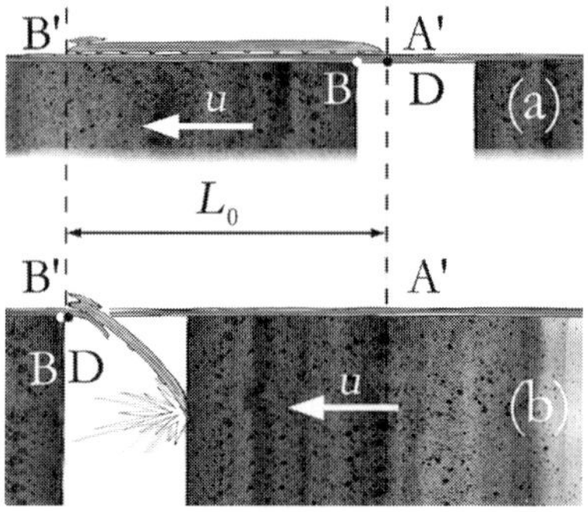

Figure 7.22 The initial (a) and final (b) moments of the train crash to be observed in Mr. Ex's reference frame (engineer's presentation). The separation point x'_u between the straight and curved part of the train (point D) does *not* coincide with B except for the very last moment of train's crash. The edge B moves to the left slower than light (with the speed u) and the separation point D moves faster than light, so that the product of their speeds is $Vu' = c^2$. It is instructive to compare this figure with Figure 7.15 (engineer's reference frame). There, both the train and the falling part of the bridge stay horizontal and remain on one (sinking) level.

$(1/4)^2 = 1/16$ of the train's proper length. Figure 7.22a depicts a moment when the train is 1/16 of its full length above the canyon. Since the gravity force between the train and Rulia is, according to (7.29), 16 times the weight of the stationary train, the load on the bridge is at its limit. This is a moment when the bridge is just about to break. The train begins to bend where the bridge begins to break. But it is not at the point B! The gap in the bridge forms instantaneously in the engineer's reference frame. But it is not one instant in Mr. Ex's reference frame! In his reference frame, the point on the bridge right under the train's front (point A′) gives in first, and it is *here* where the train starts to bend, shoving its head under the remaining part of the bridge on the right. The part on the left (which is already doomed to be broken) remains at this moment strictly horizontal and so does the train. There is thus an initial separation equal to the distance A′B (1/16 of the train's proper length) between the points D and B. Only some time later (at the zero moment $t' = 0$) will the edge B of the bridge be observed as giving in. But, by this time the edge B will reach B′, so that both points B and D will have merged only at the *rear* of the train. By the zero moment $t' = 0$ the process of train's bending in Mr. Ex's reference frame has been completed and the train is converted entirely from a rod into a hook. This is where relativity of time reveals itself in its full sway. In order for D to catch up with B at the rear of the train, it must move faster than light while the edge B is, as it should, moving slower than light with the prescribed velocity V.

The fact that the separation point moves faster than light does not by itself contradict anything, since there is *no energy transfer* associated with this motion. We will see in Chapter 11 that this kind of motion is a rather common physical phenomenon.

Now, let us describe the whole process symbolically. The time t'_D it takes the separation point to move along the train from its front to the rear is

$$t'_D = \frac{L_0}{u} = \frac{V}{c^2} L_0. \tag{7.38}$$

The time t'_B it takes the edge B to reach the train's rear is BB'/V, where BB' is the original distance between them when the bridge started to break. From the above numerical example and Figure 7.21 this distance is readily found to be

$$BB' = L_0 - \gamma^{-2}(V)L_0 = \left[1 - \gamma^{-2}(V)\right]L_0.$$

Therefore,

$$t'_B = \frac{BB'}{V} = \frac{V}{c^2} L_0 = t'_D. \quad (7.39)$$

It is thus proved that, although the edge B moves, as any physical body should, slower than light, it needs the same time to reach the rear of the train as the superluminal point D, because it has to travel a shorter distance ($BB' < L_0\}$). The boundary between the "roddish" and the "hookish" parts of the train starts from the train's front and moves to its rear faster than light. The product of these two velocities, according to (7.24), is equal to $uV = c^2$.

Ironically, the superluminal speed of point x'_V, which in Mr. Ex's opinion was a fatal flaw of the engineer's description, actually resolves the problem with "rigid" train resisting any bending force. No matter how rigid the object, the bending forces between its atoms cannot transfer information about atomic displacements faster than light. Consider a train's atom that starts falling down together with the collapsing part of the bridge. The adjacent atom finds itself also on the collapsing part and thereby already in a free fall *before* it knows that its neighbor had already started doing it. Because the process of falling down is propagating along the train faster than light, there is no time left for internal forces to respond. In this respect, the train, bent as it is in Mr. Ex's reference frame, remains physically nondeformed, so that the engineer's presentation accurately describes reality without contradicting anything.

This completes the description of a picture that the train's passenger would observe in the fleeting time interval before the crash if he or she were unlucky enough to remain there. Both physics and geometry in this description go hand in hand and see to it that there are no privileged observers, but *everybody* gets the same facts. The observed features and instantaneous physical characteristics of the phenomenon may be dramatically different in the two systems, but in either system they contrive to bring about the same result. The train that crashes in the engineer's reference frame crashes in the passenger's one. This is the way the relativity of certain physical quantities ensures the absoluteness of physical events.

7.5 The Dynamics of Relativistic Length Contraction

> *Since Einstein's 1905 paper it has been realized that Lorentz contraction is a space-time geometric effect, devoid of any dynamical content.*
>
> Anonymous reviewer

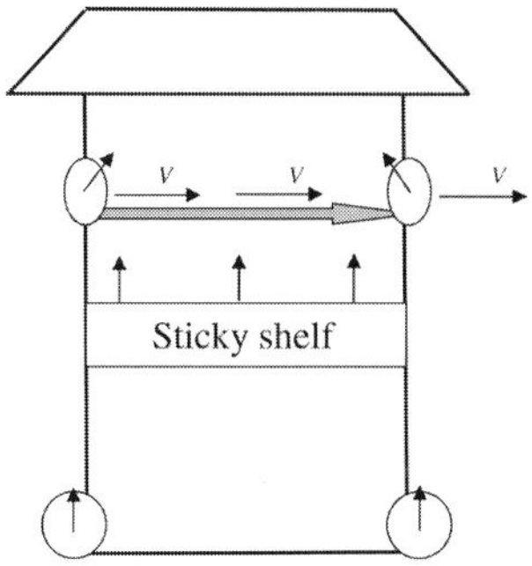

Figure 7.23 A mechanical model for stopping a rod at the moment when it is all in the shed. The rod is stopped here by a sticky shelf, approaching it from beneath. The distance between the shelf and the rod is highly exaggerated. There is practically no clearance between them left at the depicted moment. Note that in Alice's RF the sticky shelf touches all parts of the rod simultaneously. Accordingly, all her clocks (shown at the bottom of the shed) read the same time when the end points of the rod coincide with the opposite sides of the shed. In contrast, Bob's clocks (on the rod) register different times for the same events.

One can frequently read that the Lorentz contraction effect can be entirely understood as the manifestation of pseudo-Euclidean geometry of space-time. This axiomatic approach, while being powerful and elegant in problems with uniform motion, is far less straightforward when the state of motion is changing [61–64]. If we want to get a deeper insight into the nature of the so-called kinematic effects, we need to consider *transitions* between different states of motion. We will then see a much richer world of dynamical effects intimately connected with space-time geometry.

The question about dynamical aspects of relativistic kinematics in special relativity (SR) was explicitly formulated as far back as 1975 [61]. In this section, we will find the dynamics in kinematic effects by considering accelerated motions.[1)] We will study what actually happens to an object when its motion is rapidly changed. We will see the role of atomic interactions within the object in determining its final state after removal of the external forces.

Suppose that a horizontal rod of proper length $L_0 = 10$ m passes through a $l_0 = 2$ m shed (Figure 7.23). The speed of the rod is such that the Lorentz factor $\gamma(V)$ determining its length contraction

$$L = \frac{L_0}{\gamma(V)}, \qquad \gamma(V) \equiv \left(1 - \frac{V^2}{c^2}\right)^{-1/2} \tag{7.40}$$

is equal to 5, so that the rod measures only $L = l_0 = 2$ m in the rest frame of the shed. Then, at a certain moment the rod will just fit into the shed. Suppose, at this moment the rod is stopped by a very sticky shelf that has been moving upward (Figure 7.23). Immediately, an equal friction force is applied to equal parts of the rod (we assume that the heat produced thus quickly dissipates). A straightforward application of

1) We have already mentioned about a widespread misconception that SR is not applicable to accelerated motions. This misconception might have contributed to the oversimplistic view of kinematic effects in SR.

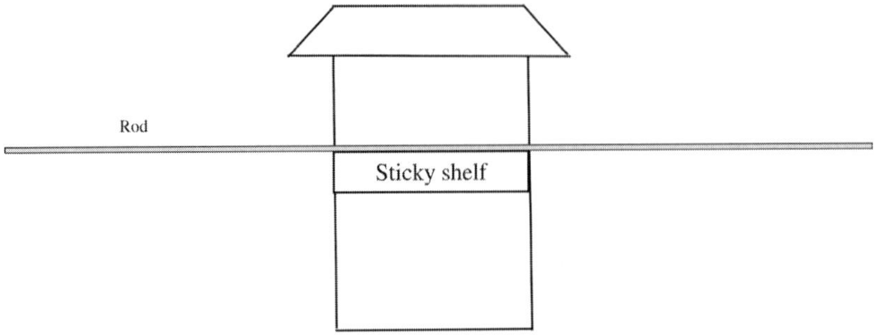

Figure 7.24 End of the stop, as one might originally expect from the formula for the length contraction. However, this outcome would be impossible if someone in the shed would stop all the parts of the rod simultaneously.

Equation (7.40) would lead us to expect that the *stopped* rod would measure its $L_0 = 10\,\text{m}$ proper length in the reference frame of the shed, which means that its edges would now each stick 4 m out of the shed (Figure 7.24).

But this would, under given conditions, contradict the dynamics of motion.

Indeed, in order for the rod to stop at once, it must be subject to a very large force in a very short time, during which each part of the rod could only progress by a negligibly small distance. How then could the ends of the rod have traveled the respective distances 4 m in the two opposite directions, as depicted in Figure 7.24? This would imply near *infinite speed* of the rod's edges and, even more bizarre, the leading edge of the rod should jump infinitely fast in the *forward* direction – opposite to the applied force. Similarly, the rear edge of the rod should jump instantaneously in the backward direction.

Thus, on the one hand, equal forces applied simultaneously to equal parts of the rod would cause them to accelerate (or decelerate) in synchrony, so that the length of the rod cannot change. But on the other hand, it *must* change if one expects it to measure its 10 m proper length.

This contradiction originates from the straightforward application of the Lorentz contraction formula to the accelerated motion and also from the sloppy use of language: assuming simultaneous braking forces, we were not careful enough to specify two crucial points:

1. Simultaneous – in what reference frame?
2. Is the braking force the only actor in the play, or there are other participants?

It turns out that the outcome of the process critically depends on its mechanism.

In the following, we will analyze the whole process as recorded by observers in two *inertial* reference frames: one associated with the shed (system A) and the other *originally* associated with the rod (system B). Accordingly, we introduce two observers: Alice in A and Tom in B.

We will first neglect the internal forces as compared with the huge braking force. Then, we will take them into account and discuss their role in the kinematic effect described by (1).

A. Braking Simultaneously in A (Stopping a Moving Rod)

Suppose that the braking forces are dominating and are applied to all parts of the rod *simultaneously in A*. We start with Alice's account of the process. Alice was pushing up the sticky shelf so that it grips with the rod and stops it precisely at the moment (zero time of Alice's synchronized clocks, Figure 7.23) when the rod is all within her shed. At this moment, she records the corresponding readings of Tom's synchronized clocks at the ends of the rod (we chose units of time so that these readings are -1 and $+1$ "units" respectively). The rod's length right *after the stop* is 2 m. This is precisely what one would expect from the dynamics of motion under the above assumptions: equal forces applied simultaneously to equal parts of the rod will cause their equal and simultaneous accelerations; this cannot change the size of the rod. Thus, our assumption of the rod slowing down uniformly (simultaneously in all its parts in A) is *logically incompatible* with the assumption of the rod retaining its *proper length* after the stop. If the moving rod stops *and* retains its proper length, then its length measured by Alice after the stop must be equal to its length measured by Tom before the stop. The length measured by Tom before the stop was 10 m. Therefore, Alice would expect to see the transition from the 2 m length-contracted moving rod to the 10 m stationary rod, as it was depicted in Figure 7.24. But this is *not* what she actually observes! This shows that our initial assumption of a constant proper length being independent of the change of motion was not generally true. The proper length of the rod may change with change of its motion. Under the considered action of braking forces the *proper* length must change from 10 m before the braking to only 2 m after the braking.

We must be very careful in our formulations here: *this* change has nothing to do with the kinematic length contraction due to relative motion. It is the *dynamic change of proper length* (actual physical deformation), which dominates over the effect of *disappearance* of length contraction owing to the stop. Accordingly, while Alice sees no change in the length of the rod, there must be dramatic change observed by Tom. If the special relativity gives a consistent description of the world, it must account for this change. There must be a physical reason for deformation of the rod in Tom's reference frame. We thus turn our attention to system B – to Tom's account of the same process.

Tom sees the shed sliding to the left down the rod. Due to the relativity of time, while the edges of the rod coincide with the opposite sides of the shed simultaneously in A, these events happen at different moments in B. The right edge of the rod coincides with the back door of the shed at -1 u ("unit") of Tom's time. The left edge coincides with the front door at $+1$ u. The time interval between the events is 2 u. Applying the Lorentz transformation for time coordinates of the two events, taking into account that both these events happen at $t=0$ in system A, and using (7.40), we can express this time interval in terms of the (initial!) proper length of the rod:

$$\Delta t' = \frac{V\,L_0}{c^2}. \tag{7.41}$$

Accordingly, the length of the shed as observed in B must be *smaller* than the length of the rod: Tom sees the moving shed contracted down to $l = l_0/\gamma = L_0/\gamma^2 = 0.4\,\text{m}$.

In addition, Tom sees the rising sticky shelf inside the moving shed. And for the same reason as discussed in Section 4.4 (relativity of time), he observes the shelf tilted to the horizontal (Figure 7.25). The events at the edges of the shelf observed *simultaneously in B* are not simultaneous in A – the one at the rear edge is later (when the shelf in A is accordingly higher) than one at the front. Therefore, Tom sees the rear edge of the shelf being higher than its front edge.

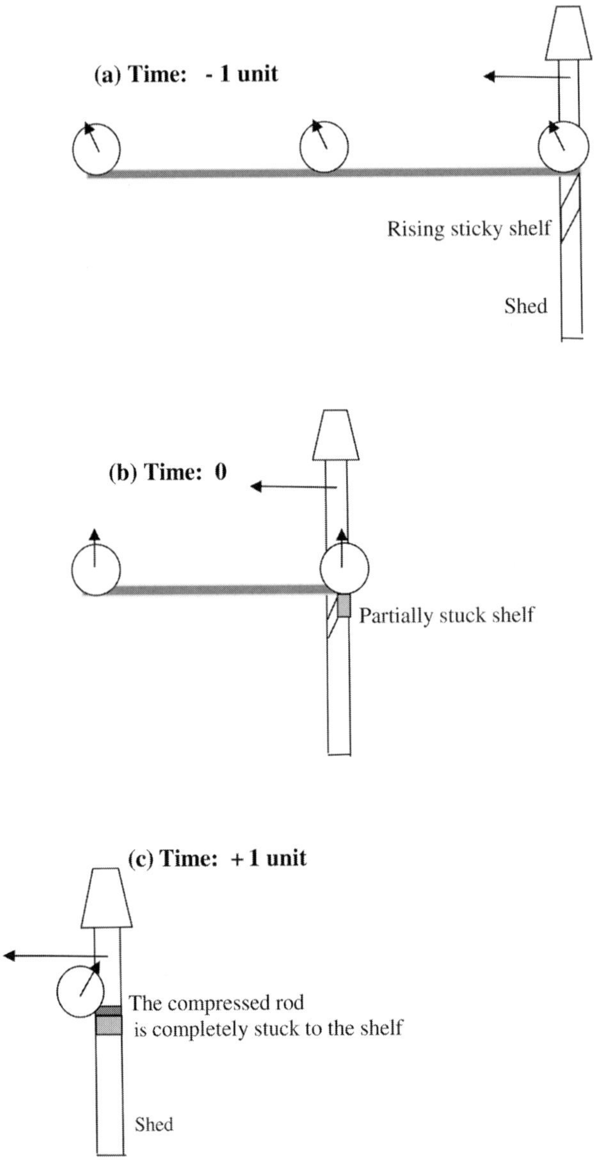

Figure 7.25

If Alice, while keeping the shelf horizontal, pushes the shelf up with a speed v in her reference frame, then it is observed in B as a "time-dilated" vertical component $v_\perp = v/\gamma$ of the shelf's velocity (another – horizontal – component $-V$ appears in B due to the motion of the shed with a speed $-V$ relative to B). Therefore, during the time interval $\Delta t'$ Tom observes a point on the shelf traveling up the distance $\Delta y' = v_\perp \Delta t'$. This is the elevation of the right edge of the shelf above the left edge. As a result, the tilt angle of the rising shelf in B is given by

$$\tan \alpha = \frac{\Delta y'}{l} = \gamma^2 \frac{V v_\perp}{c^2}. \tag{7.42}$$

At -1 u of Tom's time, the upper end of the shelf touches the right edge of the rod, sticks to it and drags it to the left (Figure 7.25a). This compresses the rod! As the rest of the shelf continues to rise, the region of compression forms and rapidly expands to the left along the rod due to the new originally relaxed material of the rod joining the motion (Figure 7.25b). The front of the expanding region can be considered as the running separation point between the two parts: the one to the left of the front – still intact – and the one to the right involved in motion together with the shed. The separation point runs down the rod faster than light. As we will see in Chapter 11, this is the kind of superluminal motion that does not contradict anything; it is just a moving boundary between two regions rather than a physical particle. Particles of the compressed part of the rod are moving (together with the corresponding part of the shelf) slower than light. The separation point outruns them because, as mentioned above, the new particles in the front join this motion as they grip with the rising part of the shelf. Because the compressed parts of the rod move toward its opposite edge, the process ends up with the rod contracting to the size of the shed (Figure 7.25c).

We emphasize that the compression wave here is *not* the shock wave. In a shock wave, the motion of a particle right behind the wave front is the cause and the motion of the adjacent particle right ahead will be the effect. In a compression wave described here, any particle of the rod changes its motion only because of the contact with the rising part of the shelf – before it knows anything about corresponding changes of an adjacent particle of the rod. The changes of state for two particles here are not in the "cause and effect" relationship. If the compression were to produce a shock wave, the latter would also lag behind the separation point, as any shock wave is subliminal. Such a wave does not form here, because the sticky shelf freezes the corresponding type of motion.

During this process we can no longer consider the rod as *one* inertial reference frame. It belongs to two different frames: one to the left of the advancing front still belongs to the system B and the one to the right of the front belongs already to system A. The separation into two systems starts from the whole rod constituting one system B and ends up with the whole rod constituting (together with the shed) system A. An important implication is that since during this transition the two parts of the rod move relative to each other, the rod *cannot be characterized as having a certain proper length*! At least not within its conventional definition as the length of an object in its rest frame, because now two parts of the object have different rest frames.

In this case, the definition of proper length of the rod should be generalized as its length in a reference frame where the *net momentum* of the rod is zero. This reference frame does not coincide with either A or B but changes from one to another during the process. Physically, this is due to the fact that the individual momenta of the two parts of the rod change with time as the separation point slides down the rod. As a result, this new reference frame (let us call it system C) turns out to be accelerating. This should come as no surprise once we consider the accelerated motion of the rod.

Since the motion of an accelerated object can be described in consecutive steps as motion in the comoving inertial frames, each with a small speed relative to the previous one, and in each such frame the rod with the zero *net* momentum consists of the two *moving* parts whose respective lengths change from frame to frame, we can no longer expect the generalized proper length to remain constant. The generalized definition of proper length should provide us with the means to watch its change as a function of time.

Since the separation point between the two parts of the rod travels the distance L_0 in time $\Delta t'$ given by Equation (7.41) its speed in B is

$$u = \frac{L_0}{\Delta t'} = \frac{c^2}{V}. \tag{7.43}$$

One can also derive this equation by considering the motion of the intersection point between the rod and the tilted part of the shelf. The velocity of the tilted part has two components: V (horizontal) and $v_\perp$ (vertical). With the first component alone, the intersection point would slide down the rod with the speed V. With the vertical component alone, the intersection point would slide down the rod with the speed $v_\perp/\tan\alpha = \gamma^{-2}c^2/V$. The total speed of the intersection point is the sum of the two contributions and yields (7.43).

The described "shelf-induced" contraction stops when the rod shrinks down to the size of the shed. The contraction is caused by deformation of the material of the rod under the external force.

This conclusion may appear false to Alice. Indeed, all particles of the rod are stopped simultaneously in her shed. The distance between any two adjacent atoms along the rod did not change in A. Therefore, it seems that there cannot be any change in the interatomic interactions, and, with heat removed, the rod cannot deform or change in any other way.

The best answer to this would be to suggest Alice to take a closer look at her experiment. She would realize that, even with heat instantly removed, what had been a regular rod is now a streak of some exotic material with unusually high density, glued to the shelf across its surface.

We can understand this outcome by taking a closer look at hitherto neglected *atomic interactions and their* role *in the process.*

Discussion: The Role of Interatomic Forces

Although it is true that one could, in principle, keep the interatomic distances constant in A during the braking, it is *not* true that interatomic forces will remain constant in this process. In relativistic dynamics, the interaction forces between two

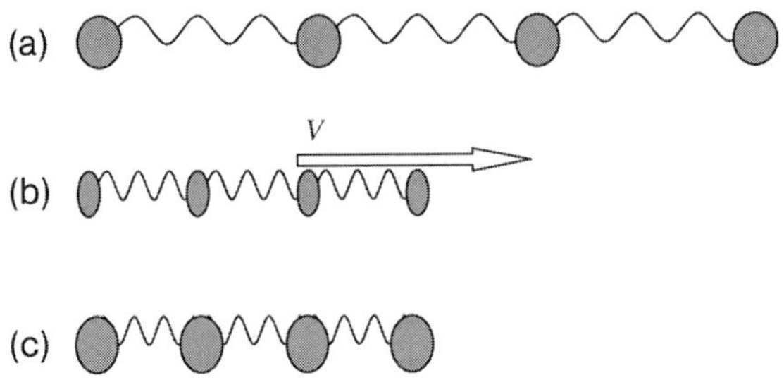

Figure 7.26 Chain of spheres connected by springs. (a) A chain in a stable mechanical equilibrium, observed in its rest frame (system B). The distance between neighboring spheres is determined by the length of nondeformed springs and is an intrinsic physical characteristic of the chain. (b) The chain in *the same* state, as observed from another reference frame A. Because it is moving relative to this frame, it is length-contracted. The distances between the neighboring spheres, as well as their longitudinal diameters are reduced by Lorentz factor $\gamma(V)$. (c) The same chain, stopped in A by braking forces applied simultaneously to each sphere. The chain is no longer in mechanical equilibrium – the reduced distance between the spheres does not correspond to the equilibrium state of the springs. The springs are compressed, and the depicted state can only be maintained by some opposing external forces. Without them, the state (c) would be unstable.

particles depend also on their velocity [29]. If the latter changes, the interaction force can also change even with the distance fixed.

The *proper* distance between two neighboring atoms in a rod is about $a_0 = 10^{-10}$ m. At this distance, each atom is in stable equilibrium (or vibrates around its equilibrium position). We can model such a state by a linear chain of equidistant particles connected by springs, as shown in Figure 7.26a. In equilibrium, the springs are neither stretched nor compressed.

According to the principle of relativity, an equilibrium state of a stationary system must be *the same* in any inertial reference frame. Therefore, Alice and Tom must each observe such a state as a stationary chain of equidistant atoms separated by the same distance a_0. Tom, initially, has a rod in such a state. Alice initially observes this state from her shed and measures the interatomic distance along the rod to be only $0.2a_0$ (Figure 7.26b). This remains consistent with the definition of equilibrium, since the rod is in a uniform motion. However, if she manages to stop the rod *without changing its length* in A, she has after the stop a *stationary* object with longitudinal interatomic distance $0.2a_0$. This does *not* correspond to an equilibrium state of a stationary object. What Alice has now is a system far from equilibrium, with all its atoms destroyed and squeezed together. The springs representing the internal forces are compressed to 1/5 of their normal length (Figure 7.26c). This corresponds to huge forces of repulsion between atoms. In a model that represents atomic forces in a solid by springs obeying the Hooke's law, the sudden removal of the shelf (disappearance of external forces) would send the system into longitudinal vibrations around its natural *proper* size. We can visualize this as a mass on a very elastic spring that is initially severely compressed down to 1/5 of its relaxed length and then released. The result will be oscillations around the equilibrium position $a = a_0$. If the vibrations are damped, the system will ultimately relax to the equilibrium size a_0, thus restoring its proper length.

Although springs only approximately model real internal forces, the conclusion that these forces act to some extent as the "keeper" of proper length does not depend on the model. As we know (recall Chapter 6), the field of a moving point charge is flattened in the direction of motion. Because the point charge has no size, this flattening is an intrinsic property of the field as such, depending *only* on the relative motion between the charge and an observer's reference frame, rather than on the shape of the charge. It can be considered as the "length contraction" of the field (Figures 6.5 and 6.6). However, if the source of the field has a finite size, it *has* to possess the same property because the Maxwell's equations are self-consistent (actual situation is more subtle because we have to consider the equilibrium conditions for a system of *moving* charges and accordingly take into account the corresponding magnetic field as well [65]. The same is true for all other forces (and quantum-mechanical probability distributions for all particles), since they all obey the relativistic equations, which are Lorentz-invariant. For a moving system, its particles and the *equilibrium distances* between them must be all Lorentz-contracted along the direction of motion [65,66] (Figure 7.27).

When the rod stops, the field of each of its particles (as observed in A) restores the shape characteristic for the stationary state and the interatomic distances necessary for the equilibrium must increase. If the particles themselves are not allowed to accordingly shift apart, their fields become strongly overlapped (Figure 7.27c), which produces huge repulsive forces. If the external forces are suddenly removed, this repulsion may blow up the rod. If external forces stay, we will have the "poststop"

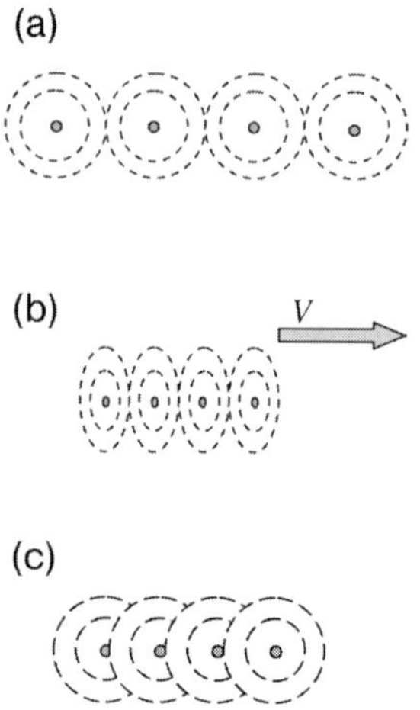

Figure 7.27 "Equipotential surfaces" of fields produced by individual point charges regularly arranged in a straight line. (a) In the rest frame of the system (the charges at the centers of spheres are kept in equilibrium by internal forces, other than the electrostatic forces). (b) In a reference frame A moving relative to the system along its length. (c) In the same frame A after the system had been stopped relative to it and "frozen" by large external forces applied simultaneously. The resulting state, if "unfrozen," will be extremely unstable.

picture observed by Alice – the rod *not* kinematically contracted but *dynamically compressed*. However, if external forces weaken sufficiently slowly, the internal forces can restore the *initial* proper length of the rod. In this respect, *the internal fields act as a memory*, keeping information about proper shape of an object.

Alice could summarize this part of the discussion in the following way. Before the stop, the rod, although length-contracted, was *not* deformed, because the atomic distances had matched the shape of the internal field produced by *moving* atoms. After the stop, even though the length of the rod did not technically change in A, the rod *was* deformed because the atomic distances no longer matched the equilibrium shape of the *stationary* internal field. The rod was not allowed to readjust to the changed shape of the individual fields of its constituents. Our concept of deformation should be refined to describe adequately this kind of process in relativistic mechanics. In particular, an actual physical deformation *does not automatically imply a change in length* in a given reference frame, and vice versa.

In relativity, a system cannot in principle be ideally rigid [29,67,68]. A rod, even when rigid in all conventional situations, behaves in the above thought experiment first as a stick of putty and some time later as a rigid body. Therefore, in cases with huge external forces acting during a very short time (less than the travel time of the shock wave between the points of interest), we can first use the model of an infinitely deformable rod (in essence, such "rod" can be represented by its two end points). Then we can try to find out how the internal forces change the results.

Whatever the model, we arrive at the description of the phenomenon that, while appearing different to different observers, is consistent for all of them and predicts the final state upon which everybody agrees. To Alice, the rod retains its initial size because the *equal* external forces are applied *simultaneously* to its equal parts. Such forces can (and do) stop the rod, but they alone cannot change its length. To Tom, the rod has been compressed by external forces acting *at different times* on different parts of the rod. Such forces both stop *and* deform the rod. Both agree that in the final state the rod is physically deformed and just fits into the shed. (In order for Tom to maintain his account unbroken, we must place him at the rear edge of the rod. After the end of the process, Tom continues his state of motion, so he literally "flies off the handle" of the rod.)

B. Braking Simultaneously in B (Accelerating Stationary Rod)

The previous analysis might prompt an assumption that the proper length would conserve if the forces were applied to the rod *simultaneously in B*, which is initially the rest frame of the rod. In this case, it will be convenient to start with the picture of the process as observed by Tom. Also, to illustrate that special relativity gives consistent description of accelerated motions regardless of accelerating mechanism, we now consider, instead of friction, another possible mechanism of braking, in which the rod is represented by two equal masses at its edges, boosted by the identical jet engines. In this scenario, Tom is originally positioned in the middle between the masses. He sees the shed moving to the left. When the shed passes Tom, both engines are switched on to accelerate the rod in the same direction (Figure 7.28). Since both actions are simultaneous in B, they boost the rod

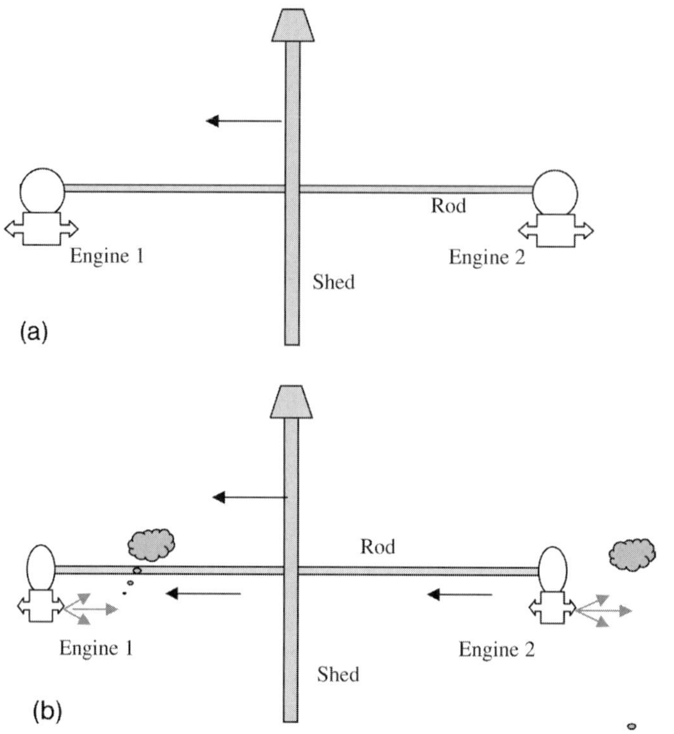

Figure 7.28 Another way to stop the rod. The forces are applied to all parts of the rod simultaneously in the system T. We use a simplified model of the rod: two end masses with the attached engines (infinitely deformable inelastic rod). (a) Immediately before the acceleration the rod is stationary in T. When the shed passes by the center of the rod, the two engines fire and instantly accelerate the rod in the direction of the shed, without changing the length of the rod. (b) The moment immediately after the acceleration. The rod is now stationary in A.

to the left without changing its length. If the engines are sufficiently powerful, the rod instantly acquires the speed of the shed. In the final state, the rod and the shed both form one single whole and the edges of the rod stick out of the shed symmetrically on both sides.

This situation is "reciprocal" to the previous one. Tom and Alice exchange their roles.

Alice had seen the moving rod stopped. Now Tom sees the stationary rod accelerated. Alice had claimed that the length of the rod did not change. Now Tom claims the same.

But there is again something wrong with this result. Tom, remaining in B, now observes the rod with its unchanged length 10 m moving together with the shed whose length in B is 0.4 m. The ratio of these lengths is 25 : 1. Since both objects are now moving as one single whole, any other observer will measure the same ratio for their lengths. Therefore, Alice (having now the rod and the shed both at rest in her system A) also sees the rod to be 25 times as long as the shed. Because the proper length of the shed is 2 m, she now measures the length of the rod as 50 m, instead of the expected 10 m.

We conclude that the rod must again have undergone *physical deformation* – this time *increase* of proper length by a factor of $\gamma(V)$.

This conclusion must be consistent with the picture of the process as seen by Alice. She first sees the length-contracted 2-m rod moving to the right. Then, she observes the ignition of the jet engines. The crucial point here is that these two events are *not simultaneous* in A (Figure 7.29). First starts the engine at the left-end mass, stopping

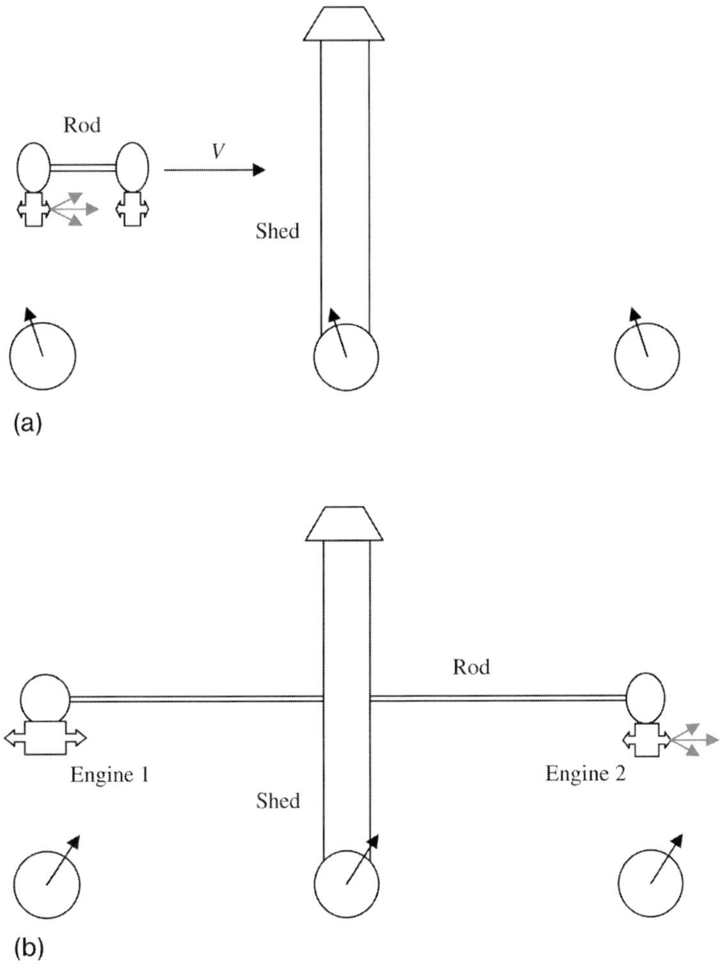

Figure 7.29 The same process as in Figure 7.28, but now observed by Alice. (a) The initial moment of stopping the rod in Alice's reference frame. The left engine stops the left end of the rod. (b) The final moment: the right engine stops the right end. Between the moments (a) and (b), Alice observes the rod being stretched from the 2-m length to the 50-m length (the drawing is not to scale).

it. The right-end mass keeps on moving, thus *expanding* the rod. The engine at the right starts some time after it had passed by Alice. When it stops, the rod and the shed form one single whole; the ends of the rod are 50 m apart and positioned symmetrically on either side of the shed.

To obtain the actual amount of the stretch quantitatively, note that the length of the rod after the stop in A is the sum of two terms: the length of the contracted rod just when the left engine fires, plus the distance traveled by the right end till the whole rod stops (Figure 7.29). The first term is $L_0/\gamma(v)$ and the second term is $V\Delta t'$, where $\Delta t' = t'_2 - t'_1$ is the time difference in A between the ignitions of the two engines. The same argument that leads to (7.41), now gives

$$\Delta t' = \gamma(V)\frac{L_0 V}{c^2}. \tag{7.44}$$

Therefore,

$$\Delta x' = \frac{L_0}{\gamma(V)} + L_0\frac{V^2}{c^2}\gamma(V) = L_0\gamma(V)\left(\frac{1}{\gamma^2(V)} + \frac{V^2}{c^2}\right) = L_0\gamma(V). \tag{7.45}$$

Suppose now that the rod is represented by a row of equidistant noninteracting point masses, each with its own engine, and all the engines fire simultaneously to stop the rod at the zero moment in B. Then, Alice would observe a succession of consecutive flashes of the engines, each stopping its corresponding mass, so that the pulse of flashes will run from the rear to the front of the rod. Since the wave starts at point x'_1 at the moment t'_1 and stops at point x'_2 at the moment t'_2 in A, the speed $u = \Delta x'/\Delta t'$ of this pulse is identical to that in Equation (7.43). Again, the pulse propagates faster than light. But, as it was the case with the running separation point in the previous section, this does not violate any laws, because it is not associated with a signal or energy transfer.

Thus, the simultaneous application of forces in B, with the atomic interactions turned off, also fails to preserve the proper length. But, contrary to the previous case, the rod now undergoes stretch.

How will this result change if there are internal forces? Let the end masses be connected by a spring. Some time after the engines stop (the masses are released), the stretched spring will start contracting. The system begins to vibrate around its equilibrium size corresponding to the relaxed spring. By definition, this size is the initial proper length of the rod. In our particular case, the vibrations will not be symmetrical with respect to the center of the relaxed rod, because the stretch exceeds the proper length (the rod must be very elastic!). Accordingly, it cannot be modeled by the Hooke's law now. Nevertheless, if the vibrations are damped, the system again will ultimately relax to its natural size, thus restoring its proper length. If, however, there are some external forces keeping the original stretch fixed, then the object uniformly accelerated from rest will remain after the boost in a physically deformed state with its *proper* length extended by a factor $\gamma(V)$. As we will see later, such forces naturally appear in rotational motion of a ring.

C. Nonsimultaneous Braking

An obvious way to preserve the proper length of the rod without using its internal interactions would be to accelerate the shed instead of the rod until they are at rest relative to each other. But this would just shift the problem from the rod to the shed.

An incremental accelerating procedure preserving the proper length of an object is described in [69]. Here, we describe a more straightforward (and accordingly more violent!) solution. We can just stop the ends of the rod at the positions, which they must have in A when the proper length *does* conserve. For instance, we could stop the rear end of the rod at 5-m mark to the left of the center of the shed and the front end at 5-m mark to the right of the center. The spacing between the marks will be the needed 10 m. Generally, for the rod with the proper length L_0, Alice would want to stop the rear and the front ends at the marks

$$x_1 = -\frac{1}{2}L_0, \qquad x_2 = \frac{1}{2}L_0, \tag{7.46}$$

respectively. But since the moving rod was length-contracted, its ends cannot pass these markings simultaneously. First, the rear end of the rod will pass the mark x_1 and we should stop it at this very moment. Only *some time later,* the front end of the rod will pass the mark x_2 and it must be stopped at this instant (Figure 7.30).

We now need to find these moments of time. When the rear point reaches the mark x_1, Alice finds Tom closer to the center of the shed by half of the contracted rod. The distance between Tom and Alice at this instant is

$$D = \frac{1}{2}L_0 - \frac{1}{2}\frac{L_0}{\gamma(V)} = \frac{1}{2}L_0\frac{\gamma - 1}{\gamma}. \tag{7.47}$$

Therefore, the moment when the rear end of the rod stops in A must be D/V. Similarly, we find that the front end of the rod must be stopped at the moment D/V. Thus, the instants of Alice's time corresponding to the marks x_1 and x_2 are, respectively,

$$t_1 = -\frac{1}{2}\frac{L_0}{V}\frac{\gamma - 1}{\gamma} \quad \text{and} \quad t_2 = \frac{1}{2}\frac{L_0}{V}\frac{\gamma - 1}{\gamma}. \tag{7.48}$$

If, instead of only two end points, we imagine the rod as a row of equidistant noninteracting particles, each stopped by its individual engine one after another in equal short time intervals, starting at t_1 and ending at t_2, then Alice could observe the pulse of engine flashes rushing from rear to front of the stopping rod. The edge of the pulse separates two parts of the rod – the one stopped and another still moving. The speed of the separation point is

$$u = \frac{x_2 - x_1}{t_2 - t_1} = \frac{L_0}{t_2 - t_1} = \frac{V}{1 - \gamma^{-1}} = \frac{c^2}{V}\left(1 + \frac{1}{\gamma}\right). \tag{7.49}$$

Again, the separation point moves faster than light. In the end, Alice sees the rod stopped in the position depicted in Figure 7.30. In the process, she observed the rod

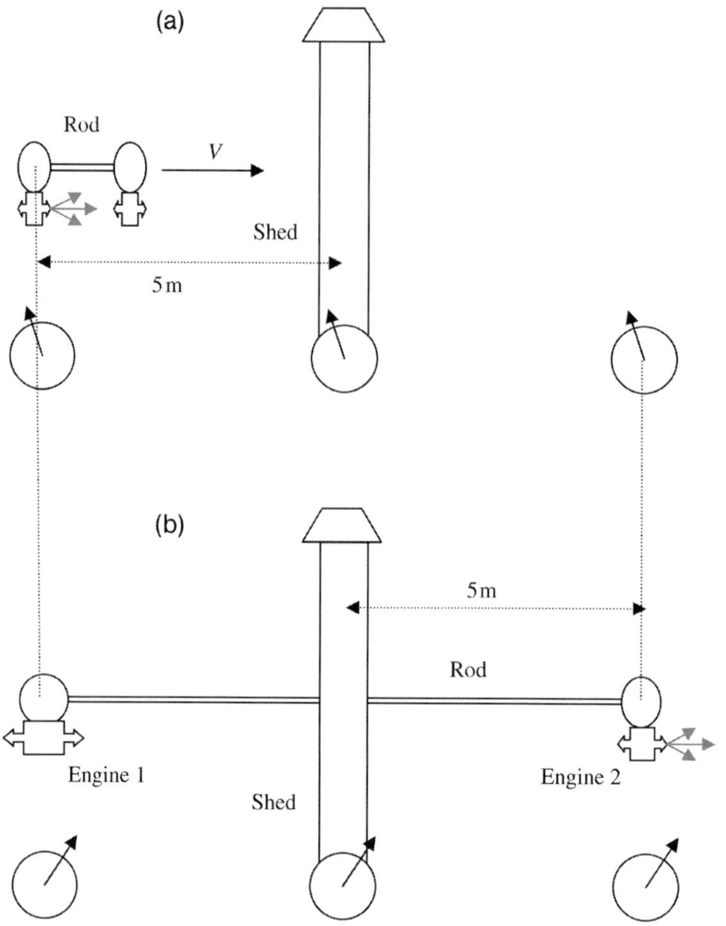

Figure 7.30 The same as in Figure 7.29, but now (a) the left engine fires and stops the left end of the rod when it is *exactly* 5 m to the left of the center, and (b) the right engine stops the right end when it is exactly 5 m to the right of the center. The final state of the rod is the same as in Figure 7.24 (the proper length is conserved), but it is reached by stopping its two ends at two different, specially timed, moments in system A.

being stretched from its Lorentz-contracted length L_0/γ to its proper length L_0. But she could only achieve this by stopping different parts of the rod *at different moments.*

How does this process look from Tom's viewpoint? Applying Lorentz transformations to coordinates x_1, x_2, t_1, t_2 (Equations (7.46) and (7.48)), we find the coordinates of the same events in B:

$$x'_1 = -\frac{1}{2}L_0, \qquad x'_2 = \frac{1}{2}L_0 \tag{7.50}$$

and

$$t'_1 = \frac{1}{2}\frac{L_0}{V}\frac{\gamma - 1}{\gamma}, \qquad t'_2 = -\frac{1}{2}\frac{L_0}{V}\frac{\gamma - 1}{\gamma}. \tag{7.51}$$

Equations (7.50) give the coordinates of the end points of the rod, measured by Tom at the moments when their engines fire. They are just what one would expect. But the moments themselves are, according to (7.51), observed in B in the reverse order: the moment t'_1 is *later* than the moment t'_2. This is another aspect of relativity of time: since the flashes are not in "cause and effect" relationship, the interval between them is spacelike and the ordering of the flashes is not invariant.

Our next question is the dynamics of the process observed in system B. Tom sees Alice and her shed moving to the left. By the time the engine on Tom's right fires, Alice has not yet passed by, because this time is *before* the zero moment. Alice's coordinate at this time is $x_A(t'_1) = -Vt'_1 = (1/2)L_0(\gamma - 1)/\gamma$, so the distance between her and the right end of the rod is

$$D_R = \frac{1}{2}L_0 - x_A(t'_2) = \frac{1}{2}\frac{L_0}{\gamma}. \tag{7.52}$$

At this moment, the right end is accelerated to the speed V – it is "hurled" into Alice's reference frame and starts moving to the left. With the left end of the rod still fixed, this means that Tom observes the rod shrinking! After Alice had passed by Tom, at the moment t'_1, the left engine fires and hurls the left end of the rod to Alice's reference frame. At this moment, Alice's coordinate is $x_A(t'_1) = -Vt'_1 = -(1/2)L_0(\gamma - 1)/\gamma$, so the distance in the system B between her and the left end is

$$D_L = x_A(t'_1) - x'_1 = x_A(t'_1) + \frac{1}{2}L_0 = D_R. \tag{7.53}$$

The resulting distance between the edges of the rod measured by Tom in the end of the process is $D_L + D_R = L_0/\gamma$. The ends of the rod are symmetrical with respect to the center of the shed. If we imagine again the rod as consisting of equidistant point particles, accelerating from rest to the speed V one after another starting from the right, then Tom would observe the compression pulse running down the rod from right to left. The velocity of this pulse would be

$$u' = \frac{x'_1 - x'_2}{t'_1 - t'_2} = -\frac{c^2}{V}(1 + \gamma^{-1}) = -u. \tag{7.54}$$

Both Alice and Tom observe the same pulse. But to Alice, it is an expansion pulse moving to the right; to Tom, it is a compression pulse moving to the left.

Tom observes that in the end of the process the rod has shrunk from its proper length L_0 down to the Lorentz-contracted length L_0/γ. Because the rod is now moving with speed V relative to Tom, he concludes that the proper length of the rod measured by Alice is the same as the one originally measured by him – it is conserved. The rod shrinks in length to retain its proper length! Crazy as it appears to be, this statement is quite consistent. It would constitute a logical paradox for a rod remaining at rest or moving uniformly in one inertial reference frame. In our case, the compression of the rod is accompanied by its *acceleration*. That the proper length here remains the same after compression is no contradiction, because the considered process of compression is relative. What is observed as contraction by Tom is observed as expansion by Alice! *One and the same system appears here to evolve in the opposite directions when viewed from two different reference frames.* The *direction of evolution* of an accelerated object can be a relative property.[2] This is due to the fact that for spatially separated parts of the system, the moments of the start and the end of their respective evolutions have opposite ordering in these systems.

D. Generalization to Arbitrary Forces

The situation considered above is a typical example of a "gedunken experiment," which is routinely used in physics to simplify the discussion. The assumption of instant and infinite forces in this "experiment" is unrealistic to the same extent as an assumption of an ideal force-free inertial reference frame with its grid of ideally rigid straight infinite rods – actually, an infinite number of such frames moving relative to each other at constant velocities. Such systems do not exist, which does not preclude us from using corresponding notion as an idealization and arriving at experimentally verifiable conclusions. We obtain Kepler's laws and Newton's law of gravitation by treating the planets and the Sun as merely mathematical points with infinite muss densities, which is factually wrong and clearly unrealistic. But it does not affect the correctness of conclusions, because under these conditions the result depends on the net masses, not their densities. Similarly, the results of the accelerated motion in the above discussions, while being sensitive to the temporal and spatial distribution of forces, actually depend on *impulse of force* $\Delta P = F\Delta t$ rather than force as such. The definition of impulse holds for an infinite force as well if $F \to \infty$ and $\Delta t \to 0$ in such a way that their product remains constant. This justifies our model with infinite and instantaneous forces. Since it is the final momentum (and thereby the final velocity) of an object that determines its longitudinal size, the resulting size after the acceleration procedure with a finite lasting force will remain the same as for an infinite but instant force with the same impulse.

This conclusion can be restated in the language of Lorentz transformations in the following way. Suppose we represent a rod by two equal end masses connected with the massless spring. We can accelerate the rod by imparting both masses with equal amount of momentum in two different acceleration procedures. In the first procedure, both masses are hit simultaneously in their initial rest frame B by infinite instant forces parallel to the connecting spring. As was shown above, in the final state both masses (and corresponding rod) will move relative to B with a speed V and the separation between them (the length of the spring) will remain the same as before. In other words, the length of the rod as measured in B will not change. However, because the rod is now *moving* relative to B, it is Lorentz-contracted, which means that its proper length as

2) This conclusion does not contradict causality, since it refers only to extended objects when the corresponding events in their different parts are separated by the spacelike intervals. The internal evolution of each single particle along its world line has the same direction for all observers.

measured in its *new* rest frame (system A) is greater by a factor $\gamma(V)$ than it had been before the acceleration. The reason for change in proper length (the actual physical deformation) becomes clear if we note that in system A the rod was initially moving and the forces were not simultaneous: the trailing mass was stopped earlier than the leading mass, which naturally resulted in the extension of the rod.

In the second procedure, both masses are subject to equal *finite* forces, which start simultaneously in B, act during equal time intervals $\Delta t_1 = \Delta t_2 = \Delta t$, and vanish simultaneously in B. Both masses will travel during this time interval equal distances $\Delta l_1 = \Delta l_2 = \Delta l$, so that in the end of the procedure the rod will move in B having the same length as before. But since it is now *moving*, it must be Lorentz-contracted, which implies that its *proper* length (again, measured in its new rest frame A) must increase by the same factor $\gamma(V)$. The physical reason for this is that in system A the masses were initially moving and the two forces neither did start simultaneously nor ended simultaneously. Because the Lorentz transformations are linear, the corresponding time intervals $\Delta t'_1$ and $\Delta t'_2$ are equal in A, but due to the relativity of simultaneity, they are *not* entirely overlapped in time as are their counterparts in B. The force on the trailing mass starts earlier than does the force on the leading mass; and it finishes earlier than does the force on the leading mass. Thus, even though both masses travel the same distance $\Delta l'_1 = \Delta l'_2 = \Delta l'$ before their consecutive stops, the trailing mass is stopped earlier than the leading mass, which naturally results in the expanding of the connecting spring, that is, the extension of the rod.

Thus, the second procedure produces *precisely the same result* as the first one. The only difference is that after the second procedure the rod as a whole will be shifted by a distance Δl in B or $\Delta l'$ in A as compared to its respective position in the first procedure. Inasmuch as we are interested only in the *length* of the rod measured in two frames, rather than in its consecutive positions, this difference is immaterial. This justifies our assumption of the infinite instantaneous forces used in the previous sections. The second procedure is clearly more realistic. But when discussing the issue for the first time in the classroom, it might obscure the underlying physics behind additional mathematical details. And the underlying physics is that an accelerated object does not generally conserve its proper length. Some exceptions exist, for example, implementing a procedure described in [69], or using the instant but infinite forces considered in part C of this section. The latter result shows that the procedure described in [69], while being the "least violent," is not the only one possible.

In the next chapter (Section 8.1), we will utilize the "violent" approach to demonstrate in the simple way the *dynamical mechanism* of deformation of a spinning disk and, accordingly, the *dynamical origin* of its specific geometry.

Summary

We can now summarize the results of the thought experiments with Alice and Tom.

1. The relativistic kinematics of accelerating objects cannot be separated from the dynamics. As is illustrated by the examples B and C and Figure 7.27, a change in

motion of particles constituting an object changes the structure of their fields and thereby the shape of the object after acceleration. The acceleration of an extended object involves interplay of various forces and the outcome critically depends on both details of the process and *physical structure* of the object. Various internal effects (elastic and shock waves, possible phase transitions, etc.) determine the actual picture of the process.

2. The size of an accelerated object cannot be uniquely determined by Equation (7.40), because the object generally cannot even be assigned the *constant* proper length.

3. The concept of deformation in relativistic mechanics is more subtle than in classical physics. Out of its two intimately linked characteristics – geometric shape *and* physical structure – the former is a relative attribute and therefore may not manifest itself in certain RF in a deformed state and vice versa, be manifest in some RF in a deformation-free state. A uniformly moving rod is technically deformed (length contraction), but physically it is deformation free, which becomes evident in its rest frame. On the contrary, as we will see later, a uniformly rotating ring, while retaining its circumference $L = 2\pi R$ in its *inertial* rest frame, is physically deformed (stretch deformation), which becomes evident in the *corotating* RF.

4. Simultaneous application of equal braking forces to equal parts of a moving rod tends to compress the rod in its original rest frame, thus decreasing its proper length.

5. Simultaneous application of equal accelerating forces to equal parts of a stationary rod tends to stretch the rod in its final rest frame, thus increasing its proper length.

6. If one stops a rod by applying braking forces to its parts at different moments timed so that its proper length remains the same after the stopping, it tends to stretch the rod in the reference frame where it had originally moved and compress it in the reference frame where it had originally rested.

7. If an object is boosted from one inertial RF to another and the boost is sufficiently gentle, the binding *internal fields and interactions* within the object tend to ultimately restore its proper length perturbed by the boost.

8. Accelerated extended objects may look quite different (and sometimes even appear to evolve in the opposite directions) in different reference frames; but each observer has a consistent description, leading to the final state upon which everyone agrees.

9. The study of *transitions* between different states of motion provides a deeper insight into the nature of the so-called kinematic effects. It shows that indicating the space-time geometry as the single basis for description of relativistic effects, may oversimplify the actual physics. It would also be against the spirit of the general relativity, according to which the space-time geometry itself is determined by the energy-momentum tensor. A close look at the relativistic kinematics shows that it has dynamical underpinnings. The "embryo" of the intimate connection between geometry and matter can be traced already to special relativity.

7.6 The Conveyor Belt Paradox

Everybody who was in a subway escalator was dealing with a conveyor belt. It is a closed ribbon made of flexible material, moved along two parallel but opposite rails by two pulleys fastened to a rigid frame MN (Figure 7.31). Denote the inertial system attached to this frame as K and the distance MN as L_0. Suppose this conveyor has been brought to motion so that the speed v of the ribbon is comparable with c. Here, we are coming at something interesting: because the ribbon is now moving relative to frame K, it undergoes the length contraction, so that its length as measured in K becomes shorter by a factor of $\gamma(v)$. But, it evidently cannot become shorter than it was initially, since the distance $MN = L_0$ remains fixed. Already this inconsistency would appear as paradox to us, had not we read the previous section. We know from the analysis therein that a moving rod in K, stopped in K simultaneously in all its parts, suffers compressing deformation, *decreasing* its proper length by a factor of $\gamma(v)$, where v is its speed before the stop. A rod accelerated from rest in K to a speed v simultaneously in all its parts suffers extending deformation *increasing* its proper length by the same factor. Of course, the rod must be elastic enough to remain whole.

In our case, a part of the conveyor that is instantly between M and N, and moving in one direction, represents such a rod. And from what was just said, it follows that its *proper* length $\tilde{L}_0$ is now equal to

$$L_0 \to \tilde{L}_0 = L_0\gamma(v). \tag{7.55}$$

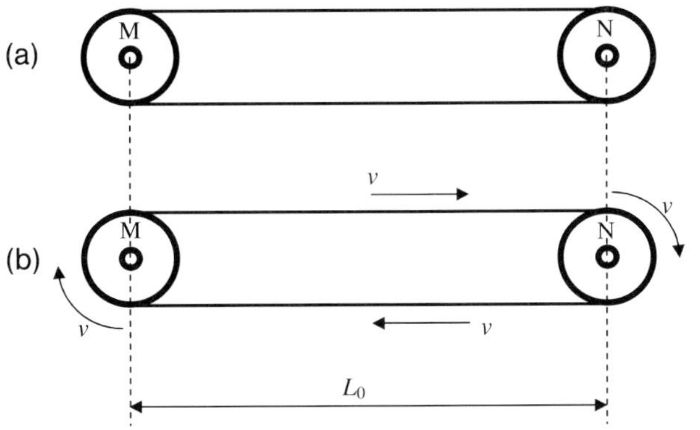

Figure 7.31 (a) Conveyor belt in its rest frame K, with engines off. (b) The same belt in K with engines on. Geometrically, there is no difference between states (a) and (b). Physically, the two states are different. In nonrelativistic physics the difference is only in that all parts of the system are stationary in (a) and moving in (b). We can call it the kinematic difference. According to relativity, the difference is much more dramatic. The system in state (b) is more massive than in (a). In addition (and of major interest for us), the belt is relaxed in (a) (gravity neglected), whereas in (b) it is stretched by a factor of $\gamma(v)$. We do not directly observe the stretch only because of the opposing effect of the Lorentz contraction.

Therefore, its Lorentz-contracted length observed in K is

$$L = \frac{\tilde{L}_0}{\gamma(v)} = L_0. \tag{7.56}$$

It thus remains precisely equal to the fixed length L_0, as is required by the given physical conditions (Figure 7.31b). If the ribbon before the boost was slightly sagging, now it will get tight. If there was no such initial supply of length, the ribbon will undergo physical deformation that can be detected by a tensor meter (stress detector). So this "paradox" is easily resolved.

If, however, we consider the situation from a reference frame K′ comoving with a straight part of the ribbon, all the above conclusions must change. Let, for instance, K′ be associated with the lower part of the ribbon, moving to the left. This part is stationary in K′ and therefore by definition of proper length as the length of a stationary object, it must have the length given by (7.55). At the same time, the frame is moving and therefore the distance between the axels M and N measured by K′-observer must shrink because of the Lorentz contraction:

$$L_0 \to L_0' = \frac{L_0}{\gamma(v)}. \tag{7.57}$$

Thus, the proper length of the belt increases after the turning on by a factor of $\gamma(v)$, whereas the length of the frame as measured in K′ decreases by a factor of $\gamma(v)$. This is, of course a physically impossible situation. By construction, the length of the straight part of the belt *must be equal* to the distance between the axels in *any* RF. This is naturally the case in K and it must, of course, hold in K′. If two objects have their lengths only instantly coincident in one RF, this coincidence is relative, since in another RF the respective end points of the two objects will not be coincident in the same instant of time. However, the *continuously lasting coincidence* of the respective end points is a physical fact, which must be the same in all RF. Therefore, these objects must have coincident lengths in all other RF. We thus came to direct contradiction.

Is there a way to resolve it? The only possibility that comes to mind is that instead of being tightly stretched, the belt must be freely sagging. But this does not help, because the fact of being either tightly stretched or sagging must also be the same in all RF, since this is an intrinsic characteristic of a physical state of the system. Geometrically, a sagging belt is curved, which is different from being straight. These two states are mutually exclusive and the corresponding experiment could determine which conclusion is correct and which is not. And this could, in turn, be used to determine which one of the two RFs we use is in an "actual motion" and which is in a state of rest. In other words, the states of motion and rest would be fundamentally different and experimentally distinguishable.

This appears to be a real paradox within the theory: applying its rules to the given situation, we come to a conclusion that contradicts its own postulate – the relativity principle (equivalence of all inertial RF).

Well, we did not apply the rules correctly. One can hear sometimes that the error lies in ignoring the parts of the belt adjacent to the pulleys and these parts do not move in straight line and therefore cannot be analyzed within the framework of special relativity. This is a wrong argument for two reasons. First, special relativity is *perfectly suited* to analyze accelerated motions (we have seen an example in the previous section and will see more in the next chapter); second, the influence of the pulleys in this problem can be made negligibly small by taking their radius to zero. Alternatively, we can retain the finite size of the pulleys but increase the distance MN between them so that contribution from their circumferences to the net length of the belt becomes totally negligible.

Where is in this case the error in our reasoning? Again, we have ignored the relativity of time!

When we considered the process from the system K′, we made an unspoken assumption that the lower half of the belt that was in K is simultaneously all under the pulleys in K′. But this totally ignores relativity of simultaneity! So, let us restart, taking a proper account of this fundamental effect.

Suppose that before the conveyor was turned on, exactly one half of the belt – say, the lower one – was painted grey (Figure 7.32a). If we consider the conveyor (still not working) from the system K′, then both halves of the belt will be Lorentz-contracted together with the frame MN by the same factor and, accordingly, will always remain equal in length (Figure 7.32b). The temporal dimension does not cause any problems in this case, since different parts of the belt do not move relative to each other.

The situation changes after the turning on. Now there is a relative motion between the belt and the frame, as well as between different parts of the belt itself. Accordingly, we have to explicitly take time into the picture.

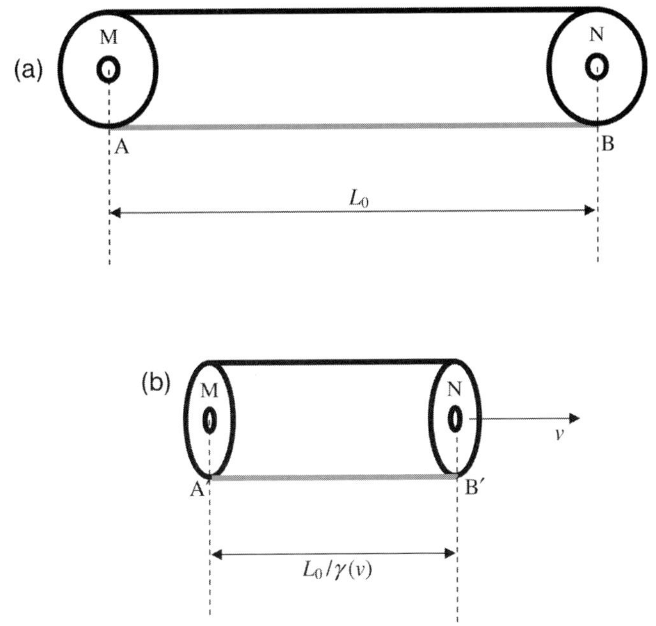

Figure 7.32 The conveyor belt, with engines off and with its top and bottom part having two distinct colors: (a) in its rest frame K; (b) the same belt in the same state as observed from system K′.

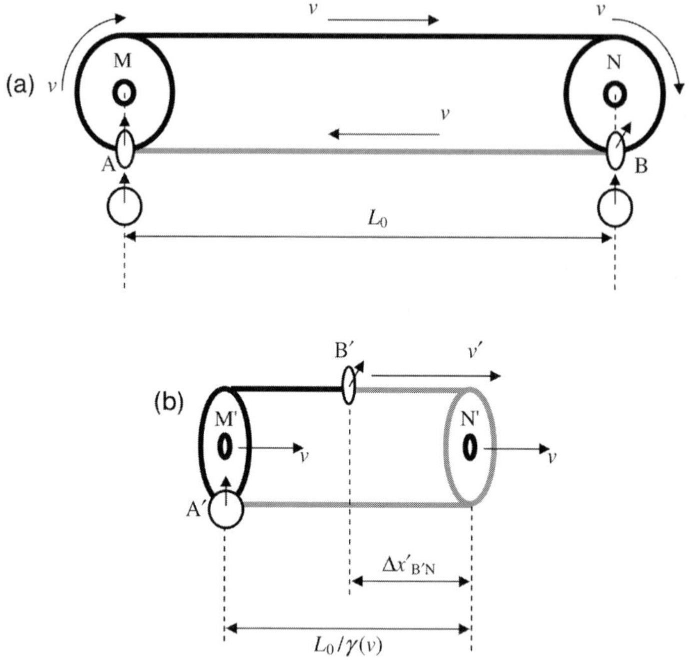

Figure 7.33

Consider the moment in K when the pink half of the belt is all at the bottom. Then, the edges A and B of this half are instantly coincident with points M and N, respectively (Figure 7.33a). Let it be the zero moment in K:

$$t_A = t_B = 0. \tag{7.58}$$

Consider at this moment the system of synchronized clocks in K′. If observed from K, the leading clock in K′ shows an earlier time than the trailing clock. Suppose we set the clocks so that the pair of clocks coincident with the event at A in both systems reads the zero time. Then, the clock in K′ instantly coincident with the event at B will show the time

$$t'_B = \gamma(v)\left(t_0 + \frac{v}{c^2} L_0\right) = \gamma(v)\frac{v}{c^2} L_0. \tag{7.59}$$

But the space of K′-observer is the continuous set of different locations at *the same moment* of his time. Then the question arises: where is the point B at the zero moment of the primed time? Obviously, since this moment is earlier than moment t'_B given by (7.59), the point B has not yet reached the position under the axel N. The axel N itself is the Lorentz-contracted distance away from M. Indeed, the distance M′N′ between the axels in K′

$$M'N' \equiv L'_0 = \frac{L_0}{\gamma(v)}. \tag{7.60}$$

The point B thus must be at this earlier moment on the upper half of the belt (Figure 7.33b). By the moment t'_B (7.59) it must be exactly under the axel N. This axel is moving in K′ together with the frame MN at a speed v. Therefore, between the moments $t' = 0$ and $t' = t'_B$ it will travel the distance

$$\Delta x'_N = v t'_B = \frac{v^2}{c^2} \gamma(v) L_0. \tag{7.61}$$

Added with (7.60), this gives the instant location of point B at the moment t'_B (but the point A will already have "turned around the corner" and be on the upper side by this moment!). Accordingly, the *net length of the painted part of the belt* in K′ is given by the sum of the last two equations:

$$L'(\text{painted}) = \frac{L_0}{\gamma(v)} + \frac{v^2}{c^2}\gamma(v) L_0 = \gamma(v) L_0 (\gamma^{-2}(v) + 1) = \gamma(v) L_0. \tag{7.62}$$

This gives the total proper distance between A and B, measured along the belt, that is, the proper length of the belt between A and B in K′. Comparing with (7.55) shows that it is, as it should be, equal to the new (stretched!) proper length $\tilde{L}_0$ after the boost. In this respect, the result is consistent with our expectations based on the relativity postulates.

We can also use another approach: at the zero moment in K′, the point B is *behind the axle* N on the upper part of the belt. According to the rule of the addition of velocities, this part is moving in K′ at a speed

$$v' = \frac{2\beta}{1+\beta^2} c. \tag{7.63}$$

By the moment $t' = t'_B$, when point B meets with axle N, it will travel the distance

$$\Delta x'_B = v' t'_B = 2 \frac{\beta^2}{1+\beta^2} \gamma(v) L_0. \tag{7.64}$$

The difference between (7.64) and (7.61) is the initial distance between B and N at the zero moment in K′:

$$\Delta x'_{BN} = \Delta x'_N - \Delta x'_B = \frac{\beta^2}{1+\beta^2} \frac{L_0}{\gamma(v)}. \tag{7.65}$$

We can consider the instant segment $\Delta x'_{BN}$ as the Lorentz-contracted length of the corresponding part of the belt, which is moving at a speed v' given by (7.63). Since $v' > v$, the degree of contraction is greater than for the frame MN. Accordingly, the *proper* length $\Delta \tilde{x}'_{BN}$ of this segment measured by the third observer sitting on the *upper* part of the belt is

$$\Delta \tilde{x}'_{BN} = \Delta x'_{BN} \gamma(v'). \tag{7.66}$$

Now, the Lorentz factor $\gamma(v')$ is given by expression (3.14) at $V = v$:

$$\gamma(v') = (1+\beta^2)\gamma^2(v). \tag{7.67}$$

Therefore,

$$\Delta\tilde{x}'_{BN} = \beta^2\gamma(v)L_0. \tag{7.68}$$

Thus, we have the instant proper length $L_0/\gamma(v)$ of the lower part of the belt between the two axles and the instant proper length (7.68) of the remaining painted part on the upper part of the belt. Adding them together will again give the same result (7.62).

Now, what about the proper length of the unpainted part of the belt? Since the proper length is an invariant, it must have the same value as was found in K. We then found that both parts of the belt have the same proper length $\tilde{L}_0$. Therefore, this must hold in K′ as well. In other words, we expect that $\Delta\tilde{x}'_{MB} = \tilde{L}_0$. It can be either measured directly in the rest frame of an object, or indirectly from another RF by measuring some related characteristics. In our case (system K′), we are not in the rest frame of part MB, so we must measure its *Lorentz-contracted* length $\Delta\tilde{x}'_{MB}$. This length, as is seen from Figure 7.33b, is

$$\Delta x'_{MB} = \frac{L_0}{\gamma(v)} - \Delta x'_{BN}. \tag{7.69}$$

In view of (7.65), this gives

$$\Delta x'_{MB} = \frac{1}{1+\beta^2}\frac{L_0}{\gamma(v)}. \tag{7.70}$$

As in the previous cases, to obtain the corresponding *proper* length $\Delta\tilde{x}'_{MB}$, we must multiply this by the Lorentz factor $\gamma(v')$, which is given by (7.67). Then, we immediately get

$$\Delta\tilde{x}'_{MB} = \Delta\tilde{x}'_{BN} = \gamma(v)L_0 = \tilde{L}_0. \tag{7.71}$$

Discussion

On the way, we obtained another interesting result. The acting conveyor belt consists of parts that are in different states of motion. For instance, in K′ the lower part of the belt is stationary, whereas the upper part is moving at a speed v' given by (7.63). Accordingly, each part has its length Lorentz-contracted with different factors. But if we use these measured Lorentz-contracted lengths to calculate the *proper* length of each part and then add them together, we obtain the total proper length of the object. This procedure constitutes an indirect measurement of the total proper length. It can be considered as a possible operational definition of such length, which does not even mention the rest frame of the object and rightly so, because determining a "rest frame" of a changing object for the purpose of direct measurement of its proper length is not always easy. To illustrate this important point, consider another version of a belt – this time an expanding elastic rod with a nonuniform mass distribution. For instance, its one part consists of heavier material than the other one. In addition, its length is a function of time and therefore we can only talk about an *instant* proper

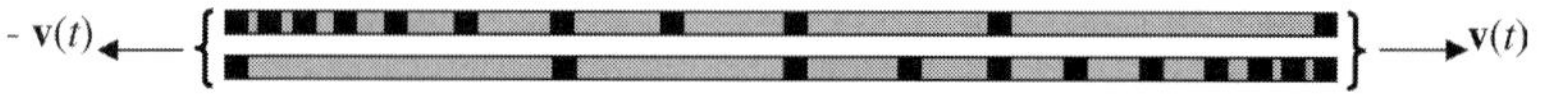

Figure 7.34 Two expanding rods remain congruent at any moment. They appear to be geometrically identical. But if they have different mass distributions (each solid square here represents a unit mass), their centers of mass do not coincide and generally are moving relative to one another. In this case, they may have different proper lengths, $L_0^{(1)}(t) \neq L_0^{(2)}(t)$. In particular, if the individual mass distributions are mirror reflections of one another, as depicted here, the center of mass of the combined system is in the middle of the figure. The proper length of the combined rod is at any moment different from those of the individual rods.

length $L_0 \to L_0(t)$. We can define it as a distance between the simultaneous positions of the rod's edges in its rest frame. Now consider the second expanding rod, which is at each moment *geometrically identical* (congruent) with the first one, but has a different mass distribution over its length (Figure 7.34). It seems obvious that the geometric congruence of both rods can be established independently of their physical properties. But on the other hand, the centers of mass of the two rods generally do not coincide and in the case of the rods' expansion, these centers are moving relative to one another. At each moment of time, the rest frame of each rod is the one instantly comoving with its respective center of mass. Since the centers are themselves moving relative to each other, so are the rest frames of the respective rods. Applying the operational procedure for the length measurement, we will obtain two generally different proper lengths for the two rods that are geometrically identical. There may be some exceptions like in a special case shown in Figure 7.34. But even in this case, the proper length of the *combined system* of the two congruent rods is different from the proper length of either rod. This example illustrates the complexity of the concept of proper length in the case of an arbitrarily changing object. And it also shows that geometry cannot be separated from physics, in particular, from mass distribution in a system.

But in the case with conveyor belt considered above, this concept can be relatively simply defined.

Based on this case, getting back to our compendium of different physical quantities in Section 5.5, we can express the properties of the proper length in the following way: proper length is (under certain conditions) additive, invariant, but not conserved.

7.7
What is Rigid in Relativity? (What Constitutes Deformation?)

The situations considered in the previous sections illustrate the subtle nature of the concepts of "rigidity" and "deformation." The horizontal train (Section 7.4) in a state of a free fall remains ideally straight as observed by the engineer and is partially straight and partially bent in an inertial RF moving horizontally with the initial velocity of the train. By all accounts, the train is recorded as geometrically deformed in the second RF, whereas in reality nothing can be in a more perfect relaxation than a train freely falling in a homogeneous field of gravity. In Section 7.5, the horizontal rod, initially moving along its length in system A and stopped by a uniformly distributed simultaneous force in this system, does not change its length as observed by Alice. By all accounts, it is geometrically nondeformed in A, whereas actually it is severely compressed. The conveyor belt in the previous section appears nondeformed in the rest frame of the conveyor, whereas each of its elements is subject to periodic stretches and compressions when bending around the pulleys. A relaxed elevator in Section 4.4 is geometrically deformed as observed by the engineer and Mr. O'Bryan, and the observed shape depends on the history of the boost.

This state of affairs is appropriately expressed by the following quotation from [68]: "Even in conventional flat space-time it is impossible to establish on an accelerated body a rigid coordinate system with a flat (Euclidean) space metric and with synchronized time. Such a coordinate system can be tied only to a uniformly moving body. With the exception of special cases (e.g., a rigidly rotating disk), any given (*accelerated* – M.F.) frame of reference will become deformed in the course of time; its geometrical properties will change, and the rates of clocks tied to it, will change."

We can strengthen this statement by noting that even a uniformly spinning disk must have undergone physical deformation during transitional period from the state of rest to the state of uniform rotation; even in the final state of *uniform* rotation, it will periodically deform when observed from any inertial reference frame moving relative to its center of mass. It looks therefore that the exception of a rigid uniformly spinning disk indicated in [68] should be explicitly restricted to a *single* inertial RF attached to the center of this disk.

In what follows, we will consider a few more examples illustrating the general rule formulated in [68]. And the first example will illustrate the point just made, about a uniformly rotating object observed from a passing inertial RF.

1. A Spinning Wheel

Suppose we have a rotating wheel of a bike and denote as K′ an inertial RF comoving with its center of mass. Let us first describe the wheel's rotation in K′. Denote its angular velocity in this frame as ω' and the plane of its rotation as $x'y'$. The position of the wheel's spoke at any moment of time is characterized by its instant angle $\varphi(t')$ with the x'-axis:

$$\varphi'(t') = \omega' t' + \varphi'_0, \tag{7.72}$$

where φ'_0 is the angle it makes with the x'-axis at the zero moment $t' = 0$ of system K′. The Cartesian coordinates of a point at a distance r' from the center on a given spoke are

$$\begin{aligned} x' &= r' \cos(\omega' t' + \varphi'_0), \\ y' &= r' \sin(\omega' t' + \varphi'_0). \end{aligned} \tag{7.73}$$

Eliminating r', we can write the equation for an arbitrary point on the spoke as

$$y' = \kappa x', \qquad \kappa \equiv \tan(\omega' t' + \varphi'_0). \tag{7.74}$$

This is the equation for a straight line passing through the center and making the angle (7.72) with the x'-axis. For a given spoke this angle changes linearly with time and at any given moment we can switch from one spoke to another by changing the parameter φ_0'. Thus, Equation (7.74) describes the entire set of spokes of the wheel spinning with angular velocity ω'. If the radius of the wheel is R, then $0 \leq r' \leq R$, and for any spoke at a fixed moment t',

$$0 \leq x' \leq R\cos(\omega' t' + \varphi_0'). \tag{7.75}$$

The wheel itself forms rotating reference frame K_0.

Consider three observers: Paul in K_0, Sam in K', and Peter in a frame K, which is moving uniformly along the x'-axis with velocity $-V$, so that the center of the wheel is sliding to the right with respect to Peter at a speed V.

What is the picture seen by each observer?

Paul sees the stationary wheel and the universe spinning about the axis perpendicular to the wheel and passing through its center. If he himself is not on the axis, then he experiences a centrifugal force tending to pull him still farther away from the axis. His system K_0 is noninertial. Accordingly, as we will see in Section 8.1, its space geometry is non-Euclidean.

Sam in system K' sees the nonrotating universe and the spinning wheel. Spokes of the wheel rotate about its center but remain straight, according to our assumption that the wheel rotates as a rigid body. The things look pretty simple for Sam.

Peter, however, sees something more interesting. To describe his observations in the most simple way, we, as usual, direct the axes of K parallel to the corresponding axes of K'. Our first prediction is that the diameter of the wheel along the x-direction will undergo the Lorentz contraction, so the shape of the wheel recorded in K is the ellipse with its minor axis along the x-direction. The two corresponding spokes (instantly horizontal in K') are shortened by the Lorentz factor. Other spokes will have their x-components shortened by the same factor, so our first expectation is to observe in K something similar to Figure 7.35.

That would be the true story, were the wheel not rotating. The rotation brings about two new effects. First, the two spokes that are now "horizontal" (are instantly alongside x) will be "vertical" (alongside y) a quarter of the period later. Therefore, their lengths will be extended from $R/\gamma(V)$ in the former position up to R in the latter. Similarly, the two spokes that are now "vertical," will be horizontal a quarter of the

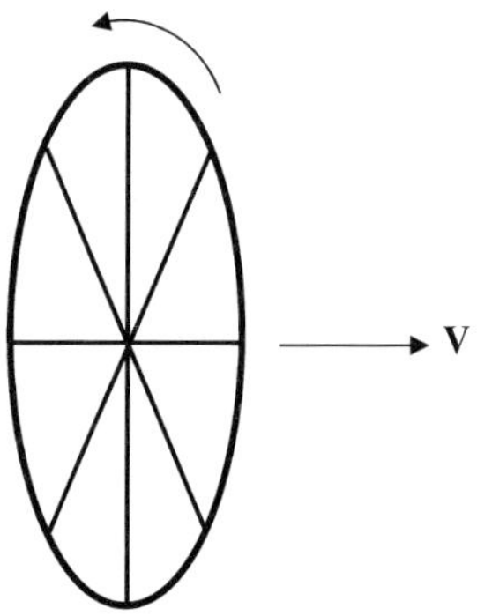

Figure 7.35 Naively expected picture of a moving and spinning wheel as observed by Peter.

period later and their lengths will be contracted from R now down to $R/\gamma(V)$ then. Thus, each individual spoke is "breathing," its length changing periodically from R to $R/\gamma(V)$ and back to R. The period of this process is determined by the time dilation effect, that is,

$$T = T'\gamma(V) = \frac{2\pi}{\omega'}\gamma(V) = \frac{2\pi}{\omega}, \tag{7.76}$$

so that the frequency

$$\omega = \frac{\omega'}{\gamma(V)}. \tag{7.77}$$

Peter in K sees the wheel spinning more sluggishly than it does in K′.

The second effect is a little more subtle. It has to do with relativity of simultaneity, which, as we have learned in Section 4.4, causes the tilt of a line moving in K′, when observed from K. In our case, each spoke is moving perpendicular to itself in K′. Let us assume that the wheel is spinning counterclockwise and consider the spoke OB, which is instantly horizontal in K′ (Figure 7.36a). Each length element of this spoke is now moving up with a speed proportional to the element's distance from the center. Therefore, in K each length element of OB is tilted downward from the horizontal by the angle given by (4.51). The tilt increases with the distance from the center owing to the increase of the corresponding velocity. For the spoke AO, which is moving down at the same moment in K′, the tilt in K is upward if viewed from A to O. The resulting instant shape of the "horizontal" diameter of the wheel in K is shown in Figure 7.36b. We see again that what is horizontal in K′ is not horizontal in K. This is because the length elements along x, which we observe at one moment of K-time, are characterized by the moments of their history that are in K′ different for different elements. Moreover, the local tilt of each element is a function of x, since its velocity depends on its position. Thus, what is straight in K′ is not straight in K. On the contrary, the spokes that are currently vertical in K′ are vertical in K, because the simultaneous

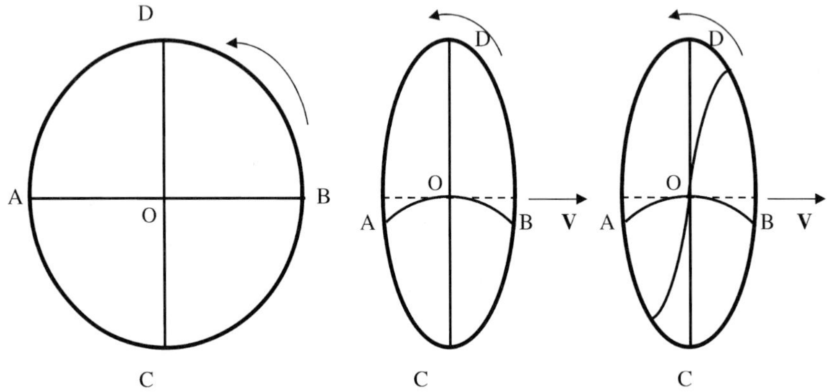

Figure 7.36 (a) The two spokes that are respectively instantly horizontal and vertical in K′; (b) the same spokes as observed in K; (c) the same as in (b) with an additional spoke making an arbitrary angle with the x-direction at the depicted moment.

events on the plane perpendicular to the direction of relative motion of the two frames are simultaneous in both frames.

The combined effect is that each spoke of the spinning wheel observed in K undergoes two kinds of periodic deformation: it is stretching and contracting and simultaneously bending and straightening with the same period. And this is happening to a system, which, by assumption, is made of rigid material! What is "rigid" then?

Before discussing the concept of rigidity in relativity, we want to describe the whole process quantitatively. To this end, we need to express the coordinates of a point (x', y') on a spoke in terms of the (x, y) coordinates:

$$\begin{aligned} x' &= \gamma(V)(x - Vt) = \gamma(V)\tilde{x}, \\ y' &= y = \tilde{y}, \\ t' &= \gamma(V)\left(t - \frac{V}{c^2}x\right) = \gamma^{-1}(V)t - \frac{V}{c^2}\gamma(V)\tilde{x}. \end{aligned} \tag{7.78}$$

Here

$$\tilde{x} = x - Vt \tag{7.79}$$

is the instant distance in K between the x-projection of point (x, y) on a spoke and the center of the wheel, and the corresponding notation is introduced for the y-projection of this point. Using (7.79), the reader can easily obtain the third of Equations (7.78). Next we plug expressions (7.78) into Equation (7.74). We see immediately that owing to time dependence of the coefficient κ in (7.74) and relativity of simultaneity, the corresponding coefficient in K will be a function of x, and thereby of $\tilde{x}$. Which means that, as we have already found from our qualitative treatment, the spokes in K are not generally straight. Quantitatively,

$$\tilde{y} = \gamma(V)\tilde{x}\tan\left\{\omega\left[t - \frac{v}{c^2}\gamma^2(V)\tilde{x}\right] + \varphi'_0\right\}. \tag{7.80}$$

Here, we have also used the relation (7.77). As to the initial angle, we have left its value expressed relative to the system K′.

Using $x' = \gamma(V)\tilde{x}$ from the first of Equation (7.78) and putting it into (7.75), we can also determine the range of change for the variable $\tilde{x}$:

$$0 \le \tilde{x} \le \frac{R}{\gamma(V)} \cos\left\{\omega\left[t - \frac{V}{c^2}\gamma^2(V)\tilde{x}\right] + \varphi'_0\right\}. \tag{7.81}$$

According to (7.81), only the values of $\tilde{x}$ that are less than the expression on the right have the physical meaning of the x-projection of the corresponding point on the spoke. Consider the limiting case when $\tilde{x}$ is *equal* to the expression on the right, and denote the corresponding value as $\tilde{x}_m$. In other words, find the appropriate root $\tilde{x} = \tilde{x}_m$ of the equation

$$\tilde{x} = \frac{R}{\gamma(V)} \cos\left\{\omega\left[t - \frac{V}{c^2}\gamma^2(V)\tilde{x}\right] + \varphi'_0\right\}. \tag{7.82}$$

Then (7.81) can be written as

$$0 \le \tilde{x} \le \tilde{x}_m(R, \omega, V, t, \varphi_0'). \tag{7.83}$$

Equations (7.80) and (7.83) (and the similar equations for negative $\tilde{x}$) give the quantitative description of the problem.

Let us consider the same two special cases as before, namely, the spokes that are, respectively, horizontal and vertical in K′. In our notations, the first case corresponds to $t = 0,\ \varphi_0' = 0$. Putting this into (7.80) gives

$$\tilde{y} = -\gamma(V)\tilde{x}\tan\left(\frac{V\omega}{c^2}\gamma^2(V)\tilde{x}\right). \tag{7.84}$$

In accordance with our qualitative prediction, this expression describes the bent line tangent to the x-axis at $\tilde{x} = 0$ and curving down from it for both positive and negative $\tilde{x}$. The spoke horizontal in K′ is bent in K. Looks pretty similar to our result with the relativistic train during its crash. The distance between the end points of this spoke is determined by (7.83) evaluated at $t = 0,\ \varphi_0' = 0$ and is less than $R/\gamma(V)$.

The second case corresponds to $t = 0,\ j_0' = \pi/2$ and yields

$$\tilde{y} = \gamma(V)\tilde{x}\cot\left(\frac{V\omega}{c^2}\gamma^2(V)\tilde{x}\right). \tag{7.85}$$

This expression appears to contradict our qualitative conclusion that the spoke instantly vertical in K′ is instantly vertical in K. Indeed, Equation (7.85) does not describe the straight line, let alone the vertical straight line. The "discrepancy" is resolved if we evaluate the range of change of $\tilde{x}$ in this case. This range shrinks down to zero, which is evident from both physical considerations and Equations (7.82) and (7.83), evaluated at given $t = 0$ and $\varphi_0' = \pi/2$. Indeed, for these parameters Equation (7.82) takes the form

$$\tilde{x} = \frac{R}{\gamma(V)}\sin\left(\frac{V\omega}{c^2}\gamma^2(V)\tilde{x}\right), \tag{7.82a}$$

which has only one physical root $\tilde{x} = \tilde{x}_m = 0$. Thus, Equation (7.85) is only valid at the zero value of $\tilde{x}$ and its right-hand side is not defined at this value. In other words, $\tilde{x} = 0$ corresponds to all physically possible values of $\tilde{y}$ between R and $-R$, that is, to the vertical segment $\tilde{x} = 0,\ -R \le \tilde{y} \le R$. Which means that the spoke CD instantly vertical in K′ is straight and instantly vertical in K.

An interesting by-product of this effect can be obtained immediately from viewing Figure 7.36b (Problem 7.11).

2. Rigid "Horizontal" Rod

Next, consider a rigid horizontal rod in K′. Let this rod move in vertical direction, first up and then down with a speed v [70]. The instant position of the rod at any moment of time is described by

$$y(x', t') = \begin{cases} v't', & t' < 0, \\ -v't', & t' > 0. \end{cases} \tag{7.86}$$

Before the zero moment the rod is moving up. It touches the x'-axis at $t' = 0$ and immediately starts moving down with the same speed. At this time the rod remains horizontal, moving perpendicular to itself.

However, the property of being horizontal is a relative characteristic, which may be different for different inertial observers even in the same locality. In another frame K, moving along the x'-axis with velocity $-V$, the rod will be tilted downward when moving up and upward when moving down [1]. Since in K′ it is moving first up and then down, in K, owing to the relativity of simultaneity, there will be a part of it moving up and the other part moving down at the same moment of time in K. These parts are accordingly tilted, one downward and the other upward, and connected at the x-axis. A straight solid rod under the described conditions is broken in K into two parts, thus forming a kink. And the kink (separation point between the two parts) will itself slide down the x-axis.

For quantitative description of the phenomenon, we again express the primed characteristics in terms of the corresponding unprimed ones using the Lorentz transformations (7.78). We get

$$y(x,t) = \begin{cases} v\left(t - \dfrac{V}{c^2}x\right), & t - \dfrac{V}{c^2}x < 0, \\ -v\left(t - \dfrac{V}{c^2}x\right), & t - \dfrac{V}{c^2}x > 0. \end{cases} \tag{7.87}$$

Here

$$v \equiv v'\gamma(v) \tag{7.88}$$

is the "vertical velocity" of the rod in system K. Equation (7.88) appears to contradict the time dilation effect, according to which the vertical component v_y of the velocity of a fixed point of the rod in K is *less*, not greater, than v' by a factor of $\gamma(V)$. Actually, there is no contradiction here, since (7.88) is the speed of the rod's element with the fixed x-coordinate. This element is *not* a fixed point on the rod, which itself is also sliding along x with velocity V. We can visualize this by imagining the intersection point P between the rod and a vertical line $x = \text{const}$ in K. As the rod moves in K, we see its *different* points successively coinciding with P (Figure 7.37). The point P is sliding along both the rod and the vertical line. And its velocity $v \neq v_y$ just because the rod is tilted in K. The tilt angle α with the horizontal is given by

$$\tan\alpha = \frac{dy}{dx} = \pm\frac{Vv}{c^2}, \tag{7.89}$$

with the sign determined according to one of the two conditions in (7.87).

Since the two slopes of the kink meet on the x-axis and at the same time are moving, the meeting point (the kink itself) must also be moving along x. We now want to find the speed of this motion. Denoting

$$\frac{c^2}{V} \equiv u, \tag{7.90}$$

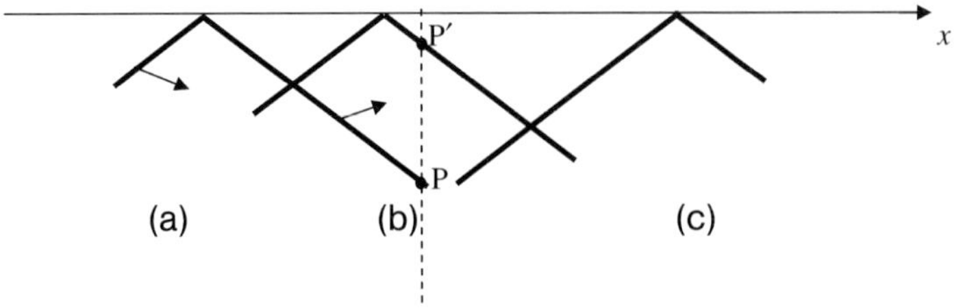

Figure 7.37 Three consecutive snapshots (a), (b), and (c) of the same straight rod, made in K. Note that two intersection points P and P′ between the rod and a vertical axis are physically different points of the rod. Note also that the points where the rod is observed broken in each snapshot are different points of the rod as well. The arrows indicate the velocities of two broken parts of the rod as observed in K.

we recast (7.87) as

$$y(x,t) = \begin{cases} v\left(t - \dfrac{x}{u}\right), & x > ut, \\ v\left(\dfrac{x}{u} - t\right), & x < ut. \end{cases} \tag{7.91}$$

This expression clearly shows that the kink is actually a "triangular wave" moving with a speed $u > c$. Figure 7.37 shows three consecutive positions of this "wave" at three different moments of time. The two parts, into which the rod breaks in K, are different at these moments. And the most interesting aspect of this effect is that, according to (7.90), the speed $u > c$. The separation point between the two parts (the kink) is moving faster than light. This does not contradict causality because this kind of motion is not accompanied by any energy transfer and therefore cannot be used for signaling.

If the rod is long enough and periodically pops up and down at constant speed between two horizontal lines x and x', then we can observe different cycles of its motion at one moment in K. The rod in this case will have many kinks (Figure 7.38).

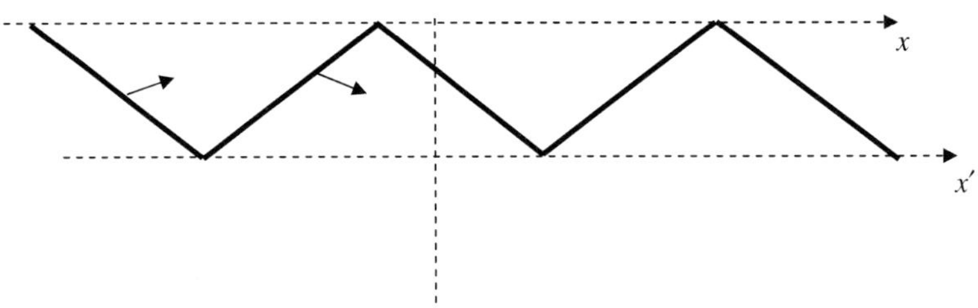

Figure 7.38

3. An Oscillating Rod

In the third situation, we consider the same horizontal rod performing harmonic oscillations up and down in K′ (i.e., along the y'-axis) with frequency ω'. Accordingly, its velocity will also be a harmonic function of time. As before, we assume the rod to be rigid in K′.

What kind of motion will be recorded in system K? Again, applying the same reasoning as above, we can predict that in K different parts of the rod will have different tilt due to the fact that we will record simultaneously in K different moments of K′-time for different parts of the rod and these parts will at these different moments have different instant velocities. With tilt being a function of x, the rod is not straight in K. Moreover, the tilt depends on time t as well, so the rod must wiggle with time. The reader can already suspect the emergence of a transverse sinusoidal wave running along the "rigid" rod!

This is precisely what relativity predicts in the given situation. The proof is very straightforward. Write down the equation of harmonic motion for the horizontal rod in K′:

$$y'(x', t') = y'_0 \sin \omega' t', \qquad v'(x', t') = \frac{dy'}{dt'} = \omega' y'_0 \cos \omega' t' \tag{7.92}$$

and plug in Equation (7.78). We will obtain

$$y(x, t) = y_0 \sin \left[\omega' \gamma(V)\left(t - \frac{V}{c^2} x\right)\right], \qquad y_0 = y'_0. \tag{7.93}$$

This can be easily recast into a more conventional form

$$y(x, t) = y_0 \sin(\Omega t - Kx) = -y_0 \sin K(x - ut), \tag{7.94}$$

where

$$\Omega \equiv \omega' \gamma(V), \qquad K \equiv \Omega \frac{V}{c^2}, \qquad u \equiv \frac{\Omega}{K} = \frac{c^2}{V}. \tag{7.95}$$

These equations describe the predicted phenomenon: as a horizontal rod performing vertical oscillations gets corresponding horizontal boost, there appear wiggles on the rod, which constitute transverse running sinusoidal wave (Figure 7.39). The wave has the following properties:

(a) If the oscillating rod is boosted to the right, the wave, according to (7.93), will run also to the right.

(b) The amplitude of the wave is equal to the amplitude of oscillations in K′.

(c) The wave frequency exceeds the proper frequency of oscillations by the Lorentz factor; we have noted already that this does not contradict anything since the wave frequency in K is determined by the oscillation frequency at a *fixed* x, while the time dilation is observed in K for a *fixed* x'.

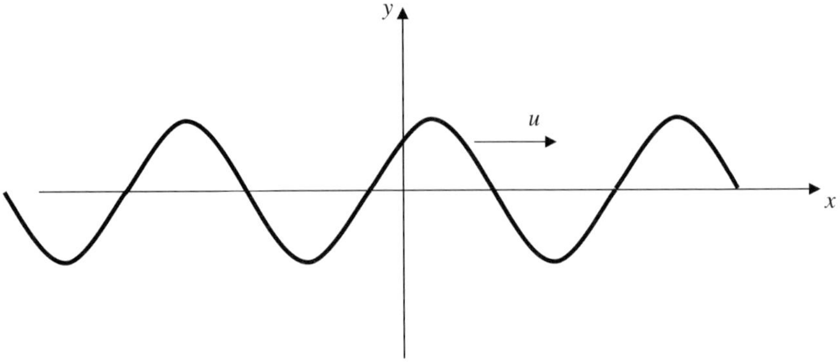

Figure 7.39

(d) The speed of the wave is greater than the speed of light in vacuum. This does not contradict anything either, since the phase velocity of a wave is not the velocity of the rod itself. The latter is just V.[3)]

Now we want to discuss the physical nature of these waves. Definitely, it is not accompanied by any physical deformation of the rod. In contrast to ordinary waves, this wave is not accompanied by the motion of different parts of the rod relative to one another. If we apply reverse Lorentz transformation to either triangular or sinusoidal wave described by (7.91) and (7.93), respectively, we will obtain straight segment moving as one piece. Alternatively, we can calculate directly the relative velocity between any two elements dl_1 and dl_2 of the rod in K. If we are careful enough to evaluate their instant velocities in K at the moments corresponding to *one and the same* moment t' in the system K′, we will again obtain zero (Problem 7.3).

Further, any wave associated with the relative motion of different elements of the supporting medium is accompanied by energy loss to heat.[4)] In our case, there is no heat generation in the rod.

In case of the triangular wave, a real rod with assumed properties would be just broken into two pieces in K. Imagine trying to make a kink in a crisp dry spaghetti and if you succeed, make the kink in addition run along spaghetti faster than light!

Precisely such an effect can be produced with our rod under the corresponding conditions. Nothing horrible happens to the rod – it does not break, not even warms up by a little bit.

A wave running along the rod in our thought experiment is a purely relativistic kinematic effect – another manifestation of the relativity of time. Its properties are determined only by the kinematic characteristics of motion (of which the most crucial is the relative speed V of the two reference frames), rather than by the physical properties of the material. If we consider any other medium instead of the rod, we will

3) More accurately, V in K stands exclusively for the velocity of center of mass of the rod. Since the rod is wiggling in K, the velocity of each of its elements becomes a function of element's position.

4) The only known exception – superfluidity – does not apply here since we have assumed the material of the rod to be rigid.

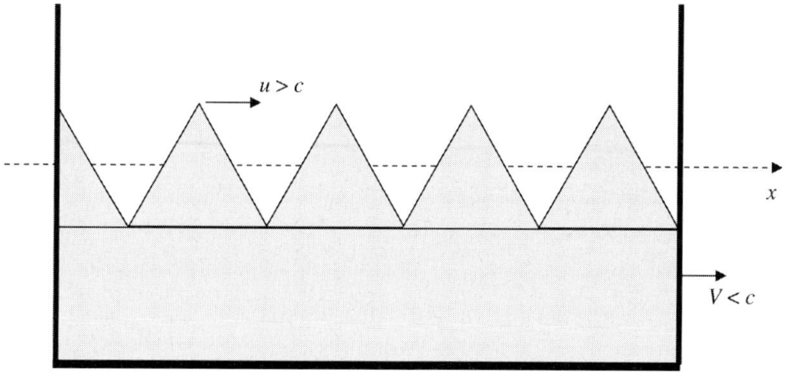

Figure 7.40

obtain the same characteristics of the wave in this medium. As an example, imagine that in situation 2 we have, instead of the rod, a horizontal water surface in K′, first steadily rising up to a certain level and then steadily sinking immediately after it. We then will obtain the same result as for the horizontal rod – the appearance of the triangular wave, but this time it will be the water wave (Figure 7.40). The running triangular kinks on the water surface are as exotic as the kinks running along dried spaghetti. They are not described by the equations of fluid dynamics. Such equations do allow us to prepare an arbitrarily shaped wave (triangular included) at some initial moment of time. But such a wave would be unstable. In contrast, the waves in our thought experiment are stable and we could observe them as long as we could create a sufficiently long water tank and maintain periodic rise and fall of water in it. Also, in contrast with a hydrodynamic wave in water, which is subliminal, the apexes in our experiment move faster than light, and water itself is partially involved into the wave motion, since it is transported with the tank at a speed V.

If instead of rising and falling of water in the tank, the tank itself would pop up and down together with fixed amount of water in it, then the bottom of the tank would form the same running zigzag wave as the water surface (Figure 7.41).

The discussed phenomenon must not be considered as a mere abstraction that cannot be observed in real everyday life. Once we find an appropriate medium capable of moving sufficiently fast, the discussed phenomenon could be as real as anything else. Do we have such a medium? The answer is yes. It is electromagnetic radiation. Imagine instead of a water surface, a wave front of a light wave, or better still, a laser pulse. Let the pulse move up and then bounce down from a horizontal reflecting surface. That would be an experimental realization of our rod or water surface moving up and then down. If the lab with this system (the reference frame K′) is moving horizontally relative to us (the residents of reference frame K), the above-described effect (the tilt of the surface in K) is precisely what has been observed routinely in astronomy and optics. It is a well-known effect – aberration of light (its relation to the relativity postulates has been described in Section 6.12). If this is a very short laser pulse, it will form an observable kink with its tip running along the

Figure 7.41 A water tank ABCD containing a fixed amount of water and bobbing up and down together with it (inertial effects are neglected).

reflecting surface. (An experimental situation closely related to this effect was discussed in Section 6.13.)

7.8
The Twin Paradox

We now turn to one of the most famous of the relativistic paradoxes – the twin paradox, or, more general, the clock paradox. This paradox can be considered as a natural extension of our previous discussions of relativity of time. Remember, we considered the time of a process observed from two different reference frames and found it to be different for each frame despite the symmetry between all inertial frames (Sections 2.5 and 2.8). We explained the difference as owing to the asymmetry of the observational procedure. In the twin paradox, a situation is considered where the observational asymmetry seems to be eliminated, because it is now one and the same pair of events that is considered in either reference frame: the departure and return of a spaceship with one of the twins aboard.

Consider first the conventional textbook version. Imagine two identical twins, Larry and Joe. Larry leaves the Earth to visit a distant star, while Joe remains at home. After a long journey, Larry returns home and turns out to have aged much less than Joe. This effect is predicted by the theory of relativity and is confirmed experimentally in a laboratory version of the situation – the experimental study of the decay rate of μ-mesons moving in a circle in a magnetic field [71]. The effect can be thought of as yet another manifestation of time dilation that we had studied earlier in Sections 2.5 and 2.8. But now the effect seems to be a real paradox because of seemingly symmetrical role of both – the twins and their respective observational procedures. According to this symmetry, Larry might claim that he could consider himself stationary and Joe moving together with the Earth first away from and then back toward Larry. In which case the same relativistic equations should give the result of Joe having had aged less than Larry. We thus come to the two mutually contradictory statements: by the time of their reunion, Larry turns out to be younger than Joe and Joe turns out to be younger than Larry. Both statements now refer to one situation when the twins are again together at one place at one time in one common reference frame.

To sharpen the paradox, let us assume that the destination of the spaceship is hundreds of light years away. In this case, there can be no way for twins' reunion unless there is some dramatic increase in human longevity, which we do not suppose to happen here. By the time the ship returns, many centuries may have gone by on Earth; all Larry's contemporaries will for a long time have been dead. But Larry may still be alive! Suppose that the ship's speed is very close to the speed of light. Then Larry's proper time interval for the journey lasting many centuries by the Earth's time may, according to Equation (2.59), be just a few years. In other words, he will be just a few years older when he returns while many hundreds of years will have passed on Earth. He will thus find himself having been "hurled" into the distant future. This is the conclusion given by the special theory of relativity applied to inertial reference frame associated with our solar system. Things might look differently from Larry's perspective if he tried to apply the same theory to his ship's reference frame. Then, he could claim that he is stationary in his ship while the solar system with the rest of the universe is moving relative to him. In which case, according to the same relativistic equations, it is *he* who will long have been dead inside the rusted ship, while Joe will be alive and kicking by the time the Earth "returns" to the ship. The paradox is thus restated as follows: when the ship returns, Larry is alive and sees Joe's ancient grave and Joe is alive and sees Larry's skeleton in the remainder of the ship. The paradox is in that both mutually excluding statements refer to the same situation with the two objects (or subjects) brought together.

We now proceed with the solution of the paradox. First, though Larry *might* consider himself as stationary, his reference frame is *not* symmetrical to that of Joe's. It is by no means inertial, because in order to return back to Earth, the ship must change its velocity, that is, it must accelerate. Larry might object that he observed *Joe* accelerated. But this dissention can be easily resolved by asking one simple question: who actually experienced the forces needed to produce the acceleration? Definitely it was not Joe. It was Larry who needed to turn the engines on to accelerate his ship toward its destination, then to decelerate it for landing, then to accelerate it this time back toward the Earth, and finally to decelerate it again for landing on Earth. The forces exerted on Larry and observed by Joe are real forces having a real source; for example, the jet stream of plasma from the ship's engines. The forces exerted on Joe (and the whole Earth) and observed by Larry are fictitious forces of inertia discussed in Chapter 1. Larry cannot identify any real physical body responsible for these forces. These forces are only due to Larry's choice of his ship as a reference frame. From this fact alone, Larry could conclude that his is not an inertial reference frame. At any single moment of their *uniform* motion neither of the twins could say with confidence who was resting and who was moving. But the whole round-trip, including the moment of turnaround, resolves the dispute. This moment makes all the difference. The symmetry originally assumed to exist between the two reference frames turns out to be an illusion. *The two systems are not equivalent.*

Second, since Larry's reference frame turns out to be noninertial, Larry must in all his calculations use the equations of the general theory of relativity.

Third, the general theory of relativity used in Larry's reference frame will give the same result as that of the special theory used by Joe: Larry will have aged less than Joe. In case of big time discrepancies, the common answer is: Larry will be alive and Joe will be dead.

Thus both possible approaches, if performed consistently, give results that are in agreement with each other and with the above-mentioned experiment with decaying muons. Both twins (if they are educated enough) know these results and thus know in advance the only possible outcome of the anticipated space odyssey: Larry will outlive Joe. They also know that there is no discrimination against Joe in this state of affairs, because Joe's biological life is not in any way physically affected by Larry's journey. Quite the contrary, it is Larry's life that has been so dramatically affected as to be extended into distant future without slightest change in his biological life span. This is just another, and very sound, manifestation of the relativity of time.

We will now turn to a detailed description of the phenomenon and its possible observation. We will consider here an idealized situation to see how we can get the same answer for both parties using only the special theory of relativity. Imagine a distant star M 2000 orbited by a planet that according to recent data might harbor extraterrestrial life. The star is 300 light years away from our solar system. Larry starts a journey to the star in a spaceship that accelerates practically instantaneously (this is our idealization!) up to the 0.999 of the speed of light, then moves uniformly till he reaches the star, then turns also instantaneously and rushes back to Earth with the same speed. As in our previous discussions, we neglect technical details such as motions of the Earth and the other planet, the energy needed to accelerate the spaceship, the rate of energy supply to accelerate it that fast, and so on. These details, important as they are, will not change one essential feature of the anticipated journey: its dramatic effect on Larry's proper time.

We will try to understand the origin of the time discrepancy between the twins' life spans by considering radiocommunication between the two reference frames. The Earth and the ship communicate by sending regular radio signals. Suppose they arrange to send each other one signal a year. Then each will know the other's age by just counting the total number of signals received between the moments of departure and return of the ship. Since time is a continuous variable, while the signals come in discrete lumps, let us divide each year into 10 equal time intervals and mark each interval by sending a weaker signal; the number of such signals will give us the corresponding number in the first decimal place after the integral number of years; for instance, if we received 5 strong signals followed by only 2 weak signals, we can write the total number of signals as 5.2, which will correspond to 5.2

years of the sender's time. We can follow this procedure to mark ever-smaller decimal fractions of the year. Thus, the total number of signals in our treatment can be fractional, which enables us to specify time by the number of signals to an arbitrarily high precision.

Now, we can proceed with quantitative treatment of the problem. Assume our solar system and another star M to be stationary. Let L_0 be the distance between them measured in their common inertial reference frame S_0. A spaceship starts from Earth at a speed v and moves to M. While moving to M the ship represents inertial system S. Upon reaching M, the ship turns and moves with the same speed but in the opposite direction and therefore it now represents another inertial reference frame S′. The time it takes for a round-trip is

$$T_0 = 2t_0, \quad \text{where we denoted } t_0 \equiv \frac{L_0}{v}. \tag{7.96}$$

The proper time of the ship can be found in two different but equivalent ways: you can find the contracted distance L between the star M and the Sun in the rest frame of the ship, divide this distance by v and double the result for a round-trip time:

$$L = \gamma^{-1}(v)L_0, \qquad T = 2\frac{\gamma^{-1}(v)L_0}{v} = 2\frac{t_0}{\gamma(v)}. \tag{7.97}$$

Or, you can apply Equation (2.58) for the time dilation to (7.96):

$$T = \gamma^{-1}(v)T_0 = 2\frac{\gamma^{-1}(v)L_0}{v} = 2\frac{t_0}{\gamma(v)}. \tag{7.98}$$

Thus, simple treatment within the framework of the special theory of relativity gives us unambiguous result that the proper time of the spaceship (7.98) is less than that of the Earth (7.96) by a factor $\gamma(v)$. This factor can be made arbitrarily large for relativistic motions, so the proper time of the ship can be made arbitrarily small. Some people think that the equations of the special theory of relativity cannot be applied to a spaceship in this kind of problem because the ship's motion is not inertial. This is a common misconception. The special theory of relativity can be applied to accelerated motions as well as to uniform ones; the only condition is that the system of reference where this motion is being considered must be inertial, if we want a straightforward treatment. Insofar as we consider the ship's motion relative to an inertial reference frame S_0 associated with the solar system, our treatment is not only valid but also pretty simple. The difficulties may arise when we try to apply the equations of special relativity *in* the reference frame of the ship, which does not move inertially all the time. In the following treatment, we will avoid these difficulties by applying the equations of special relativity to the ship twice: first, when the ship moves uniformly toward the star M (the inertial system S), and second, when the ship moves toward the Earth (the inertial system S′).

With this in mind, we are now going to obtain the same results (7.96) and (7.98) by collecting the counts of radio signals received by Earth's and ship's detectors from their respective senders.

The number of signals from the ship received on Earth is the sum of the low rate and high rate successions of signals. The low rate signals are from the receding ship and the high rate signals are from the approaching ship. If the proper frequency is $f_0 = 1$ S/y (signal/year), then the low and high rates are, respectively,

$$f_L = \sqrt{\frac{1-\beta}{1+\beta}}, \qquad f_H = \sqrt{\frac{1+\beta}{1-\beta}}, \qquad \beta \equiv \frac{v}{c}. \tag{7.99}$$

The time T_L on Earth for the incoming low rate signals consists of two intervals: the time $T_1 = t_0$ during which these signals are produced by the receding source and the time $T_2 = L_0/c = \beta t_0$ needed for the last of these signals to travel the distance L_0 between the Earth and the source. So $T_L = T_1 + T_2 = (1+\beta)t_0$. The amount of low rate signals is thus

$$N_L = \sqrt{\frac{1-\beta}{1+\beta}} \cdot T_L = t_0(1+\beta)\sqrt{\frac{1-\beta}{1+\beta}} = \frac{t_0}{\gamma(v)}. \tag{7.100}$$

The time T_H of high rate signals coming to Earth from the ship is just the time interval between the moment of the first such signal's arrival and the moment of the ship's return to Earth:

$$T_H = \frac{L_0}{v} - \frac{L_0}{c} = t_0(1-\beta). \tag{7.101}$$

The corresponding number of high rate signals is

$$N_H = \sqrt{\frac{1+\beta}{1-\beta}}\, T_H = \frac{t_0}{\gamma(v)}. \tag{7.102}$$

The total number of signals from the ship and, accordingly, the number of years passed there is

$$N = T = N_L + N_H = 2\frac{t_0}{\gamma(v)}. \tag{7.103}$$

This result is identical with (7.98).

In a similar way we calculate the number of signals from Earth received on the ship. During the first part of the journey, the ship's detector counts low rate signals; during the second part, which starts immediately after the turn, the ship receives high rate signals. Both parts of the journey take the same time $(1/2)T = \gamma^{-1}(v)t_0$. Therefore, the total number of low and high rate counts in the ship's detectors is

$$N_0 = \frac{t_0}{\gamma(v)}\left(\sqrt{\frac{1-\beta}{1+\beta}} + \sqrt{\frac{1+\beta}{1-\beta}}\right) = 2t_0 = T_0, \tag{7.104}$$

which is identical with Equation (7.96). Thus both the theoretical prediction of the time dilation effect and the described thought experiment with signal exchange give the same result: the Earth's history and the ship's history evolving during their separation are characterized by different times. The ship's time is less than the

Earth's time. This property of split histories does not depend on the reference frame. The ship's crew observing the history of Earth from their rushing outpost gets the same reading T_0 for Earth's time as do historians or physicists on Earth. The latter get the same reading T for the ship's round-trip proper time as do the ship's clocks. And in everybody's account T turns out to be less than T_0 by the same amount. The time discrepancy between the two split and then reunited histories is their common physical characteristics.

There is, however, something missing in this account. It describes *how* all participants of the experiment get a common result for T_0 and then a common result for T. It does not explain *why* the results turn out to be the way they are. The attentive reader may have noticed an asymmetry in the counting procedure that was employed without much comment. When counting low and high rate signals coming *to Earth from the ship*, we said that low rate signals kept arriving for much longer time than high rate signals. When counting the signals coming *to the ship from Earth* we said that both low and high rate signals had been received within the two *equal* time intervals. Why is it so?

The observers on Earth remain in place when the ship is on its way to the star M. If we had not calculated the ship's progress in advance, we would not know when the ship reaches its destination. Actually, we have no other means to even know whether the ship has reached it at all, except the radio signals coming from the ship. But these signals need time to travel from the ship to Earth: zero time for the very first signal (sent at the start) and a very large time L_0/c for the last low rate signal. The time between the arrivals of the first and the last low rate signals is thus extended by this time interval L_0/c. When we receive high rate signals from the returning ship, the last high rate signal is received immediately (the ship is already here), while the first high rate signal is received only L_0/c years after the moment it was sent. Thus, the time between the arrivals of the first and last of high rate signals is *shortened* by the same time interval. As you think of it, you find both processes to be just manifestations of the Doppler effect, which works for the first and the last signal in a succession the same way it does for the two neighboring signals.

The crucial question is: why does the same not apply to the ship's reference frame? There, you remember, we found *the same* time intervals for the successions of low rate and high rate signals from Earth. Does not the ship's crew have to wait till the last low rate signal and the first high rate signal from Earth reached the ship? The answer is: it would have to, *had it remained in the same reference frame S*. But as the ship jerks back upon reaching the star M, it is equivalent to jumping from one reference frame S moving away from the Earth to another reference frame S′ moving toward the Earth. The moment the ship performs the jump, its detectors already start rushing toward the Earth (making thereby Earth rush toward the ship). This *immediately* changes the extended succession of low rate signals incident on the ship into compressed succession of high rate signals. The transition from low rate to high rate signals in the ship's reference frame occurs the moment the ship reaches the star M, while on Earth it occurs long time after this event. You remember, we discussed the difference between the Joe's and Larry's motions in our original example? We had emphasized that no real forces act on Joe whereas there are real forces exerted on Larry. It is these forces that cause the ship to jump and thereby change so dramatically its proper time. It is here where the crucial difference between the two systems comes into play.

It is instructive to extend this line of reasoning in the following way. Suppose, Larry's spaceship consists of two parts: one occupied by Larry and the other carrying a sophisticated robot named Glen. The moment the spaceship reaches the star M, these parts separate and while the compartment occupied by Larry turns back, the capsule with Glen keeps on moving away from the Earth as before. Thus, Glen remains all the time in one inertial reference frame S. Larry now starts sending out pairs of simultaneous signals – one toward the Earth and the other in the opposite direction toward Glen. Glen keeps on collecting the signals coming from the Earth and from Larry. Our question is: what will be Glen's conclusions after he compares Joe's and Larry's times at the moment of their meeting.

Glen initially determines the times of Joe and Larry during the first part of the trip – from Earth to star M. These times are, respectively,

$$t_{\text{Larry}}^{(1)} = \frac{t_0}{\gamma(v)} \quad \text{and} \quad t_{\text{Joe}}^{(1)} = \frac{t_{\text{Larry}}^{(1)}}{\gamma(v)} = \frac{t_0}{\gamma^2(v)} = t_0(1-\beta^2). \tag{7.105}$$

According to Glen's measurements, the time $t_{\text{Larry}}^{(1)}$ during the first half of the trip is *greater* than the time $t_{\text{Joe}}^{(1)}$, that is, Larry was aging faster than Joe! No surprise, if you recall that at this stage Larry is stationary in the ship whereas Joe is moving away and therefore undergoes the time dilation. But in this case, why does not the same apply to the second stage, when Joe is moving toward Larry?

We have analyzed this question in Larry's reference frame and found the explanation in Larry's jump from reference frame S to reference frame S′. But Glen does not make any jump – he remains in the same inertial reference frame as in the first stage. In this frame, Joe keeps on receding from Glen, Larry turns back and is now moving away from Glen faster than Joe. Accordingly, his aging rate will slow down dramatically and drop far below Joe's aging rate. As a result, by the moment he returns to Earth, he will be younger than Joe. Let us look into Glen's calculations at this stage.

Larry's speed v' relative to Glen at this stage is composed of his speed v relative to Earth and the Earth's speed v in the same direction relative to Glen. Applying the law of addition of velocities to this case yields

$$v' = \frac{2v}{1+\beta^2}. \tag{7.106}$$

Denote Glen's time between Larry's jump and his catching up with Joe as t_{Glen}. During this time Larry covers the distance $v' t_{\text{Glen}}$ and Joe travels the distance $v t_{\text{Glen}}$. The difference between these distances is the initial spacing between Larry and Joe at the moment of jump, which is the Lorentz-contracted distance $L_0/\gamma(v)$. Thus, we have the equation

$$(v' - v)t_{\text{Glen}} = \frac{L_0}{\gamma(v)}. \tag{7.107}$$

Solving this for t_{Glen} and using (7.11), Glen obtains

$$t_{\text{Glen}} = \frac{t_0}{\gamma(v)}\frac{1+\beta^2}{1-\beta^2} = t_0\,\gamma(v)(1+\beta^2). \tag{7.108}$$

It is now a straightforward matter to determine the Joe's proper time $t_{\text{Joe}}^{(2)}$ for the second stage of Larry's trip. Since Joe is moving with speed v in Glen's RF, we have

$$t_{\text{Joe}}^{(2)} = \frac{t_{\text{Glen}}^{(2)}}{\gamma(v)} = \frac{t_0}{\gamma^2(v)}\frac{1+\beta^2}{1-\beta^2} = t_0(1+\beta^2). \tag{7.109}$$

Similarly, we find Larry's proper time $t_{\text{Larry}}^{(2)} = t_{\text{Glen}}/\gamma(v')$. Applying the rule (3.17) to our case gives

$$\gamma(v') = \frac{1+\beta^2}{1-\beta^2},$$

so that

$$t_{\text{Larry}}^{(2)} = \frac{t_0}{\gamma(v)} = t_{\text{Larry}}^{(1)}. \tag{7.110}$$

During the first stage of the trip, the rate of Larry's aging in Glen's RF was the highest possible, because Larry was stationary in this frame. In the second stage, Larry's aging rate was the lowest possible in these conditions because Larry's speed v' is now greater than v. However, according to (7.108), the second stage itself lasts much longer – Larry has to catch up with receding Earth rather than simply cover the initial distance between him and Joe. Both competing effects totally balance each other, so that Larry's proper times turn out to be equal for both stages. The important thing is that comparing with (7.109) shows that $t_{\text{Larry}}^{(2)} \ll t_{\text{Joe}}^{(2)}$. Just as we had expected, this time Larry is aging much slower than Joe because he is moving faster and, accordingly, his proper time is much shorter than that of Joe.

Finally, we sum up the proper times of each participant during two stages to get

$$t_{\text{Joe}} = t_{\text{Joe}}^{(1)} + t_{\text{Joe}}^{(2)} = 2t_0, \qquad t_{\text{Larry}} = t_{\text{Larry}}^{(1)} + t_{\text{Larry}}^{(2)} = 2\frac{t_0}{\gamma(v)}. \tag{7.111}$$

The result obtained by Glen is exactly the same as we got considering two other reference frames. The proper times are invariant characteristics depending only on the path between the two events in space-time, not on reference frame used.

The whole phenomenon becomes crystal clear if we draw the world lines of both systems (Figure 7.42) and of the radio signals they use. Alternatively, we can draw these lines using Glen's capsule as the stationary RF (Problem (7.12)). In both cases, we see that the world line of Earth is straight as a laser beam, whereas the world line of the ship's compartment carrying Larry consists of two different segments of equal "lengths." The separation point C on the Earth's world line between the low rate and high rate signals from the ship is shifted from the middle of the line toward the

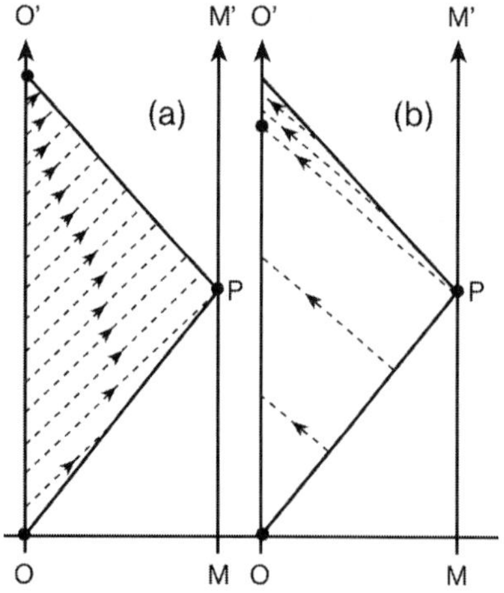

Figure 7.42 Space-time diagram of Larry's round-trip. OO′ – the world line of Earth between the events O (Larry's departure from Earth) and O′ (his return to Earth). OPO″ is the world line of the spaceship between the same events. MM′ is the world line of the planet system M. Dashed lines are the world lines of radio signals between the Earth and the ship. (a) Signals from the Earth to the ship; (b) signals from the ship to the Earth.

future. There is not much room left for high rate signals, which results in relatively small net amount of signals received from the ship. On the ship's world line, the transition point between two successions of signals from Earth is at the vertex. The high rate signals arrive during the same time as the low rate ones, which results in large total number of signals. Accordingly, a large time is recorded for Earth and a small time for the ship by the moment they meet again. Topologically, it is due to the vertex in the ship's world line, that is, to the singular point with an infinite curvature. This is a manifestation of a general law: out of the two lines between the two points (events) in Minkowski's world, the one with larger curvature has the shorter "length." Accordingly, an object moving along the line with a larger curvature has a shorter proper time between the two events.

This stands in a sharp contradiction with what we actually see in Figure 7.42. The path OPO′ is longer, not shorter, than the straight line OO′. It is because we see the *geometrical* path, in whose length the contributions from the horizontal and the vertical components of the segment OP, say, come with the same sign (Pythagorean theorem $\mathrm{OP}^2 = \mathrm{OM}^2 + \mathrm{MP}^2$). The reader should recall that the world line being discussed here is the *kinematic* path in Minkowski's world. Segment MP represents *proper time,* not the distance. Contributions from the temporal and the spatial components of the interval in space-time come, as we have learned in Chapter 4, with the *opposite* signs. The *kinematic* length of the OPO′ is smaller than that of OO′. Because the kinematic length (the space-time interval) is absolute (invariant under Lorentz transformations), this relationship between OPO′ and OO′ is true for *all* observers. Physically, the kinematic length of the OPO′ is Larry's proper time (multiplied by c) during his journey (recall Section 4.1). Thus, the proper time of anyone making a round-trip relative to an inertial reference frame is always less than the proper time of the one remaining in place in this system. This statement is the shortest and the clearest explanation of the "twin paradox."

... We conclude the section by turning our attention to something that has been half hidden in the intricacies of the above analysis and that actually forms its most essential part: *the time travel is possible*. This exciting conclusion follows in the most straightforward way from the comparison of the world lines OO′ and OPO′ in Figure 7.42. The space traveler comes back to find himself in a distant future that is extended far beyond the lifetime of all his former contemporaries on Earth. He thereby becomes a time traveler. He has shifted from his epoch to another one. Such a possibility is not a speculation. It is a scientific prediction confirmed by experiment. The above-mentioned μ-mesons circling in a magnetic field provide a perfect laboratory model of time travel. Suppose you put a stationary μ-meson somewhere in the circular path and call this meson A. Call the circling meson B and draw a space-time diagram for A and B (Figure 7.43). Its only difference from the diagram in Figure 7.42 is that the world line of B forms a helix. Because of its curvature, the kinematic length of the helix is less than the world line of A between their two consecutive meetings. Accordingly, the proper time of B is less than that of A.

Now suppose that A and B are born simultaneously. But if B is moving fast enough, it makes a full cycle within its lifetime while A will have decayed long before the cycle is completed. We call it the time dilation for B, but actually it is the time travel: B finds itself in a distant future; its former "contemporary" had long ago decayed. The time travel is a real physical phenomenon.

The theory of relativity gave more than just a prediction of this phenomenon. It has shown practical means for its realization for humans. The means to travel in time (toward the future) is to travel in space in a closed loop with a speed close to c. It is very difficult technically but possible in principle. A couple of decades before the first landing on the Moon, only few believed that this would ever happen. Incredible as it may seem, the time travel for humans may be just a question of time.

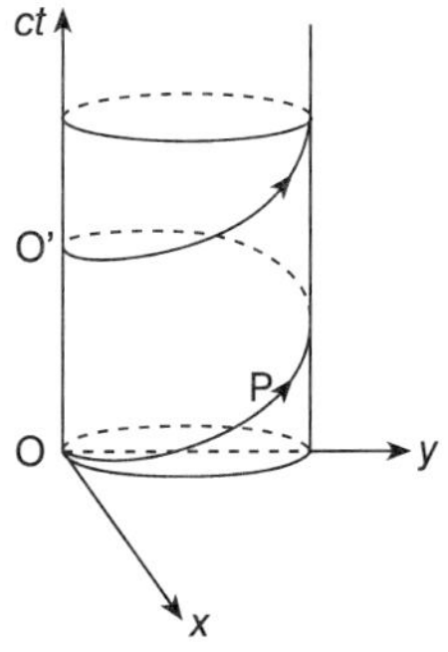

Figure 7.43 The world lines of mesons A (OO′) and B (OP). Meson B is moving in a circle in plane XY and meson A is resting at a point O on this circle. The kinematic length OPO′ is shorter than the length OO′. This means that the proper time of B between two successive meetings with A is less than that of A. If B is moving fast enough, it will find A to have decayed long before B completes one cycle, while it will take many cycles for B to decay.

7.9 The Three Friends' Paradox

Imagine two spaceships B and C, distance x apart, piloted by the astronauts Bob and Cathy, respectively. A space station A with astronaut Alan aboard is positioned exactly between the ships (Figure 7.44). At the zero moment of time, Alan sends out two laser pulses toward B and C. Since Alan is positioned midway between them, the time it takes the light signal to reach either astronaut is

$$t_0 = \frac{1}{2}\frac{x}{c}. \tag{7.112}$$

Immediately after the astronauts receive their respective signals, they set their clocks to the reading t_0 and start moving with a speed v toward each other. These two events (the onset of motion of both spaceships) occur simultaneously in A. The time it takes either astronaut to reach Alan is

$$\Delta t_B = \Delta t_C = \frac{1}{2}\frac{x}{v}. \tag{7.113}$$

So the moment of time in A when all three friends are instantly coincident is

$$t_A = t_0 + \Delta t_B = t_0 + \Delta t_C = \frac{x_0}{2}\left(\frac{1}{c} + \frac{1}{v}\right). \tag{7.114}$$

This is the reading of Alan's clock at this moment. Now, the reading of Bob's clock can be easily evaluated by noticing that its initial reading is equal to (7.112) by the arranged experimental setup and the additional time interval $\Delta t'_B$ spent in motion is less than (7.113) (the time dilation effect) by the corresponding Lorentz factor

$$\Delta t'_B = \frac{\Delta t_B}{\gamma(v)} = \frac{x}{2v\gamma(v)}. \tag{7.115}$$

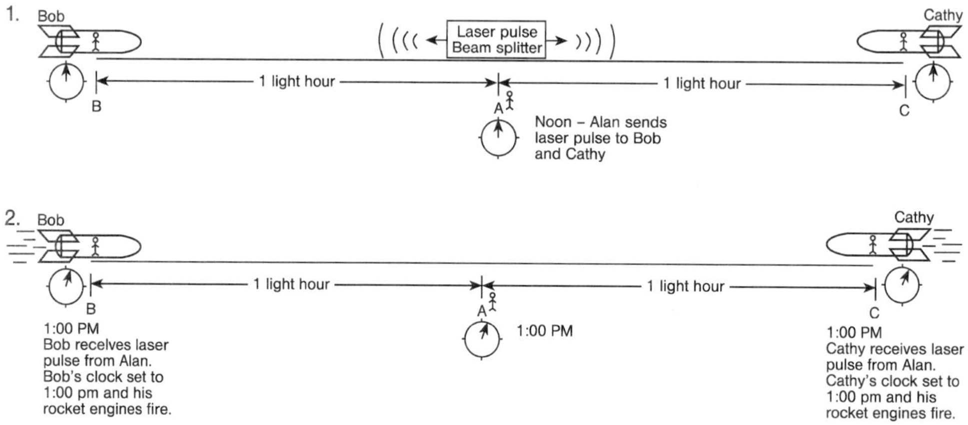

Figure 7.44 (From Ref. [72]).

Thus,

$$t'_B = t_0 + \Delta t'_B = \frac{x}{2}\left(\frac{1}{c} + \frac{1}{v\gamma(v)}\right). \tag{7.116}$$

By symmetry, the same result obtains for Cathy's clock

$$t'_C = t'_B \tag{7.117}$$

at this moment. Both moving clocks read the same time, which is less than the reading (7.114) of Alan's clock.

The paradox appears when we try to describe the same process from Bob's (or Cathy's) perspective. In Bob's RF (frame B) it is Cathy's clock, not his, that undergoes the time dilation. Therefore, the symmetry (7.117) between Bob's and Cathy's clocks observed in A breaks down in B.

One could evaluate the corresponding asymmetry quantitatively in three steps. First, write the expression for the speed v' of Cathy's spaceship relative to B: since in the original frame A Cathy and Bob move toward each other with the same speed v, their *relative* speed is

$$v' = \frac{2v}{1+(v^2/c^2)}. \tag{7.118}$$

Second, write the Lorentz factor corresponding to this relative speed (this can be easily derived):

$$\gamma(v') = \gamma^2(v)\left(1+\frac{v^2}{c^2}\right). \tag{7.119}$$

Finally, using this factor, write the expression for the dilated time interval of Cathy's clock, corresponding to the interval $\Delta t'_B$ in frame B:

$$\Delta t''_C = \frac{\Delta t'_B}{\gamma(v')} = \frac{1}{2}\frac{x}{v\gamma^3(v)(1+(v^2/c^2))}. \tag{7.120}$$

As a result, the reading of Cathy's clock in B at the moment of coincidence with Bob's clock will be

$$t''_C = t_0 + \Delta t''_C = \frac{x}{2}\left(\frac{1}{c} + \frac{1}{v\gamma^3(v)(1+(v^2/c^2))}\right) \neq t''_B (= t'_B). \tag{7.121}$$

In contrast with the result (7.117) obtained by symmetry in Alan's frame of reference, it is now NOT the same as the reading (7.116) of Bob's clock, because here there is no symmetry between Bob's clock sitting in one place in B and Cathy's clock *moving* relative to B. But, on the other hand, the result obtained in A, including Equation (7.116), must hold for *all* observers, because this result is a statement about the individual readings of the three clocks at their *instant coincidence*, which constitutes an event in space-time, and all three readings form an intrinsic, and thereby an absolute, characteristic of this event.

Thus, we have arrived at the paradox [72]. In essence, this is a more complicated version of the clock paradox.

Just as in conventional version, the origin of this paradox is in our deeply rooted perception of absolute time. Namely, in the above calculation we have used an unspoken assumption that in Bob's RF the two events – the onset of motion of both spaceships – occur simultaneously, as they do in Alan's RF. This assumption is wrong. Let us denote these events as E_B and E_C. Then, in the frame A

$$t(E_B) = t(E_C) = t_0. \tag{7.122}$$

But since these events are spatially separated by the distance x in A, they are not simultaneous in system B, which at the moment t_0 is already moving relative to A. Namely, the event E_C happens earlier than event E_B (see Equation (2.55) and Figure 2.6). To find the time discrepancy between these events in B, we have to extend Bob's RF from the immediate vicinity of his spaceship at least as far as to include Kathy's ship as well. Then Equation (2.55) will tell us that in the system B, Cathy starts moving *earlier* than Bob by the time

$$\Delta \tilde{t}_B = \gamma(v)\frac{v}{c^2}x. \tag{7.123}$$

In other words, Cathy's spaceship with her clock being instantly set to t_0 starts moving toward Bob at the moment

$$t'_0 = t_0 - \gamma(v)\frac{v}{c^2}x \tag{7.124}$$

of Bob's time – a while before Bob's moment of start. We can imagine the instantaneous boost of Bob's spaceship as its rapid transfer from system A into a "relativistic train" (system B) moving to the right with a speed v, and this transfer instantly changes Bob's set of simultaneous events from those simultaneous in A to those simultaneous in this train. This also shows the physical interpretation of the product $\gamma(v)x$ in Equation (7.123): x being the distance between the initial positions x_B and x_C of Bob's and Cathy's clocks B and C is also the distance in A between the two *moving* clocks on the train that are instantly coincident respectively with B and C; therefore, $\gamma(v)\,x$ is the *proper* distance between these two moving clocks. In other words, it is the initial distance between Bob and Cathy in Bob's new RF. Now, the speed of Cathy's ship toward Bob is given by Equation (7.118). Therefore, the time in B it will take Cathy to reach Bob is

$$\Delta t'_B = \frac{\gamma(v)x}{2v}\left(1+\frac{v^2}{c^2}\right). \tag{7.125}$$

Adding this to the initial time t'_0 in (7.124) gives

$$\tilde{t}'_B = \frac{1}{2}\frac{x}{c} - \gamma(v)\frac{v}{c^2}x + \gamma(v)\frac{x}{2v}\left(1+\frac{v^2}{c^2}\right). \tag{7.126}$$

After simple algebra this reduces to

$$\tilde{t}'_B = \frac{1}{2}x\left(\frac{1}{c} + \frac{1}{v\gamma(v)}\right) = t'_B. \tag{7.127}$$

Bob's time at the moment of meeting is the same as in Alan's RF – no surprise, this is the proper time between the same pair of events and as such it is an invariant. I leave it for the reader to show that the reading of Alan's clock in B is greater than (7.127) by exactly the same amount as in system A (Problem 7.6). No surprise either, the difference in readings of the two instantly coincident clocks is an absolute characteristic of an event – it is the same for all observers.

Now we want to figure out the reading of *Cathy's* clock at the moment of meeting as observed in B. Alan has recorded the identical readings of Bob's and Cathy's clocks at the moment of their instant coincidence. As already mentioned, such coincidence constitutes an event and its intrinsic characteristics (such as the individual readings) must be the same for all observers. Therefore, we expect that Bob will also observe the readings of both clocks to be identical.

Here is the proof: Cathy's clock starts ticking with the reading set at t_0. The time in B it takes Cathy to reach Bob is given by (7.125). But this is time read by the clocks *stationary* in B. Since Cathy's clock is moving relative to B, it will show the *time-dilated* interval:

$$\Delta t'_{\mathrm{C}} = \frac{\Delta t'_{\mathrm{B}}}{\gamma(v')}. \tag{7.128}$$

Using (7.125) and (7.119) gives

$$\Delta t'_{\mathrm{C}} = \frac{1}{2}\frac{x}{v\gamma(v)}. \tag{7.129}$$

Finally, adding this to the initial reading, we obtain

$$t'_{\mathrm{C}} = t_0 + \Delta t'_{\mathrm{C}} = \frac{1}{2}x\left(\frac{1}{c} + \frac{1}{v\gamma(v)}\right) = t'_{\mathrm{B}}. \tag{7.130}$$

As we had expected, this is identical to the reading (7.127) of the Bob's clock – despite the time dilation. The identity is maintained by nonsimultaneity (7.123) and (7.124), which in this case exactly balances the time dilation effect. In frame B, Cathy's clock is ticking slower than Bob's; but it is set to and starts ticking from the reading t_0 accordingly *earlier* than Bob's clock.

There is a subtle point here that frequently prevents even the experienced reader from realizing the role of nonsimultaneity in resolving this paradox: it is the condition that both Bob's and Cathy's clocks are initially set to *the same* reading t_0. The fact that the initial *readings* are set to be *the same* may be misinterpreted as an indication that the *events* of setting are themselves simultaneous in all reference frames. This is, of course, not true. This kind of confusion frequently occurs in situations when the same clocks are transferred between different reference frames. Therefore, it is worth repeating: in the given problem, in system B, Cathy's clock starts ticking from the same reading as Bob's clock, but it does it earlier than Bob's one.

The similar reasoning from Cathy's perspective will give the same, paradox-free, result.

Thus, all three friends, observing the instant coincidence of the three clocks from their respective reference frames, agree upon the following fact: Bob's and Cathy's clocks show at this moment *the same reading*, and this reading shows an earlier time than the reading of Alan's clock.

Problems

7.1 We face a dark wall at a distance d in front of us. At $t = 0$ the wall flashes brightly for an instant and then gets dark again. We can describe this by saying that each point of the wall becomes for an instant a source of light radiated uniformly in all directions.
(a) What will we see?
(b) What will a short exposure snapshot show at some $t > 0$?
(c) Describe your answers quantitatively.

7.2 Suppose that a long horizontal rod oscillates in vertical direction in the inertial frame K′. The oscillation amplitude is a, the oscillation speed u is constant, only undergoing abrupt reversals at the equidistant moments of time.
(a) Describe qualitatively the shape of this rod as seen from an inertial RF K moving horizontally along the length of the rod with a speed V relative to K′.
(b) Derive analytical expression describing motion of the rod as observed in K.

7.3 Consider the *harmonic* oscillations of the rod from the previous problem, with oscillation frequency ω.
(a) Find relative velocity in K of its two elements at x_1 and x_2 considered at the same moment of time in K.
(b) Do the same for two elements considered at the same moment of time in K′. Explain the results.

7.4 On May 1, 2005, a rocket carrying an astronaut Alvin starts from Earth for a round-trip to a star M. The speed of the rocket is $0.8c$. Simultaneously, another rocket with Alvin's twin brother Dan starts to a stellar system R 61. The speed of Dan's rocket was classified at that time. The astronauts' twin sister Mary remains on Earth. Upon reaching their destinations and completing their very short research programs there, they return to Earth on the same day. Upon arrival, when all three twins get together to celebrate their reunion, Alvin finds himself 15 years younger than Mary and 5 years older than Dan.
(a) Draw the space-time diagram with the world lines of Mary and her twin brothers.
(b) When (what year) did the spaceships return to Earth?
(c) What was the speed of Dan's spaceship?
(d) How far away from Earth (in light years) are the star M and the system R 61, respectively? What are these distances in km?

7.5 On May 1, 2005, a rocket carrying an astronaut Alvin starts from Earth for a round-trip to a distant star M. The speed of the rocket is $0.95c$. Simultaneously, another rocket with Alvin's twin brother Dan starts to a stellar system R 61. The speed of Dan's rocket is 0.8c. The astronauts' twin sister Mary remains on Earth. Many years later, both spaceships complete their missions and return to Earth on the same day. Upon arrival, when all three twins get together to celebrate their reunion, Dan finds himself 20 years younger than Mary and 5 years older than Alvin.

(a) With the same assumptions as in the previous problem, are the conditions of this problem self-consistent? If not, how should we change the assumptions to make the conditions consistent?

(b) Draw the space-time diagram with the world lines of Mary and her twin brothers and find the kinematic lengths (or proper times) of all their respective branches.

(c) When (what year) did the spaceships return to Earth?

(d) How far away from Earth (in light years) are the star M and the system R 61, respectively? What are these distances in km?

7.6 In Section 7.9, find the reading of Alan's clock *in Bob's* RF at the moment of the three friends' meeting.

7.7 Suppose you measure the relativistic length contraction effect on a very thin horizontal rod of proper length l_0, using the procedure shown in Figure 7.7. The rod is moving along its length with a speed V. Assume that the diffraction effects can be neglected. A short illuminating pulse is incident vertically and has the length δz. How short must this length be to achieve the accuracy $\delta l/l = 10^{-2}$, where l is the measured length and δl is the fuzziness caused by δz?

7.8 A very thin dart is shot so that it moves at $v = 280\,000$ km/s. During its flight it is illuminated from the side by a short laser pulse whose front is wider than the length of the dart. How short (in time) must the pulse be if you want the shadow on the screen to represent the length of the dart to an accuracy of 5%?

7.9 A moving clock ticks slower. Geometrically, its measured period T' can be considered as the projection of its proper period T onto the observer's temporal axis ct'.

A moving rod contracts. Would it be correct to say that its measured contracted length l' is the projection of its proper length l onto the observer's respective spatial axis x'?

7.10 (a) Derive Equation (7.41).

(b) Derive Equation (7.44).

7.11 Show that the center of mass of a spinning wheel observed from a moving system K does not coincide with its geometrical center (the direction of motion lies in the wheel's plane). Do it

(a) by direct inspection of a few properly chosen spokes

(b) by considering the wheel's mass distribution in K.

7.12 In Section 7.8, sketch a space-time diagram of Larry's round-trip and his radiocommunications with Joe, using Glen's RF.

8
Miracles of a Spinning World

"Curiouser and curiouser!"
Lewis Carroll, *Alice's Adventures in Wonderland*

In this chapter, we will study accelerated systems with nonzero angular momentum. This may be a merry-go-round, a planet orbiting its star, an electron in circular motion in an external magnetic field, or just a spinning disk. All these systems display a plethora of interesting new phenomena.

We will discuss some of them.

8.1
The Ehrenfest Paradox

The phenomenon we consider in this section is known as the Ehrenfest paradox [73,74]. In its initial formulation, the paradox appears when we consider the rim of a spinning disk and try to apply special relativity (SR) for finding its proper length. The paradox emerges when we assume that the relation (2.25) between the Lorentz contracted and the proper length of a rod uniformly moving along the direction of its length is applicable also to the length of a circle rotating in its plane. We can confirm this assumption by a pretty sound argument, similar to those used in the previous sections. Namely, even though the whole disk does not constitute an inertial reference frame (RF), we can introduce for each of its small area elements a comoving inertial RF, in which this specific element is instantly stationary. Of course, these RFs will be different for different elements of the disk. But each individual length element is Lorentz contracted in the laboratory frame (system S) if the element is oriented along the local direction of motion, or remains equal to its proper length if the element is oriented perpendicular to this direction.

Consider the radius of the spinning disk. All its length elements are the same as in the rest frame K of the disk, since they are all oriented perpendicular to the direction of their motion. Therefore, the length of the whole radius measured by the stationary observer in S as the sum of its constituting length elements will remain equal to its proper length.

Special Relativity and How it Works. Moses Fayngold

ISBN: 978-3-527-40607-4

Now consider the circumference of the spinning disk. Each length element of this circumference, observed from S (which is a *nonrotating* reference frame attached to the center of the disk), undergoes Lorentz contraction. Denote the radius of the disk as R and its angular velocity as Ω. Then the corresponding Lorentz factor for each element of the circumference is $\gamma(\Omega R)$. Accordingly, the whole circumference L measured as the sum of lengths of its individual elements is less than its proper length Λ by this Lorentz factor:

$$L = \frac{\Lambda}{\gamma(\Omega R)}. \tag{8.1}$$

But here a thoughtful reader may ask – hey, wait a minute, how can this be true if the circumference in question is in all its parts and at all moments of time coincident with the *stationary* circumference of the same radius? In geometry, such coincidence (or congruence – recall Section 7.6 and Figure 7.34) is one of definitions of equality! Therefore, we appear to have come to a logical contradiction: on the one hand, the proper length of a *stationary* ring congruent with the ring of the rotating disk is

$$L = 2\pi R \tag{8.2}$$

and putting this into (8.1) yields

$$\Lambda = 2\pi R \gamma(\Omega R). \tag{8.3}$$

On the other hand, precisely because both rings are congruent, there must be just

$$\Lambda = L = 2\pi R, \tag{8.4}$$

as for the nonrotating disk.

This contradiction results from our deeply ingrained convictions that Equation (8.2) holds in a system S′ corotating with disk and the properties of time in a rotating system are the same as in an inertial RF. Well, both these convictions are unjustified.

The correct logic is this: of course, the circumference of the rotating disk measured by an inertial observer is shorter than its proper circumference measured by a resident of the disk; it must be equal to $2\pi R$ for the stationary observer and greater than $2\pi R$ for the disk resident, as in Equation (8.3). These two requirements appear to be mutually exclusive, but they are not, because the circumference of the spinning disk is a relative characteristic, just as is any length. The part $\Lambda = L$ of Equation (8.4) is wrong and the *proper* length Λ, that is, length measured in frame S′, must be greater than $2\pi R$ by the corresponding Lorentz factor. This means that the disk must be in a state of a complex deformation such that the plain of the disk is *no longer a Euclidean plain.*

Before moving further, let us get a visual feeling of a non-Euclidean surface by examining two examples (Figures 8.1 and 8.2).

In Figure 8.1 we see a familiar spherical surface. The shortest line between two points A and B on the sphere is the arc of the big circle (meridian) passing through the two points. The length of the arc is

$$L_\theta = R\theta, \tag{8.5}$$

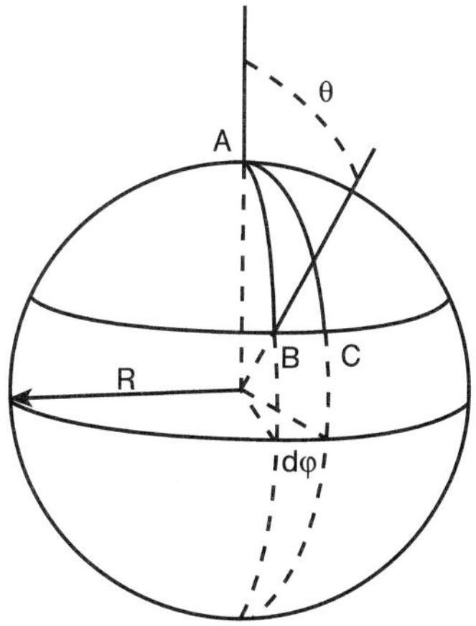

Figure 8.1

where R is the radius of the sphere and θ is the angular distance between points A and B. Let one of the points (say, point A) be a "North Pole" of the sphere. Now draw another circle through point B so that this circle lies in the plane perpendicular to the polar axis. This circle is a "parallel." Using spherical coordinates, we find for the length element of this parallel,

$$dl = R \sin\theta \, d\varphi, \tag{8.6}$$

where φ is the azimuthal angle. The total length of the parallel is

$$\Lambda = 2\pi R \sin\theta. \tag{8.7}$$

In a special case when $\theta = 90^\circ$, the parallel becomes the equator and its length becomes equal to $2\pi R$.

The arc AB (8.5) plays the role of the radius of the circle (8.7) centered at A. Denoting this radius as R_θ, we have

$$R_\theta = L_\theta = R\theta. \tag{8.8}$$

Now imagine a two-dimensional resident of the surface initially believing that his world is flat. This resident will find the geometry of his two-dimensional world rather unusual. For instance, a rocket launched in a certain direction and moving straight along this direction (i.e., along the corresponding meridian in Figure 8.1) farther and

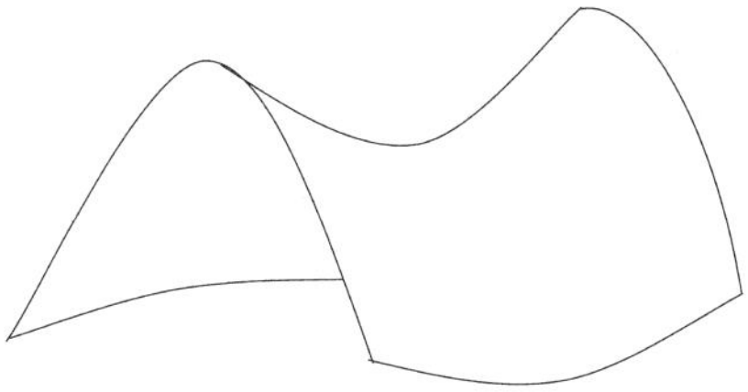

Figure 8.2

farther away from the observer will eventually return to the launching pad from the opposite side (this is directly related to the fact that the resident's world is finite even though it has no boundaries); a meridian (intersection between the sphere and a plane passing through its center) plays the role of a straight line (geodesic) on the sphere; any two points A and B on the sphere can be connected by a straight line in two different ways; any two straight parallel lines (meridians) eventually converge and intersect; the sum of the angles of a triangle defined by points A, B, and C is greater than 2π, and it approaches 2π only in the limit of infinitesimally small or narrow triangle; any circle has two different interiors, with different (diametrically opposite) centers O and O′ and accordingly different radii. Both centers are the opposite poles of the sphere and are equally legitimate. We can even generalize Equation (8.8) to embrace both cases:

$$R_\theta = \begin{cases} R\theta, & \text{center O,} \\ R(\pi - \theta), & \text{center O}'. \end{cases} \tag{8.9}$$

Also, combining (8.8) and (8.7), the curious resident finds that the ratio of a circle to its radius is

$$\frac{\Lambda}{R_\theta} = 2\pi\frac{\sin\theta}{\theta} \leq 2\pi. \tag{8.10}$$

The kind of geometry with these properties is known as geometry with positive curvature (Riemann geometry).

Another example is a hyperbolic surface (a "saddle," Figure 8.2). The total area of this surface is infinite, any two parallel lines eventually diverge, the sum of the angles of a triangle is less than 2π, and the ratio of the length of a circle to its radius is greater than 2π. This kind of geometry is known as geometry with negative curvature or Lobachevsky's geometry. Sometimes it is referred to as a hyperbolic geometry – after the name of a surface on which it is realized.

Comparing these two examples with the case under study, we see that the surface of a spinning disk has negative curvature – it is described by Lobachevsky's geometry.

Now we can discuss the connection between the geometrical and the dynamic properties of rotating objects.

To simplify the discussion, let us focus on the rim of the disk (circular ring of radius R) rather than considering the whole disk. The ring is rotating with angular velocity Ω about its symmetry axis perpendicular to its plane.

How can a rotating ring with radius R, if it is coincident with the stationary one, be equal to it (i.e., have length $2\pi R$) in the stationary frame, but have the length greater than $2\pi R$ for an observer sitting on the ring? Does not this observer see the same congruence between the two disks?

We can understand the situation better in the framework of our results in the previous chapter, as another case of the dynamics of relativistic length contraction. A good way to do it is to start from a stationary ring, bring it to rotation, and watch how the change of state of motion affects the geometry.

So imagine two identical concentric rings in one plane P in an inertial frame S. Consider both rings (one of which is scheduled for rotation) as composed of small elements of arc of length dl. In the list of scheduled operations we include, first, the rotational boost of one of the rings from S to a corotating system S′ and then measuring its length in either of the systems.

We want to take special care to ensure that the boosting process does not introduce any deformations to the boosted ring as observed in system S. We explicitly name the system S since we know already that a process that is deformation-free (in the narrow sense of geometric deformation) in one reference frame is not deformation-free in another, and by the initial condition the *boosted ring retains its length in S*. To satisfy this condition, we boost the ring by applying equal tangential forces to its equal parts d*l* *simultaneously in S*. Such operation, as we know, does not change either the size of an element d*l* or the shape of the whole boosted ring as observed in S, even though the ring is now rotating with linear velocity $v = \Omega R$. It is natural to call the boosted ring S′ – the same as the corotating system. For the S-based observer, the ring S′ after the boost remains coincident with ring S – precisely as stated in the initial conditions. Accordingly, the length measurement of the boosted ring performed by the S-observer gives the result $L' = 2\pi R$.

We have completed the first two stages in our list of scheduled things. Let us go to the next stage on the list – the observation made by resident of the rotating ring.

We have said above that the ring S′ after the boost, as observed by its resident, must be in a state of some complex deformation. Now we will trace out the nature and origin of this deformation. Let us introduce an *inertial* RF comoving with an arbitrarily chosen element dl' and call it also S′. Since the element dl' actually is a part of rotating ring, which is a noninertial system, it will only for a split second be comoving with this inertial frame. But if we manage to understand what happens within this infinitesimally short time interval, we will get an insight into the nature of the whole phenomenon. So let us employ two observers: Paul in S′ and Sam in S. As Paul watches the whole process, he sees initially (before the boost) the element dl' moving to the left and Lorentz contracted, since before the boost this element is a part of system S:

$$dl'(\text{initial}) = \frac{dl}{\gamma(v)}, \tag{8.11}$$

where $v = \Omega R$. As has already been pointed out, we consider the boost *instantaneous in S*: the two instant and equal forces applied simultaneously (in S) to the end points of the element d*l*. In S′, these two events are not simultaneous – they are separated by the time interval

$$dt' = \gamma(v)\frac{v}{c^2}dl. \tag{8.12}$$

As we know, the S′-clock instantly coincident with the leading edge Q of the element PQ = d*l* right about to be boosted by Sam shows earlier time than the clock at the trailing edge (as observed from S, Figure 8.3). Therefore, when Paul (for whom the edge Q is trailing!) observes the boost, he sees first the trailing edge of the element stopped relative to him, while the opposite (leading) edge P keeps on moving with the velocity *v*. This stretches the element! The amount of stretch is

$$\delta' = v\,dt' = \gamma(v)\frac{v^2}{c^2}dl. \tag{8.13}$$

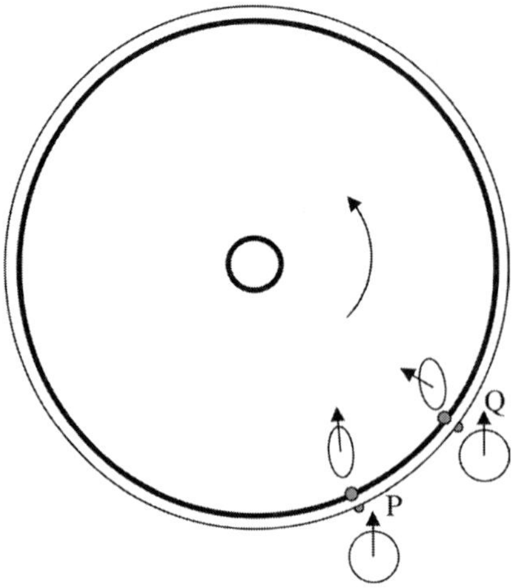

Figure 8.3

As in Section 7.5, we can represent the element by the two equal point masses at its two respective edges connected by a spring. The spring will keep on stretching until the opposite mass will be instantly stopped, thus transferring the entire element to system S′. The resulting new length of the element in S′ will be the sum of initial length (8.11) and the amount of stretch (i.e., the distance traveled by the opposite edge before it is stopped in S′):

$$\mathrm{d}l'(\text{final}) = \mathrm{d}l'(\text{initial}) + \delta' = \gamma(v)\left(\frac{v^2}{c^2} + \gamma^{-2}(v)\right)\mathrm{d}l' = \gamma(v)\mathrm{d}l. \tag{8.14}$$

Since the element dl was chosen arbitrarily, this result holds for all other elements and thereby for the whole circumference of the disk. This proves the result (8.3), but now we can see its dynamical origin! Formally, we can obtain this result by integrating (8.14) over the circumference L:

$$\Lambda(R) = \oint \mathrm{d}l'(\text{final}) = \gamma(v)\oint \mathrm{d}l = \gamma(v)L = 2\pi R\gamma(\Omega R). \tag{8.15}$$

This result shows that the kinematic (or, as many say, geometric) effect has dynamical underpinnings.

Both observers – Sam and Paul – came to the same conclusion that the rings S and S′, while having the same radius R and being coincident in S, have nevertheless different proper lengths. The ring S is in the state of equilibrium and has "normal" length $L = 2\pi R$; the S′-ring is in a state that is physically deformed. This deformation is not geometrically manifest in S. However, when measured in S′ by Paul, the ring turns out to be longer than L with its length described by Equation (8.15). In particular, if the angular velocity is such that the speed of the rim of the disk approaches that of light, then, according to (8.15), its circumference becomes infinitely long, even if the radius remains small!

Both observers agree on the results, but give different explanations to it. For Sam, the S′-ring has to have its proper length greater than L by the Lorentz factor in order to be coincident with ring S due to the Lorentz contraction. For Paul, the ring S′ is longer than S ($\Lambda > L$) by the corresponding Lorentz factor because it had been stretched due to nonsimultaneous application of the corresponding boosting forces during the acceleration. The fact that Λ is greater than $2\pi R$ is explained by Paul as a manifestation of the non-Euclidean geometry in a rotating system, and this geometry is, in turn, the result of physical deformation.

Exercise 1

When Alice and Tom learned about weird properties of a spinning disk, they both came up with an idea: what if they buy a very small (and accordingly cheap!) plot of land, install a platform on it, build a small house on the platform, and then bring the platform into a rapid rotation? Will not the area of the house dramatically increase due to circumferential extension of the rotating system? They were especially excited by the fact that circumferential extension can be made arbitrarily large at sufficiently rapid rotation. Then they could sell the house for a much higher price and, which is more important, patent the idea!

Is the idea viable?

They e-mailed to Paul, asking what is the area of his disk, and meanwhile performed the following calculation.

Solution

Denote the sought-for area as A'. The area of a disk can be written as

$$A' = \int_0^R \Lambda(r)\mathrm{d}r, \tag{8.16}$$

where $\Lambda(r)$ is the proper length of a circle of radius r. In our case,

$$\Lambda(r) = 2\pi r\,\gamma(\Omega r). \tag{8.17}$$

It follows

$$A'(R) = 2\pi\int_0^R r\,\gamma(\Omega r)\mathrm{d}r = 2\pi\int_0^R \frac{r\mathrm{d}r}{\sqrt{1-(\Omega^2/c^2)r^2}}. \tag{8.18}$$

This integral can be easily taken by change of variable $((\Omega/c)r = \sin\chi)$. Integration yields

$$A'(\Omega R) = 2\pi\frac{c^2}{\Omega^2}\left(1-\sqrt{1-\frac{\Omega^2}{c^2}R^2}\right), \quad \Omega R \le c. \tag{8.19}$$

Consider two special cases.

(a) The disk is so small or spinning so slowly that

$$\Omega R \ll c. \tag{8.20}$$

In this case, expanding the square root in (8.19) into Taylor series and retaining only the first two terms, we obtain

$$A'(\Omega R) \xrightarrow[\Omega R \ll c]{} \pi R^2. \tag{8.21}$$

This is familiar expression for the area of a stationary disk – precisely what one would expect in a nonrelativistic limit. In terms of Alice and Tom financial project – no gain. The project could work only in the relativistic domain. So they turned their attention to the opposite limit.

(b) The speed of the disk's edge approaches c:

$$v_R \equiv \Omega R \to c, \quad \text{so that} \quad \Omega \to \frac{c}{R}. \tag{8.22}$$

In this case the general expression (8.19) reduces to

$$A'(\Omega R) \xrightarrow[\Omega R \to c]{} 2\pi R^2. \tag{8.23}$$

This is twice the area of a stationary disk of the same radius.

Soon after our friends had completed their calculations, they received an experimental confirmation from Paul. Paul's disk was ultrarelativistic, that is, its circumference was moving with a speed very close to the invariant limit c. And the measured area of the disk exceeded its area before the boost nearly by a factor of 2. This is a dramatic difference from the Euclidean high-school geometry, but still not as large as our friends had expected, given that the circumferential dimension under these conditions can, according to (8.15), be made arbitrarily long. However, when they thought over these results, they realized that such dramatic elongation must be observed by Paul only within a very thin *peripheral* annulus of the disk, whereas the central region has, for any r not too close to R, almost normal area πr^2. So even the contribution from the peripheral area is finite and the net area turns out to be restricted by (8.23). When Alice and Tom weighted this very modest gain against the expenses needed to spin their would-be house on the platform to subluminal velocity of the edge, they dropped the idea.

But, while feeling satisfied that Alice's and Tom's calculations are consistent with his measurements, Paul gets embarrassed when trying to consider the whole problem the other way around. Namely, consider now the *stationary* ring S congruent with ring S′ and try to evaluate its length *from Paul's viewpoint*. This immediately raises two questions. First, can Paul apply the same reasoning as Sam (i.e., use the Lorentz contraction effect) to explain the correlations between the two rings from the viewpoint of *his* RF? And second, if he can, will the result be in agreement with the previous conclusion – that his S′-ring has a greater proper length than Sam's S-ring?

The immediate answer to both questions seems to be no. Let us start with the second one. Imagine yourself sitting together with Paul on the ring S′. The first thing you feel is the inertial (centrifugal) force pushing you away from the center, but it is balanced by some other force necessary to keep you in place. The ring S′ is stationary with respect to you and has the proper length Λ, while the ring S is spinning around the center with the angular velocity $\Omega' = \Omega$. Its instant length measured by you is equal to Λ, since it appears at any moment to be congruent with S′. On the other hand, by assumption, you can, at least in your immediate locality, apply the rules of special relativity according to which each element of ring S undergoes the Lorentz contraction and is therefore shorter in your frame S′ than its proper length dl. Or, which is the same, the *proper* length dl of the element must be greater than its Lorentz-contracted length $d\tilde{l}$ measured by you. Since the choice of the element is arbitrary, this holds true for all the elements and thereby for the whole length of ring S. But the proper length of ring S is L and its whole Lorentz-contracted length measured by you must, by congruence, be equal to Λ. We thus come to conclusion that

$$L = \gamma(\Omega R)\Lambda > \Lambda, \tag{8.24}$$

in flat contradiction with our previous conclusions and result (8.15), according to which $\Lambda > L$. Thus, it turns out that there are two different lines of reasoning for S′-observer, leading to two contradicting results. This contradiction, arrived at by *the same observer*, shows that at least one line of reasoning (we suspect, the second one) must be wrong. According to some treatments, this is a consequence of the fact that system S′ is noninertial and therefore the rules of SR are not applicable to it.

This conventional explanation, just dismissing the second line of reasoning and thereby the accompanying paradox, does not explain anything – it simply shoves the problem under the rug.

Actually, the SR has a full authority over rotating systems as well, at least when applied to a circumference of a rotating disk or to a very narrow ring [75]; it describes correctly the corresponding phenomena and gives their complete, sometimes really surprising, explanation.

The situation in case provides a good illustration to this.

We want to measure the length of S in S′. Each element of S is moving relative to S′. Recall the crucial requirement that in length measurement of a moving object, its leading and trailing edges must be marked simultaneously in a system where we perform the measurement. In the given case, this is the system S′. Since the object (ring) is a closed loop, its leading and trailing edges coincide. Therefore the requirement that their positions be marked at the same moment of time seems to be satisfied automatically, at least when we measure the length of the *whole ring*.

That would be true were the system S′ inertial. Since it is not, the situation is much more subtle. Look at Figure 8.4. At a certain moment in the *stationary* system all stationary clocks read the same time. On the disk, each clock ahead of the previous one reads an earlier time. I emphasize – the members of each pair of clocks on the moving disk *are* synchronized in Paul's reference frame. However, if we apply the same procedure until we return to the original clock on the disk, we realize that this clock should read two different times at once. (Go back to the chain of moving clocks in Figure 2.6 in Chapter 2 and try to imagine what happens if you wrap the chain around a circle!) We see that the clock synchronization procedure carried out along a

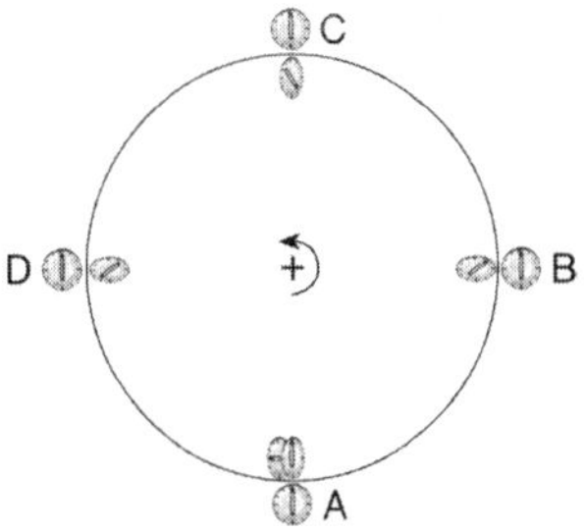

Figure 8.4 The system of clocks on a rotating disk. If we synchronize the clocks going step by step around the circle, then we realize that it is impossible to assign only one moment of time to an event. A, B, C, D: synchronized stationary clocks.

closed loop in a rotating system will allocate two (and, by iteration, an infinite number!) different times to the same event!

Let us calculate the time lag $\Delta t'$ for the ring of radius R. Apply Lorentz transformation,

$$dt' = \gamma(v)\left(dt - \frac{v}{c^2}dl\right) \tag{8.25}$$

to two close events at the edges of the element dl on the ring L. Integrating along the whole ring gives

$$\Delta t' = \oint dt' = \gamma(v)\left(T - \frac{v}{c^2}L\right). \tag{8.26}$$

Here, as before, L is the proper length of the stationary ring; T is the time interval between the two events at the same point of the ring. Since we consider all corresponding events at one moment of time in S, we have $dt = 0$ and $T = 0$. Setting $dt = 0$ in (8.25) and accordingly $T = 0$ in (8.26) gives

$$dt' = -\gamma(v)\frac{v}{c^2}dl \tag{8.27}$$

and

$$\Delta t' = -\gamma(v)\frac{v}{c^2}L. \tag{8.28}$$

As we know (Section 2.6), Equation (8.27) is the syncronization condition for the S′-clocks as observed from S. As to (8.28), it shows that applying the synchronization procedure to the set of clocks along a *closed* loop gives (at least!) two different times to the same event on the loop, with the time discrepancy (or time lag) given by the right-hand side of the equation.

Since $v = \Omega R$ and $L = 2\pi R$, Equation (8.28) can be recast in the form

$$\Delta t' = -\gamma(v)\frac{v}{c^2}L = -2\pi\gamma(\Omega R)\frac{\Omega}{c^2}R^2 = -2\frac{\gamma(\Omega R)}{c^2}\Omega A, \tag{8.29}$$

where A is the area of the ring in S. The time lag is proportional to the angular velocity Ω of rotation and the area A enclosed by the loop. In principle, it can be generalized to the loops of arbitrary shape, but we here will consider only a circle around the rotational center.

Now we are in a position to complete the discussion. In order for the length measurement of the ring S to be simultaneous in S′, we need to take account of the time lag (8.29). Thus, the total length of the ring S in S′ will consist of the two contributions: its Λ-congruent length and the additional distance $v\Delta t'$ traveled by a point on the ring during the time lag $\Delta t'$:

$$L' = \gamma(v)L + v\Delta t' = \gamma(v)L - \frac{v^2}{c^2}\gamma(v)L = \frac{L}{\gamma(v)}. \tag{8.30}$$

The quantity L' is precisely the Lorentz-contracted length of ring S as measured in S′. However, this result is obtained only if (1) Paul marks the position of a point on ring S at two different moments of his time separated by the time lag $\Delta t'$ and then (2) subtracts the distance between the two consecutive markings from the proper length Λ of his ring S′. This subtraction can be considered as a correction for the fact that both moments of marking are equally legitimate temporal labels of an event in Paul's frame. Crazy as it may seem, this procedure constitutes a length measurement of a ring S in system S′ and gives an operational definition of this length as measured in a spinning world [76]. Note that this procedure has not been chosen specially to obtain the desired result, but is obtained by consistently applying the rules of SR. And result (8.30) that follows is itself consistent with these rules.

From this result Paul concludes that the *proper* length of the ring S is $\gamma(v)L' = L$, as it is, indeed, measured by Sam.

Thus, the second line of reasoning by Paul, if corrected for the time lag, also gives the same result as before and is consistent with the conclusion made by Sam.

We can also obtain these results from another perspective – that of the space-time diagrams (Figure 8.5). Let us start with the stationary ring S – draw its instant position at the zero moment of system S. We will get the circle of radius R centered at point O (the origin) of the horizontal plane $(0, x, y)$. The continuous set of its instant positions at later moments forms a cylindrical surface – the *worldsheet* of the ring. You can also think of it as the set of the world lines of all points of the ring, which form generatrices of the cylindrical surface. One of these generatrices (ct-axis) is the world line of Sam. Note that by this construction we chose the axis passing through the event M on the ring rather than event O at its center as the ct-axis. This axis represents temporal dimension of space-time, while the circle of the ring represents one of its spatial dimensions (closed onto itself).

Let us now turn to Paul and consider a plane tangent to the cylinder at line MM′ (plane $x = -R$, Figure 8.5). In this plane, the axes (ct', x') would represent inertial system S′ if Paul were moving in straight line MN. But Paul is sitting on the spinning ring, so, accordingly, the initially plane sheet (ct', x') must be wrapped around the symmetry axis over the same cylindrical surface that forms the worldsheet of the ring S (Figure 8.6). The helix ct' is the world line of Paul, the helix x' is the former x'-axis in Figure 8.5 (or, which is the same, line of simultaneity in S′, or, which is the same, the

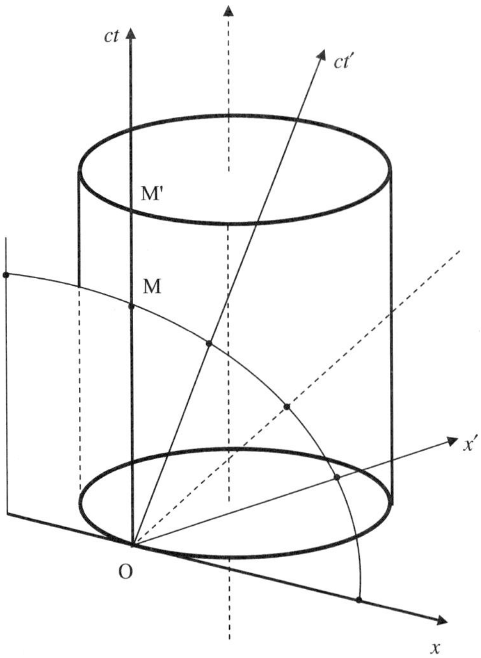

Figure 8.5

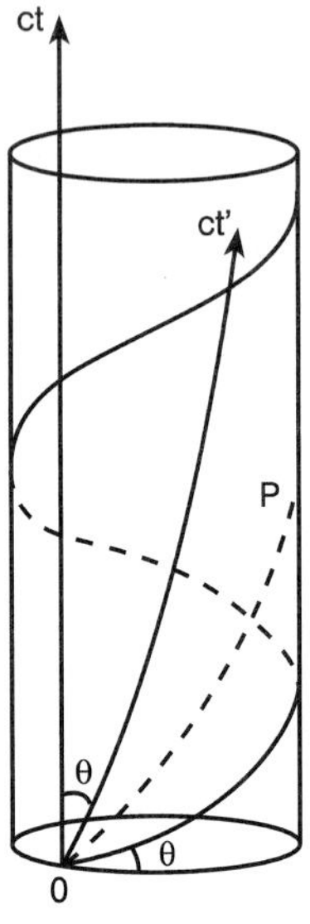

Figure 8.6

world line of a fictitious particle moving infinitely fast along the ring in S′); and the helix OP is the world line of a photon moving in a circle along the ring. Altogether we have the worldsheet of the spinning ring S′. The worldsheets of both rings are mapped onto the same cylindrical surface. This is another case of plotting the space-time diagrams representing two reference frames in relative motion, on one graph. The difference from the case discussed in Chapter 4 is that here one RF is rotating, rather than moving uniformly.

And this difference is crucial. Note that the helices representing the basis axes ct' and x' of system S′ have *different pitch*. Therefore, there will be a later time $t > 0$ in S (and thereby $ct' > 0$ in S′), when the x'-line intersects with the ct'-line the second time! Actually, in case of uniform rotation, they intersect periodically infinite amount of times! Since the x'-line is, by the original definition, the line of simultaneity in K′ (now – in S′), and the ct'-line is, by the same definition, Paul's world line (Paul, by the initial arrangement, is sitting on the edge of the spinning disk), this means that the space and time are literally intertwined: the two helices representing them in Figure 8.6 are mixed together as opposed to being just tilted in case of two *inertial* reference frames!

We can now use the space-time diagrams to get another derivation of the time lag lying at the heart of the Ehrenfest's paradox. The x'-line, by its original definition, is the line $ct' = 0$. In other words, it intersects the time axis at the origin. But then, lo and behold, it intersects the ct'-line again at some other point with $ct' > 0$. Similarly, the ct'-line is the line $x' = 0$. Thus, the intersection point of the two lines marks an event $(0, 0)$ in space-time, which is chosen as an origin of coordinate system. Since the two lines intersect again, one and the same event at the origin has at least two distinct times $t' = 0$ and $t' \neq 0$. And since the choice of an origin is arbitrary, this startling result is true for any event in a rotating world.

The problem now is to find this next closest moment of time t' assigned to the same event. We use the fact that the moments t' in S′ are related to moments t of Sam's time. So it is sufficient to find the moment ct of Sam's time, corresponding to intersection. Once this is done, we can use the relation $t \leftrightarrow t'$ to find the corresponding ct'.

Since Paul is at $x' = 0$ of S′, and the origin of S′ is moving relative to S, the relation is determined by the time dilation effect:

$$ct = \gamma ct'. \tag{8.31}$$

Let θ be the angle between ct and ct', so that $\tan\theta = v/c \equiv \beta$. The equation of Paul's world line in terms of S-coordinates is

$$ct = \tan\left(\frac{\pi}{2} - \theta\right)x = \frac{x}{\tan\theta} = \frac{x}{\beta}. \tag{8.32}$$

The equation of line of simultaneity of system S′ is

$$ct = (\tan\theta)x = \beta x. \tag{8.33}$$

After we have wrapped the x–ct plane around the symmetry axis of the ring, the worldsheet of the x-axis becomes the worldsheet of the ring and x becomes $R\varphi$, where

φ is the azimuthal angle of a point on this sheet. The two equations (8.32) and (8.33) for two coordinate lines of system S′ take the form

$$ct = \frac{R\varphi}{\beta}, \tag{8.34}$$

and

$$ct = \beta R \varphi. \tag{8.35}$$

They obviously intersect at the origin ($ct = \varphi = 0$). Since the second curve (the simultaneity line of S′) has a smaller slope, by the moment when they meet again at some $\varphi \neq 0$, it will make one turn more around the symmetry axis than the first curve, that is,

$$\frac{R}{\beta}\varphi = R\beta(\varphi + 2\pi). \tag{8.36}$$

(Actually, we can add $2\pi m$ with integer m instead of 2π, since the process reiterates.)

Solving for φ, we find that this happens at

$$\varphi = 2\pi\beta^2\gamma^2. \tag{8.37}$$

Putting this into (8.34) gives

$$ct = 2\pi R\beta\gamma^2. \tag{8.38}$$

In view of (8.31), we finally obtain

$$ct' = \frac{ct}{\gamma} = 2\pi R\beta\gamma \tag{8.39}$$

or, since $\beta = v/c = \Omega R/c$,

$$t' \equiv \Delta t' = 2\gamma\frac{\Omega}{c^2}A, \tag{8.40}$$

where $A = \pi R^2$ is the (Euclidean!) area of the disk.

Suppose Paul is sitting on the rim of the disk and ct' is his world line. Clearly, the intersection O′ of this line with x' is an event of Paul's biography different from O and it has accordingly assigned a later moment given by (8.40). But since it is connected with O by the world line of a particle with $v = \infty$ in Paul's RF (i.e., by simultaneity line in this frame!) it can be assigned the time $t' = 0$ as well. Both moments are equally legal! By reciprocity, the event O can also be assigned the earlier time equal to negative of (8.40). And by iteration, all multiple integers of (8.38) would be also legal. And since the choice of reference event O is arbitrary, the obtained conclusion holds for any event on the spinning disk. And, finally, since ct' is the time elapsed from the zero moment from which we count time in this example, the obtained equation is identical (at $m = -1$) to (8.28) for the time lag $\Delta t'$.

A skeptical reader may still have a certain feeling of dissatisfaction. However, justified as such feeling may be, it comes from the totally unusual character of the resulting conclusions, rather than from any logical inconsistency. Nothing seems to be more remote from the common world of our intuition than the notion of the two distinct but equally legitimate moments of proper time characterizing a given event. But the above analysis provides us with the legal operational procedures within the formalism of the special relativity to obtain a coherent, logically consistent description of the whole process for all observers.

As a by-product of our search we made a remarkable new discovery: in a spinning world, time is not only relative, but it is also not even single-valued! Even with the proper synchronization procedure, we cannot establish a single common time globally, that is, for the whole rotating system. And we obtained this result in two different ways, first, by integrating local Lorentz transformations, and second, by analyzing the corresponding space-time diagrams. In the following two sections we will arrive at the same result from the third, and quite different, perspective.

8.2 Circumnavigations with the Atomic Clocks

More than half a century after the birth of the theory of relativity, the manifold story about the twin paradox (or more generally, the clock paradox) took on a new twist. By that time, the advances in experimental physics had led to the development of new types of clock – so-called atomic clocks, whose tickings were periods in radiation emitted by some atoms in the optical transitions between specified states with different, sharply defined energy levels (a more detailed description of optical transitions is given in Chapter 10). The precision and accuracy of the atomic clocks by far exceed those of any other clocks based on known macroscopic phenomena – mechanical, electromagnetic, or astronomical. For instance, the reported inherent drift of hydrogen maser clock is less than 1 part in 10^{14} [19]. This allows us to extend the testing ground of the special theory of relativity into the realm of nonrelativistic velocities. Using an atomic clock, one can detect tiny changes of its proper time caused by motion with nonrelativistic speed, like that of satellites and even jet planes, by comparing it with the proper time of a similar clock that remained in place. Such an experiment had been suggested, designed, and carried out by Hafele and Keating and the crews of commercial flights in early 1970s [19,20]. The experimental scheme was pretty simple. Four cesium beam atomic clocks were flown on regular commercial jet flights around the Earth twice – once eastward and once westward. In both cases the clocks' speed was the same – equal to typical jet speed about 300 m/s. After one circumnavigation the flown clocks were compared with the identical reference clocks at the US Naval Observatory. The results, within the margin of the experimental and computational errors, confirmed the prediction of the theory for a given situation. Before presenting the results, it is worthwhile to ask: what kind of prediction should we expect for this case?

According to analysis in Section 7.8, we would expect that both flying clocks lose a certain amount of their proper time as compared to the reference clock that remained stationary. And inasmuch as they move with equal speeds, the lost amounts must be equal for the eastbound and westbound clocks.

Now, both statements here are wrong. The experiment showed that not only there is no symmetry in the lost proper time of the flown clocks, but also there is even no loss of proper time in one of them. The west-flying clock turned out to have *gained* proper time against the reference clock! These results appear to be in flat contradiction with the unambiguous prediction of the theory in the previous section.

Therefore, when the results were published, many people have considered them as the experimental refutation of the theory of relativity. Here is a typical comment of the opponents of the theory of relativity:

> The theory of relativity clearly predicts that the time dilation should be equal for both flying clocks if they move with equal speeds. But in fact the westward flying clock showed no time dilation at all – quite the contrary, it gained time against the reference clock. So there is a conflict between theoretical prediction and observation. If the observations made by Hafele and Keating are correct, we can consider the theory of relativity to have been empirically refuted.

Now, where is the fallacy of these arguments?

It is in that the reference clock on Earth is confused with the stationary clock in an inertial reference frame. According to the theory, the loss in time, which is independent of the direction of flight, will be observed relative to a stationary clock in an *inertial reference frame*. In the experiment we discuss, the reference clock was stationary *relative to the rotating Earth*, which is by no means an inertial reference frame (Figure 8.7). There is an innate asymmetry between eastward and westward directions in a rotating system, which is imposed by the given direction of its rotation. This asymmetry can be described in three different but equivalent ways: first, on the qualitative level, one can draw and analyze a space-time diagram for the world lines of all clocks (including the imaginary stationary clock on the extension of Earth's rotational axis in the inertial reference frame); second, one can consider all the clocks involved (reference clocks included!) as nonstationary clocks moving with different velocities and accelerations in an inertial reference frame, and apply the correlations of the special theory; third, one can use the known solution to Einstein's equations of the general theory of relativity for a gravitating and rotating mass representing the Earth and apply it to clocks moved around the mass in circular orbits. We will use the first two options and start with the simplest one – the space-time diagram.

To understand things better, consider first a hypothetical situation of a nonrotating Earth. The world lines of all clocks involved are shown in Figure 8.8a. Here the cylindrical surface represents the worldsheet of the Earth's equator. The vertical generatrix OO' of the cylinder represents the world line of the reference clock. In the absence of rotation the reference clock is stationary and inertial. The two symmetric helices represent the world lines of the flying clocks. As was the case in the previous sections, these world lines have the same curvature if the clocks fly at the same speed. They must accordingly have a smaller kinematic length than the line OO'. Because the kinematic length of a world line represents the proper time of an object tracing out this line, the proper times of the flying clocks after one circumnavigation are both

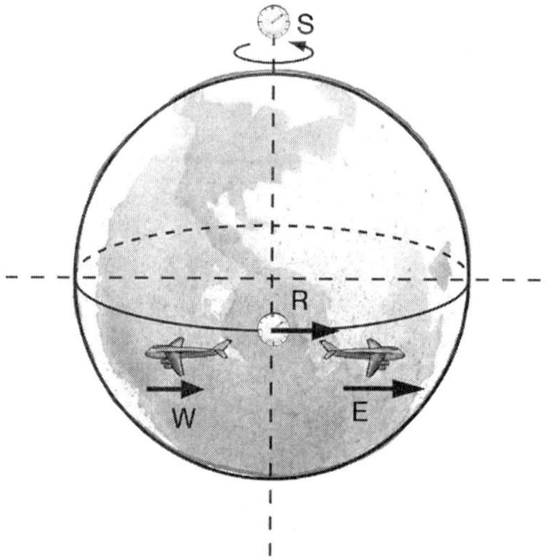

Figure 8.7 The rotating Earth in the inertial (nonrotating!) reference frame. S – stationary clock; R – reference clock on the Earth's equator; E and W– the eastbound and westbound clocks, respectively; arrows represent the speeds of the clocks. To the stationary observer, the westbound plane appears to fly backwards because a typical speed of a commercial flight is less than rotational speed of Earth at the equator.

less than the proper time of the reference clock. If their speeds are equal, the loss of the proper time is the same for eastward- and westward-flying clocks – in total accord with the prediction of the theory of relativity for the nonrotating Earth.

Now we turn to the real rotating planet Earth. Because of rotation, the reference clock at the Naval Observatory is no longer an inertial clock! Its world line is accordingly also a helix and has now to be considered on a par with the world lines of the two other clocks. We can still consider the system of the three clocks within the framework of the special theory of relativity. All we have to do is to compare their readings with those of a stationary clock in an inertial reference frame.

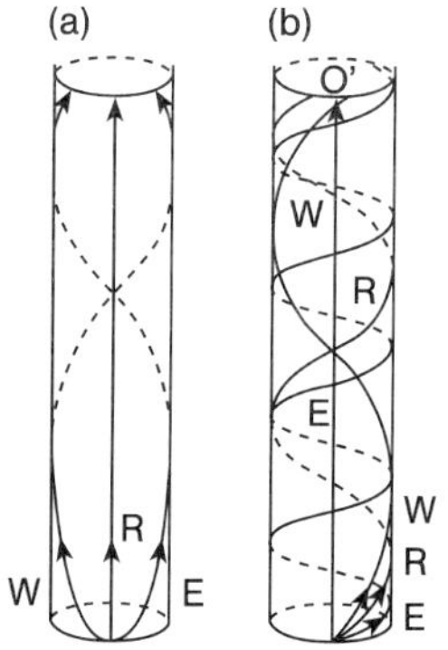

Figure 8.8 The world lines of all the clocks involved (a) for nonrotating Earth; (b) for rotating Earth. In (b), the E-clock is moving faster than the R-clock and its world line is curved more; this clock ages slower than the R-clock. The W-clock has the least curved world line; this clock ages faster than R-clock. All three moving clocks age faster (to different degrees) than the stationary clock S, whose world line OO′ is straight.

We will then immediately see that the world lines of all clocks fall neatly into two quite different categories (Figure 8.8b). The world line of the stationary inertial clock on the Earth's axis will be straight as a laser beam, whereas the world lines of all other clocks (including the reference clock on the ground) will be twisted and curved into helices. In terms of time intervals it means that the world line with no curvature corresponds to the maximum possible interval of the proper time between a *given pair of events* (the start and the finish of the circumnavigation trip, both signaled to the inertial clock). In other words, we first consider a very special case when all the clocks involved start at one point A and then meet all at some other point B of space-time. Then the difference in proper times of all the clocks at the meeting point will be determined by the curvatures of the corresponding world lines. In this case, the world lines of the remaining three sets of clocks correspond to smaller proper times, that is, *all three sets of the experimental clocks run slower than the stationary clock in the inertial reference frame*, in direct accordance with the prediction of the theory of relativity. Now, among these three experimental sets, the world line of the clock flying westward has the least curvature, which corresponds to the greatest proper time for this category. The world line of the reference clock on the ground has greater curvature (and smaller proper time) because this clock has greater acceleration than the westward-flying clock. And, finally, the eastward-flying clock has the greatest curvature and the smallest proper time per one circumnavigation. So there *must be an asymmetry* in the time readings in this situation. The actual flying clocks had, of course, traveled at some altitude (more than 10 000 feet) above the reference clock and had accordingly experienced a different gravitational field strength than the reference clock. The gravitational field slows down the evolution of an object. This effect can be rigorously described by the general theory of relativity, which lies beyond the scope of this book. The fact that the flying clocks had all been at a higher altitude caused a corresponding increase in their rate, which adds an additional contribution to the asymmetry in changes of proper time. This contribution partially accounts for the fact that the readings of the reference clocks are much closer to those of the eastward-flying clocks than to those of the westward-flying clocks. Even this fact has a full and clear explanation on the quantitative level. Thus the theory of relativity had predicted a new subtle and remarkable phenomenon that was first pointed out by Hafele and Keating and soon after found experimental confirmation.

Now we will treat the Hafele–Keating experiment quantitatively. We present here a simplified version of the analysis, considering an experimental model in which all sets of clocks had all the time been at the same altitude. But even so, the real situation is more complicated than the special case considered above; generally, as it was in the real situation, all the clocks do not meet again at the same point of space-time. As we will see soon, the westbound clock will return to the reference clock earlier than the eastbound clock. But, hopefully, the special case mentioned above will give the reader a gist of what is going to happen in a more general case when we treat the problem relativistically.

The special theory of relativity is equally well equipped for considering *all* possible kinds of motion – uniform or accelerated, rectilinear or in arbitrarily curved lines, subluminal or superluminal. The calculations are most straightforward if the motion to be discussed is considered relative to an inertial reference frame. This is precisely what Hafele and Keating do in their papers [19,20]. The *nonrotating* system attached to the center of the Earth, *is* inertial to a high degree of accuracy, because the centripetal acceleration associated with the Earth's orbital motion around the Sun is about $6\times 10^{-3}\,\mathrm{m/s^2}$, that is, less than one thousandth of the acceleration due to gravity on Earth. And if, in addition, we take into account that this system is in a state of a free fall in the gravity field of the Sun, then, according to Einstein's principle of equivalence (which had been shown experimentally to be correct to the accuracy 10^{-12} [3]), this system is inertial for all practical purposes.

We said that the situation at hand is nonrelativistic in that all the clocks involved, as well as all the points on the rotating Earth, move much slower than light. But even in a nonrelativistic situation, we can come to fundamentally different conclusions depending on whether we treat the problem according to Newtonian mechanics or according to relativistic mechanics. And the motions in a rotational system that we are considering provide a good illustration. To see it, let us first find the results in a Newtonian approximation.

Place the reference clock at some spot on the Earth's equator. Let Ω be the angular velocity vector of the Earth's rotation,[1)] and R be its equatorial radius. Relative to the inertial (nonrotating!) frame K, which we can consider stationary, the reference clock moves with the speed $v_R = \Omega R$. Now suppose two planes with the clocks departing simultaneously from the same spot and with the same speed v relative to the reference clock – one eastward and the other westward. Call the respective clocks "clock E" and "clock W." Relative to the inertial frame, the planes move with the speeds

$$v_E = v_R + v, \qquad v_W = v_R - v. \tag{8.41}$$

Now find the time t_E for one circumnavigation of the E-clock. To do it, we notice that *relative to the stationary frame* K, the E-clock makes precisely one rotation more than the reference clock (if you outrun your competitor on a racetrack, next time you catch up with him you will make one round more than him). Therefore, the distance $v_E t_E$ traveled by the E-clock will be one equator longer than that $v_R t_E$ of the reference clock by the time they meet again:

$$(v_R + v)t_E = 2\pi R + v_R t_E. \tag{8.42}$$

It follows

$$t_E = \frac{2\pi R}{v}. \tag{8.43}$$

The same reasoning applied to the clock W shows that since it is moving slower than the reference clock relative to the inertial reference frame, by the time they meet

1) We neglect here a very slow (1 cycle in about 26 000 years) precession of the Earth's rotational axis.

again it will have traveled one equatorial length less than the reference clock, so we can write

$$(v_R - v)t_W = v_R t_W - 2\pi R, \qquad t_W = \frac{2\pi R}{v} = t_E. \tag{8.44}$$

The travel times are the same for both types of clocks if they move relative to Earth with the same speed.

The Earth-based observer can comment with a smile: "You could have saved your time if you had asked me. I would have told the result to you immediately without any calculations. In Newtonian physics the time measured in a reference frame is the same in all other reference frames, therefore the times ((8.43) and (8.44)) found for the inertial frame must by definition be the same as the time of the Earth-based observer. Such an observer would obtain this result immediately without "looking out" beyond the Earth."

I agree. And now, before doing the *relativistic* treatment of the same problem, I want to follow this advice and ask: "What outcome would one expect in this case?"

We know that the relativistic mechanics is more subtle than Newtonian, but *in this case* the result seems obvious. If I sit in a spot somewhere on the equator and dispatch two planes – one eastward and one westward, and the planes move with the same speed relative to me, they will surely circumnavigate the Earth in the same time by my clock and so after circumnavigation they will arrive simultaneously.

Well, this is wrong! We will now show this rigorously.

Let me remind again, that in special relativity, if we want to obtain the correct results in the most straightforward way, we need to consider the events relative to an *inertial* reference frame. So, we must first find the speeds of the flying clocks *relative to the nonrotating frame K attached to the center of the Earth*. For this frame, we have, by applying, instead of (8.41), the *relativistic* rule for addition of velocities (Section 3.1):

$$v_E = \frac{v_R + v}{1 + (v_R v/c^2)}, \qquad v_W = \frac{v_R - v}{1 - (v_R v/c^2)}. \tag{8.45}$$

Accordingly, Equations (8.42)–(8.44) for the travel times measured by the *stationary* (inertial!) observer in K will now take the form

$$\frac{v_R + v}{1 + (v_R v/c^2)} t_E = v_R t_E + 2\pi R, \qquad \frac{v_R - v}{1 - (v_R v/c^2)} t_W = v_R t_W - 2\pi R. \tag{8.46}$$

Solve these equations for t_E and t_W:

$$t_E = \frac{2\pi R}{v} \frac{[1 + (v_R v/c^2)]}{[1 - (v_R{}^2/c^2)]}, \qquad t_W = \frac{2\pi R}{v} \frac{[1 - (v_R v/c^2)]}{[1 - (v_R{}^2/c^2)]}. \tag{8.47}$$

Now, just as we have done in the previous section for the rim of a spinning disk, we can introduce here the *proper* length of the Earth's equator as measured by people on Earth:

$$\Lambda = \gamma(v_R)L = 2\pi R\gamma(v_R), \tag{8.48}$$

where $L = 2\pi R$ is its length measured "from the outside" – that is, in the reference frame K. Also, we introduce "the proper period":

$$\tau_\nu \equiv \frac{\Lambda}{\nu}. \tag{8.49}$$

As is seen from its definition, "the proper period" is the time one would expect to read on the reference clock when an object departing from it with a speed ν returns after circumnavigation along the circumference of length Λ. In the astronomical language, it would be a period of an object on a circular orbit. Even if the spatial geometry of a rotating world were non-Euclidean, one would still expect for the period of an object with speed ν on a circular orbit of radius R the "obvious" value τ_ν given by Equation (8.49). Very soon we will see the reason why this term is taken into the quotation mark.

When expressed in terms of Λ and τ_ν, Equation (8.47) can be written more simply as

$$t_{\mathrm{E,W}} = \tau_\nu \gamma(\nu_\mathrm{R})\left(1 \pm \frac{\nu_\mathrm{R}\nu}{c^2}\right). \tag{8.50}$$

At this point we introduce a new term: the *coordinate time*. This is the time measured by an inertial observer not participating in a studied motion. In our case it is the time of a process as measured in K. Then we can say that t_E and t_W in the above equation are the *coordinate times* for the eastward and westward circumnavigations, respectively. The coordinate times for the orbital motions in the opposite senses turn out to be different (the "+" sign for E and the "−" sign for W circumnavigation in Equation (8.50)).

Now we can find the corresponding *proper* times of the reference clock for the two circumnavigations. According to our general result in Section 2.7, the proper time of a moving clock is related to the coordinate time read by the inertial system of clocks in K by the corresponding Lorentz factor depending on the clock's speed.

The speed of the reference clock is ν_R. Therefore, its proper time $\tau_\mathrm{E}^{(\mathrm{R})}$ between the departure and return of the E-clock and the proper time $\tau_\mathrm{W}^{(\mathrm{R})}$ between the departure and return of the W-clock are given by

$$\tau_\mathrm{E}^{(\mathrm{R})} = \frac{t_\mathrm{E}}{\gamma(\nu_\mathrm{R})} = \tau_\nu\left(1 + \frac{\nu_\mathrm{R}\nu}{c^2}\right), \tag{8.51a}$$

$$\tau_\mathrm{W}^{(\mathrm{R})} = \frac{t_\mathrm{W}}{\gamma(\nu_\mathrm{R})} = \tau_\nu\left(1 - \frac{\nu_\mathrm{R}\nu}{c^2}\right). \tag{8.51b}$$

Let us stop here and formulate our first conclusion: not only the coordinate times, but also the travel times (time intervals between the departure and return of a plane) *measured by the reference clock* are different for the eastward- and westward-bound planes moving with *the same* speed relative to this clock! Contrary to our naive expectations, the "proper period" introduced in (8.49) is *not* what will be actually measured by the reference clock. Actually, there are two different periods for any circular orbit around the center, depending on the sense of revolution. The orbit with the angular velocity parallel to Ω (the angular velocity vector of the rotating world) has a period greater than τ_ν and the same orbit with the angular velocity antiparallel to Ω has a period less than τ_ν.

Where does this difference come from? The times were equal when we considered the problem in the framework of Newtonian mechanics and derived Equations (8.41)–(8.44). It is only because we have switched from the Newtonian rule (8.41) of addition of velocities to the relativistic rule (8.45) that the difference in times crept in. And the reason is that by rule (8.41) the velocities of the flying clocks remain symmetrical with respect to the speed v_R (one greater and the other less than v_R by the same amount), whereas by rule (8.45) this symmetry is broken. We see that the resulting difference in the travel times is a *purely relativistic effect*.

Thus, in the rotating world, not only space geometry is unusual, but also the temporal characteristics of orbital motions become quite weird. This is another illustration of our statement in the previous section that time and space are both "intertwined" in a rotating world.

Now we can ask, what are the *proper* times of the circumnavigating clocks themselves?

Let us start with case E: the coordinate time for one circumnavigation in the eastward direction is given by the first equation (8.47). The speed of the E-clock in the inertial frame K is v_E. Therefore *its* proper time τ_E is

$$\tau_E = \frac{t_E}{\gamma(v_E)} = t_E\sqrt{1 - \frac{v_E^2}{c^2}}, \tag{8.52}$$

where v_E and v_W are given in (8.45). Now using our results (3.16) from Section 3.1, we can rewrite this in the form

$$\tau_E = \frac{t_E}{\gamma(v_E)} = \frac{t_E}{\gamma(v_R)\gamma(v)[1 + (vv_R/c^2)]} = \frac{\tau_v}{\gamma(v)}. \tag{8.53}$$

In a similar way, we find the proper time of the W-clock:

$$\tau_W = \frac{t_W}{\gamma(v_W)} = \frac{t_W}{\gamma(v_R)\gamma(v)[1 - (vv_R/c^2)]} = \frac{\tau_v}{\gamma(v)} = \tau_E \tag{8.54}$$

This is another surprising result. From the nonrelativistic perspective, it is quite natural: if the two travelers, Ann and Bob, start moving simultaneously with equal speeds $v_A = v_B = v$ from a certain point on the equator, one eastward and the other westward, then, after one circumnavigation, they must return to the starting point simultaneously, and accordingly, having equally aged. But according to relativity (Equations (8.50) and (8.51)), they will not return simultaneously either by coordinate time or by the time of the reference clock. So in the relativistic domain, we already got used to something crazy (from the classical viewpoint) popping up from our calculations. What has popped up here, however, is indeed crazy, but now from relativistic viewpoint, since it is natural from the classical viewpoint! Whereas the two proper times $\tau_E^{(R)}$ and $\tau_W^{(R)}$ of the reference clock, as well as the two *coordinate* times t_E and t_W for the motion of the E-clock and W-clock, are different (which was shown above to be, under given conditions, the correct relativistic result) – the *proper* times of the traveling clocks upon return turn out to be equal! Or, using our example with Ann and Bob, Ann and Bob start simultaneously from a spot on the equator, one eastward and the other westward, and move equally fast with respect to Celia who remains on that spot. They complete their respective circumnavigations and return to Celia at different times by her clock: if Bob, say, was traveling westward, then he returns first and Ann will then be the second. Once we accept this crazy result, we already expect that Ann will have aged more than Bob. However, *this* is not the case! Both travelers, by their inner clocks, have aged equally upon their return.

We can roughly understand this in the following way. It takes a longer coordinate time for one circumnavigation of the eastbound clock than it does for the westbound clock. The proper time is shorter than the coordinate time by the corresponding Lorentz factor. The eastbound clock is moving faster in the inertial frame than the westbound clock. Accordingly, its Lorentz factor is greater than that of the westbound clock. To get the corresponding proper time, the longer coordinate time for the E-clock is divided by the greater Lorentz factor, and the shorter coordinate time for the W-clock is accordingly divided by a smaller Lorentz factor. This produces a mechanism that compensates for the coordinate time discrepancy. But the corresponding characteristics (coordinate times and the Lorentz factors) are not in proportion to the respective velocities v_E and v_W. Therefore, it seems really surprising that the compensation is so finely tuned as to produce the equal final outcomes for the proper times. Well, this is another surprise of the spinning world.

Now we are able to estimate the *experimentally observed quantity* – the relative time offset between a traveling clock and the reference clock:

$$\delta_E \equiv \frac{\tau_E - \tau_E^{(R)}}{\tau_E^{(R)}}, \qquad \delta_W \equiv \frac{\tau_W - \tau_W^{(R)}}{\tau_W^{(R)}}. \tag{8.55}$$

Before doing any calculations, we can make some qualitative predictions simply by comparing the world lines of the respective clocks. From Figure 8.8b, the world line of the E-clock has a greater curvature than that of the reference clock. Accordingly, it has a shorter kinematic length.

In other words, we will always have

$$\tau_E^{(R)} > \tau_E \tag{8.56}$$

and therefore δ_E must be always negative. This is just a generalized version of the clock paradox: out of two clocks moving between the events O and O′, the one that is moving (and accordingly accelerating) faster follows the world line with a greater curvature and accordingly spends less of its proper time. As to the W-clock, we cannot make a universal prediction for the reasons that will very soon become clear.

Now back to the equations. Combining Equations (8.51), (8.54), and (8.55), we obtain

$$\delta_{E,W} = \frac{1}{\gamma(v)[1 \pm (v_R v/c^2)]} - 1. \tag{8.57}$$

As we had expected, the offsets are different. For the case at hand, retaining only the terms of the order of c^{-2}, we get (you have to do some calculus to check it (Problem 8.4))

$$\delta_E = -\frac{v(2v_R + v)}{2c^2}. \tag{8.58}$$

As for the clock W, notice that the coordinate time of its circumnavigation, as well as all other expressions for the W-clock, can be obtained from the corresponding equations for the E-clock by just changing the sign of v. Therefore, we can obtain the offset for the W-clock in terms of v, v_R directly from (8.58):

$$\delta_W = \frac{v(2v_R - v)}{2c^2}. \tag{8.59}$$

As we have already predicted, for the E-clock the offset is negative, that is, it loses a certain fraction of its proper time with respect to the proper time of the reference clock. We have already explained it by noting that the E-clock moves relative to the stationary frame faster than the reference clock and therefore "ticks" slower. It turns out to have aged less when the clocks meet again. Now we obtained this result quantitatively, using the equations of Special Theory of Relativity (STO).

For the W-clock, the sign of the offset depends on its speed. If the speed is greater than $2v_R$, the offset is also negative and this clock will also have aged less than the reference clock when they meet again. We can explain this if we notice that at a speed exceeding $2v_R$ the W-clock is moving westward *faster* relative to the stationary frame than the reference clock is being carried eastward by the Earth's rotation. Again, the clock that moves faster ages slower. If the speed v is much larger than $2v_R$, so that we can neglect the rotational motion, then Equations (8.58) and (8.59) tell us just what one would expect for this case – that both traveling clocks will have aged less than the reference clock by approximately the same amount when they meet again. However, if the speed v is less than $2v_R$, then the W-clock moves slower relative to the stationary system than the reference clock, and will after one circumnavigation have aged more than the reference clock. This will produce a positive offset.

(At speeds somewhere between v_R and $2v_R$, the W-clock is still moving westward relative to the stationary frame, albeit slower than the eastward motion of the reference clock. If the speed v of the W-clock is equal to v_R, it "cancels" the effect of the Earth's rotation and the W-clock stands still in the stationary reference frame. In this case it has the maximal aging rate. If its speed is less than v_R, it is moving eastward relative to the stationary frame.)

For typical conditions of international jet flights (altitude about 10 km and a speed $v \cong 300$ m/s) the theoretical prediction (including the above mentioned contribution from the gravitational effect) was that the E-clock should have lost 40 ± 23 ns, and the W-clock should have gained 275 ± 21 ns compared with the reference clock. The experimental data showed that the E-clock actually lost 59 ± 10 ns and the W-clock gained 273 ± 7 ns.

We see that the experimental results stand in quantitative agreement with the predictions of the theory of relativity if we take a proper account of the motion of the reference clock. As Hafele and Keating put it, "These results provide an unambiguous empirical resolution of the famous clock "paradox" with macroscopic clocks." So, paraphrasing the above-quoted critic of STO, if we believe in the observations made by Hafele and Keating, we can then believe in Einstein's theory of relativity even more firmly than ever before.

8.3
Surprises of the Rotland

In this short section, Rotland will stand for a "rotating land." It may be any extended spinning object, including our planet. We have seen that accurate time measurements with clocks flying around the Earth, counterintuitive as they seem to be, confirm the relativistic predictions. Here, we want to notice that the obtained results concerning the two proper times (8.51) of the reference clock provide us with yet another way to find the time lag discussed in the previous section.

Equations (8.51), although obtained for a special problem of the Earth circumnavigation, are quite general and hold for *any* speed v. It may be even superluminal speed and we will see later that such velocities are as legal as the subluminal ones, and do not by themselves contradict anything. So let us consider, instead of the planes, two fictitious particles moving along the circumference with infinite speed relative to the *reference clock* and accordingly set $v = \infty$ in Equation (8.51). As we know, the world line of a particle with infinite velocity along certain direction coincides with the spatial axis in this direction in a given reference frame, or, which is the same, with the corresponding line of simultaneity. Therefore, we come again to the apparently paradoxical result that the temporal and the spatial axes of the rotating reference frame intersect more than once. Equations (8.51) give us the two corresponding moments of proper time of the reference clock at these intersections:

$$\begin{aligned} \tau_{\mathrm{E}}^{(\mathrm{R})}(\infty) &= \lim_{v\to\infty} \frac{t_{\mathrm{E}}}{\gamma(v_{\mathrm{R}})} = \lim_{v\to\infty} \frac{\Lambda(v_{\mathrm{R}})}{v}\left(1+\frac{v_{\mathrm{R}}v}{c^2}\right) = \frac{v_{\mathrm{R}}}{c^2}\Lambda(v_{\mathrm{R}}) \\ \tau_{\mathrm{W}}^{(\mathrm{R})}(\infty) &= \lim_{v\to\infty} \frac{t_{\mathrm{W}}}{\gamma(v_{\mathrm{R}})} = \lim_{v\to\infty} \frac{\Lambda(v_{\mathrm{R}})}{v}\left(1-\frac{v_{\mathrm{R}}v}{c^2}\right) = -\frac{v_{\mathrm{R}}}{c^2}\Lambda(v_{\mathrm{R}}) \end{aligned} \quad . \tag{8.60}$$

This is identical to the expressions (8.18) for the time discrepancy (v there is v_{R} here) for the same event (intersection between the axes). Indeed, once the fictitious particle is moving infinitely fast, it is expected to return to the reference clock after the circumnavigation at the same moment of time as its departure. However, Equations (8.60) show the finite proper time read by the reference clock between the departure and return of such a particle. Moreover, this time turns out to be different for different particles: later than the moment of departure for the E-particle, and even more bizarre, earlier than the moment of departure (negative sign of $\tau_{\mathrm{W}}^{(R)}(\infty)$!) for the W-particle. Both these times are as legitimate as the zero moment $\tau = 0$ and are accordingly represented together with this moment, on the same footing, as the temporal characteristics of the event. The difference between them is the time lag discussed in Section 8.1.

And yet there remains one point mentioned above in passing (in comment to Equations (8.42–8.44)) which seems to constitute a real paradox. Here it is. We have proved that the two objects launched with *the same* speed in the opposite directions must have different circumnavigation times read by the reference clock. But this implies that they must have had *different* speeds. Indeed, imagine that you are sitting somewhere on the equator with the reference clock. At a certain moment of time (call it the zero moment), you record two planes departing with equal speeds – one eastward and one westward. If they maintain equal speeds during the whole flight, then how can it be that they return after circumnavigation at different times by your clock? Restate the argument. Suppose that the two planes take off simultaneously, fly around the Earth in the opposite directions, and return back to the same airport at different times. Will you say that the planes had the same speed? The likely answer is: "Not, unless I am crazy". And you confirm your answer with a simple calculation. If the radius of the Earth is R, then the length of the Earth's equator measured by an inertial observer is $L = 2\pi R$. As we know already, its proper length measured by people on Earth is $\Lambda = \gamma(\Omega R)L = 2\pi R\gamma(\Omega R)$. The speed of the E-clock is $\Lambda/\tau_R^{(E)}$ and the speed of the W-clock is $\Lambda/\tau_R^{(W)}$. Using Equations (8.47) and (8.51) we easily obtain

$$\tilde{v}_E = \frac{v}{1 + (v_R v/c^2)}, \qquad \tilde{v}_W = \frac{v}{1 - (v_R v/c^2)}. \tag{8.61}$$

(Here, we use tildes to distinguish v_E and v_W from the corresponding velocities measured in the stationary inertial frame.)

Thus, if the travel times are not equal, we obtain travel speeds that are *not equal*. This result appeals to our common sense but apparently contradicts the initial condition, according to which the speeds of both flying clocks relative to Earth are *the same*. The statement about the equal speeds resulting from unequal travel times over the same distance seems absolutely crazy. The equivalent statement that it takes different times for planes with equal speeds to fly around the Earth must therefore be also crazy. As we mentioned in the comments to Equation (8.47), this result only appears when we switch from Newtonian mechanics to relativistic mechanics. But now we obtained Equation (8.61) also using relativistic mechanics. Does this mean that the theory of relativity contains contradictions after all? If so, this would be the death sentence for the theory. Or does this mean that all our calculations have been fundamentally flawed? The whole situation appears to be not even a paradox, but just a huge nonsense. How can one reasonably explain this nonsense?

We will look for the answer in the next section.

8.4 Photons' Races in a Centrifuge

Here, we will discuss another interesting and important application of the general description of orbiting in rotating world. Consider a disk with a circular path of length L around its center O. Pick a point P on the path and place two detectors there.

Suppose we launch two photons (considered as localized particles) from this point in the opposite directions along the path. Think of the path L as a long corridor with the ideally reflecting walls. A photon launched down the corridor will be forced into a circular motion by the outer wall. Therefore each photon will circumnavigate the closed path and come back at P where it will be detected. Since both photons travel the same distance, and the speed of light does not depend on direction, both photons will return at the same time $T_0 = L/c$ and the detectors will fire simultaneously.

Now complicate the matter. Bring the whole platform with the pathway L into rapid rotation about its center O. Call it an optical centrifuge. Launch again two photons from P in two opposite directions along the path and wait for their return back at P after one circumnavigation around the path (Figure 8.9). What would you expect to see? Will the two photons return at P simultaneously or not?

The answer to this question has been known for a pretty long time, after a French physicist Sagnac had carried out the corresponding experiment [76, 77]. The two photons no longer return back at the same time. The one launched in the direction of rotation arrives at P later than the one launched in the opposite direction.

How can we explain this result in terms of what we have learned?

According to the theory of relativity, the speed of light is a universal constant. It is the same in all directions and does not depend on the motion of its source. Therefore, each photon will start from P with the same speed c independent of direction even though the source P itself is moving. If we mark the position of P at the moment of the emission, both photons will return back to this position simultaneously. But the emitter itself (carrying also the detectors) is no longer there – it has progressed to another position P′ due to the rotation of the platform (Figure 8.9b). The photon that had been launched in the same direction has not yet reached this point; the photon launched in the opposite direction, however, has already passed this point and had accordingly been detected. Thus, the photons cannot arrive back at the emitter at the same time. Once the platform starts rotating, the clockwise and counterclockwise motions in a circular path are no longer equivalent.

We will start again with the arrival times. And as before, it is easier to do it from the viewpoint of the stationary observer. Look again at Figure 8.9. Call the photon launched from P in the direction of the platform's rotation, the east photon. The photon launched in the opposite direction will be called the west photon.

As is seen from the figure, the west photon will meet with the detector again when it is at position P′. The east photon will hit the detector at a later time, when the detector will be at a position P″. The same treatment as in the previous section gives for the arrival times measured by a stationary observer:

$$T_{\mathrm{E}} = \frac{2\pi R/c}{[1-(v_{\mathrm{R}}/c)]} = \frac{T_0}{[1-(v_{\mathrm{R}}/c)]}, \quad T_{\mathrm{W}} = \frac{2\pi R/c}{[1+(v_{\mathrm{R}}/c)]} = \frac{T_0}{[1+(v_{\mathrm{R}}/c)]}, \tag{8.62}$$

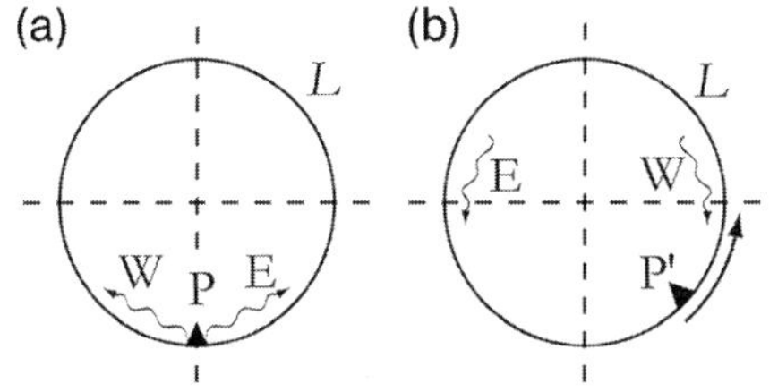

Figure 8.9 A circular racetrack (optical centrifuge) with a photon emitter and two detectors. The emitter launches two photons in the opposite directions ("east" and "west"), while the detectors record their return time. The photons do not complete their respective cycles simultaneously.

where T_0 is the travel time along the loop for the stationary platform. These expressions for coordinate times follow directly from the general equation (8.47) if we set there $v = c$ and change notation $t \to T$.

We see that $T_W < T_0$ – no surprise, the west photon has traveled the shorter length between its emission and absorption. The time T_E, on the other hand, is greater than T_0 because the east photon has to travel a longer distance in the stationary system between its emission and absorption.

Consider now the same process from the viewpoint of the disk inhabitant Paul. He is sitting near P and watches the motion of photons in his frame. Both photons travel the same distance along the circular loop. The length of the loop measured by Paul is its proper length given by $\Lambda = L_\gamma(v_R)$ and a proper time of his clock is related to corresponding time T in the stationary system by $\tau = T/\gamma(v_R)$, so that $\tau_E^{(R)} = T_E/\gamma(v_R), \tau_W^{(R)} = T_W/\gamma(v_R)$. It follows from (8.62) that, just as we have found in the general case, in the optical centrifuge, too, the clock at the detector measures two different circumnavigation times for the east and west photons:

$$\tau_E^{(R)} = T_0\sqrt{\frac{1+\beta_R}{1-\beta_R}}, \qquad \tau_W^{(R)} = T_0\sqrt{\frac{1-\beta_R}{1+\beta_R}}, \tag{8.63}$$

where $\beta_R \equiv v_R/c$.

Exercise 2

Find *the photons'* proper times, using the same general equations (8.53) and (8.54) and explain the results.

Solution

Our qualitative prediction before using any equations is that both photons' proper times must be zero, simply because a photon's world line has the zero kinematic length and, accordingly, zero proper time. Another way to express the same thing: we know from the twins' paradox that for a traveler making a loop with a speed approaching that of light, the corresponding aging time goes to zero. And the photons do not age at all!

Quantitatively, the Lorentz factor for a photon is infinite, so dividing the expressions (8.62) by such a factor gives zero for both photons. The same follows from Equations (8.53) and (8.54) by setting there $v = c$. Our general conclusion in Section 8.2 was that the proper circumnavigation times are equal for both travelers. As we see, this remains true when the travelers are photons, but now these equal proper times have the fixed zero value.

Now, in the spirit of the concluding remark in the previous section, Paul can define the average speed of the photon circumnavigating the platform as the ratio $\Lambda/\tau^{(R)}$. Applying this definition to our case and using (8.63) gives

$$c_E = \frac{\Lambda}{\tau_E^{(R)}} = \frac{c}{1+\beta_R}, \qquad c_W = \frac{\Lambda}{\tau_W^{(R)}} = \frac{c}{1-\beta_R}. \tag{8.64}$$

Equation (8.64) can be also obtained directly from the general expressions (8.61) in the previous section as a special case when $v = c$.

The photons' speeds (8.64) found by Paul from his measurements of their travel times are different from the speed of light! The west photon appears to move faster than light; if the rotation of the disk is rapid enough, such a photon travels arbitrarily fast. The east photon, on the other hand, appears to move slower than light and on a sufficiently rapidly rotating disk it can slow down to half the speed of light.

So far these results pertain to the average speed of the photons. But we apparently cannot avoid the conclusion that the same must be also true for the local speed: it must be smaller than c if the light moves in the direction of rotation of the disk and greater than c if light moves in the opposite direction. The results of the Sagnac experiment seem to give unambiguous proof to it. Here is the argument presented by Paul:

> One of the goals of Physics is to draw meaningful conclusions from experiments. As a disk-based observer I do not have to think much about its rotation and how it may affect the phenomena. I have to do the measurements. If the measurements tell me that both photons arrive back at P simultaneously, I conclude that the light propagates in two opposite directions with the same speed. If the oppositely traveling photons that were emitted simultaneously do not return to me simultaneously, I have to conclude that they travel at different speeds. Now, I did perform the experiment and I see that when there is no rotation, the photons that were emitted simultaneously in the two opposite directions, return simultaneously. I accordingly interpret this as another confirmation of Einstein's postulate about the constancy of the speed of light. However, when I repeat the experiment during rotation of the disk, the photons do not return simultaneously. The only conclusion I can draw from it is that the speed of light in a rotating system is different in different directions. And this must be true not only for the average speed, but also for local speed in any location.

This argument by Paul seems very strong indeed: since the conditions are the same at any point along the photon's circular path around the center, the velocity measured locally must be also the same at any point, and thus be equal to its average over the whole circumference. Therefore, the Sagnac experiment can be considered as a direct measurement of the local speed of light separately for one and the other direction.

And yet, the last conclusion would be wrong. It does not follow from the Sagnac experiment! This experiment by itself only shows that apart from the local speed c that describes the rate of photon motion from one point to another, one can also introduce two other speeds characterizing *complete cycles* of clockwise and counterclockwise motions of light around the center of a rotating system. These new speeds are different from the local speed c because the procedure of their measurement *is different* from the local measurement. To show the difference, I will first present a purely physical argument in the form of a thought experiment, and then the arguments based on the definition of measurement and on the concept of time in rotating systems.

Consider on the rim of our disk an element ΔW with the segment of arc AB (Figure 8.10). According to Paul's interpretation of Equation (8.64), the local speeds of light in the directions $A \rightarrow B$ and $A \leftarrow B$ differ from one another. Let now the radius of the disk increase and its angular velocity decrease in such a way that the product ΩR and thereby the speeds c_E and c_W in Equation (8.64) remain constant. The reference frame formed by the local region ΔW, with ever increasing accuracy approximates the inertial reference frame in the limit $R \rightarrow \infty$, $\Omega \rightarrow 0$. (One can make sure of it by noticing that centripetal acceleration $a = \Omega^2 R = (\Omega R)\Omega = \text{const}\cdot\Omega$ of the element ΔW under given condition goes to zero.) It follows then that Paul's conclusion about different local speeds of light in different directions has to be true for this inertial reference frame. And due to the principle of relativity, it then has to be true in general!

To see the physical consequences of this, let us continue our thought experiment. In Figure 8.10 you see a spaceship traveling alongside with ΔW with the same speed v as the rim of the disk, that is, $v = \Omega R$. Assume that $\Omega R = 0.999\,999\,99c$, so that c_W is nearly infinite and c_E is nearly one half of c. The spaceship moves in a straight line and therefore represents an inertial reference frame. The disk radius R is so huge that even the region ΔW that is small with respect to the whole disk is still huge by our standards. It is so huge that our spaceship is comoving together with it for a considerable time required to perform our experiment and all this time they practically touch each other. A crew member Sam, which is an old friend of Paul, comes out to meet his friend. They greet each other and start discussing problems of common interest. As they do it, they run together on cosmic roller blades specially designed for space strolls and able to accelerate to nearly the speed of light. They are running shoulder by shoulder – Sam on a sideboard of his spaceship and Paul along the rim of his disk – you remember, their platforms nearly touch each other. Suppose they start running in the direction from A to B. Nothing seems to be in the way of this run. For instance, Sam can step on Paul's platform without even noticing it, since the spaceship and the region ΔW are to the highest accuracy at rest relative to each other. But the moment Sam's roller blades speed him up to half the speed of light, Paul yells:

- Hey, slow down, I cannot accompany you any farther!
- Why?
- Because I cannot outrun light and it propagates with the speed of only $0.500001c$ in this direction.
- Come on, I also see this beam – it behaves quite normally.
- Yes, in your reference frame, but things are different on my disk.

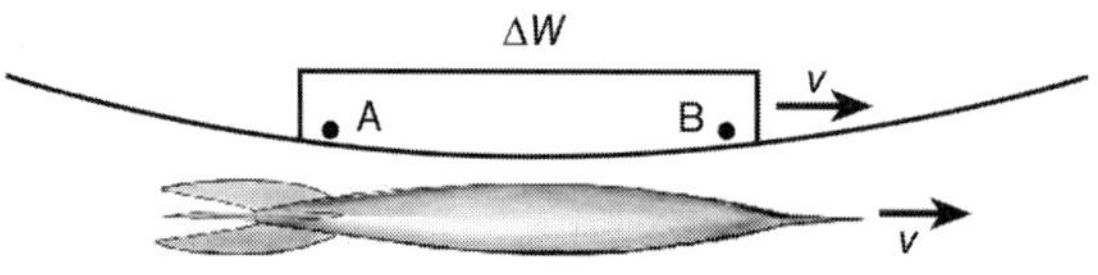

Figure 8.10

- What are you talking about? There is no difference between our systems! They are stationary relative to each other. For all practical purposes they form one common reference frame. I do not know what is going on elsewhere on your disk, but this spot does not in any way differ from my spaceship. We are in the same conditions. But if you mind going farther this way, let us turn back.

They turn back, accelerate up to nearly the speed of c, and now Sam yells at Paul:
- Hey, slow down, I cannot follow you farther!
- What's wrong?
- I cannot move faster than light, my blades are at their limit!
- Come on, we are just crawling like the wretched photons in your spaceship, but here the speed of light is $10\,000\,000c$ – we can frolic to our heart contents!

And Paul disappears from Sam's view – he crosses the light barrier without violating anything: even though he now runs much faster than c, it is still less than his limit of $10\,000\,000c$!

Now, if Paul is right in his conclusions, the above scene could be in principle possible. But this would flatly contradict the laws of Nature in the domain where they are established beyond any doubt. We have to conclude that we *cannot* identify the speeds c_E and c_W as the two possible local speeds of light. The local speed of light is a single fixed invariant quantity. The speeds c_E and c_W are the two different values of some variable quantity $c(\Omega r, \hat{n})$, which depends on locality, direction $\hat{n}$ of photon's motion, and angular velocity of the disk.

What then is the difference between them?

The best answer to this question would probably be the inspection of corresponding space-time diagrams. To measure the *local* speed, we mark, as usual, two events M and N along the trajectory of an object – its departure from one point and arrival at another point. Since each point is equipped with its own clock, the local speed of light is determined from the segment of the photon world line between the two world lines of these clocks (Figure 8.11a). To measure the speed c_1 or c_2 in Sagnac experiment, we wait for the photon to return to *the same* clock after one circumnavigation. The resulting speed is in this case determined from the segment of the world line of this clock within one pitch of the helix representing the photon's world line (Figure 8.11b). In the more familiar terms of three-dimensional space – the local speed of light is measured along a small segment with two end points, whereas the speeds c_1 and c_2 are measured from two consecutive readings of only one clock on the closed loop. The fundamental difference between these two procedures is seen, for instance, from the fact that the former admits also the speed measurement for the "geodesic" photon moving in a straight line tangential to the rim of the disk, whereas the latter is unable to do this in principle. No surprise that the two fundamentally different procedures of measurement give generally different results – they actually measure two different physical characteristics. Only for a stationary disk do these characteristics coincide. Generally, the speeds c_E and c_W in Sagnac experiment have nothing to do with local speed of light.

From Figure 8.11 we can also clearly see why the "global" speed $c_{E,W}$ turns out to be double-valued. If the photons were moving along the rim of a *stationary* disk, the

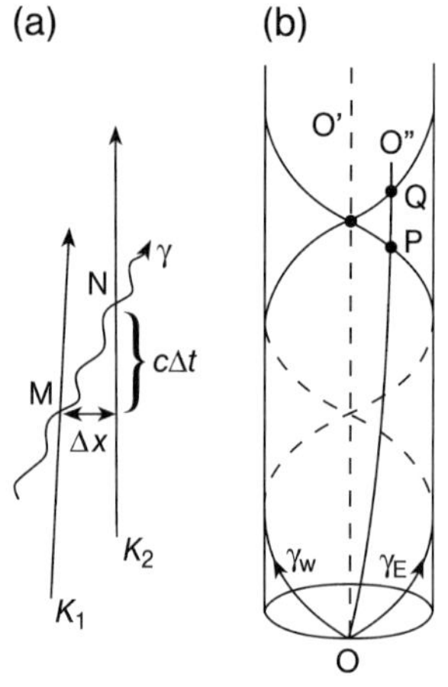

Figure 8.11 The diagrams for local and "global" speed measurement on a closed racetrack. (a) *local* speed of an object can be measured as a slope of its world line MN between the world lines K_1 and K_2 of the two measuring clocks Δx meters apart, (b) *global* speed of an object (photons included) is determined by the world line of a *single* measuring clock between the departure and arrival of the object and the circumference of the track. If the centrifuge is stationary, the clock's world line is vertical OO′ and the segment between the departure and arrival of the object with constant local speed will be the same for the east and west directions. However, if the centrifuge is rotating, the world line of the measuring clock is twisted (OO″) and accordingly intersects with the world lines of the east and west photons at *different* points (P and Q). This corresponds to different arrival times and thereby different global speeds of the two photons.

world line of the recording clock would be the vertical line OO′ – the generatrix of the cylindrical surface representing "the world pipe" of the rim. In this case the intersection points of OO′with the world lines of the east and west photons coincide, which would yield the same travel time for both photons. In other words, the "global" and local characteristics of motion coincide in this special case. In case of rotation, the helices representing the world lines of the photons do not change because the local speed of light does not change. The world line of the clock, however, is now also twisted into a helix and intersects with the photons' lines already at different points, which produces the difference between the travel times and thereby the different "global" speeds c_E, c_W.

Therefore if Paul asks, "Why does it take different times for a photon to travel along the same circle in opposite directions?" – the answer would be – first, because the world line of the measuring device (the clock P) is twisted by the rotation of the disk; and second, this twist is mismatched with those of the photon lines. Were the photon lines affected by the rotation in the same way, as is the line of the clock, they would remain symmetrical *with respect to this twisted line*, and all three would again intersect at the same point. This is precisely what happens in nonrelativistic mechanics when we use the Newtonian law of addition of velocities. Recall our nonrelativistic treatment of the circumnavigating clocks in Section 8.2 – Equations (8.41–8.44).

In that case, you remember, we obtained the same value for the travel times in the east and west directions (and accordingly for the local and global speeds). It was not for nothing, after all, that we had derived apparently unnecessary nonrelativistic Equations (8.41)–(8.44). They now clearly illustrate that had the photons behaved in the same nonrelativistic fashion, they would not display any time discrepancy either. Because the Sagnac experiment *does* record the discrepancy, it can be considered as yet another evidence of the relativistic nature of photons; and the measured magnitude of this discrepancy corresponds to the relativistic limit, when the photon speed added with any other speed (for instance that of the disk's rim) remains unchanged. In other words, the Sagnac experiment, which on the face of it appears to disprove Einstein's assertion about the invariance of the speed of light, gives an additional proof that the *local* speed of light is one and the same in *any* reference frame, rotating included!

Reversing this argument, we can get yet another perspective of the Sagnac experiment: the appearance of two additional speeds c_E and c_W indicates the rotation of the system. Thus, not only can we detect such rotation mechanically without "looking out" (recall Chapter 1), but also we can do it optically. From this viewpoint, the Sagnac experiment can be considered as an optical analog of Faucault's famous experiment with a pendulum [31].

The above analysis pertains not only to the speed of light – it has a general character. As an example, suppose you launch a photon in the E-direction on the disk spinning counterclockwise, and a particle in the W-direction. Then, at sufficiently large Ω or R, the particle returns to the starting point earlier than the photon (Problem 8.11). You then are again tempted to conclude that the particle's speed $v > c$. But then you decide to take a more conservative stand, and check it again, this time launching the photon in the same W-direction as the particle. Then, sure enough, the photon will return earlier than the particle.

As another example, we illustrate the difference between the local and "global" speeds for the case opposite to that of light: the stationary particle. So imagine a particle stationary in K and sitting close to the rim of the rotating disk. Its local speed relative to Paul is $v_L = \Omega R$. Let us now try to find this speed using the procedure analogous to Sagnac's experiment, that is, divide the circumference length $\Lambda = 2\pi R\gamma(\Omega R)$ of the circle in K' by the time interval $\tau = T/\gamma(\Omega R)$ on the clock P between its two consecutive meetings with the particle (T is the rotation period of the disk). The result will be

$$v_G \equiv \frac{\Lambda}{\tau} = \frac{2\pi R\gamma(\Omega R)}{T/\gamma(\Omega R)} = \frac{2\pi R}{T}\gamma^2(\Omega R) = \frac{v_L}{1-(\Omega^2 R^2/c^2)} \neq v_L. \tag{8.65}$$

This result is a special case of the general equation (8.58) in Section 8.2, when the westbound object moves relative to a rotating system with speed $v_R = \Omega R$ and therefore remains still in the stationary inertial system.

Suppose that we are unaware of the above analysis and naively believe that result (8.58) yields the local speed of the particle. But upon closer inspection we will realize that the speed v_G has nothing to do with the local speed and at sufficiently large ΩR it can become infinite – and this for a particle resting in K!

The speeds v_L and v_G in this example, as in the previous ones, have different physical meaning (Figure 8.12). The local speed is determined from the readings of the two synchronized clocks in K′ at the end points of the segment MN of the world line of the particle. The global speed is determined by the proper time of the one pitch of the helix representing the world line of one clock K′. There is no reason to expect any equivalency between these two essentially different experimental procedures.

The following analysis of connections between the local and global speeds leads to another astounding discovery in the rotating wonder-world: there is no such thing there as one time for the whole space – even in one reference frame.

Let us apply the measurement procedure for the local speed of light to our case. The segment of the photon trajectory is along the circle of radius R (Figure 8.4). In an inertial (nonrotating) system K associated with the center of the disk, the spatial separation between the events is $\mathrm{d}l$ and the time interval between them is $\mathrm{d}t$. Since the events are on the photon's world line, $\mathrm{d}l/\mathrm{d}t = \pm c$, depending on the direction of the photon's motion along the segment $\mathrm{d}l$. The signs $\pm$ relate to this direction. The speed itself is c regardless of direction. Now, perform the same measurement in the role of Paul – the observer on the rotating disk. If Paul measures the local speed, the procedure he uses has nothing to do with the Sagnac experiment. He also marks the end points of the directed segment $\mathrm{d}l'$ and the moments of the time interval $\mathrm{d}t'$ between the same events and then calculates the ratio $\mathrm{d}l'/\mathrm{d}t'$. The result is given by the Lorentz transformation:

$$c' = \frac{\mathrm{d}l'}{\mathrm{d}t'} = \frac{\mathrm{d}l - \Omega R \mathrm{d}t}{\mathrm{d}t - (\Omega R/c^2)\mathrm{d}l}. \tag{8.66}$$

Since $\mathrm{d}l/\mathrm{d}t = \pm c$, the last equation reduces to $c' = \pm c$.

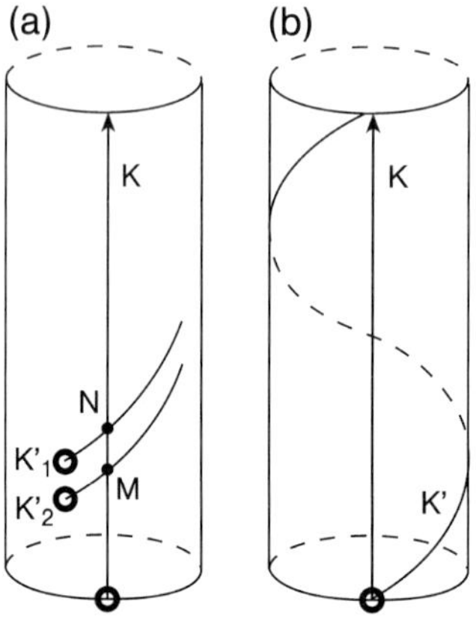

Figure 8.12 The local and global speeds of a particle measured in a rotating system K′ in a special case when the particle sits still in the stationary system. (a) The local speed is determined by the difference in readings of *pair* of clocks K'_1 and K'_2 at the intersections M and N with the world line of the particle and the spatial separation between these clocks in K′. (b) The global speed is determined by the kinematic length of the world line of a single clock K′ and the circumference of the track.

How does this relate to the Sagnac experiment? Note that all the conditions at all points of the considered circular path are the same. Therefore for each pair of close points along the path we can use inertial reference frame K, with the direction of motion only slightly tilted with respect to that of neighboring locality. Because the Lorentz transformations are linear and contain in our case the same parameter ΩR, their times dt' and distances dl' just add up algebraically, so we can apply the previous equation to an arc of finite length l' and to corresponding finite travel time t' and obtain the same result.

$$c' = \frac{l'}{t'} = \pm c. \tag{8.67}$$

On each stage of this process we can use the "integral" relation (8.67) that will always give $|c'| = c$ for both directions of the photon motion.

The situation changes radically when we apply (8.67) to the whole rim of the disk so that the end points of the arc merge together at the opposite side of the rim, producing a complete circumference. In this case we can still use formally Equation (8.67) and get the result $c' = \pm c$. But now this result does not correspond to the real physical situation. The local times of the two events in the photon's life (its departure from and return to the detector) are now measured by *the same* clock. This by itself would be OK were not the system rotating. As we have learnt in Section 8.1, in a rotating system we cannot uniquely allocate one time to a point using synchronization procedure around the closed loop. Moreover, according to the general result (8.27), we can even drop the requirement that the loop be circular and centered around the rotational axis. Any closed loop will produce the same effect of multivaluedness, with only numerical difference.

Peculiar properties of "global" time in rotating systems are accompanied by peculiar properties of space. First of all, Equation (8.65) tells us the same thing that we have mentioned in the previous section about the Earth's equator: that the rim of the rotating platform is not the same as the rim of the stationary platform. This has no practical consequences for Earth because the difference is negligible in this case, but generally the effect may be important. Due to the Lorentz factor, the circumference Λ increases with the rotational speed of the disk. The disk radius, on the contrary, does not change. Therefore the ratio of the circumference to the radius on the rotating disk is greater than 2π. What can this mean? Only one thing: the geometry of a rotating system is not the Euclidean geometry we learned in school! What we call space is now curved so that some of the axioms of the "regular" space no longer work there. Strange as this appears to be, it can still be understood if we refer to the space-time diagram. Consider such a diagram for a point P on the rotating disk (Figure 8.13). Its world line is a helix. What we perceive as space is the set of events simultaneous in a given reference frame. For the observer in this frame all three spatial axes are perpendicular to the time axis, which is the world line of a stationary particle in this frame. Unable to represent all three spatial axes on two-dimensional sheet of paper, Paul limits himself to two axes x and y in the plane of the disk. He draws them perpendicular to the world line of P. In our RF this surface is tilted with respect to Paul's world line (his time axis; it makes the angle (4.19) with this line (axis)). But

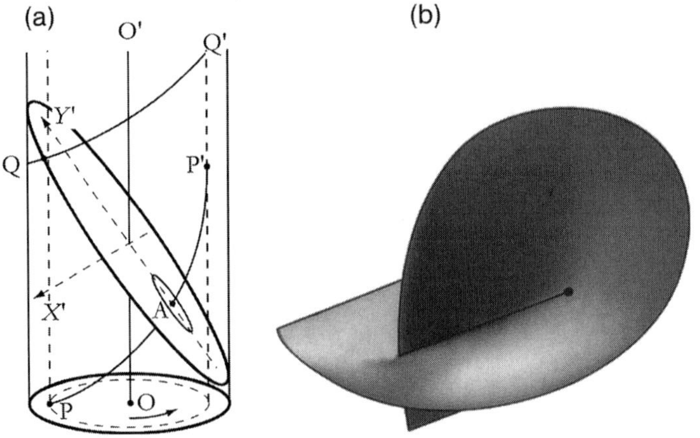

Figure 8.13 (a) The plane $X'Y'$ is perpendicular to the world line PP′ of a disk particle. Therefore its vicinity around intersection point A represents a small area on the rotating disk. However, this plane is not perpendicular to the world lines of other disk particles (for instance, line QQ′ or line OO′ of the center of the disk). Corresponding elements $\tilde{O}$ or $\tilde{Q}$ cannot represent *spatial* areas around points O and Q. (b) If we twist the plane to make it perpendicular to the world lines of all the disk particles, its geometry departs from Euclidean geometry and the surface becomes self-intersecting.

this world line is neither worse, nor better than the world line of any other point on the disk. Therefore the plane defined by the axes x, y must be also perpendicular (in Paul's RF!) to the world line of any other point on the disk. In an inertial reference frame we have no problem in satisfying this requirement, because the world lines of all the points of this frame are represented by straight lines all parallel to each other; so a plane xy perpendicular to one of them is automatically perpendicular to all the rest. Now, try to perform the same trick here together with Paul! You cannot! Because all the world lines of the particles on the rotating disk (except for that of its center) are twisted, a plane perpendicular to one of them will not be perpendicular to another (Figure 8.13a). Accordingly, the events on such a plane are not simultaneous – they do not form the space of the disk. The only way for Paul to make the plane perpendicular to the world lines of all the particles of his disk is to twist the plane accordingly, so it is no longer a plane from our viewpoint (Figure 8.13b). But it is a plane for Paul, with the distinction that it is no longer a Euclidean plane!

The considered effect of the twist is only manifest along a line of finite length. It has no effect on a local speed of light. The same is true about the whole space of the system. We define the space as a three-dimensional "hypersurface" perpendicular to the world lines of the clocks stationary in a given system. Rotation, while retaining the same structure of four-dimensional space, disrupts the topological connectivity of the three-dimensional space. But it is well known that the change in topological properties of a space (in our case – space-time) does not by itself affect its local properties. Therefore, synchronization of clocks and setting common simultaneity turns out to be impossible for the whole space if the system is rotating, but it remains always possible in any local region – in accordance with the postulate of the constancy of the local speed of light.

Applying this to the concluding part of the previous section, we see that there was no contradiction between Equation (8.61) and the initial condition that both flying clocks had the same speed. The speeds v_E and v_W found there are the "global" speeds characterizing complete cycles of motion of the clocks, whereas v is the local speed of the clocks. Both types of speed are *different* characteristics of motion in a rotating system and both can be used for the full description of such motion.

Thus a close examination clarifies the "paradoxes" associated with the Sagnac experiment and rotational motion and reveals some of its subtleties.

8.5 The Thomas Precession

The phenomenon we want to consider here is closely related to the material discussed in Section 4.4. It is intimately linked with relativity of direction. But it is manifest in its full sway in processes involving orbital motion, when a vector with preserved direction is transported along a closed path and returns to its starting point. As a mechanical model we can consider a gyroscope installed on the rim of a rotating disk. In the microworld it may be an electron orbiting a nucleus with the zero net magnetic moment, so that the only force on the electron is Coulomb's force. This force does not exert any torque on the electron, so the electron's intrinsic angular momentum (spin), and thereby its direction, is conserved. Classically, we can consider such an electron as a microscopic gyro, whose axis is being continuously transferred along its orbit. And yet, in all such cases, after one complete circumnavigation the direction of the gyro's axis as measured in the inertial system S attached to the atom turns out to be changed by a certain angle depending on the gyro's orbital speed.

Any change in direction of rotational axis of a spinning body is called precession. This effect is well known in astronomy and the best-known example is the precession of the polar axis of our planet. Another well-known case is precession of a spinning top. In all these examples, the precession is caused by external forces producing a torque on an object.

The precession we are going to discuss here has a fundamentally different origin. In contrast to the above examples, it is not caused by any torques on the spinning object. Together with other relativistic phenomena, it is the direct consequence of the relativity of simultaneity.

In any physical situation, a direction can be represented by a rod lined up with this direction. Recall our previous results from Section 4.4: two rods, moving with different velocities $\mathbf{v}_1$ and $\mathbf{v}_2$ and parallel in a reference frame K, are generally not parallel in another reference frame K′. For instance, the deck of the aircraft carrier and the floor of the rising elevator in it are parallel to each other for Mr. O'Bryan, comoving with the carrier, and not parallel for the engineer sitting on the seashore. Another aspect of this phenomenon is that orientation of a rod boosted to a velocity $\mathbf{v}$ in a process preserving its direction depends on path of the boost in the velocity space.

Preservation of direction means, in particular, that if such a path is a curved line, that is, the boost was a succession of rod transfers from one instantly comoving RF to another with a different direction of motion, the direction of the rod was maintained in each such transfer. In other words, it has to be the same in each comoving RF. And, miraculously, if we do two such successions of boosts, carefully performing the preservation procedure, by two different paths but both leading to the same final velocity **v**, the resulting directions of the rod upon achieving this velocity may be different. This leads to another astounding conclusion: if we accelerate the rod along a *closed path* in the velocity space, so that in the end it is in the same state of motion as in the beginning, its *orientation* will be *not* the same as before, even though it had been thoroughly preserved throughout the whole operation.

Consider a situation that can be relevant to this case: a rod whose center of mass is performing a uniform circular motion around a fixed center, but the rod itself is not rotating. The last statement requires careful clarification: the rod is not rotating in its own rest frame, that is, in an instantly comoving inertial frame. We can imagine the rod being transferred parallel to itself from one such RF to another. Or else, we can imagine an observer (call him Paul) sitting on the rotating disk and having a uniform rod whose center can freely rotate on a fixed support (Figure 8.14). The rod is made to point in the direction of a distant galaxy M, which happens to be in the plane of the disk. Since the whole world is observed by Paul in state of rotation around the symmetry axis of his disk, so is the galaxy. Since the rod is pointing in the direction of this galaxy, it is also rotating *as observed by Paul*. This does not contradict our condition, since this rotation of the rod is observed in the noninertial RF rigidly connected with the disk, and thus it merely reflects the disk's rotation. In contrast, in any *inertial* RF instantly comoving with the rod (at any given moment of time this is a *different* RF!), the rod is observed pointing toward the same galaxy M, and this means the same fixed direction common to all these RFs, since these frames do not rotate. But if we refer all these observations to *one and the same* inertial RF K attached to the center of the disk, the result will be different. The orientation of the rod at any given moment of time will generally not coincide with its initial orientation. If initially it

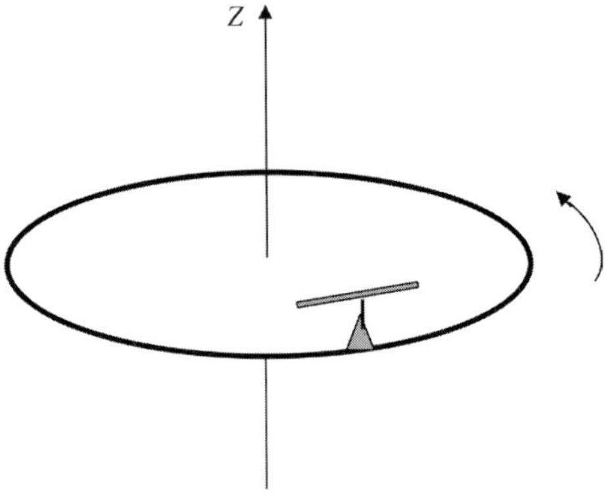

Figure 8.14

was found pointing toward the galaxy M, then upon one cycle it will, while remaining in the same plane of rotation, point to a different direction in this plane. This is a direct consequence of the results found in Section 4.4 and reformulated in the beginning of this section. When the center of the rod completes one cycle and returns to its initial position, its velocity vector accordingly completes a closed circular path in the velocity space. As a result, the orientation of the rod will change for the observer in the frame K (Figure 8.15).

Let us now describe the whole process quantitatively. In contrast with our initial approach where we considered a special case of a rod parallel or perpendicular to direction of its initial motion, here we must consider a general case where this direction is an arbitrary variable [2,78].

We want to find the change of this variable when the center of the rod traces out a circle in configuration space while the rod's orientation is kept fixed in each instantly comoving inertial RF. In other words, the rod is being continuously transported from one inertial RF to another in such a way that in each such transport the two consecutive orientations of the rod are aligned.

We will first approximate the circle traced out by the center of the rod by an n-sided polygon and then take the limit $n \to \infty$. Then the whole process can be considered as a succession of uniform motions of the rod's center along each side of the polygon (Figure 8.16).

Consider what happens in transition from one such side to the next. Let K be the rest frame of the polygon. We start with the rod moving along the bottom side of the

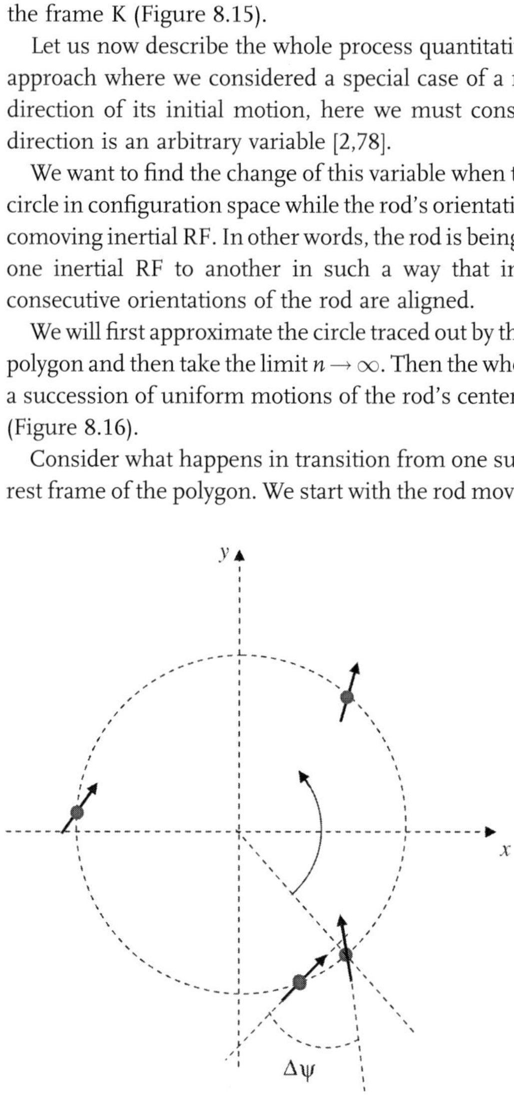

Figure 8.15 Thomas precession. Depicted is a particle with an intrinsic angular momentum (spin), orbiting around a fixed center (e.g., an electron orbiting a nucleus). After completing one cycle, the direction of spin turns out to be rotated through an angle $\Delta\psi$ even if there is no torque on the particle.

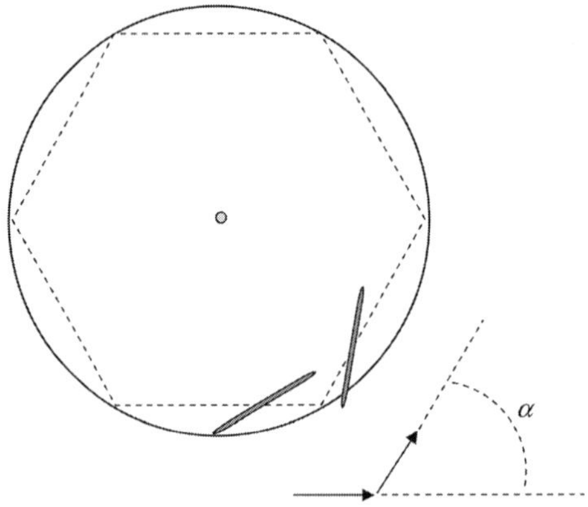

Figure 8.16

polygon (inertial system K′). Upon reaching the vertex, the rod is transported (remaining parallel to itself) to an inertial system K″ moving along the next side of the polygon. In other words, K′ and K″ are the corresponding comoving frames (or the rest frames of the rod), respectively, before and after the transport. It is convenient to represent two consecutive orientations of the same rod before and after the transition by two different rods A′B′ (rod 1) and A″B″ (rod 2). By the initial conditions, in this kind of transition (the parallel transport), the two rods must be lined up in K′. But, as we already know, their orientations may be different as observed in K (Figure 8.17).

We want now to find this difference. Just as we have done in Section 4.4, write the equation of the rod A′B′ as

$$y - y_1 = k_1(x - x_1), \qquad k_1 \equiv \tan\theta_1. \tag{8.68}$$

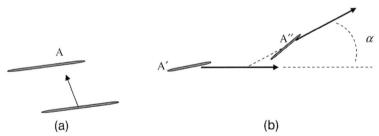

Figure 8.17 Parallel transfer of a rod between two different RFs K′ and K″. Arrows indicate the corresponding velocities. (a) Two rods representing two consecutive orientations of the same rod A (in its initial rest frame K′) before and after transition K′ → K″. The rods are parallel in K′. (b) The same rod A before and after transition as observed in the stationary frame K. The two orientations are not parallel.

Here x_1, y_1, just as x_0, y_0 in Equation (4.54), are the coordinates in K of a fixed point on the rod. We choose y_1 so that it is on the rod's intersection with the bottom of the polygon

$$x_1 = vt, \qquad y_1 = 0. \tag{8.69}$$

Equation (8.68) then takes the form

$$y - k_1 x = -k_1 vt, \tag{8.70}$$

where v is the orbital speed of the rod.

Similarly, we write the equation of the rod 2:

$$y - y_2 = k_2(x - x_2), \qquad k_2 \equiv \tan\theta_2. \tag{8.71}$$

This rod is moving along the side 2 of the polygon, therefore, if we make the point x_2, y_2 of the second rod coincident with point x_1, y_2 of the first rod at $t=0$,

$$x_2 = v_x t, \qquad y_2 = v_y t, \tag{8.72}$$

where

$$v_x = v\cos\alpha, \qquad v_y = v\sin\alpha, \tag{8.73}$$

then Equation (8.71) will take the form

$$y - k_2 x = -(k_2 v_x - v_y)t. \tag{8.74}$$

Now we express these equations in terms of coordinates of system K′

$$\begin{aligned} t &= \gamma\left(t' + \frac{v}{c^2}x'\right), \\ x &= \gamma(x' + vt'), \\ y &= y'. \end{aligned} \tag{8.75}$$

Then after some simple algebra we find that the equation of the rod 1 as observed in K′ is

$$y' = \frac{k_1}{\gamma}x'. \tag{8.76}$$

We could arrive at this equation immediately if we imagine that the rod 1 is a diagonal of the rectangle with sides along x and y, respectively, and noticing that this rectangle is Lorentz contracted in K by a factor of γ.

Now, putting the expressions (8.75) into the equation of the rod 2, we have

$$y' - \gamma k_2(x' + vt') = -\gamma(k_2 v_x - v_y)\left(t' + \frac{v}{c^2}x'\right),$$

or, after a simple rearrangement,

$$y' - \gamma\left[k_2\left(1 - \frac{vv_x}{c^2}\right) + \frac{vv_y}{c^2}\right]x' = \gamma\left[k_2(v - v_x) + v_y\right]t'. \tag{8.77}$$

Setting here, say, $t' = 0$ (i.e., considering the rod 2 at one moment of time in K′) gives

$$y' = \gamma\left[k_2\left(1 - \frac{vv_x}{c^2}\right) + \frac{vv_y}{c^2}\right]x'. \tag{8.78}$$

We are now asking, what must be the difference between k_1 and k_2 (i.e., the difference between the angles θ_1 and θ_2 in K) in order for both rods to be perfectly aligned in K′?

Setting the coefficients in the right-hand sides of Equations (8.76) and (8.78) equal to one another and using (8.73) gives the condition

$$k_1 = \gamma^2\left[\left(1 - \frac{v^2}{c^2}\cos\alpha\right)k_2 + \frac{v^2}{c^2}\sin\alpha\right] \tag{8.79}$$

or

$$\tan\theta_1 - [\gamma^2(1 - \beta^2\cos\alpha)]\tan\theta_2 = \beta^2\gamma^2\sin\alpha. \tag{8.79a}$$

Thus, we have found that if the two rods are aligned in K′ (i.e., the rod is transported from K′ to K″ parallel to itself), then they have different directions in K and the difference is described by Equation (8.79).

Next, we take the limit $n \to \infty$, that is, $\alpha = 2\pi/n \to 0$. In this limit, $\cos\alpha \to 1$, $\sin\alpha \to \alpha$, and the above equation takes the form

$$\tan\theta_2 - \tan\theta_1 = -\beta^2\gamma^2\alpha. \tag{8.80}$$

But

$$\tan\theta_2 - \tan\theta_1 \equiv \tan(\theta_2 - \theta_1)(1 + \tan\theta_2\tan\theta_1).$$

In the limit $\theta_1 \to \theta_2 \to \theta$ this gives

$$\tan(\theta_2 - \theta_1)(1 + \tan^2\theta) \equiv \frac{\tan(\theta_2 - \theta_1)}{\cos^2\theta} \to \frac{(\theta_2 - \theta_1)}{\cos^2\theta}. \tag{8.81}$$

Finally, combining this with (8.80) and redenoting α as $d\alpha$, we obtain

$$\theta_2 - \theta_1 = -\beta^2\gamma^2\cos^2\theta\, d\alpha. \tag{8.82}$$

Our next impulse is to write $\theta_2 - \theta_1 = d\theta$, but this would be wrong. The difference $\theta_2 - \theta_1$ is the change in the angle between the rod and the fixed side x' of the polygon. It is therefore equal to the change $d\psi$ of rod's orientation in the fixed RF K:

$$\theta_2 - \theta_1 = d\psi. \tag{8.83}$$

After completing the transition process we switch from side x' to x'', that is, reset the reference direction by $d\alpha$. Only after that will we have the total change of the angle θ between the rod and its velocity vector. In other words,

$$d\theta = d\psi + d\alpha. \tag{8.84}$$

We are interested in $d\psi$. Putting (8.83) in (8.82) gives

$$d\psi = -\beta^2\gamma^2\cos^2\theta\, d\alpha. \tag{8.85}$$

This is the misalignment in K between the initial and final orientations of a vector transported parallel to itself around a single vertex of a polygon at $n \to \infty$. Note again, the expression "parallel to itself" does not contradict the word "misalignment" since the former refers to the comoving frame K′, whereas the latter refers to frame K. We can talk interchangeably about two different vectors existing simultaneously or two consecutive orientations of the same vector before and after its parallel transport from K′ to K″. In either case the two vectors (or two orientations of the same evolving vector) are aligned in K′ and misaligned in K. In sharp contrast with Newtonian mechanics, alignment (or, which is the same, parallelism) of two vectors moving with respect to one another is a relative characteristic.

Dividing (8.82) by the time interval dt (the time it takes to go around one corner – to make one transfer) gives

$$\omega_T = -\beta^2\gamma^2\cos^2\theta\,\omega_{\mathrm{Or}}. \tag{8.86}$$

Here, $\omega_{\mathrm{Or}} = \mathrm{d}\alpha/\mathrm{d}t$ is the angular frequency of our rod's (or any other object's) orbital motion and $\omega_T = \mathrm{d}\psi/\mathrm{d}t$ is the angular frequency of the corresponding rotation of the rod. Since the rod just represents a spinning object, ω_T is angular velocity of spin precession.

We obtained the basic equation describing a remarkable phenomenon – the Thomas precession. It is precession of the spin of an object caused entirely by object's orbital motion, not by any torque.

Note that ω_T is an *instantaneous* angular velocity, which may be different for different points of the orbit. According to (8.86), it depends on angle θ between the spin and instantaneous orbital velocity. However, since it is proportional to the *square* of the cosine function, it has the same sign throughout the whole orbital period T. Therefore, the incremental rotations of spin accumulate rather than cancel each other out and produce a nonzero effect after each complete cycle, as depicted in Figure 8.15.

Now we can estimate the corresponding total angle of rotation $\Delta\psi$ quantitatively. Consider the most simple case of a uniform circular motion with $\beta \ll 1$. In this case $\gamma \to 1$ and Equation (8.86) reduces to

$$\omega_T = -\beta^2\cos^2\theta\,\omega_{\mathrm{Or}}, \tag{8.87}$$

with constant ω_{Or}. Since the orbital motion is slow, we have $\omega_T \ll \omega_{\mathrm{Or}}$ and the net effect must be small. We can imagine the spin vector turning through a very small angle as the position vector of the object itself makes one complete rotation through the angle 2π, that is, $\Delta\psi \ll 2\pi$. Therefore, the angle θ varies practically at the same rate as α over the whole range $0 \le \theta \le 2\pi$ as the object completes one cycle, and we will have in this approximation

$$\omega_{\mathrm{Or}} = \frac{\mathrm{d}\alpha}{\mathrm{d}t} = \frac{\mathrm{d}\theta}{\mathrm{d}t}. \tag{8.88}$$

The resulting rotation then is obtained as

$$\Delta\psi = \int_0^T \omega_T(t)\,\mathrm{d}t = -\beta^2\int_0^{2\pi}\cos^2\theta\,\mathrm{d}\theta = -\pi\beta^2. \tag{8.89}$$

The Thomas precession can be observed in some atomic spectra. It has to be distinguished from precession due to torque on the electron's magnetic moment exerted by magnetic field of the nucleus; there is also magnetic field produced by the orbital motion of the electron itself, and it interacts with the electron spin as well (spin–orbit interaction). In a hydrogen atom precession of the electron spin due to such magnetic interactions is in the opposite sense and twice the rate of the Thomas precession. Both effects are relativistic (recall that magnetic field itself is a relativistic phenomenon!), but are opposing one another. The net effect is only half of the frequency corresponding to the magnetic field alone.

In the general case of arbitrary velocities the calculations are more involved [78] and give the result

$$\Delta\psi = -2\pi(\gamma - 1). \tag{8.90}$$

It is easy to see that it reduces to (8.89) in the limit of small β. However, in the opposite limit of velocity range the Thomas precession turns out to dominate over the orbital motion: the precession frequency is much higher than the orbital frequency. Such cases can, in principle, be observed in some astrophysical objects, like a spinning satellite orbiting a neutron star or a black hole.

Problems

8.1 Suppose you are in a Disneyland.
   (a) Explain why we do not observe unusual phenomena predicted by SR for a rotating system when we are riding on a merry-go-round.
   (b) Determine the fractional change of the circumference length and the area of the merry-go-round as a function of its radius R and angular velocity Ω.
   (c) Estimate these changes numerically for $R = 5$ m and $\Omega = 1$ rad/s.

8.2 Three runners on a stadium – Ann, Bob, and Celia – start simultaneously from the same mark. Celia is running at a speed v_R, Ann is trailing her at a speed $v_R - v$, and Bob is leading at a speed $v_R + v$. Here v is the speed determined relative to Celia. The radius of the circular track is R. In nonrelativistic approximation, find the meeting time of the runners. Will Ann meet again with Celia at the same time as Celia will meet with Bob?

8.3 Consider the same situation when relativistic effects are significant.

8.4 Derive approximation (8.58) in Section 8.2.

8.5 Derive approximation (8.59) by direct calculation.

8.6 Find the offsets for $v = 300$ m/s and $v_R = \Omega_E R_E$, where Ω_E is the angular velocity of the Earth's rotation and R_E is the Earth's radius.

8.7 A flying clock departs from an airport on the Earth's equator and returns after one circumnavigation around the Earth in its equatorial plane. During circumnavigation, the clock was moving east at a constant speed of 460 m/s relative to Earth.

(a) With Earth's radius $R = 6370$ km, calculate the speed of the airport's reference clock relative to the nonrotating frame associated with the center of the Earth.

(b) Find the time offset between the proper times of the flying clock and the reference clock when both clocks are compared after circumnavigation.

8.8 The equatorial radius of the neutron star in Crab nebulae is $R = 10$ km. The rotational speed of an equatorial point of the star is $v = 250\,000$ km/s. You observe the star from your spaceship that is kept stationary just above the surface of the spinning star on its equator. Your spaceship is a part of an inertial (nonrotating) reference frame K. Neglecting gravitational effects, find

(a) The rotational period T in K and the corresponding "proper" period of the star.

(b) The local speed of your spaceship relative to the surface of the star.

(c) The corresponding "global" speed of your spaceship.

(d) Explain result (c) and the discrepancy between (c) and (b).

8.9 A flying clock departs from a point on the rim of a rotating disk of radius $R = 5$ m. The clock is moving at $v = 300$ m/s around the rim in the direction of its rotation. After one circumnavigation, the time offset between the flying clock and the reference clock sitting at the starting point of the rim was found to be $\delta = 10^{-13}$. Find the angular velocity of the disk.

8.10 You use a beam splitter to send two ultrashort laser pulses in two opposite directions around the rim of a doughnut-shaped space station of radius $R = 60$ m. After having traveled over the same distance (the circumference of the station) with the same speed the pulses return with a time lag 10^{-12} s. You conclude from this fact that your station is rotating and, being an expert in relativity, you even determine its rotation rate. What is this rate (the angular velocity of the station)?

8.11 Suppose that an E-photon is emitted at a point on the rim of a disk of radius R, rotating with an angular velocity Ω. Simultaneously, a massive W-particle is emitted from the same point at a local speed v. Assuming that both emitted particles remain on the circumference during their flight, find the condition at which the massive particle returns to the initial point earlier than the photon.

8.12 Imagine two identical disks that are touching one another and rotating about their respective parallel symmetry axes with the same angular velocity but in the opposite senses. They carry each a rod similar to that shown in Figure 8.14. At the onset of disks' rotation, both rods point in the same direction (in the fixed inertial RF K) in the plane of the disks. Will they point in the same direction when they meet again after one complete cycle?

9
Theory of Relativity is the Theory of Absoluteness

Invariants are diamonds.
Do not throw away diamonds!

E. Taylor, J. A. Wheeler, *Exploring Black Holes*

9.1
What Is Relative and What Is Absolute

All the material in the previous chapters helps better understand the actual status of the theory of relativity. In this short but important chapter we will discuss this status and the discussion will also include some questions of semantic. The semantic as such may seem something secondary. But frequently, if not properly addressed, an ambiguous semantic may cause serious misconceptions in understanding basic principles. This is especially true for such domain of human knowledge as the theory of relativity, frequently dealing with phenomena far remote from our everyday life experience.

Ironically, the word "relativity" in the name of the theory, while reflecting properly one of its most important aspects, leaves in obscurity the equally and maybe even more important aspect from which the name itself has originated.

This aspect is absoluteness!

The basic postulate of relativity, named "the principle of relativity," is that all laws of Nature are absolute – they are the same for all inertial observers (and the general relativity removes the restriction to only inertial reference frames). But in this case, which RF to board is merely a matter of choice, taste, or coincidence – not a matter of principle. It is a mere coincidence that we have studied the world basically from a RF named the Earth. Now we have an overwhelming amount of *direct* observations of the world from other RFs – trains, planes, spaceships, and so on, including an RF called the Moon. And the world operates in the same way from all of them. This implies that a state of rest or motion is a relative property of a system – it depends on choice of a RF. And it is this *relativity of states of motion*, which stems from *absoluteness of natural laws* that gave name to a theory upholding this absoluteness.

Physically, if we switch from one RF to another, all observed phenomena will run according to the same rules that we found in the first RF. This means that the basic

Special Relativity and How it Works. Moses Fayngold

ISBN: 978-3-527-40607-4

equations describing the world must be the same in both systems. The property of an equation to remain unchanged when expressed in the coordinates of another RF is called covariance. Since the transformations of coordinates, under which this covariance holds, are the Lorentz transformations, the corresponding property is called Lorentz covariance and the equations with this property are Lorenz covariant.

The current trend in physical literature is to use the term "Lorentz invariant" or just "invariant" instead of "covariant" for the corresponding equation or expression.

As we know, the covariance (or invariance) of an equation does not necessarily mean that an actual value of a physical quantity figuring in this equation is the same in both RFs. If this quantity is a variable, such as velocity, position, and so on, then its value specified, say, by a given initial condition, may be different in the two different RFs (in this case the numerical values specifying condition itself will also be different in another RF, see the example discussed in Section 2.2). A quantifiable physical characteristic whose numerical value, when measured simultaneously from two different RFs turns out to be different, is called the relative characteristic (the word "simultaneously" here has exact meaning since it refers to the two measurements made on the same object, for instance, the same car). A characteristic, whose value is the same for all observers, is called the invariant characteristic or just the invariant. The speed of light in vacuum far from lumps of matter is already known example. Another example is the electric charge of an object.

Thus, we have the world described by Lorentz covariant equations and numerous physical characteristics of the world's objects described by physical quantities or variables. Most of these quantities are relative and some of them are invariant. And there is one, which is neither totally relative nor totally invariant! We will now turn to its discussion.

9.2 The Speed of Light and the Invariant Speed

... Most modern physicists interested in SR
emphasize this invariance of the speed of light,
rather than the idea that the speed of light is a limiting speed.

C. Pickover, *Time. A Traveler's Guide*

Relativity has turned many apparently well-established notions upside down.

It has shown that time, which had always been known as the absolute quantity, is a relative quantity.

On the contrary, and ironically, the speed, which has been the best-known *relative* quantity, turned out to be only relatively relative, since it contains within its range

$$0 \le v \le \infty \tag{9.1}$$

a certain "sacred" finite value c, which is invariant!

We can express this invariance using a simple geometrical construction. Represent all theoretically possible speeds as points on a "speed axis." Then the

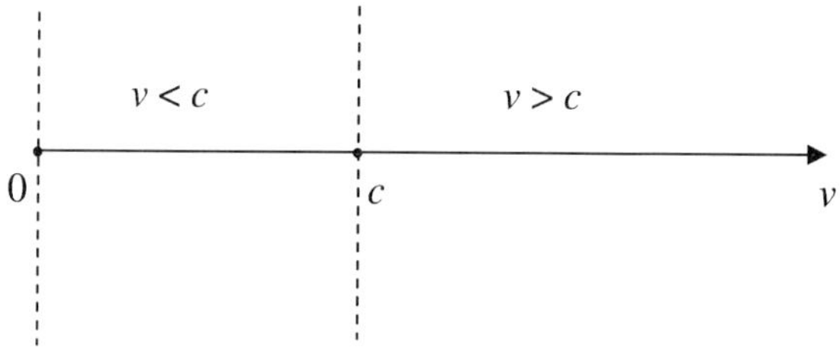

Figure 9.1 Velocity space in one dimension.

speed c will also be represented by a point on this axis (Figure 9.1). This point is singled out from the rest by its immobility. Indeed, any speed different from c changes under the Lorentz transformations along the direction of motion and therefore the point representing such a speed changes its position on the axis. However, the point corresponding to the speed c remains in place. This can be considered as a geometrical representation of the physical fact – the invariance of the speed of light in vacuum.

From this representation one can deduce another important property of the speed of light. As we have found from the rule of addition of velocities, no speed less than c can be made equal to c. If at some moment of time an object is moving slower than light, then no matter how many Lorentz transformations we apply (no matter how hard and how long we accelerate the object), the representing point can approach the point "c" infinitesimally closely, but will never merge with it. (Otherwise, one could by a succession of reverse Lorentz transformations change the position of the point "c," which would contradict its basic property – immobility.) It follows that the value of "c" is not only a fundamental constant, but also the unattainable limit for all other speeds. The speed of light turns out to be a barrier that not only cannot be crossed, but also it cannot even be attained by any object with a nonzero rest mass.

More generally, we can consider velocity rather than just speed, and represent it as a point in a 3-velocity vector space (Figure 4.18). In this case, an object moving in an arbitrary direction with a speed c will be represented by a point C on a spherical surface of radius c. Any subluminal object (i.e., an object moving at $v < c$) will be represented by a point P within this surface. A superluminal object (an object moving at $v > c$) will be represented by a point $\tilde{P}$ outside the surface. In all three cases, the position of a point is uniquely specified by its radial distance v from the origin and its two angular coordinates θ and φ for its polar angle and azimuth, respectively. A Lorentz transformation representing the switch to another RF will generally change both – radial distance and angular coordinates of the point. However, a point originally inside the sphere will always remain inside and a point outside will always remain outside. A point C on the surface itself ($v = c$) can only change its angular position but not its radial coordinate and will thus remain on the surface. In other words, any Lorentz transformation or their combination maps the interior of the sphere onto itself and its exterior onto itself; the spherical surface is mapped onto itself. Thus, this surface can be considered as an impenetrable boundary between subluminal and superluminal worlds. This boundary is absolute – it has the same shape and size for all observers.

Will it then be legitimate for us (the subluminal observers!) to ask what is there beyond the boundary? Does it make sense to discuss superluminal velocities?

We will consider these questions in Chapter 11.

Now I want to discuss the invariant radius of the sphere – the fixed "solid rock" in the ocean of elusive relativity.

The invariance of the value $v = c$, which forms one of the foundations of relativity, is frequently expressed in the formulation: "The speed of light does not depend on speed of its source."

This formulation is technically correct, but it may be conceptually misleading. It strongly implies, but does not explicitly require, the absence of ether. If a beginning reader takes this formulation literally, within the narrow domain of its meaning, he or she can miss the point. Indeed, in fluid dynamics, the speed of sound in a medium does not depend on speed of its source either. What, then, is the difference between the acoustic wave in a medium and the light wave in vacuum?

According to nonrelativistic physics, there was no essential difference between them, since light was also believed to be a wave in an elastic medium – ether, and its speed, while being independent on state of motion of its source, was believed to depend on state of motion of an observer. According to relativity, there is no ether and the speed of light in vacuum does not depend on state of motion of the observer. Thus, we come to another formulation:

The speed of light in vacuum is the same for all observers.

This is a far better formulation. It answers the above question and it includes the previous formulation involving *the source* of light. If one observer is sitting on the source of light, while another one is moving relative to it, then the source of light is moving relative to the latter observer and the second formulation automatically includes this case, while avoiding any reference to a source. At the same time, it is much more general since it includes the cases when both observers are moving (although differently) relative to the source. One might argue that independence of the source's motion is sufficient to get the same value for the second observer. Strictly speaking, this is not necessarily so. Imagine a world in which one observer measures the speed of light equal to c_1 regardless of the motion of the source and another observer moving with respect to the first one will measure the speed of light equal to $c_2 \neq c_1$, also regardless of motion of the source. It is true that in this world, the speed of light might also depend on direction. But even so, for each fixed direction, the speed of light may be different for different observers, while being, for each of them, totally independent of motion of the source. In this world, we may have an infinite set of different speeds of light, one for each observer, each depending also on direction, but each one, for each fixed direction, totally independent of motion of the source. And if, in addition, we take into account dispersion, then even for a fixed direction, the speed of light would also depend on its frequency, while, for a fixed frequency and direction, remaining totally independent on motion of its source. What a nightmare! But this nightmare is precisely the description of acoustic waves in a medium (Problem 9.1). The speed of acoustic wave is independent of speed of the source, but it is *not* invariant. If we replace such a wave with light, the first formulation, within its semantic domain, will leave the nightmare for light as well.

The second formulation eliminates this nightmare. It is more general and conveys the main point – the *invariance* of the speed c.

Discussion

Having thus promoted the second formulation, I want now to present an argument against it.

The point of this argument is that the expression "the speed of light in vacuum" in the second formulation should be replaced with the expression "there exists an invariant speed c." In other words, it would be better to eliminate any reference to light in relativistic postulates.

If we insist on use of the word "light," then, strictly speaking, the second formulation is generally wrong. There are cases when the light is moving in vacuum and yet its speed is *not* the same for all observers. Automatically, it is not equal to c for any observer. And these cases are well known, they are taught in colleges to students majoring in electrical or computer engineering, let alone in physics. This is a phenomenon known as the evanescent light, occurring, for instance, in total internal reflection of light incident from glass onto the glass–air interface. In this effect, even though all the incident light energy is reflected back into glass, there is a light wave penetrated into air too. It carries energy *along* the interface and exponentially decays with distance from it. We have studied this effect in some detail in Section 6.12. Here, it is sufficient to say that the phase velocity of such wave is less than c. So if we just pump the air out, then here we are dealing with light spreading along the interface but on its vacuum side, that is, just in vacuum. And its speed is *not* equal to c. And accordingly it is *not* the same for all observers.

We could argue that this is actually not a free light wave since it is bound to the surface. This is true, but, on the other hand, it retains potential for becoming free again. For instance, if we take another piece of glass and put it close to the first interface and parallel to it, the evanescent wave will resurrect in the second glass as a full-fledged plane light wave (Figure 9.2a). This effect is known as frustrated total internal reflection. And if, in addition, we make the second glass wedgelike, then the resurrected wave will reemerge in vacuum behind it as a free plane wave (Figure 9.2b).

There is another way for an evanescent wave to become an exact free wave, which does not even require the presence of a second glass. Just let it go along the surface until it reaches the corner (Figure 9.3), which eventually always happens in the real world. At the corner, in order to remain attached to the glass surface, the evanescent wave has to make a sharp turn and, ironically, precisely by doing this, it detaches from it. The mechanism of detachment is another beautiful example of relativity at work and can be intuitively understood if we note that, as the wave makes a turn, its translational motion converts into rotational motion, with the corner as a rotational axis. At a certain distance from the axis the phase velocity of the wave reaches the value of c and starting from this point it becomes free wave – plane, cylindrical, or

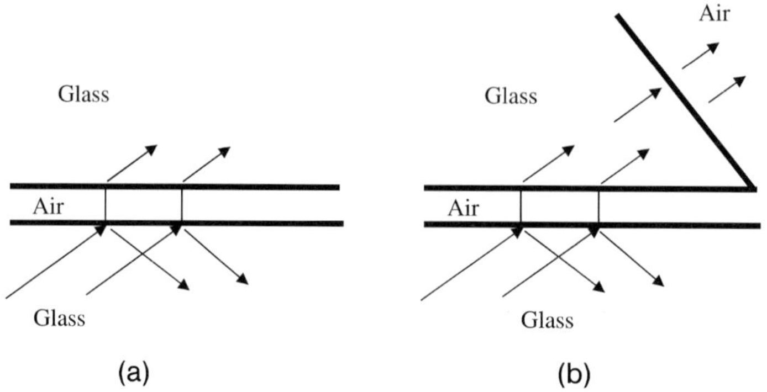

Figure 9.2 Schematic for conversion of evanescent light (in air) into "regular" light (a plane wave) (a) in the second glass or (b) in air again. The arrows represent directions and intensities of the corresponding plane waves.

spherical depending on geometry of the experiment. This is another case of the effect discussed in Section 6.11.

In view of these phenomena, once the evanescent wave retains the potential to become a free wave again, we should rather exempt the evanescent light by rephrasing our formulation to something like this: "Light in vacuum very far from all material objects moves with speed that is the same for all observers." But here we can ask, how far should it be to make "very far?" After all, we carry out our experiments while surrounded by material objects. And in the vicinity of the so-called critical angle, the evanescent wave may decay so slowly that it will be practically indistinguishable from the plane wave. Since the concept of a plane wave is itself only

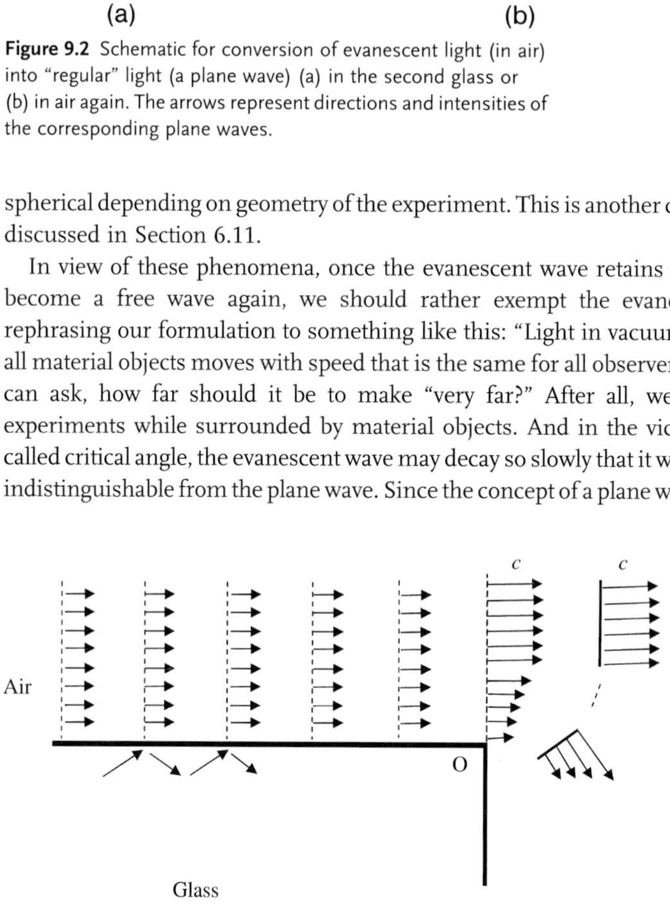

Figure 9.3 Another possible mechanism of conversion of evanescent wave into a plane (or diverging) wave. Arrows indicate velocities. The wave right above the glass is the evanescent wave, moving to the right slower than light, with its amplitude exponentially decaying along the vertical direction. When going around the comer O, the wave front acquires rotational component to its motion leading to the increase of the phase velocity. At a certain critical distance from O the velocity reaches the value of c, at which point the motion of farther areas of the front becomes again purely translational (more accurately, radial, diverging from O). This part of the front is detached from the glass and propagating with the speed c.

an idealization, even the revised second formulation may be a rather slippery ground. It is especially dangerous in view of the fact that the concept of light waves in free space may be an extraordinarily accurate approximation in all respects *but its speed*. For, even though its speed $\beta = v/c$, say, in an evanescent state, may be short of 1 by a ridiculously minute value 10^{-100}, it can be Lorentz transformed to zero, that is, such light can be stopped in the same free space by a succession of boosts. The difference $1 - \beta = 10^{-100}$ is not good enough for such a wave to be a good plane wave in vacuum! In terms of Figure 9.1, just a proximity of a point to c, no matter how close, is not sufficient for such a point to become the "sacred" one.

There may be other situations, when light in vacuum is not attached to any boundary and yet its speed is significantly different from c. Recall, for example, the case considered in Section 5.5 of a single photon in a superposition of two different states of motion. To such a photon, we cannot even assign a definite velocity – we can only speak about its expectation value. The photon in such state has a nonzero rest mass and accordingly can only be characterized by an average speed, which may, depending on conditions, have any value between c and zero.

Therefore, it would be much simpler to avoid any reference to light and just say: "According to relativity, *there exists an invariant speed, with numerical value c.*"

This invariant value c is a fundamental constant of Nature. Of course, its experimental measurement does involve light, but, as any measurement, it also involves an unavoidable, although very small, experimental error. An invariant speed, as a theoretical concept, emerges as a generalization, or, in a way, a sublimation of experiments and observations on which it is based.

The last formulation would be the best formulation of the corresponding relativity postulate. It conveys its true meaning, while avoiding possible ambiguities associated with mentioning light. All the more so that gravitational waves and, most probably, neutrinos, as well as possibly some other yet undiscovered particles can also claim the same status as light.

Thus, light, which played such an important role in the development of the relativity theory, is demoted by relativity to one of the objects of Nature that can, under specified conditions, have the invariant speed.

Therefore, in many cases throughout this book, I use the term "the invariant speed" instead of "the speed of light in vacuum."

9.3
Space-Time Intervals

Another important invariant that we have studied is, of course, the space-time interval. Here, we only rehash some of its properties, emphasizing some of the possible pitfalls.

An interval is an invariant characteristic of a pair of events in space-time. Geometrically, it is the norm of the 4-vector connecting these events. Recall that intervals fall into three distinct categories: timelike, null or lightlike, and spacelike, depending on sign of the vector's dot product with itself. Consider first a timelike interval

$$s_{12}^2 = c^2(t_2 - t_1)^2 - \mathbf{r}_{12}^2 > 0. \tag{9.2}$$

Since the value of s_{12} is absolute, we can use *any* RF for its calculation. Naturally, we choose one where the corresponding expression is the simplest possible. And this is a RF where the spatial part $\mathbf{r}_{12} = 0$. Physically this means that it is most convenient to observe such an interval from a RF where both events forming it happen in one place. Accordingly, both events are on the temporal axis of this RF. Setting in (9.2) $\mathbf{r}_{12} = 0$ gives

$$s_{12} = c(t_2 - t_1). \tag{9.3}$$

The value of the interval is uniquely determined by the time lapse between the events in this RF. The amount of this lapse,

$$\tau_{12} = t_2 - t_1 = \frac{s_{12}}{c}, \tag{9.4}$$

is called the *proper time* between the events.

By its definition, the proper time is a relativistic invariant. If we are interested in the proper time between two events separated by a timelike interval, but for whatever reason cannot board the corresponding RF for direct measurement, we can calculate it as the absolute value (the norm) of the interval in a system where the two events are observed:

$$\tau_{12} = \frac{1}{c}\sqrt{(t_2 - t_1)^2 - \mathbf{r}_{12}^2}. \tag{9.5}$$

Suppose now that the two events are separated by a spacelike interval, that is,

$$s_{12}^2 = c^2(t_2 - t_1)^2 - \mathbf{r}_{12}^2 < 0. \tag{9.6}$$

Then we can find a RF where both events happen simultaneously and the interval reduces to pure distance r_{12} between them. In this RF, we have $t_2 = t_1$ and get

$$r_{12} = |\mathbf{r}_{12}|. \tag{9.7}$$

The distance r_{12} is called the *proper distance* between the two events. Generally, the proper distance is the norm

$$\lambda_{12} = \sqrt{r_{12}^2 - c^2(t_2 - t_1)^2}. \tag{9.8}$$

Note that the *proper distance* between two events is generally *not* the same as the *proper length* of an object whose end points happen to be respectively coincident with these events. Consider a solid rod of constant proper length l_0. If you are in the rest frame K_0 of the rod and you want to measure its length, you can do it by first marking its end points. And it is *not* necessary that you mark them simultaneously in K_0. You can mark one end now (at a moment t_1) and the other end later (at a moment $t_2 > t_1$) in K_0 and then quietly measure the distance between the marks. We can even consider such measurement as a possible operational definition of proper length. From the viewpoint of the experimental physics, the requirement that the marks be made simultaneously is redundant for a stationary object with constant shape and size and can in this case be removed from such definition. Since the rod is stationary in K_0, the distance between the marks is the proper length of the rod regardless of the time lapse between the two markings. On the contrary, it is *not* the proper distance between the marking events if the marks are not made simultaneously in K_0. Moreover, we can make the time separation between the markings so big that the corresponding interval becomes the timelike interval. Consider a stationary rigid rod of 1 m proper length. In its rest frame, we mark its left end now and its right end 1 million years from now. The interval between the markings is definitely timelike, in which case it cannot even be assigned a proper distance. There is no such thing as a proper distance for a timelike interval. And yet we can measure the distance between the marks and, low and behold, we obtain exactly 1 m, which is, according to definition, the proper length of the rod.

Consider now a reciprocal situation: let the rod move with velocity $\mathbf{v}$ relative to a reference frame K and let its length be along the direction of its motion. We want to measure the proper length of the rod while remaining in K. We can do it, again, by marking the end points of the rod and then measuring the distance between the marks; but now, since the rod is moving, it is absolutely essential that the instant positions of its leading and trailing edges be marked simultaneously in K. In this case, the spatial separation $r_{12} = l$ between the marks is, by definition, the *proper distance* between the marking events, but it is *not* the proper length of the rod. Indeed, the described procedure constitutes the length measurement of the *Lorentz-contracted* rod; the rod's *proper* length l_0 (assuming its speed is known) is obtained by multiplying the measured distance l by the corresponding Lorentz factor: $l_0 = \gamma(v)l \neq r_{12}$.

Thus, the proper distance and the proper length are both relativistic invariants. But they, generally, describe different characteristics of a process or an object. The proper distance relates to a pair of events in space-time, which are connected by a spacelike interval; the proper length describes geometrical properties of a material object observed in its rest frame.

The two invariants become identical in some special cases. Consider, for instance, an expanding balloon, so that its diameter is increasing with time. Suppose that you observe this process from the rest frame of the balloon, that is, from the inertial frame attached to its center of mass. You see two ants sitting quietly at the end points of the diameter. Then suddenly the ants start running. When you process your data, you find that both ants started running simultaneously in your RF. In this case, the proper distance between these two events is also the instant proper size of the balloon.

We can use the invariance of an interval for the most simple description and resolution of the twin paradox (Section 7.8). Recall that the paradox appears when we try to compare the proper times of two twins, of which one is stationary, and the other travels back and forth. Since the proper time is just the norm of the corresponding interval measured in units of time, we can describe this process in terms of the two respective paths (timelike world lines) taken by the twins between the events of their parting and meeting. This allows us to easily generalize the process to include the case when the traveling twin moves along an arbitrarily curved world line (Figure 4.20). We know from Section 5.7 (Equation (5.144)), as well as from the analysis of the twin paradox (Section 7.8), that this twin's motion is not free – he must be subject to an external force causing his world line to curve. Consider now the difference in the two world lines in terms of their kinematic lengths rather than in terms of their 4-velocities. We will prove that the length of the straight world line is greater than length of any other world line connecting the same points. In terms of proper time: the proper time of the clock moving with constant 4-velocity is longer than the proper time of a clock moving with varying 4-velocity between the same events. In other words, *the path a free object takes between two events in space-time is the one for which the object's aging is maximal.* This is known as the principle of extremal aging [79]. It is totally equivalent to the principle of the least action considered in Section 5.7 and to the earlier result found in Section 4.6.

The proof is very easy if we utilize the fact that the norm of an interval between any two events is relativistic invariant. Since the total kinematic length of an arbitrarily curved path is the sum of the corresponding small intervals, which are invariants, the whole length is absolute as well. Consequently, it is independent of choice of RF in which it is measured or calculated. Therefore, we can choose the one in which the calculation is the simplest possible. In our case, it is the rest frame of the freely moving twin (and accompanying clock). According to (9.3), its kinematic length in this frame takes the form

$$s_{MN} = \int_M^N ds = c\int_M^N dt = c\,\tau_{MN}. \tag{9.9}$$

The kinematic length of the second twin traveling between the same events along a curved world line with a variable speed is given by

$$\tilde{s}_{MN} = \int_M^N d\tilde{s} = c\int_M^N \sqrt{1-\frac{v^2(t)}{c^2}}dt \quad < s_{MN}. \tag{9.10}$$

No matter what is the dependence $v(t)$, the curved path followed by the second twin is shorter than s_{MN} and, accordingly, he ages less than the uniformly moving or stationary twin:

$$\tilde{\tau}_{MN} < \tau_{MN}. \tag{9.11}$$

Summary

Here are the rules the Nature uses for sorting out and executing actual motion from all logically possible ones. Out of all possible paths between the two points in space-time, an object follows the one requiring the least possible action. The path requiring the least possible action has the greatest kinematic length and takes the longest proper time. The more twisted your path in space-time is, the slower you age. Hence, the relativistic prescription for those who want to remain young: do not stay in one place. Get subject to varying forces and move in a wiggling line!

9.4
Relativistic Doppler Effect II

The material of this section provides us with another illustration of how special relativity works in describing phenomena associated with nonuniformly moving objects.

... I am sitting on an *arbitrarily* moving source S (say, a laser gun) emitting light with frequency ω. This source emits a photon. Since the photon's frequency is measured in the rest frame of the source S (with recoil negligibly small), we denote it $\omega = \omega_S$.

My partner is sitting on a detector (system D) arbitrarily moving relative to the source. Its instant location at any moment t is determined by position vector $\mathbf{r} = \mathbf{r}(t)$. A time t after the emission, the photon is recorded by the detector. From my viewpoint, the recorded frequency must be determined by the detector's velocity relative to me *at the moment of detection.*

From my partner's viewpoint, the detected frequency was determined by the relative velocity between us *at the moment of shooting.*

Who is right? Both are, because both events – that of emission and that of detection – are equally involved in the process and the outcome is determined by the relative velocity between S and D at the respective moments of time. This can be illustrated by a space-time diagram with S and D in arbitrary (not necessarily uniform!) states of motion (Figure 9.4). It is clearly seen from the diagram that both observers say the same thing in different languages. The relative velocity they mention is a relative velocity between S at t_S and D at t_D. Both individual velocities can be evaluated in some inertial RF K at the corresponding moments of time in this frame. From this

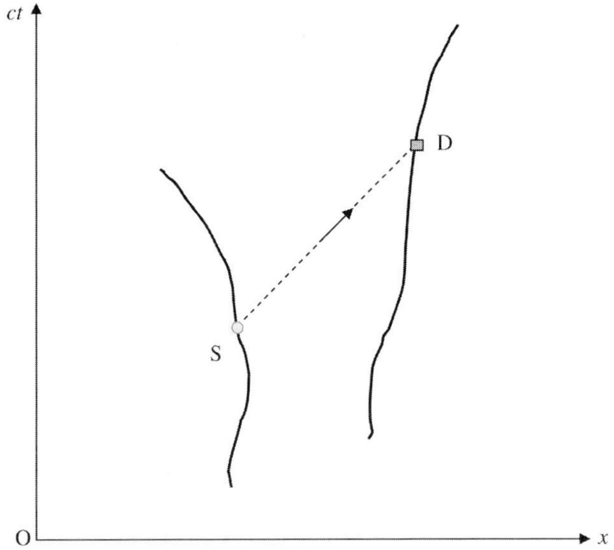

Figure 9.4

RF, we see that the moments of emission and detection are actually the end points of the world line of the photon connecting these events. Thus, the corresponding operational procedure determining it would be, first, choosing an arbitrary inertial RF K; second, introducing an auxiliary *inertial* RF K_S instantly comoving with system S at the moment t_S of K-time; third, introducing another auxiliary *inertial* RF K_D instantly comoving with system D at the moment t_D of K-time; and finally, calculating the relative velocity between the auxiliary frames K_S and K_D. If we have found $\mathbf{v}_{rel}(t_S)$ in the instant RF of the detector (frame D), then we will have $\mathbf{v}_{rel}(t_D) = -\mathbf{v}_{rel}(t_s)$ in the instant rest frame of the source (frame S). In other words, when I say that the recorded frequency is determined by the relative velocity of my partner at the moment of detection, it implies relative velocity with respect to me. But since my motion is generally not uniform, this relative velocity with respect to me must be specified by my instant velocity at the moment of emission.

Because I refer all the events to my reference frame S, the possible time dependence of its velocity in K is totally obscured in my mind. And it appears to me that it is only my partner's state of motion that determines the frequency of the photon he records.

Similar reasoning shows that for my partner, the frequency of the photon recorded by him is determined entirely by the motion of the source at the moment of emission. This can be considered as another aspect of the clock paradox, when *both* clocks are in accelerated motion.

In such situation, the best way to describe the process and find the relation between the two frequencies ω_S and ω_D is to find an *invariant characteristic* of the process and express the observed frequencies in terms of this characteristic.

It turns out that such an absolute characteristic does exist. It is a dot product $up = u_j p^j$ of two 4-vectors: 4-velocity u of an object and 4-momentum p of the photon. We have

$$up = u^0 p^0 - \mathbf{u} \cdot \mathbf{p}. \tag{9.12}$$

Recall that

$$u = (u^0, \mathbf{u}) = \gamma(v)\left(1, \frac{\mathbf{v}}{c}\right), \tag{9.13}$$

where $\mathbf{v}$ is the velocity of the object (Equation (5.5b)), and

$$p = (p^0, \mathbf{p}) = \hbar\left(\frac{\omega}{c}, \mathbf{k}\right) = \frac{\hbar\omega}{c}(1, \hat{\mathbf{n}}), \tag{9.14}$$

where ω is the photon's frequency, $\mathbf{k} = (\omega/c)\,\hat{\mathbf{n}}$ is its wave vector, and $\hat{\mathbf{n}}$ is the unit vector along the propagation direction (Equations (5.22), (5.31)–(5.34)). Combining all this, we obtain an invariant expression

$$up = u^0 p^0 - \mathbf{u} \cdot \mathbf{p} = \frac{\hbar\omega}{c}\gamma(v)\left(1 - \frac{\mathbf{v} \cdot \hat{\mathbf{n}}}{c}\right). \tag{9.15}$$

It is very precious, indeed. As an invariant, it has the same value regardless of the reference system used. Therefore we can use the one in which the expression (9.15) is the simplest and, accordingly, its physical meaning is immediately seen. This is the rest frame of the object. In this frame, $\mathbf{u}=\mathbf{v}=0$, $\gamma(v)=\gamma(0)=1$ and the expression (9.15) reduces to

$$up = u^0p^0 - \mathbf{up} \Rightarrow \frac{\hbar}{c}\omega_R. \tag{9.16}$$

Thus, the dot product of 4-velocity of an object and 4-momentum of a photon is, up to a constant coefficient $\hbar/c$, the photon's frequency $\omega=\omega_R$ *in the rest frame of the object.*

What is it good for?

It enables us to compare the frequencies of the same photon as observed from different reference frames. Because such frequency can be encoded in an expression that is an invariant, we can apply it to any moving object at any moment of time, regardless of how this object had been moving before or is going to move after this moment. In other words, we can apply it to an *arbitrarily* moving object. This means that we can use it to answer the question posed in the beginning of this section, without using coordinates, their transformations, and so on.

In our case, the two different RFs (two related objects) are the source and the detector, respectively, connected by the photon's world line. We want to compare the photon's frequency ω_S in the rest frame of the source at one end of this line to its frequency ω_D with respect to detector at the other end of the line. All we need to do is to read Equation (9.16) backward and apply it twice – to the object S (the source, $\omega_R=\omega_S$) and to object D (detector, $\omega_R=\omega_D$), respectively. Using (9.15), we obtain

$$\frac{\hbar}{c}\omega_S = u'_S p = \frac{\hbar\omega}{c}\gamma(v'_S)\left(1-\frac{\mathbf{v}'_S\cdot\mathbf{n}}{c}\right) \tag{9.17}$$

and

$$\frac{\hbar}{c}\omega_D = u_D p = \frac{\hbar\omega}{c}\gamma(v_D)\left(1-\frac{\mathbf{v}'_D\cdot\hat{\mathbf{n}}}{c}\right). \tag{9.18}$$

Since the 4-velocity u of an object is generally a function of time, in any RF we have to evaluate it at the moment of time specifying the corresponding end point of the photon's world line. Thus, in (9.17) we evaluate the dot product at the moment of emission and in (9.18) we evaluate it at the moment of detection. Hence the primes in (9.17) – they indicate that we, as has already been mentioned above, evaluate v_S and accordingly u_S at the retarded moment of time when the photon was emitted by the source. But in either case, the corresponding expression is relativistic invariant. Since we want to compare the values of these two invariants, let us consider their ratio

$$\frac{\omega_D}{\omega_S} = \frac{u_D^0p^0 - \mathbf{u}_D\cdot\mathbf{p}}{u_S^0p^0 - \mathbf{u}_S\cdot\mathbf{p}} = \frac{\gamma(v_D)[1-(\mathbf{v}_D\cdot\hat{\mathbf{n}}/c)]}{\gamma(v_S)[1-(\mathbf{v}_S\cdot\hat{\mathbf{n}}/c)]}. \tag{9.19}$$

This ratio is, of course, also an invariant. Therefore, the expression (9.19) will give the same result regardless of the RF used for its calculation.

Now we have a powerful mathematical tool to answer the questions posed in the beginning of the section. Let us start with the trivial situation, when both – the detector and the source – are inertial and stationary with respect to one another. In this case, the recorded frequency must be the same as the emitted one. Indeed, using the common rest frame of both devices as the computational basis, we have $\mathbf{v}_S = \mathbf{v}_D = 0$ and $\gamma(v_S) = \gamma(v_D) = 1$, and we obtain, as expected, $\omega_D = \omega_S$. We will obtain the same result if we use any other inertial RF, not necessarily the rest frame of the system. In any such RF, both – the source and detector – are moving, but since their state of motion is the same, we still have $\mathbf{v}_S = \mathbf{v}_D$ and the fraction on the right in (9.19) is equal to one. Thus, a resident of such moving RF will be able to predict the outcome of the corresponding experiment performed in the *rest* frame of the system.

Consider now the most general case – the source and the detector are each in an *arbitrary* state of motion and the states are different. Take the reference frame of the detector (frame D) as our computational basis. In this frame $v_D = 0$, $\gamma(v_D) = 1$, and the Equation (9.19) immediately gives

$$\omega_D = \frac{\omega_S}{\gamma(v'_S)\left[1 - (\mathbf{v}'_S \cdot \hat{\mathbf{n}}'/c)\right]}. \tag{9.20}$$

This is a famous expression describing relativistic Doppler effect.[1)] The primes here are to emphasize that we must evaluate in frame D the instant velocity of the source (and the corresponding direction from the source to the detector) at the retarded moment of the photon emission (Problem 9.9).

Consider a few special cases.

I. The source is moving in a straight line directly toward or away from the detector (the "longitudinal" Doppler effect). In this case $\mathbf{v}'_S \cdot \hat{\mathbf{n}} = \pm v_S$ and Equation (9.20) reduces to

$$\omega_D = \frac{\omega_S}{\gamma(v'_S)\left[1 \mp (v'_S/c)\right]} = \omega_S\sqrt{\frac{1 \pm \beta'_S}{1 \mp \beta'_S}}, \qquad \beta_S \equiv \frac{v'_S}{c}. \tag{9.21}$$

(Very interesting! The *longitudinal* Doppler effect for a photon observed in two different RFs is described by the same equation as the *transverse* force in two such frames (Equation (5.91)).

II. The source is moving past the detector and emits a photon from the point of the closest approach exactly toward the detector (Figure 9.5a). In this case $\mathbf{v}'_S \cdot \hat{\mathbf{n}}' = 0$ and we obtain

$$\omega_D = \frac{\omega_S}{\gamma(v'_S)}. \tag{9.22}$$

When the line of sight (the path of the incoming photon) is perpendicular to the trajectory of the source, the detected photon's frequency is decreased by the

1) There are a few different ways to derive the relativistic expression for the Doppler effect for the case of *inertial* motion of S and D. All of them are pretty straightforward in this special case (see Section 6.10 and [1]). The relativistic invariants used here demonstrate the ability of SR to describe phenomena in *non* inertial RF.

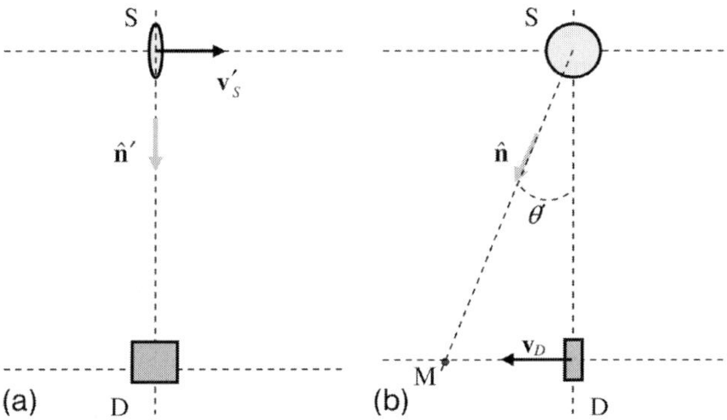

Figure 9.5 (a) Case II observed in system D. (b) The same case observed in system S. In this system, the angle θ is such that the photon must hit the detector at point M.

corresponding Lorentz factor. This is a direct manifestation of the time dilation (TD) effect: the oscillations in the *moving* source, which are responsible for the emitted light, are slowed down in the rest frame of detector as compared with their *proper* frequency ω_S measured in the rest frame of the source. That is, again, a moving clock ticks slower! Since in the optical spectroscopy (area of optics studying the light emission and absorption by matter) the frequency decrease is associated with the shift of the corresponding color to the red end of the spectrum, it is often referred to as the redshift.

In order to illustrate the invariance of the expression (9.19), let us consider the same situation with the computational basis switched from system D to system S. In system S, the detector is moving uniformly past the stationary source, which emits a photon at the moment of the closest approach in such a direction that the photon will hit the detector, as does a sniper aiming ahead of a flying target in order to hit it (Figure 9.5b). Setting in (9.19) $v_S = 0$ gives

$$\omega_D = \omega_S \gamma(v_D)\left(1 - \frac{\mathbf{v}_D \cdot \hat{\mathbf{n}}}{c}\right). \tag{9.23}$$

On the face of it, the result seems quite different from (9.22). But examining Figure 9.5b we see that

$$\mathbf{v}_D \cdot \hat{\mathbf{n}} = v_D \sin\theta' = \frac{v_D^2}{c}, \tag{9.24}$$

so the expression in parenthesis in (9.23) reduces to $\gamma^{-2}(v_D)$ and, since $\mathbf{v}_D = -\mathbf{v}'_S$, the result is identical to (9.22).

III. The source is on the rim of a disk of radius R, spinning with angular velocity Ω. The detector is at the disk's center (Figure 9.6a). In this case, the outcome can be immediately predicted from the initial version of situation II. Recall that in the Doppler effect only an instantaneous velocity of the source at the moment of emission, *not* acceleration of the source, determines the frequency of the

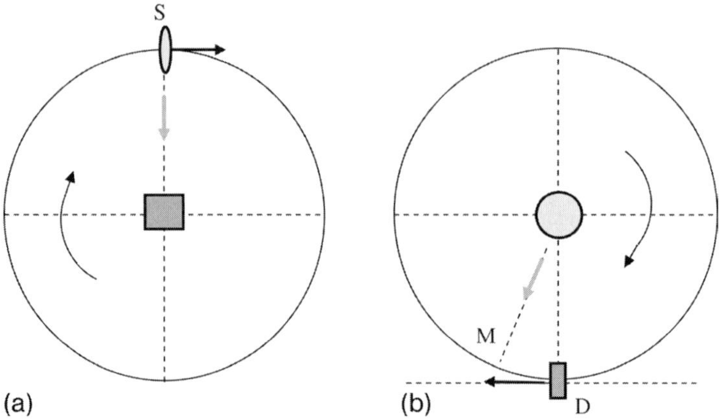

Figure 9.6 Doppler effect on a spinning disk. (a) Observed in system D; (b) observed in system S. In this system, the angle θ is such that the photon must hit the detector at point M.

corresponding detected photon. Therefore, it is immaterial whether the source keeps on going with the same velocity or orbits around the detector after the emission. The detected frequency is the same as in case II. Moreover, this time we do not even have to bother about how to determine the retarded moment t': the light emitted toward the center at *any* moment from *any* position of the source makes 90° with its instantaneous velocity at that moment. Therefore, we can drop "the prime" label in this case. Also, we can now express the detected frequency in terms of the angular velocity of rotation, so the result (9.22) can be written as

$$\omega_D = \frac{\omega_S}{\gamma(\Omega R)}. \tag{9.25}$$

But while the result is the same, its interpretation is quite different for the disk's resident Paul: since Paul is spinning together with the disk, the source is stationary with respect to him and Paul cannot interpret the decrease in the observed frequency as the time dilation due to motion of the source. So how will Paul explain the observed effect? We will postpone its discussion until the last chapter.

IV. Paul was very intrigued by the fact that the spectral lines of all chemical elements from the sources positioned off-center are redshifted with respect to their corresponding *proper* frequencies. So he decided to inverse the experimental conditions: to relocate with his detector a distance r away from the center and observe the frequencies from the source positioned at the center (Figure 9.6b). When he completed the measurements, all the measured frequencies turned out to be blueshifted with respect to the corresponding proper frequencies. This result is also predicted by our general expression (9.19). We can see it immediately if we again utilize the relativistic invariance of this expression – its independence on RF. This time the most convenient choice of RF is the rest frame of the source. Setting in (9.19) $\mathbf{v}_S = 0$ gives

$$\frac{\omega_D}{\omega_S} = \gamma(v_D)\left(\frac{1 - \mathbf{v}_D \cdot \hat{\mathbf{n}}}{c}\right). \tag{9.26}$$

This expression is identical to (9.23), but the result it produces is different. The difference is in the fact that now the detector is moving in a circle (Figure 9.6b) rather than along the tangential to it; and, since all the photons from the source are moving in radial directions, we have at the moment of detection $\mathbf{v}_D \cdot \hat{\mathbf{n}} = 0$ and Equation (9.26) reduces to

$$\omega_D = \omega_S \gamma(v_D) = \omega_S \gamma(\Omega r). \tag{9.27}$$

Now the detected frequency is *greater* than the proper frequency by the Lorentz factor.

Note that when the systems S and D were moving uniformly along the parallel lines (case II, Figure 9.5), the photon exchange at the closest approach between them was symmetric with respect to both observers. In all cases, we have for the relative velocities at the respective moments the same relation $\mathbf{v}_D = -\mathbf{v}'_S$; but in the case of uniform motion of both objects the corresponding rest frames are physically equivalent, whereas generally they are not. Accordingly, in the former case the detected photon would be redshifted regardless of the position of the detector, whereas in the latter case the redshift is observed only when the detector is closer to the center than the source and the blueshift otherwise. And again, it is difficult, if not impossible, to explain the blueshift in case IV as Doppler effect for the disk's resident Paul, since there is no relative motion between S and D in the RF rotating with the disk.

But, even though we cannot *explain* in familiar terms within special relativity (SR) the photon's frequency shift between the events of emission and detection on a rotating disk, the theory still gives the correct prediction of the effect itself. This is a spectacular demonstration of the power of SR: first, contrary to the widely spread misconception, it describes correctly the phenomena in noninertial RF; and second, it provides us with accurate quantitative predictions even in cases when explanation of these phenomena lies beyond its conceptual framework. In these cases the theory itself suggests possible ways of its generalization, as was the case with the appearance of the general relativity. The way it does this enables us to fully appreciate the power of the relativistic invariants used here.

9.5
Relative Is Real

It is physics, not language, that really matters

Brian Greeny, *The Elegant Universe*

Extremely important as the absolute aspect of the world is, it should not be accepted at the cost of its relative aspect. For many people the word relative (observer-dependent) is the synonym of "subjective." They confuse relative with apparent or even illusory. For them, whatever falls into category "relative" (i.e., any experience not shared by *all* observers) ceases to be real.

Actually, few things can be farther apart than "relative" and "subjective."

Subjective mostly refers to an opinion, idea, or interpretation. Relative refers to an actual physical characteristic, observation, or measurement.

Subjective means depending on a person's background, moral values, life experience, state of mind, taste, mood, or just fleeting whim. Accordingly, subjective can be as far from real as you can get.

Relative as observer-dependent means depending on observer's *state of motion*, not on his or her state of mind. State of motion is an objective characteristic of a physical system. Accordingly, relative, while being not absolute, is totally objective – it is a *real face of the world observed from a certain angle.*

Unfortunately, the relativistic effects such as Lorentz contraction(LC) or time dilation are described in some texts in a language that implies that they are merely *appearing* to an observer. For example, "A moving rod appears to be shortened . . ." or "A moving clock appears to tick slower . . .." Such formulations produce an impression that a moving rod actually retains the same fixed length in *any* RF and only seems to be shortened. In other words, the Lorentz contraction is a sort of illusion imposed on the observer, rather than a real physical phenomenon, even though it can be experimentally observed and measured. Sadly, even some professional physicists share this confusion between relative and illusory and quote the Einstein's original paper on SR [80]: "A rigid body that, measured in a state of rest, has the form of a sphere, therefore has in a state of motion viewed from the stationary system the form of an ellipsoid of revolution . . .. Thus, whereas the Y and Z dimensions of the sphere (and therefore of every rigid body of no matter what form) do not *appear* modified by the motion, the X dimension *appears* shortened . . ."

Those who take this literally fall victim of the language with its ambiguities. In physics, the citations cannot be taken out of the physical context. The same pertains to the Einstein's unfortunate expression "appears shortened" in the English translation of his earliest work on relativity. The above quotation illustrates how dangerous the magic of words can be when words are put thoughtlessly before physics and then used as interpretation of physical phenomena. It was this uncritical use of the word by a few who were misled into thinking of Lorentz contraction as something merely apparent, that inspired John Bell to write his remarkable paper "*How to teach Special Relativity*" [24].

For those who build their picture of the world only from quotations of the masters, I would suggest to consider another quotation from a later work by Einstein and Infeld [81]:

"It follows from the Lorentz transformation that a moving stick contracts in the direction of the motion and the contraction increases if the speed increases . . ."

A moving stick does not appear to contract, it just contracts! Einstein had enough time since 1905 through 1938 to think about clear presentation of relativity to educated laymen for whom the quoted book was intended and enough opportunities to return the word "apparent" back into the text in any of his latest works, if he had thought it appropriate.

Here are a few more quotations:

> *We find that a meterstick measures shorter in the ratio* $\gamma^{-1}(V)$: 1 when moving with the velocity V past the system in which the observation of length is being made, than when measured in a system in which it is at rest. . . . This result . . . is to be regarded as entirely *real* one, which except for experimental difficulties could be verified by direct observation . . ."
>
> (R. C. Tolman [82])

> *It is natural to define the length l* of the rod relative to S as the difference between *simultaneous* coordinate values of the end points . . . A body, which moves with a velocity v relative to an arbitrary inertial system S *is* contracted in the direction of its motion . . .. Lorentz contraction . . . is a *real* effect observable in principle by an experiment.
>
> (C. Moeller [31])

These authors took special pain to warn the reader that the loosely used word "appear" in some descriptions of the relativistic effects should not be understood literally.

As we have seen in Section 7.1, an observation of a fast moving sphere could give yet *another proof* of the reality of the length contraction effect. For, were not a moving sphere *really* flattened in the direction of motion, the optical aberration and the light retardation effects would result in the optical image of the sphere that would be *extended* in the direction of motion. It is only a combination of the two equally real but opposing effects – length contraction, on the one hand, and light aberration and time delay – on the other hand, that produce together the undistorted image of the moving sphere.

The Lorentz-contracted length of a moving object is obtained from measurements *after all corrections have been made* for all obscuring effects (including different time propagations of light from different parts of the object). And the result turns out to be less by a factor of $\gamma(V)$ than the length measured by the observer sitting on the object. This is a real physical phenomenon because it can be observed and confirmed by using different experimental techniques, such as direct marking of the instantaneous positions of the leading and trailing points of the object simultaneously in the given frame, optical measurement described in Section 7.1, or measuring the time it takes the object to pass by one fixed clock in an RF.

Some people think that relative effects or characteristics are not real because they are not observed by *all* observers. According to this logic, even such an experience as a sunrise is not a real effect, because what is observed by you as the sunrise from one position on Earth is observed simultaneously as the sunset by someone else from another position and not observed at all (say, midnight of local time!) from a third one ("simultaneously" here means at the same moment of global time in an inertial RF associated with the center of the Earth, not local geographic time.)

I wonder, if such an interpreter of relativity smashes a car against a concrete wall while driving at 80 miles/h, would he or she still insist that nothing has really happened because the high velocity that has caused the crash is merely a relative quantity and therefore not real? Would such an observer still refer to the rest frame of the car, where the car was stationary and that was what “really mattered?”

The reader can also come across a statement that, unlike the time dilation experiments (recall the muon decay discussed in Sections 2.8 and 7.6), the Lorentz contraction effect has never been observed. This is downright wrong. As we have seen in Sections 7.1 and 7.2, there are more than one different experimental procedures (and accordingly, several different but equivalent operational definitions) of Lorentz contraction.

Here, I want to refer to an actually performed experiment that has provided the first direct and unambiguous observation of the Lorentz contraction effect. It is the Michelson experiment!

Historically, the Lorentz–Fitzgerald contraction was proposed especially to account for the negative result of the Michelson experiment. Actually, as has already been mentioned, this contraction is the result of motion of the interferometer relative to *any* inertial system whose observer may watch the experiment and try to explain its outcome. Thus, in the rest frame of the interferometer the Michelson experiment gives the negative result because there is no ether and space is isotropic, so the turning of the device cannot affect the outcome. In the frame where the interferometer is moving the same result can only be explained as the consequence of the length contraction of the arm that is in the direction of motion.

> ...*just as the Michelson–Morley experiment can be regarded as a direct test of the Lorentz contraction, the Kennedy–Thorndike experiment*[2] can be regarded as a direct test of time dilation.
>
> (R. C. Tolman [82]) (see also [34])

As shown Section 2.8, the Lorentz contraction can be considered as a *direct consequence* of the time dilation and therefore any experimental observation of the latter is thereby observation (albeit indirect) of the former. And since the time dilation effect is being routinely observed in the experiments with atmospheric muons discussed in Section 2.8, or in the lab experiments with unstable particles in high-energy physics, we may say that the LC effect is also an experimentally observed phenomenon. Actually we can show more – that the two effects are totally equivalent.

Exercise

Show that the time dilation follows as a consequence from the Lorentz contraction.

Solution

Suppose that two observers – Homer and Ulysses – are sitting on the opposite ends of a rod. Let the rod have a proper length L_0 and move with a speed V relative to my frame. Let Ulysses with his clock sit at the front of the rod and Homer with his one at the rear. Both clocks are synchronized in the system of the rod.

2) The reader can find a very clear description of the Kennedy–Thorndike experiment in [2].

I watch the approaching rod and set my clock so that it reads the zero time when Ulysses passes by me and so is set the Ulysses' clock. I also have a team of assistants positioned equidistantly down the way and provided with synchronized clocks. Let L be the length of the moving rod in my frame. It means that when Ulysses passes by me at the zero time by my clock, Homer is L m away from me. That of my assistants who observes Homer passing by him at the zero moment by his (and my!) clock, raises the flag to mark the instant position of Homer passing by at this moment, and the subsequent measurement confirms that the distance between us was L m. Thus, we satisfy the requirement that the length of the rod moving in my frame be measured as the distance between its end points marked simultaneously in this frame.

Just as Ulysses is passing by me, a stationary muon is produced in a special device that I have. The muon lives its proper lifetime and decays at a moment t by my clock exactly when Homer passes by. Because Homer was L m away from me at the zero time, and moving with a speed V, I know that $t = L/V$. According to the initial requirement, none of us knows anything of the TD effect at this stage of the experiment, but we all know about the length contraction and therefore agree that the length of the rod in my system is $L = L_0/\gamma(V)$, or, which is the same, $L_0 = L\gamma(V)$. This means that Homer is γL m away from Ulysses in the frame of the rod. Homer observes the muon born near Ulysses. Because Homer's clock is synchronized with that of Ulysses, it reads the zero time at that moment. Then by the time the muon (together with me) passes by Homer and decays, Homer's clock will read

$$t' = \frac{\gamma L}{V} = \gamma t. \tag{9.28}$$

That is, the lifetime of the muon is t s in my frame (where the muon is stationary) and γt s in the frame of the rod (where the muon is moving). This is time dilation effect. It follows from the length contraction just as the length contraction follows from it. If we have two distinct statements A and B such that A implies B and A follows from B, then these two statements are equivalent. Thus, the TD effect and LC effect are equivalent implications of the relativistic postulates. Both effects are the sides of the same coin, either one as real as the other. Therefore, as stated above, the experimental observation of one is also observation of the other.

Discussion

Some people, while tacitly admitting the reality of the longitudinal contraction of a moving object, say that this reality is somehow inferior to that observed in the *rest frame* of an object. Such a viewpoint cannot be consistently maintained. First, the fact that the world reveals itself in two distinct aspects – some of its features are absolute and some are relative – does not mean that one of these aspects is more real (within its domain) than the other. Second, an object's shape (or other characteristics) in its rest frame is not necessarily the simplest possible. Third, the rest frame of an object is merely one of infinite number of different frames, which are, according to the principle of relativity, all equivalent. The fact that a certain RF can be more convenient for our observations of a system does not make the properties of that system more real when observed from this frame than when observed from elsewhere.

Consider, for example, a stationary spherical (in its rest frame K) mass M bombarded by an ultrarelativistic point particle. In the rest frame K_0 of the *whole* system we would have the point mass and an ellipsoid approaching each other. This picture is not simpler or "truer" than the one observed from the original frame K.

As another example, consider a straight "horizontal" rod oscillating up and down in a long train (system K_1) moving with a speed V relative to the platform K. When observed from the platform, the rod is at each instant a continuous but wiggling line (e.g., zigzag) described in Section 7.7. The adepts of the supremacy of the rest-frame concept may say that this line is somewhat an illegal description of the rod. "If you want to observe its *true* shape, board the train!" – is their verdict. Well, once in the train, you will indeed observe the rod straight; but the rod, while remaining horizontal, will still pop up and down and may display different kinds of behavior depending on its acceleration program. For instance (recall Section 7.5), it will, due to the Lorentz contraction, have its *thickness* periodically shrink and expand as an accordion, if the acceleration program conserves its *proper* thickness; or it may retain its thickness in K_1 if it is accelerated simultaneously in *all* parts of its cross- section as well as of its length in K_1. But in this case its *proper* thickness will change periodically. In either case even the frame K_1 is still not good enough. Then we may be advised to jump onto the rod itself and cling to it if we want to observe its "*true*" shape. But even if we do so, we can still find the rod's proper thickness periodically changing if *all* its parts are accelerated simultaneously in K_1. We then face a dilemma. On the one hand, we have the rod periodically changing only its position but retaining fixed shape and size when observed from the rest frame of the train. On the other hand, we have the same rod with its center of mass fixed, but changing periodically its proper thickness when observed from its own rest frame. In which of these two frames does the rod exhibit its "true" shape?

Now, suppose that while pondering over this question, we receive a message that the rod is not the only one object to be studied and we have to include another rod residing outside our train. We look outside and see the whole universe whizzing up and down, accordingly swelling and shrinking in its size until we get dizzy. Eventually we jump off the rod back into K_1. And immediately we notice another train (system K_2) moving relative to K with the same speed V but in the opposite direction and also containing a similarly oscillating rod in it. If the rods' oscillations are out of phase (i.e., the center of mass of the whole system is stationary in K), then our initial platform K is the rest frame of the whole system. According to the logics upholding the rest-frame supremacy, now each wiggling rod as a part of the whole system must be "legal" for the platform-based observer. So the wiggling shape of the first rod is illegal when considered alone, but legal in the company of the second wiggling rod moving in the opposite direction. Or will we be required to jump first onto one rod and observe *only* it and then jump onto the other rod and focus on it *only*?

All this may set us wondering: if each object shows its "true face" only in a certain single RF that must be specified for this object only, then how come that we have been able to learn anything reliable about the world?

According to logics of relativity, *any* RF can provide the observer with equally true picture of the universe and *any* of its part. It is true that some of the RFs (e.g., inertial ones) may offer a view that is simpler to interpret. A RF associated with the Sun turned out to be much more convenient and revealing more directly the laws of planetary motions than the RF associated with the Earth. But this does not mean that the scientific observations from the Earth have not been true or real.

A good geometric analogy illustrating this point (and frequently used to illustrate Lorentz contraction) is, for instance, a rectangular block and its possible various cuts. If I cut the block along a plane parallel to one of its faces, I can measure the angle between the edges of the cut, find it to be 90° and call it the proper angle. It can represent the proper length of a rod measured in its rest frame. If I cut the same block along a plane that is not parallel to either one of its edges, the obtained cut will not be a rectangle, but it is as real as the first cut. The angle between the edges of the second cut may be 45° (and represent the length of the same rod when in motion), but none of the angles is less real than any other.

The actual Lorentz contraction can be inferred by considering similar cuts in space-time instead of three-dimensional space in the way considered in Sections 4.2 or 7.4.

The essence of relativity is the revelation that geometry is intimately connected to dynamics.

It was the question what makes space-time geometry the way it is that led Einstein to the general theory of relativity, according to which the geometry is itself determined by the dynamics of energy–momentum distribution. Getting back to SR, one could ask, also together with Einstein, what is time and space but the readings of clocks and metersticks (at least operationally)? And if so, since clocks and metersticks are dynamical systems, what determines their observed behavior?

Let us summarize this part. Any observed characteristic, corrected for possible distorting effects (friction, aberration, time delay, etc.) and confirmed by other independent experiments, is real regardless of whether it is shared by observers in different state of motion or not. If, in addition, the same result is obtained by *all* observers regardless of their state of motion, we are dealing with an *absolute* or *invariant* property (e.g., an interval between two events in space-time, or the norm of any 4-vector, the trace of a second-rank tensor, etc.).

Whatever is absolute is real, but the opposite is not necessarily true. A real experimental result may or may not be invariant. In the latter case it is relative. Relative has nothing to do with subjective or apparent. The "apparent" is what cannot be confirmed by other observations or experiments even in the same reference frame. The Lorentz contraction is not in this category. While being relative, it can be measured by using various experimental procedures in the same RF, which all give the same result.

Similarly, the relativistic mass $m = m_0\gamma(v)$, while being a relative characteristic, is as real as you can get, which becomes evident in the inelastic collisions, or when you try to bend a particle's trajectory, for example, that of a point charge in an external magnetic field. Recall our example with colliding particles in Section 5.5. What can be more real than new particles born in a collision from the relativistic mass of the initial particles? Are these new particles our illusion? Their origin is relative (the relativistic mass of particle 1 in the rest frame of particle 2, or vice versa; but it is relativistic mass in either case). The adepts of supremacy of the rest mass can argue that the new particles come from the rest mass of the colliding system. This is true; but the rest mass itself originates from the *relativistic* masses of its constituents; recall that the rest masses are not additive, but the relativistic masses are. This makes the concept of relativistic mass convenient for explanation of the origin of the rest mass of a compound system without appealing to its rest energy, all the more so that the rest energy itself can be merely the sum of relativistic energies of the constituents.

We will illustrate this by an example that can be called the "kinetic energy paradox." The kinetic energy of a system, being additive, is the sum of kinetic energies of its parts

$$K = \sum_{j=1}^{N} K_j, \tag{9.29}$$

where N is the number of parts. On the other hand, the kinetic energy is determined by Equation (5.20) as the difference between the relativistic energy of the system and its rest energy:

$$K = \mathcal{E} - \mathcal{E}_0 = \mathcal{E} - M_0 c^2. \tag{9.30}$$

Let us apply this to the simplest possible system – the two photons of the same frequency ω and with opposite momentums. According to (9.29), the total kinetic energy of this system is doubled energy of one photon, that is,

$$K = \mathcal{E} = 2\mathcal{E}_1 = 2\hbar\omega. \tag{9.31}$$

The rest mass m_0 of each photon is zero, but the rest mass M_0 of the whole system is not:

$$M_0 = \frac{\mathcal{E}_0}{c^2} (\neq 2m_0). \tag{9.32}$$

Since we are in the rest frame of the system (net momentum is zero!), its total relativistic energy $\mathcal{E}$ is equal to its rest energy $\mathcal{E}_0$ and therefore according to Equation (9.30)

$$K = 0. \tag{9.33}$$

Thus, according to definition (9.30) the total kinetic energy is zero, whereas according to (9.29) it is not zero.

The explanation of this "paradox" is that the two definitions of the kinetic energy are not equivalent. The first of them (9.29) gives the *total* kinetic energy, whereas the second one – *only* kinetic energy associated with the motion of the center of mass of the system. No surprise that in our case, when the center of mass is stationary, this part of kinetic energy is zero. The remaining part of kinetic energy of the system is hidden in its rest mass and is represented by Equation (9.29) but not by (9.30). According to (9.30) and (9.31) the rest mass of the system can be conveniently written as

$$M_0 = 2m_1, \tag{9.34}$$

where $m_1 = \mathcal{E}_1/c^2$ is the relativistic mass of each photon. As we have seen in Sections 5.3 and 5.5, relativistic mass is a meaningful characteristic describing the dynamics of motion at high speeds. Here, this concept is useful in showing the real origin of the rest mass of a system. In contrast with the rest mass, which is invariant, the individual relativistic masses m_1 are relative characteristics equal to each other only in the rest frame of the system. But they are no less real than the rest mass. At least for a system of noninteracting particles, it is much more convenient to think of its rest mass as built up from the individual *relativistic* masses that are additive than from the individual rest masses, which are not.

The rest mass is an important characteristic of a particle or a system. But this does not mean that we should sanctify a particle's mass when measured in its rest frame and vilify it when measured in any other frame. Actually, it is the latter measurements that are practical in most cases. Consider a prerelativistic definition of mass as a ratio of an external force to the resulting acceleration. As we know from Section 5.4, relativity imposes an additional requirement that for a moving particle, the force must be perpendicular to its velocity. This requirement is automatically satisfied by the Lorentz force on a charged particle. Therefore, for charged particles at least, the whole procedure is easily executable and is routinely used in high-energy physics. In most cases, the newly born particles emerge from the high-energy collisions with a speed close to c. In such cases, the relativistic mass of a particle can be easily measured. Suppose that $\mathbf{v} \perp \mathbf{B}$ and we know the particle's charge q. Then, having a photograph of the particle's track, say, in a bubble chamber, we find the force from $f = qvB$ and acceleration from $a = v^2/R$. Then we evaluate the mass as $m = f/a = qBR$ by measuring R and setting $v \cong c$. If $\mathbf{v}$ is not perpendicular to $\mathbf{B}$, the procedure is a little more involved, but also executable.

The measurement of the *rest mass* under the same conditions would be much more complicated. For instance, we could also measure, in addition to $\mathcal{E} = mc^2$, the momentum p and then determine m_0 from (5.23). But the measurements of m and p must in this case be much more accurate than in case just described for measurement of m only, and m_0, unlike m, is *computed* rather than directly measured according to the original definition of mass.

This example shows that relativistic invariants fall into two distinct categories: (1) operational invariants and (2) computational invariants.

Operational invariant is a quantity that can be determined by a universal operation, that is, by an experimental procedure executable in any RF and any such procedure gives the same result in all of them.

Computational invariant is a quantity such that it can be computed by a universal algorithm from other, noninvariant, quantities measured in an arbitrary RF.

Examples of an operational invariant are the invariant speed c and electrical charge q. One and the same procedure measuring c in one RF will give the same result in another RF. Similarly, any experiment for measuring charge of an isolated system gives the same result regardless of the state of motion of the system or measuring device. One of the most compelling evidences of the charge invariance is the comparison of the net charge of an atom in its ground state and its highly excited state with high orbital momentum of the excited electron (the so-called Rydberg's atom). This kind of experiment is very sensitive to invariance violation and it shows that in both states the net charge of the atom is zero, which implies that the electron's charge is invariant.

In contrast, the basic experimental procedure for measuring *mass* of an object as measure of its inertness gives different results in different RFs. But its value measured in the rest frame can be also computed from other quantities measured in an arbitrary RF. Thus, this value stands out as an absolute characteristic of the system, but only computationally, not operationally. In other words, the invariant m_0 as measured from an *arbitrary* RF actually describes the inertness of the object in its *rest* frame, not in the frame where the measurement is performed. In this respect, it can also be called the conditional invariant. The same is true for the proper length and proper time.

But mass, as well as time and space, as such, are all relative characteristics of a system or process.

Thus, even invariants, diamonds as they are, come in two distinct varieties. Other (relative) characteristics, while being not diamonds, can be compared, using the same language, with graphite or with coal. But even these latter forms, albeit less precious, are no less real than any diamond. Different as they are for us, from Nature's viewpoint all are essentially carbon.

Summary

The two postulates of relativity are *physical* statements about very general patterns of behavior of dynamic systems. According to the first postulate, the laws of Nature are absolute (invariant) – they are the same in all RFs; according to the second postulate, there exists an absolute (invariant) speed.

The theory based on these postulates describes the geometry and underlying physics of space-time. It describes features of reality that are relative and features that are absolute and shows relations between them. In particular, the absoluteness of natural laws makes all RFs equivalent and thereby the notions of state and motion only relative concepts. The existence of the invariant speed makes space and time relative characteristics – two different but intimately related faces of space-time.

Problems

9.1 Consider a medium, in which an acoustic wave, for a certain frequency, has a velocity $\mathbf{u} = u\hat{\mathbf{n}}$, where the unit vector $\hat{\mathbf{n}}$ determines the propagation direction. Let an observer move through such a medium with velocity $\mathbf{v}$. Find an expression for velocity $\mathbf{v}'$ of acoustic wave relative to this observer, in terms of $\mathbf{u}$ and $\mathbf{v}$, assuming that u and v are both small with respect to c.

9.2 In the previous problem, find the *relativistic* expression for velocity $\mathbf{v}'$.

9.3 In Problem 9.2, find the frequency of the acoustic wave as a function of direction, using the relativistic Doppler effect.

9.4 Consider a photon passed through a beam splitter described in Section 5.5. The photon is in a superposition of two equally weighted states of motion, whose directions make 90° with one another. Find the average speed of this photon.

9.5 A rod with a nonuniform mass distribution has most of its mass concentrated near the edges, so it can be represented by the two end points with the rest masses $m_1^{(0)}$ and $m_2^{(0)}$. The masses are flying apart with the speeds v_1 and v_2 from their respective initial positions x_1 and x_2.
(a) Find the proper distance between the masses as a function of time.
(b) How will you determine the *proper length* of the rod?

9.6 Suppose you observe two parallel elastic rods. They are both expanding but remain geometrically identical – their respective end points coincide during all the time of your observation. However, their mass distributions are different: the first rod has its right edge twice as massive as the left one, whereas the second rod has its left edge three times as massive as the right one. Do these congruent rods have the same proper lengths?Explain your answer.

9.7 Suppose that you are sitting on an inertial detector and the motion of the source is described as $\mathbf{r}_S = \mathbf{r}(t)$ in your RF. Find the condition determining the retarded moment t' of the photon emission for a photon arriving at the detector at a moment t.

9.8 Paul is sitting at the center of a disk of radius R spinning with angular velocity Ω.
(a) Describe the motion of a photon emitted from the periphery of the disk and heading toward its center when observed by Peter from an inertial frame K attached to the center of the disk. Is this motion radial for Paul?
(b) Sketch the spatial trajectory of this photon relative to the spinning disk.
(c) Find the equation describing this trajectory in terms of the disk's polar coordinates r, $\tilde{\varphi}$.

9.9 Suppose we have a photon emitter S and detector D both positioned on the periphery of a disk of radius R. The angular separation between S and D is Φ. Let the disk rotate with an angular velocity Ω. What is the frequency shift between the emitted and detected photon? Give your answer based on
(a) symmetry considerations;
(b) use of the relativistic invariant (9.19);
(c) sketch the whole process, indicating the instant positions of S and D at the moment of emission t_S and moment of detection t_D, and the world line of the photon between the corresponding events.

9.10 Consider a system of the two photons with velocities $\mathbf{v}_j = c\hat{\mathbf{n}}_j$ and relativistic masses m_j defined as $m_j = \varepsilon_j/c^2 = \hbar\omega_j/c^2, j = 1, 2$.

(a) Show that the sum of these masses is equal to $M \equiv m_1 + m_2 = M_0\gamma(v_c)$, where M_0 is the rest mass of the system and $\mathbf{v}_c$ is its center-of-mass velocity.
(b) Find M and M_0 in the limiting case $\hat{\mathbf{n}}_1 \to \hat{\mathbf{n}}_2 \to \hat{\mathbf{n}}$, where $\hat{\mathbf{n}}$ is a unit vector specifying direction.

9.11 Derive the expression for relativistic mass of a particle of a charge q moving in a homogeneous magnetic field $\mathbf{B}$. The invariant mass of the particle is unknown. The particle's velocity $\mathbf{v}$ makes an angle θ with $\mathbf{B}$ and its magnitude is close to c. The particle traces out a helix of radius R. Find the expression for particle's relativistic mass in terms of B, R, and θ.

10
Relativity at Work III: Quantum Mechanics

Einstein said that if Quantum Mechanics is right, then the world is crazy. Well, Einstein was right. The world is crazy.
Daniel Greenberger

10.1
Basic Ideas of Quantum Mechanics

Quantum mechanics (QM) emerged in the early twentieth century from attempts to explain some properties of the blackbody radiation and atomic spectra. It had soon become clear that the classical physics was unable to account for those properties. Classical predictions of behavior of atomic systems and their interactions with radiation contradicted the experiments. Moreover, the mere existence of atoms seemed to be a miracle in the framework of the classical physics. According to the classical picture ("planetary" model) based on Rutherford's famous experiments, an atom is a system of electrons orbiting around the nucleus like the planets around the Sun (hence the name of the model). This model looked simple and very attractive. Why, indeed, could not the electrons travel around the atomic nucleus like planets orbiting the Sun or like a planet's satellites around the planet? All the more so that the interaction law (the attraction inversely proportional to the square of the distance) is mathematically the same in both cases. And how compelling it would be from the philosophical point of view: the big is just an upscale of the small and vice versa.

Well, according to the same classical physics such an atom cannot exist. And the reason is very simple. Here are the electrons in a Rutherford's atom (Figure 10.1). It seems they are in far more favorable conditions than, for instance, the Earth's artificial satellites, which eventually spiral downward owing to a small drag force in the upper atmosphere. In contrast, the electrons appear to be totally free of any dissipative forces such as drag or friction: there is nothing but themselves in the space around the nucleus. They should circle and circle around with no hindrance. Ideal planets with no energy losses on the way!

This conclusion overlooks a fundamental fact that each electron carries the electrical charge and, accordingly, its own electromagnetic (EM) field. This must convert them from ideal planets to the downspiraling satellites. According to the electromagnetic theory (recall Section 6.10), if an electron is moving with constant velocity, its field just

Special Relativity and How it Works. Moses Fayngold

ISBN: 978-3-527-40607-4

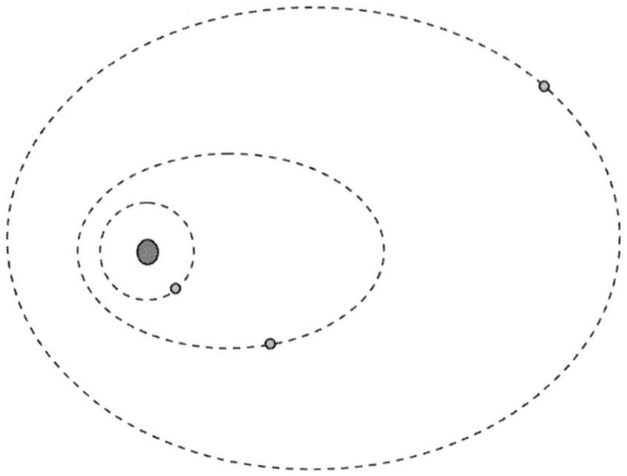

Figure 10.1

follows this motion, remaining "rigidly" attached to its "master." If, however, the electron changes the state of motion – for example, it accelerates – its field is getting partially detached. This is precisely what should happen with an orbiting electron – an orbital motion is an accelerated motion. Accordingly, the orbiting electron must be loosing its energy, which is radiated away together with the "detached" part of its electromagnetic field. The classical atomic electron must, in a way, move in an atmosphere of its own radiation field and it must loose energy because of the "radiative friction" in this atmosphere. As a result, very soon (in about 10^{-8} s) all the electrons, having emitted a blend of electromagnetic waves of different frequencies, must fall onto the nucleus and the Rutherford's atom would cease to exist.

It seemed that the classical physics had come to a hopeless dead end. On the one hand, the Rutherford's experiments had shown that his planetary model is the only one possible. On the other hand, it turned out to be impossible.

According to the ongoing experiment carried out by Nature itself, atoms are stable. And in cases when they do radiate (say in collisions or after an optical excitation), the corresponding spectrum is discrete: on a dark screen or photographic film there forms a set of distinct spectral lines (Figure 10.2). An atom of each chemical element has its own unique discrete spectrum.

This clash between the classical predictions and the experiment shows that something was badly wrong in the classical picture of the electrons moving along their paths within an atom.

The correct and consistent theory including the description of quantum-mechanical phenomena had required no less radical departure from some classical concepts than it had been the case at the birth of relativity theory.

First, to explain the electromagnetic spectrum of a body in thermal equilibrium, Max Planck had to postulate that the energy of atoms interacting with radiation of frequency ω could only change by discrete portions $\varepsilon = \hbar\omega$ [83]. Then Einstein noted that other phenomena in the light–matter interactions, such as the so-called photo-effect (electron emission from the illuminated surface of some materials), could be correctly explained if electromagnetic radiation itself actually consisted of indivisible portions (quanta of radiation) [84]. For radiation with frequency ω, the

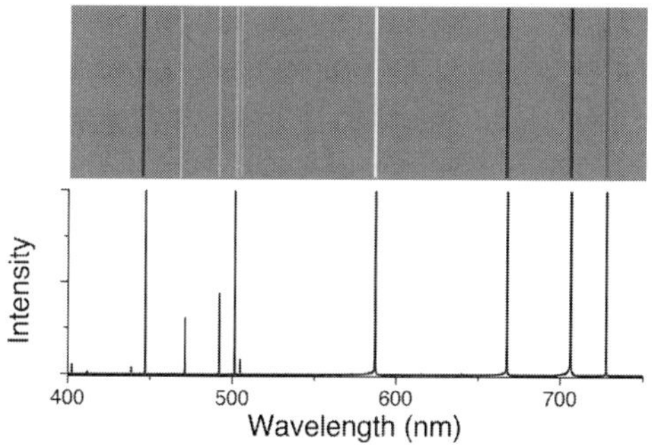

Figure 10.2 Top: Visible spectrum of a Helium lamp; Bottom: schematics of the corresponding optical transitions for a Helium atom. (Courtesy Andrei Sirenko, Department of Physics, NJIT)

energy and momentum of such portion are given by

$$\mathcal{E} = \hbar\omega, \qquad \mathbf{p} = \hbar\mathbf{k}. \tag{10.1}$$

Here, $\omega = 2\pi f$, where f is the number of cycles per second (actual frequency),[1] and $\mathbf{k}$ is the wave vector known from classical wave mechanics and electrodynamics: it points in the direction of propagation of the wave and has a magnitude equal to $2\pi/\lambda$, where λ is the wavelength (recall Section 6.9).

The crucial new characteristic here is the coefficient $\hbar$ – the so-called Planck's constant. Just as special relativity has introduced the invariant speed c as a universal constant of nature, quantum mechanics has introduced another universal constant – the Planck's constant (quantum of action)

$$h = 2\pi\hbar = 6.626\,068\,76 \times 10^{-34}\ \mathrm{J\,s}. \tag{10.2}$$

As we know, in relativity, c plays the role of an invariant scaling parameter in the velocity space. If velocity $\mathbf{v}$ of a particle satisfies $v \ll c$, then its motion can be described accurately by the laws of Newton; otherwise, the relativistic physics must take over.

Similarly, the Planck's constant is another scaling parameter, representing the least possible (indivisible) action. If the action S of a one-particle system, evaluated classically, satisfies $S \gg h$, then the classical approximation may be accurate; otherwise, quantum mechanics takes over. Thus, the constant h determines the domain of applicability of the classical description.

Later Niels Bohr developed these ideas in his theory of atom. Soon after it had become clear that the Rutherford's model needed radical changes, Niels Bohr came up with a new model of a hydrogen atom [85]. To account for stability of atoms, Bohr just had postulated that there must be certain selected orbits in the atomic space, on which the electrons do not radiate. As long as the electrons are moving in these stable orbits, the atom remains in a stationary state. The only opportunity for the atom to radiate (or

1) Note that we use here the term "frequency" for both f and $\omega = 2\pi f$. The correct term for ω is, of course, the "angular frequency," but in the physicists' jargon the word "angular" is frequently dropped.

absorb) energy is to do it in a quantum leap – jumping from one stationary state to another. At these jumps the atom changes its energy discontinuously, by one discrete portion. Accordingly, it radiates one lump of the electromagnetic energy of a fixed frequency. These single lumps or portions of energy are nothing else but the original quanta first suggested by Einstein (and later named "photons" by Lewis); the energy of the radiated photon is equal to the energy difference between the atomic initial and final states. The radiation frequency is connected with the energy difference by

$$\omega = \frac{\mathcal{E}_2 - \mathcal{E}_1}{\hbar} = \frac{\mathcal{E}}{\hbar}, \tag{10.3}$$

which is just the combination of Equation (10.1) and the original Planck's suggestion about the allowed discrete energies of material particles. If passed through a prism, such portion of light all refracts by one fixed angle determined by its frequency. Accordingly, a sharp line rather than a fuzzy band will appear on the screen. This is precisely what is observed in the atomic emission spectra shown in Figure 10.2.

A reverse process is also possible: an atom can absorb a photon with a frequency tuned to an allowed optical transition and jump to a state with higher energy. This transition is called excitation.

An atom does not usually stay long in an excited state. In about a few billionths of a second, it returns to its ground state, after radiating a photon of the same frequency. This gives a simple and natural explanation of yet another law, first formulated by Kirchgoff, that atoms of a given element radiate light of the same frequencies as they absorb.[2)]

Despite its success in explaining some atomic spectra, the Bohr's model left open the basic question: what is it that allows an electron to live on certain privileged orbits and what is the criterion for selecting such orbits? In other words, what is the nature of the so-called atomic stationary states and what determines their characteristics?

A breakthrough in answering this fundamental question is associated with the names of de Broglie, Schrödinger, Born, and Heisenberg.

Louis de Broglie suggested that if light waves can, under certain conditions, exhibit corpuscular characteristics, then the symmetry requires that the reverse should also be true – all particles must possess wavelike characteristics and the relation between the wave and particle aspects must also be described by Equation (10.1). Therefore, the motion of a free particle with energy $\mathcal{E}$ and momentum $\mathbf{p}$ is, according to de Broglie, represented by a wave [86]

$$\Psi(\mathbf{r}, t) = \Psi_0\, \mathrm{e}^{i(\mathbf{kr} - \omega t)} = \Psi_0\, \mathrm{e}^{i((\mathbf{p}/\hbar)\mathbf{r} - (\mathcal{E}/\hbar)t)}. \tag{10.4}$$

One of motivations for de Broglie's work was the notion that the wave nature of the electrons' motion in an atom would *naturally* admit the emergence of discrete states. Indeed, as we know from mechanics, only waves of certain frequencies can survive in a bound volume. A well-known example is the fundamental frequency and its harmonics on a vibrating string.

2) The reason why the excited atomic states are not exactly stationary is the electron interaction with the so-called vacuum fluctuations. This is a purely quantum-mechanical effect, although some of its results (e.g., the spontaneous emission) can be described by the classical picture of radiative losses briefly outlined above.

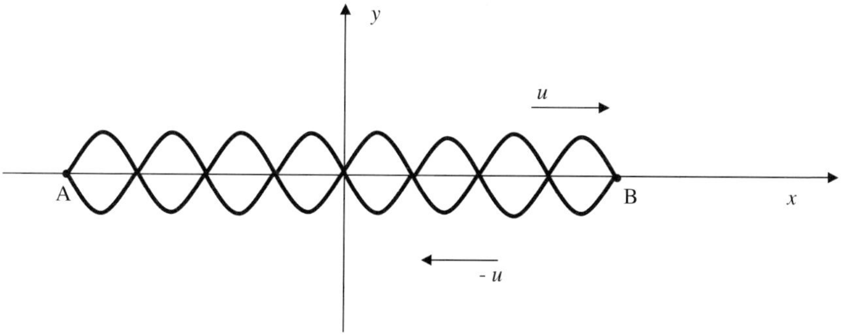

Figure 10.3

The waves in a bounded volume are subject to the process of "natural selection": only the fittest can survive. The fittest are those whose half wavelength (or its integral multiple) fits into the size of the available volume along the direction of propagation of the wave (this condition weeds out inappropriate directions as well). The best-known example is the phenomenon of the standing waves on a string with the fixed ends. In this simplest case, there are only two possible directions of propagation – those along the string; and we can observe a discrete set of possible waves, such that 1, 2, 3, ..., or any other integral number of half wavelengths fits into the length of the string. The word "standing" means that the wave motion here is not accompanied by the energy flow as there are two symmetric sets of waves running in the two opposite directions, so there is no net energy transfer (Figure 10.3).

In case when the path of the wave is closed onto itself, forming a loop, only an integral number of waves can fit into the path. Imagine a snake biting its own tail and wiggling. If all adjacent wiggles are equal, then obviously only an integral number of them will fit into the length of the snake. Similarly, if you hit a ring, it will start vibrating, with only integral number of waves running around (Figure 10.4). For a ring of radius a all possible wavelengths are given by

$$\lambda_n = \frac{2\pi a}{n}, \qquad n = 1, 2, 3, \ldots . \tag{10.5}$$

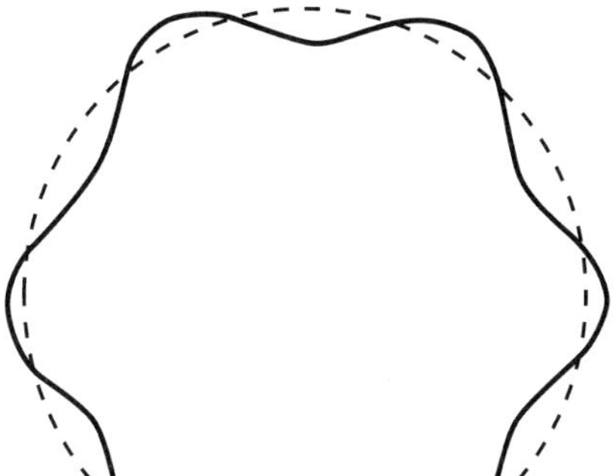

Figure 10.4

In view of (10.1) this can be written as

$$ap_n = n\hbar, \tag{10.6}$$

which is a selection rule for possible observed momenta in an appropriate experiment for the corresponding momentum measurement. And there is more to it: because the particle is orbiting the center, it has also a nonzero angular momentum **L**. The product on the left is the component of this angular momentum onto the axis perpendicular to the classical orbital plane. Denoting this axis as z, we have

$$L_z = m\hbar \tag{10.7}$$

with integer m. We switched notation here from n to a more conventional m used in the description of the orbital momentum; note also that (10.7) is more general than the special case (10.6) from which it was derived. Thus, the selection rule for the wavelengths produces the selection rule for the angular momentum of a system. It also produces a similar rule for the frequencies because for any medium the frequency f of a wave is related to its length by

$$\lambda f = \frac{\omega}{k} = u, \tag{10.8}$$

where u is the characteristic wave velocity in the medium. Therefore, any selection rule for the allowed wavelengths (so-called "quantization") automatically results in the corresponding quantization for frequencies. Thus, de Broglie's insight, expressed in relations (10.1) and (10.4), automatically leads to the quantization of some physical characteristics such as angular momentum, as well as the *energy* of a bound system. This is one of the crucial points where quantum mechanics comes in.

Exercise 1

As an example let us apply this to one of the simplest quantum-mechanical system: the hydrogen atom. We start with the classical expression for the electron's energy in the Coulomb's field of the nucleus:

$$\mathcal{E} = \frac{p^2}{2m_e} - \frac{e^2}{a}. \tag{10.9}$$

Here, m_e is the electron's mass, a is radius of its orbit, and e^2 is the shorthand for $q^2/4\pi\varepsilon_0$. From the expression for the centripetal force

$$\frac{m_e v^2}{a} = \frac{e^2}{a^2}, \tag{10.10}$$

we find[3)]

$$\frac{e^2}{a} = \frac{p^2}{m_e} = -2\mathcal{E}. \tag{10.11}$$

3) This relation is a special case of the so-called "virial" theorem [87], relating average kinetic or potential energy of a system to its total energy.

Now, we say that there is the de Broglie wave associated with the electron and use the same reasoning as before (namely, that only those λ that satisfy (10.5) are allowed within the atomic space) to relate the electron's momentum p to the radius of its orbit by (10.6). Combining then two equations (10.11) with (10.6) and eliminating a yields

$$\mathcal{E} = -\frac{m_e e^4}{2\hbar^2 n^2}. \tag{10.12}$$

This is known equation for the discrete energy levels of a hydrogen atom. Together with Bohr's rules (10.3) connecting energy levels with the radiation frequency for corresponding optical transitions, it explains quantitatively the observed spectrum of atomic hydrogen.

Let us summarize these examples. According to the classical picture of the world, all characteristics specifying the state of a physical system are continuous variables. According to quantum mechanics, some of them can take on only certain discrete values. For instance, the energy of any localized system bound to a certain finite region of space can only assume certain discrete values determined by the binding forces. An electron in the Coulomb's field of a proton can have only discrete energies when it is bound by the proton ($\mathcal{E}_n < 0$, $n = 1, 2, 3, \ldots$), thus forming a hydrogen atom; and it has a continuous set of energies when ionized ($\mathcal{E} > 0$).

We talk of the possible energy values (or the energy levels) of a system as its spectrum. Thus, the energy spectrum in quantum mechanics can be discrete, or continuous, or combination of both.

On the other hand, the angular momentum of any object can only take on the discrete values. As an example, consider the electron spin (intrinsic angular momentum unrelated to any orbital motion of the electron). This variable is discrete and, moreover, it has only two possible projections onto an arbitrary chosen direction (call it the z-direction), namely,

$$s_z = \begin{cases} +(1/2)\hbar, \\ -(1/2)\hbar. \end{cases} \tag{10.13}$$

The top line in (10.13) is frequently referred to as "spin up" and the bottom as "spin down."

The wave aspect of matter exhibited in a stationary state explains also why an electron in such states does not radiate continuously as it should according to the classical physics. A wave is something quite different from a tiny ball rolling down an orbit. Behaving as a wave, an electron is "smeared out" over the whole orbit, as does a wave on an elastic ring, which is vibrating in *all its parts at the same time*. The electron is spread out in the atomic space as music in a concert hall. Therefore, the physicists nowadays rarely say "an electron orbit." They prefer to say "an electron configuration," "electron shell," or "electron cloud" even when it is about *only one* electron. In a state with definite energy, spinning of such a cloud around the nucleus would produce only a steady current loop, which does not radiate.

Depicted in Figure 10.5 are a few different stationary states of an electron in a hydrogen atom. In each such state the electron indeed forms a sort of a "cloud"

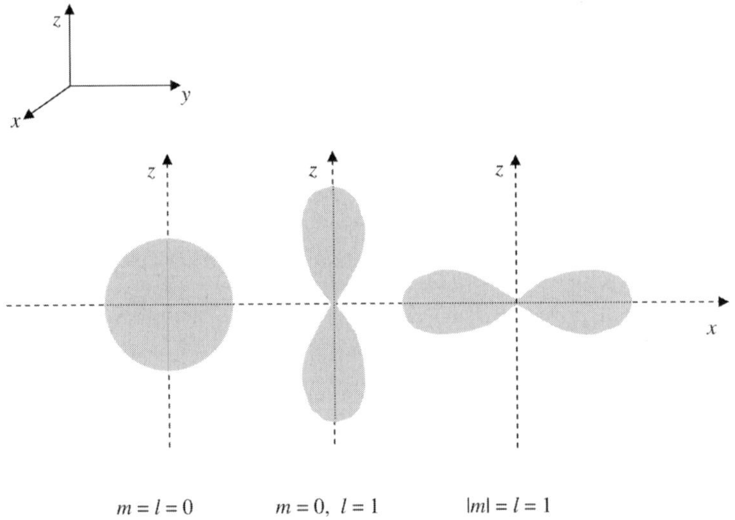

Figure 10.5 A few different configurations of the probability cloud formed by a single electron in a hydrogen atom. The integers l, m are the quantum numbers determining the magnitude and the z-projection of electron's orbital angular momentum.

around the nucleus. And as music in a hall, reflecting from the walls and chairs, can sound in some places louder than in others, so the electronic "cloud" in an atom can be denser at some points than at others.

We will discuss some of these features in more detail later and now focus on other fundamental implications of de Broglie's idea.

Equations (10.1) describe a very strange (from the classical viewpoint) symbiosis of concepts considered to be mutually exclusive: on the left we see the energy and momentum of a corpuscular object; on the right we see the characteristics of a periodic process – the frequency and the wave vector of propagating monochromatic wave. The equations tell us that knowing one type of characteristics determines the other, which means that both aspects – one related to *discrete things* and the other to a *periodic process* undulating in space and time – are relevant characteristics of the same entity.

Thus, these equations express the universal connection between the waves and particles, the "wave–particle dualism" of the world. According to this dualism, the wavelike or particlelike behavior as such would give us only an incomplete image of an object, working fine in some experimental conditions, but leaving it to its counterpart to take over in others. For instance, if we do find an electron at some place in an experiment especially designed for a position measurement, we always find the whole electron there, not a part of the electron. This is a corpuscular behavior. On the other hand, there is no way of indicating specific position of the electron *between* such measurements without actually performing a new one. All evidence indicates that the electron between the position measurements does exist in all locations of an extended region at once, rather than in a well-defined single place. This is a wave behavior. We express this by saying that a microscopic object is recorded as a particle but propagates as a wave. These two properties act as mutually complementary characteristics of reality (Bohr's complementarity principle).

From this duality emerges, like an exotic plant from a seed, the chain of properties that exhibit dramatic departure from some "sacred" concepts of classical physics and all together form the basis of quantum mechanics. In some aspects, this departure seemed to be so irrevocable that even Einstein, who was, together with Planck, the initiator of the quantum physics, later refused to accept its most "outrageous" features (as an interesting historic fact, we will note that Einstein had received the Nobel Prize in physics not for his theory of relativity but for his pioneering contribution to the quantum theory of radiation). The same skeptical attitude was shared by Schrödinger who refused to accept the probabilistic interpretation of the wave function and its apparently paradoxical implications.

Below we will outline the basic quantum-mechanical concepts that superseded the classical picture of the world.

(I) The moment we assumed that a particle can be represented by a wave and be treated as such, there naturally arises a question – *what is it* that is waving? Ironically, the answer to this question also starts with Einstein. According to the electromagnetic theory, the light intensity is proportional to the square of the amplitude of the corresponding electromagnetic wave. On the other hand, if light is to be considered as a bunch of particles, then its intensity must stand in proportion to the number of particles per unit volume (i.e., concentration) in given locality. It follows then, according to Einstein, that the square of the wave amplitude must be a measure of concentration of the light particles. And according to de Broglie this has to be generalized to particles of any kind.

Well, what if we have only one particle? Clearly, we cannot speak of concentration (in statistical terms) when there is only one particle. The answer to this fundamental question was suggested by Max Born [88]. According to Born, the square modulus of the particle's wave amplitude at a certain point in space is *probability* per unit volume (probability density) to find this particle in the vicinity of this point. This is the famous probabilistic interpretation of the wave function. It has become the part of a theory with an astounding explanatory and predictive power.

But does it really explain what is waving there? Can we say that it is the probability amplitude? To get a deeper understanding, we need to consider the quantum-mechanical probability quantitatively and in more details.

If a quantum-mechanical state is described by a function of position $\Psi(\mathbf{r}, t)$ (called a wave function in $\mathbf{r}$-representation), then, according to Born, the expression

$$d\mathcal{P}(\mathbf{r}) = |\Psi(\mathbf{r})|^2 d\mathbf{r} \tag{10.14}$$

gives the probability $d\mathcal{P}(\mathbf{r})$ to find the particle in a small volume $d\mathbf{r} = dx dy dz$ containing the point $\mathbf{r}$. Accordingly, the expression

$$|\Psi(\mathbf{r})|^2 = \frac{d\mathcal{P}(\mathbf{r})}{d\mathbf{r}} \equiv \rho(\mathbf{r}) \tag{10.15}$$

represents the *probability density* of finding the particle at **r**. Thus, the wave function $\Psi(\mathbf{r}, t)$ describes the probability distribution associated with the state Ψ.

According to the probability theory, the probability to find the particle in a finite region of space of volume V will be given by the sum (in our case of continuous variable – by the integral) of the probability density taken over this volume:

$$\mathcal{P}(V) = \int_V d\mathcal{P}(\mathbf{r}) = \int_V |\Psi(\mathbf{r})|^2 d\mathbf{r}. \tag{10.16}$$

If the volume V embraces all positions allowed for the particle under given conditions (e.g., if the particle is locked in a box, then V should be the volume of this box; if it is free, then V should be infinite), then the probability in (10.16) is 1: the particle is certainly to be found somewhere within the available volume. In such cases, we do not indicate the integration region and, assuming that $\Psi(\mathbf{r})$ is square integrable, can write

$$\int |\Psi(\mathbf{r})|^2 d\mathbf{r} = \int \Psi^*(\mathbf{r})\Psi(\mathbf{r}) d\mathbf{r} = 1. \tag{10.17}$$

Similarly, if we represent the state as a function of another characteristic, say energy, $C = C(\mathcal{E})$ (the wave function in $\mathcal{E}$-representation), then its square modulus

$$|C(\mathcal{E})|^2 = C^*(\mathcal{E}_j)C(\mathcal{E}_j) = \mathcal{P}(\mathcal{E}_j) \tag{10.18}$$

will give the probability to find the particle in a state with energy $\mathcal{E}_j$ (we assume here that the energy spectrum is discrete; in the case of a continuous spectrum, Equation (10.18) gives the probability density, that is, $d\mathcal{P}(\mathcal{E})/d\mathcal{E}$). The net probability to find the particle in any one of all available energy states is given by

$$\sum_j \mathcal{P}(\mathcal{E}_j) = \sum_j C^*(\mathcal{E}_j)C(\mathcal{E}_j) = 1. \tag{10.19}$$

The normalization condition (10.17) assumes that $\Psi(\mathbf{r}, t)$ is squarely integrable. In some (idealized) situations this assumption is not satisfied. For example, the de Broglie wave (10.4) describing a state with sharply defined $\mathbf{p} = \mathbf{p}'$ is not squarely integrable. The integral of $|\Psi|^2$ with Ψ from (10.4) diverges. This by itself does not constitute any problem for the theory, as was just mentioned, the state (10.4) with sharply defined $\mathbf{p} = \mathbf{p}'$ in the continuum of states is only an idealization never realized in an experimental situation. Any real situation involves a group of waves (wave packet) within a narrow region $\Delta\mathbf{p}$ such that all the monochromatic components of the group reinforce each other (interfere constructively) within a finite region of space $\Delta\mathbf{r}$ and the resulting perturbation falls off very fast (the monochromatic components interfere destructively) outside of this region, so the resulting wave function is squarely integrable. But even in the idealized case (10.4) of a free particle with only one sharply defined momentum, the normalization condition can be reasonably generalized owing to the fact that the amplitude of such wave $\Psi'_0 \neq 0$ only for a given $\mathbf{p} = \mathbf{p}'$ and is zero for all other $\mathbf{p}$. Similarly, the amplitude of another monochromatic wave with $\mathbf{p} = \mathbf{p}''$ is $\Psi''_0 = \text{const} \neq 0$ only for $\mathbf{p} = \mathbf{p}''$ and is zero for all other $\mathbf{p}$. Then the integral of the product $\Psi''^*(\mathbf{r})\,\Psi'(\mathbf{r})$ over the whole space is zero if $\mathbf{p}'' \neq \mathbf{p}'$ and is infinite if $\mathbf{p}'' = \mathbf{p}'$. In fact, such integral has the properties of the δ-function first introduced by Dirac [89]. Therefore, the monochromatic function (10.4) can be normalized so that

$$\int \Psi^*_{p'}(\mathbf{r},t)\Psi_{p''}(\mathbf{r},t)d\tau = \delta(\mathbf{p}'-\mathbf{p}''). \tag{10.20}$$

This type of the normalization condition holds for an arbitrary observable κ with continuous spectrum. If the spectrum is discrete (e.g., when we use states with sharply defined energies, as in (10.19)), but we still describe each state in **r**-representation, that is, use $\Psi_j(\mathbf{r}) = \Psi_{\mathcal{E}_j}(\mathbf{r})$ instead of $C(\mathcal{E}_j)$, then the normalization condition takes the form

$$\int \Psi^*_{\mathcal{E}_j}(\mathbf{r})\Psi_{\mathcal{E}_k}(\mathbf{r})d\mathbf{r} = \delta_{jk}. \tag{10.21}$$

This naturally reduces to (10.17) when $j = k$.

The conditions (10.20) and (10.21) tell us, among other things, that the integral of product of the wave functions of two distinct states is always equal to zero. The complete set of functions satisfying these conditions is known in mathematics as the orthonormal set. In many respects, such a set is analogous to a set of the orthonormal vectors in vector algebra.

(II) If even a single particle is a wave, it must be able to do what any wave can do. A wave can split into two or more waves, for instance, by passing through a screen with two or more slits. Or, it can split into two parts when passing through a beam splitter (recall the Michelson experiment or Figure 4.5 in Section 4.4). Then the split parts can propagate independently of one another. The wave is now in a *superposition* of two distinct states with different **k**. And vice versa, the different parts of the wave can combine together where they overlap, producing the interference pattern. This may be precisely the kind of interference formed, say, in Michelson–Morley experiment we had discussed in Chapter 1. Now we can say that *each single photon*, as a smallest part of an electromagnetic wave, must still be able to do the same thing. A single photon can take both paths in the Michelson interferometer and then interfere *with itself* on the screen. It still lands on the screen as a particle, producing a single spot, but only at places where its interference with itself is not totally destructive. If it is totally destructive (the two split parts of the wave, when recombining, turn out to be out of phase, so the resulting amplitude is zero), the particle cannot land in the corresponding place. It has a tendency to hit primarily the spots where the parts of the associated wave interfere constructively (the amplitude of the resulting wave is maximal).

If all other particles are also waves, they should exhibit similar behavior in similar situations. In other words, de Broglie's hypothesis predicts the possibility of all wave phenomena – superposition, diffraction, and interference – for particles as well. Such prediction was experimentally confirmed (Figure 10.6).

The most striking part of these confirmed predictions is the possibility for a single particle to be simultaneously in two or more states that classically are mutually exclusive. Thus, the results of the electron diffraction in the double-slit experiment can be only explained if we accept that the electron before hitting the screen does indeed take both available paths at once. Any attempt to peep in and catch it in action

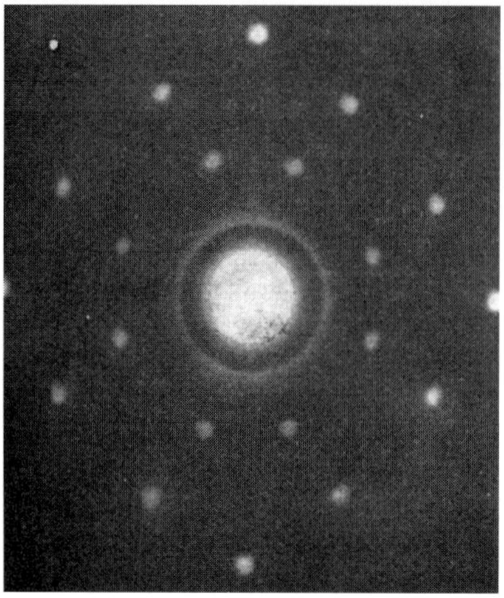

Figure 10.6 Neutron diffraction.

results in the whole electron being found in one or the other path only, but this destroys the interference!

Superposition can be formed not only by waves with different phases (resulting, for instance, from different paths of arrival), but also with different frequencies. The resulting wave will no longer be monochromatic and may have an arbitrary shape. And vice versa, an arbitrary shape can be represented as a combination of monochromatic waves with different frequencies. Mathematically, this corresponds to Fourier expansion of an arbitrary function:

$$\Psi(x) = \sum_j C(\omega_j) e^{i(k_j x - \omega_j t)}, \qquad \omega_j = j\omega_0, \quad j\text{-integer}. \tag{10.22}$$

A well-known example is again a vibrating string. It may vibrate with only one of the set of allowed frequencies, producing a pure tone. Its instant shape is then a sinusoid. But it also can vibrate with a few different frequencies at once and all of them can be recorded with the corresponding set of tuning forks. This is a superposition of states with different frequencies. The corresponding shape in the resulting state cannot be represented by only one sine or cosine function.

Now, if a particle is a wave, it can also be in a superposition of states with different frequencies.

But, as definite frequency corresponds to a definite energy via (10.1), it follows that a single particle can be simultaneously in states with different energies. Mathematically

$$\Psi(\mathbf{r}) = \sum_j C(\mathcal{E}_j) \Psi_j(\mathbf{r}) e^{-i\omega_j t}, \qquad \omega_j = \frac{\mathcal{E}_j}{\hbar}. \tag{10.23}$$

The resulting state cannot be characterized by a sharply defined energy.

(III) The above examples illustrate the fundamental role of the concept of a wave function in the description of a quantum-mechanical system. In particular, they demonstrate how efficient the concept of waves is in understanding and quantitative description of the orbital momentum and energy quantization, as well as superposition of states. However, an attempt to answer the question "The waving of what is represented by the wave function?" shows its rather subtle and even somewhat elusive physical nature. The probability amplitude Ψ_0 of the monochromatic de Broglie wave (10.4) definitely does not wave. The exponential factor in (10.4) does "wave," but it is merely a complex function of $\mathbf{r}$ and t, which by itself cannot be experimentally observed for a free particle in an empty space. This factor does not seem to correspond to an undulating physical entity like a water surface on a lake or electric field in a light wave. The latter may be described by the *real part* of an expression similar to (10.4), while the correct description of a *light quantum* (the photon) uses the whole complex function (10.4) in its entirety.

It is easy to show, using the nonrelativistic limit of QM, that the wave function of a particle must be complex in its entirety for a fundamental reason: it not only stores *all* the information about the particle's state at a given moment, but also determines its evolution for all future moments. In other words, knowing $\Psi(\mathbf{r}, 0)$ determines $\Psi(\mathbf{r}, t)$. This can only be the case if $\Psi(\mathbf{r}, t)$ is a solution of differential equation of the first order in time:

$$i\hbar \frac{\partial \Psi}{\partial t} = \hat{H}\Psi. \tag{10.24}$$

Here,

$$\hat{H} = -\frac{\hbar^2}{2m}\nabla^2 + U(\mathbf{r}) \tag{10.25}$$

is a differential operator called Hamiltonian operator, which for a free particle (the potential energy $U(\mathbf{r}) = 0$)) reduces (up to a constant) to the Laplacian ∇^2. Equation (10.24) is Schrödinger's famous equation. In contrast to the d'Alambert equation that, for a monochromatic wave, can be equally well satisfied with $\sin \omega t$, $\cos \omega t$, or an exponential function $e^{\pm i\omega t}$, the first-order (in time) wave equation can only be satisfied by an exponential function of the latter type, which is an explicitly complex function.

The abstract nature of the wave function becomes even more evident if we consider a *system* of particles. If we have N particles, then the wave function describing the whole system is, in the case of $\mathbf{r}$-representation, a function of at least $3N$ variables:

$$\Psi(\tilde{\mathbf{r}}) = \Psi(\mathbf{r}_1, \mathbf{r}_2, \ldots, \mathbf{r}_N, t) = \Psi(x_1, y_1, z_1; x_2, y_2, z_2; \ldots, x_N, y_N, z_N; t). \tag{10.26}$$

According to Born's interpretation, its square modulus gives the probability density to find one particle at $\mathbf{r}_1$, another particle at $\mathbf{r}_2$, ..., and the Nth particle at $\mathbf{r}_N$ at the moment t. The probability itself will be given by

$$d\mathcal{P}(\tilde{\mathbf{r}}, t) = |\Psi(\tilde{\mathbf{r}}, t)|^2 d\tilde{\mathbf{r}}. \tag{10.27}$$

Here, $\tilde{\mathbf{r}}$ can be considered as a $3N$-dimensional vector with $3N$ components

$$\tilde{\mathbf{r}} = (\mathbf{r}_1, \mathbf{r}_2, \ldots, \mathbf{r}_N) = (x_1, y_1, z_1; x_2, y_2, z_2; \ldots; x_N, y_N, z_N), \tag{10.28}$$

and $d\tilde{\mathbf{r}}$ is the corresponding volume element:

$$d\tilde{\mathbf{r}} = d\mathbf{r}_1\, d\mathbf{r}_2 \ldots d\mathbf{r}_N, \tag{10.29}$$

with each subelement determined, like in (10.27), by

$$d\mathbf{r}_j = dx_j\, dy_j\, dz_j, \quad j = 1, 2, \ldots, N. \tag{10.30}$$

In the simplest possible case, when the particles do not interact with and are totally independent from one another, the above probability, as the probability of independent events, will be the product of the individual probabilities to find each particle at a respective location (we drop the time dependence here):

$$d\mathcal{P}(\tilde{\mathbf{r}}) = d\mathcal{P}(\mathbf{r}_1, \mathbf{r}_2, \ldots, \mathbf{r}_N) = d\mathcal{P}_1(\mathbf{r}_1)\, d\mathcal{P}_2(\mathbf{r}_2) \ldots d\mathcal{P}_N(r_N). \tag{10.31}$$

Accordingly, the probability amplitude will be the product of the individual amplitudes:

$$\Psi(\tilde{\mathbf{r}}) = \Psi(\mathbf{r}_1, \mathbf{r}_2, \ldots, \mathbf{r}_N) = \Psi_1(\mathbf{r}_1)\Psi_2(\mathbf{r}_2) \ldots \Psi_N(\mathbf{r}_N). \tag{10.32}$$

If we want in all these examples to consider $\Psi(\tilde{\mathbf{r}})$ as a wave, it will be a wave in an abstract $3N$-dimensional space.

We can come to the same conclusion from a slightly different perspective. It can be shown that the wave function satisfies the continuity equation (see Appendix A):

$$\frac{\partial \rho}{\partial t} + \vec{\nabla} \cdot \mathbf{j} = 0, \tag{10.33}$$

where ρ is the probability density (10.15), and

$$\mathbf{j} = \frac{i\hbar}{2m}(\Psi\, \vec{\nabla}\Psi^* - \Psi^*\, \vec{\nabla}\Psi) \equiv -\frac{\hbar}{m}\, Im(\Psi \vec{\nabla} \Psi^*) \tag{10.34}$$

is the probability flux density.

If we multiply the ρ and $\mathbf{j}$ by the mass of a particle $\rho_m \equiv m\rho$, $\mathbf{j}_m \equiv m\mathbf{j}$, then the obtained new quantities ρ_m and $\mathbf{j}_m$ have a very simple interpretation as the mass and flux density of matter in a given state, which satisfy the corresponding equations of fluid dynamics. Similarly, multiplying ρ and $\mathbf{j}$ by a particle's charge q, we obtain the electric charge and current density in a given state. In either case, the corresponding densities satisfy the continuity relation identical with (10.33). These relations express the conservation laws for the mass and electric charge, respectively, and can be considered as an additional argument for probabilistic interpretation of the wave function. But again, the simple view of the amplitude Ψ as something whose square modulus represents certain density in space becomes far less straightforward for a *system* of particles. In this case, both expressions (10.33) and (10.34) are functions of many variables satisfying the continuity equation in an abstract multidimensional space rather than in the real space.

So it is better to consider the wave function just as a mathematical expression storing all information about the physical state of a system, and accordingly, we can (and frequently will) call it the "state function" as well.

(IV) It follows from I and II that physical attributes of a system can be represented by operators. For example, the energy or momentum quantization considered above can be obtained mathematically as a result of the solution of a certain differential equation for a wave function satisfying the corresponding boundary conditions. In the case of energy, the corresponding equation for a particle in a stationary environment

$$\hat{H}\Psi(\mathbf{r},t) = \mathcal{E}\Psi(\mathbf{r},t) \tag{10.35}$$

has solutions

$$\Psi(\mathbf{r},t) = \Psi(\mathbf{r})\,e^{-i\frac{\mathcal{E}}{\hbar}t}, \tag{10.36}$$

satisfying these conditions only for a special set of the allowed energies $\mathcal{E}$. Equation (10.35) is time-independent Schrödinger's equation, where $\hat{H}$ is the Hamiltonian operator, which for a particle in a potential field, is defined in (1.10.25) (from now onward we will denote the operators in **r**-representation by capped symbols). As the same wave function on the right of (10.35) is multiplied by the corresponding energy, the Hamiltonian represents energy of the system in the given state.

As another example, consider the momentum operator in representation, in which the state function of a system is explicitly expressed as a function of position **r** (**r**-representation). This must be an operator that, when applied to a state function (10.4) representing a state with sharply defined momentum **p**, will produce the same function multiplied by **p** (Problem 10.1). Such an operator is

$$\hat{\mathbf{p}} = -i\hbar\vec{\nabla}, \tag{10.37}$$

where $\vec{\nabla}$ is the well-known del-operator (gradient) (4.116).

It is worthwhile to show the origin of operators in a more general way, which illustrates the operational structure of QM. Consider first an arbitrary state $\Psi(\mathbf{r})$ in **r**-representation (we drop here the possible t-dependence). According to probabilistic interpretation, $|\Psi(\mathbf{r})|^2$ is the probability density to find the particle (originally prepared in the state $\Psi(\mathbf{r})$) in the vicinity of point **r**. As the Nature, according to Born's interpretation, is intrinsically probabilistic, another measurement on an identical particle originally in the same state $\Psi(\mathbf{r})$ may find it at another location $\mathbf{r}'$. Because of the variability of the outcomes in measurement of the same characteristics in the same state, we generally characterize the state by the *mean* of this characteristic. We thus can introduce the mean position (an average $\langle \mathbf{r} \rangle$) in a given state:

$$\langle \mathbf{r} \rangle = \int \mathbf{r} \mathcal{P}(\mathbf{r}) d\mathbf{r} = \int \Psi^*(\mathbf{r}) \mathbf{r} \Psi(\mathbf{r}) \, d\mathbf{r}. \tag{10.38}$$

Similarly, we can find the mean momentum in the same state. If we use, as before, the same **r**-representation of the state, then the mean momentum is determined by the expression (Appendix B):

$$\langle \mathbf{p} \rangle = \int \Psi^*(\mathbf{r}) \hat{\mathbf{p}} \Psi(\mathbf{r}) \, d\mathbf{r}, \tag{10.39}$$

where $\hat{\mathbf{p}}$ in the integrand is precisely the operator determined by (10.37)!

Thus, to find the mean *momentum* when the state is expressed in terms of *position* (i.e., in **r**-representation), we use the same formula (10.39) as for the mean position but with **r** replaced by an operator $\hat{\mathbf{p}}$. Since the use of this operator in (10.39) determines the mean momentum in the corresponding state, it is natural to name it the *momentum operator*. It can be considered as the "position representative" of observable momentum and this representative (**r**-representative) is identical with (10.37). Thus, (10.37) defines the momentum operator in **r**-representation.

Generally, the mean value of an arbitrary observable L in a state expressed in **r**-representation is given by

$$\langle L \rangle = \int \Psi^*(\mathbf{r}) \hat{L} \Psi(\mathbf{r}) \, d\mathbf{r}, \tag{10.40}$$

where $\hat{L}$ is the L-operator in **r**-representation.

The operators representing physical observables must satisfy two important requirements.

First, they must be linear; if Ψ is a superposition $\Psi = c_1\Psi_1 + c_2\Psi_2$ with coefficients c_1 and c_2, then the application of the linear operator to Ψ must be again the linear superposition:

$$\hat{L}\Psi = \hat{L}(c_1\Psi_1 + c_2\Psi_2) = c_1\hat{L}\Psi_1 + c_2\hat{L}\Psi_2. \tag{10.41}$$

Second, they must be self-conjugate, which means the following. As seen from the example (10.37), an operator $\hat{L}$ may be a complex expression. But for it to represent a physical observable, which is a real number, the whole expression on the right of (10.40) must be real. This imposes a specific requirement on the quantum-mechanical operators representing physical observables: the complex conjugate of the expression on the right of (10.40) must not change the expression. More generally, for any two state functions $\Psi_1(q)$ and $\Psi_2(q)$, the operator $\hat{L}(q)$ must satisfy the condition

$$\int \Psi_1^*(q)\hat{L}(q)\Psi_2(q)dq = \int \Psi_2(q)\hat{L}^*(q)\Psi_1^*(q)dq. \tag{10.42}$$

The operators satisfying this condition are called self-conjugate, self-adjoint, or Hermitian operators. Applying this to the case $q \to \mathbf{r}$, $\Psi_1 = \Psi_2 = \Psi$, it is easy to check that the right-hand side of (10.40) is, indeed real, if the operator $\hat{L}$ is Hermitian. It is also easy to prove that the momentum operator (10.37) is Hermitian in the class of the square-integrable functions (Problem 10.2).

Now, suppose you are a statistician studying the population of a certain country or a coach of a basketball team. In the first case, you may be interested not only in the average longevity of the population, but also in the individual longevities' spread around the average. In the second case, you will be interested not only in the average tallness of your players, but no less important in its distribution among individual players. Similarly, if we measure an observable L in a system described by a state function $\Psi(\mathbf{r})$, we may be interested not only in the mean value of L but also in how the individual measurement results are spread around this value. The measure of the spread of the variable L around its mean is given by the *mean-square deviation* $\langle(\Delta L)\rangle^2$:

$$\langle(\Delta L)\rangle^2 \equiv \langle(L-\langle L\rangle)^2\rangle. \tag{10.43}$$

We will call its square root the *variance*:

$$\mathrm{Var}\{L\} \equiv \Delta L \equiv \sqrt{\langle(\Delta L)^2\rangle} = \sqrt{\langle(L-\langle L\rangle)^2\rangle}. \tag{10.44}$$

If a certain *quantum-mechanical* quantity L is an observable and is accordingly represented by an operator $\hat{L}$, then an individual deviation of such observable from its mean, as well as the square of this deviation, are also observables. Accordingly, they can be represented by the respective operators

$$\Delta\hat{L} \equiv \hat{L}-\langle L\rangle \tag{10.45}$$

and

$$\hat{V}\{L\} \equiv (\Delta\hat{L})^2 = (\hat{L}-\langle L\rangle)^2. \tag{10.46}$$

We leave it to the reader to prove (Problem 10.4) that if $\hat{L}$ is Hermitian, then the operators $\Delta\hat{L}$ and $\hat{V}\{L\}$ are also Hermitian.

The quantum-mechanical mean-square deviation is given by (10.40) with the operator $\hat{L}$ replaced by $(\Delta\hat{L})^2$:

$$\langle(\Delta L)^2\rangle = \int \Psi^*(q)(\Delta\hat{L})^2\Psi(q)dq. \tag{10.47}$$

It is important to distinguish between the classical and quantum-mechanical deviation. In classical physics, the deviation results from natural experimental errors accompanying each measurement. Such errors cannot be eliminated entirely, but they can be made very small as compared to the measured value itself. In other words, fractional deviation (and accordingly, variance) $\Delta L/L$ can be, at least in principle, made negligible.

The situation is fundamentally different in quantum mechanics. Generally, quantum-mechanical variance cannot *in principle* be reduced below a certain level, characteristic for a given state, even under ideal conditions when all possible experimental errors are totally eliminated so that classical variance would be zero (later we will see examples illustrating this point).

There are, however, *special* states for any given quantum-mechanical observable, in which this observable has a sharply defined value and accordingly the zero variance. The mathematical condition selecting these states is a requirement that the corresponding variance (10.47) be zero.

As the operator $\Delta\hat{L}$ is Hermitian, we can write this requirement in the form

$$\int \Psi^*(q)\Delta\hat{L}\,\Delta\hat{L}\,\Psi(q)dq = \int \Delta\hat{L}^*\Psi^*(q)\,\Delta\hat{L}\,\Psi(q)dq = \int |\Delta\hat{L}\,\Psi(q)|^2 dq = 0. \tag{10.48}$$

It follows immediately that in the corresponding states, $\Delta\hat{L}\,\Psi(q) = 0$ or, in view of (10.45) and the fact that $\langle L\rangle = L$ in a state with sharply defined L,

$$\hat{L}\Psi(q) = L\Psi(q). \tag{10.49}$$

Thus, the application of the $\hat{L}$-operator to the state function of a state with the zero variance in L is equivalent to just multiplying this function by a number. Such a function is called an eigenfunction of the corresponding operator and the number L is the eigenvalue of this operator. Equation (10.35) determining the *energy* eigenfunctions and eigenvalues is a special case of Equation (10.49).

If $q \to \mathbf{r}$ (coordinate representation), then $\hat{L}\,(q) \to \hat{L}(\mathbf{r})$ may be, like in (10.37), a differential operator acting on $\Psi(\mathbf{r})$. We see that $\Psi(\mathbf{r})$ is in this case a solution of a linear differential equation.

An interesting question is what operator represents position-variable – the position-vector $\mathbf{r}$?

Evidently, in $\mathbf{r}$-representation it is just $\mathbf{r}$. This is an example of somewhat tautological statement: operator of any observable in its own representation is the observable itself.

Exercise 2

Find the eigenfunctions of the $\mathbf{r}$-operator in $\mathbf{r}$-representation.

Solution

An obvious (but wrong) answer would be: any function. Indeed, if $\hat{\mathbf{r}} = \mathbf{r}$, then

$$\hat{\mathbf{r}}\Psi = \mathbf{r}\Psi = \mathbf{r}\Psi. \tag{10.50}$$

Here, the equation for eigenfuctions becomes an identity. This solution is wrong because $\mathbf{r}$ in its both parts is a variable, whereas in the true equation for the eigenfunctions, $\mathbf{r}$ on the right must be a specific eigenvalue, *not* the variable. This is like in programming: there is a variable name and there is a specific value of this variable. To distinguish between them, denote the eigenvalue as $\mathbf{r}'$. Then the correct equation for an eigenfunction takes the form

$$\mathbf{r}\Psi = \mathbf{r}'\Psi. \tag{10.51}$$

The only function satisfying this equation is delta function

$$\Psi_{\mathbf{r}'}(\mathbf{r}) = \delta(\mathbf{r}-\mathbf{r}'). \tag{10.52}$$

(V) Another manifestation of superposition of states is indeterminacy of some physical characteristics of a system. According to the "classical" vision of the world, all characteristics specifying a physical system can be determined simultaneously to an arbitrary high precision; if we fail to do so, it is only because of our incompetence or imperfection of our devises. In quantum mechanics, certain pairs of quantities, such as position and momentum of a particle, cannot *in principle* be determined simultaneously to an arbitrarly high accuracy. We can see immediately the origin of this principle from the de Broglie postulate in the following way. Imagine a particle in a state with definite momentum p. It is represented by the de Broglie wave (10.4). According to the Born interpretation of the wave function, the corresponding probability distribution is uniform throughout the whole space. The particle cannot be ascribed a distinct position – it is totally indeterminate. Suppose now that we have performed an accurate position measurement and found a particle at a point $\mathbf{r}'$. This new state has been created by the measurement and can be described by a function that is infinite at $\mathbf{r}'$ and zero elsewhere (Dirac's delta function $\delta(\mathbf{r}-\mathbf{r}')$ – an eigenfunction (10.52) of position operator). The delta function can be represented by an infinite number of the de Broglie waves with different $\mathbf{p}$, which reinforce each other in one place (at $\mathbf{r}'$), producing through constructive interference a huge splash there, and interfere destructively canceling each other altogether elsewhere. In other words, this state is a superposition of states with all possible momenta and it cannot be attributed only one distinct momentum.

Thus, in a state with definite momentum, a particle has no definite position, and in a state with definite position it has no definite momentum. Generally, the more accurately is determined one of these attributes in a certain physical state, the less accurately, *in this state*, is determined the other one. Quantitatively, if we express the indeterminacy of position x in terms of the above-discussed variance Δx and the corresponding indeterminacy in the x-component of momentum p_x as the corresponding variance Δp_x, then these indeterminacies satisfy the requirement

$$\Delta x\, \Delta p_x \geq \frac{1}{2}\, \hbar. \tag{10.53}$$

This is the Heisenberg so-called uncertainty relationship – probably the second (after $\mathcal{E} = mc^2$) most famous equation in physics. It tells us that the product of Δp_x and Δx cannot be less than one half of $\hbar$; so, the more accurately is determined, say, the coordinate x, the less accurate (better to say more indetermined) is the x-component p_x of momentum and vice versa.

Variables (observables) a and b that cannot be measured simultaneously are called *incompatible*. The corresponding operators $\hat{A}$ and $\hat{B}$ representing them do not commute: their successive application to a function gives different result depending on ordering, that is,

$$\hat{A}\hat{B}\Psi \neq \hat{B}\hat{A}\Psi \quad \text{or} \quad (\hat{A}\hat{B}-\hat{B}\hat{A})\Psi \neq 0. \tag{10.54}$$

The opposite is also true: if two linear Hermitian operators representing different variables do not commute, the corresponding variables are incompatible. The examples of incompatible variables are p_x, x; p_y, y; p_z, z. On the contrary, s_z and z, or x and p_y are pairs of compatible variables.

Introducing a new operator called commutator

$$\hat{C} \equiv [\hat{A}, \hat{B}] \equiv \hat{A}\hat{B} - \hat{B}\hat{A}, \tag{10.55}$$

we can formulate the general uncertainty principle in terms of variances of the corresponding observables:

$$\Delta a\, \Delta b \geq \frac{1}{2} |\langle \hat{C} \rangle|. \tag{10.56}$$

This constitutes the Robertson–Schrödinger theorem: if the operators $\hat{A}$ and $\hat{B}$ representing two observables a and b do not commute (their commutator $\hat{C} \neq 0$), then the corresponding observables are incompatible and the product of their variances Δa and Δb cannot be less than the expectation value of the commutator $\hat{C}$.

I want to caution you here about the word "uncertainty" frequently used for such characteristics as Δx or Δp_x, or generally, for variances Δa and Δb of any two incompatible variables a and b. It may produce an impression that the actual values do exist, but they are only unknown to us probably owing to some subtle flaw in the communication between the real world and the observer; and maybe this uncertainty will be eliminated someday when we find out the nature of this flaw. Actually, as far as we know today from the overwhelming amount of scientific evidence, Δp_x and Δx are *real indeterminacies* in position and momentum of a system, rather than observer's uncertainty about them. These indeterminacies are the objective characteristics of the system and therefore it would be more appropriate to call the uncertainty principle the "indeterminacy principle."[4)]

(VI) Even more fascinating result of superposition of states is found in a *system* of particles.

We will consider here the simplest possible example of only two particles. Suppose that each particle has a spin (intrinsic angular momentum) $s = 1/2$. Then each one of them can exist in one of the two possible states characterized by the spin component along a fixed direction z: either with spin up or with spin down. In this way the rule (10.7), requiring that the spin projections onto an axis form discrete set with "period" $\hbar$, is satisfied (you can find more about spin in Appendix F).

Let us denote the corresponding state function of particle 1 as Ψ_1 and the state function of particle 2 as Ψ_2. Here, we will be interested only in spin state and accordingly will drop in the following equations the position and time dependence.

4) Historically, the German "unbestimmheit" in the first formulation of the principle means "indeterminacy." The term "uncertainty" in the English formulation of this principle is the result of a sloppy translation.

Then the possible spin states of particle 1 can be written as

$$\Psi_1 = \begin{cases} |\uparrow\rangle_1 \text{ (spin up)}, \\ |\downarrow\rangle_1 \text{ (spin down)}, \end{cases} \tag{10.57}$$

and similarly for the second particle, with index 1 replaced by 2. Each one of the states on the right side of Equation (10.57) is the eigenstate of the spin operator (more accurately, of the operator s_z of the spin projection onto the z-direction). Of course, either particle taken separately can also be in a superposition of these two states:

$$|\tilde{\Psi}_\alpha\rangle = c_{\alpha\uparrow}|\uparrow\rangle_\alpha + c_{\alpha\downarrow}|\downarrow\rangle_\alpha, \qquad \alpha = 1, 2. \tag{10.58}$$

The expansion coefficients here are the probability amplitudes describing the same state $|\tilde{\Psi}_\alpha\rangle$ in s_z-representation and satisfying the normalization condition (10.19): $|c_{\alpha\uparrow}|^2 + |c_{\alpha\downarrow}|^2 = 1$.

But the state (10.58) is not the kind of superposition that will interest us now. We will be interested in a *combined* state of the whole system. There can be many such states. Consider one of them, specified by the condition that the net spin of the system is zero. In this case, the projection of the *net* spin on any axis is, of course, zero. But a single particle can only have one out of the two projections described in (10.57), neither of which is zero. The only possibility for the *net* projection to be zero is that spin of one particle is opposite to the spin of the other. How can we describe such a state? Well, the rule (10.32) suggests that it should be the product of the individual amplitudes, that is, $\Psi_1\Psi_2$. But this is for the case when the particles are independent. Our two particles, even if they do not interact, and may be far apart, are not independent. They are correlated by the requirement that their spin directions are opposite to one another. So if the particle 1 has its spin up, then the particle 2 must have its spin down and vice versa.

Therefore, if we specify Ψ_1 as $|\uparrow\rangle_1$ then Ψ_2 must be specified as $|\downarrow\rangle_2$. Then the wave function of the whole system may be the product

$$|\Psi_{12}\rangle = |\uparrow_1\rangle|\downarrow\rangle_2. \tag{10.59}$$

But our specification is not as detailed as to specify the spin direction of either particle. It only specifies the correlation between these directions. Each particle knows that its spin must be the opposite to the spin of its partner, but neither of them knows its actual spin direction. The particle 1 might as well have its spin down, in which case the particle 2 must have its spin pointing up. In this case, the wave function of the system would be

$$|\Psi_{21}\rangle = |\downarrow_1\rangle|\uparrow\rangle_2. \tag{10.60}$$

If a system can be in either of these two states, then, according to the superposition principle, it can also be in both states at once. So the most general state satisfying the above requirement can be written as

$$\Psi = a|\Psi_{12}\rangle + b|\Psi_{21}\rangle = a|\uparrow\rangle_1|\downarrow\rangle_2 + b|\downarrow\rangle_1|\uparrow\rangle_2, \tag{10.61}$$

where the amplitudes a and b describe the weight of the corresponding state in the superposition. Neither particle in this state can be characterized independently of its partner. Their properties are entangled with one another even when they are far apart in space. Accordingly, neither of them has its distinct state function. Only the whole system has. The corresponding state (10.61) is called the entangled state and the whole phenomenon – the entanglement. We will discuss some interesting implications of this phenomenon in the following chapter (Section 11.10).

(VII) In classical statistical physics we often use the phase space to describe an ensemble of particles. For one particle moving in one direction the corresponding phase space is two-dimensional (2D): one dimension representing x and the other one representing p. At each moment the physical state of the particle is uniquely defined by its position (x, p) on the *phase diagram*. As time evolves, both position and momentum change and the point on the diagram representing the particle traces out a trajectory. In a well-known example of an oscillator, the trajectory is an ellipse (Figure 10.7).

Thus, the complete classical description using the phase space requires an additional dimension – let us call it the p-dimension – for each spatial dimension.

Since in QM the particle cannot have simultaneously definite p_x and x, its position on the diagram is fuzzy, filling out a patch, whose area, according to the indeterminacy principle, cannot be less than $(1/2)h$ (we could call it the quantum of phase space (Figure 10.8)). The patch can have any shape. For instance, it can be stretched into an infinitely narrow (and accordingly, infinitely long) stripe, say, parallel to the x-axis at $p = p_0$; this would correspond to the above example – the state with definite

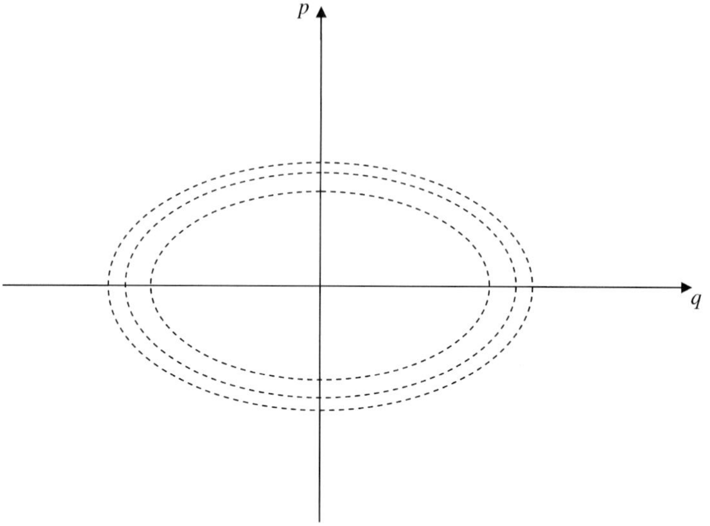

Figure 10.7 Trajectories of a harmonic oscillator in the phase space. The size of each ellipse is a continuous function of oscillator's energy $\mathcal{E}(p, q)$, where p and q are coordinate and the corresponding momentum, respectively.

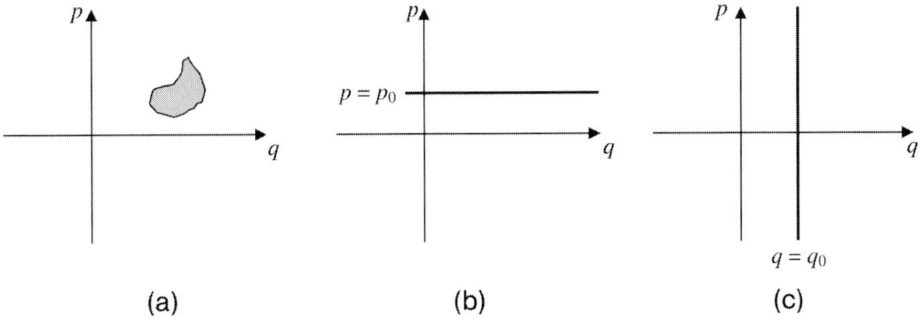

Figure 10.8

momentum p_0 and totally undefined position $-\infty < x < \infty$. Or, it could be stretched into similar vertical stripe at $x = x_0$; that would correspond to the state with definite position and totally undefined momentum $-\infty < p < \infty$. These two extremes can be considered as the examples of the so-called "squeezed" states (the term comes from the analogy with the toothpaste rapidly squeezed out of a tube). Another possibility is for the patch to form a closed loop. This situation applies to a particle performing a periodic motion along a certain direction q. Classically, for each position q, the particle would have a definite momentum $p(q)$ along this direction and its state would be represented by the ellipse on Figure 10.7 ($q = x$). In this case, the classical action for one period can be written as [87]

$$S = \oint p(q)\mathrm{d}q. \tag{10.62}$$

Geometrically, expression (10.62) gives the area enclosed by the loop and its value is uniquely determined by the oscillator's energy. As in the classical picture the latter is a continuous parameter, so is the area within the loop. Thus, it is a continuous function of energy

$$S = I(E). \tag{10.63}$$

Quantum-mechanically, however, this area can only be a multiple of h (Figure 10.9). This results in the energy quantization for any system performing periodic motion (the Bohr–Sommerfeld quantization condition [87–90])

$$I(E) = \oint p(q)\mathrm{d}q = \left(n + \frac{1}{2}\right)h, \qquad n = 0, 1, 2, \ldots. \tag{10.64}$$

But the additional p-dimension on the phase diagram remains linearly independent of the x-dimension, so the phase space remains two-dimensional.

(VIII) In classical physics, knowing all characteristics describing uniquely the system, for instance, the exact position and momentum of a point particle at some moment of time, as well as the forces acting on it, was sufficient to determine in all details its future behavior. In other words, the sequence of all future states of a system – its evolution – is predetermined by its current state and the environment it is in.

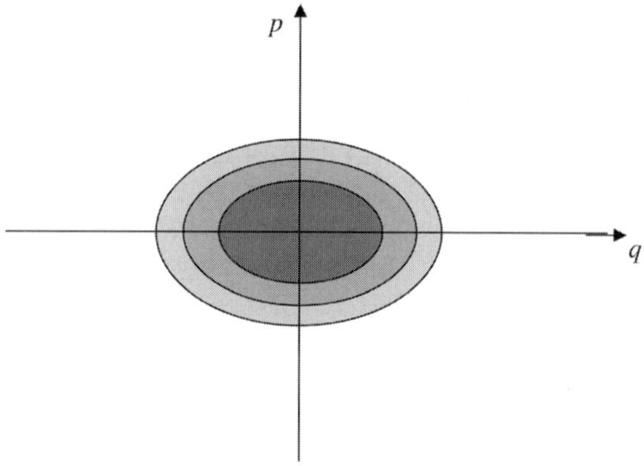

Figure 10.9 Trajectories of a quantum oscillator in the phase space. They look like those in Figure 10.7, but have quite different meaning. The oscillator with a fixed energy is described by the whole area of the corresponding ellipse, rather than by the ellipse itself. Also, in contrast with the classical physics, the area of an ellipse can only change by a fixed quantity h (is quantized). Since there is one-to-one correspondence between area and energy, together with area the energy is also quantized.

This statement, known as Laplacian determinism, seemed to be as fundamental as the concept of the absolute space and time.

Inherent in the classical determinism is the predictability of the future behavior of a system. A soccer player is able to estimate beforehand the trajectory of a soccer ball from its instant position and velocity and act accordingly. Astronomers can predict the future positions of celestial bodies (in particular, solar and lunar eclipses) with astounding accuracy.

According to quantum mechanics, the deterministic (predictable) description of the world is restricted. The state function of a system changes *continuously* with time between measurements and this change is described by the Schrödinger equation. Thus, evolution of the state function of a system can be exactly predicted – it is deterministic. But the wave function as such is not a dynamic variable – it only carries information about probability of measurement results of the relevant dynamical variables. Therefore, generally, only *probability* of a certain event (or a certain outcome of a measurement) can be predicted, not the event (the outcome) itself. According to (10.56), a system may even not possess any definite value of a certain characteristic, the exact knowledge of which, from classical viewpoint, may be absolutely essential for description of its evolution. This shows that the unpredictability is intimately linked with the indeterminacy principle.

As we will see below, the wave function (and the state of a system it describes) while changing continuously between measurements can change abruptly and unpredictably (collapse to another quite different state) in the process of measurement. The act of measurement and its exact outcome is not described by the Schrödinger equation. Many people, including Einstein, Schrödinger, and Bohm, regarded this as the evidence of incompleteness of quantum theory. However, all attempts to find a deeper reality, a sort of hidden parameters that eventually determine the measurement results, have failed. Moreover, as has been shown by Bell [24], the existing theory, on the one hand, and all its known modifications incorporating hidden parameters, on the other, lead to quite different predictions that can be checked experimentally. And all the experiments carried out up to now have unequivocally confirmed the conventional QM, according to which the world is intrinsically probabilistic. Chance turned out to be an irreducible element of reality.

(IX) One of the most striking features of QM is the role of observation. In classical physics, a system possesses its individual characteristics before and independent of any measurements; the latter (if carried out properly) can only reveal these characteristics to the observer or verify their expected values without perturbing the system. From this viewpoint, a measurement, while being important means of verifying or discovering the reality, cannot by itself change the reality; it can only serve as a "witness" or "recorder" of the reality.

In quantum physics, the measurement plays much more fundamental role because it can "create" a new reality by dramatically changing the state of a quantum system subject to measurement. This property has much to do with the existence of the minimal quantum of action (10.2). Indeed, any measurement (local or nonlocal!) involves interaction between the measured system and measuring device. Classically, this interaction can be made arbitrarily small, thus preserving the measured characteristic from undesired perturbations. According to quantum mechanics, however, the action associated with the interaction energy cannot be less than h; therefore, if the system itself is sufficiently small, the significant perturbations cannot be avoided and, accordingly, the exact outcome of measurement generally cannot be predicted. For instance, an electron in an atom has an indeterminacy in its position $\Delta q \cong a$, where a is the atomic size. One could ask *where exactly* the electron is in the atom at a given moment of time, but this would be a meaningless question unless and until one performs a position measurement, by, say, illuminating the atom with gamma photons and recording the characteristics of a scattered photon. The recording could, in principle, narrow down the indeterminacy from a to a far less value. However, in the process of such scattering, the electron would undergo uncontrollable exchange of momentum with the photon, which would radically change its initial state. It could even get from the photon a "kick" sufficient for it breaking loose of the atom altogether (ionization). Thus, we will get a state with a far more "finely" defined instant position of the electron at the moment of measurement, but this happens at the cost of a radical change of the very physical state of the system we wanted to measure. The atom will be destroyed (more politely, ionized), and the indeterminacy in the electron's momentum will, according to (10.56), sharply increase. This is in stark contradiction with the classical picture (the planetary model) mentioned in the beginning of this section, according to which the electrons orbit the atomic nucleus along definite trajectories with the values of both – position and momentum – being the exact functions of time.

Consider another example of measurement. Suppose, a moving particle with the initial momentum $\mathbf{p}$ passes through a very narrow slit of a width a. This can constitute a position measurement: the particle must have been in a state with $\Delta x \approx a$ at the moment of passing. This creates a new momentum state, with a certain spread in the x-component of momentum around $p_x = 0$. Suppose, we place a screen at distance z from the slit and the particle eventually lands at a point of the screen at distance x from its center, which is just opposite the slit. This may provoke us to think that at the moment when the particle was passing through the slit, when it was localized within the region of the size a, it has accordingly acquired the transverse component of momentum $p_x \approx p(x/z)$. If $a \ll x, z$, this can be considered as a very accurate measurement of the transverse momentum, that is, $\Delta p_x \ll \hbar/a \approx \hbar/\Delta x$. It looks like refutation of Heisenberg's indeterminacy.

The flaw in this reasoning is in the assumption that the particle after passing the slit was following toward its destination along the classical path making an angle $\theta = \arctan(x/z)$ with its initial direction of motion. Actually, the narrow slit plays the role of a source of a diverging spherical (or cylindrical, depending on length of the slit) wave; and the particle leaves the slit in a superposition of plane de Broglie's waves forming in their entirety such diverging wave. In other words, the particle recedes from the slit in all directions at once and if we calculate the corresponding variance in Δp_x, we will find that it exceeds the minimum $\hbar/a$, in total accord with the indeterminacy principle. Each one of all virtual directions followed by the particle has a certain chance to actualize. One of these chances materializes when the particle experiences shock from collision with the second screen. Its wave function collapses from the superposition of infinite number of waves (i.e., from diverging wave illuminating the whole screen) to a singularity (δ-function) determined by the position of the landing point. In other words, the collision creates a new state, which is an eigenstate of the position operator, not a momentum operator. The corresponding experiment does not constitute a momentum measurement for a particle "squeezing" through the slit.

As we mentioned above, any experimental setup for position measurement "creates" a new state with definite position, but this destroys the wavelike aspects of the object.

(X) In accordance with the indeterminacy principle, the physical state of a system can generally be completely specified by a more abstract entity than its directly observable characteristics; this entity can be considered as a vector. As such a vector specifies a physical state, we can call it a state vector. All physically possible state vectors reside in an abstract vector space – the so-called Hilbert space. Each dimension of the Hilbert space represents a specific state in which corresponding variable of interest has a sharply defined value. We can draw some analogy from our real space. In this space, we frequently use a triplet of vectors $\hat{\mathbf{x}}, \hat{\mathbf{y}}, \hat{\mathbf{z}}$ as a basis for a Cartesian coordinate system. Accordingly, $\hat{\mathbf{x}}, \hat{\mathbf{y}}, \hat{\mathbf{z}}$ are called the basis vectors. Similarly, we can call the physical states of a *quantum system* chosen for the description of its properties, the basis states. By analogy with a triplet of *basis vectors* $(\hat{\mathbf{x}}, \hat{\mathbf{y}}, \hat{\mathbf{z}})$, used in our physical space, we call these states basis states and use the same term for the mutually perpendicular unit vectors representing these states.

Thus, the degree of freedom associated with the spin variable is, in the case of an electron, described by the two-dimensional Hilbert space – one dimension for spin up and the other for spin down. If we also want to add the possible momentum states for a more complete description of the electron, we need to add the corresponding new dimension (new basis vector linearly independent of all others) for each possible momentum, to the original Hilbert space. Since the number of distinct momentum states of an electron is infinite, the resulting new Hilbert space is not just multidimensional; it has an infinite number of dimensions! Moreover, in the case of the continuous spectrum, all possible dimensions of the Hilbert space form an indenumerable set!

We might as well represent the electron's state in terms of its position, rather than momentum. Then, even in case of only one dimension (say, only x-direction) in real space, each point for electron's residence on the x-axis is a distinct physical state and should be represented by an independent eigenvector; accordingly, we have again as many dimensions of the associated Hilbert space, as the amount of points on the line x!

Just as in the case of momentum, all points of a line x form an indenumerable set. This means that the corresponding Hilbert space is not only infinitely dimensional (∞-D, in our notations), but indenumerably dimensional!

Do not even try to visualize such a space in its "wholeness." We can make only a few steps in this direction. We may attempt to imagine an eigenvector associated with a point $x = x_\alpha$ on the line as a unit vector "sticking out" from this point in some direction. But, as the directions for unit vectors representing each its respective point must be not just different, but all mutually perpendicular, our ordinary 3D space can only "accept" three such vectors, each corresponding to one out of three different arbitrarily chosen points on the line (Figure 10.10a). We can make it look more similar to familiar 3D Cartesian coordinate system with the triplet of unit vectors with common origin as the basis of the system, if we bring all three vectors to one origin. In doing this, we do not lose any information about the locations of the points being represented by these vectors, since each vector carries the corresponding label (Figure 10.10b). For instance, if $x_2 = 5.3$, then the basis vector $\hat{\mathbf{x}}_2$ sticking out of its new position at the origin in Figure 10.10b knows that it represents an eigenstate corresponding to the eigenvalue $x = 5.3$ of the position operator.

For each new point on the line we have to add a new direction perpendicular to all three already employed, and this is beyond our ability to depict things graphically. It is because, generally, the Hilbert space is complex, and so are its basis vectors, so we cannot visualize even one of them as a single directed segment representing an ordinary vector.[5)]

But even though we cannot visualize multidimensional (and complex) vector spaces, we can understand them as abstract spaces with the same rules of operation as those in familiar spaces of 1, 2, and 3 dimensions.

5) Strictly speaking, the Hilbert space is defined for square integrable functions. As the spectrum of eigenvalues of the position operator is continuous, its eigenfunctions are not square integrable in the ordinary sense. The normalization condition for them is given by (10.19).

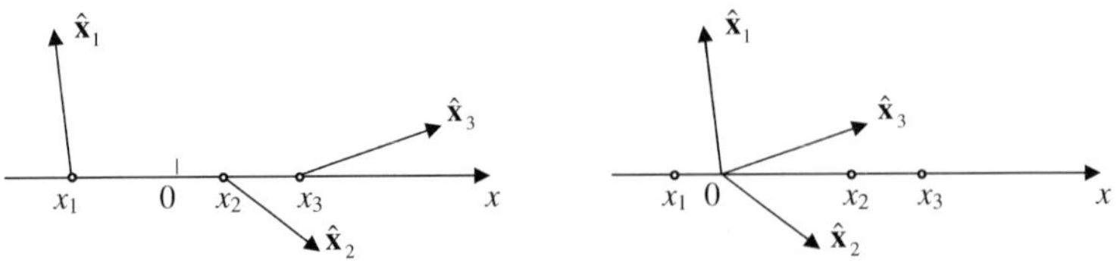

(a) (b)

Figure 10.10 A subspace of the Hilbert space for three eigenfunctions of the position operator, corresponding to the eigenvalues x_1, x_2, and x_3. (a) Points $x=x_1$, $x=x_2$, $x=x_3$ on the x-axis are different eigenvalues of the $\hat{\mathbf{x}}$-operator. The corresponding eigenstates can be represented graphically by mutually perpendicular unit vectors $\hat{\mathbf{x}}_1, \hat{\mathbf{x}}_2, \hat{\mathbf{x}}_3$ sticking out each from its respective point. (b) All three vectors are brought to a common origin by parallel translations, which do not change the vectors and thereby the represented states.

(XI) If you are still in doubt how can a line forming only a 1D space give rise to ∞-D Hilbert space, ask yourself: what do we plot along a basis direction of our ordinary space and along, say, a basis "x'-direction" of the corresponding Hilbert space?

Plotted along the basis x-direction of the ordinary space is the geometrical projection of a position vector $\mathbf{R}$ onto the x-axis. Plotted along the x'-direction of the corresponding Hilbert space is the "projection" $a_{x'}$ of the state vector $|\Psi\rangle$; but, in addition, it has a physical meaning that makes a fundamental departure from classical intuition: the square modulus of this projection gives only the *probability* to find, upon the appropriate measurement, the particle described by a state vector $|\Psi\rangle = \psi(x)$ (Equation (10.2)) at a point $x = x'$. This reflects the fact that the world, on the atomic level, is intrinsically probabilistic. In case when the considered states are discrete (the corresponding characteristic specifying the system has a discrete spectrum), we can find the total probability by summing all these individual probabilities. Geometrically, the result is square of the length of the state vector (Pythagorean theorem in multidimensional space!). Physically, the result is the net probability to find the system in any of its allowed states. As the system is certainly in one of its allowed states, this net probability is, of course, the certainty, that is, it is equal to 1. Thus,

$$|a(n)|^2 \equiv \mathcal{P}(n) \leq 1; \quad \sum_n |a(n)|^2 = |\langle\Psi|\Psi\rangle|^2 = 1. \tag{10.65}$$

The state vector in this case turns out to have always the fixed length equal to one. In other words, its tip cannot depart from the surface of multidimensional sphere of the unit radius. It can only slide over this sphere, "scanning" its surface.

If the characteristic in question has a continuous spectrum (say, a space coordinate q), then we can only talk about some finite probability $d\mathcal{P}(q)$ to find it within a small interval $(q, q + dq)$. By analogy with mechanics of continuous medium we can define the corresponding probability density $w(q)$ as

$$w(q) \equiv \frac{d\mathcal{P}(q)}{dq}. \tag{10.66}$$

In these cases the probabilistic interpretation says that the square modulus of a wave function in the corresponding representation gives the probability density to find the system at q.

So the variable plotted along the "direction" x' is $a_{x'}$, whose modulus, according to (3), can have any value between 0 and ∞, depending on state ψ. The variable plotted along some other "direction," say, x'', would be $a_{x''}$, also depending on state ψ and generally with the same range of change as $a_{x'}$.[6] Again, keep in mind that here we consider, for instance, $a'_x = \psi(x')$ as a variable depending *not* on x' (x' is fixed as an eigenvalue of the position operator), but on physical state of the system. For example, consider different eigenstates $\psi(x')$, $\psi'(x')$, $\psi''(x')$ of an operator other than the position operator or different possible solutions of the Schrödinger equation corresponding to different boundary conditions or even the same solution of the time-dependant Schrödinger equation, but just evolving with time. In all these cases, the state vector $\psi(x')$ evaluated for the same x' spans through the corresponding Hilbert space, changing thereby its component $a_{x'}$ along the x'-direction of the Hilbert space.

So far we have considered a state vector $\psi(x)$ only as a function of coordinate (time not counted). This may be a complete description for a spinless particle. Suppose now that we have a spin 1/2 particle, say, an electron. Then, apart from describing the particle with respect to the position variable, we must also include the spin variable. We know that the spin variable for the electron has only two eigenvalues, corresponding to two possible spin components onto a chosen direction (call it z-direction): $s_z = +1/2$ and $s_z = -1/2$. Question: what is the dimensionality of the corresponding Hilbert space? Answer: 2∞ (twice the indenumerable set of dimensions corresponding each to its respective x and to one out of two of its respective spin states at each x.) In other words, you can find an electron at each point being in one out of the two different definite spin states: either with its spin up or with its spin down. In this case the position operator has *two* different linearly independent eigenstates, both corresponding to the same eigenvalue of this operator. Hence, the number of dimensions of the corresponding Hilbert space doubles.

In the classical phase space, the direction p_x is perpendicular to the x-direction. Therefore, if we consider a state with definite p (the p-eigenstate), it is tempting to think that the corresponding direction must be "orthogonal" to the Hilbert space representing the "x-eigenstates" considered above; and if we want to include it into the picture, we must expand the previous Hilbert space by adding a new direction perpendicular to all other preexisting directions. But this would be wrong. Indeed, according to the general relation (10.34), each p-eigenstate can be represented as an expansion over different "x-eigenstates":

$$\psi_p(x) = \frac{1}{\sqrt{2\pi}}\, e^{ikx} = \frac{1}{\sqrt{2\pi}} \int e^{ikx'} \delta(x - x')\, dx'. \qquad (10.67)$$

6) You may wonder why the amplitudes a are allowed here to go beyond $|a| = 1$, if the state vector is to be unitary. The answer is that since the variable x is continuous, the corresponding expansion is an integral, not the sum, and $|\psi|^2$ is the probability density, rather than just probability. The probability density may be arbitrarily big, hence the range of variable a_x may be infinite. If this variable were discrete ($a_x \rightarrow a_n$, with n integer), its modulus would range between 0 and 1.

But $\delta(x-x')$ is the eigenstate of the position operator x corresponding to the eigenvalue x'. Here, the whole set of $\delta(x-x')$ consists of distinct eigenstates for different x' and e^{ikx} are the expansion coefficients. If a vector can be represented as a superposition of the eigenvectors of a given Hilbert space, this vector is itself an element of this space. The only thing that distinguishes it from the vectors used in the expansion is that this vector is generally not along any of the used eigenvectors (if it were, it would be an x-eigenstate, not p-eigenstate). Thus, all states with definite p (the p-eigenstates) also reside in the Hilbert space formed by the x-eigenstates (and vice versa). The only difference is that the corresponding bases $\hat{\mathbf{x}}, \hat{\mathbf{x}}', \hat{\mathbf{x}}'', \ldots$ and $\hat{\mathbf{p}}, \hat{\mathbf{p}}', \hat{\mathbf{p}}'', \ldots$ are rotated with respect to each other, so that no p-eigenstate is along an x-eigenstate, and vice versa.

Summarizing, we have arrived at rather counterintuitive conclusion that for any additional variable *compatible* with the x-variable, there should be the set of additional new dimensions in the Hilbert space, for it to represent completely the state of the system. For any additional variable *incompatible* with the x-variable, there are no additional dimensions in the Hilbert space. All the eigenstates of the additional variable are already within the existing Hilbert space. The same Hilbert space thus describes all x-eigenstates *and* all p_x-eigenstates, as well as the eigenstates of all other observables incompatible with x.

10.2 Relativistic QM Indeterminacy

Relativistic description of quantum world involves a major revision of some basic concepts such as the indeterminacy principle. According to QM, certain pairs of observables such as the position and momentum of a particle cannot have simultaneously definite values. This does not preclude, however, a *single* observable from having a definite value in certain states. For instance, we can, in principle, measure a particle's position with infinitely high accuracy – provided no momentum measurements are performed at the same time. And vice versa, we can prepare a particle in a state with sharply defined momentum. The possibility to measure each dynamical variable separately to an arbitrarily high accuracy at any moment of time is of paramount importance for all nonrelativistic QM. Only because of such possibility one can introduce the concept of a wave function, which is a basic concept in its theoretical structure. As we know, the square modulus of the wave function $\Psi(q)$ determines probability $\mathcal{P}(q)$ (or its density, in case of continuous q) of the corresponding value of q resulting from an appropriate measurement. Evidently, the necessary premise for such concept is the physical possibility of arbitrary fast and accurate measurement of q; without such possibility the concept of QM probability $\mathcal{P}(q, t)$ as a function of q and t would be meaningless.

The existence of the invariant speed c, which is also the upper limit for speeds of all known material objects, imposes new fundamental restrictions on possibilities of measurement of dynamical variables [92–94]. Consider again the uncertainty relation

$$\Delta p \, \Delta x \geq \hbar. \tag{10.68}$$

This can be recast in the relation between Δp and the indeterminacy Δt of the particle's time of arrival at the detector or, equivalently, the time interval necessary for complete measurement of momentum p. Indeed, using

$$\Delta x \approx v \Delta t \tag{10.69}$$

yields

$$v \, \Delta p \, \Delta t \geq \hbar. \tag{10.70}$$

Here, v is the average velocity of the corresponding wave packet. Also, using (10.69) and the relation

$$\Delta \mathcal{E} = v \, \Delta p, \tag{10.71}$$

which holds in both nonrelativistic and relativistic domains, we obtain

$$\Delta \mathcal{E} \, \Delta t \geq \hbar. \tag{10.72}$$

It is necessary to emphasize here the fundamental difference between the relations (10.70) and (10.72) and the indeterminacy relation (10.68) for the coordinate and momentum. In the latter, the symbol Δ means the degree of indeterminacy of both – momentum and coordinate – in one state. In contrast, the meaning of Δt in the last equation is just the time duration of performed measurement of p or $\mathcal{E}$. It is *not* indeterminacy of time itself for the corresponding state of an object. As an example, consider a measurement of an excited energy level $\mathcal{E}$ in a system by determining the frequency $\omega = (\mathcal{E} - \mathcal{E}_0)/\hbar$ of the emitted photon in optical transition to the ground-level $\mathcal{E}_0$. In such an experiment, there is a certain expectation time (called the *lifetime* of the given excited state) – the average time we have to wait for the emergence of the expected photon after the preceding excitation. It is this lifetime that determines a characteristic time interval Δt for the described measurement. Consequently, it defines the indeterminacy $\Delta \mathcal{E}$ in the corresponding energy $\mathcal{E}$ through the relation (10.72) (this indeterminacy is frequently called the "level's width"). In spectroscopic measurements, this transforms into the "line width," referring to the spectral width of a line representing the measured frequency ω. A sharply defined level (small $\Delta \mathcal{E}$) has an accordingly high longevity (big Δt) and vice versa.

Another manifestation of the "time–energy" relation (10.72) is already mentioned phenomenon of "vacuum fluctuations." Even in the absence of all matter (more accurately, in a state with lowest possible energy), vacuum is not a pure emptiness. On a sufficiently short timescale all kinds of particles can be born literally out of nothing, but only to return almost immediately back to nothing. These ephemeral entities are called the *virtual particles* and the corresponding timescale (the lifetime of an emerged object) depends on its energy. The greater energy implies greater $\Delta\mathcal{E}$ and accordingly smaller Δt. A virtual proton–antiproton pair has much lower life expectancy than a similar pair of electron–positron since the former is much more massive.

Thus, vacuum turns out to be a dynamic system that can, and does, interact with observed objects and influences their behavior. It is this interaction that causes instability of the so-called "stationary" excited states. Actually, they are all only quasistationary. There are many other observable manifestations of vacuum fluctuations.

Now let us turn to Equation (10.70). They say that if we want to perform a sufficiently accurate measurement of momentum of a system during sufficiently short time (i.e., we want to prepare an object with small Δp or/and $\Delta\mathcal{E}$ during a short time Δt), we need the sufficiently high group velocity of the object. That posed no problem in nonrelativistic theory, according to which there is no restrictions on velocities, so v might be arbitrarily high. However, the existence of the invariant limit c changes things dramatically. Setting $v = c$ in (9.2.3) gives the maximal possible accuracy (minimal possible Δp) for the momentum measurement within a given time interval Δt

$$\Delta p_{\min} \cong \frac{\hbar}{c\,\Delta t}. \tag{10.73}$$

Thus, relativity renders it impossible to carry out arbitrarily accurate *and* fast measurement of momentum.

To see clearly the difference between the nonrelativistic and relativistic situations, consider the momentum measurement by means of a diffraction grating. We place the grating in the way of de Broglie's wave associated with a studied particle (actually, an ensemble of particles all in one state and passing through the grating all together or one at a time). Then the diffraction pattern on a distant screen (the distribution of the individual landings on the screen) tells about the wavelength of a particle in a studied state (i.e., *before* passing through the grating). Knowing the wavelength, we determine the momentum through (10.1). Now, suppose we are in a rush and want to complete the measurements as soon as possible. We cannot make it shorter than the time Δt it takes the wave packet to pass through the grating. But if the world is nonrelativistic so that there is no upper limit to objects' speeds, this time can be made arbitrarily short if we are operating with the extremely fast particles, even when the packet representing them is very broad in space. If, in addition, the mass of a particle is vanishingly small, so that $p = mv$ remains finite, there will be no limitations on either measured momentum itself or accuracy of its measurement. Thus, we could, while remaining within restriction (10.68), measure *momentarily* any momentum with any accuracy.

In the real (relativistic) world, this is impossible due to the invariant limit c. If a system is in a state with sharply defined p, it will have totally undefined position; that is, $\Delta x \to \infty$, the wave packet "degenerates" into the infinitely lasting monochromatic wave permeating the whole universe, and accordingly it would take eternity to measure its momentum accurately.

The same holds for the energy of the system. In particular, if the system has a sharply defined energy, this state lasts forever and it would take us forever to measure its energy with the accuracy it deserves. This example also illustrates that the quantity Δt here has nothing to do with "time indeterminacy"; in particular, a system can have a definite energy at any definite moment of time. Landau used to express this by saying "I can measure the energy and look at the watch." The amount Δt in (10.70) and (10.72) refers to system's lifetime or the time it takes to prepare the system or measure its energy.

The same restrictions are imposed by relativity on the coordinate measurement. Starting again from general relation (10.68), rewrite it now as

$$\Delta q \geq \frac{\hbar}{\Delta p}, \tag{10.68a}$$

and use again (10.71). We will get

$$\Delta q \geq \frac{\hbar v}{\Delta \mathcal{E}}. \tag{10.74}$$

But according to relativity, v cannot exceed c; for this upper limit, the inequality reads

$$\Delta q \geq \frac{\hbar c}{\Delta \mathcal{E}}. \tag{10.75}$$

Suppose now that we are in the rest frame of the particle. In classical world, this would mean that the particle itself is at rest. In quantum-mechanical world, this only means that the particle's probability cloud is at rest, only its *average* momentum is zero, *not* its variance Δp. Accordingly, the variance of energy is not zero either:

$$\Delta \mathcal{E} \neq 0. \tag{10.76}$$

Moreover, according to (10.75), the more accurate is the position measurement the greater becomes indeterminacy in the particle's energy. At sufficiently accurate position measurement the energy indeterminacy becomes of the order of magnitude of the particle's rest energy, that is,

$$\Delta \mathcal{E} \approx m_0 c^2. \tag{10.77}$$

Putting this into (10.75), we find

$$\Delta q \geq \frac{\hbar}{m_0 c}. \tag{10.78}$$

What happens if we try to determine position with even higher accuracy? In this case, the energy indeterminacy exceeds the rest energy. According to what we have learned in Section 5.5, the energy is massive! If we have enough energy, we can cook up the corresponding amount of the stuff – specifically, new particles. In our case, when the energy indeterminacy exceeds $2m_0c^2$, there appears a nonzero chance of the particle–antiparticle production. This would result in the emergence of a system, quite different from the single particle in the initial state.

We thus find that the quantity $\hbar/m_0c$ is a characteristic scaling parameter for a particle with given rest mass m_0, determining the minimal possible linear dimension of its localized state in space. This parameter is called the *Compton wavelength* for the reasons that will become clear later in this chapter.

In a reference frame (RF) where the particle is moving and has the relativistic energy $\mathcal{E}$, the same argument leads to the conclusion that the minimal possible position indeterminacy compatible with the notion of a lasting (self-identical) single particle is given by

$$\Delta q \geq \frac{\hbar c}{\mathcal{E}}. \tag{10.79}$$

In the ultrarelativistic limit we have (Section 5.5)

$$\mathcal{E} = \sqrt{m_0^2c^4 + p^2c^2} \xrightarrow[m_0 \to 0]{} pc, \tag{10.80}$$

that is, the energy is nearly proportional to momentum. Putting this into (10.80) and recalling (10.1) and (10.2), we get

$$\Delta q \geq \frac{h}{p} = \lambda. \tag{10.81}$$

This means that the minimal possible position indeterminacy in the ultrarelativistic limit cannot be less than de Broglie's wavelength of the particle. In particular, this is always true for the photons, since for them $m_0 = 0$. Therefore, the notion of the photons' coordinates is meaningful only when the characteristic size of a system where they move by far exceeds their wavelength. But this corresponds to the classical limit of the "ray optics," representing light as particles moving along classical trajectories – light rays. Generally, when the wavelength is not small with respect to the characteristic size of a system, the notion of coordinates of a photon becomes meaningless.

Thus, we are led to the conclusion that in a consistent relativistic QM theory the notion of coordinates as dynamic variables loses its meaning – there is no known operational procedure for their instant and accurate measurement. For the same reason, the momentum of a particle cannot retain its original meaning either. In relativistic QM, the notion of momentum or energy can only apply to free particles; in this case $\mathcal{E}$ and p are conserved variables, lasting forever and therefore they can be measured to arbitrary high accuracy. Thus, the only observable variables that can be meaningfully defined turn out to be the characteristics of *free* particles (such as 4-momentum, polarization, etc.). Therefore, in the relativistic QM it is convenient to use momentum representation.

A typical process illustrating such situation is scattering, when the particles converge from infinity, engage into interactions ("collide" in the everyday language), and then disengage and fly apart to become free again, generally in states other than before the collision. Accordingly, we can associate the dynamical characteristics with the initial conditions of such a system, when its constituents are still free and about to enter into interactions and then with a final state, when the constituents (albeit possibly different than in the beginning) can again be considered as free. The initial and final states are usually specified by the momenta and energies of the corresponding particles in the respective states. In this case, the basic requirement to a consistent relativistic quantum theory is that, for each known type of interaction, it must give working algorithms to calculate the transition probabilities connecting the initial and final states. For example, consider the annihilation of a particle and antiparticle (denote them as p^+ and p^-, respectively). They can disappear and produce instead a few photons of gamma radiation. Schematically,

$$\mathrm{p}^+ + \mathrm{p}^- \rightarrow \sum_{i=1}^{N} \gamma_i, \quad N \geq 2 \tag{10.82}$$

where N is the number of resulting photons (most frequently, this number is just 2). In this process, the initial system changes beyond recognition, but relativistic quantum electrodynamics gives prescription for calculating probabilities for each possible transition without describing what happens in between – a detailed description of the process in time turns out to be as illusory as classical trajectories in quantum mechanics.

As another interesting consequence, there is no longer any reliable criterion for deciding whether a given particle is elementary or complex. We cannot make a judgment on this question without considering the process of interactions between the particles and such a process, as we have learnt, cannot be described in all details in terms of dynamical variables only.

We know now quite a few particles called elementary. For instance, we can name electrons, protons, neutrons, muons, pions, kaons, neutrinos, and so on. They are characterized by their respective rest masses and can have a fixed electrical charge (e.g., zero or one electron charge). Now, consider one of them, say, a neutron. A free neutron is unstable, with an average lifetime of about 11 min. It decays into a proton, electron, and antineutrino. Can we say, on the basis of this information, that a neutron actually consists of these three particles – products of its decay? In other words, can we visualize a neutron as a system of a proton, electron, and antineutrino temporarily bound by some attractive force into a system called a neutron? Such a possibility seems very feasible. After all, we know about a hydrogen atom consisting of one proton and electron as its building blocks. Why cannot the same particles be building blocks of a neutron?

The relativistic quantum theory says no and explains why. When we say that a building consists of bricks, this statement has a meaning because a brick within a building is the same as a single brick. In particular, a single brick is much smaller than a building and remains smaller when it constitutes a part of the building.

Similarly, when we say that an atom consists of elementary particles like protons and electrons, this statement is meaningful, if the constituting particles are smaller and simpler than the atom harboring them. This condition is satisfied for the atom and the electrons and protons as its parts. The Compton wavelength of an electron, determining its most compact possible localization, is about 10^{-13} m. The Compton wavelength of a proton is about 5×10^{-17} m. The size of an atom, on the contrary is, as we have calculated in the previous section, about 10^{-10} m. Thus, the atom provides a lot of free room for the electron's residence, about 10^9 volume of the electron's most compact state, and even more so for the proton, about 10^{18} volume of its minimal localized size. This is why we can say that an atom as a system consists of the protons, neutrons, and electrons in the same sense as the building consists of bricks.

Consider now a neutron as a harboring system for the proton, electron, and neutrino. The size of a neutron is 10^{-15} m – much smaller than minimum possible localization of the electron. The electron cannot be squeezed into the neutron without being totally destroyed as an independent entity. Such an attempt would just result in production of the electron–positron pairs and may be other particles.

How then can we explain the emergence of the proton, electron, and antineutrino at the moment of neutron's decay? The relativistic quantum theory describes this by saying that the neutron as such *disappears* and the three other particles appear instead. Moreover, under some conditions where there is enough ambient energy, a particle considered as "elementary" can "decay" into heavier particles usually considered as composite. For instance, a proton within a nucleus can change into a neutron, positron, and neutrino. In terms of our everyday life experience, this looks as if, say, a car suddenly disappears, after producing an elephant, Cinderella, and a beam of light. But weird as it appears, such transformations are not just possible, they are routine phenomena on the subatomic scale. Recall our previous reference to an exotic hypothesis that all observed universe had emerged from collision of two sufficiently energetic "elementary" particles. Highly speculative as it is, it is feasible within the framework of the modern physics. All conservation laws are satisfied, so such process does not contradict any known laws of physics, except, maybe, our intuition. But we already learned more than once that our intuition is not always a reliable guide in the relativistic and quantum world. Relativity, through its mass–energy relation, makes it possible to produce the universe out of two particles, and the Big Bang might have been no more than just their collision. Would it make much sense to say that these original two particles were elementary if all known world has (or could have) emerged out of them?

Summary

Symbiosis of relativity and QM forms the basis of spectacular phenomena in the microworld.

QM brings in indeterminacy which makes the world fuzzy – much more fuzzy than it looks when observed on the macrolevel. This indeterminacy admits, by virtue of Equation (10.72), brief violations of energy conservation, and relativity through mass–energy relation uses the "illegal" energy to produce short-lived particle–

antiparticle pairs. Although the borrowed energy is almost immediately returned to vacuum, so no violations are observed on the macroscale, the mere existence of such processes "behind the scene" makes vacuum a shimmering active entity influencing many observed phenomena.

10.3 Relativistic Wave Equations

Relativistic postulates determine basic features of equations of QM describing particles' states and their evolution. A student who even very thoroughly studied the nonrelativistic QM, frequently emerges from the course under the impression that all QM is essentially described by Schrödinger's equation. Nothing can be farther from the truth. It turns out that in relativistic QM, particles of different nature are described by different wave equations. Some of them, for example, those for particles with nonzero rest mass, reduce to Schrödinger's equation in the nonrelativistic limit. Generally, the correct relativistic equations for particles and their interactions fall into two different categories. One of them describes particles with an integer spin (bosons) and the other describes particles with half-integer spin (fermions). The solutions of equations describing bosons transform as the integer rank tensors; the tensor rank is equal to spin quantum number. For instance, the solutions for the zero spin particles transform as a zero rank tensor, that is, they are either scalar or pseudoscalar functions. The solutions describing spin-one particles transform as a first rank tensor and accordingly they are the vector functions.

Similarly, the equations for the fermions have solutions transforming as half-integer rank tensors. Such tensors are called spinors. The simplest fermions have spin 1/2 (e.g., electrons, nucleons – protons and neutrons, neutrinos, and their antiparticles). Accordingly, their wave functions are spinors of rank 1/2, which transform as a tensor of rank 1/2. For this reason, they are sometimes even called "half vectors."

To avoid possible confusion, let us introduce a generic name for an equation accurately describing the states of an unidentified particle for all possible spins and energies – a "relativistic wave equation." In this book, we consider the relativistic equations describing massive particles: those with spin 0 and those with spin 1/2.

10.3.1 The Klein–Gordon Equation

Let us recall the way the Schrödinger equation can be obtained. We can start with the classical nonrelativistic expression for the energy of a particle in an external potential field $U(\mathbf{r},t)$

$$\mathcal{E} = \frac{p^2}{2m} + U(\mathbf{r}, t), \tag{10.83}$$

and then replace $\mathcal{E}$ and $\mathbf{p}$ by the corresponding operators:

$$\mathcal{E} \to i\hbar \frac{\partial}{\partial t}, \qquad \mathbf{p} \to -i\hbar \vec{\nabla}. \tag{10.84}$$

Now we will have instead of (10.83) the energy operator on the left and its equivalent (Hamiltonian) on the right in position representation. Finally, we apply them to a wave function $\Psi(\mathbf{r}, t)$ and here we are with the Schrödinger equation (A1).

Now, if we want to use the same procedure to obtain a *relativistic* equation, we must start with the corresponding relativistic expression for the energy. Immediately, even for a most simple case of a free particle ($U = 0$), we come across a stumbling block – according to (5.5.21), the energy is not a rational function of momentum. If we use (10.84) in the relativistic expression for energy, we will get

$$i\hbar \frac{\partial}{\partial t} = \pm\sqrt{m_0^2 c^4 - \hbar^2 \vec{\nabla}^2}. \tag{10.85}$$

There arises a question how to interpret the operator on the right – the square root of a sum of a constant (square of the particle's rest mass) and the Laplacian operator. A natural way to do it would be to go back to (5.5.21) and expand the right-hand side into a Taylor series:

$$\sqrt{m_0^2 c^4 + p^2 c^2} = m_0 c^2 + \frac{1}{2} \frac{p^2 c^2}{m_0 c^2} - \frac{3}{8} \frac{p^4 c^4}{(m_0 c^2)^3} + \cdots. \tag{10.86}$$

Then replacing $\mathcal{E}$ and $\mathbf{p}$ with the corresponding operators and acting on a function $\Psi(\mathbf{r})$ will yield

$$i\hbar \frac{\partial \Psi}{\partial t} = \left(m_0 c^2 + \frac{1}{2} \frac{\hbar^2 \vec{\nabla}^2}{m_0} - \frac{3}{8} \frac{\hbar^4 \vec{\nabla}^4}{m_0^3 c^2} + \cdots \right) \Psi. \tag{10.87}$$

We have obtained a partial differential equation of infinite order in spatial coordinates. Leaving aside the infinite complexity of such an equation, finding its exact solution describing a real physical situation would require an infinite number of the boundary conditions – fixing the values for all orders of the partial derivatives in x, y, z and their combinations of the type

$$\frac{\partial^N \Psi(\mathbf{r})}{\partial x^{2k} \partial y^{2l} \partial z^{2m}} \Big|_A, \qquad 2k + 2l + 2m = N, \tag{10.88}$$

evaluated on the boundary A around the volume of interest. Even if the world were indeed described by such equation, we should forlorn all hope to solve it exactly and thereby learn something about the behavior of the corresponding system in view of the infinite amount of information necessary to specify the boundary values. But, still more important, even if we use a sufficiently accurate approximation by leaving only a finite number of terms in (10.87), the truncated version of (10.87) will not be Lorentz covariant; accordingly, the solution accurate in one RF may be not accurate in another. Therefore, Equations (10.85) and (10.87) do not give meaningful or satisfactory description of the related phenomena.

We have another option to obtain sensible wave equation: we can, instead of (5.2.21), use the mass–energy relation in the form of (5.2.20), in which the invariant term m_0c^2 is isolated on one side. Then we will get the wave equation for a free particle

$$\left(\nabla^2 - \frac{\partial^2}{c^2\partial t^2}\right)\Psi = \lambda_C^{-2}\Psi, \qquad \lambda_C \equiv \frac{\hbar}{m_0 c}, \tag{10.89}$$

where λ_C is the corresponding Compton wavelength of the particle. The Lorentz invariance of this equation is immediately seen, since its left-hand side is just the d'Alambertian operator, whose form is covariant under the Lorentz transformations; therefore, its application to Ψ results in a tensor quantity of the same rank as on the right. In particular, if Ψ is a scalar, then we will have a scalar, that is, the zero rank tensor, on both sides of the equation.

Equation (10.89) was obtained almost at the same time by Klein, Gordon, and Fock [95–97]. It is usually called the Klein–Gordon equation.

This equation can be conveniently written in the explicitly four-dimensional form if we introduce the co- and contravariant components of the 4-momentum operator

$$p_j \to \hat{p}_j \equiv -i\hbar\frac{\partial}{\partial x^j}; \quad p^j \to \hat{p}^j \equiv -i\hbar\frac{\partial}{\partial x_j}. \tag{10.90}$$

Then (10.89) takes the form

$$\hat{p}_j\hat{p}^j\Psi = -m_0^2c^2\Psi, \tag{10.91}$$

or, recalling our notations in Section 5.1, we can also write this in the form

$$(\Box^2 + \lambda_C^{-2})\Psi(\mathbf{r}, t) = 0, \tag{10.89a}$$

where

$$\Box^2 \equiv \frac{\partial^2}{\partial x_i \partial x^i}. \tag{10.89b}$$

The wave function here can be considered as the function of a 4-vector of position in space-time:

$$\Psi(\mathbf{r}, t) = \Psi(x), \quad x = (x^0, \mathbf{r}) = (x^0, x^1, x^2, x^2). \tag{10.92}$$

The simplest solutions of this equation are the de Broglie waves describing states with definite energy and momentum:

$$\Psi_\mathbf{k}(\mathbf{r}, t) = \Psi_0(\mathbf{k})\, e^{i(\mathbf{k}\cdot\mathbf{r} - k_0 t)}.$$

Here, $\mathbf{k} = \mathbf{p}/\hbar$, $k_0 = \pm\sqrt{k^2 + \lambda_C^{-2}}$ and $\Psi_0(\mathbf{k})$ is the one-component amplitude that can be a function of $\mathbf{k}$. The mathematically allowed values of the 4-vector (k_0, $\mathbf{k}$) satisfy the condition

$$k_0^2 - k^2 = \lambda_c^{-2} > 0,$$

and thus lie on the twofold hyperboloid of the revolution in the 4-momentum space, similar to the situation discussed in Section 5.1. The top fold of the hyperboloid corresponds to the particles' states with positive energies and the bottom fold would correspond to the states with negative energies.

Even though infinitely more simple than (10.87), the Klein–Gordon equation is fundamentally different from the Schrödinger equation: it is the *second-order* partial differential equation with respect to time. This causes the corresponding drastic difference in solutions describing the behavior of the system. To determine the system's evolution described by $\Psi(\mathbf{r},t)$, we must know the initial conditions. In contrast to the Schrödinger equation, which is of the first order in time, and accordingly requires as an initial condition the value of $\Psi(\mathbf{r}, 0)$ at the initial moment $t=0$, now we must know in addition the value of its time derivative $(\partial\Psi/\partial t)|_{t=0}$, which is independent of Ψ. This in turn results in a profound difference in the interpretation of such concepts as probability density and probability flux density. But before going into this, we need to consider the behavior of the wave function itself under the general Lorentz transformations, including rotations.

The expression on the left in (10.91) can be interpreted as the square of the length of the 4-vector. As this is an invariant under the Lorentz transformations, it follows that under these transformations, the wave function can only be multiplied by a factor of modulus 1:

$$\Psi(x) \to \Psi'(x') = \kappa\Psi(x), \qquad \kappa = e^{i\alpha}, \tag{10.93}$$

with a real constant α. In the case of *continuous* transformation (like any rotations in space-time), the only possible value for κ can be $\kappa = 1$, because in the limit of the identical transformation $x' = x$ there must be $\Psi'(x') = \Psi(x)$.

The spatial reflections were described in Section 4.7. They are determined by

$$\mathbf{r} \to \mathbf{r}' = -\mathbf{r}, \qquad t' = t. \tag{10.94}$$

Applying such transformation twice leads to identical transformation, that is, after the second reflection we have the initial state restored:

$$\Psi(\mathbf{r}, t) \to \Psi'(-\mathbf{r}, t) \to \Psi(\mathbf{r}, t). \tag{10.95}$$

However, using (10.92) and (10.93) we get

$$\begin{aligned} &x = (\mathbf{r}, t) \to x' = (-\mathbf{r}, t),\\ &\Psi(\mathbf{r}, t) \to \Psi'(-\mathbf{r}, t) = \kappa\Psi(\mathbf{r}, t), \quad \text{(first reflection)},\\ &\kappa\Psi(\mathbf{r}, t) \to \kappa\Psi'(-\mathbf{r}, t) = \kappa^2\Psi(\mathbf{r}, t). \quad \text{(second reflection)}. \end{aligned} \tag{10.96}$$

Thus, we have

$$\Psi(\mathbf{r}, t) \to \Psi'(-\mathbf{r}, t) = \kappa^2\Psi(\mathbf{r}, t) = \Psi(\mathbf{r}, t). \tag{10.97}$$

This can only be the case if $\kappa^2 = 1$, that is,

$$\kappa = \pm 1. \tag{10.98}$$

If $\kappa = 1$, then

$$\Psi(\mathbf{r}, t) \to \Psi'(-\mathbf{r}, t) = \Psi(\mathbf{r}, t). \tag{10.99}$$

If $\kappa = -1$, then

$$\Psi(\mathbf{r}, t) \to \Psi'(-\mathbf{r}, t) = -\Psi(\mathbf{r}, t). \tag{10.100}$$

In the first case, the wave function is a scalar, in the second a pseudoscalar. They behave equally under spatial rotations and proper Lorentz transformation. However, under mirror reflections (inversion) the scalar function remains the same, whereas the pseudoscalar function changes sign.

As mentioned above, the behavior of a wave function under the spatial inversion and rotations in space-time is intimately connected with the type of wave equation it satisfies and, accordingly, with the intrinsic physical characteristics of the particles it describes. In particular, the transformation properties are related to the spin of a particle. The Klein–Gordon equation and its solutions – the scalar and pseudoscalar wave functions – describe particles with the zero spin and nonzero rest mass.

As we have learned in Section 10.1 (see also Appendix A), the Schrödinger equation (between the position measurements) describes the probability distribution that can flow in the configuration space like a fluid. Accordingly, it satisfies the continuity Equation (A10) similar to that of the fluid and expresses the conservation of probability [90, 91]. In relativistic case, the situation becomes much more subtle. We know already that the concept of probability density as an exact continuous function of position generally loses its meaning for high energies due to the production of new particles. We therefore can expect that the concept of quantum-mechanical probability in the configuration space may not even be defined consistently in the relativistic domain. Let us see whether this is true.

To this end, we perform on (10.89) exactly the same procedure as in Appendix A: multiply (10.89) by Ψ^* and subtract from the result its complex conjugate. We will get

$$\frac{\partial \varsigma}{\partial t} = -\vec{\nabla} \cdot \mathbf{s}, \tag{10.101}$$

where

$$\varsigma = \frac{\hbar}{2im_0c^2}\left(\Psi\frac{\partial \Psi^*}{\partial t} - \Psi^*\frac{\partial \Psi}{\partial t}\right) = \frac{\hbar}{m_0c^2}\mathrm{Im}\left(\Psi\frac{\partial \Psi^*}{\partial t}\right) \tag{10.102}$$

and

$$\mathbf{s} = \frac{i\hbar}{2m_0}(\Psi\,\vec{\nabla}\Psi^* - \Psi^*\,\vec{\nabla}\Psi) = \frac{\hbar}{m_0}\,\mathrm{Im}(\Psi\vec{\nabla}\Psi^*). \tag{10.103}$$

Introducing a 4-vector

$$s = (c\varsigma, \mathbf{s}), \tag{10.104}$$

we can rewrite (10.101) in a four-dimensional form

$$\Box \cdot s \equiv \frac{\partial s^i}{\partial x^i} = 0. \tag{10.105}$$

Now we can analyze and interpret the obtained expressions. The expression (10.103) for the flux density looks exactly the same as in the case of the Schrödinger equation and in principle, could be interpreted in the same way. However, the expression (10.102) is *not* the square modulus of the wave function. Accordingly, there is no guarantee that it will always be definitely positive. In addition, since the wave function is now the solution of the equation with the second-order time derivative, the values of Ψ and its first-time derivative $\dot{\Psi}$ are mutually independent. As a result, the quantity ζ *can* take on any value, positive or negative! It therefore cannot be interpreted as probability, since the latter can be only nonnegative. Once it is not probability, the corresponding vector **s** is not probability flux. This is mathematical manifestation of the physical fact following from QM and special relativity (SR) that we cannot localize a particle within a sufficiently small region of space-time due to the possibility of production of new particles. In other words, the number of particles is no longer an integral of motion (a conserving quantity). Still, this by itself is not sufficient to explain the possibility of *negative* values of ς. Let us think how can we interpret such behavior? There is only one possibility: to associate ς with the electric charge density, which can also be of both signs. But in this case the described particles must have the electric charge q. Thus, the relativity postulates predict the possibility of existence of the electrically charged massive particles with the zero spin. They can come in three varieties: with $q, -q$, or zero charge. And indeed, we know that such particles do exist. Among them there are so-called π-mesons, or just *pions*, having $|q| = |q_e|$ and the rest mass m_0 about 270 of the electron rest mass m_e. We now know that they are carriers of the so-called strong force responsible for the strong (or nuclear) interactions between the nucleons – protons and neutrons. They are described by a pseudoscalar wave function.

Another sort of particles in the same category are the K-mesons (*kaons*). They also have the zero spin and electric charge $q_K = \pm q_e$, or 0, but are about 3.5 times more massive ($966 m_e$) than pions.

Of course, the corresponding charge density must depend on q. Therefore, for our interpretation to reflect adequately the actual properties of the system, we must include the charge q into its description as a factor to both – ς and **s**. Mathematically, this is equivalent to merely multiplying Equations (10.101) or (10.105) by q. But the moment we do it, the terms involved acquire the simple physical meaning as the electric charge density $\rho = q\varsigma$ and the corresponding current density $\mathbf{j} = q\mathbf{s}$. The possibility that ς can be of either sign reflects the fact that particles of either sign can occur. Thus, relativity demands that the zero-spin particles with nonzero rest mass be carriers of electric charge and establishes the connection between the wave function of such particles and the corresponding charge distribution:

$$\rho = q\varsigma = \frac{q\hbar}{2im_0c^2}\left(\Psi\frac{\partial\Psi^*}{\partial t} - \Psi^*\frac{\partial\Psi}{\partial t}\right) = \frac{q\hbar}{m_0c^2}\,\mathrm{Im}\left(\Psi\frac{\partial\Psi^*}{\partial t}\right), \tag{10.106}$$

and

$$\mathbf{j} = q\mathbf{s} = \frac{iq\hbar}{2m_0}(\Psi\,\vec{\nabla}\Psi^* - \Psi^*\,\vec{\nabla}\Psi) = \frac{q\hbar}{m_0}\,\mathrm{Im}(\Psi\,\vec{\nabla}\Psi^*). \tag{10.107}$$

Note that in the nonrelativistic quantum theory we also deal with the electric charges. They enter the corresponding expressions for Hamiltonian and multiply the probabilities in exactly the same way as we have seen in the last two equations. However, in the Schrödinger's mechanics the probability density has a meaning of its own regardless of charges and the latter are just introduced into the picture "by hand" when we need to calculate the corresponding characteristics such as energy of electrical interactions, currents, their magnetic moments, and so on. By contrast, in the relativistic theory of the massive spinless particles, the corresponding expression (10.102) cannot be interpreted as a local probability density and the charges are *demanded* by the theory to account for the possibility of negative values in this expression.

The charge diversity of the particles described by the Klein–Gordon equation can be considered as an additional (discrete!) degree of freedom. Using this terminology, we can say that in the nonrelativistic theory of a spinless particle there is only one state for its free motion with a fixed momentum **p**. The relativistic theory gives for such state three different solutions corresponding to three possible values of the particle's charge. These three values define the domain of the new degree of freedom.

The extra degrees of freedom associated with the massive zero spin particles can also be described by a set of two differential equations of the first order in time for the two functions, which is equivalent to the single Klein–Gordon equation of the second order for one function.

Let us introduce two new functions χ and φ so that

$$\varphi+\chi=\Psi, \qquad \varphi-\chi=i\,\frac{\lambda_C}{c}\,\frac{\partial\Psi}{\partial t}, \qquad \lambda_C\equiv\frac{\hbar}{m_0c}, \tag{10.108a}$$

and accordingly

$$\varphi=\frac{1}{2}\left(\Psi+i\,\frac{\lambda_C}{c}\,\frac{\partial\Psi}{\partial t}\right), \qquad \chi=\frac{1}{2}\left(\Psi-i\,\frac{\lambda_C}{c}\,\frac{\partial\Psi}{\partial t}\right). \tag{10.108b}$$

Since the solution Ψ and its time derivative $\partial\Psi/\partial t$ for a second-order differential equation are, generally, mutually independent, so are the functions φ and χ. Putting (10.108a) into (10.89) produces the set of two equations for these functions (Problem 10.12)

$$\left.\begin{aligned} i\hbar\,\frac{\partial\varphi}{\partial t}&=-\frac{\hbar^2}{2m_0}\,\nabla^2(\varphi+\chi)+m_0c^2\varphi\\ i\hbar\,\frac{\partial\chi}{\partial t}&=\frac{\hbar^2}{2m_0}\,\nabla^2(\varphi+\chi)-m_0c^2\chi \end{aligned}\right\}. \tag{10.109}$$

It is easy to verify that the system of Equations (10.109) is equivalent to (10.89) (Problem 10.13).

Equations (10.109) can be written in a more simple form if we consider φ and χ as the two components of function Ψ. This allows us to represent Ψ as a single-column matrix

$$\Psi = \begin{pmatrix} \varphi \\ \chi \end{pmatrix}. \tag{10.110}$$

Now we introduce four matrices

$$\hat{\sigma}_1 \equiv \begin{pmatrix} 0 & 1 \\ 1 & 0 \end{pmatrix}, \quad \hat{\sigma}_2 \equiv \begin{pmatrix} 0 & -i \\ i & 0 \end{pmatrix}, \quad \hat{\sigma}_3 \equiv \begin{pmatrix} 1 & 0 \\ 0 & -1 \end{pmatrix}, \quad \hat{I} = \begin{pmatrix} 1 & 0 \\ 0 & 1 \end{pmatrix}. \tag{10.111}$$

The first three of them were first introduced by Pauli to describe the electron spin (Appendix C). The fourth one is the unitary matrix.

Using these matrices, we can rewrite the set of Equation (10.109) in the matrix form as one single equation

$$i\hbar \frac{\partial \Psi}{\partial t} = \hat{H}\Psi. \tag{10.112}$$

for unknown matrix function Ψ defined in (10.110). The matrix Hamiltonian here is

$$\hat{H} = (\hat{\sigma}_3 + i\hat{\sigma}_2)\frac{\hat{\mathbf{p}}^2}{2m_0} + m_0 c^2 \hat{\sigma}_3. \tag{10.113}$$

Equation (10.112) looks like the Schrödinger equation, but here this similarity is illusory, since each component φ and χ constituting Ψ, actually satisfies the Klein–Gordon equation.

An interesting question is how to perform transition from the Klein–Gordon equation to a nonrelativistic limit. For such transition, one must keep in mind that when we talk about energy in nonrelativistic physics, we do not include the rest energy into it. Therefore, for the transition to be meaningful, we can write the relativistic wave function for a special case of a stationary state in the form with the rest energy $\mathcal{E}_0$ explicitly singled out:

$$\Psi(\mathbf{r}, t) = \Psi_0(\mathbf{r})e^{-i(\mathcal{E}/\hbar)t} = \Psi_0(r)e^{-i((\mathcal{E}_0 + \mathcal{E}')/\hbar)t} = \Psi_0(r, t)e^{-i(m_0 c^2/\hbar)t}. \tag{10.114}$$

Here, $\mathcal{E}'$ is the rest of the system's energy (*not* its rest energy!) and

$$\Psi_0(\mathbf{r}, t) \equiv \Psi_0(\mathbf{r})e^{-i(\mathcal{E}'/\hbar)t}. \tag{10.115}$$

In a nonrelativistic limit we have

$$\mathcal{E}' \ll \mathcal{E}_0 = m_0 c^2, \tag{10.116}$$

and therefore

$$i\hbar \frac{\partial \Psi_0(\mathbf{r}, t)}{\partial t} = \mathcal{E}'\Psi_0(\mathbf{r}, t) \ll m_0 c^2 \Psi_0(\mathbf{r}, t). \tag{10.117}$$

Thus, in this limit we can write, in view of (10.116)

$$-\hbar^2 \frac{\partial^2 \Psi}{\partial t^2} = (\mathcal{E}_0 + \mathcal{E}')^2 \Psi = (\mathcal{E}_0 + 2\mathcal{E}_0 \mathcal{E}' + \mathcal{E}'^2)\Psi \approx (\mathcal{E}_0 + 2\mathcal{E}_0 \mathcal{E}')\Psi. \tag{10.118}$$

Putting this into Klein–Gordon equation and using the relation (10.118) for the "auxiliary" wave function $\Psi_0(\mathbf{r}, t)$ yields

$$i\hbar \frac{\partial \Psi_0(\mathbf{r}, t)}{\partial t} = -\frac{\hbar^2}{2m_0} \vec{\nabla}^2 \Psi_0(\mathbf{r}, t). \tag{10.119}$$

This is the Schrödinger equation. We can thus conclude that the Klein–Gordon equation in the nonrelativistic limit reduces, as it should, to the Schrödinger equation satisfied by the auxiliary wave function (10.115). This property also holds for nonstationary states. We can also verify that in the nonrelativistic limit the auxiliary wave function gives correct expressions for the probability density and probability current density and can thus be interpreted as the probability amplitude in this limit.

Summary

When the particles have, apart from translational, also additional degrees of freedom corresponding to discrete variables, we can represent the wave function as a single-column matrix. In the case of spinless particles, the additional degree of freedom is associated with the charge variable. For charged particles, this variable has two values corresponding to "+" or "–" charge. For charged particle with nonzero spin, the additional degree of freedom also includes possible spin projections onto some fixed reference direction. In the case of spin 1/2, there are two such components and accordingly the wave function must have four components – two for the possible value of the charge and two for the possible value of the spin projection. We will consider such cases in the next two sections.

10.3.2
Dirac's Equation

Detailed studies of the atomic spectra led to the discovery of a new characteristic of an electron, which later came to be known as the electron spin (intrinsic angular momentum). It turned out, however, that this angular momentum cannot be interpreted in terms of our intuitive classical picture of a rotating body, at least, for two reasons.

First, if an object possesses properties characteristic of the rotation, it must, according to common sense, be spatially extended; and initially an electron was expected to have some size; but actually it turned out to be, by all accounts, a point particle and it does not make much sense to speak of a point spinning around itself.

Second, even if the electron had an extended charge, the equatorial speed of its rotation necessary to produce the observed angular momentum must exceed the speed of light (Problem 10.14). Therefore, no consistent mechanical model satisfying relativistic requirements can explain the electron spin in terms of the rotation of an extended object.

Thus, it would be meaningless to try to imagine or visualize this property as actual rotation (spinning) of the electron around itself.

We can try to represent the electron spin schematically just as a point charge with an arrow attached to it, pointing along the direction of the angular momentum. But even that would not represent adequately the actual nature of the spin, as according to QM, an angular momentum, vector as it is in classical physics, does not have a definite direction in space owing to indeterminacy relation between its components. It would be more accurate to represent the spin by a conical surface with its symmetry axis along the reference direction taken to determine one out of the three spin components. Then the height of the cone could represent the exact value of this component, while the diameter of its basis represents the indeterminacy in the two other components.

Thus, the electron spin does not imply anything of our classical notion of a rotating material body. This may appear embarrassing. As Max Born [98] put it, "The idea of spin without something actually spinning seems to be rather abstruse. But one should remember that there are other examples of such abstractions; for instance, the theory of relativity has deprived the ether, the carrier of electromagnetic waves, of all properties of ordinary matter, so that one has to speak about vibrations without having anything material which vibrates."

It is even more so in the concept of the de Broglie wave, which exists and propagates through empty space without any material substance actually waving.

The translation of the spin concept into the language of wave mechanics may follow the idea of the spin carried by a photon in the language of the electromagnetic waves. In this case, we associate the photon spin with the property of a plain EM wave to be right- or left-circularly polarized. Similarly, we can regard the spin as a sort of polarization of the de Broglie wave.

Dirac has formulated a relativistic equation that describes accurately the behavior of the spin 1/2 particles. It turns out that it not only does this, but it also does much more than this. Namely, the resulting equations do not require the spin to be preliminary introduced "by hand." Quite the contrary, the existence of spin $(1/2)\,\hbar$ follows automatically from the requirement of the Lorentz invariance. In addition, the same requirement predicts the existence of antiparticles. Thus, both – the existence of spin as well as the particle–antiparticle symmetry – are purely relativistic effects.

Dirac's work was not in line with the Klein–Gordon equation. It showed that we can construct the relativistic wave equation in such a way that its solution could lead to the continuity equation with a positive definite probability density ς, defined just as square modulus of Ψ, as in Section 10.1. Then we can write for ς and the corresponding local charge density for a given particle

$$\varsigma = \Psi^*\Psi, \qquad \rho = q\varsigma = q\Psi^*\Psi. \tag{10.120}$$

It then follows from both the charge and the probability conservation that

$$\int \Psi^* \Psi \mathrm{d}\tau = 1 \tag{10.121}$$

and

$$\frac{\mathrm{d}}{\mathrm{d}t}\int \Psi^* \Psi \mathrm{d}\tau = \int \left(\frac{\partial \Psi^*}{\partial t} \Psi + \frac{\partial \Psi}{\partial t} \Psi^* \right) \mathrm{d}\tau = 0. \tag{10.122}$$

Now let us consider further implications of this requirement in relativistic domain. If Ψ is to satisfy the Klein–Gordon equation, then at some moment of time taken as the initial moment, the wave function and its time derivative must be independent of one another. But in this case we can, at least for this moment, determine, say, the time derivative $\partial \Psi / \partial t$ everywhere within the integration volume so that the integral (10.122) will not be zero. Thus, the expression (10.120) for the probability density and the following requirement (10.122) turn out to be in logical contradiction with requirement that Ψ be a solution to the Klein–Gordon equation. If we want a description based on (10.120), then it is necessary that the time derivative of Ψ be determined by the Ψ itself at any given time. This can only be the case when Ψ satisfies a first-order differential equation with respect to time. But this requirement brings us back to the apparently insurmountable difficulty with the square root of an operator, discussed in Section 10.3.

Here comes the crunch. What cannot be done with one (scalar or pseudoscalar) function Ψ, perhaps can be done if Ψ is a multicomponent quantity, say, tensor of a certain rank, that is, $\Psi \to \Psi_n$? Relativistic covariance requires that if a wave equation is linear in $\partial/\partial t$, it has to be linear in spatial derivatives as well. Thus, the problem reduces to the question, whether there exists a linear operator containing only first-order spatial derivatives, such that, when acting upon tensor Ψ_n, it produces the same result as the nonlinear operator (10.86)?

Actually, we are already familiar with a similar situation in the classical electromagnetism. The d'Alambert's Equation (6.134) for the 4-potential A is a second-order equation, whereas the Maxwell equations (6.180) for the electromagnetic field tensor F_{jk} are the first-order equations. But this gain in lowering the order of equation is inseparable from a certain loss.

Namely, in the EM theory, we have four uncoupled components (first-rank tensor A_j) in the d'Alambert's equation, but six components (the antisymmetric second-rank tensor F_{jk}) in the system of six coupled Maxwell's equations. An increase in the number of components and coupling between them is the price we pay for a decrease in their order.

In the current case, if the starting point is the Klein–Gordon equation for a scalar (zero rank tensor) Ψ, then we can expect in its linearized form a certain amount of linear but coupled equations for a vector quantity $\Psi \to \Psi_j$. And, invading the unknown territory, we cannot even say in advance the dimensionality (number of components) of this vector.

Let us now formulate the problem quantitatively.

If Ψ consists of N components, we can represent it as a column matrix

$$\Psi = \begin{pmatrix} \Psi_1 \\ \Psi_2 \\ \cdots \\ \cdots \\ \Psi_N \end{pmatrix}. \tag{10.123}$$

In this case, the operators acting on Ψ must be the matrix operators. There must be four such matrix operators – one for each spatial derivative. The elements of these matrices may be complex numbers. Denoting these matrices as $\hat{\gamma}^{(j)}$, $j = 0, 1, 2, 3$, we can write

$$\hat{\gamma}^{(j)} = \begin{pmatrix} \gamma_{11}^{(j)} & \gamma_{12}^{(j)} & \cdots & \gamma_{1N}^{(j)} \\ \gamma_{21}^{(j)} & \gamma_{22}^{(j)} & \cdots & \gamma_{2N}^{(j)} \\ \cdots & \cdots & \cdots & \cdots \\ \gamma_{N1}^{(j)} & \gamma_{N1}^{(j)} & \cdots & \gamma_{NN}^{(j)} \end{pmatrix}, \qquad j = 0, 1, 2, 3. \tag{10.124}$$

Now, write the d'Alambertian operator in the original Klein–Gordon equation in the form

$$\left(-\frac{\partial^2}{\partial x_0^2} + \nabla^2\right)\Psi = \lambda_c^{-2}\Psi. \tag{10.125}$$

According to (10.125), the quantity λ_c^{-2} can be considered as an eigenvalue of the d'Alambertian operator, corresponding to a particle with the rest mass m_0. Next step: we try to define a plausible expression for the square root of the d'Alambertian such that it be a *linear* operator:

$$\sqrt{-\frac{\partial^2}{\partial x_0^2} + \nabla^2} = i\hat{\gamma}_0 \frac{\partial}{\partial x_0} + \hat{\gamma}_\alpha \frac{\partial}{\partial x_\alpha}. \tag{10.126}$$

Nonsensical as this equation looks, its square makes perfect sense:

$$\begin{aligned} -\frac{\partial^2}{\partial x_0^2} + \nabla^2 = \left(i\hat{\gamma}_0 \frac{\partial}{\partial x_0} + \hat{\gamma}_\alpha \frac{\partial}{\partial x_\alpha}\right)^2 &= -\hat{\gamma}_0^2 \frac{\partial^2}{\partial x_0^2} + \hat{\gamma}_\alpha^2 \frac{\partial^2}{\partial x_\alpha^2} \\ + i(\hat{\gamma}_0\hat{\gamma}_\alpha + \hat{\gamma}_\alpha\hat{\gamma}_0)\frac{\partial}{\partial x_0}\frac{\partial}{\partial x_\alpha} &+ (\hat{\gamma}_\alpha\hat{\gamma}_\beta + \hat{\gamma}_\beta\hat{\gamma}_\alpha)\frac{\partial}{\partial x_\alpha}\frac{\partial}{\partial x_\beta}. \end{aligned} \tag{10.127}$$

But for this to hold, the matrix gamma operators must satisfy the following requirements:

$$\left\{\hat{\gamma}_j, \hat{\gamma}_k\right\} \equiv \hat{\gamma}_j\hat{\gamma}_k + \hat{\gamma}_k\hat{\gamma}_j = 2\delta_{jk}. \tag{10.128}$$

Already at this stage we notice a remarkable similarity with the properties of the Pauli matrices.

Next step: we need to determine the number N. It can be shown (Appendix D) that the least number satisfying all the conditions is $N=4$. The corresponding matrices are all Hermitian, so they can represent physically measurable quantities, that is, be their operators. They are called the *gamma matrices* or the Dirac's matrices and can be composed of the Pauli matrices:

$$\hat{\gamma}^{(0)} = \begin{pmatrix} I & 0 \\ 0 & -I \end{pmatrix}, \qquad \hat{\gamma}^{(1)} = \begin{pmatrix} 0 & \sigma_1 \\ \sigma_1 & 0 \end{pmatrix},$$

$$\hat{\gamma}^{(2)} = \begin{pmatrix} 0 & \sigma_2 \\ \sigma_2 & 0 \end{pmatrix}, \qquad \hat{\gamma}^{(3)} = \begin{pmatrix} 0 & \sigma_3 \\ \sigma_3 & 0 \end{pmatrix}. \tag{10.129}$$

Here, each element of the matrices in (10.129) is itself a (2×2) matrix.

Since the Pauli matrices are the spin operators, so Dirac's matrices can be considered as such as well. The difference is that Pauli had introduced his matrices into the theory "by hand" to describe the previously discovered spin properties of an electron. In Dirac's theory, the electron spin is obtained automatically from the requirement that the corresponding wave equation be relativistic covariant. Suppose one does not know anything about spin, one has not read the reports about its discovery. Then one could have predicted its existence and basic properties from relativity principles applied to the quantum wave equations.

Once we found matrix operators to be 4×4 matrices, we must ascribe four components to the wave function Ψ. Only in this case, we will have four linked equations for four unknown functions Ψ_j.

As seen from the way it was derived, the fact that Ψ has four components, has nothing to do with the fact that our space-time is four-dimensional. In particular, the four components Ψ_j do not transform as a four-vector under Lorentz transformations. As mentioned before, they form a half-integer rank tensor, called bispinor or Dirac spinor. If, instead of $N=4$, we introduce the matrices of higher dimensions, this would not change the formal structure of the theory, except that it would describe hypothetical particles with half-integer spin higher than 1/2.

The linearized relativistic equation is of the first order in all derivatives and has explicitly covariant form:

$$\hat{\gamma}^{(j)} \frac{\partial}{\partial x_j} \Psi = \frac{i}{\lambda_c} \Psi, \tag{10.130}$$

or, in terms of the 4-momentum operator,

$$\hat{\gamma}^{(j)} \hat{p}_j \Psi = \frac{\hbar}{\lambda_c} \Psi. \tag{10.131}$$

This, as expected, can be written in the form similar to Schrödinger's equation if we first write

$$i\hbar \frac{\partial \psi}{\partial t} = (-i\hbar c \hat{\gamma}^{(\alpha)} \nabla_\alpha + m_0 c^2 \hat{\gamma}^{(0)}) \Psi, \qquad \nabla_\alpha \equiv \frac{\partial}{\partial x_\alpha}, \tag{10.132}$$

and then introduce the Hamiltonian

$$\hat{H} \equiv -i\hbar c\hat{\gamma}^{(\alpha)}\,\nabla_\alpha + m_0 c^2\hat{\gamma}^{(0)} = -ic\hat{\gamma}^{(\alpha)}\hat{p}_\alpha + m_0 c^2\hat{\gamma}^{(0)} = -ic\hat{\gamma}\cdot\hat{\mathbf{p}} + m_0 c^2\hat{\gamma}^{(0)}, \tag{10.133}$$

so that

$$i\hbar\,\frac{\partial\Psi}{\partial t} = (-ic\hat{\gamma}\cdot\hat{\mathbf{p}} + m_0 c^2\gamma^{(0)})\Psi = \hat{H}\Psi. \tag{10.132 a}$$

Here, $\hat{\gamma}$ is the matrix "vector" operator in 3-space:

$$\hat{\gamma} \equiv (\hat{\gamma}^{(1)}, \hat{\gamma}^{(2)}, \hat{\gamma}^{(3)}). \tag{10.134}$$

These are different forms of Dirac's famous equation. In all these forms, the wave function Ψ is a 4-component quantity – Dirac's bispinor, so that actually any of the equations is the system of four linked equations for four unknown functions Ψ_j. To write this system explicitly for all four components Ψ_j, we need to use the rules of matrix multiplication (Problem).

Now we can derive the continuity equation for the probability distribution in space-time. To simplify the discussion, we will restrict to the case of a free particle. Following the procedure outlined in Section 10.1, write the expressions for the Hermitian conjugate of Dirac's equation. In doing this, use the fact that the Hermitian conjugate of the product of two matrices $\hat{A}$ and $\hat{B}$ is

$$(\hat{A}\hat{B})^+ = \hat{B}^+\hat{A}^+ \tag{10.135}$$

(Problem 10.17). We will get the complex conjugate of the Dirac's equation in the form

$$-i\hbar\,\frac{\partial\Psi^+}{\partial t} = -c\hat{\mathbf{p}}\Psi^+\cdot\hat{\gamma}^+ + m_0 c^2\Psi^+\hat{\gamma}^{(0)+}. \tag{10.136}$$

But since all gamma operators are Hermitian, both Equations (10.132) and (10.137) can be written as

$$\begin{aligned} i\hbar\,\frac{\partial\Psi}{\partial t} &= (c\hat{\gamma}\cdot\hat{\mathbf{p}} + m_0 c^2\hat{\gamma}^{(0)})\Psi, \\ -i\hbar\,\frac{\partial\Psi^+}{\partial t} &= -c\hat{\mathbf{p}}\Psi^+\cdot\hat{\gamma}^+ + m_0 c^2\Psi^+\hat{\gamma}^{(0)+}. \end{aligned} \tag{10.137}$$

Multiply the first of these equations by Ψ^+ on the left, the second by Ψ on the right, and then subtract the products. We get

$$i\hbar\left(\Psi^+\,\frac{\partial\Psi}{\partial t} + \Psi\,\frac{\partial\Psi^+}{\partial t}\right) = c[\Psi^+\hat{\gamma}\cdot\hat{\mathbf{p}}\Psi + (\hat{\mathbf{p}}\Psi^+\hat{\gamma})\Psi]. \tag{10.138}$$

The parentheses on the right side of this equation mean that the momentum operator is acting only on the function Ψ^+. The whole expression in the brackets can be brought to the form

$$c[\Psi^+\hat{\gamma}\cdot\hat{\mathbf{p}}\Psi+(\hat{\mathbf{p}}\Psi^+\,\hat{\gamma})\Psi]=\vec{\nabla}\cdot(\Psi^+\hat{\gamma}\,\Psi). \tag{10.139}$$

Then we obtain

$$\frac{\partial\varsigma}{\partial t}=-c\vec{\nabla}\cdot\mathbf{i}, \tag{10.140}$$

where the quantity

$$\varsigma=\Psi^+\Psi=\psi_j^*\,\psi^j \tag{10.141}$$

is the probability density, and

$$\mathbf{i}=\Psi^+\hat{\gamma}\Psi \tag{10.142}$$

represents the probability flux density.

Thus, the wave function Ψ in Dirac's equation admits, as it does in Schrödinger's theory, probabilistic interpretation. But the structure of the corresponding expressions for the probability and its flux in terms of the wave function is more complicated than it is in Schrödinger's equation for a spinless particle. It is more similar to that in Pauli's original formulation, but contains twice as many terms. The relativistic description doubles the number of terms in the expression for the probability density. What is the physical meaning of the extra terms?

To answer this question, we must consider the solutions for Dirac's equation. Again, we limit to the case of a free particle. To simplify the equations, denote $\hat{\gamma}^{(0)}\equiv\hat{\beta}$. Look for a solution of Dirac's Equation (10.137) describing a stationary state

$$\Psi(r,t)=\Psi_0\,e^{-i(\mathcal{E}\mathbf{kr})}. \tag{10.143}$$

Putting this into (10.137) we will get the equation for $\Psi_0\equiv\tilde{W}$

$$\mathcal{E}\tilde{W}=(c\hat{\gamma}\cdot\mathbf{p}+m_0c^2\hat{\beta})\,\tilde{W}. \tag{10.144}$$

Now, since $\tilde{W}$ is a 4-component quantity as is Ψ, we can represent it as a column matrix

$$\tilde{W}=\begin{pmatrix}w_1\\w_2\\w_3\\w_4\end{pmatrix}=\begin{pmatrix}W\\W'\end{pmatrix}, \tag{10.145}$$

where W and W' are the short hands (called "spinors") for the two column matrices

$$W=\begin{pmatrix}w_1\\w_2\end{pmatrix},\qquad W'=\begin{pmatrix}w_3\\w_4\end{pmatrix}. \tag{10.146}$$

Putting this into (10.145) and taking account of the representation of the gamma matrices in terms of the Pauli matrices gives

$$\mathcal{E}W = c\hat{\sigma}\cdot\mathbf{p}W' + m_0c^2W, \qquad \mathcal{E}W' = c\hat{\sigma}\cdot\mathbf{p}W - m_0c^2W'. \tag{10.147}$$

For this system of homogeneous equations to have a nonzero solution, its determinant must be zero:

$$\begin{bmatrix} \mathcal{E} - m_0c^2 & -c\hat{\sigma}\cdot\mathbf{P} \\ -c\hat{\sigma}\cdot\mathbf{p} & \mathcal{E} + m_0c^2 \end{bmatrix} = 0 \tag{10.148}$$

It follows

$$\mathcal{E}^2 - m_0^2c^4 = c^2(\hat{\sigma}\cdot\mathbf{p})^2. \tag{10.149}$$

Using the properties of Pauli matrices it is easy to show (Problem 10.19) that $(\hat{\sigma}\cdot\mathbf{p})^2 = \mathbf{p}^2$. Therefore, we come to the already known relation between the energy and the momentum of a particle

$$\mathcal{E}^2 = m_0^2c^4 + p^2c^2. \tag{10.150}$$

It shows that to each given momentum there correspond two different possible states of a particle – one with positive energy and the other with negative energy. For each sign of the energy, the system (10.147) has a distinct solution for bispinor $\tilde{W}$. Let us denote these solutions as $\tilde{W}_1$ for the positive-energy states and $\tilde{W}_2$ for the negative-energy states. Then

$$\tilde{W}_1 = \begin{pmatrix} a_1 \dfrac{cp}{\mathcal{E} - m_0c^2} \\ b_1 \dfrac{cp}{\mathcal{E} - m_0c^2} \\ a_1 \\ b_1 \end{pmatrix}, \qquad \tilde{W}_2 = \begin{pmatrix} a_2 \\ b_2 \\ a_2 \dfrac{cp}{\mathcal{E} + m_0c^2} \\ b_2 \dfrac{cp}{\mathcal{E} + m_0c^2} \end{pmatrix}. \tag{10.151}$$

Here, a_1, b_1 and a_2, b_2 are the constants that cannot be determined from Equations (10.147) since the equations are homogeneous. They can be determined from the normalization condition for the probability.

Comparing the last equation with (10.145) allows us to rewrite these solutions directly for the spinors W and W'. For the positive-energy solution, the spinors W and W' are

$$W_1 = \frac{cp}{\mathcal{E} - m_0c^2}\begin{pmatrix} a_1 \\ b_1 \end{pmatrix}, \qquad W'_1 = \begin{pmatrix} a_1 \\ b_1 \end{pmatrix}. \tag{10.152}$$

For the negative-energy solutions we will have

$$W_2 = \begin{pmatrix} a_2 \\ b_2 \end{pmatrix}, \qquad W'_2 = \frac{cp}{\mathcal{E} + m_0c^2}\begin{pmatrix} a_2 \\ b_2 \end{pmatrix}. \tag{10.153}$$

The physical meaning of these solutions becomes clearer if we consider the limiting cases. In the nonrelativistic limit we have $p \approx m_0v$, and

$$\begin{aligned} \mathcal{E}-m_0c^2 &\approx \frac{p^2}{2m_0} \approx \frac{m_0v^2}{2} \quad \text{for } \mathcal{E}>0, \\ \mathcal{E}+m_0c^2 &\approx -\frac{p^2}{2m_0} \approx -\frac{m_0v^2}{2} \quad \text{for } \mathcal{E}<0. \end{aligned} \tag{10.154}$$

Therefore, in this limit we have

$$\begin{aligned} W_1 &\approx 2\,\frac{c}{v}\begin{pmatrix} a_1 \\ b_1 \end{pmatrix} \gg W'_1, \qquad \mathcal{E}>0, \\ W'_2 &\approx 2\,\frac{c}{v}\begin{pmatrix} a_2 \\ b_2 \end{pmatrix} \gg W_2, \qquad \mathcal{E}<0. \end{aligned} \tag{10.155}$$

If out of the two spinors constituting a bispinor, one is much smaller than the other, it can be neglected and Dirac's bispinor reduces to a dominating spinor introduced by Pauli. Thus, in the nonrelativistic limit the general solution describing bispinor reduces to only one spinor W_1 in the case of the positive-energy solution and W'_2 in the case of the negative-energy solution.

Accordingly, the elements a_1, b_1 in (10.151) are the probability amplitudes to find a positive-energy particle (associated with an electron) in a state with its spin up or down, respectively. Similarly, the elements a_2, b_2 are the probability amplitudes to find a negative-energy particle (associated with a positive-energy positron) in a state with its spin up or down, respectively.

In the ultrarelativistic limit $|\mathcal{E}| \gg m_0c^2$ (regardless of the sign of energy), we have both spinors – primed and unprimed – roughly of the same magnitude. This corresponds to the fact that in performing a high-accuracy position measurement, we are equally likely to find either a spin 1/2 particle or its antiparticle, or even a combination of both. Note that an antiparticle can be found even if initially we had a low-energy state with one particle only. As we know, a sufficiently accurate position measurement automatically supplies the particle with extra energy, thus making it relativistic. This, in turn, increases the probability of particle–antiparticle generation while interacting with measuring system (say, high-energy gamma radiation). And it may so happen that the generated antiparticle will pop up in our device instead of the original particle. Such an outcome, totally exotic from classical viewpoint, is a routine event in relativistic microworld. It is this process that is described by the probability amplitudes (10.152).

10.4
Relativistic Doppler Effect III: A Child of Relativity and Quantum Mechanics

The Doppler effect can be best explained by a picture of propagating wave from a moving source, which directly appeals to our common sense and everyday life experience. But for the result to be general we must in the end take into account the relativistic effect of time dilation – the decrease of observed frequency of the source

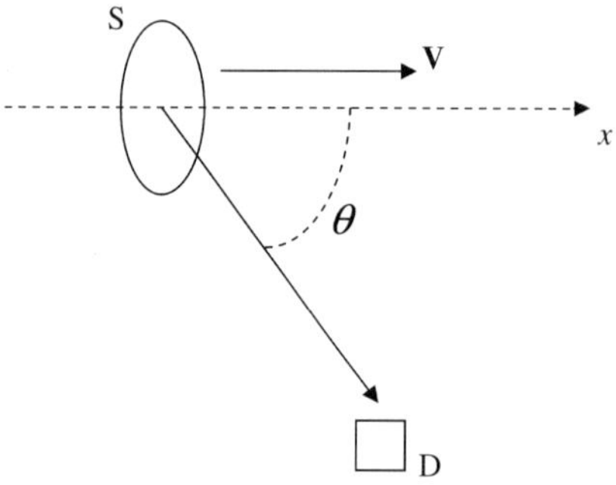

Figure 10.11 The geometry of the Doppler effect: S, source; D, detector.

due to its motion:

$$\omega = \frac{\omega_0}{\gamma(V)}, \tag{10.156}$$

where ω and ω_0 are the observed and the proper frequency, respectively, V is the speed of the source, and $\gamma(V)$ is the corresponding Lorentz factor. Then for the frequency ω' of waves arriving from the source to the observer one gets the expression

$$\omega' = \frac{\omega}{(1-(V/c)\cos\theta)} = \frac{\omega_0}{\gamma(V)(1-(V/c)\cos\theta)}, \tag{10.157}$$

where θ is the angle between the directions of motion of the source and the wave (Figure 10.11). Since ω' here is the frequency recorded by detector, $\omega' = \omega_D$, this equation is identical to Equation (6.228). In particular, at $\theta = 0$ (an approaching source) Equation (10.157) describes the increase of the wave frequency (the blue shift), at $\theta = \pi$ (a receding source) we have the decrease of frequency (the red shift), and at $\theta = 90^\circ$ a less prominent decrease in frequency owing entirely to the motion of the source regardless of its direction (pure time dilation effect).

The same Doppler effect can be described more formally as the result of Lorentz transformation of 4-vector $(\omega/c, \mathbf{k})$, where $\mathbf{k} = k_x, k_y, k_z$ is a "normal" three-dimensional wave vector. Let ω_0, k_x, k_y, k_z be the components of the 4-vector in the rest frame of the source, the corresponding primed quantities be the components of this vector in the rest frame of the observer, and the x-axis be directed along the relative motion of the source and the observer. Then, according to the Lorentz transformation,

$$\omega_0 = \gamma(V)(\omega' - Vk'_x). \tag{10.158}$$

and, since $k'_x = k' \cos\theta = (\omega'/c)\cos\theta$, we have $\omega_0 = \gamma(V)\omega'(1 - V\cos\theta/c)$, or

$$\omega' = \frac{\omega_0}{\gamma(V)(1-(V/c)\cos\theta)}, \tag{10.159}$$

which is Equation (10.157).

Now, according to quantum mechanics, the radiation of light of a given frequency is a discrete process in which indivisible light particles – photons – are emitted by a luminous object losing a strictly specified amount of energy with each emission. The lost "quantum" of energy is carried away by the emitted photon and is equal to $\hbar\omega$. But because of the Doppler effect, the frequency and thereby energy of the emitted photon will experience a blue or red shift if the emitting object is moving respectively toward or away from the observer. On the other hand, the emitted energy is only associated with the energy difference between the initial (excited) and final (let us call it ground) state of the emitting system. These states and thereby their energies $\mathcal{E}_e$ and $\mathcal{E}_g$ as well as their difference $\mathcal{E}_e - \mathcal{E}_g$ are determined by the *intrinsic* properties of the system only. It is therefore not immediately clear, how, within the energy conservation law, the photon's energy can be uniquely determined by $\mathcal{E}_e - \mathcal{E}_g$, and at the same time depend on direction of emission? In other words, how does the quantum-mechanical mechanism of light emission incorporate into the picture of Doppler effect?

On the qualitative level, the answer to the question is that the photon's energy, while being determined by the difference $\mathcal{E}_e - \mathcal{E}_g$, is not just equal to $\mathcal{E}_e - \mathcal{E}_g$. Two important factors affect the relation between $\mathcal{E}_e - \mathcal{E}_g$ and $\hbar\omega$: the relativistic increase of mass of a moving source and its recoil when shooting out a photon.

The first factor alone may at first appear to be sufficient to account for a blue shift. The total energy of a moving source (let it be an atom) is its rest energy times the Lorentz factor. Thus, the total energy of a moving *excited* atom is

$$\mathcal{E}'_e = \sqrt{\mathcal{E}_e^2 + p_e^2 c^2} = \gamma(V)\,\mathcal{E}_e, \tag{10.160}$$

where

$$p_e = \frac{\mathcal{E}_e}{c^2}\gamma(V)\,V \tag{10.161}$$

is the atom's momentum. Similarly, the total energy of the moving atom in its *ground* state is also increased by the same factor, so that neglecting the recoil effects we can state that the energy difference between the excited and the ground state for the atom in motion is (Figure 10.12)

$$\mathcal{E}'_e - \mathcal{E}'_g = \gamma(V)(\mathcal{E}_e - \mathcal{E}_g). \tag{10.162}$$

This energy is larger than the energy $\mathcal{E}_e - \mathcal{E}_g$ of the optical transition in the atom's rest frame and therefore the energy of the emitted photon is accordingly larger. This explanation, however, is not satisfactory. It is not in quantitative agreement with the amount of the blue shift (setting $\theta = 0$ in Equation (10.157) gives for the photons emitted in forward direction the blue shift *larger* than the relativistic energy difference (10.61)). And, even worse, it stands in flat contradiction with (10.157) in the case of the backward radiation, when the energy of the red-shifted photons is *less* than $\mathcal{E}_e - \mathcal{E}_g$.

To account for these discrepancies, we must also consider the effects of atom recoil in the emission process. These effects are well known in our everyday life world. If you dive head first from the bow of a rushing motorboat, you kick the boat in the backward direction, which *decreases* its speed and kinetic energy (the boat recoils back). In response, you receive the equal kick from the boat in the forward direction,

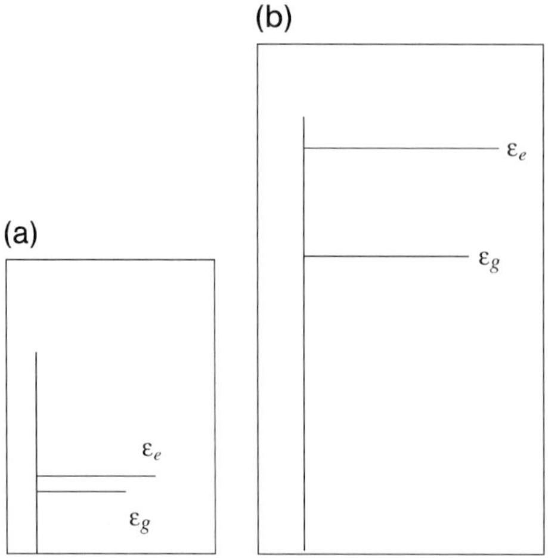

Figure 10.12 Ground and excited states (not to scale) for (a) stationary atom and (b) moving atom.

which *increases* your kinetic energy. As a result, your energy is *larger* than the initial kinetic energy you had had immediately before diving, as a result of moving together with the boat. The additional energy comes from the kinetic energy lost by the boat. If you dive from the stern of the rushing boat, you kick the boat in the forward direction, which increases its speed and kinetic energy (the boat recoils forward). The boat responds by kicking you in the backward direction, which decreases your initial kinetic energy. The energy thus lost by you is gained by the boat.

If we apply the same reasoning to the photons quitting the moving atom, we will obtain similar results. The photon emitted in the forward direction will gain an additional energy at the cost of kinetic energy of the atom. This explains why the blue shift is even larger than the Lorentz-increased energy of the optical transition (10.61). The photon emitted in the opposite direction will lose some energy to the atom, which will receive from it a kick in the forward direction. This energy loss is even larger than the energy increase caused by the Lorentz factor. As a result, the red-shifted photon is even less energetic than its sister emitted by the stationary atom.

To make this crude analogy exactly accurate, we must describe the process rigorously, taking into account the conservation of both energy and momentum of the system.

We start with the rest frame of the excited atom. The atom decays into its ground state and the photon. The photon has an energy $\hbar\omega_0$ and momentum $\hbar\omega_0/c$. Since the original atom was stationary, the atom in the ground state recoils back with the same momentum

$$\Delta p = \frac{\hbar\omega_0}{c}. \tag{10.163}$$

Therefore, its energy is $\sqrt{\mathcal{E}_g^2 + \Delta p^2 c^2}$. Note that it is *not* just $\mathcal{E}_g$ and accordingly the photon's energy is not just $\mathcal{E}_e - \mathcal{E}_g$. The law of conservation of energy states that

$$\mathcal{E}_e = \sqrt{\mathcal{E}_g^2 + \Delta p^2 c^2} + \hbar\omega_0. \tag{10.164}$$

Using the conservation of momentum (10.162) and solving (10.163) for $\hbar\omega_0$, we get

$$\hbar\omega_0 = \frac{\mathcal{E}_e^2 - \mathcal{E}_g^2}{2\mathcal{E}_e} = \frac{\mathcal{E}_e + \mathcal{E}_g}{2\mathcal{E}_e}(\mathcal{E}_e - \mathcal{E}_g). \tag{10.165}$$

Only in a special case $\mathcal{E}_e - \mathcal{E}_g \ll \mathcal{E}_e$ (when we can use the approximation of an "infinitely heavy" atom) does this expression reduce to $\mathcal{E}_e - \mathcal{E}_g \cong \hbar\omega_0$.

Consider now a "primed" reference frame where our excited atom moves with velocity V before it decays. Its energy and momentum before the emission are, respectively,

$$\sqrt{\mathcal{E}_e^2 + p^2c^2} = \mathcal{E}_e\gamma(V) \tag{10.166}$$

and

$$\mathbf{p} = \frac{\mathcal{E}_e}{c^2}\mathbf{V}\gamma(V). \tag{10.167}$$

The atom emits a photon in a direction $\mathbf{n}$, which makes an angle θ with $\mathbf{p}$ (Figure 10.13). The photon's energy is $\hbar\omega'$ and momentum is $\hbar\mathbf{k}' = \hbar(\omega'/c)\mathbf{n}$. After the emission, the atom is in its ground state and has the momentum $\mathbf{p}' = \mathbf{p} - \hbar\mathbf{k}'$ and the energy $\mathcal{E}_g' = \sqrt{\mathcal{E}_g^2 + (\mathbf{p} - \hbar\mathbf{k}')^2 c^2}$.

The law of conservation of energy states that $\mathcal{E}_e\gamma(V) = \mathcal{E}_g + \hbar\omega'$. This gives the equation

$$\mathcal{E}_e\gamma(V) = \sqrt{\mathcal{E}_g^2 + (p^2 - 2p\hbar k'\cos\theta + \hbar^2 k'^2)c^2} + \hbar\omega'. \tag{10.168}$$

Putting here (10.167) for p and ω' for $k'c$, we obtain the relation between $\hbar\omega'$, the energies of the atom's ground and excited states, and the emission angle θ, which after some algebra simplifies to

$$\hbar\omega' = \frac{\mathcal{E}_e^2 - \mathcal{E}_g^2}{2\mathcal{E}_e\gamma(V)(1 - (V/c)\cos\theta)}. \tag{10.169}$$

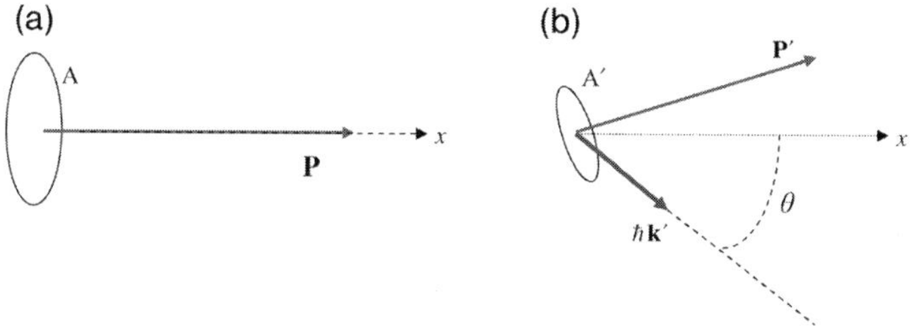

Figure 10.13 A moving atom (a) before the photon emission, state A, and (b) after the emission, state A′.

But according to (10.164), this can in turn be expressed in terms of the proper frequency ω_0 as

$$\omega' = \frac{\omega_0}{\gamma(V)(1-(V/c)\cos\theta)}. \tag{10.170}$$

This is the original Equation (10.157) for the Doppler effect.

10.5
Field–Particle Interactions: The Compton Effect

In this section, we will consider another case of the field–particle interaction – the Compton effect. This phenomenon also involves a few problems of general physical interest, such as the question of Newton's third law in relativity, and the description of scattering of EM waves by a point charge from classical and QM perspectives.

We know that in relativity, the interaction between two objects does not always obey Newton's third law (one of the simplest cases is the interaction between a stationary and a moving point charges – recall Section 6.1). The same appears to be the case when a photon interacts with a charged particle. Let us first consider the corresponding interaction classically. In this case instead of the concept of photons, we use a concept of a continuous EM wave packet. It acts on a charged classical particle with a force. The particle (let it be an electron) responds by exerting reciprocal force on the EM field. Already at this stage it seems strange to think of a force exerted on the field – the field is considered by many as a mere abstraction used only because it is convenient for describing interaction between the charges. In static situation, we sometimes think about electric field around the charges as a kind of mediator transferring the forces between them, rather than something existing in its own right. In such cases, we say that a charge q_1 (not its field) exerts a force on charge q_2 and the latter, in turn, exerts the force on the former.

But then we turn to the field description and say that it is the external *field* at the corresponding location that exerts force on q_2. In this case, in the spirit of the Newton's third law, we have to ask whether q_2 exerts a force on the field in this locality? If somebody says no, we get surprised and ask: hey, wait a minute, how about the Newton's law? But then we recall that this law does not always hold even for the charges, or current elements for that matter, like in the examples in Sections 6.1–6.3. In this case, why should we expect equal action–reaction between a charge and the field?

Even more so when the field cannot be identified as being due to a specific charge, like, for instance, field of an incident EM wave. Since its primary source is unknown, we are more prone to consider this field as an independent entity existing in its own right. Does this field exert a force on a charge? Definitely, because the charge in this field begins to oscillate and we can even calculate the amplitude and frequency of this oscillation if we know the characteristics of the incident wave. Does the charge, in turn, exert a force on the incident field? This time, if somebody says no, we are no longer surprised since we are already prepared to the fact that the action–reaction law does not always hold between the two participants only, but requires a broader picture. And the answer seems to be correct in view of the experimentally known fact that the incident plain wave does not change its direction in the electric field of a single fixed charge.

The same question can also be recast into quantum-mechanical language. We can ask how come that a photon as a particle of the EM field, exerts a force on a charge, but itself does not experience any force in the charge's field since it is electrically neutral? Again, we can say that there is no contradiction here because we know that the action–reaction rule may not hold between two interacting objects. But in both cases – classical and quantum-mechanical – the answer may be wrong.

Let us start with the classical case. Consider a monochromatic, linearly polarized, plane wave incident on a stationary electric charge. The charge starts oscillating in the alternating electric field of the wave, and having acquired speed, it also responds to the wave's magnetic field. These oscillations are the charge's way to respond to the field's action and thereby produce a reaction.

Well, what *is* the reaction?

An oscillating charge is a microscopic antenna. As any working antenna, it is, in turn, radiating the EM field of its own. Microscopically, this happens because the oscillating charge has acceleration. We know (Section 6.11) that this produces EM radiation propagating away from the charge in the form of a diverging spherical wave. This secondary radiation appears as a *scattered wave*. The resulting field is no longer a single incident wave. It is a superposition of the incident and scattered waves – it has changed. This change is the result of the charge's reaction – if you wish, it is a manifestation of the action–reaction law. The field does experience the influence from the charge. We can describe this influence by saying that the charge exerts a force on the field. On the face of it, this seems at odds with the fact that the scattered wave spreads around practically in all directions, whereas one would expect that the field gets a "kick" in a certain single direction. This expectation is wrong in classical case, because what we perceive as a classical wave is the result of many photons moving and oscillating in total synchrony and the scattered wave is the macroscopic average of scattering of billions and billions of individual photons. If we want to talk reasonably of a kick and maybe of energy exchange between the charge and the field, we must narrow down to a single photon and, in addition, take care to provide with the experimental environment ensuring the photon momentum measurement – for example, put the photon detectors around the scatterer (the charge).

In the considered situation, when the charge was initially at rest, one expects the scattered frequency to decrease and interpret this as energy decrease of the field. However, in the classical picture the scattered frequency in this case is considered to be equal to the incident frequency. This discrepancy is only apparent one, because in classical picture we implicitly assume the charge oscillating around its initial fixed position. In reality, the charge will start gradually drifting along the direction of incidence, that is, away from the oncoming wave. As a result, the wave's frequency as seen by the charge decreases and so does the reradiated frequency.

This is only a very crude description. At a closer look one can find subtle correspondence between the classical and quantum description of this process [99].

Consider the electron oscillation under the incident monochromatic wave

$$\mathbf{E} = \mathbf{E}_0\, \mathrm{e}^{i\,(\mathbf{k}\cdot\mathbf{r}-\omega t)}, \qquad \mathbf{B} = \mathbf{B}_0\, \mathrm{e}^{i\,(\mathbf{k}\cdot\mathbf{r}-\omega t)}, \qquad \mathbf{B}_0 = \frac{\hat{\mathbf{n}} \times \mathbf{E}_0}{c}, \tag{10.171}$$

where the unit vector $\hat{\mathbf{n}}$ is along the direction of incidence.

According to classical theory, the electron starts oscillating with the same frequency ω. Oscillating electron radiates waves of the same frequency (in the reference frame where the center of oscillation is stationary). The radiated field carries energy, but its net momentum is equal to zero because of the symmetry with respect to the axis of the oscillating dipole moment and the mirror symmetry with respect to the plane perpendicular to this axis and passing through the center of oscillations. A very interesting situation: the radiated waves carry away the energy initially taken by the oscillating electron from the incident wave, but the net momentum of the scattered wave is zero owing to the wave's symmetry. In the first approximation, the electron borrows the energy from the incident plane wave to immediately return it in the form of the scattered wave, but by doing this it separates the incident energy from the associated momentum and absorbs the momentum. This momentum must therefore transfer to the electron itself. We can describe this as the radiation pressure exerted by the incident wave on the electron. In the next approximation, we must take into account the relatively small amount of the incident energy that inevitably must be acquired together with momentum by the initially resting electron. This comes as the correction to the first approximation: we now say that not all of the borrowed energy is returned to the EM wave. The EM radiation in its final state has less energy than in the initial state. If the radiation is represented by only one photon, then this photon, apart from being scattered, also loses a certain fraction of its initial energy and thereby its associated frequency must decrease and the wavelength accordingly decrease.

Summarize this part:

In the classical language of forces, we are talking first only about the electric force exerted by the EM wave on the electron. Then, we take into account the magnetic force which appears the moment the electron starts moving under the electric force. According to relativistic rule (6.2.9) following from the Lorentz transformation of forces, the magnetic force accelerates the electron *along* the direction $\hat{\mathbf{n}}$ of the incident wave. The resulting motion of the electron must therefore be a superposition of two oscillations: one transverse along $\mathbf{E}$ and the other longitudinal along $\hat{\mathbf{n}}$. If we now take an average of the magnetic force over one complete cycle, taking into account the longitudinal oscillations, there remains a nonvanishing residue along $\hat{\mathbf{n}}$. Therefore, in the *third* approximation of classical description, there appears the electron *drift* along the direction of incidence, that is, the actual motion is a superposition of three motions: two oscillations and one relatively slow drift along $\hat{\mathbf{n}}$. It is this drift that is the result of momentum transfer from the incident field to the electron.

In the corresponding equation of motion, we have the Lorentz force acting on a charge in the incident EM wave. This force, in view of (10.171), can be written in 3D form as

$$\mathbf{f} = q_e(\mathbf{E}_0 + \mathbf{v} \times \mathbf{B}_0)e^{i(\mathbf{k}\cdot\mathbf{r}-\omega t)} = q_e\left(\mathbf{E}_0 + \frac{\mathbf{v}}{c} \times (\mathbf{E}_0 \times \hat{\mathbf{n}})\right)e^{i(\mathbf{k}\cdot\mathbf{r}-\omega t)}, \tag{10.172}$$

or, using the known identity for double cross product:

$$\mathbf{f} = q_e\left(\mathbf{E}_0 + \left(\frac{\mathbf{v}}{c}\cdot\hat{\mathbf{n}}\right)\mathbf{E}_0 - \left(\frac{\mathbf{v}}{c}\cdot\mathbf{E}_0\right)\hat{\mathbf{n}}\right)e^{i(\mathbf{k}\cdot\mathbf{r}-\omega t)}. \tag{10.172a}$$

In the nonrelativistic case, when the speed of the particle remains significantly less than c, the first term on the right dominates the expression, so in this approximation, for the linearly polarized light, we can consider the established stationary motion of the particle as oscillations along the direction of polarization. Accordingly, the second term vanishes. The remaining third term represents a relativistic correction, which describes the component of force along the direction of incidence. As is clearly seen from the derivation, this additional force comes from the magnetic field acting on the moving charge. If we find the solution of Equation (10.172) including the terms with v/c, in the RF where the charge *on the average* is stationary and find the average force over the period, then we find that the average electric force is zero, but the average magnetic force (represented by the third term on the right in the last equation) gives a small but nonvanishing residue along the direction of incidence.

Let the x-axis run along $\hat{\mathbf{n}}$-direction and the y-axis along the $\mathbf{E}$-direction. Then, the first approximation (neglecting the magnetic term) gives

$$m_0\ddot{\mathbf{r}} = q\mathbf{E}_0\,e^{-i\omega t}. \tag{10.173}$$

The special solution of this equation with the initial conditions $\mathbf{r} = \dot{\mathbf{r}} = 0$ at $t = 0$ is

$$\mathbf{r}^{(1)}(t) = -\frac{q}{m_0\omega^2}\,\mathbf{E}_0\,e^{-i\omega t}, \tag{10.174}$$

and the corresponding velocity

$$\mathbf{v}^{(1)} = \dot{y}(t)\hat{\mathbf{y}} = i\,\frac{q}{m_0\omega}\,\mathbf{E}_0\,e^{-i\omega t}. \tag{10.175}$$

Next step: we try to solve the whole Equation (10.172) by putting there the velocity $\mathbf{v}^{(1)}$; although still approximate, this equation will be already much more accurate than (10.173)

$$\mathbf{f} = q(\mathbf{E} + \mathbf{v}^{(1)} \times \mathbf{B}) = q\left(\mathbf{E}_0\,e^{-i\omega t} + i\,\frac{q}{m_0\omega}(\mathbf{E}_0 \times \mathbf{B}_0)e^{-2i\omega t}\right). \tag{10.176}$$

In this approximation, the second term on the right, while adding nothing to the transverse oscillations, will produce oscillation along the $\hat{\mathbf{n}}$-direction (the longitudinal, or the x-oscillations), so it may be convenient to rewrite (10.176) in components:

$$\begin{aligned} m_0\ddot{y} &= qE_0\,e^{-i\omega t}, \qquad y(t) = -y_0\,e^{-i\omega t}, \qquad y_0 \equiv \frac{q}{m_0\omega^2}\,E_0; \\ m_0\ddot{x} &= i\,\frac{q^2E_0^2}{m_0\omega c}\,e^{-2i\omega t}, \qquad x(t) = -\frac{1}{4}\,iky_0^2\,e^{-2i\omega t}, \qquad k = \frac{\omega}{c}. \end{aligned} \tag{10.177}$$

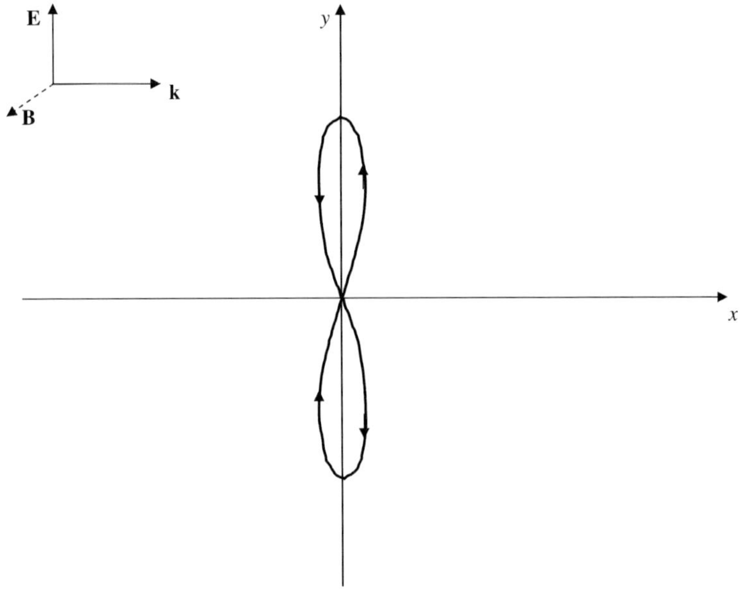

Figure 10.14 Trajectory of a point charge under the incident plane monochromatic EM wave. It represents 2D harmonic oscillator with frequencies' ratio 1 : 2.

The imaginary unit here indicates that the longitudinal oscillations are phase-shifted with respect to the transverse ones by 90°; we also see that their frequency is twice the transverse frequency. And, for the nonrelativistic oscillations $ky_0 = (qE_0/m_0c\omega) \ll 1$, so their amplitude is much smaller than that of transverse oscillations (the used approximation itself is based on this fact). But the oscillation frequency, as seen from (10.177), is twice the "transverse" frequency ω. The trajectory of the resulting x–y oscillations can be represented by number 8, squeezed in horizontal direction (Figure 10.14).

Now, if we take the average force (both electrical and magnetic) over one complete cycle for only transverse oscillations, we will get zero, but due to the longitudinal oscillations, the x-coordinate is no longer zero. Accordingly, in averaging the x-component of the magnetic force, we may still use only the y-component of oscillations, but must take account of the fact that according to (10.171), the incident field is a function of x. Then for the x-component of the force we will have

$$f_x = q(\mathbf{E} + \mathbf{v}^{(1)} \times \mathbf{B})_x = q\left(i \frac{q}{m_0\omega} (\mathbf{E}_0 \times \mathbf{B}_0)\hat{n} e^{2ikx(t)} e^{-2i\omega t} \right). \tag{10.178}$$

with $x(t)$ from (10.177).

Averaging over the period gives

$$\langle f_\parallel \rangle = \langle f_x \rangle = \frac{q^2}{2\pi m_0 c} \frac{(ky_0)^4}{\omega} E_0^2 = \frac{1}{2\pi} \left(\frac{qE_0}{m_0 c\omega} \right)^5 qE_0. \tag{10.179}$$

If we know the interaction time Δt, this would allow us to calculate the acquired momentum as

$$\Delta p = \langle f_{\parallel} \rangle \Delta t. \tag{10.180}$$

But we do not know this time.

We can evaluate the acquired momentum in the following way. Using the results (6.201) and (6.231) in Sections 6.8 and 6.11, we can find the intensity (radiated power) of the dipole radiation [34]

$$I = \frac{2}{3} \frac{q^2 \omega^4 y_0^2}{(4\pi\varepsilon_0)^2}. \tag{10.181}$$

Putting here y_0 from (10.177) gives

$$I = \frac{2}{3} \frac{q^4}{4\pi\varepsilon_0 m_0^2} E_0^2 = \frac{1}{3} \frac{q^4 \mu_0}{\pi\varepsilon_0 m_0^2} S = \frac{1}{3} \frac{\mu_0^2 q^4 c^2}{\pi m_0^2} S, \tag{10.182}$$

where S is the magnitude of the Poynting vector $\mathbf{S}$. Equation (10.182) gives the amount of energy $\Delta\mathcal{E}/\Delta t$ per unit time borrowed by the electron from the incident wave (to return almost immediately in the form of the scattered wave). Whatever the fate of the borrowed energy, there was a momentum $\Delta p = \Delta\mathcal{E}/c$ associated with it. This momentum goes to the electron. The momentum per unit time is the force. Thus, the longitudinal force we were looking for, can be simply found, using (10.181) as

$$f_{\parallel} = f_x = \frac{\Delta p_x}{\Delta t} = \frac{\Delta\mathcal{E}}{c\Delta t} = \frac{2}{3} \frac{q^2 \omega^4 y_0^2}{4\pi\varepsilon_0}. \tag{10.183}$$

The ratio of the scattered power to the incident intensity is the *cross section* of the whole process:

$$\sigma = \frac{1}{3} \frac{q^4}{\pi m_0^2 c^2}. \tag{10.184}$$

A word of caution may be timely. It appears as if it is the magnetic force that is responsible for the momentum and thereby energy transferred to the electron. This is an illusion. We know that magnetic force does no work, since according to the Lorentz force law, it is always perpendicular to a particle's velocity. The appearance of nonzero work done by the magnetic force comes here from the fact that we determined only the longitudinal component of the magnetic force on the electron. This component of magnetic force is much greater than its transverse component, since their ratio is in proportion to the ratio of transverse and longitudinal components of the electron's velocity or, which is the same, to the ratio of the height of the 8-figure traced out by the electron to its width in Figure 10.14:

$$\frac{f_{\parallel}}{f_{\perp}} = \frac{v_{\perp}}{v_{\parallel}} = \frac{8(\parallel)}{8(\perp)}, \quad \text{so that} \quad |f_{\parallel} 8(\perp)| = |f_{\perp} 8(\parallel)|. \tag{10.185}$$

On the other hand, it is seen from the same figure that the products themselves in (10.185) have the opposite sign. Therefore, if we calculate the net work done by the magnetic force on the electron, it will be zero, as it should:

$$W_{\text{magn}} = f_{\parallel} 8(\perp) + f_{\perp} 8(\|) = 0. \tag{10.186}$$

Well, in this case what force is responsible for the accelerating electron along the x-direction? The electric force is. It accelerates the electron along the y-direction and the role of the magnetic force is only in turning part of the acquired velocity vector by 90°, from the y-direction to the x-direction. Hence the appearance that it is the magnetic force that does the work.

Here, we arrived at the possibility to discuss the frequency of the scattered wave. As mentioned above in our qualitative description, the charge starts drifting along the residual average "magnetic" force that we have called the radiation pressure force. There appears a nonvanishing component of the charge's motion, along the direction of incidence. As a result, the charge sees the redshifted frequency of the incident radiation. In the RF comoving with the charge, the reradiated light is accordingly redshifted as compared to the incident frequency in the laboratory RF. But for the observer in this RF (for us) the reradiated light is emitted by the *moving* source and thus is, in turn, Doppler-shifted. Since the charge radiates (in classical description) in all directions (except the oscillation axis), the Doppler shift is different for different directions (see Sections 9.4 and 10.5 on relativistic Doppler effect). In particular, it is blueshifted in the forward direction, redshifted in the backward direction, and slightly redshifted (pure time dilation effect!) in the transverse direction. The correct calculation of the resulting frequency comes as a combination of both stages. Let us estimate the result quantitatively.

Suppose that an initially stationary electron has acquired a drift velocity v_d after being exposed to incident monochromatic wave (10.171) during a time interval T. Then in an instantly comoving inertial RF (system K′), the incident frequency seen by the charge, and accordingly the proper frequency of its oscillations, is given by Equation (10.159) of the previous section with $\theta = \pi$. Indeed, the incident wave can be considered as emitted by a distant source with proper frequency ω; this source is stationary in K, but receding from the charge in K′. The charge feels the wave emitted by this source in the direction opposite to its motion. Thus, the secondary radiation frequency in K′ is

$$\tilde{\omega} = \frac{\omega}{\gamma(v_d)(1+\beta_d)}, \tag{10.187}$$

where $\beta_d \equiv v_d/c$. In system K′ this frequency of secondary radiation is the same in all directions. In K, the source of radiation is drifting and the wave diverging from it is considered as a scattered wave. Its frequency depends on the direction and is described by the same Equation (10.159), with the distinction that now θ is a variable – it is the angle between the scattered and incident wave. Thus, we have

$$\omega' = \frac{\tilde{\omega}}{\gamma(v_d)(1-\beta_d \cos\theta)}. \tag{10.188}$$

Combining thc last two equations gives

$$\omega' = \omega \frac{1-\beta_d}{1-\beta_d \cos\theta}. \tag{10.189}$$

Usually, instead of frequency change, the corresponding equation is expressed in terms of the wavelength change. Accordingly, we can rewrite (10.189) as

$$\lambda' = \lambda \frac{1-\beta_d \cos\theta}{1-\beta_d} = \lambda\left(1 + \frac{2\beta_d}{1-\beta_d} \sin^2 \frac{\theta}{2}\right). \tag{10.190}$$

The change of the wavelength is

$$\lambda' - \lambda = 2\lambda \frac{\beta_d}{1-\beta_d} \sin^2 \frac{\theta}{2} \xrightarrow[\beta_d \ll 1]{} 2\lambda\beta_d \sin^2 \frac{\theta}{2}. \tag{10.191}$$

In particular, at $\theta = 0$ (forward scattering) there is no change. This should come as no surprise, since the receding electron in K′ sees and accordingly radiates the redshifted frequency, but we in K see the receding electron radiating in the same direction and accordingly blueshifted. The two opposite shifts cancel each other out and we see no shift in this direction. The forward scattering is always coherent. In contrast, the backscattered radiation experiences the maximal red shift since both shifts are of the same sign in this direction and reinforce each other. In the transverse direction, we have intermediate shift. If $\beta_d \to 1$ (the particle has accelerated nearly to the speed of light by receiving sufficiently large energy from the field), then the change of the wavelength becomes infinite, that is, $\Delta\lambda \to \infty$.

Classically, the frequency change must be a variable even for a fixed direction θ, since it also depends on the acquired drift velocity v_d. In high-energy physics, we usually describe motion in terms of the particle's energy rather than its speed. Therefore, it is reasonable to express in (10.191) the speed v_d in terms of the final energy of the particle associated with its drift. We can do it by using Equation (5.28).

A thoughtful reader may ask another question: what about the double oscillation frequency of the electron along the x-direction? The electron oscillating along any direction is a dipole oscillating in this direction. Therefore, there has to be radiation with corresponding frequency spreading symmetrically around the x-direction. According to Equation (10.181), relating radiated power to the oscillation amplitude, the intensity of this radiation will be much less than the intensity of radiation of frequency ω, since the amplitude of longitudinal oscillations is much smaller than the amplitude of transverse oscillations.

Thus, we have predicted the emergence of an additional component of radiation and even some of its properties. This radiation, however, is not reflected in (10.189). Why?

The answer is that we just skipped it when we derived Equation (10.189) from the Doppler effect. We were only focused on the primary frequency ω, since it gives the dominating contribution to radiation. But we could as well repeat the whole line of reasoning to obtain the similar expressions for the double-frequency component.

However, we did not do it for the following reason. The expressions (10.171)–(10.176) are classical and we know that classical wave (10.171) is represented by a huge (theoretically, infinite) number of the incident photons. From this viewpoint, the frequency doubling is a nonlinear optical effect corresponding to a very high intensity in the incident wave. When we switch to quantum-mechanical description with one incident photon at a time, this effect cannot occur since it requires many photons. But in the case of high light intensity with many photons at a time, this effect can, in principle, be observed in quantum domain as well. It seems to contradict the law of conservation of energy, since in the quantum-mechanical description double frequency corresponds to twice the energy of scattered photon, while the energy of each incident photon is only $\mathcal{E} = \hbar\omega$. Nevertheless, there is no contradiction here. Simply at high intensity there appears a nonzero probability of the double-photon processes, specifically double-photon absorption and scattering. In our case, the electron can absorb two photons at once, but then reemit the acquired energy with only one photon of double frequency. This would be a double-photon Compton effect.

For a single photon, the moment we have introduced this concept by quantizing the EM field, we return from the concept of long-range interaction to the concept of a local interaction. On the one hand, a photon in a state with definite momentum is simultaneously everywhere; on the other hand, its interaction with a point charge happens at the location of the charge. The single photon after the interaction can again be described by a monochromatic wave. This wave can be spherical. This describes a situation of scattering only – not recording the scattered photon in a detector. The spherical wave is a continuous superposition of plane waves scattered each in its own distinct direction. The amplitude of each such wave represents the probability amplitude to find the photon in a detector placed in the corresponding direction. Once the photon is found in the detector, we know that it has chosen to have been kicked by the electron in this direction. More accurately, its wave function has collapsed to a delta function $\delta(\mathbf{r}-\mathbf{r}')$, where $\mathbf{r}'$ is the position of detector. That the photon turned out to have veered in this direction is the manifestation of a force exerted by the charge on this photon. The electric neutrality of the photon has not prevented it from either acting on a charge or experiencing the reaction of the charge.

We gave an intuitive qualitative description of the important phenomenon – the Compton effect, playing the important role in both – microphysics and macrophysics, on a cosmological scale. It is worthwhile therefore to consider some of its features quantitatively. In this treatment we will use both – quantum-mechanical and relativistic rules governing the effect.

Quantum-mechanically, we can go to Equation (10.191). Writing there $v_d = p/m_0c$, we note that according to QM, the product $p\lambda = h$. Therefore, (10.179) takes the form

$$\lambda' - \lambda = 2\frac{h}{m_0 c}\sin^2\frac{\theta}{2}. \qquad (10.192)$$

This is the original formulation of the Compton effect. The quantity h/m_0c is the Compton wavelength, already familiar to us from the previous section.

Now, after so lengthy derivation of Equation (10.192) describing the Compton effect, it may be instructive to derive it directly from the conservation of relativistic

4-momentum. We have for the 4-momentum of the initial state (stationary electron and incident photon)

$$m_0c^2 + \hbar\omega = \sqrt{m_0^2c^4 + p_e^2c^2} + \hbar\omega', \qquad (10.193)$$
$$\mathbf{p} = \mathbf{p}' + \mathbf{p}_e$$

Here, $\mathbf{p} = \hbar\mathbf{k}$ and $\mathbf{p}' = \hbar\mathbf{k}'$ are the momenta of the incident and scattered photon, respectively; $\mathbf{p}_e$ is the momentum of the electron after the collision with the photon. Thus, on the left of the two equations are, respectively, the total energy and 3-momentum of the system before the reaction; on the right, the same characteristics after the reaction. Using $\omega = kc$, $\omega' = k'c$, we rewrite this as

$$m_0c + \hbar(k - k') = \sqrt{m_0c^2 + p_e^2},$$
$$\mathbf{p}_e = \mathbf{p} - \mathbf{p}' \Rightarrow p_e^2 = \hbar^2(k^2 - 2\mathbf{k}\cdot\mathbf{k}' + k'^2) = \hbar^2(k^2 - 2kk'\cos\theta + k'^2). \qquad (10.194)$$

Here, θ is the angle between the directions of the incident and scattered photons (the scattering angle).

Squaring the first equation and eliminating p_e by using the second one, we easily obtain after simple algebra

$$k - k' = \frac{\hbar}{m_0c}kk'(1 - \cos\theta) \qquad (10.195)$$

or

$$\frac{1}{k'} - \frac{1}{k} = \frac{\hbar}{m_0c}(1 - \cos\theta). \qquad (10.196)$$

But since $k = 2\pi/\lambda$, this is equivalent to (10.192).

The Compton effect, first discovered by Compton in the lab, plays the important role on the cosmological scale. One of the examples is some anomalies in the cosmic background radiation (CBR) – an afterglow of the Big Bang. A crude explanation of this phenomenon is as following. Imagine a sealed box filled with hot gas in thermal equilibrium with the accompanying radiation. If the walls of the box are made movable and elastic so that we can expand its volume, then the hot gas inside will cool down and the radiation, remaining in the equilibrium with gas, will experience the redshift. The latter may be also explained as a manifestation of the Doppler effect – decrease of light frequency after reflection from a receding wall. Our universe is analogous in this respect to such an expanding box. The hot radiation released in the Big Bang remains in thermal equilibrium with the other cooling contents of the expanding universe and cools down as well, which means that its spectrum corresponds to ever decreasing average temperature of the universe. This is manifest in the redshift as well. In this case we can understand the redshift as the manifestation of the Doppler effect: the background radiation is light from extremely distant (and accordingly hot since what is observed now is their earlier moments!) objects receding from us. An equivalent visualization: imagine standing waves of a type described in Section 9.1, but this time the length of the circumference itself is expanding. This will cause the corresponding extension of the wavelength of each standing wave fitting into the circumference. Now imagine that the circumference is an image of the expanding universe and the standing waves on it represent the waves of the background radiation filling its space. The increase of wavelengths is equivalent to the redshift and cooling down of radiation. This helps visualize the connection between the cosmological expansion and the background radiation spectrum. The spectral curve of radiation obtained from numerous observations [100] corresponds to a pretty cool environment of our average backdrop – the temperature of about $T = 2.7$ K.

However, more accurate measurements carried out later found a noticeable deviation from the equilibrium state [101]. Namely, the radiation intensity on the "blue" end of the spectrum turned out to be higher than it follows from thermal equilibrium. In other words, the number of the high-frequency photons is greater than expected from the above analogy. It looks like there is some process beyond our simplified model of the photons in otherwise empty expanding space. And this process must produce extra high-energy photons.

Such process does exist. It is the so-called inverse Compton effect. Its essence is pretty simple. According to relativity, a process observed in one RF may look pretty different when observed from another RF. This is true, of course, for the Compton effect as well. Imagine that we observe the previous process of the photon scattering by an initially stationary electron from an RF where this electron is moving toward the incident photon. You can obtain such an RF by boarding a spaceship moving away from the incident photon. In this spaceship, the photon's frequency will be significantly redshifted because of the Doppler effect, whereas the electron is initially moving toward the incident photon. After they collide, the electron comes out slowed down – part of its initial kinetic energy has "changed hands" – it has transferred to the photon. Accordingly, the photon after the collision is in a scattered state with its frequency *higher* than before. This situation is very similar to our example with the two colliding billiard balls in Section 5.6. There, you remember, the energy transfer was from ball A to B in one RF and from B to A in another RF. Now we have the same situation with the balls A and B replaced by a photon and electron, respectively.

A similar result will obtain if we imagine our initial conditions in this section just reversed in time: initially the electron and the photon are both moving in collision courses and in the final state the electron is stopped, with all its kinetic energy having gone to the photon, whose frequency after the collision accordingly increases. Hence, the name of the process following from this analogy: the reversed Compton effect. Once the direct effect is possible, so is its time-reversed counterpart.

This possibility actualizes in the real world on the global scale. There is an unimaginably huge amount of low-energy photons in the frequency band corresponding to micro- and radiowaves. On the other hand, there is a huge supply of high-energy charged particles (so-called cosmic rays, mostly the electrons and protons) generated in such powerful processes as supernova explosions, collapses of massive stars forming the neutron stars and black holes, galaxy collisions, and so on. Such a high-energy particle may bump into a low-energy, low-frequency lazy photon and give it a hearty kick, changing the photon from radiowaved entity into visible, UV, X-ray, or even a gamma-ray photon. These processes, far from thermal equilibrium in the current, relatively old and accordingly cold universe, act as an additional generator of "young" energetic photons. The estimations of the excess of such energetic photons correspond to the amount of the available high-energy particles in cosmic rays piercing the interstellar and intergalactic space.

Thus relativity, together with QM, allows us to understand and explain in simple terms yet another observed global-scale phenomenon.

Problems

10.1 Derive the expression for the momentum operator $\hat{\mathbf{p}}$ in the coordinate representation.

10.2 In **p**-representation, write the expressions for
(a) the momentum operator;
(b) the position operator.

10.3 Show that the operator $-i\hbar\vec{\nabla}$ is Hermitian.

10.4 Prove that if $\hat{L}$ is Hermitian, then the operators $\Delta\hat{L}$ and $\hat{V}\{L\} \equiv (\Delta\hat{L})^2$ are also Hermitian.

10.5 Express $\langle \Delta L^2 \rangle$ in terms of $\langle L^2 \rangle$ and $\langle L \rangle^2$.

10.6 Prove that the spin operator and the position operator commute.

10.7 Find the general expression for the commutator of the two operators: $\hat{x}_\alpha$ and $\hat{p}_\beta$ ($\alpha, \beta = 1, 2, 3$ where 1, 2, 3 stand, as usual, for x, y, z, respectively).

10.8 On the basis of the solution of the previous problem find the general expression for the product of variances $\Delta x_\alpha \Delta p_\beta$ from the Robertson–Schrödinger theorem.

10.9 Assume that the algorithm outlined in Section 10.1 for obtaining the Hamiltonian can be used to find other operators representing the corresponding observables known in classical physics.
(a) Use this algorithm to find the expression for operator of the angular momentum $\hat{\mathbf{L}}$ in **r**-representation;
(b) Find the commutation relations between the components $\hat{L}_\alpha$, $\hat{L}_\beta$ of this operator;
(c) Find the commutation relations between $\hat{L}^2$ and $\hat{L}_\alpha$;
(d) On the basis of the solution for (b), use the Robertson–Schrödinger theorem to find the indeterminacy relations for variances ΔL_α, ΔL_β and interpret the results.

10.10 Consider a nonrelativistic particle of mass m bound to the origin by a restoring force $f = -kx$ (a 1D harmonic oscillator).
(a) Write the equation of motion of the particle;
(b) Determine its oscillation frequency.

10.11 For the same conditions as in the previous problem find
(a) The coordinate and momentum of the particle as functions of time;
(b) Determine the relation between $x(t)$ and $p(t)$;
(c) Sketch the dependence $p(x)$ for a few different values of the total energy. You will obtain the corresponding phase diagrams. What do the obtained curves look like?

10.12 Derive Equations (10.109).

10.13 Prove that (10.109) is equivalent to (10.89).

10.14 Derive the nonrelativistic expressions for the probability and probability current densities in terms of auxiliary function $\Psi_0(\mathbf{r}, t)$.

10.15 Assuming an electron to be a spinning solid sphere of radius $r_0 = 10^{-15}$ m, estimate the necessary equatorial speed needed to produce spin $(1/2)\hbar$.

10.16 Using the rules of matrix multiplication, write explicitly the system of four equations for all four components of Dirac's bispinor Ψ_j.

10.17 Using the definition of a complex conjugate operator and the rules of matrix multiplication, prove that $(\hat{A}\hat{B})^+ = \hat{B}^+\hat{A}^+$.

10.18 Prove Equation (10.139) for the expression in the brackets for the flux density in Dirac's equation:

$$\Psi^+(\hat{\boldsymbol{\alpha}} \cdot \vec{\nabla})\Psi + (\vec{\nabla}\Psi^+ \cdot \hat{\alpha})\,\Psi = \vec{\nabla} \cdot (\Psi^+ \hat{\alpha}\Psi).$$

10.19 Show that $(\hat{\sigma} \cdot \hat{\mathbf{p}})^2 = p^2$.

10.20 Estimate the maximal possible lifetime of particle pairs that wink into existence and annihilate in the ongoing process of vacuum fluctuations.
(a) For two UV photons with frequency $\omega \cong 4 \times 10^{15}\,\text{s}^{-1}$.
(b) For an electron–positron pair.
(c) For a proton–antiproton pair.

10.21 Consider the following naïve picture of a fleeting episode in vacuum fluctuations: a particle–antiparticle pair has emerged at a certain point and its constituent particles immediately start flying apart. Since their mass–energy is destined to go back to vacuum, they must soon disappear and the physical mechanism for it is their mutual annihilation. But for them to annihilate they must meet again at some point and this seems impossible since they are flying apart.

Explain this "paradox" and confirm your explanation by evaluating the characteristics involved in the process – the particles' rest mass, possible speed, lifetime, and wavelengths.

10.22 In classical (non-QM) relativistic theory a charged particle can spiral to the Coulomb's center of attraction. Determine the critical impact parameter b_c in terms of particle's angular momentum L. Can such spiraling happen according to QM?

(*Hint*: Apply quantization of angular momentum and the indeterminacy principle).

11
Relativity and Causality

The past does not equal the future.

Anthony Robbins

11.1
Space-Time and Causality

After the appearance of the special theory of relativity, many people came to believe that speeds exceeding the speed of light in vacuum are impossible in principle. Sometimes, one can come across a reference to Einstein himself or some other authority as saying that no motions faster than the invariant speed c are possible. Even if Einstein or anyone else might have said something to this effect, the corresponding references are irrelevant to actual scientific contents of the theory. In this chapter we will discuss these contents.

Relativity as such does not say anything about possibility or impossibility of superluminal motions. The only statement it makes in this respect is that among all possible speeds there exists one (e.g., that of light in vacuum), which is invariant. This has been somehow misconstrued into quite different statement that this speed c, apart from being invariant, is also the upper limit of *all* speeds and no motions faster than c are possible.

Physicists routinely deal with various kinds of speeds exceeding c. These speeds are perfectly and consistently described within the framework of relativity. All we know now is one limitation: the value of c is the upper limit for the speeds of material bodies or of the processes that could be used for transmission of a signal. And this limitation is imposed by the principle of causality, not by relativity theory.

By the term "signal" we mean the transmission of a certain amount of energy that carries information about an event at a point $\mathbf{r}_1$ at the moment t_1 and can change the state of a certain physical system at a point $\mathbf{r}_2$ at the moment t_2. According to what we know today from experiments, the nature of things is such that for all motions of this type

$$|\mathbf{v}| = \frac{|\mathbf{r}_2 - \mathbf{r}_1|}{t_2 - t_1} \leq c. \tag{11.1}$$

Special Relativity and How it Works. Moses Fayngold

ISBN: 978-3-527-40607-4

And nevertheless, motions with $v \geq c$ also occur in nature. This does not contradict the above statement (11.1), since the latter type of motion cannot be used for a *signal transfer* from $\mathbf{r}_1$ to $\mathbf{r}_2$. The corresponding point (or, more generally, sphere) representing this motion in the velocity space (recall Sections 3.4 and 9.1) is not associated with the velocity of a real object. A point in the velocity space lying beyond the light barrier, that is, a point $v > c$, cannot be obtained from subluminal velocities in the framework of the regular Lorentz transformations. In this chapter, we will consider a number of examples of superluminal motions, including the superluminal tunneling of a light pulse in a specially prepared optical "barrier."

There are situations, when superluminal motions of physical objects may appear in the phenomena associated with strong gravitational fields. As we will see in the last chapter, gravitation can be considered as manifestation of curvature of space-time caused by a stationary or moving matter. The superluminal phenomena in a curved space-time are described by Kip Thorne in his book, *Black Holes and Time Warps* [102].

We had said before that relativity is consistent with causality principle. But so was the Newtonian nonrelativistic mechanics. Moreover, the Newtonian mechanics seemed to be an absolutely perfect match with classical causality. There was no serious ground for any discord in this respect for two reasons: first, there were no restrictions whatsoever within Newtonian paradigm on the velocity of signal transfer since any velocities, no matter how high, were allowed (recall that nonrelativistic limit formally is obtained at $c \to \infty$, so any speed would remain less than c and would thereby be legal). Second, the time is absolute in this limit, so the ordering of events is the same in all possible reference frames (RFs).

The situation is more subtle in relativity. With the discovery of the invariant speed, and with time stripped of its absolute status, the symbiosis of relativity and causality needs much more careful examination. We will be concerned mostly with different types of superluminal velocities and their relation with causality in the relativistic domain.

The apparently simple concept of velocity turned out to be not that simple after all. A few quite different velocities can be associated with the same process, for example, phase velocity, group velocity, signal velocity, and front velocity.

The first two types of velocity can take on any value [1, 5] and it would not contradict anything. But there is one velocity – that of a signal transport – that does not exceed c in any observations. The same is true for the front velocity. The relation between the front velocity and the invariant speed will be discussed in Section 11.7.

The special status of the signal velocity is attributed to the fact that signal (and thereby energy) exchange carry out the *causal connections* between spatially separated events. To see how the ban on superluminal signal transfer between causally connected events emerges from the existing theory, we have to discuss *causality* – one of the most important scientific concepts.

In the physical world, no event is isolated from others. One of the most important manifestations of causality is that the world's events always influence one another in a certain way. Namely, for any event (the *effect*) it is always possible to find at least one other event that has brought it into being – its *cause*. (There is one remarkable exception that does not fall into this scheme: the Big Bang, which has brought into being our universe. The Big Bang can be considered as the ultimate cause for everything in existence today; but what caused the Big Bang itself, or whether it had any cause at all, remains a murky issue at the moment of writing this book.).

All observable events are governed by a fundamental principle: the cause precedes the effect. We call this principle the retarding causality (an effect occurs later than its cause). We introduce here this principle as an *additional element* in the description of the world. This additional element, combined with relativity, restricts the speed of any interactions transferring a signal. Let us see how it works.

Suppose that the signal velocity u can take on *any* value and consider an event A at a point $\mathbf{r}_A$ at moment t_A; let A cause another event B to happen at a position $\mathbf{r}_B$ and moment t_B. Draw the x-axis through points $\mathbf{r}_A$ and $\mathbf{r}_B$. Then, the y- and z-coordinates of the two events are zero and the positions of the events are characterized by their x-coordinates x_A and x_B, so that separation between the events will be $\Delta x \equiv x_B - x_A$.

According to the principle of retarding causality, B happens later than A, that is,

$$\Delta t \equiv t_B - t_A > 0. \tag{11.2}$$

Since the events are connected with the signal traveling at a speed u, we have

$$\Delta x = u\Delta t. \tag{11.3}$$

Consider now another system K′, moving uniformly along the x-direction with velocity V. What time interval between the same events will be measured by an observer in K′? Assuming the axes x', y', z' in K′ running parallel to x, y, z and using Lorentz transformations (2.52), we have

$$\Delta t' \equiv t'_B - t'_A = \gamma(V)\left(\Delta t - \frac{V}{c^2}\Delta x\right). \tag{11.4}$$

In view of (11.3), the last equation gives

$$\Delta t' = \gamma(V)\left(1 - \frac{uV}{c^2}\right)\Delta t. \tag{11.5}$$

Thus, $\Delta t'$ is proportional to Δt. The factor $\gamma(V)$ is, according to its definition, positive for all $V < c$. As to the second factor $(1 - (uV/c^2))$, it can generally have any sign, depending on signal velocity u. But according to the principle of causality, if A is the cause of B, this relation between the two events is invariant. The causality relation for a considered pair of events must hold in *all* reference frames.

Therefore, there must be $\Delta t' > 0$. As is clearly seen from (11.5), this means that the factor $(1 - (uV/c^2))$ must always remain positive irrespective of the relative velocity V between the reference frames. And this can only be the case if the signal velocity u does not exceed c. Indeed, if it were possible for events A and B to be connected by a superluminal signal, that is, $u > c$, one could always find a reference frame K′, moving relative to K with a speed

$$V > \frac{c^2}{u}, \tag{11.6}$$

for which the factor $uV > c^2$, that is, $1 - (uV/c^2) < 0$, and accordingly, $\Delta t' < 0$. This means that for the pair of causally connected events A and B the effect would be observed before its cause.

So here is the logical chain restricting the speed of causal interactions: the existence of the invariant speed requires relativity of time; relativity of time makes it possible for a succession of events to be different in different reference frames: an event A can precede B in a reference frame K and follow B in a reference frame K′ (recall, for instance, the phenomena discussed in Sections 7.4 and 7.5). However, if A and B are *causally connected*, then, according to retarding causality, their ordering must be the same for all observers, *despite* relativity of time. This requires the speed of any causal interaction between them not to exceed c. If this requirement is not met, then the time ordering of A and B can be reversed for an observer in some other reference frame K′. In the framework of the above reasoning, this would be a violation of causality. To prevent this from happening, it seems to be necessary to exclude the possibility of superluminal signals.

11.2 Tachyons and Tardyons

... a pleasing symmetry exists between tardyons and tachyons,
each of which stays on its own side of the speed-of-light barrier.
C. Pickover, *Time. A traveler's Guide*

Once we realized that special relativity (SR) admits superluminal motions, we can explore the corresponding new domain. In this section, we will discuss hypothetical superluminal particles – tachyons. We will find a striking symmetry between the world of tachyons and our conventional physical world.

First, let us introduce new terms. If we have given a name to the superluminal particles, it is reasonable to do the same for the subluminal particles. Using he same arsenal from ancient Greek, physicists have dubbed regular well behaved subliminal particles the *tardyons* (the English words *retardation, retarded* stem from this root.) Thus, all the particles that can possibly exist in the universe, fall into three different categories according to their speed: tardyons ($v < c$), photons and gravitons ($v = c$), and tachyons ($v > c$).

It is easy to see that these three categories correspond to the three different domains of space-time (Section 4.1). Pick an event O in space-time. Let it be a reference event. Consider all possible world lines passing through the event O. *Each* line can be a four-dimensional path of a particle. The world lines of tardyons are all timelike and feel out the interior of the light cone with the apex at event O. The world lines of tachyons are spacelike and fill out the exterior of this cone. The world lines of photons (and gravitons) are isotropic (have zero kinematic length) and form the generatrices of the light cone.

Next, let us look again at some basic concepts of relativistic kinematics – the kinematics of tardyons and photons. We will see how easily it can incorporate the tachyons.

We start with the basic expression for the interval ds between two close events on the world line of a particle moving with a speed v:

$$\mathrm{d}s^2 = c^2\mathrm{d}t^2 - \mathrm{d}r^2 = \frac{c^2\mathrm{d}t^2}{\gamma^2(v)}. \tag{11.7}$$

Recall that according to its definition (2.25), the Lorentz factor γ is positive for the tardyons, infinite for the photons, and imaginary for the tachyons. Therefore, the interval ds is real for a tardyons and for the photons it is zero, just as one would expect from the definition of the photon's world line. As far as tachyons are concerned, the value of ds turns out to be imaginary. But this does not by itself preclude tachyons from existence, because the interval is not directly measurable quantity. It is only the mathematical expression formed from dt and dr, which are all real for all kinds of particles.

The same can be said about the components of 4-velocity of a tachyon. Although the magnitude of 4-velocity, according to its definition (see Equations (5.1) and (5.2)), is always equal to 1 for *any* kind of particle, the components of 4-velocity for tachyons

$$u_0 = \gamma(\tilde{v}); \qquad u_\alpha = \frac{\tilde{v}_\alpha}{c}\gamma(\tilde{v}), \quad \alpha = 1, 2, 3, \tag{11.8}$$

in view of the corresponding Lorentz factor, turn out to be imaginary. But this does not mean that these components cannot be observed. Unlike usual coordinates and velocities, as well as energy and momentum, the components of *4-velocity* are not directly measurable quantities. But they can be measured indirectly, by computing from directly measured components of $\tilde{\mathrm{v}}$, which are always real.

The product of the rest mass by c and by 4-velocity gives the 4-momentum (Equation (5.12)). For an ordinary particle

$$p_j = m_0 c u_j, \quad j = 0, 1, 2, 3. \tag{11.9}$$

For a tachyon

$$\tilde{p}_j = \tilde{m}_0 c \tilde{u}_j. \tag{11.10}$$

Since the tachyon's momentum must be an observed quantity, it has to be real. With $\tilde{u}_j$ imaginary, this implies that the tachyon's invariant mass has to be imaginary:

$$\tilde{m}_0 \equiv i m_0. \tag{11.11}$$

All the components of 4-momentum have a simple physical interpretation. Setting in (11.10) $j = 0$, one has

$$\tilde{p}_0 = \tilde{m}_0 c \gamma(\tilde{v}) = \frac{\tilde{\mathcal{E}}}{c} \tag{11.12}$$

and for the spatial components $\alpha = 1, 2, 3$

$$\tilde{p}_\alpha = \tilde{m}_0 \tilde{v}_\alpha \gamma(\tilde{v}) = \tilde{m}\tilde{v}_\alpha. \tag{11.13}$$

As we see, the zeroth component is (as is the case with tardyons) just the energy divided by c, and the spatial components are just the components of the regular relativistic 3-momentum. Thus, the components of the 4-momentum for a tachyon have the same meaning as the components of the 4-momentum for a tardyon.

On the basis of this interpretation, we obtain the relation between the energy and the momentum of a tachyon

$$\frac{\tilde{\mathcal{E}}^2}{c^2} - \tilde{\mathbf{p}}^2 = \tilde{m}_0^2 c^2 \tag{11.14}$$

in the same way as we did in Equation (5.23) for a tardyon. Note again that while the tachyon's rest mass $\tilde{m}_0$ is imaginary (it cannot be brought to rest in *our* space), its energy and momentum are real (and therefore measurable) physical quantities.

We see from (11.14) that, while the 4-momentum of a tardyon is timelike (has real norm $m_0 c$), the 4-momentum of a tachyon is spacelike. This is just another way to say that tardyons move always slower than light and reside in the interior of the light cones, while tachyons, if in existence, move faster than light and reside in their exteriors.

For a photon ($m_0 = 0$), the 4-momentum is isotropic (has a zero length). This is a natural result of the fact that the photons' world lines form the generatrices of light cones.

These statements are expressed analytically in three simple relations:

$$\left.\begin{array}{l} \mathcal{E} > pc, \\ \mathcal{E} = pc, \\ \tilde{\mathcal{E}} < \tilde{p}c, \end{array}\right\} \tag{11.15}$$

for a tardyon, a photon, and a tachyon, respectively.

We can now express the speed of a particle in terms of its energy. Using the general definition (5.27), we have

$$v = \frac{d\mathcal{E}}{dp} = \frac{p}{\mathcal{E}}c^2 = c\sqrt{1 - \frac{m_0{}^2 c^4}{\mathcal{E}^2}} < c, \tag{11.16a}$$

$$\tilde{v} = \frac{d\tilde{\mathcal{E}}}{d\tilde{p}} = \frac{\tilde{p}}{\tilde{\mathcal{E}}}c^2 = c\sqrt{1 - \frac{\tilde{m}_0^2 c^4}{\tilde{\mathcal{E}}^2}} > c. \tag{11.16b}$$

Now, we want to depict Equations (5.23), (11.14) graphically and compare the graphs for different kinds of particles. To get rid of the inessential details, consider the case when the vectors $\mathbf{p}$ and $\tilde{\mathbf{p}}$ are collinear and the *absolute values* of masses are equal (m_0 has the same value for both kinds of particle). Then, the 3-vectors of momentum are reduced to only one component along their common direction and the equations take the form

$$\mathcal{E}^2 - p^2 c^2 = m_0^2 c^4, \tag{11.17a}$$

$$\tilde{\mathcal{E}}^2 - \tilde{p}^2 c^2 = -m_0^2 c^4. \tag{11.17b}$$

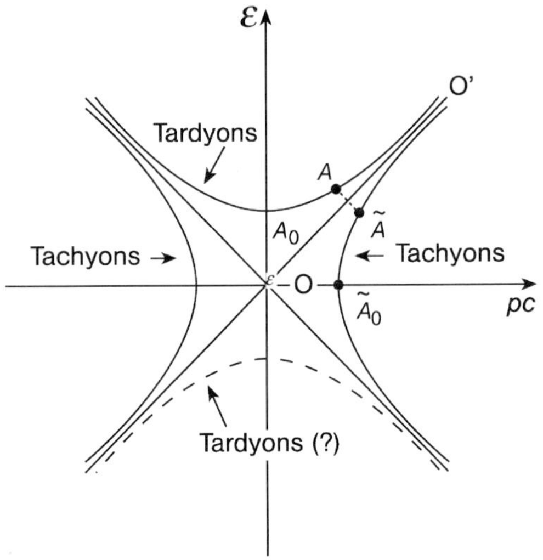

Figure 11.1 $\mathcal{E} \leftrightarrow p$ relations in the momentum space.

In a coordinate system where the energy and momentum of a particle are plotted along the corresponding axes (Figure 11.1), Equations (11.17) describe a hyperbola whose slope at each point (for each pair of variables p and $\mathcal{E}$) determines the speed of a particle with given values of the variables. For a regular tardyon (the branch of the hyperbola in the upper part of the plane), this slope is everywhere less than c. No matter how much we increase the energy and momentum of the particle, the corresponding point in the plane will slide along the curve ever farther from the origin and the speed of the particle, although approaching ever closer the speed of light, will remain less than c.

Of a special interest is the point, where $p = 0$. It corresponds to a particle at rest with the minimal possible energy. This minimum, as is seen from (11.17a) at $p = 0$, is equal to $\mathcal{E}_0 = m_0 c^2$. It is just the rest energy of the particle.

If the rest mass of the particle is zero (a photon or graviton), so is its rest energy. It means that if you try to stop such a particle, you are left with nothing. Particles of this kind do not exist at rest. The corresponding Equation (11.17a) splits into two simpler equations

$$\mathcal{E} - pc = 0, \tag{11.18a}$$

$$\mathcal{E} + pc = 0, \tag{11.18b}$$

that is, $\mathcal{E} = \pm pc$. They describe a hyperbola, degenerated into two intersecting straight lines passing through the origin. Physically, they correspond to the propagation of photons, whose energies, as we know, satisfy the Equations (11.18). The straight lines (11.18) lie on the generatrices of the light cone in the "momentum space" and are the asymptotes of the hyperbolas (11.17).

In geometry, the set of hyperbolas described by the equation $x^2 - y^2 = \text{constant}$ consists not only of the curves for which the constant is positive or zero but also includes the curves for which the constant is negative. A pair of such curves is shown in the same Figure 11.1 – they are the right and the left branches of a hyperbola, with their apexes on the *pc*-axis. These branches are described by Equation (11.17b) and correspond to tachyons. Thus, by allowing tachyons to exist, we give a physical meaning to this group of curves in the hyperbolas family, that is, introduce an element completing the picture to the full symmetry.[1)]

The hyperbolas of this group, as is clearly seen from Figure 11.1, have at any point a slope (i.e., $d\tilde{\mathcal{E}}/cd\tilde{p}$) *larger* than 1. But the difference from tardyons is not restricted to this distinction only. The dependence of the tachyon's energy and momentum on its speed is also dramatically different from that of the tardyon. Namely, the energy and momentum of a tachyon *decrease* with the increase of its speed. Look at a branch of the hyperbola corresponding to tachyons. As a point on this branch slides away from its apex, both $\tilde{\mathcal{E}}$ and $\tilde{p}$ increase, while the slope of the curve approaches 1, that is, the speed of the tachyon decreases, approaching *c*. Thus, to slow down a tachyon with $\tilde{\mathcal{E}} > 0$, one has to pump it with additional energy, and to speed up the tachyon one has to subtract energy from it!

Such behavior seems at first paradoxical. But if we give it more thought, this behavior is natural and even necessary in the world of superluminal velocities. The term "superluminal" means that the speed of light is the *lower limit* for this type of particle. To approach this limit, the particle must slow down and for the limit to be unattainable, the slowdown must require unlimited energy supply. This "paradoxical" behavior of tachyons ensures the symmetry of the light barrier: no matter from which side an object approaches the barrier (the speed of a particle approaches *c*), this is accompanied with an unlimited growth of energy and momentum of the object.

The curves $\tilde{\mathcal{E}}(v)$ and $\tilde{p}(v)$ illustrate another weird feature in the behavior of tachyons: as the tachyon is accelerated to an infinite speed, its energy does not just decrease but goes to zero independently of its proper mass $\tilde{m}_0$ while its momentum remains finite and goes to $m_0 c$.

The speeds $v = 0$ and $v = \infty$ can be considered as symmetrical with respect to *c* in that either of them is maximally remote (in corresponding domain) from the light barrier. We can say that the speed $v = \infty$ plays the same role for a tachyon, as the speed $v = 0$ for a tardyon. However, if we compare the energy and momentum of the tardyon at $v = 0$ with the energy and momentum of the tachyon at $v = \infty$, we will notice another peculiarity. While for a tardyon

$$\mathcal{E}(v = 0) = m_0 c^2, \quad p(v = 0) = 0, \tag{11.19a}$$

for an "equivalent" tachyon with "symmetrical" speed $\tilde{v} = \infty$ one has

$$\tilde{\mathcal{E}} = 0, \quad \tilde{p} = m_0 c. \tag{11.19b}$$

1) In fact, the full symmetry is not achieved even in this case, because the tardyons with negative energies corresponding to the lower branch of the hyperbola are unknown. The related topics have been discussed in Section 10.4.

The quantities $\mathcal{E}$ and p change roles: the energy of the tachyon behaves more like momentum and the momentum behaves more like energy. This is a natural consequence of the "reinterpretation" of the meaning of the temporal and spatial components for tachyons, as discussed in Section 11.3.

We call the particle with $v = 0$ the stationary particle. The tachyon with the infinite speed also deserves a special name. It has been named the *transcendent* (or *transient*) tachyon. The properties of transcendent tachyons are very unusual. Such a tachyon, tracing out the whole space in an instant, is observed at all points of its trajectory at once. But this observation lasts only an infinitesimally small moment, because owing to the infinite speed of the transcendent tachyon, it emerges and momentarily disappears simultaneously at all points on its track. Therefore, a rather strange phenomenon is observed in the corresponding reference frame: at first there is nothing there and then there suddenly appears and momentarily disappears an infinitely long rigid "rod" consisting of the tachyon "smeared out" along its whole length.

In this respect, a stationary tardyon and a transcendent tachyon act as certain kind of antipodes: the former stays at one point in space throughout the whole time; the latter stays for only an instant at all points of a spatial line. The former has zero momentum and the finite energy $\mathcal{E} = m_0c^2$, while the latter has the zero energy and final momentum $\tilde{p}c = m_0c^2$. The former is represented by an apex A_0 of the hyperbola (11.17a) and the latter by the point $\tilde{A}_0$ of the hyperbola (11.17b), which is symmetrical to A_0 with respect to a photon line OO'. The equations

$$\tilde{\mathcal{E}} = pc \qquad \text{and} \qquad \tilde{p}c = \mathcal{E} \tag{11.20}$$

are the analytical expression of this symmetry. In some respect, the resting tardyon and the transient tachyon of the same mass m_0 are the symmetrical counterparts of one another.

But there is more to it! If we take a closer look at Figure 11.1, it hints at an obvious generalization of this symmetry. Suppose we pick up a tardyon with some *arbitrary* speed $v < c$. It will be represented by a point A with corresponding "coordinates" $\mathcal{E}$ and pc on the upper branch of the hyperbola in Figure 11.1. Can we find a tachyon symmetrical to this tardyon? The graph suggests the positive answer. It would be the tachyon represented by a point $\tilde{A}$ symmetrical to A with respect to the asymptote (generatrix) OO'. The "coordinates" of this point are related to coordinates $\mathcal{E}$ and pc by the same Equation (11.20). The corresponding vectors $(\mathcal{E}, pc)$ and $(\tilde{\mathcal{E}}, \tilde{p}c)$ satisfy the equation

$$\mathcal{E}\tilde{\mathcal{E}} - p\tilde{p}c^2 = 0, \tag{11.21}$$

which is equivalent to (11.20). In the geometry of Minkowski's world, the expression $\mathcal{E}^2 - p^2c^2$ determines the square of the 4-vector $(\mathcal{E}, pc)$ and thereby its magnitude (kinematic length). Similarly, the expression $\mathcal{E}\tilde{\mathcal{E}} - p\tilde{p}c^2$ determines the scalar product of the two different vectors. In our case, this product is equal to zero. As we know from geometry, it means that the two vectors are perpendicular. The fact that the vectors OA and $O\tilde{A}$ in Figure 11.1 do not look mutually perpendicular is caused by inadequacy of the used graphical representation: we have to represent the relations of the *pseudo-Euclidean* geometry on the ordinary Euclidean plane. For the pseudo-Euclidean space, we say that these vectors are *dual* to one another. Let us also call the tardyon and symmetrical tachyon represented by mutually dual vectors on the diagram in Figure 11.1, mutually dual. The property of two particles to be mutually dual is Lorentz invariant. If this property is found for a pair of particles in one inertial reference frame, then due to Lorentz invariance of the scalar product of the representing vectors, it will be found in any other inertial reference frame.

Now, here is an interesting question: how are the speeds of mutually dual tardyon and tachyon related to each other? According to the general definition of speed (11.16), we have

$$\tilde{v} = \frac{\tilde{p}}{\tilde{\varepsilon}}c^2, \qquad v = \frac{p}{\varepsilon}c^2, \tag{11.22}$$

so that

$$v\tilde{v} = \frac{p\tilde{p}}{\varepsilon\tilde{\varepsilon}}c^4 = c^2, \tag{11.23}$$

or

$$\tilde{v} = \frac{c^2}{v}. \tag{11.24}$$

A few times, on different occasions, we have obtained this same relation between two velocities and now we have it again. Recall, for instance, that the same relation connects the phase and group velocities of the de-Broglie waves of a free particle.

Thus, the speed of a tachyon dual to a given tardyon is equal to the phase velocity of the de-Broglie wave associated with this tardyon. This allows us to suggest that the concept of a tachyon is not only logically possible within the framework of the special theory of relativity, but also has some physical meaning. As we think of it, there appears an impression that we have caught a glimpse of something deep. But today we can only say that *if* the tachyons do exist, then tardyons and tachyons can come in dual pairs whose characteristics are described by Equation (11.21) or by the equivalent equations

$$\tilde{m}_0 = im_0 \qquad \text{and} \qquad \tilde{v} = \frac{c^2}{v}. \tag{11.25}$$

We want to emphasize once again that the transition from one member of a dual pair to another described by the transformations (11.25) cannot be realized by continuous change from v to $\tilde{v}$ or vice versa through the light barrier. We have already shown this using the law of conservation of energy (the transition through the light barrier would require an infinite energy). It will be instructive to show the same thing using the law for addition of velocities.

The impossibility of reaching the barrier from its subluminal side has already been shown in Chapter 3. It is clearly seen from expressions (3.39) there that if one of the two input speeds is c, the output does not depend on the second speed. If we admit the possibility for superluminal reference frames consisting of tachyons, then the speed of light relative to such reference frames would also be constant and equal to c.

Consider now a tardyon and a tachyon, whose velocities are collinear and differ from c by the same amount δv so that

$$v = c - \delta v, \qquad \tilde{v} = c + \delta v. \tag{11.26}$$

Their relative speed, according to rule (3.7), is

$$\tilde{v}' = \frac{\tilde{v} - v}{1 - (v\tilde{v}/c^2)} = 2\frac{c^2}{\delta v}. \tag{11.27}$$

Let the tardyon speed up and the tachyon slow down, so that $\delta v \to 0$, and their speeds approach from the opposite sides their common limit – the speed of light. If we plot the two velocities as points on the velocity axis, then corresponding points get closer and closer to each other, to eventually merge together at the common point representing c. One would be tempted to imagine the corresponding two particles eventually moving together at one common speed – the speed of light. However, their *relative* speed, given by Equation (11.27), will go to infinity! Here, the impossibility of crossing the light barrier from either side is manifest in the most dramatic and impressive way.

When the two particles approach the light barrier from opposite sides, they must remain in different realms separated by the barrier. If you sit on the tardyon, the tachyon on the other side of the barrier should move relative to you faster than light – its relative speed must exceed c, no matter how close to the light barrier you both are. If the special theory of relativity is logically consistent, it must meet this requirement. And this is precisely what it does – with an astounding efficiency: the infinite relative speed in the limit $\delta v \to 0$ is definitely larger than c!

On the other hand, if the tardyon with the same speed $v = c - \delta v$ is moving *toward* the tachyon with a speed $\tilde{v} = c + \delta v$, their relative speed *decreases*; applying rule (3.7) to this case yields

$$\tilde{v}' = \frac{\tilde{v} + v}{1 + (v\tilde{v}/c^2)} = \frac{c}{1 - (\delta v^2/2c^2)} < c + \delta v. \tag{11.28}$$

The reader can check that this inequality holds for all $\delta v < c$. It tells us that the relative speed, while remaining larger than c, is *smaller* than the tachyon speed in the initial reference frame. This result, although not so dramatic as the previous one, still contradicts our intuition, according to which, if I go 5 km/h to meet a friend who runs toward me at 10 km/h, our relative speed will increase to 15 km/h, rather than decrease.

Notice, that the speed of the transcendent tachyon relative to a stationary tardyon is infinite. On the other hand, we know that such a pair is a special case of mutually dual particles. In this connection, there arises an interesting question: what is the relative velocity between two dual particles in the general case? We can get an answer if we recall that duality between the two particles is an invariant property. Paradoxical as it may sound, the same is true for the value of relative velocity, by mere definition of this quantity (to measure *relative velocity* of the two objects, *any* observer has to transfer to the rest frame of one of the objects, recall Chapter 3).

The relative velocity *between two fixed objects* is a computational invariant, using the language of Chapter 9.

Therefore, the relative velocity between an arbitrary tardyon and its dual tachyon will not change, nor will they stop being dual to each other, if we switch to the rest frame of this tardyon. By doing this we come back to the special case of the tachyon dual to the stationary tardyon. But such a tachyon is transient, it moves with an infinite speed. Because the relative speed of two *fixed* objects is invariant, it must have been infinite in the original reference frame also.

We can get the same result directly from the law of addition of velocities. Putting in the first equation of (3.22) mutually dual velocities v and $\tilde{v} = c^2/v$, we will obtain

$$\tilde{v}' = \frac{\tilde{v} - v}{1 - (v\tilde{v}/c^2)} = \infty. \tag{11.29}$$

Reversing this argument, we can give another definition of dual particles: two particles with equal absolute values of the rest mass are mutually dual if their relative speed is infinite. Indeed, the infinite relative speed requires that the denominator of the above expression be zero, from which there immediately follows $v\tilde{v} = c^2$ – the definition of dual particles.

Another interesting question is: what is the relative velocity between the two tachyons? Once we admit that such particles can, at least in principle, exist, then we could, at least in principle, admit a reference frame and clocks connected with each of them (probably made of the same kind of particle.) That would allow one to measure velocity of one tachyon relative to another. Let the tachyons move in one direction with the speeds $\tilde{v}_1$ and $\tilde{v}_2$. The relativistic law of addition of velocities applies equally well to *any* velocity – subluminal or superluminal. Applying it to two superluminal velocities $\tilde{v}_1 = c + \delta_1$ and $\tilde{v}_2 = c + \delta_2$, $\delta_1, \delta_2 > 0$, gives

$$\tilde{v}_{12} = \frac{\tilde{v}_2 - \tilde{v}_1}{1 - (\tilde{v}_1\tilde{v}_2/c^2)} = \frac{\delta_2 - \delta_1}{-(\delta_1 + \delta_2)/c + \delta_1\delta_2/c^2} < c. \tag{11.30}$$

Let us make the tachyons move in the opposite directions ($\tilde{v}_2 = -\tilde{v}_1$) to increase their relative velocity. Then, the same addition law gives

$$\tilde{v}_{12} = c\frac{2 + (\delta_1 + \delta_2)/c}{2 + (\delta_1 + \delta_2)/c + (\delta_1\delta_2/c^2)} < c. \tag{11.31}$$

Thus, any two tachyons move relative to one another with a speed less than the speed of light! In other words, they behave like tardyons with respect to each other!

Let us allow our imagination to run wild for a moment. Then, based on the obtained results, we can speculate that tachyons could form conglomerates of particles, planets, stars, galaxies, and so on, similar to those in our world. Maybe, they can even produce intelligent life? Maybe, someone wrote a book there hypothesizing our existence and calling *us* the tachyons?

Summary

In one and the same space-time there can exist two equivalent worlds. The objects of either world move relative to each other slower than light. But the relative velocity of two objects belonging to different worlds is always more than c. No object can by continuous change of speed transfer from one world to another. The worlds are impenetrable. They are separated by the impenetrable barrier – the invariant speed – the speed of photons and gravitons. The latter particles form the third world, all the particles of which move relative to all other particles with the same fundamental speed *c*.

There emerges a picture so complete in its symmetry that one would start wishing that tachyons really exist!

Hypothesis about tachyons can be seriously considered only under condition that they could be observed. This condition is equivalent to an assumption that tachyons can interact with the known matter – tardyons, photons, and gravitons. We consider some of these interactions in the next two sections.

11.3
The Tolman Paradox

There was a young lady named Bright
Whose speed was faster than light;
She started one day
In the relative way
And returned on the previous night.

Punch, December 19, 1923 or A.H. Reginald Buller

Suppose we reversed the time.

The result would be a very unusual world, where people would live their lives backward, first emerging from their graves, then changing into babies and returning into their mothers' wombs. The amount of disorder in such a world would decrease, and the amount of order would increase. According to thermodynamics, which studies subtle connections between the observable macroscopic phenomena and the motions of the constituent microparticles, the probability of such a world is zero. But in all other respects, this reversed world, for all its apparent weirdness, would be subordinated to laws that are intrinsically consistent. It would follow the rule of cause and effect. The only difference is that as compared to our usual world, the cause and effect switch roles. What is the cause in our world is the effect in the described one, and vice versa. For instance, the cause for a cup of tea jumping onto the table would be its self-assembling from the splinters on the floor, absorbing moisture from it and collecting heat, part of which would accumulate into kinetic energy. Although some of the laws of nature appear to be turned inside out, causality not only conserves but, strangely enough, even retains its retarding character. This is due to the fact that simultaneously with the reversal of time the cause and effect *change* roles.

Most of the laws of nature are invariant with respect to the time reversal. This means that, unlike the macroscopic world, which seems different and strange when run backward, in the microworld of single particles there is often no difference between the direct and reversed flow of time.[2)]

2) There are some phenomena associated with the so-called CP violation [103], where this statement must be generalized: with time-reverse things would run as usual, if simultaneously we perform the mirror reflection and change signs of all charges. The laws of nature are invariant under this combined transformation (CPT theorem) [104].

For example, let an excited atom A_1 radiate a photon at a moment t_1 and return to its ground (normal) state. The emitted photon gets absorbed by another atom A_2 at a later moment t_2 and causes its transition from the ground state to the excited state. Clearly, the cause of the excitation of the atom A_2 was the photon emission from the atom A_1. Let us now reverse the process. Then we will first observe the photon emission by atom A_2 due to its optical transition to the lower state at the moment t_2. This will cause the excitation of the atom A_1 due to absorption of the photon at the moment t_1. Because time is reversed, the moment t_1 now occurs *later* than the moment t_2, so that again the cause precedes the effect. Despite the reverse of time, there is nothing unusual in the resulting process.

In contrast, the reversing of macroscopic phenomena seems unusual; but the laws of nature remain self-consistent, because synchronously, the cause and effect also change their roles. Ordinarily, if a hare is shot dead by a hunter, the hunter's shot is the cause and happens earlier in time, while the death of the hare is the effect and happens later. In the time-reversed world, the dead hare would suddenly resurrect, with the bullet emerging out of it and then this bullet, moving backward, would whack into the barrel of the hunter's gun. One would now call the first event (the emission of the bullet from the hare) the cause and the second one (the "absorption" of this bullet by the gun) the effect. This *reinterpretation* of the cause and the effect saves the principle of the retarded causality in the time-reversed world.

Let us now apply a similar trick to the problem of superluminal signals. Suppose that our atoms exchange superluminal signals instead of photons. Let such a signal be represented by a fictitious superluminal particle (tachyon). Imagine observing such a particle emitted by an atom A_1 at a moment t_1 and then absorbed by another atom A_2 at a later moment t_2 in an inertial reference frame K. It is clear that the first event is the cause of the second. But we know already that for a superluminal signaling the interval between the corresponding events is spacelike, and one can always find an inertial reference frame K′, in which the time ordering of the events changes. This seems to contradict retarded causality. However, one can avoid this contradiction in a way similar to the one described for the time reverse, but in a more limited sense: one might reinterpret the cause and effect *only* for the events along the spacelike interval and its end points, when their time ordering is reversed under the corresponding Lorentz transformation (i.e., when we transfer to another reference frame moving sufficiently fast). The agent moving from A to B and causing some change in B would be observed from the other reference frame as moving from B to A and causing corresponding change in A.

"What's the big deal," one might think. "This is a familiar effect, I often see it during driving, when I happen to outpass a pedestrian strolling along the sidewalk. Relative to my car, the pedestrian then appears to move in the opposite direction."

But this would be a false analogy. If the pedestrian first crossed the 6th Street and then the 7th Street, from your car you will see her doing this in the same succession. You will *not* see her crossing the 7th Street first and the 6th Street after that, no matter how fast you drive.

The situation with a superluminal particle is totally different. You do not (and cannot) outrun such a particle. And yet you can see its motion literally in reverse, that is, crossing the 7th Street first and only then the 6th Street. This is a purely relativistic effect, when the two events are interchanged *in time* for an observer in another reference frame.

We now can describe some implications of the above properties of the spacelike trajectories on a macroscopic scale. Imagine that tachyons do exist and people have learned how to manufacture superluminal bullets out of them. Imagine that the hunter Tom fires such a bullet and kills a hare. Because the bullet is superluminal, these two events (the shot and resulting death of the hare) are connected by a spacelike interval.

Now consider the same process from the viewpoint of Alice flying by in a spaceship. Traveling in a spaceship does not produce any *global* time reverse of the type described in the beginning of this section, so Alice will observe Tom's and the hare's lives in their normal course. In Alice's reference frame, just as in Tom's, Tom first aims, then shoots; the hare first grazes, then dies. And yet in the shooting episode she will see something strange.

Here is her account.

> I flew by and watched a hare frolicking on a forest meadow. Then all of a sudden the hare dropped dead. A bullet burst out of it and zipped away with a stupendous speed. Then I saw my friend Tom hunting. His behavior was a little weird. He noticed the hare and took good aim at it as if the hare were not dead. At this moment the bullet from the hare struck Tom's gun right in the barrel and a tiniest fraction of a second later Tom pulled the trigger. Then he ran to see what happened to the hare. It appears to me from what I saw that the hare died by itself and produced that horrible bullet aimed at Tom, and the recoil of Tom's gun was the effect of this event.

As you compare Alice and Tom's accounts, you will see obvious contradictions between them. Tom insists that he has fired first and killed the animal with his bullet. His shot was the cause and the hare's death was the effect. Alice witnessed that the hare died first and its death was accompanied by the emergence of the bullet that caused the recoil of Tom's gun.

Who is right?

Both are, because the time ordering of the events separated by a spacelike interval is *relative*, and so may be the designation –which event is the cause and which is the effect. Alice's reinterpretation of what is the cause and what is the effect is logically consistent and helps save the principle of retarded causality. By using the reinterpretation, the principle holds in either reference frame.

Such "reinterpretation principle" had been suggested by Bilaniuk, Deshpande, and Sudarshan [105, 106]. Its essence can be described in the following way. The authors had noticed that the same condition $V\tilde{v} > c^2$ that swaps tachyon's past and future automatically reverses the sign of its energy and momentum. They suggested that in all such cases we should make an additional sign reverse of the two latter dynamical quantities. In other words, together with reinterpreting the cause in K as the effect in K′ and vice versa, we should reinterpret the negative tachyon in K′ as positive (the sign of the momentum will then be reversed automatically). So, if we do

$$\tilde{\mathcal{E}}' \Rightarrow -\tilde{\mathcal{E}}' > 0 \tag{11.32}$$

then, automatically,

$$\tilde{p}' \Rightarrow -\tilde{p}' < 0. \tag{11.33}$$

Physically, this means that in another RF for which the Lorentz transformations result in the sign change of time, the absorption of negative tachyon coming to a detector from the future is actually observed as the emission of positive tachyon moving from the detector and (in time) to the future.

The possibility of such reinterpretation would mean that superluminal communications do not by themselves contradict retarded causality. One can therefore speculate about the possibility of the existence of the superluminal particles and superluminal communications.

It would be much more difficult for Alice to explain why Tom's aiming and triggering his gun was so remarkably accurately timed with the arrival of the bullet from the hare. In Alice's reference frame, the triggering of the gun is not the cause of the bullet having flown into it. Nor is it its effect. The two events are obviously correlated in both RFs, but while this is a causal correlation for Tom, it appears to Alice to be noncausal. Alice can still find causal explanation of the strange correlation she observes, if she turns to Tom's records of the two events.

Isn't it strange that causality between two events can be manifest in one FR but not in the other? Well, relativity of time also seemed strange a century ago. The basic requirement for a concept or a theory to be considered as viable is its logical consistency. The tachyon hypothesis fails this test, if we treat them in all other respects as ordinary particles (tardyons) and consider a two-stage tachyon exchange. We will immediately run into a logical contradiction (the Tolman paradox [82]).

Consider a simple scenario. There are two RFs, K and K′, sliding with respect to one another with a speed V along the common x-axis. We assume, as usual, that their sets of clocks are synchronized and set so that the local clocks at their respective origins read $t = t' = 0$ when the origins instantly coincide. At this moment, an observer Peter at the origin of system K shoots a tachyon at his friend Sam in K′, L meters away from the origin. According to the preliminary arrangement, Sam shoots back right after Peter's tachyon hits his detector. Tolman has shown that in such two-stage exchange the secondary tachyon can return to the origin of K *before* Peter has shot the primary one. The second tachyon, a *response* to the first one, arrives at an *earlier* moment $t < 0$, that is, before the first shot. If this secondary tachyon disables Peter's equipment, then Peter cannot shoot his original tachyon, thus precluding the appearance of the secondary one. So we have come to a nonsensical dilemma: if Peter shoots, he is disarmed beforehand and therefore he does not shoot. If he does not shoot, his devise remains functional and he shoots.

This paradox is called sometimes the time loop because the world lines of the shooter and the two tachyons form a closed loop. The emergence of this kind of loop can be seen from the space-time diagram of the process (Figure 11.2). To simplify the

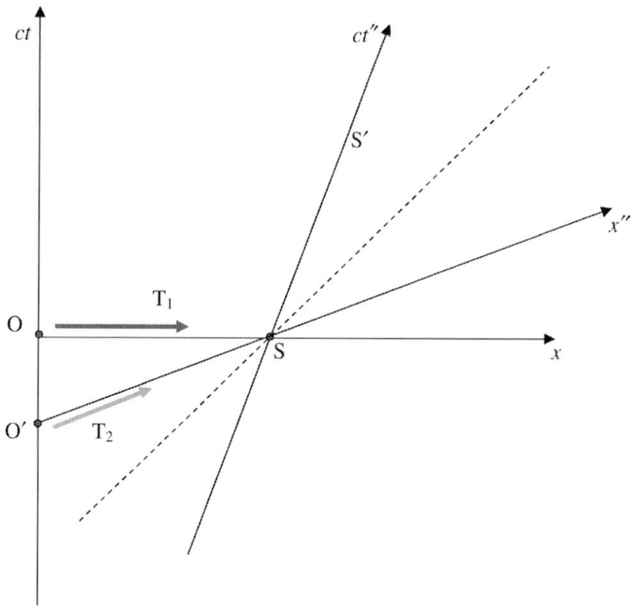

Figure 11.2 The time loop. *ct* and *ct″* are the world lines of Peter (system K) and Sam (system K′), respectively. OS is the world line of the primary tachyon T_1 shot by Peter. Since this tachyon's speed is infinite in the system K, its world line is coincident with the *x*-axis of K. SO′ is the world line of the secondary tachyon T_2 shot back at Peter by Sam. Since this tachyon's speed is infinite in the system K′, its world line is coincident with the *x″*-axis of K′. And since it is shot back at Peter (i.e., in the negative *x″*-direction), this world line intersects with *ct*-axis (and accordingly hits Peter's emitter) at O′ *before* the emission of the primary tachyon. Peter observes this event as *spontaneous* emission of a tachyon toward Sam. The directions of both tachyons *as observed by Peter* are shown by arrows.

diagram, consider a special case when the speed of both primary and secondary tachyons is infinite relative to their respective emitters. In this case, the world line of the first tachyon coincides with the *x*-axis (simultaneity line with the origin in K). Accordingly, it intersects Sam's world line at point S, which is *below* the x'-axis (the line of zero time in K′ passing through O parallel to O′S and not shown in the Figure). In other words, Sam sees the primary tachyon *before* its emission at $t = t' = 0$. This is already unusual, but by itself does not involve any contradiction. It would be just another illustration of the known fact that the time ordering of two events separated by a spacelike interval can be reversed in another RF.

Since the energy of this tachyon in K′ is also negative (Problem 11.2), then, according to reinterpretation principle, this process is actually observed by Sam as spontaneous emission of a tachyon with $\mathcal{E} > 0$ from his detector *before* the zero time and its absorption in Peter's emitter *at* the zero time. This means that Sam observes the primary tachyon moving after the emission toward Peter with finite speed.

By arrangement, right after this emission Sam shoots the secondary tachyon at Peter with infinite speed in K′. On the diagram, the world line of this tachyon is represented by the line SP parallel to the x'-axis (simultaneity line with the event S in

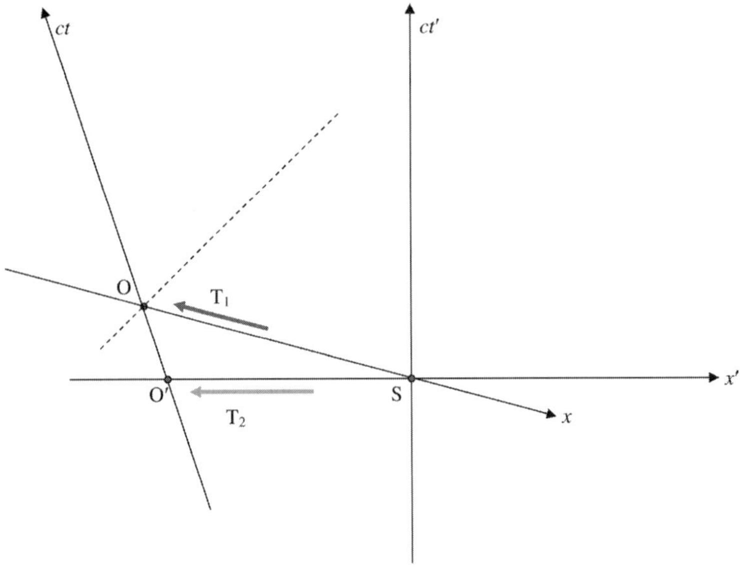

Figure 11.3 The time loop. The same process as in Figure 11.2, but now as observed by Sam. As before, ct'' and ct are the world-lines of Sam (system K′) and Peter (system K), respectively. SO is the world-line of the primary tachyon T_1 shot by Peter. Since this tachyon's speed is infinite in system K, its world-line is coincident with the x-axis of K. SO′ is the world-line of the secondary tachyon T_2 shot back at Peter by Sam. Since this tachyon's speed is infinite is system K′, its world-line is coincident with the x''-axis of K′. Since T_1 is shot by Peter at Sam (that is, in the positive x-direction), its world-line intersects with ct''-axis (and accordingly hits Sam's emitter) at S right before the emission of the secondary tachyon. Sam observes this event as *spontaneous* emission of a tachyon towards Peter. The directions of both tachyons *as observed by Sam* are shown by arrows.

K′). The intersection point P with Peter's world line (oblique projection of event S onto the time axis ct) determines the moment of arrival of the secondary tachyon at the origin. We see that this moment is *before* the initial event O that "triggered" the whole loop. Thus, Peter records the secondary tachyon before he shoots the first one. The energy of this tachyon in K, obtained from the Lorentz transformation, is also negative. Accordingly, Peter sees this as *spontaneous* emission of a positive tachyon toward Sam with finite speed (Figures 11.2 and 11.3).

Thus, each observer sees only emissions. Peter first sees the spontaneous emission of the secondary tachyon from his device at a moment $t_P < 0$. This tachyon starts moving toward Sam at finite speed. At the zero moment Peter shoots his "original" (primary) tachyon with infinite speed, which reaches Sam at the same moment as the spontaneously emitted tachyon (Figure 11.2).

Sam, in his instant location, records almost simultaneous emissions of two tachyons from his device at point S at a moment $t'_S < 0$. Both tachyons start toward Peter with different speeds. One is infinitely fast and reaches Peter at the same moment t'_S by Sam's clocks. The other one is moving slower (although faster than light) and reaches Peter at the zero moment just when Peter passes through the origin O (Peter is moving relative to Sam at a speed $V' = -V$).

Let us now do it quantitatively.

When the primary tachyon reached Sam, the origins of both systems coincided and Sam was a distance L away from the origin in K. Therefore, his coordinates in K at that moment were:

$$t_S = 0; \qquad x_S = L. \tag{11.34}$$

The corresponding K′-coordinates:

$$t'_S = \gamma(V)\left(t_S - \frac{V}{c^2}x_S\right) = -\gamma(V)\frac{V}{c^2}L, \qquad x'_S = \gamma(V)L. \tag{11.35}$$

At this moment, Peter, moving to the left away from Sam, has the x'-coordinate

$$x'_P = -Vt'_S = \gamma(V)\frac{V^2}{c^2}L. \tag{11.36}$$

The secondary tachyon has infinite speed in K′, therefore it hits Peter at the same moment $t'_P = t'_S$ and at the same position (11.36):

$$t'_2 \equiv t'_P = t'_S = -\gamma(V)\frac{V}{c^2}L; \quad X'_2 \equiv X'_P = \gamma(V)\frac{V^2}{c^2}L. \tag{11.37}$$

In Peter's RF, the K-coordinates of recording the secondary tachyon are

$$\begin{aligned} x_2(P) &= \gamma(V)(x'_2 + Vt'_2) = 0; \\ t_2(P) &= \gamma(V)\left(t'_2 + \frac{V}{c^2}x'_2\right) = \gamma^2(V)\frac{V}{c^2}L\left(\frac{V^2}{c^2} - 1\right) = -\frac{V}{c^2}L. \end{aligned} \tag{11.38}$$

For Sam, the corresponding event is the tachyon absorption; for Peter, it is the tachyon emission.

The whole process, as observed in K, is a spontaneous emission of a tachyon from Peter's device at the origin at a moment $t < 0$ *preceding* the start of the whole project, and then the emission of another tachyon at $t = 0$ (preplanned start of the project). The second (actually, primary) tachyon is moving faster and catches up with the first (actually, the secondary) one precisely at Sam's device, where both are absorbed.

The same process is observed in K′ as almost simultaneous emission of the two tachyons from Sam's device; one (the primary) is emitted spontaneously and zips toward Peter with a finite (albeit superluminal) speed; the other (the secondary one) is emitted intentionally a split second later by following the initial agreement (Sam is educated enough to recognize the spontaneously emitted tachyon as the reinterpreted primary one coming from Peter's future). The secondary tachyon zips toward Peter with infinite speed and therefore hits and disables his device *before* the primary one is absorbed by it.

Such a situation contradicts the imperative condition that Peter's device is to be destroyed by the secondary tachyon and therefore Peter's equipment cannot produce (in Sam's view, absorb) the primary one.

The contradiction we have come to could be much more dramatic and make the world the total mess. Imagine that someone named Sig sent a superluminal signal to a distant partner on a space mission who, following the initial arrangement, immediately responded by sending similar on-coming signal. That secondary signal arrives back to Earth a few centuries before the moment the primary signal was sent and kills Christopher Columbus before he embarked on his historic voyage. As a result, the American continent was not discovered in 1492. The whole human history had taken a different turn, so that Sig has not been born and no signal was sent . . ..

But in this case, . . .

I encourage the reader to ponder on what the world would look like in this case – in case superluminal signaling were possible.

Summary

If the tachyons exist and can be used for superluminal signaling, this would bring in irresolvable logical contradictions, of which, we believe, the Nature is free. It follows then that either tachyons do not exist or they cannot be harnessed for superluminal communication.

11.4
Tachyons and Causality

> *If tachyons are one day discovered,*
> *the day before the momentous occasion*
> *a notice from the discoverers should appear in newspapers*
> *announcing:*
> *"Tachyons have been discovered tomorrow."*
>
> Paul Nahin, *Time Machines*

There is a widely spread misconception that condition $v > c$ for any object or process automatically involves causality violation. As we will see, the detailed analysis in each individual kind of processes shows such concerns to be ungrounded.

In the book [1], an argument has been presented against the possibility of superluminal signaling using tachyons. The argument was based on the fact that on the quantum level the Cerenkov radiation of photons or gravitons by a tachyon is a random process. In each such individual process, the tachyon is being kicked off its classical path. Each such kick is observed as a spontaneous event, whose exact moment is totally unpredictable, and so is the direction of the kick suffered by the tachyon. As a result, the tachyon's path can be considered as a combination of its classical motion under the drag force, onto which is superposed the random Brownian-like motion. This motion renders the tachyon unsuitable for a controlled message or information transfer.

But even if this process were by itself totally sufficient to save causality, is it the only possible mechanism? Suppose we are not interested in the fine details of the tachyon's trajectory, or the tachyon itself is so massive that its ever spreading jittering around the classical trajectory becomes negligible and only the averaged path, that is, its motion under the classically calculated drag force is essential in determining the final result. Since this *average* motion, while being predictable, remains superluminal, will not it violate the causality?

Here, we take the analysis one step further to answer this question. We will discuss two entirely relativistic mechanisms saving causality for a motion on a macroscopic scale and discuss the conditions necessary for their full effectiveness. The first mechanism is the change of tachyon's spectrum $\mathcal{E}(p)$ and thereby $\omega(k)$ in such a way that it becomes impossible to localize the tachyon at a point. In other words, a tachyon cannot in principle be considered as a point particle. The second mechanism is the relativistic length contraction effect that, when sufficiently well within the superluminal domain, starts working "in reverse" – as relativistic length extension.

We will start with formulating the most essential features of tachyon's behavior that follow from relativistic postulates.

1. Tachyons have an irreducible effective size a_0 [107].

2. Tachyons' Lorentz factor is imaginary:

$$\gamma(v) = i\tilde{\gamma}(v) = i\left(\frac{v^2}{c^2} - 1\right)^{-1/2}, \quad v > c, \tag{11.39}$$

   and so are its "proper" characteristics – mass $\tilde{m}_0 = im_0$, longitudinal size $\tilde{a}_0 = ia_0$, and, if unstable, its proper lifetime $\tilde{\tau}_0 = i\tau_0$. The corresponding characteristics observed by us are represented by real numbers and can be obtained in measurements using conventional experimental procedures [1, 10, 11]. Figure 11.4 shows the graph of the Lorentz factors $\gamma(v)$ and $\tilde{\gamma}(v)$ in the subluminal and superluminal domains, respectively. We see that whereas $\gamma(v)$ (at $v < c$) is always greater than or equal to 1 (at $v = 0$), its counterpart $\tilde{\gamma}(v)$ in (11.39) is greater than 1 in the range $c < v < \sqrt{2}c$ and less than 1 in the range $v > \sqrt{2}c$.

3. The longitudinal size of a tachyon undergoes Lorentz contraction (or extension), depending on its velocity according to (11.39) (Figure 11.5):

$$a = \frac{\tilde{a}_0}{\gamma(v)} = \frac{a_0}{\tilde{\gamma}(v)}. \tag{11.40}$$

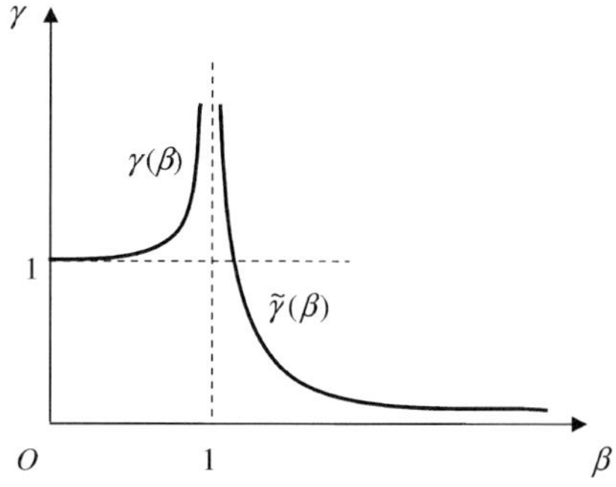

Figure 11.4 The Lorentz factor $\gamma(\beta)$ and its superluminal counterpart $\tilde{\gamma}(\beta)$ versus β.

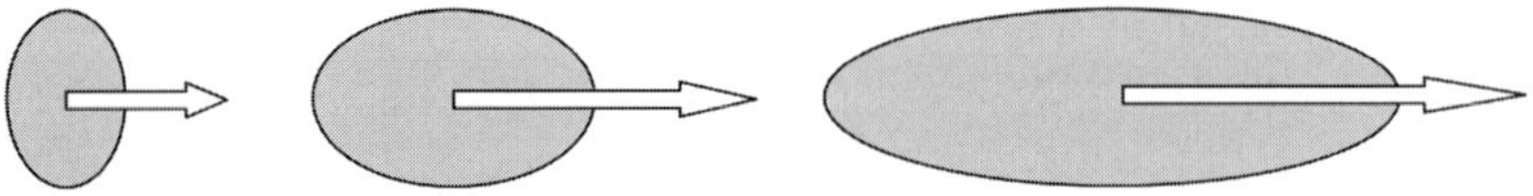

Figure 11.5 Instant shapes of a tachyon at different speeds. As the tachyon is being accelerated beyond $v_c = \sqrt{2}c$, the Lorentz contraction changes to Lorentz extension, which rapidly increases as $v \to \infty$. The arrows represent the tachyon's velocity: (a) $v < v_c$; (b and c) $v > v_c$.

4. A tachyon cannot be a scalar particle. Each tachyon has an intrinsic vector **e** built into it, similar in its transformational properties to that of a dipole moment vector [1, 10, 11].

5. Even electrically (and magnetically) neutral tachyons emit Cerenkov radiation in the form of gravitational waves at a rate

$$\frac{d\mathcal{E}}{dt} = \mathbf{f} \cdot \mathbf{v}, \tag{11.41}$$

where **v** is the tachyon's instant velocity and **f** is the radiative force. It plays the role of a drag force on the tachyon and its value depends on tachyon's structure [108].

6. In all RF where the tachyon's velocity **v** is parallel to **e**, the force **f** is along the direction of tachyon's motion and has a constant magnitude. This force accelerates the tachyon by decreasing its momentum and energy.

7. On the microscopic level, tachyons, as well as tardyons, obey the rules of QM.

Because the tachyon's proper mass is imaginary, $\tilde{m}_0 = im_0$, the universal energy–momentum relation (5.22), when applied to a tachyon, takes the form

$$\left(\frac{\mathcal{E}}{c}\right)^2 - p^2 = \tilde{m}_0^2, \quad \text{that is,} \quad \left(\frac{\mathcal{E}}{c}\right)^2 - p^2 = -m_0^2. \tag{11.42}$$

The generalized expression for a 4-momentum of an object, embracing all objects – subluminal (tardyons), superluminal (tachyons), and luminal (photons, gravitons, and maybe other yet undiscovered zero-mass particles) can be written as

$$\left(\frac{\mathcal{E}}{c}\right)^2 - p^2 = \begin{cases} m_0^2 c^2, & v < c, \\ 0, & v = c, \\ -m_0^2 c^2, & v > c. \end{cases} \tag{11.43}$$

The corresponding curves $\mathcal{E}(p)$ for all three cases are shown in Figure 11.1.

The relations (11.44)–(11.46) below, applicable to the tachyons, follow from (11.4) and (11.5):

$$\mathcal{E} = mc^2 = \tilde{m}_0 \gamma(v) c^2 = m_0 \tilde{\gamma}(v) c^2 = \frac{m_0 c^2}{\sqrt{\beta^2 - 1}}, \tag{11.44}$$

$$p = mv = \frac{\mathcal{E}}{c^2} v = \frac{m_0 c}{\sqrt{\beta^2 - 1}} \beta, \tag{11.45}$$

$$v = \frac{p}{\mathcal{E}} c^2 = \frac{d\mathcal{E}}{dp}. \tag{11.46}$$

Now we consider the implications of these rules.

Start with rule 1. While it is possible, in principle, to localize a tardyon at a point by an appropriate position measurement, no known experimental procedure can do the same with a tachyon. Classically, a tachyon is characterized by a finite size a_0 and quantum-mechanically, at any possible position measurement, its wave function can only collapse down to this characteristic size and not a single bit less. This imposes the corresponding restrictions on the "arrival time" of a tachyon at a detector. Namely, the tachyon arrival at the detector cannot be specified by an exact moment t'. Its irreducible size produces the corresponding indeterminacy Δt_{ar} in time of its arrival at any detector (or its interaction time with any point particle):

$$\Delta t = \frac{a_{\parallel}}{v}, \tag{11.47}$$

where $a_{\parallel}$ is its dimension (11.40) along the direction of motion and v is its instant velocity at the detector's location.

Similarly, there is indeterminacy in time of the tachyon's emission:

$$\Delta t_i = \frac{a_{\parallel}^{(i)}}{v_i}, \tag{11.48}$$

where $a_{\parallel}^{(i)}$ and v_i are, respectively, its longitudinal size and instant initial velocity in the emission event.

Therefore, when we calculate the time necessary for signal transfer with a specific tachyon, we must correct for the corresponding indeterminacy. Since tachyon is an indivisible particle, we can consider it as being emitted only after it is completely outside the emitter; and we can consider it as absorbed by a detector only when its rear gets completely inside the detector. Accordingly, the moment of its emission is later by the amount (11.47) than the emergence of its center and the moment of its absorption is later by the amount (11.46) than the arrival time of its center to the absorber. Altogether, the effective time it takes for a complete signal transfer by a tachyon, is

$$t_{\text{eff}} = t_c + \Delta t_i + \Delta t, \tag{11.49}$$

where t_c is the time the *tachyon's center of mass* takes to travel between the emitter and detector.

An equivalent (and more simple) approach is just to say that if the tachyon is to carry a signal, then in view of the above argument, the signal's speed will be determined by the *speed of tachyon's rear,* rather than by the speed v of its center of mass.

The problem thus reduces to determining the rear's speed v_r.

Suppose the tachyon is moving along the positive x-direction, like in Figure 11.5. If x is the instant coordinate of the tachyon's center, then the instant position of its rear is

$$x_r = x - \frac{a_0}{\tilde{\gamma}(v)}. \tag{11.50}$$

We assume here that for an *accelerated* tachyon (and tachyons do not know motion free of acceleration) its instant longitudinal size is determined only by the instant Lorentz factor of its center.

Taking the time derivative of (11.50) determines the rear's speed $\beta_r = v_r/c$:

$$\beta_r = \beta\left[1 - \frac{r_0}{c}(\beta^2 - 1)^{-1/2}\dot{\beta}\right], \tag{11.51}$$

where $\dot{\beta} \equiv d\beta/dt = c^{-1}dv/dt \equiv \dot{v}/c$ is acceleration of the tachyon's center of mass. To find it, we need a quantitative description of the tachyon's motion under the constant drag force. Combining Equations (11.41) and (11.46), we see that for the case $\mathbf{e}||\mathbf{p}$ the relativistic equation of motion is

$$\frac{dp}{dt} = f. \tag{11.52}$$

The equation for a tachyon is the same as known relativistic equation of motion for a tardyon. But keep in mind that in case of a tachyon, the drag force, while decreasing its momentum, increases its velocity (recall Section 11.2). A tachyon is accelerated by the drag force!

Let us express the tachyon's motion in terms of its acceleration a. Since in our case the velocity and force are collinear, we can apply the rule (5.62b), with "tachyonic" Lorentz factor. In view of the above comment about the relation between the direction of force and acceleration, we can write

$$f = -m_0\tilde{\gamma}^3(\beta)a = -m_0(\beta^2 - 1)^{-3/2}a. \tag{11.53}$$

Combining this with (11.51) at $\overset{\circ}{\beta} = \frac{a}{c}$ gives

$$\beta_r = \beta\left[1 - \chi(\beta^2 - 1)\right] = (1 + \chi)\beta - \chi\beta^3, \quad \chi \equiv \frac{r_0|f|}{m_0c^2}. \tag{11.54}$$

It is easy to see that in the region $\beta > 1$, that is, in all "tachyon's domain," its rear's velocity is less than its center's velocity. Moreover, at sufficiently large β, the tachyon's trailing edge can have zero speed or even move in the direction opposite to velocity $\mathbf{v}$ of the tachyon's center! This is another manifestation of the fact that at sufficiently high speeds ($v > \sqrt{2}c$) the Lorentz contraction is converted into its antipode, Lorentz extension. This effect can be easily visualized by examining Figure 11.5. If an accelerated superluminal object is moving faster than $\sqrt{2}c$, then according to (11.41), it undergoes relativistic length extension. Acceleration, through increasing v, increases its longitudinal dimension. At sufficiently large v the object is extending even faster than it is moving, so that eventually its rear starts moving in the opposite direction.

However, while being less than β, the value of β_r generally can still be greater than 1. Causality must preclude such situations.

It is easy to see that $\beta_r(\beta) = 1$ at $\beta = 1$ regardless of the value of χ. The function $\beta_r(\beta)$ has a maximum in tachyon's domain (at $\beta > 1$) if $\chi < 1/2$. This maximum exceeds 1, but it shifts to the boundary $\beta = 1$ and decreases down to 1 as the value of χ approaches 1/2. Thus, the causality is not violated if we require that

$$\chi = \frac{1}{2}. \tag{11.55}$$

We can express this requirement in terms of the drag force. Such force depends on specific details of tachyon's structure. The simplest model is the one of a uniformly charged sphere (i.e., tachyon's proper shape reconstructed from observations in subluminal RF [10, 11]). For such model Jones [108] has obtained the expression

$$f = \frac{9}{8} \frac{q^2}{4\pi\varepsilon_0 a_0^2}. \tag{11.56}$$

Since the actual tachyon shape and charge distribution may be different from the accepted model, we will write here a coefficient κ instead of 9/8, considering it as an adjustable parameter. Thus, for an electrically charged tachyon,

$$f_E = \kappa \frac{q_E^2}{4\pi\varepsilon_0 a_0^2}. \tag{11.57}$$

We can write analogous expressions for a tachyon, carrying magnetic charge q_M (hypothetic magnetic monopole [44,45]):

$$f_M = \kappa \frac{\mu_0 q_M^2}{4\pi a_0^2}. \tag{11.58}$$

Here, we use the units of magnetic charge, in which the "magnetic Coulomb's law" for magnetic field $\mathbf{B}$ has the form

$$\mathbf{B} = \frac{\mu_0}{4\pi} \frac{q_M}{r^2} \hat{\mathbf{r}}. \tag{11.59}$$

For an uncharged tachyon (neglecting possible effects of general relativity), we can obtain the similar expression for a gravitational drag force, treating tachyon's mass as the "gravitational" charge. In this case, however, we have to take into account the total relativistic mass of the tachyon, since all of it produces the gravitational field responsible for Cerenkov radiation:

$$f_G = \kappa \frac{G m_0^2 \tilde{\gamma}^2(v)}{a_0^2}. \tag{11.60}$$

In contrast to the electromagnetic (EM) drag force, the gravitational drag force turns out to be velocity-dependent.

We will limit our discussion to the case of electric or magnetic charges.

Let a tachyon carry the same amount of electric charge as an electron, that is, $q_E = q_e$. Going back to (11.57) and using (11.54), we get

$$\chi = \kappa \frac{q_e^2}{4\pi\varepsilon_0 a_0 m_0 c^2}. \tag{11.61}$$

Now, we make an additional (but quite natural) assumption: we set a_0 equal to the corresponding Compton wavelength $a_0 = \hbar/m_0 c$. It follows then

$$\chi = \kappa \frac{q_e^2}{4\pi\varepsilon_0 \hbar c} = \alpha\kappa = \frac{1}{2}. \tag{11.62}$$

Here, α is the famous fine structure constant

$$\alpha \equiv \frac{q_e^2}{4\pi\varepsilon_0 \hbar c} \approx \frac{1}{137}. \tag{11.63}$$

Then, the requirement $\chi = 1/2$ uniquely determines our adjustable parameter κ:

$$\kappa = \left(\frac{1}{2}\right)\alpha^{-1} \approx 68. \tag{11.64}$$

This is much greater than the value 9/8 corresponding to the model of a uniformly charged sphere. If this result is true, it would indicate that actually most of the tachyon's charge is concentrated close to the center, forming a dense pointlike core similar to the atomic nucleus within an atom. The charge density falls off rapidly with the increase of radial distance, but never becomes exact zero. This "fine structure" of a tachyon is consistent with the absence of the long wave range in its spectrum [1, 107].

Now, we come to the final and most interesting result. We will show that the same parameter κ, once adjusted to the value (11.64), determines the "superluminal analogue" of the electric charge quantization first introduced by P. Dirac from quite different considerations.

Consider a tachyon carrying *magnetic* charge q_M. Applying the same treatment to this case, we get from causality nonviolation:

$$\chi = \kappa \frac{\mu_0}{4\pi} \frac{q_M^2}{a_0 m_0 c^2} = \kappa \frac{\mu_0}{4\pi} \frac{q_M^2}{\hbar c} = \frac{1}{2}. \tag{11.65}$$

If κ is the same for electric or magnetic tachyon, then it follows from Equations (11.62) and (11.65)

$$\frac{q_M}{q_e} = \sqrt{\frac{1}{\mu_0\varepsilon_0}} = c. \tag{11.66}$$

We can work out from here an equivalent relationship between the electric and magnetic charges. Namely, writing

$$\frac{q_M}{q_e} = \frac{q_M q_e}{q_e^2} = \frac{q_M q_e}{4\pi\varepsilon_0 \alpha \hbar c} = c, \tag{11.67}$$

we obtain

$$q_M q_e = \frac{4\pi}{\mu_0}\alpha\hbar. \tag{11.68}$$

The relations (11.66) and (11.68) look similar to Dirac's famous conditions for a subluminal monopole [109], although the numerical values are different. Namely, the Dirac's relations in the same units are

$$\frac{q_M}{q_e} = \frac{1}{2}\frac{n}{\alpha}c, \quad \text{or} \quad q_M q_e = 2\pi n \frac{\hbar}{\mu_0}, \quad n = 0, 1, 2, \ldots \tag{11.69}$$

The numerical difference is in the fine-structure factor. Also, in Dirac's conditions at least one of the charges is explicitly quantized. These differences are natural, because the corresponding relations follow from quite different requirements. The relations (11.66) and (11.68) must hold for the tachyons and follow from causality nonviolation. The relations (11.69) hold for tardyons and follow from quantum-mechanical quantization conditions [44, 45, 109].

It should be stressed that the relations (11.66) and (11.68) have been obtained under certain assumptions about the tachyon's structure mentioned in the beginning and without taking into account the contribution from gravitational Cerenkov radiation. Although this contribution may be very small for a charged tachyon with a proper mass comparable with those of the known tardyons, it may be significant or even dominating for supermassive tachyons. For a neutral tachyon without dipole or higher electrical or magnetic moments, the gravitational radiation would be a sole actor in the play. But in this case, too, causality must impose its requirements quite similar to the one considered here.

Summary

The superluminal speed of a tachyon by itself does not necessarily involve superluminal signaling. The considered analysis demonstrates how QM and Relativity can conspire to prevent it from happening. QM imposes a minimal size on tachyon's length and thereby makes a signal transfer complete only when the *whole* tachyon crosses the "finish line." Relativity, trough the length extension effect, makes sure that the resulting time of arrival remains within the causal limit.

11.5 Wave Packet Propagation and Position Measurements

Here, we consider a possibility of the superluminal signaling via position measurements in QM. Suppose we have a particle in a state $\Psi(\mathbf{r}, t)$. Its probability distribution at $t = 0$ (taken as the initial moment) is shown in Figure 11.6a and at this moment we perform a position measurement. If the measurement is sufficiently accurate, it may locate the particle that was initially (at $t \leq 0$) described by the wave function $\Psi(\mathbf{r}, t)$, within a very small region of space around a certain point $\mathbf{r}'$ (Figure 11.6b). In the

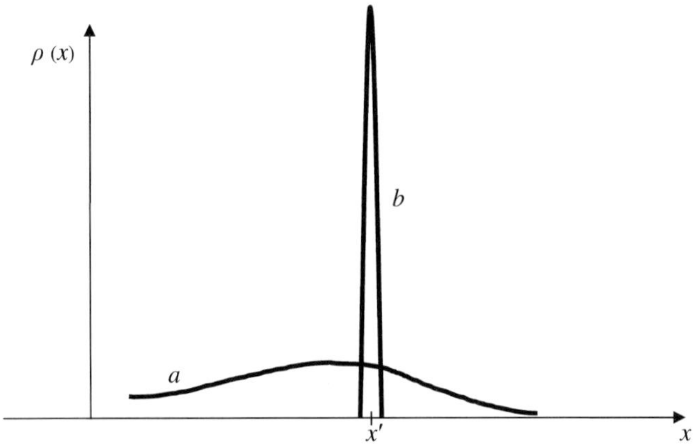

Figure 11.6 An instantaneous probability distribution ρ (x): (a) before the position measurement; (b) immediately after the position measurement.

process of measurement, the probability distribution in Figure 11.6a practically instantly collapses to an extremely narrow configuration in Figure 11.6b. Even for a single electron with initially undefined position (i.e., in a state described by a plane wave embracing the entire universe) its wave function collapses instantaneously at the moment of measurement to a well-defined wave packet within a small region of space. Such a collapse or reduction of the wave packet from initially infinite to a finite size occurs with an infinite speed (imagine something shrinking within a twinkling of an eye from the size of the visible universe down to the size of an atom!). And there is no contradiction with the theory of relativity in it, because already prior to the act of measurement there was a nonzero chance for the electron to happen to be at precisely the same atom where it was found to be after the measurement. And on the contrary, even though there was a much greater chance for it to find itself some place within the Andromeda nebulae, it was only a chance, not a certainty. One cannot therefore describe the collapse of the wave function in terms of cause and effect, for instance, as the convergence of a certain compressible fluid whose tractable parts occupy at any moment a well-defined place in space. Therefore, these quantum-mechanical phenomena, subtle as they are, do not in any way undermine the foundation of the theory of relativity.

Consider now a more provocative situation, illustrated in Figure 11.7. Now the initial configuration right before the measurement is described (also in **r**-representation) by a very narrow wave packet. The state right after the measurement is represented by a wave packet shown in Figure 11.7b. The narrow wave packet has jumped from position a to another position b a finite distance apart. In the everyday language, the particle has changed its residence from compartment a to compartment b in practically no time. Is it not the evidence that the wave packet representing the particle was moving between positions a and b with almost infinite speed?

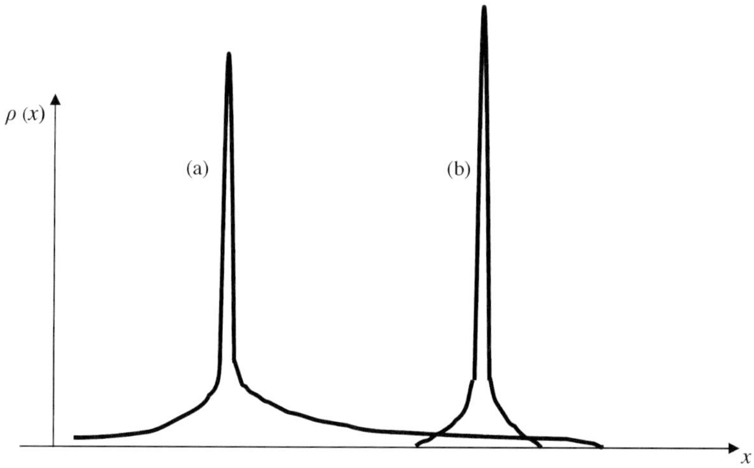

Figure 11.7 Position measurement in a localized state. Instantaneous probability distributions ρ (x): (a) before the position measurement; (b) immediately after the position measurement.

The answer is no. As we have stressed in Chapter 10, the state function is evolving deterministically *between* the measurements. The jump of the packet from a to b occurred *during* the measurement as its result.

Even though b is far from a, the situation shown in Figure 11.7 indicates the preexisting, albeit small, but nonzero probability of the particle to show up at b in the appropriate measurement. Therefore, when it happened to be found there, this is merely actualization of a preexisting potentiality. It does not require the intermediate motion and subsequent invasion of the wave packet a to location b. Nor *is* there any such motion. Rather, the measurement *disrupts* a possible smooth motion of a, which might, indeed, have been occurring before $t=0$, and the preexisting state instantly disappears. Instead, the measurement creates a new state represented by a new packet that instantly pops up at b. The possibility of such creation was guaranteed by a preexisting nonzero probability density at b. Because in the given case this probability was small, it *looks* like the result of the packet's motion from a to b, but it is actually the collapse of the wave function at an unlikely place with low *original* probability density.

Another way to state this is that an evolution of state $\Psi(\mathbf{r}, t)$ is, as just mentioned, described by QM only *between* the measurements. The event of measurement itself (in particular, the collapse of the wave function to a new configuration at a position measurement) is not described by it. One of manifestations of this fact is that, in contrast to the deterministic (predictable!) evolution of $\Psi(\mathbf{r}, t)$ between the measurements, the new location b of *collapsed* state was totally unpredictable. If it were the result of smooth motion of packet a, then it could have been predicted in advance *with certainty*, but this is not the case when we perform a measurement. Just before the measurement, the wave function $\Psi(\mathbf{r}, t)$ tells us only about probability to collapse at a position $\mathbf{r}=\mathbf{r}'$, not about the position $\mathbf{r}'$ itself at which it happens to collapse. That the

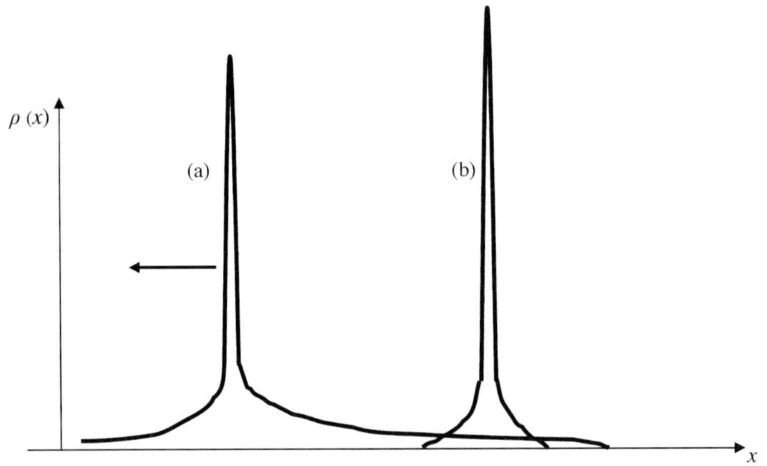

Figure 11.8 Position measurement in a localized state. Instantaneous probability distributions ρ(x): (a) before the position measurement; (b) immediately after the position measurement. It is the same as in Figure 11.7, but now with the additional information that the object (wave packet) before the measurement was moving to the left. Therefore in (b), it is found behind its earlier position.

corresponding localization of the particle at $\mathbf{r}'$ is *not* the result of packet's motion is evident from the fact that sometimes the particle can be found at $\mathbf{r}'$ even if the initial packet had been moving away from $\mathbf{r}'$ (Figure 11.8)!

Suppose now that we have prepared a particle in a totally localized state, that is, a state represented by a wave function which is *exactly* zero outside a small region $\Delta\mathbf{r}$ at $\mathbf{r} = \mathbf{r}'$ (Figure 11.9). Now "release" the particle from its temporary trap prepared by the measurement and perform a new position measurement immediately after that. Suppose we found it this rime at $\mathbf{r}''$ far away from $\mathbf{r}'$. According to our initial conditions, the probability density $|\Psi(\mathbf{r}'', 0)|^2$ was exactly zero before the measurement. Therefore, in this case there remains only one possible explanation of the particle's emergence at $\mathbf{r}''$: it is due to the motion or the spread of the initial wave packet. In other words, now the two subsequent states are in cause and effect relation. The initial presence of particle at $\mathbf{r}'$ and the subsequent shift and/or spread of its wave packet from this location is the cause. The subsequent discovery of the particle at $\mathbf{r}''$ far from $\mathbf{r}'$ is the effect. This is consistent with the fact that evolution of a wave packet between the measurements is described by the corresponding wave equation.

Figure 11.9 shows the case with $\Delta x' < \Delta x''$. In the opposite case, the particle's appearance at $\mathbf{r}''$ is still an unpredictable outcome. It has its cause, the propagation of the initial packet, but it is unpredictable because the corresponding probability was not equal to 1; there were lots of other competing locations with emerging nonzero probability density; and the evolving wave function contained only relative likelihoods of finding the particle in one of these locations. Causality does not necessarily mean exact predictability!

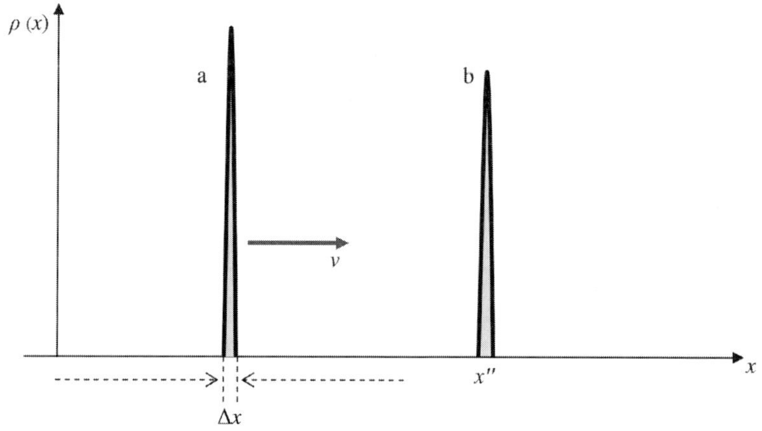

Figure 11.9 Position measurement in a localized state with sharply defined boundary. Instantaneous probability distributions $\rho(x)$: (a) the distribution at $t=0$ (before the position measurement); (b) the distribution at $t=t_0>0$ (immediately after the position measurement made at $t=t_0$). The function (a) is exactly zero outside the region Δx and (b) emerges in a place where the particle certainly could not be found at $t=0$. In this case, the appearance of (b) can only be attributed to the initial pulse propagation between the moments $t=0$ and $t=t_0$.

As seen from this example, the necessary condition for strong causal connection between the two described events is the requirement that $\Psi(\mathbf{r}, 0)=0$ outside the volume element $d\tau'$ around $\mathbf{r}'$. Since it is nonzero within $d\tau'$, there has to be a point (more generally, a surface S) somewhere between $\mathbf{r}'$ and $\mathbf{r}''$ where either the function itself or its derivative of a certain order with respect to $\mathbf{r}$ is discontinuous. Thus, the surface S forms a sharply defined border between the two corresponding regions. A function with such properties is *not analytical* with respect to variable $\mathbf{r}$. If, in addition, the function also depends on time, then the surface S forming the border between the regions can be moving. In this case, it is called the wave front or just front.

The appearance of the particle in a place where there was initially no chance to find it can only be the result of the particle's arrival at this place from elsewhere due to the motion of the corresponding wave front. It turns out then that the signal transfer can be associated with motion of such front.

Thus, the relevant question to ask would be: what is the propagation speed of the wave front? Can it be superluminal?

This question has two aspects – experimental and theoretical. We will discuss both aspects in the following sections.

11.6
Superluminal Quantum Tunneling

Another example of the superluminal motion may be found in the experiments with the quantum tunneling. Quantum tunneling is a remarkable phenomenon, in which a particle passes through a region, which is totally forbidden classically, because in this region its potential energy would exceed its total mechanical energy.

Here is a simple example. Imagine that you are driving a car with its engine off. On a straight road and without friction you could still keep on moving at a constant speed. But if there is a hill ahead, then your car, while going uphill, will gain potential energy at the cost of its kinetic energy and will accordingly slow down. If the hill is high enough, then your car will eventually stop at some point on the slope. At this point your motion will reverse from going uphill to rolling back downhill. The hill becomes an insurmountable barrier and the point where you instantly stop before turning back can be called the turning point. Graphically, this point is the intersection of the horizontal line representing the total energy of you car and the curve representing its potential energy as a function of position (Figure 11.10). We say that this curve represents a potential barrier.

All particles with the total energy exceeding the top of the barrier will merely slow down as they pass, only to regain their kinetic energy on the back side of the barrier; for those with the total energy lower than the top, the barrier cannot be passed. The particle is reflected from it at the turning point. The position of this point depends, as seen from Figure 11.10, on barrier's shape and on total energy of the particle.

For a barrier with one top and for a particle with a fixed total energy lower than the top, there are two turning points: one for a particle approaching the barrier from the right and the other one for an identical particle with the same energy approaching from the left. The situation is in this case analogous to that of the Section 5.6, with the distinction that in Section 5.6 (Equation (5.107)) the region of a stable radial motion constitutes *potential well* and, accordingly, the regions *outside* of the range $r_{min} < r < r_{max}$ are classically inaccessible, whereas here the range $x_1 < x < x_2$ *between* the two points x_1 and x_2 is classically inaccessible. In both cases,

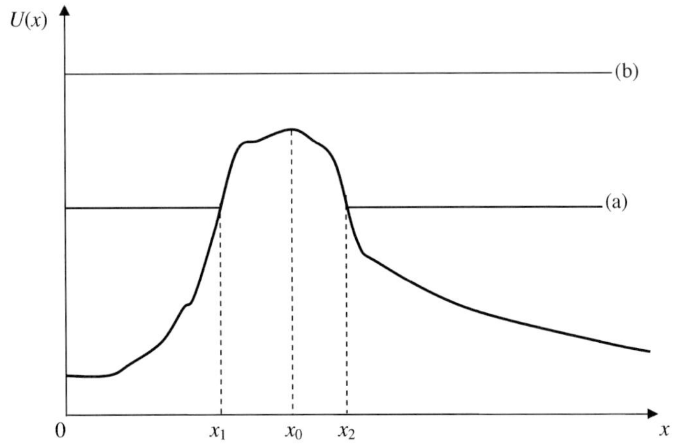

Figure 11.10 The graph of potential energy $U(x)$. (a) $\mathcal{E} < U(x_0) = U_{max}$. The region $x_1 < x < x_2$ is classically inaccessible; the barrier cannot be passed; (b) $\mathcal{E} > U(x_0)$. The barrier can and will be passed.

the total energy of the particle is less than its potential energy in the inaccessible regions, which means that within such regions the classical particle would have negative kinetic energy and, accordingly, imaginary velocity. As in the case (5.107), this is immediately seen in the nonrelativistic limit, where the kinetic energy is $mv^2/2$, so that, for instance, within the barrier one would have

$$K = \mathcal{E} - U = \frac{1}{2} m_0 v^2 < 0, \quad x_1 < x < x_2. \tag{11.70}$$

In a relativistic case, using the corresponding expressions (5.116)–(5.118), we obtain the same result (Problem 11.3).

Quantum mechanics (QM) changes all that. A free particle in a state with definite energy may have definite momentum and therefore, according to the uncertainty principle, totally indeterminate position (its wave function is smeared out uniformly over the whole space). Therefore, introduction of a barrier can only change probability distribution for the particle to be found on this or that side of the barrier; but it cannot restrict the particle's motion to only one side (unless the barrier is infinitely high or has an infinitely wide top above the particle's energy).

In a more realistic case, the particle's momentum is not sharply defined. Accordingly, its wave function (and thereby its probability cloud) is no longer smeared out uniformly over the whole space but rather will form a packet of a finite width representing the degree of indeterminacy in the particle's position. Now, we can ask an interesting question: what happens if, after a position measurement, the particle (its wave packet) is found entirely within the barrier (Figure 11.11)? In this case we know the particle is all within the region where it cannot be found classically.

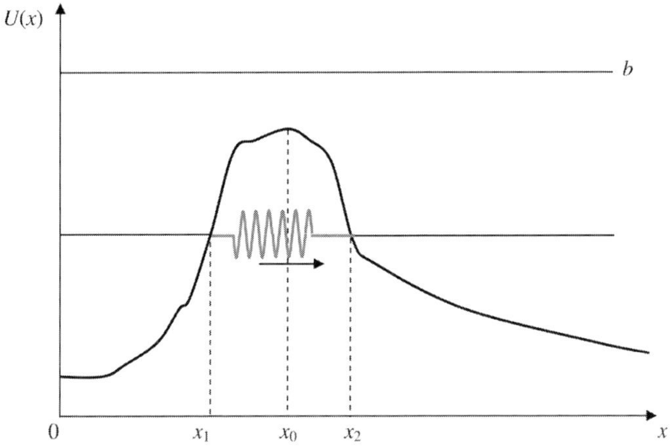

Figure 11.11 The particle is entirely within the barrier. This automatically changes its total energy from a sharply defined discrete value into an energy band, which gives a nonzero probability to find, in an appropriate measurement, the particle's energy above the top of the barrier. This, in turn, gives the corresponding probability for the particle to emerge on the side of the barrier opposite to one from which it had come. Actually, the position of the wave packet representing the particle is not essential for this conclusion.

Accordingly, we might expect that the particle will have the imaginary velocities. Does this conclusion constitute a paradox?

In quantum physics, the answer is no – for two reasons.

First, the moment the particle's position is restricted within the barrier (i.e., its position indeterminacy has been reduced down to $\Delta x \leq |x_2 - x_1|$), it automatically introduces the corresponding indeterminacy $\Delta p \geq h/\Delta x$ into its momentum and thereby into its energy:[3)]

$$\mathcal{E} \rightarrow \mathcal{E} \pm \frac{1}{2}\Delta\mathcal{E}. \tag{11.71}$$

Owing to this indeterminacy there is a chance to find in the particle's energy spectrum an energy that is higher than the top of the barrier; this is sufficient for the particle to have may be a small but nonzero chance to pass over the barrier "legally," with its velocity remaining a real quantity.

Second, the wave aspect of matter *allows* for an imaginary velocity and thereby momentum, because according to the de-Broglie relations, this would only mean the imaginary wavenumber, and the latter is well known in the wave theory as indicating attenuation (or amplification) of the wave. Indeed, set the wavenumber k in (10.4) to be imaginary by writing $k \rightarrow i\kappa$, with κ real, and you will obtain the exponentially decaying (or increasing, depending on sign!) probability amplitude (for more detailed description, see Appendix E, G).

There is another aspect of the indeterminacy, which, already at this stage, provides with some insight into the nature of the superluminal tunneling. Namely, as we found in Section 10.2, the indeterminacy $\Delta\mathcal{E}$ in the wave packet's energy is inseparably linked with the indeterminacy Δt of time of the packet's arrival at the detector:

$$\Delta t \geq \frac{h}{\Delta\mathcal{E}}. \tag{11.72}$$

This uncertainty sets the lower limit for the average time spent by the particle within the barrier. In typical experimental situations discussed below, $\Delta t \sim 10^{-13}$ s. For the tunneling to be superluminal, the particle must spend inside the barrier less time than it would take light to travel the barrier's width a; and for time comparison to be unambiguous, the corresponding time difference $t_2 - t_1 = a/c - t_u$ must be greater than the indeterminacy (11.72). Combining these two conditions gives

$$\frac{a}{c} - t_u > \Delta t. \tag{11.73}$$

3) We must draw a subtle distinction here from a stationary state with sharply defined energy in an external potential field. In such state, even though the total energy $\mathcal{E}$ is determined exactly, its constituents – the particle's potential and kinetic energy (and thereby its momentum) are undetermined (this is reminiscent of a system with its *net* spin component exactly known (Section 10.1), while the individual spins of its constituents are not determined). The position measurement destroys the initial stationary state and introduces the indeterminacy (11.71) into the *total* energy $\mathcal{E}$.

If this difference lies beyond the uncertainty range, there must be a specific mechanism for a statistically significant difference in times of arrival. Such mechanism may be the reshaping of the wave packet similar to one considered in Ref. [1]. In this case, we will have apparent superluminal tunneling.

Thus, a particle with energy sharply defined and lower than the barrier, can pass through the barrier without invoking any contradictions because at no time is it entirely within the barrier. On the other hand, a particle, even when entirely within the barrier and with its *average* energy again lower than its top, can still pass through the barrier without any contradictions because now its energy spectrum has a part going beyond the top of the barrier. By this trade-off the Nature always avoids the above paradox [90]. This allows us to represent the particle's passing through the barrier while having the energy (or its average) below the barrier's top as if there were a tunnel cut through the barrier, hence the (very appropriate) name "tunneling" for this effect. Of course, in reality there is no hole or tunnel cut through the barrier, the latter is not affected by the passage of a particle. But, since such a passage is allowed (although with probability less than 1) quantum-mechanically, it is *as if* there were some virtual tunnels in the barrier. In terms of the everyday language, in a case when the possibility of tunneling is actualized, it is as if a solid object banged into a concrete wall and instead of bouncing back, emerged on the opposite side, both the object and the wall remaining intact. A good description of such tunneling can be found in the Lewis Carroll's immortal story "Through the Looking Glass" [111].

It is important to stress again that quantum-mechanical *possibility* of tunneling is intimately linked with the *impossibility* to predict an individual outcome. In the classical world, the situation is strictly deterministic: if you know *exactly* the initial conditions, you can predict the individual outcome with certainty. In case of passing a barrier, if the particle's energy is above the top (and there is no dissipation), it will pass. If the energy is below the top, it will not. In contrast, quantum mechanics gives *any* particle (even with a very low energy) a chance to pass (in our jargon, to tunnel), but this is only a chance. Whether the particle will tunnel or not in a given single trial is totally unpredictable. This unpredictability is *not* the result of some incompleteness of our knowledge about the system. It may be as complete as possible. As already discussed in the previous sections, it is the nature of the initial conditions themselves that renders exact predictions generally impossible. This is what we mean when we say that quantum-mechanical determinism (which gives far more accurate description of Nature) is more restricted than classical one. It has nothing to do with elimination of determinism, which some claim is inherent in quantum mechanics. The chance (the probability) of tunneling is an objective characteristic of a given system, which can be accurately calculated and tested. It is an exact function of given conditions. If a barrier is tall or (and) wide and the particle's energy is low, the chance of tunneling is small. If the barrier is thin or (and) low and the particle's energy is only slightly below the top, the probability is close to 1. And, even though quantum mechanics does not predict an individual outcome, its probabilistic predictions can be with high accuracy confirmed statistically. If we solve the quantum-mechanical equation for a system with given conditions and find the tunneling probability to be, say, 0.43, it enables us to *predict* that on the massive scale, 43% of the particles will

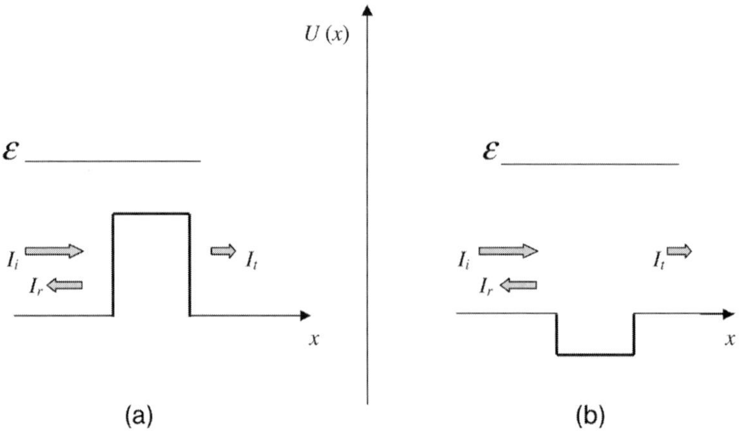

Figure 11.12 (a) A particle with energy $\mathcal{E} > U_{max}$, which classically would pass above the barrier, actually has a chance to be reflected from it. (b) If, instead of the barrier, we have a potential well, there is still a nonzero probability for the particle to be reflected from it. Arrows represent intensities of the incident (I_i), reflected (I_r), and transmitted (I_t) flux, respectively (recall that we can speak about flux even in the case of only one particle). If there is no absorption in the region of the barrier or well, we have $I_r + I_t = I_i$.

tunnel through this barrier. If the particles are, say, the photons, and you perform an experiment with a laser pulse containing billions and billions of them, then a screen behind the corresponding optical barrier will, in accord with prediction, be illuminated with 43% of the initial intensity.

There is another remarkable aspect of the tunneling effect – it has an "antipode" associated with it by symmetry: just as a particle with energy below the top of the barrier is allowed to tunnel through it, a similar particle with its energy above the top is, under certain conditions, allowed to be reflected from it (Figure 11.12a). Moreover, even if, instead of a barrier, we have a potential well, there may be a nonzero chance for the particle to get reflected from the well (Figure 11.12b) – a totally crazy situation from the classical viewpoint!

Now, the tunneling particles we want to discuss here are the photons. Therefore, it is natural to start with a question: what constitutes a barrier (or a potential well) for a photon? It is tacitly assumed sometimes that, since the photon has no charge but only a *relativistic* mass owing to its energy–momentum

$$m = \frac{\mathcal{E}}{c^2} = \frac{p}{c} = \frac{\hbar\omega}{c^2},$$

and this energy–momentum interacts gravitationally with other lumps of matter, the photon can only have potential energy in the gravitational field. But, as a particle of the electromagnetic field, the photon is a carrier of the corresponding electromagnetic force and it interacts with matter electromagnetically. Manifestations of this interaction are bending of a light ray in the atmosphere, as well as its reflection and refraction on the surface of a medium. Another manifestation is the Compton effect discussed in Section 10.6. Therefore, the motion of a photon in a homogeneous

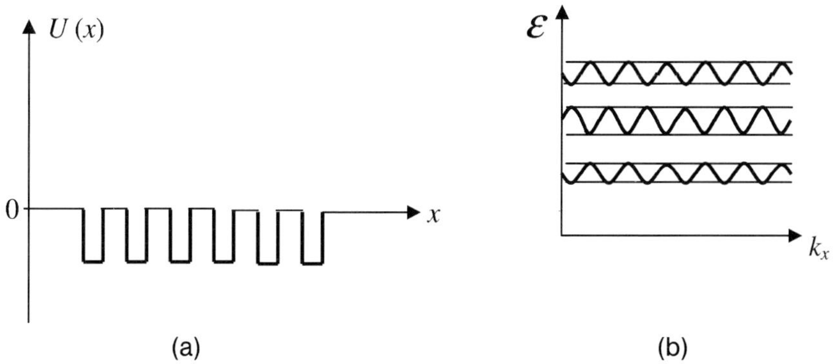

Figure 11.13 (a) The Kronig–Penni potential; (b) the allowed energy bands. Energy in an allowed band is a periodic function of wavenumber k_x for a particle moving in such field.

medium with index of refraction n can be described as the motion of a massive particle in an effective potential field U.

It turns out that for a medium with $n > 1$ (such as glass or water), the photon's "potential energy" is $U > 0$, so actually this constitutes a potential well (Appendix E). But in the professional jargon it is also called a barrier and the passing through it the tunneling. Another type of an optical quantum barrier for a photon may be an array of alternating layers with different optical properties (Figure 11.13).

Figure 11.14 shows the wave function of a particle with fixed energy interacting with a barrier (see more details in Appendix E).

Once we can manufacture an optical barrier in more than one way, there appears another interesting question: for the particles that happen to go through the barrier, what is the transition time?

The question formulated in this form is rather ambiguous and can result in confusion. Before asking about a numerical value of transition time, we must define the concept itself in terms of an operationally executable procedure. It turns out that there are a few possible definitions.

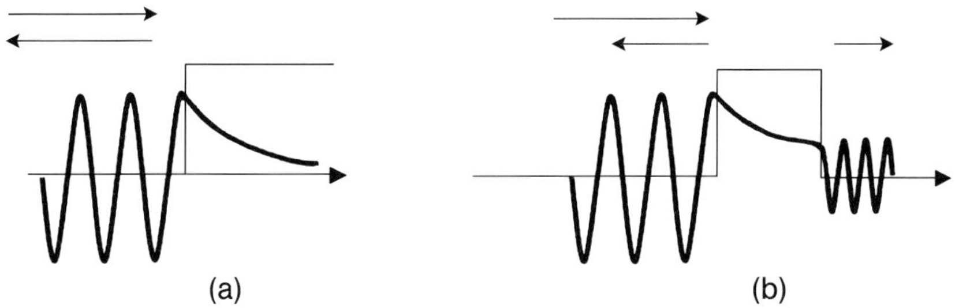

Figure 11.14 (a) An infinite-width barrier; the wave function on its left is a superposition of the incident and reflected waves of equal amplitudes. Within the barrier, the wave function exponentially decays; there is no transmission. (b) The finite-width barrier; there is a transmitted wave on the right. On the left, the amplitude of reflected wave is less than that of the incident wave. Overall, there is a nonzero flow of energy to the right.

We will start with the most straightforward procedure that comes to mind: we just prepare one-photon state, throw this photon toward the barrier, record the moment of time t_1 when it reaches the entrance x_1, and in case the photon emerges from the opposite side x_2, we record the moment t_2 of its exit, then find the corresponding time difference $\Delta t = t_2 - t_1$. We can define this difference as the transit time. And then, we may divide the width of the barrier $\Delta x = x_2 - x_1$ by this time to find also the speed of the photon $u \equiv \Delta x/\Delta t$ when inside the barrier.

Well, since the photon is a quantum particle, such procedure would not work. The more accurately is measured photon's position and moment of time at the entrance the more violently is disrupted its initial state of motion, so that the corresponding second measurement at the exit will not tell us much about the "would be" crossing time.

This unsuccessful attempt illustrates that in an operational definition an account must be taken of the wave aspect of the tunneling particle. We can do it by visualizing a particle as a wave packet similar to a "bump" in Chapter 7 of the book [1]. The shape of a bump is nothing else but the graph of the corresponding probability distribution of finding a photon at the corresponding position in the proper position measurement.

There still remain other possibilities. We consider a few of them.

1. First, we can define the average transit time as the time difference between the moment of arrival of a photon to *detector* (i.e., the total time of motion) and the time it takes to travel along the path outside the barrier. Since in both cases the time of arrival is determined only within the range (11.72), the difference between them will be, at best, determined with the same accuracy (provided there is no spreading or reshaping of the packet inside the barrier.) In other words, the average transit time can be measured only to the accuracy given by (11.72).

2. We can determine the time the *top of the bump* takes to cross the barrier. The top has at each moment a definite position, which can be calculated if we know the spectrum of the bump and the optical characteristics of the barrier. We can also find the corresponding time experimentally, by measuring the distribution of arrival times of the individual photons in many trials and taking the result recorded most frequently. The most frequent result corresponds to the most probable location of a photon, which is at the top of the bump. Under certain conditions, as we know, the top may move faster than light, but this kind of motion does not violate causality.

3. In all experimental situations, the bump representing a photon's wave packet, has a sharply defined leading edge. It is the motion of this edge that actually constitutes a signal. Its instant position is very difficult to measure experimentally, because it corresponds to the vanishingly small probability of the photon's location; however, it is exactly defined at any moment and can be calculated from the spectrum. We can define the transit time as the time it takes the edge to cross the barrier and the speed of the photon within the barrier as the speed of the edge. But this will not give any new information, since, as we will see in the next section, this speed cannot exceed the speed of light in vacuum.

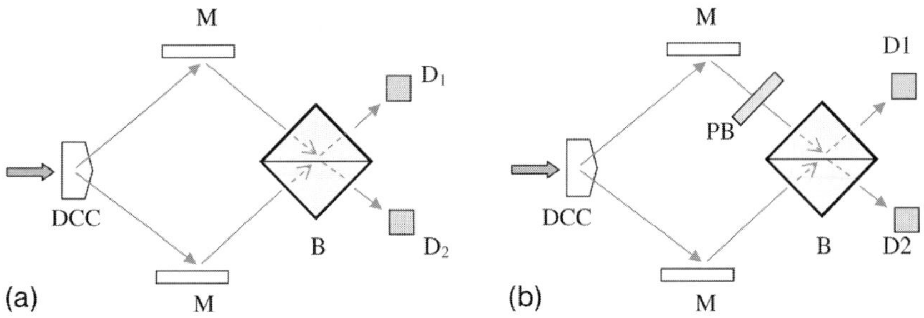

Figure 11.15 Schematic representation of the superluminal quantum tunneling (SQT) experiment. DCC – down-conversion crystal; M – mirrors; B – beam splitter; D_1, D_2 – detectors; PB – the plate creating *potential barrier* for one of the photons. Solid arrow – the incident UV-photon; thin arrows – the respective paths of the two daughter-photons born in the crystal. (a) Stage 1 – matching the paths; (b) Stage 2 – measuring the tunneling time.

Now we are in a better position to discuss a real experimental situation [12, 13]. The working procedure was much more complicated and designed so as to avoid the *direct* measurement of the photon's traveling time within the barrier. The experimenters had used two twin photons produced simultaneously as the products of decay of the initial more energetic (with higher frequency) photon in a crystal. This nonlinear process is known as spontaneous parametric down conversion [113]. Typically, in this process one UV photon produces two IR photons, which start from their birthplace in two different directions. The photons are then directed by an optical system to a beam splitter, after which they wind up in one of the two different detectors (Figure 11.15). The experiment consists of the two basic stages. On stage 1 the two photons are sent along the two legs whose lengths can be adjusted to ensure equal travel distances for both photons *without* the barrier. This can be done by observing the statistics of photons' arrivals at the detectors.

Each photon can be transmitted through or reflected by the beam splitter. Therefore, there are four possibilities: both photons are transmitted; both photons are reflected; photon 1 is transmitted while photon 2 is reflected; photon 1 is reflected while photon 2 is transmitted. As seen from Figure 11.15, first two possibilities lead to the photons winding up in the different detectors. The last two possibilities lead to both photons hitting one and the same detector, that is, finishing their respective journeys in one common state. If the photons arrive at the beam splitter at different times, then they act as independent particles and, accordingly, all four possibilities have equal probability. If the photons arrive at the beam splitter simultaneously, then they are strongly coupled and choose out of four possibilities those that provide one common final state for both. These are only one of the last two possibilities. In this case, only one out of the two detectors clicks, indicating a simultaneous arrival of both photons (more accurately, the rate of coincident clicks of both detectors is the lowest). Thus the stage 1 is tinkering with the two paths until their lengths become equal, which is manifest in the fact that in all trials only one out of the two detectors clicks, indicating a simultaneous arrival of both photons.

Once it is established that two path lengths are equal, there comes stage 2: placing a barrier in one of the two paths. Then any time discrepancy in the two photons' arrival can only be attributed to the quantum tunneling through the barrier. The researchers indeed found such discrepancy. This was evident from the fact that the coincidence rates (clicking of both detectors) were no longer at a minimum. The only possible explanation was that now the photons did not arrive at the beam splitter simultaneously. And a remarkable fact was that the only way to restore the minimum (simultaneous arrivals of both photons) was to *lengthen* the path containing the barrier (or shorten the empty one). This clearly showed that it took less time for the photons to traverse the barrier than the same distance in air. In other words, the tunneling was superluminal.

Now we, while skipping the otherwise interesting details, will focus on the way the statistical data on superluminal tunneling were collected. As mentioned above, the wave packet of each one of the two racing photons is represented by a bump. The initial bumps are identical and leave the starting gate simultaneously (Figure 11.16a). Let us call the photon traveling along the free path the "bump 1" and its sister having to deal with the barrier the "bump 2."

When the second bump reaches the barrier, it splits into two smaller secondary bumps: one that turns back (is reflected) and one that keeps on moving forward (is transmitted). The relative sizes of the secondary bumps represent the probability of the corresponding outcomes. The reflected bump in Figure 11.16b is bigger because in the actual experimental setup used in Refs [12, 13], the reflection was much more likely then transition. Keep in mind that in no way the splitting of the bump 2 means the splitting of the corresponding photon. Bumps represent the *probability distribution* for a given photon and the emergence of the two secondary bumps out of the primary one only means that there appear two likely regions where the photon can be found and that these regions are moving apart from one another. Any attempt to find an "actual" whereabouts of the photon by measuring its position will discover a single photon in only one of the secondary bumps (whose area instantly increases up to that of the primary bump), while the other one instantly disappears. For instance, if we found the photon in the secondary bump that is moving forward, we know for certain that the photon has been transmitted and with the same certainty we can tell that we will not find it on the reflected side (the corresponding probability, big as it might have been before the measurement, has instantly vanished).

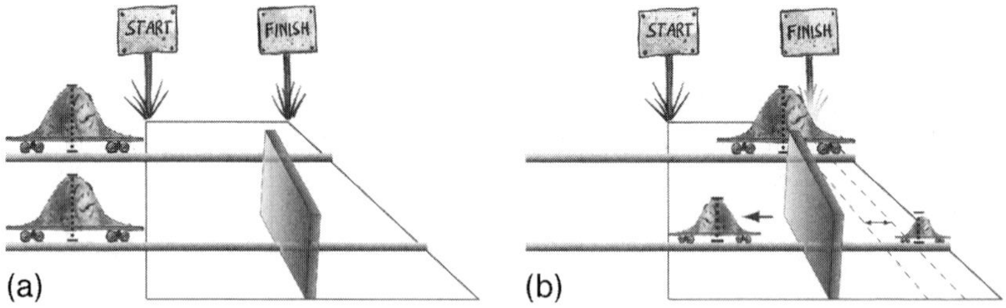

Figure 11.16

We now want to analyze this situation in more detail. In all cases when the photons had tunneled through the barrier, it was observed that the peak of the corresponding secondary bump arrived at the detector *earlier* than the peak of the bump 1 that traveled all in air. The peak is the most likely position of the photon. Therefore, experimentally, a tunneling photon is most likely to reach the "finish line" before the photon traveling in air. On Figure 11.16b depicting the final stage of the race, the top of the bump representing the tunneling photon leads the top of its feely moving counterpart by a distance indicated by double arrow. This means that the top of the wave packet (the maximum of the position probability distribution) travels faster within the barrier than it does in air. In other words, since the speed in air is (for all practical purposes) equal to c, the transmitted photons appear to tunnel faster than light. But this does not violate causality, since the peak's propagation does not transfer a signal. Its superluminal velocity may result from a reshaping effect considered before. To observe the rate of *signal* transfer, we must record the arrival times of the leading edges of the bumps. What does this observation show?

As mentioned above, the edge represents low probability and accordingly, a photon is rarely observed there. Since we are talking about the leading edge, it is represented by the photons arriving first. Thus, we have to focus on these rare occasions of the first arrivals. Instead of comparing the *most frequent* arrival times for the two photons, we must compare the *earliest* moments of arrival for both paths – one with and one without the barrier. We can reformulate the problem in the following way: does the presence of a barrier decrease the arrival time of the *earliest* photons? If yes, that would mean violation of causality as we know it today. But the experiment says no. The leading edge of bump 2 arrives no earlier than that of the bump 1. That's it! The tunneling may be superluminal, but it remains causal.

There remains an interesting question what is the mechanism of reshaping that results in the superluminal tunneling of the bump's peak. Obviously, it is different from the mechanism considered in [1]. One of the ways to get the answer in the current case can be the direct solution of Maxwell's equation for EM waves in the corresponding periodic structure, finding the dispersion relation showing how the frequency depends on wavenumber and then finding the group velocity. But whatever the mechanism, it does not affect the basic result: it does not violate causality.

11.7 Wave Front Propagation

Here, we discuss a fundamental question about the wave front velocity. Consider a running group (impulse) such that there is no perturbation ahead of it. The corresponding wave function (or functions, if there are more than one), considered at a certain point in space, is zero at all times before a certain moment. We can call this moment the arrival time of the corresponding wave packet. Then the separation point between unperturbed and perturbed regions is called the wave front. This separation point is running together with the whole pulse and we want to find out how fast it runs. This is a very important question since we know that it is a discontinuity on the wave profile associated with a signal, carrying new information, and such a signal must move no faster than the invariant speed. In other words, a signal can be transported only by a discontinuity in a pulse whose profile is represented by a nonanalytic function. Such a function is either discontinuous itself, or has a discontinuous derivative. And the point of discontinuity is, according to our definition, a wave front.

As we know, there is a very interesting reciprocal relation between a fine structure of a function and its Fourier transform: to shape out the point of discontinuity, we need a continuous set of infinitesimally short monochromatic waves, that is, waves with infinite k. In other words, the k-spectrum of any nonanalytic function is unbounded. Therefore, we expect that the front speed v_F must be equal to the phase velocity of the infinitesimally short waves, that is,

$$v_F = \lim_{k \to \infty} u(k). \tag{11.74}$$

We will prove a fundamental theorem stating that this is, indeed, the case [15, 16].

Note that the actual measurement of the front velocity may be a tricky problem. To identify its instant position, we need to identify the field at the edge where it starts from zero (Figure 11.9). If the field remains a continuous function of position, then in the vicinity of the front it is close to zero and its detection at this moment will accordingly require an infinite sensitivity of a measuring device.

We will consider a one-dimensional case of a pulse in a homogeneous medium. As we know (Section 10.2 and 10.3), a wave function of any particle can be represented by a system of N components $\psi_1, \psi_2, \ldots, \psi_N$, which satisfy the corresponding wave equation or system of N linear homogeneous equations of the first order:

$$\sum_{m=1}^{N} \left(a_{jm} \frac{\partial \psi_m}{\partial t} + b_{jm} \frac{\partial \psi_m}{\partial x} + c_{jm} \psi_m \right) = 0, \quad j = 1, 2, \ldots, N. \tag{11.75}$$

The equations may be of higher order, but we can always denote the first order derivatives as independent functions, which will satisfy the first-order Equation (11.75) (recall the situation with Maxwell's and the Klein–Gordon equations, Section 6.8 and 10.4).

Note that the time variable enters (11.75) on equal footing with the spatial variables. In other words, the equations describing the wave are Lorentz covariant.

Consider a solution as a monochromatic wave

$$\psi_m(x, t) = A_m e^{i(kx - \omega t)}. \tag{11.76}$$

Putting this into (11.75) yields the system of algebraic equations determining the amplitudes A_m:

$$\sum_{m=1}^{N} (\omega a_{jm} - k b_{jm} + i c_{jm}) A_m = 0, \quad j = 1, 2, \ldots, N. \tag{11.77}$$

Since the system is homogeneous, a nontrivial solution $A_m \neq 0$ can only exist if its determinant is zero:

$$\text{Det}\left[a_{jm}u - b_{jm} + i\frac{c_{jm}}{k}\right] = 0. \tag{11.78}$$

Here,

$$u = u(k) \equiv \frac{\omega}{k} \tag{11.79}$$

is the *phase* velocity, whose dependence on k (or ω) is determined by Equation (11.78). Since this dependence produces what we call the dispersion of waves, Equation (11.78) is called the dispersion equation.

Now we can ask how is all this related to the *front* velocity v_F?

Motion of the front is represented by a world line F on the (ct, x)-plane (Figure 11.13a). By our definition of wave front, this line is singled out by the condition that on one side of it all the amplitudes A_m are equal to zero, whereas on the other side at least one of them is nonzero (in this respect the function describing the wave profile must look similar to a Green's function, see Appendix F, Figure F2).

Now, we can show that if the wave dispersion is given by the function $u(k)$, then the front of any pulse is moving at the speed

$$v_F = u(\infty). \tag{11.80}$$

We will make use of the Cauchy theorem: suppose all ψ_m are given on a curve Q of the (ct, x)-plane; then all $\psi_m(x, t)$ are uniquely determined on all the plane by their values on this curve. An important feature of this theorem is that it may have some exceptions: there exists one or more special curves, C, in the plane, such that setting the values $\psi_m(C)$ on them does not determine a unique solution in the rest of the plane. Such curves are called the *characteristics* of the differential Equation (11.75).

Here, we have arrived at the crucial point: the curve F representing the motion of the front must be a characteristic, that is, $F \in C$. Indeed, by definition, all $\psi_m(F) = 0$ on the line F. But for these values there are *two* different solutions: first is the pulse in question and the second is the trivial solution $\psi_m(x,t) \equiv 0$. Thus, the solution is not uniquely determined and therefore F is a characteristic.

The question now arises: how can we find characteristics?

Consider a curve on the (ct, x)-plane with the segment Q containing a point $A(ct_0, x_0)$ (Figure 11.17). The values $\psi_m(Q)$ are given on the whole segment Q. We will denote the values of ψ_m at point $A(ct_0, x_0)$ as $\psi_m^{(0)}$.

Expand the sought-for solution in Taylor series:

$$\psi_m = \psi_m^{(0)} + \left(\frac{\partial \psi_m}{\partial t}\right)_0 (t - t_0) + \left(\frac{\partial \psi_m}{\partial x}\right)_0 (x - x_0) + \cdots. \tag{11.81}$$

We need to find the expansion coefficients, starting from the first derivatives. We can do it, moving along the segment Q, since the functions ψ_m are determined on it.

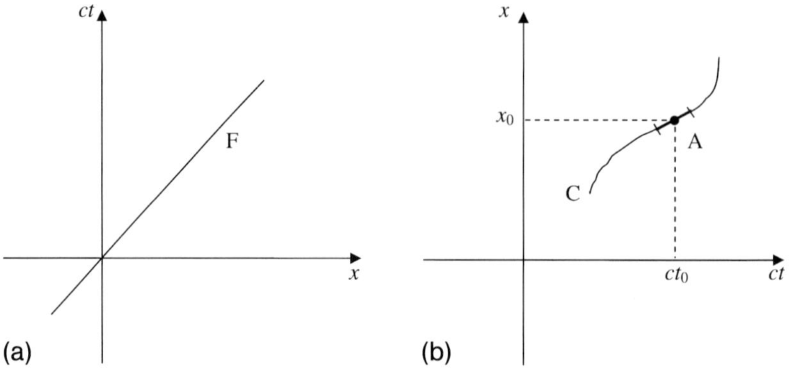

Figure 11.17 (a) The world line of a point on the front; the direction of the motion of the front is taken as the *x*-direction; (b) one of the characteristics of the differential Equation (11.4).

So let us shift along *Q* starting from point A. We can find the perfect differential along the way:

$$d\psi_m^{(0)} = \left(\frac{\partial \psi_m}{\partial t}\right)_0 dt + \left(\frac{\partial \psi_m}{\partial x}\right)_0 dx. \tag{11.82}$$

Introducing the slope $s = dx/dt$ of curve *Q* at A, we can rewrite (11.82) as

$$\left(\frac{\partial \psi_m}{\partial t}\right)_0 = \frac{d\psi_m^{(0)}}{dt} - \left(\frac{\partial \psi_m}{\partial x}\right)_0 s. \tag{11.83}$$

Note that s is known from the equation of the curve containing the segment *Q* and the full derivative $d\psi_m^{(0)}/dt$ is known only when we move along this segment.

Now we can find the partial derivatives figuring in (11.83). Putting (11.83) into (11.75), we first eliminate $(\partial \psi_m/\partial t)_0$ and obtain the equation for $(\partial \psi_m/\partial x)_0$:

$$\sum_{m=1}^{N} \left\{ (-a_{jm}s + b_{jm}) \left(\frac{\partial \psi_m}{\partial x}\right)_0 + \left(a_{jm} \left(\frac{d\psi_m^{(0)}}{dt}\right) + c_{jm}\psi_m^{(0)} \right) \right\} = 0. \tag{11.84}$$

This system of linear inhomogeneous equations has a unique solution if its determinant is nonzero. In this case, we first determine $(\partial \psi_m/\partial x)_0$ and then determine $(\partial \psi_m/\partial t)_0$ from Equation (11.83). Reiterating the process, we can then obtain all higher order derivatives of ψ_m at A, that is, find the coefficients for Taylor expansion of ψ_m.

But our initial goal was to find characteristics, among which there is line *F* for the front motion. To this end, we now require the opposite – that the determinant of system (11.84) *be* zero, that is,

$$\mathrm{Det}[a_{jm}s - b_{jm}] = 0. \tag{11.85}$$

In doing this, we switch from known s on the given curve Q to yet unknown s determining *N* local slopes of the sought-for characteristics. We know they determine characteristics, because the requirement (11.85) precludes the existence of unique solutions of (11.84), and the absence of uniquely determined solutions is the definition of a characteristic. Since the line F is a characteristic, one of the N solutions of (11.84) for s is the s_F, that is, the speed u_F of the front. Finally, comparing (11.85) with (11.78), we see that $v_F = u(\infty)$, since (11.78) becomes (11.85) when $k \to \infty$. This completes the proof.

The obtained result is of paramount importance. It says that all we need to find the front velocity is the velocity of very short monochromatic waves in a given medium. The latter velocity is relatively easy to determine from the index of refraction $n(\omega)$ as a function of ω or k. We do not even need to know this index for *all* ω – it is sufficient to know it only for very high frequencies or, which is the same, for high k. And, as we will see later from a very simple argument, for the EM waves in any medium the refraction index at high enough frequencies is just equal to one. Which means that the front velocity $v_F = c$. And this, in turn, means that even when we have a pulse running faster than light, the actual electromagnetic signal moves with the invariant speed c and not a single bit faster.

11.8 What Constitutes a Signal?

The examples considered above are only a small part of numerous experiments on superluminal group velocities. As an example, we can mention very interesting experiments by Nimtz [114].

Below we discuss the question whether superluminal group velocities contradict relativistic causality, which states that a signal cannot be communicated faster than the invariant speed.

To answer this question, we need to find out whether a wave packet or pulse with superluminal group velocity carries a superluminal signal. In other words, we ask about the relation between group velocity and signal velocity. This will naturally help us answer a more general question: what constitutes a signal?

The superluminal group velocities are not something entirely new. Actually, the analysis of the refraction index (Appendix F and G) *predicts* the possibility of group velocities exceeding the invariant speed. This analysis shows that the index of refraction is a complex function

$$n(\omega) = n_r(\omega) + i n_i(\omega), \tag{11.86a}$$

whose real part determines the phase velocity and imaginary part determines the attenuation of the EM wave. Since $k = (\omega/c)n$, the same is true for the propagation number k.

The general definition of group velocity is

$$v_g = \frac{d\omega}{dk_r}. \tag{11.86b}$$

For a medium with refraction index n we have used Equation (G31) (Appendix G)

$$v_g = \frac{c}{n_r + \omega\,(dn_r/d\omega)}. \tag{11.87}$$

We see that the group velocity is determined not only by n_r, but also by the sign and magnitude of the derivative $dn_r/d\omega$. In cases when we are close to the absorption line, the denominator of fraction (11.87) can be very small and the group velocity may become $v_g > c$.

For the sake of simplicity, we consider the expression obtained in the Appendix G for the two-level system, that is, a system with only one optical transition $\omega_{12} = (\varepsilon_2 - \varepsilon_1)/\hbar$:

$$n(\omega) \cong 1 + \frac{q_e^2 f_{12}}{2m_e\varepsilon_0}\frac{N_1 - N_2}{\omega_{21}^2 - \omega^2 - 2i\gamma\omega}. \tag{11.88}$$

Denoting

$$Q \equiv \frac{q_e^2 f_{12}}{2m_e\varepsilon_0}(N_1 - N_2). \tag{11.89}$$

we obtain in the case of a diluted medium the expressions for real and imaginary parts of n:

$$n_r(\omega) = 1 + Q\frac{\omega_{21}^2 - \omega^2}{(\omega_{21}^2 - \omega^2)^2 + 4\gamma^2\omega^2}; \qquad n_i(\omega) = Q\frac{2\gamma\omega}{(\omega_{21}^2 - \omega^2)^2 + 4\gamma^2\omega^2}. \tag{11.90}$$

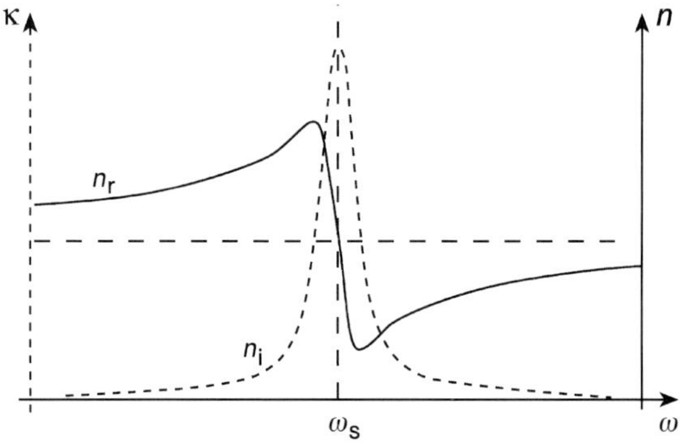

Figure 11.18 Two optical properties of a medium – the coefficient of absorption $\kappa = n_i$ (dashed line), and refractive index n_r (solid line) are plotted against frequency w. All three quantities are expressed in arbitrary units. The value of w_s is the frequency at which the absorption reaches its peak. Within the absorption band (the frequency range with high absorption), the index of refraction undergoes a rapid change. In mathematical terms, the derivative dn/dw and thereby dk/dw becomes very large in this region. According to Equation (11.87), this corresponds to small group velocities.

The graph of function $n_r(\omega)$ is shown in Figure 11.18. We see that the sign and magnitude of the slope close to the absorption center $\omega = \omega_{21}$ determine the value of the group velocity. Close to the center of the absorption linc this simplifies to

$$n_r(\omega) \approx 1 + \frac{Q}{2\omega_{21}} \frac{\omega_{21} - \omega}{(\omega_{21} - \omega)^2 + \gamma^2}; \qquad n_i(\omega) \approx \frac{Q}{2\omega_{21}} \frac{\gamma}{(\omega_{21} - \omega)^2 + \gamma^2}. \tag{11.91}$$

Denoting $\Delta = \omega - \omega_{21}$ and putting this into (11.87), we obtain the following expression for the group velocity close to narrow absorption line ($\Delta \ll \omega_{21}$):

$$v_g = \frac{c}{1 + \frac{Q}{2}(\Delta^2 - \gamma^2)/(\Delta^2 + \gamma^2)^2}. \tag{11.92}$$

Consider a few special cases.

A. $Q > 0$ ($N_1 > N_2$): the medium is in, close to, or not outrageously far from thermodynamic equilibrium. Then

$$v_g = \begin{cases} \dfrac{c}{1 - Q/2\gamma^2} > c, & \Delta = 0 \\ > c, & |\Delta| < \gamma \\ < c, & |\Delta| > \gamma \end{cases}. \tag{11.93}$$

B. $Q < 0$ ($N_1 < N_2$): the medium is inversely populated (the absolute temperature $T < 0$), that is, very far from thermodynamic equilibrium, utterly unstable. In this case,

$$v_g = \begin{cases} \dfrac{c}{1 + |Q|/2\gamma^2} < c, & \Delta = 0 \\ < c, & |\Delta| < \gamma \\ > c, & |\Delta| > \gamma \end{cases}. \tag{11.94}$$

In case B, to observe the superluminal pulse propagation, the central frequency of the wave packet must be sufficiently detuned from the center of the absorption (now amplification) band. The physical reason for it that, as seen from Figure 11.18, in case B the curve $n_r(\omega)$ is the mirror reflection of that in case A. Accordingly, $dn_r(\omega)/d\omega > 0$, and Equation (11.92) for the group velocity gives $v_g < c$ for all $|\Delta| < \gamma$; it can give $v_g > c$ outside the band, where $|\Delta| > \gamma$.

In most of the experiments, the superluminal pulse propagation can be represented by a pair of snapshots shown in Figure 11.19. We see from this figure that the snapshot b at a moment $t_0 > 0$ can be actually constructed from the for-runner – the far-ranged leading slope of a at the earlier moment $t = 0$ rather than from the *whole* snapshot a at the earlier moments. How can that be? Answer: because the original function describing the shape a is analytic. We know from the Taylor expansion of an analytical function $f(x)$ that any even infinitesimally small region $\Delta x \to 0$ of this

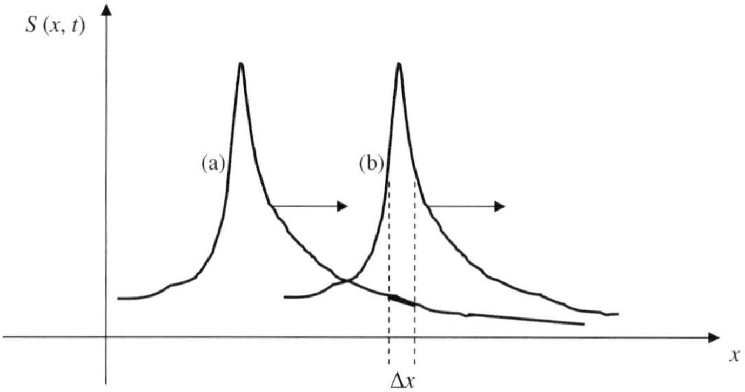

Figure 11.19 A possible shape of a superluminal pulse is represented by analytical function $S(x, t)$ at two different moments of time (shown here is a special case with no pulse distortion). (a) Instantaneous shape as a function of x at the moment $t = 0$; (b) as a function of x at a later moment $t_0 > 0$. The small segment Δx of the curve (a) at $t = 0$ contains full information about the rest of this curve, including its top. Therefore, the actual arrival of this top at $t = t_0$ does not bring in any new information.

function (knowing $f(x)$ within any region $\Delta x \to 0$, Figure 11.19) is sufficient to completely reconstruct $f(x)$ in all range of its existence. Indeed, if we know $f(x)$ within $x_0 - (1/2)\Delta x < x < x_0 + (1/2)\Delta x$, we can find $f(x)$ and all its derivatives $f^{(l)}(x_0) = d^l f(x)/dx^l$ at x_0. And knowing them enables us to find $f(x)$ everywhere:

$$f(x) = \sum_{l=0}^{\infty} \frac{1}{l!} f^{(l)}(x_0)(x - x_0)^l. \tag{11.95}$$

Thus, any arbitrarily small region of a pulse with analytic shape $f(x)$ stores all the information about the whole pulse. It is as if a small patch of a painting contained the information about the whole painting. This could, in principle, be the case, if the paint distribution over the canvas, in terms of color and saturation, were an analytic function of position. But in this case it would hardly be a piece of art.

An optical hologram can serve as another example. Its small area element already contains all information necessary for reconstruction of the whole picture if the hologram is broken. Using this language, we can compare a superluminal pulse described by an analytical function with a holographic plate. As each small part of a hologram contains "the whole thing," so does a small part (highlighted segment in Figure 10.19) of such a pulse.

The analogy can be extended in terms of energy. Each splinter of a hologram encodes practically all information about the whole imaged object; in this respect, all patches are the same and the rest of the hologram may seem to be redundant. However, it may be necessary in terms of the *brightness* – each part, adding nothing in terms of new information, adds up to the net brightness of the resulting image. Similarly, a small part of the pulse may be enough in terms of the encoded information, but the whole pulse may nevertheless be necessary to get a noticeable response of the recording system.

Discussion 1

Using the "time language" (the temporal succession of events within a fixed locality), we can compare a sensitive detector that can predict the arrival of the top already by recording its far-advanced leading slope with a weather-sensitive person who can predict a storm a couple of days before its arrival. Nobody would say that this person receives a superluminal signal from the coming storm. The person just receives the information from rheumatic pain or headache caused by preceding changes in atmospheric pressure, humidity, or other relevant weather changes associated with the approaching storm.

Using the "holographic" language, we can consider in more details the above-mentioned example with a good painting. Unlike a hologram, the different patches of the painting do not reiterate.[4)] In order for a painting to be a piece of art, it must be impossible to reconstruct the image on the whole canvas from the image on a small patch of the canvas. In this respect, physics provides us with at least one criterion necessary to define what constitutes art. I doubt that art can be rigorously defined in its totality, but at least one condition necessary for something to be art can be defined.

To see this condition clearly, consider one more example. Paleontologists have reconstructed the shape and functions of many extinct species only from a few fossilized bones they managed to find. They were able to do it only because there is a connection between the structure of an organ or bone and its function, as well as connections between different parts of an organism. Those creatures were natural product of evolution – they did not constitute art (we are talking here about *human* art).

Thus, one of the distinctions between paleontology and art is that, by contrast with the extinct species, it is impossible to reconstruct a piece of art from only one small part of it. A photograph of only one square inch of a painting by Rembrandt would be absolutely insufficient to reconstruct the whole painting. The color and intensity distribution over this painting are *not analytic.* On the other hand, the Malevich's "Black Square," widely acclaimed as a masterpiece, is analytic (holomorphous) almost over the whole area of the painting, except for the square's border. Just a few kilobytes of information about this border would be sufficient to reconstruct the whole painting from its small patch. It is true that on the border itself, as if to compensate for the scantiness of information, it is violently nonanalytic – it is discontinuous already in the zero-order derivative of the intensity and color distribution over the canvas.

This comparison helps us outline a distinction between a true piece of art and a pretender. A true piece of art is unpredictable in all its parts. It is beaming with signals conveying hundreds of kilobytes of quantified information, which affects our emotions without saying it explicitly. By contrast, pretenders have in all, at best, a few kilobytes, but so explicitly exposed that it literally flies into your face.

4) This statement should be used with care. While the rest of a hologram merely reiterates the contents of its small patch, the patch itself may contain lots of information, in which case it would carry a signal. On the other hand, a painting may have reiterating elements as a part of the artistic design.

In this respect, the above-discussed superluminal pulses, in terms of predictability of the whole pulse from its small region, are similar to an optical hologram or to a coming storm – they should be ranked even below the pretenders. As is the case with a hologram, it is possible to build up the pulse's whole shape from only a small part of its leading slope. Therefore, recording its top does not give us any new information above what had already been recorded in the for-runner. And, therefore, the arrival of the bump ahead of schedule (earlier than $t_c = x/c$, where x is the separation between the source and receiver of the pulse) does not constitute a signal.

Summary

A pulse whose shape is represented by an analytical function (a typical example is the Gaussian pulse), whether this pulse is subluminal or superluminal, does not carry a signal, since the arrival of its top does not bring any new information.

Discussion 2

Of course, the actual situation is more subtle. The top of the bump is much higher than its far-advancing slope and, accordingly, transports more power. Where does this power come from? In the case discussed in Ref. [1], when the medium had itself been pumped up with energy beforehand, the acceleration of the top of the bump beyond the light barrier could be understood as the result of "skimming" the medium by the pulse's leading slope. This reshapes the pulse owing to its growing leading slope. Eating away the energy of the trailing slope is another possible mechanism leading to effective pulse acceleration beyond the light barrier.

In some experiments, the superluminal pulses with fixed shape have been observed. What is the energy of the coming top built up from in such situations? The only possible answer is that, in this case, it comes from the top's previous position at $t' = t - x/v_g$. But, since this energy, after passing through $x = 0$ shows up at x sooner than it would take the light in vacuum to travel this distance, we have a superluminal *energy transfer*. Is this not equivalent to signal transfer? Even if it does not transfer any new information, should not we extend our definition of a signal onto the cases when just energy is transferred between places, even without accompanying information? After all, we can use this energy for changing the preexisting state at a given locality.

A possible answer is: in vacuum, yes. But it does not happen in vacuum! The discussed superluminal transportation has been observed in a medium, so let us discuss this case more carefully. Suppose our detector is not sensitive enough to see the leading slope of the coming pulse and reacts only to high intensity in the pulse's top region. Then, it seems that the information in the advanced slope cannot be used to reconstruct the whole shape with the top before the top itself arrives "in person." In this case, we could not avoid the conclusion that we have, indeed, recorded a superluminal signal.

This argument, however, would not be exactly accurate. A detector eventually responds to a certain amount of absorbed energy, not to the instant power. Precisely because the intensity in the leading slope is much lower than at the center of the pulse, its passing time is accordingly longer. The net amount of energy received by the detector from the slope and necessary for it to respond may be comparable with and in some cases even greater than the net amount in the top's region. But its accumulation time is so long that it clicks when the top is almost here. Such a click may be the response to the energy flux that has been streaming through it long before the top's arrival, rather than to the top itself. Therefore, we cannot say that whatever triggered the detector was transported by the top.

If, however, we make the detector so sensitive that even a very low slope is already sufficient for detection, then, by the same token, it may be sufficient to extract all the information there in the whole pulse. Taking account of this fact brings us back to the necessity of including the for-running slope into full consideration.

Another related issue is the noise in a detector. A detailed discussion of these important questions can be found in a very interesting book by Milonni [18].

...So, what constitutes a signal?

From what we saw, the necessary requirement is that a signal from a system A received by a system B must include an observable agent propagating from A to B. The arrival of this agent at B must change the physical state of B in a way that *could not have been predicted* before a certain moment of time t'. The traveling agent must contain the new information in such a form that it is not available before the moment t'. This critical moment of time, such that there is *nothing* before it but something after it, can be called the *time of arrival*. Quantitatively, the profile representing the agent must have no precursor that could be felt before this moment. A function of this type is a special case of nonanalytic functions. So, mathematically, one of the essential characteristics of a signal must be its *nonanalyticity*. In this respect, the response functions (or, which is the same, the temporal Green's functions) considered in Appendix F, are signal carriers, – they are nonanalytic!

So far, we have discussed a classical signal. In the case of light, this means a signal involving an astronomical number of photons, so that we remain in classical domain dealing with predictable expectation values. How would this change when we dim the pulse so that it contains only one photon, or, more generally, one particle?

This question takes us to the study of the possibility of superluminal signaling via position measurements in QM. We already discussed the corresponding situation in Section 11.6 and we found that the criteria for a wave packet (representing a single particle) to be a signal carrier are the same as for a classical pulse: it has to have a sharply defined front.

We can thus summarize this part: true signal must be associated with the wave front defined as a discontinuity point in the profile of the pulse.

Now we can turn to the final question: what is the propagation speed of this point. If it turns out to be faster than the invariant speed, the causality would be violated.

We have a powerful and at the same time simple tool to answer this question: it is the fundamental result obtained in the previous section that the speed of the front is just the limit of phase velocity at $\omega \to \infty$. All we need is to determine in each case the phase velocity $\omega/k\,(\omega)$ at high frequencies.

Let us do it! Suppose, we have a QM wave packet initially prepared in a state shown in Figure 11.20.

We want to find the propagation speed of the edge of the packet. The beauty of the fundamental theorem in Section 11.7 is that we do not need to figure out the evolution

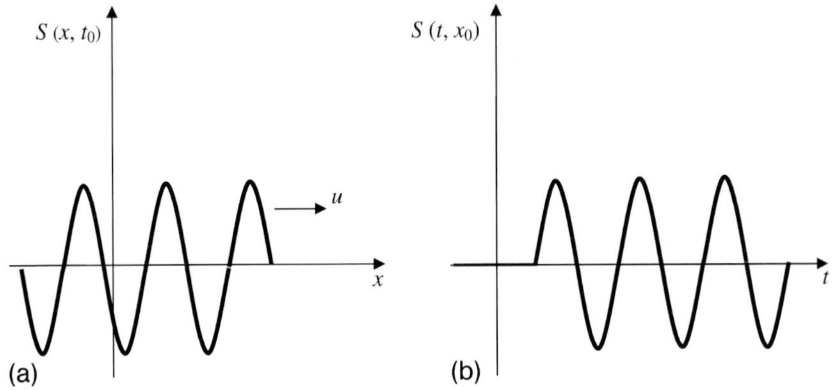

Figure 11.20 A possible shape of a signal-carrying perturbation $S(x, t)$: (a) instantaneous shape as a function of x at a moment $t = t_0$; (b) as a function of time t at a fixed point $x = x_0$.

of the packet, which may be an extremely difficult problem. All the information we need is the mass of the particle.

Start with a nonrelativistic case of a free scalar particle. The phase velocity of associated de Broglie's wave is

$$u = \frac{\omega}{k} = \frac{\mathcal{E}}{\hbar k}. \tag{11.96}$$

For a free nonrelativistic particle the energy $\mathcal{E}$ is its kinetic energy:

$$\mathcal{E} = K = \frac{\hbar^2 k^2}{2m}, \qquad \omega = \frac{\mathcal{E}}{\hbar} = \frac{\hbar k^2}{2m}. \tag{11.97}$$

Putting this into (11.96) gives

$$u = \frac{1}{2}\frac{\hbar k}{m}. \tag{11.98}$$

Now, we see from (11.97) that infinite ω corresponds to infinite k. Therefore, it is sufficient to consider the last expression for infinite k. This gives us the result

$$u_s = \lim_{k \to \infty} u(k) = \infty. \tag{11.99}$$

If we manage to prepare the particle's initial state so that it has a sharp edge of the type shown in Figures 11.9 or F2, then the next moment the particle could be found in another galaxy. And the basic requirement – nonanalyticity – necessary for this process to constitute a signal is satisfied. Have we come to a contradiction with the theorem we had discussed? The answer is no. The only thing the theorem says is that the signal moves with the speed corresponding to infinitesimally short waves. It does not regulate this speed itself. So there is no contradiction with the mathematics of the theorem.

Is there any *physical* contradiction – does the result contradict causality? The answer is no. We are in the nonrelativistic domain that allows any speeds! The result we obtained only shows that the expression (11.97) we used for the energy loses its validity at high frequencies. We can even estimate the order of magnitude of the frequencies or corresponding wavelengths at which the nonrelativistic description should be abandoned. Recalling Section 10.2, we see that we should switch to the relativistic description a way before k reaches the value corresponding to the Compton wavelength. So we can take

$$k < k_c = \frac{2\pi}{\lambda_c} = 2\pi\frac{mc}{\hbar}. \tag{11.100}$$

To be on the safe side, let us take as the maximal valid value $k = k_c/\pi$. Putting this into (11.98), we immediately obtain

$$u_s = c. \tag{11.101}$$

Of course, this is somewhat of a trick – I had known in advance what to put into (11.98) to get this result. Nevertheless, the result is valuable in that it shows what we would get if we had from the very beginning used the correct relativistic expression for the particle's energy.

So let us now consider the same initial shape, but now use the *relativistic* relation between ω and k:

$$\omega = \frac{1}{\hbar}\sqrt{m_0^2c^4 + \hbar^2c^2k^2}, \quad \text{so} \quad u = \frac{\omega}{k} = c\sqrt{1 + \frac{1}{\lambda_c^2 k^2}}. \tag{11.102}$$

Here, the Compton wavelength from the very beginning explicitly enters the expression. According to this expression, large frequencies also correspond to large k. At $k \to \infty$ we immediately get the exact result (11.101) – this time without any tricks!

Consider now optical experiments with superluminal group velocities. As such, they do not contradict anything, since a group velocity is not necessarily the signal velocity. As to the signal velocity, it needs the profile with a front – with a sharp edge similar to that in Figure 11.9. The speed of the edge is given by the expression

$$u_f = u_s = \lim_{k\to\infty} u(k) = \lim_{\omega\to\infty} \frac{c}{n_r(\omega)}. \tag{11.103}$$

Using Equation (11.90) we see that the real part of the refraction index is going to 1 at high frequencies. The front of an EM wave in any medium propagates with the invariant speed.

Physically, the last result could have been predicted from the very beginning. We know that the optical properties of any medium, including its ability to modify the wave propagation speed, arise from the oscillations of the induced dipoles. Each oscillating particle in this process has a finite mass and thereby finite inertia. At high enough frequencies the atomic oscillators do not have time enough to follow rapidly changing acting EM force. At $\omega \to \infty$ they just stop responding to the incident waves and the medium for such waves becomes as good as pure vacuum.

Summary

We can now summarize the whole discussion by formulating the fundamental difference between a signal and a group velocity.

The group velocity of a wave packet or pulse, according to its definition (11.86) and (11.87), has a physical meaning only when we have a group of waves within a narrow frequency band. The corresponding functions in the configuration space are called functions with restricted spectrum [115]. Graphically, such a function is very narrow in the frequency space and broad in the configuration space. It is clear from this description, as well as from the graphical representation, that the corresponding wave packet can have a very long for-runner in the form of a leading slope.

By contrast, a signal must have a discontinuity, that is, a sharp boundary in configuration space. This automatically implies unlimited range in frequencies, so while it may be narrow in configuration space, it is infinitely broad in the frequency space. In this case, the group velocity loses its meaning as a characteristic of the group, since it may have quite different values for different frequencies. But, even more to the point, at high enough frequencies, as we know, the phase velocity in that region of spectrum cannot exceed the invariant speed. Accordingly, in this limit we have

$$\mathrm{d}\omega/\mathrm{d}k \underset{k\to\infty}{\longrightarrow} \omega/k \to c.$$

Thus, even if we want to stick to the notion of the group velocity for such pulse, it is actually equal to the maximal signal velocity in free space – to the invariant speed.

Therefore, if someday an experiment would show the velocity of sharply discontinuous front exceeding the invariant speed, it would be a new scientific revolution.

11.9 Direct Signaling Into the Past

... Tom had been fascinated by the reports of observed pulses with superluminal group velocities; but then came disappointment when he read other research papers showing that these group velocities were useless for faster-than-light communications.

Tom has learnt that in order for a group of waves to form a signal one needed to concoct a pulse with a discontinuity (the wave front) necessary to send a new message and any such front requires an unbounded range of frequencies. He also learnt that the front cannot travel faster than the invariant speed. Farewell to round-trip time travels and superluminal communications!

After a few days of intense reading and thinking Tom came up with an idea that he thought could get around the relativistic ban on superluminal signaling. His project seemed as powerful and promising as it was simple. It was simple both conceptually and experimentally. Tom's idea was to turn the vice into a virtue. Instead of trying to get the whole bunch of frequencies to form something that cannot outrun light anyway, he suggested to select a very narrow band, ideally only one frequency $\omega = \omega_0$, out of the bunch and let this frequency alone travel in empty space, while blocking all the others. Tom believed that this could easily be done by making a filter that would absorb nearly the whole spectrum, except for only one very narrow spectral line centered on a selected frequency ω_0. If such a line could be made extremely narrow, it would be just one monochromatic wave. But a monochromatic wave is filling the whole space and lasting forever – its undulations extend into the future and into the past! All we need for the signaling is to extract this frequency from all the rest. This seems simple. Just find a plate of material opaque for the light in all ranges of spectrum except for only one frequency ω_0 for which it would be transparent and then place it in the way of a laser pulse. Suppose we have prepared a very sharp pulse – the graph of its intensity versus time looks like a sharp spike. In the absence of the filter, the spike would move without any for-runners, because all its constituent Fourier components interfere destructively, totally cancelling each other everywhere outside the current position of the pulse (the way such cancellation works is described in many sources, e.g., [115]). So there is no perturbation ahead of the pulse until its front arrives at a given point and this happens no earlier than the time t_c necessary to travel to this point with the invariant speed. An important thing here is that *all* the constituent frequencies are needed for the total cancellation of all perturbations before the front. If at least one of them is exempt, that would immediately lead to a nonzero perturbation. The same will happen if only one of them is left, whereas all the rest are removed. In the first case, the one removed frequency was balancing all the rest; in the second case, all the rest were balancing the one selected. Both cases are, of course, equivalent and both seem suitable for Tom's goal. So Tom's idea could work both ways. If, for whatever reason, it was difficult to manufacture a filter transmitting only one frequency while absorbing all the others, then the "reciprocal" filter transmitting all the light except for only one frequency that was absorbed, could do the job as efficiently.

Now suppose that we do have such a filter. Let a sharp laser pulse (with a nonanalytic profile) hit it at the zero moment of time. Let the position of the filter be the origin of our coordinate system with the x-axis along the direction of the motion of the pulse. Since there is absolutely no intensity ahead of the front, then, *without* the filter, it would be impossible to know anything about the pulse before it comes, so an observer (Alice) positioned on the other side of the origin will be unaware of the pulse before the moment $t=0$; and if she is at a distance x from the origin, she will not receive any signal before the moment $t=t_c=x/c$. Alternatively, if she records the pulse arrival at $t=0$, the pulse must have been emitted *before* that moment at $t=-x/c$.

But here comes Tom and places his filter at the origin. The filter must change the whole situation. No frequency except for ω_0 passes beyond it, so the wave with the frequency ω_0, once on the other side, sees no competition and keeps on running without any interference. Moreover, once this wave is monochromatic, it must not only run forever in the future, but also must have had existed all the time in the past! Alice would have known about Tom's action *before* the action itself, regardless of how far away from the origin she is (Figure 11.21).

In terms of signaling, this is much better than just sending a superluminal signal. For an observer sending such signal from the system K, it would move into the future, faster than light but still into the future. We know that it could be observed before the moment of its emission only by another observer in different system K′; and K′ cannot be just another system – it must satisfy the condition $V>c^2/u$, where u is the superluminal velocity of the signal in K and V is the relative velocity of the two frames. And even when all these conditions are met, the signal detection in K′ is observed as a spontaneous acausal event – the emission without any appropriate prehistory. By

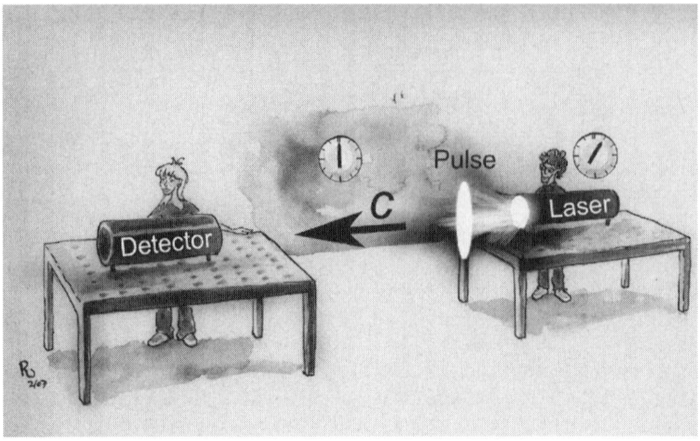

Figure 11.21

contrast, in Tom and Alice's experiment, the signal appears to be sent directly into the past, because Alice can detect the light from the filter long *before* the filter is hit by the pulse – and all this while being in *the same* system K as Tom. In addition, the signal does not appear to be a spontaneous emission from Alice's detector – it keeps coming *toward* her from Tom's filter all the time, so its cause must be in that filter, but this cause starts acting (for both Alice and Tom) *after* Alice observes its effect. In principle, the incoming radiation from the filter can go into a far more distant past than Alice and Tom's time – back to pyramids, dinosaurs, and so on (Figure 11.22).

Tom became even more excited when he compared his idea with descriptions of travels into the past using hypothetical tachyons. His idea seemed more attractive than that of tachyon signaling for a few reasons.

First, to send a tachyon signal into the past from the rest frame K of the emitter, we actually need at least two tachyons and two different reference fames. The auxiliary frame K′ must have a tachyon detector and emitter and harbor an observer with very specific instructions as to the emission moment, direction, and speed of the secondary tachyon. Recall Section 11.3 with the Peter–Sam thought experiment

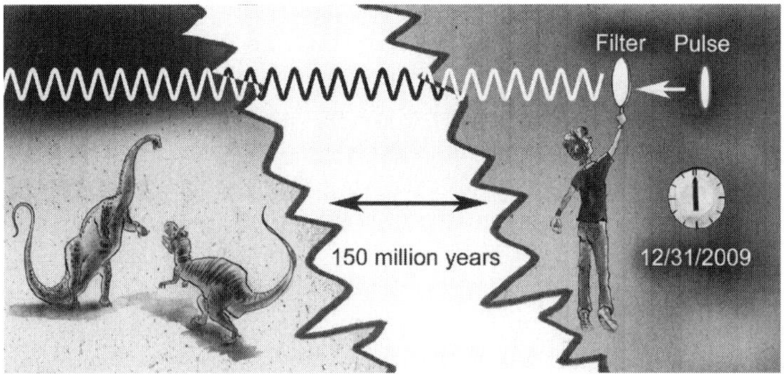

Figure 11.22

illustrating the Tolman paradox. Only via employing the K′-observer, the secondary tachyon would be observed by the K-observer in the past, but again only as a spontaneously emitted particle with no prehistory justifying such an emission. By all accounts, it would not be a *direct* signaling into the past. All these complications are absent in Tom's proposal.

Second, the mere possibility of using tachyons for controlled faster-than-light signaling was shown to be, at best, problematic [1, 108, 116]. In addition, another argument against such possibility was presented in Section 11.4 of this book.

Third, the tachyons up to now remain purely hypothetical objects.

Alice and Tom have decided to discuss Tom's idea with their friends, including Peter, Paul, and Sam. After discussion, the team came to the conclusion that the idea, beautiful as it looks, would not work.

And the reasons, as you think a little more about it, are almost obvious. Even if a filter suggested by Tom could exist, the wave transmitted by it is evidently not a signal. By suggestion, it must be monochromatic, but as we know such wave does not convey any information. Imagine yourself sitting somewhere on another planet, or on Earth some thousand years BC and observing a monochromatic wave passing by from a certain direction. Even if you start looking in that direction, will you see anything? Nothing except for a steady uniform gleaming in the field of vision. Let us alter the condition: the filter passes a narrow band of frequencies. Such band may represent only a pulse whose profile is described by an analytic function. A pulse with such a profile, as we know, does not carry a signal either. For transmitted light to carry a signal, its coordinate profile must be nonanalytic. The moment you said this, you see immediately the fundamental flaw in Tom's initial idea: the Fourier spectrum of a nonanalytic function has unbounded range! We started with the wanted ideal filter passing only one frequency and wound up with the filter passing infinite range of them!

Then maybe we should consider a symmetric alternative mentioned above – the filter *transmitting* all frequencies except for only one. In this case, we would have on the transmitted side plenty of ingredients for constructing a profile with a sharp edge necessary to produce a signal. Alas, this will not work either. An unbounded spectral range is only necessary but not sufficient condition for constructing an edge. If the pulse with this spectral range needs only one additional monochromatic wave to be cancelled (recall that all the frequencies *including* ω_0 produce nothing ahead of the initial pulse), then its net effect is just one monochromatic wave complementary to that of frequency ω_0. Again, this is as far from a signal as you can get.

But there is still something in Tom's idea that, it seems, can be justified.

Even though the transmitted wave, being monochromatic, would not constitute a signal, it would still be a process extended well into a distant past and possibly acting on some objects in that past long ago before its cause (which is the incidence of the laser pulse onto the filter).

Alas, this argument makes things even more serious. Recall the Tolman paradox: any signal into the past is a potential source of logical contradiction. Tom's filter leads to even stronger statement: *anything*, be it a signal or not, sent into the past can produce logical contradiction. We said that Tom's monochromatic wave is not a signal since it does not carry information. And yet it creates an impossible situation. Indeed,

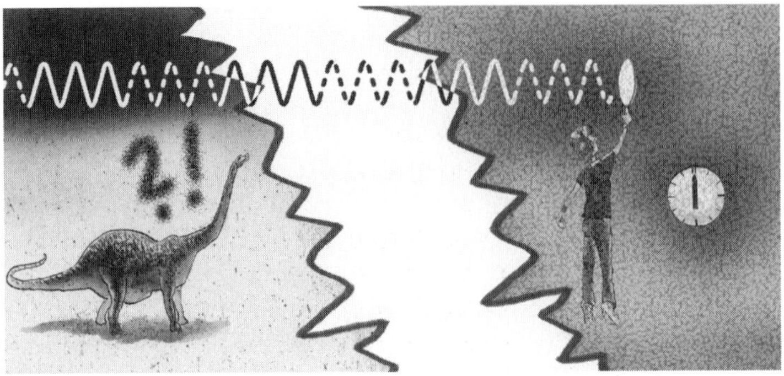

Figure 11.23

imagine yourself in a distant past, observing a permanent radiation with an extremely narrow frequency band. This radiation comes from Tom's filter. But Tom's filter has not been in place from time immemorial. Nor has the laser been shooting pulses right from the moment of the Big Bang. Neither Tom nor even his distant ancestors have been born yet, and of course, there are no lasers in the distant past.

Then a relevant question is what does this wave originate from?

The only possible answer is: the radiation you observe now, eons before Tom's birth, exists only because it is predestined that Tom in distant future will place his magic filter in front of a laser pulse.

But what if it turns out that it is impossible to create such a filter? Then, Tom cannot be predestined to perform such an action and the radiation you thought you observe does not exist (Figure 11.23).

So do you or do you not observe this radiation?

This can be considered as an utter ambiguity, but still not a direct contradiction.

However, the situation can be "promoted" up to the status of a real paradox.

For instance, the frequency ω_0 may be such that it affects in a certain way the metabolism of certain bacteria abundant billion years ago. As a result, the whole evolution of life on Earth takes entirely new direction and the human race does not appear. Then there is no laser pulse with magic filter in front and there is no resulting radiation. But then the evolution goes in its known track, producing Tom with his filter.

We can summarize such a situation as follows.

If Tom uses his filter, the evolution of life changes, the human race does not appear, and there is neither Tom nor a filter to use. If the filter is not used, the life evolves as known to produce Tom and technology, and the filter is used.

We have a self-contradictory situation similar to the Tolman paradox.

A consistent description of the real world requires that there must be something in nature precluding situations of the type just described. A filter with desired properties must be physically impossible. Quoting Milonni, "*It must...be impossible to have a perfect filter, one that absorbs one frequency without affecting any other frequency components of an input signal.* Any realizable filter must produce phase shifts in other Fourier components in such a way that they interfere destructively for all $t < 0$, so that, in fact, there is no output before any input [18]."

Now the question is what specifically forbids such a filter?

To answer it, we must go to Appendix E and F and consider the implications of the fundamental properties of a response function (known as the Green's function) discussed there. We will find that its ω-representation $G(\omega)$ considered as a function of complex variable ω is holomorphic (analytic) in one of the complex semiplanes. Now, it turns out that this leads to a specific connection between the real and imaginary parts of $G(\omega)$ on real ω-axis. At this stage, it may appear to be of only mathematical interest. But we know that polarization of the medium, which determines its optical properties, is the medium's response to an acting EM field. In fact, we have even expressed the medium's permittivity and refraction index in terms of the response function $G(\omega)$ of the corresponding oscillators (Equation (F44)). Therefore, we expect that these optical characteristics must have all the properties of a response function.

So consider such a function as described in Appendix F. It is holomorphic in one of the semiplanes ω, depending on whether it is retarded or advanced Green's function. Let, for instance, it be holomorphic in the upper semiplane (the ω-representation of the advanced Green's function, $\gamma < 0$). Then, it can be represented in this region ($\mathrm{Im}\,\omega > 0$, Figure 11.24) by Cauchy's formula:

$$G(\omega) = \frac{1}{2\pi i}\left[\int\limits_{C} \frac{G(\omega')}{\omega' - \omega}\,\mathrm{d}\omega' + \int\limits_{C'} \frac{G(\omega')}{\omega' - \omega}\,\mathrm{d}\omega'\right]. \tag{11.104}$$

The second integral here is zero, since ω is outside the contour C′. Now, letting the radii of semicircles C and C′ go to infinity, we notice from (11.104) that $G(\omega)$ falls off fast enough at $\omega \to \infty$ for the contributions from these semicircles to vanish in this limit. The integration thus can be reduced to that over two straight lines C and C′ (Figure 11.9b). This is all very well, but we are interested in real ω used in the description of physical processes. To get ω real, let us now take both lines C and C′ to the real axis, *retaining ω between them.* As a result, we obtain

$$G(\omega) = \frac{1}{\pi i} P \int\limits_{-\infty}^{\infty} \frac{G(\omega')}{\omega' - \omega}\,\mathrm{d}\omega'. \tag{11.105}$$

Here P denotes the Cauchy principal value of the integral. The frequency ω is already real and is a simple pole on the axis of integration.

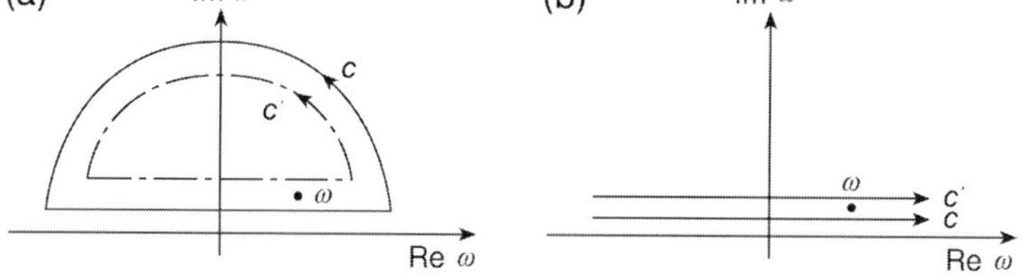

Figure 11.24

Now write down (11.105) separately for the real and imaginary parts of the response function:

$$G(\omega) = G_r(\omega) + iG_i(\omega). \tag{11.106}$$

Because of the imaginary factor in (11.105), we get

$$G_r(\omega) = \frac{1}{\pi} P \int_{-\infty}^{\infty} \frac{G_i(\omega')}{\omega' - \omega} d\omega'; \qquad G_i(\omega) = -\frac{1}{\pi} P \int_{-\infty}^{\infty} \frac{G_r(\omega')}{\omega' - \omega} d\omega'. \tag{11.107}$$

We can get the same result for the advanced response function as well (do it! see Problem 11.9).

The obtained equations are called the dispersion relations (not to be confused with dispersion equation between ω and k!). They are mathematically equivalent to either one of the two conditions:

$$G(\tau) = 0 \text{ at } \tau < 0 \quad \text{or} \quad G(\tau) = 0 \text{ at } \tau > 0. \tag{11.108}$$

Thus, the three different features of the response function $G(\omega)$:

1. to obey the dispersion relations;
2. to represent in frequency–space the Green's function $G(\tau)$ equal to zero on one of the half-lines τ;
3. to be holomorphous;

are, in essence, only one feature expressed in three different languages.

In practical applications, it is more convenient to consider only positive frequencies. Accordingly, the dispersion relations can be rewritten in the form involving integration only over positive ω. We know that a real input produces real output (e.g., real force produces real displacement). According to Equation (F10), this can be the case only if the Green's function $G(\tau)$ is real. Since $g(\omega)$ is its Fourier transform, this means that

$$G^*(\omega) = G(-\omega). \tag{11.109}$$

This converts into the additional (local) separate relations for the imaginary and real parts of $g(\omega)$:

$$G_r(-\omega) = G_r(\omega), \tag{11.110}$$

$$G_i(-\omega) = G_i(\omega). \tag{11.111}$$

Using this, we can write

$$G_r(\omega) = \frac{1}{\pi}\left[P \int_{-\infty}^{0} \frac{G_i(\omega')}{\omega' - \omega} d\omega' + P \int_{0}^{\infty} \frac{G_i(\omega')}{\omega' - \omega} d\omega' \right] = \frac{2}{\pi} P \int_{0}^{\infty} \frac{\omega' G_i(\omega')}{\omega'^2 - \omega^2} d\omega'. \tag{11.112}$$

Similarly, we have for G_i:

$$G_i(\omega) = -\frac{2}{\pi} P \int_{0}^{\infty} \frac{\omega\, G_r(\omega')}{\omega'^2 - \omega^2} d\omega'. \tag{11.113}$$

Now we are prepared to discuss physical implications. We know (Appendix F and G) that permittivity is a collective response of the medium to the applied field and, as such, it has all the properties of $G(\omega)$. It follows that the same dispersion relations must hold between the real and imaginary part of ε (and the same for μ). And, in view of the general relation $n^2 = (\mu\varepsilon/\mu_0\varepsilon_0)$, the index of refraction also turns out to be complex and obeys the dispersion relations [18]

$$n(\omega) = 1 + \frac{1}{\pi i} P \int_{-\infty}^{\infty} \frac{n(\omega') - 1}{\omega' - \omega} d\omega', \tag{11.114}$$

$$n_r(\omega) = 1 + \frac{1}{\pi} P \int_{-\infty}^{\infty} \frac{n_i(\omega')}{\omega' - \omega} d\omega', \tag{11.115}$$

$$n_i(\omega) = -\frac{1}{\pi} P \int_{-\infty}^{\infty} \frac{n_r(\omega') - 1}{\omega' - \omega} d\omega'. \tag{11.116}$$

We also have the symmetry relations for n similar to (11.109)–(11.111); this allows us to rewrite (11.109) and (11.110) in the form

$$n_r(\omega) = 1 + \frac{2}{\pi} P \int_{0}^{\infty} \frac{\omega' n_i(\omega')}{\omega'^2 - \omega^2} d\omega', \tag{11.117}$$

$$n_i(\omega) = -\frac{2}{\pi} \omega P \int_{0}^{\infty} \frac{n_r(\omega')}{\omega'^2 - \omega^2} d\omega'. \tag{11.118}$$

Now, what are the physical implications? Very simple: recall that the real part of n is the "traditional" refraction index determining the phase velocity of light in a given medium; the imaginary part determines the absorption rate. Once they are related to one another by Equations (11.117) and (11.118), you cannot set them by hand to desired values independently. In particular, if you attempt to manufacture a medium with desired absorption properties in a wide region of spectrum, it automatically will determine the corresponding transmittance in this region.

Let us now look at Tom's original proposal from this viewpoint. Tom wanted a filter that would totally absorb all the light except for only one frequency ω_0. This means that absorption coefficient and thereby n_i (ω) must be infinite in all range of the EM spectrum except for ω_0. Equation (11.117) then tells us that the refraction index for all frequencies, *including the desired frequency* ω_0, will be infinite! Indeed, applying (11.117) to desired $\omega = \omega_0$, we obtain

$$n_r(\omega_0) = 1 + \frac{2}{\pi} P \int_{0}^{\infty} \frac{\omega' n_i(\omega')}{\omega'^2 - \omega_0^2} d\omega' \underset{n_i \to \infty}{\longrightarrow} \infty. \tag{11.119}$$

What does this mean physically? A plate with an infinite n_r will totally reflect the wave with the selected frequency ω_0, instead of transmitting it. Tom's perfect filters are as impossible as perpetual motion machines. The latter ban is because of the conservation of energy; the former one is because of the causality.

Talking about the energy, we can also look at the whole problem from another angle, involving the time–energy relation (10.72). Since in the discussed case the frequency is sharply defined, so must be the photons' energy. Thus, we have $\Delta\mathcal{E} = 0$. It follows then that the mean deviation Δt must be infinite. But this quantity has a meaning of the indeterminacy in the time of arrival of a photon to a detector. In our case, the detector is a dinosaur, or a living cell of a plant, or a bacterium about 60 million years ago. If the indeterminacy of the arrival time is infinite, so must be the average expectation time for any of them to absorb the photon in question. This means that such absorption does not take place 60 million years ago as Tom had originally thought; at best it happens in our time – after, not before, Tom has used his hypothetical filter. And note that this argument does not even use the impossibility to manufacture such a filter!

Note also another important and interesting detail. The ban for signaling into the past follows equally from retarded or from advanced causality. In advanced causality the effect precedes the cause. But even in this case the principle makes the communications of the type considered here impossible. In the time-reversed world, you cannot get a laser pulse bursting out from a filter illuminated by monochromatic light coming from Alice.

11.10
The Mystery of Quantum Telecommunication

We discuss here a possibility of information exchange between the particles in an entangled state described in Section 10.1. Such exchange appeared to be an inevitable conclusion from the nonlocal correlations between the results of measurements performed on the particles of an entangled pair. These results, on the face of it, appear to demonstrate the existence of superluminal (and even instantaneous) communication between the distant objects. The phenomenon is a manifestation of the so-called quantum nonlocality and is so impressive and difficult to accept that it appears to be on the verge of mystical. It has been widely discussed from the early days of quantum mechanics up to this day. Its most characteristic features have been confirmed in experiments by Aspect and Grangier [117].

As we have seen so far, none of the discussed superluminal motions can be used for faster-than-light communication. Here, we want to show that this applies as well to the manifestations of nonlocality demonstrated in the mentioned experiments. The demonstrated nonlocality is just another aspect of the probabilistic nature of the wave function describing the behavior of physical objects (see Section 10.1).

Let me first describe a situation that appears to suggest the existence of instant communication between the entangled particles. To make the example more specific, imagine that two high-energy photons (γ -quanta) move from the opposite directions toward each other. There is a chance that when they collide they both disappear and produce instead a pair of particle and antiparticle. Suppose this chance has been materialized and they produce an electron–positron $(e + p)$ pair.

Suppose that the initial photons were so polarized that their total angular momentum was zero. Then the total spin of the system $(e + p)$ must also be zero. It means that the arrows representing the individual spins of the two particles must point in the opposite directions. If, for instance, the electron is in the state with its spin up $|e\uparrow\rangle \equiv |\uparrow\rangle_1$, then the positron must have its spin down $|p\downarrow\rangle \equiv |\downarrow\rangle_2$. The combined state of the pair would then be $|\uparrow\rangle_1|\downarrow\rangle_2$ (Here and below the first arrow in the double is for the electron and the second one – for the positron.).

Let us call this state A. Were the electron to have its spin down $|e\downarrow\rangle \equiv |\downarrow\rangle_1$, then the positron by necessity would have its spin up $|p\uparrow\rangle \equiv |\uparrow\rangle_2$. The combined state of the system of two particles would then be $|\downarrow\rangle_1|\uparrow\rangle_2$. Call this state B. Since in either state the individual arrows point in the opposite directions, the total spin of the pair in both cases is zero, as it should be. Nevertheless, the states A and B are physically different, because the individual spins of either particle are different in the two cases. For instance, the electron's spin in state A points up, whereas in state B it points down.

As we have assumed, the initial condition at the moment of the creation of the pair did not specify spin direction of either particle. Then all we can say about the physical state of the pair can be expressed in the two statements:

1. The electron has its spin either up or down.
2. If the electron spin happens to point down, then the positron spin has to be up, and vice versa.

The state function describing such a state is given by Equation (10.61):

$$\Psi = a|\Psi_{12}\rangle + b|\Psi_{21}\rangle = a|\uparrow\rangle_1|\downarrow\rangle_2 + b|\downarrow\rangle_1|\uparrow\rangle_2. \tag{11.120}$$

As we know from Section 10.1, either of the entangled particles in the superposition (11.120) shares a certain part of its individuality with the other one. Neither of them has a physical state of its own, because its state is not separated from the environment (in our case, from the state of its counterpart).

The basic property of an entangled particle is that one or several of its characteristics depend on analogous characteristics of another particle. Measurements of any such characteristics in one particle would accordingly change the corresponding characteristics of another. After the measurement, the particles will be disentangled with respect to this characteristic. To disentangle our system $(e + p)$, we must perform the measurement of an individual spin. Suppose we measure the spin of the electron and the measurement has shown that electron spin is up. Then, according to our basic statement about the properties of the pair, it is immediately known that the spin of positron must be down. Thus, the measurement of the spin of the electron also determines spin of the positron. The resulting state of the system is A. The system has collapsed from being in state A and state B at once, to being only in state A. If the measurement shows that the electron spin is down, it automatically determines the spin of positron to be up. The corresponding state of the whole pair is now state B. The wave function has collapsed from superposition of A and B to B only.

Things can be understood better through comparison. To better understand the nature of the entangled state (11.120), compare it with another possible state of the electron–positron pair. Consider again a situation when the direction of an electron spin is not specified by the initial physical conditions, but this time it does not depend on spin direction of any other particle. Then, it has a state of its own – a superposition of the two basic states, one with its spin up and another with its spin down. Denoting this superposition as Ψ_e and assuming it, for simplicity, to be equal-weighted, we can write

$$\Psi_e = |\uparrow\rangle_1 + |\downarrow\rangle_1. \tag{11.121}$$

(The superposition (11.121) of the two arrows is *not* the same as a product, say, $|\uparrow\rangle_1|\downarrow\rangle_2$, figuring in (11.120). Both arrows in superposition pertain to the same particle, whereas in the product one arrow pertains to the electron and another one to the positron.)

Similarly, we can consider a positron whose spin variable can be measured without disturbing other particles. Such a positron also can be described by its individual wave function Ψ_p. If the spin direction of the positron has not been determined, this wave function can be in a superposition of two basic states – spin up and spin down, similar to (11.121):

$$\Psi_p = |\uparrow\rangle_2 + |\downarrow\rangle_2. \tag{11.122}$$

We can consider the system of the electron and the positron described by expressions (11.121) and (11.122) as one physical system. This system will be represented by a wave function

$$\tilde{\Psi} = \Psi_e\Psi_p = (|\uparrow\rangle_1 + |\downarrow\rangle_1)(|\uparrow\rangle_2 + |\downarrow\rangle_2). \tag{11.123}$$

Carrying out multiplication, we obtain

$$\tilde{\Psi} = |\uparrow\rangle_1|\uparrow\rangle_2 + |\uparrow\rangle_1|\downarrow\rangle_2 + |\downarrow\rangle_1|\uparrow\rangle_2 + |\downarrow\rangle_1|\downarrow\rangle_2. \tag{11.124}$$

Apart from the familiar products $|\uparrow\rangle_1|\downarrow\rangle_2$ and $|\downarrow\rangle_1|\uparrow\rangle_2$, describing the $(e+p)$ states with opposite (antiparallel) individual spins, we also see here the states $|\uparrow\rangle_1|\uparrow\rangle_2$ and $|\downarrow\rangle_1|\downarrow\rangle_2$, describing two particles with parallel spins – both up or both down. Physically, the presence of the additional terms means that there is *no correlation* between the particles – they have not been born together and are generally independent from one another. Measuring the spin of the electron bears no effect on the state of positron, and vice versa. *Both spins are to be measured independently*, and their measurements can produce any possible outcome. In the state (11.124), there is an equal 25% chance for any one of the four possible outcomes of measurement.

The entangled state (11.120), on the contrary, describes rigid correlation between the particles – their spins must always come up opposite, so that the spin measurements can only produce the antiparallel doubles and *measurement of only one spin automatically gives the result for another one.*

Now, let the members of the entangled pair fly apart, so that some time later they are far away from one another. But, if they do not interact with anything else, they keep their common entangled state Ψ. Much later, even though they may be separated by millions of light years of space, they remember their initial conditions. Neither particle knows the direction of its spin until the measurement is performed. And yet either particle knows that this direction, whatever it might turn out, must be opposite to that of its partner.

Now, here comes the crunch. Suppose that our electron is now on Earth and its twin antiparticle is in another galaxy on planet Rulia. Let Alice measure the spin of the electron on Earth and find it pointing up. Immediately it becomes certain that the spin of the positron on Rulia must point down. If Tom, whose curiosity has taken him as far away as Rulia, performs the measurement there (which is no longer necessary), the measurement will only confirm this fact. It seems as if some agent has instantaneously transferred to Rulia the information about the measurement outcome on Earth, which helped the positron decide upon the direction of its own spin.

In 1937 Einstein, Podolsky, and Rosen (EPR) published their analysis of this kind of phenomena. Their intention was, *first*, to demonstrate that a particle's characteristic can, in principle, be measured without performing an actual measurement on it and, *second*, to point out that such measurement demonstrates an instantaneous physical action ("*spooky action at a distance*," as Einstein has called it), which would be incompatible with special relativity. The authors had interpreted these results as an indication that theoretical description of the world given by quantum mechanics was incomplete (the EPR paradox).

The first of these statements (regarding the possibility of the "interaction-free" measurements) is true. The second one (regarding incompleteness of QM) is false.

But before moving further and explaining why it is false, let us analyze a possible objection to the interpretation of this thought experiment. Suppose that someone on Earth prepared two bottles of wine. One bottle is Burgundy and another is Chardonnet. He prepared one bottle for himself and the other for a friend on Rulia. It does not matter who gets which wine, so the sender just puts one of them into his fridge and another one into the cargo spaceship due Rulia, without looking, and then goes to hibernation for a few million years. After a long trip, the spaceship arrives at its destination. The Earth-based physicist wakes up completely unaware which wine is on which planet. He opens the fridge and immediately knows that the wine on Rulia is Chardonnet.

How could he in an instant get information about an object in another galaxy?

The answer is very simple in this case. The earthling did not receive any signal from Rulia. The only signal he got is the one from his own fridge. The signal has changed his knowledge about the fridge's contents from total uncertainty (50:50 chance of the wine there being of a definite kind) to complete certainty (100%) that it is Burgundy. Together with preliminary information about the two bottles he had, this enables him to conclude with the same dead certainty that the wine on Rulia is Chardonnet.

The act of observation in this case did not (and could not) physically change the type of wine in either package. The Burgundy that was taken out from the fridge had been Burgundy before the observation. The Chardonnet on Rulia had been

Chardonnet long before it arrived there. We can distinguish here between the physical state of the observed object and the physical state of the observer. Only the latter has changed in the act of observation. There was no distant communication in this case.

This might tempt some readers to draw the same conclusion about the experiment with the entangled particles. But such a conclusion would be wrong. The situation with our pair of particles is fundamentally different from the two bottles with wine described above. When Alice on Earth performs the experiment and finds her particle having spin up, this does *not* mean that the particle had had its spin up before the experiment. No! The particle undergoes a dramatic change of its physical state in the process of measurement. It converts from the entanglement with its distant partner into the disentangled state of its own. In the former state the particle, even though separated from its partner by the vastness of space, did not have its full identity totally independent of the partner's. Their identities remained intimately shared. In the final state, each particle has its own full identity and can be described by a wave function of its own, independent of the rest of the world.

Thus, the measurement made on Earth changes instantaneously the situation not only on Earth, but also on Rulia (and vice versa). In the language of the wave functions, we can say that the wave function of the whole entangled system instantly collapses into one of the two possible independent wave functions:

$$\Psi = a|\uparrow\rangle_1|\downarrow\rangle_2 + b|\downarrow\rangle_1|\uparrow\rangle_2 \Rightarrow \begin{cases} \text{either} & |\uparrow\rangle|\downarrow\rangle \\ \text{or} & |\downarrow\rangle|\uparrow\rangle. \end{cases} \tag{11.125}$$

It appears that some physical agent does indeed carry the information about the change on Earth and this communication occurs instantaneously, changing immediately the situation on Rulia. It does look like we face a quite new phenomenon – a superluminal (instant) quantum telecommunication.

And yet this conclusion would be wrong. Relativity is not violated, because the changes of states we discuss are inherently statistical. By its original definition, the event N can be considered as causally affected by the event M, if the change in M changes an observable property of N from one uniquely defined value to another uniquely defined value; for instance, if the influence from M changes the spin of N from up $|\uparrow\rangle_N$ to down $|\downarrow\rangle_N$. In our case, however, the original spin direction of the positron on Rulia was not sharply defined. The positron was in a state with an indefinite direction of spin. Suppose that $|a|^2 = |b|^2$. Then, there was from the very beginning a 50% chance to find its spin up and a 50% chance to find it down. Therefore, in any individual measurement, when Tom finds the positron's spin pointing up, he can always say, "So what?" Indeed, if we knew with certainty that the spin before the measurement was down and now find it up, we could interpret this change as the effect of some outside agent. But when there had already been a 50% chance to find it up, and we do find it up, why should we attribute it to some external influence? We would rather say that this outcome is just the actualization of a preexisting potentiality.

Thus, no individual outcome in the above-described type of measurement can be the evidence of superluminal or any other telecommunication. If the measurement results are statistical by nature, the only way we can make sure that the instantaneous telecommunication does take place is to find the difference in *statistical distributions* of measurements at N with and without corresponding measurements at M.

Following this program, we must perform the *set* of individual measurements in two different conditions and compare their results.

Condition 1. Prepare a big ensemble of (e + p) pairs all in the same state Ψ (11.120). For each pair, send the electron to Earth and the positron to Rulia. Measure only the positron spin on Rulia without disturbing its counterpart on Earth and record the results.

Condition 2. Prepare again the same ensemble. But now, for each pair, measure first the electron spin on Earth and immediately after the electron spin on Rulia and record the results.

For both conditions, we prepare the pairs of entangled particles in the same state. For simplicity, consider the special case when the probabilities of both spin states are equal, that is, $|a|^2 = |b|^2 = 1/2$.

Next, we compare the records obtained for the two conditions. Here they are:

1. The results for condition 1: the measurement outcomes are distributed uniformly – 50:50. Within the margin of statistical fluctuations, half the positrons collapse to the state with spin up and another half collapse to the state with spin down.
2. The results for condition 2: the measurement outcomes are distributed uniformly – 50:50. Within the margin of statistical fluctuations, half the positrons collapse to the state with spin up and another half collapse to the state with spin down.

The net results are identical. The measurements on Earth *do not cause* any changes on the collective experimental data for measurements on Rulia. The whole experiment does not reveal any evidence whatsoever of any communications or signaling between the distant objects.

This result constitutes the contents of the Eberhard theorem [6], according to which the expectation values for the characteristics of two objects separated by a spacelike interval are totally independent of one another.

At the same time, if we compare in case 2 the *individual* results for members of each pair, rather than only the averaged data, we obtain the absolutely irrefutable evidence of the long-range (nonlocal) correlations between individual events. For each pair, if the electron spin on Earth collapses to the state up, the positron spin on Rulia, measured nearly at the same moment, collapses to the state down. If the electron on Earth is found in the state down, the positron on Rulia is found in the state up. The positron appears to know instantaneously what happened to its counterpart millions of light years away. And yet it does not prove the existence of a signal exchange between them, because in each case the positron (or we) can say that the correlation is a mere coincidence. Its chance was 50%, so there is nothing extraordinary that it happens. But why do these coincidences happen 10, 20, . . ., 1000 times in a row? Well, this is really weird, the chances of n coincidences in a row are small indeed $(1/2)^{-n}$, but they remain finite for any finite n.

These nearly absolutely improbable multiple coincidences that, from the classical viewpoint, become totally improbable as $n \to \infty$, reveal a new physical phenomenon – nonlocal quantum correlations (or quantum nonlocality). They have no classical analogue and show that a quantum system can keep certain correlations between its parts even when these parts are separated by voids of space. These correlations are manifest in the fact that the spatially separated parts of an entangled system collapse simultaneously to correlated states even when the measurement is performed only on one part. As you think more of it, this collapse seems no more (and no less) surprising than the collapse of the wave function for a single particle from the state of the uniform and infinitely large probability cloud to the state of the infinite density at a single point surrounded by emptiness with no clouds.

Of course, the situation described above is only a thought experiment. The real experiments are much more difficult, because it is extremely difficult to maintain the "purity" of the system – to protect its parts from random perturbations. Such perturbations (or fluctuations) from the surroundings destroy the quantum correlations (quantum coherence) within the system and the farther apart its constituents the harder it is to preclude them from "decoherence." But, despite numerous difficulties, the experiments [117] have been successfully carried out. They had been performed on a much smaller scale than the thought experiment considered here, but their idea and results are essentially the same. They showed the ability of quantum systems to keep a memory about common past. They did not show in either individual or collective experimental data any evidence of faster than light communications.

The whole situation also shows how tricky Nature is. She seems to tease us. In the above experiments, she appears to blatantly violate the relativity postulates, but with each new trial attempting to catch her in action she comes up free of any violations.

11.11 The "No Cloning" Theorem

We have realized that quantum teleportation using a pair of entangled particles does not constitute any signal transfer, let alone a superluminal communication. The basic reason for this was that for Tom finding his electron in a certain spin state does not tell anything about the result of Alice's measurement because of the principal unpredictability of any single quantum event. Let me use an analogy. If Tom receives a telephone call from Alice and hears her voice in the receiver and in the conversation she mentions a few details that nobody but them two can know, then Tom knows that the call is indeed caused by Alice having dialed his number at the other end of the line. No other event in the world could have caused this call.

When Tom measures his electron and finds its spin pointing up, it had, at best, only 50% chance of being caused by Alice's measurement of her particle's spin. Tom has no other means to know whether Alice has even performed the measurement in the first place. And even without her doing anything, Tom's electron might emerge from the measurement with spin up of its own accord. So even if the electron has, before the measurement, been a member of an entangled pair, the observed value of its spin can hardly be considered as a message from the other member. And there is another aspect to this. When Alice sends a message using certain carrier, she encodes the contents of her message by "imprinting" it into the carrier in a certain way. When she uses EM waves as a carrier, the "imprints" are the corresponding modulations of the wave's frequency or amplitude. When she wants to use an entangled system, there is no way for her to encode her message into a spin state of the particle she measures, since the measurement outcome is totally unpredictable. It is not under her or anybody else's control.

Thus, we have trouble at both ends of the communication line. Tom cannot consider the output of his measuring device as a message and Alice cannot convert her intended message into a controllable input.

In the language of information theory, the two distinct electron spin states (up or down) cannot be used by Alice as one bit (zero or one) of information sent and by Tom as one bit of information received.

In 1982, Nick Herbert published a paper [4] describing an important variation of the above-discussed experiments. Suppose that instead of using the two distinct spin states in one representation (the z-axis) as components of one bit, Alice chooses two distinct representations (x-axis or z-axis) as such components. In other words, both Alice and Tom make an arrangement that when Alice uses an experimental setup measuring her electron's spin along the z-direction, it corresponds to "zero" of a bit; and when she uses a setup measuring spin along the x-direction, it corresponds to "one" of a bit. In contrast to individual spin up or spin down states, the use of the corresponding x- or z-setup is totally under Alice's control. Tom, on the other hand, keeps on measuring his electron's spin only along the z-direction. Of course, the z-directions of both observers must coincide. To receive information from Alice, Tom must have the means to determine from his measurement results what choice Alice has made in each given case.

Let us first evaluate different possible outcomes in case when each partner uses only one electron of an entangled pair. Suppose Alice performs the series of spin measurements along the z-direction. We know that despite a correlation between the outcomes of her and Tom's results (spin up or spin down), it does not convey any information to Tom because even without such correlation there was 50–50 chance for either outcome. But now Tom's concern is to distinguish between *Alice's* two choices, rather than between the electron's choices. So the question now is as following: if Alice switches from the z-measurement to x-measurement, will it in any way affect the distribution pattern of Tom's measurement results? If yes, the superluminal communication is possible; if no, you can turn to watching the *Star Trek* for consolation.

So let Alice do switch to spin measurements along the x-direction. Let us call the result with the spin along the positive x-direction "spin right" and the outcome with the spin along the negative x-direction "spin left." Using the same notations as in (10.1) and in the previous section (Dirac's notations), we can write

$$\begin{aligned} |\rightarrow\rangle &= \text{spin right;} \\ |\leftarrow\rangle &= \text{spin left.} \end{aligned} \tag{11.126}$$

Before moving further, we must determine how must such states be written in the z-representation? If the electron spin is along the x-direction, its z-component is totally undetermined. Both possibilities – spin up and spin down – are potentially present with equal weights. In other words, a definite spin component along x is a quantum-mechanical superposition of spin up and spin down states. And since the right and left states are different, the corresponding pairs of superposition amplitudes for these two cases must be also different:

$$|\rightarrow\rangle = \frac{1}{\sqrt{2}}(|\uparrow\rangle + |\downarrow\rangle), \tag{11.127}$$

$$|\leftarrow\rangle = \frac{1}{\sqrt{2}}(|\uparrow\rangle - |\downarrow\rangle). \tag{11.128}$$

Suppose that in one of such measurements Alice got spin right. Immediately, Tom's electron state collapses to spin left and becomes, accordingly, described by Equation (11.128). But Tom measures spin only along the z-direction. Equation (11.128) predicts that in such measurement, there is 50:50 chance for his electron to collapse either into up, or into downstate. The probability distribution for one electron is exactly the same as in the case when Alice performs the z-measurement, although for a different reason. In the case of x-measurement, Tom's electron is in equal-weighted *superposition* of upstate and downstate; in the case of z-measurement, his electron is either in definite upstate, or in definite downstate, but Tom does not know which state is materialized until he performs the measurement. After a succession of such measurements, he will obtain a 50:50 outcome distribution, because such was the distribution of Alice's results. The experimental pattern is the same regardless of subtle difference in physical mechanism. Therefore, Tom cannot distinguish between the choices made by Alice within this experimental scheme.

Let me repeat the same thing somewhat differently: in case of the z-measurement from the x-state, the electron has no definite z-component before Tom's measurement; in the case of the z-measurement from the z-state, the electron already is in a definite z-state and the measurement is only necessary in order for Tom to learn this state. In other words, Tom's measurement from an x-state *creates a new state,* while measurement from a z-state only confirms preexisting state.

Nick Herbert's idea was to modify the experiment in such a way that could utilize the described subtle difference. The way to do it was as follows. Suppose, in each single trial Tom, before testing his electron, makes many exact copies of it. Note, he copies his electron's state (makes clones) without himself knowing this state. Call it "cloning of unknown state." There seems to be nothing unusual in such procedure – the automatic video cameras work autonomously taking images of people, objects, and events without our knowledge of those people and objects, or our participation in the events.

What are we going to achieve with such cloning?

Suppose Alice has chosen the z-measurement, with the result $|\uparrow\rangle$. Immediately, Tom's electron collapses to the state $|\downarrow\rangle$. Before measuring this state, Tom makes lots of its clones, all in the same state $|\downarrow\rangle$. Only after that he starts his measurements, passing one electron at a time through his apparatus. Since he measures a value of conserving variable in a state in which this variable is already defined, the measurement will only confirm the preexisting result. All of his cloned electrons will be found in the same state $|\downarrow\rangle$. Since Tom knows that all of them are exact clones of his original electron, he knows that electron was in the state $|\downarrow\rangle$ *before* his measurement. This could only be the case if the Alice's electron was found in the state $|\uparrow\rangle$, which means that Alice has performed the z-measurement. In the language of the information theory, Tom receives 1 bit of reliable information.

Let now Alice choose to make the x-measurement. She re-orients her spin detector from the z- to x-direction and finds her electron in the state $|\rightarrow\rangle$. Immediately, Tom's entangled electron collapses to the state $|\leftarrow\rangle$. Before measuring this state, Tom again makes lots of its exact copies. Each copy is in the same state $|\leftarrow\rangle$ as the original electron. In the z-representation, this state is described by superposition (11.128) of the "up" state and "down" state. Therefore, for each copy there is a 50:50 chance to be found in either of the two possible z-states. In which state each electron is going to be found is totally unpredictable, and not known in advance even to the electron in question, let alone to Tom. Tom passes all these electrons again through his apparatus, one at a time, and finds in the end a 50:50 distribution between the two possible z-states. This is quite different from the previous result, when all the electrons were found in only one of the two z-states. Tom immediately concludes that this time all his electrons, including the original one, had initially been in an x-state, which means that Alice must have performed the x-measurement. Again, Tom receives 1 bit of reliable information. Thus, since each kind of Alice's measurements is associated with a possible value of 1 bit, the net result of the described experiment is 2 bits of meaningful information. Since the change of the state of two entangled electrons, no matter how far apart, happens instantaneously at least in one RF, we now must have superluminal communication.

The question is whether such scheme will work. Since its crucial part is cloning, this reduces to the question whether it is possible to clone an unknown state. The answer to this question was given in 1982 [7, 8]. It says that for a wide class of quantum systems exact cloning is impossible. It is forbidden by the requirement that QM operators must be linear. In the case of our interest an operator in question is a mathematical expression describing an algorithm or a process possessing the following property. Suppose, we have a particle in an unknown state $|\Psi\rangle$ and an object in a state $|\Xi\rangle$. Considered as one combined system S, they can be described by a state function $|\Psi\rangle|\Xi\rangle$. But this state may be nonstationary, especially if the parts of the system interact with one another. As a result, within a time interval t the system S may evolve from the initial state $S(0) = |\Psi\rangle|\Xi\rangle$ to a certain final state $S(t)$. The corresponding outcome can be considered as a result of the action of a certain operator $\hat{U}(t)$ called evolution operator:

$$S(t) = \hat{U}(t)S(0). \tag{11.129}$$

It is related to the Hamiltonian operator of the system, but here we will only use the fact that it is linear.

Now, imagine a situation when the final state is $S(t) = |\Psi\rangle|\Psi\rangle$. This means that the system has evolved from one particle in state $|\Psi\rangle$ and an object in state $|\Xi\rangle$ to two particles in the same state $|\Psi\rangle$. We say that in this case the operator $\hat{U}(t)$ has copied the state $|\Psi\rangle$ onto a system $|\Xi\rangle$. This would constitute the cloning of the initial state – precisely what we need for superluminal signaling. Actually, we need "a little" more – the possibility to clone an *arbitrary* state, since in the Alice–Tom experiment described above it was essential that Tom could clone both – the state $|\leftarrow\rangle$ of his electron after Alice had performed the x-measurement and the state $|\downarrow\rangle$ after she had performed the z-measurement.

Is it possible? The answer is no. Here is the proof.

Represent the state $|\Psi\rangle$ as a superposition over the eigenstates $|\Psi_i\rangle$ with unknown amplitudes a_j

$$|\Psi\rangle = \sum_j a_j|\Psi_j\rangle. \tag{11.130}$$

Copying this state means that

$$\hat{U}(t)|\Psi\rangle|\Xi\rangle = |\Psi\rangle|\Psi\rangle = \left(\sum_j a_j|\Psi_j\rangle\right)^2. \tag{11.131}$$

But, if $\hat{U}(t)$ is to be a universal copier able to clone *any* state, then it must also be able to clone any single eigenstate in expression (11.130), that is,

$$\hat{U}(t)|\Psi_j\rangle|\Xi\rangle = |\Psi_j\rangle|\Psi_j\rangle. \tag{11.132}$$

Let us apply again the evolution operator to $|\Psi\rangle$ as we did in (11.131), but now making use of the property (11.132) and linearity of $\hat{U}(t)$. In this case, we obtain

$$\hat{U}(t)|\Psi\rangle|\Xi\rangle = \hat{U}(t)\left(\sum_j a_j|\Psi_j\rangle\right)|\Xi\rangle = \sum_j a_j|\Psi_j\rangle^2. \tag{11.133}$$

The right-hand side of this equation is not the same as in (11.131). The requirements (11.131) and (11.132) are incompatible and therefore our assumption about universal ability of the evolution operator for cloning was wrong. If it can clone all the eigenstates in (11.130), it cannot clone their superposition, and vice versa. Recall that an electron with its spin, say, to the right is a superposition of the spin up and spin down states and you will immediately see that the procedure we have described in the Alice–Tom experiment was not executable.

A similar example is known in applied optics: we can make a laser with a special filter such that it amplifies, say, vertically polarized light (i.e., clones vertically polarized photons); but the corresponding orientation of the transmission axis of the very same filter makes it impossible to amplify the horizontally polarized light.

Discussion

The no-cloning theorem was interpreted by many as a decisive argument against the possibility of superluminal signaling. However, actual situation turned out to be more subtle. For instance, S. J. van Enk [118] has pointed out that such signaling does not necessarily violate the no-cloning theorem. On the other hand, it is easy to see that there are loopholes in the theorem itself, but they do not disable the ban imposed by it.

Indeed, the proof (11.131) and (11.133) applies only to the cases when there is no preexisting information about the amplitudes a_j. In the special case when we know that all $a_j = 0$ but one, that is,

$$a_j = \delta_{jk} \tag{11.134}$$

(but we do not know k), Equations (11.131) and (11.133) are equivalent and the no-cloning theorem actually proves that copying such a system is mathematically allowed, even though the state of the system (the number k) may be unknown. In other words, there must be at least one exception to the no-cloning theorem, namely, when a system is known to be in one of the mutually orthogonal states, which are the eigenstates of a known operator. Thus, (11.134) constitutes the simple analytical condition for the breach in the no-cloning theorem.

The subtle point here is that even though the ban seems to be lifted in this case (the cloning is mathematically possible), it still remains impossible physically. This is evident from the above example with the laser amplification: suppose I know that the photon to be cloned in my laser has a distinct polarization along one of the axes, x or z, but I do not know which one specifically. To prevent generation of irrelevant photons, and for the cloning to be exact, I need the corresponding filter. But since I do not know the exact polarization state of the incident photon, I do not know how to orient my filter for it to work!

Summary

In one of the most fascinating phenomena – quantum entanglement – we observe very strong correlations in the behavior of the entangled particles even when they are separated by a great distance. This effect seems very promising for the superluminal transfer of information and appears to take us to the realm of superluminal communication. But again the laws of Nature conspire to prevent this violation of causality.

Problems

11.1 Peter shoots a tachyon with an invariant mass m_0 and an infinite speed.

(a) What is the energy of this tachyon in Peter's RF?

(b) What is its energy in a frame moving in the same direction with a speed V?

11.2 Using Equation (11.68) find the possible value of the tachyon's magnetic charge q_M.

11.3 Show that the region $x_1 < x < x_2$ where $U > \mathcal{E}$ is classically inaccessible in the relativistic case as well as in non-relativistic one.

11.4 Derive the solution of Equation (F17) from the symbolic relation (F18) by using the corresponding expansions.

11.5 Find the equation for the Green function of an $\hat{L}$-operator by applying this operator to expression (F25).

11.6 Find the Green's function for an operator describing damped harmonic oscillations.

11.7 For the conditions in the previous problem, find
(a) the Green's function in ω-representation;
(b) the Green's function in the coordinate representation by taking the Fourier transformation of the result (a).

11.8 Find the advanced Green's function of a damped oscillator, corresponding to a negative damping factor γ (this is *not* the Lorentz factor).

11.9 Find the dispersion relation between the real and imaginary parts of the refraction index for the case of the *advanced* response.

11.10 Suppose you have an exotic filter with the extremely narrow transmission band $\Delta\omega \cong 10^{-18}\,s^{-1}$. You illuminate it with a sharp laser pulse. Estimate the indeterminacy in the time of arrival of a transmitted photon from this pulse to a detector.

12
Applied Relativity

12.1
Relativistic Jet Propulsion: A "Star Trek" Cruiser

Imagine that you are fishing offshore and your boat engine gets dead. The weather is calm, but you have no oars. What would you do to get back home?

If you know some physics, the answer is obvious. You can throw away the load of fishes you had caught, one by one, throwing each fish as far as you can in the direction away from the shore. Due to the conservation of the net momentum of your loaded boat, the momentum acquired by each thrown fish will be compensated for by the opposite momentum acquired by the rest of the system (the boat with the remaining fishes). The boat will gradually accelerate (within the limits imposed by the drag force of the water) toward the shore and with luck, you can get home.

The process you have used to achieve your goal is an illustration of the principle of jet propulsion. It uses the same mechanism that causes recoil of a gun firing a shot. Each fish (or each bullet, for that matter) represents an incremental portion of hot gas ejected through the nozzle of a jet engine; the speed of the fish relative to the boat (or the nozzle speed of the bullet) represents the nozzle velocity of the jet gas.

After this short introduction, it is easy to describe jet propulsion motion quantitatively. We want to find the amount of fuel necessary to accelerate the rocket (or whatever remains of the rocket after the fuel has been consumed) to a desired final speed.

Let V be the speed of the jet stream relative to the rocket's nozzle and

$$\mu = -\frac{\mathrm{d}M}{\mathrm{d}t} \tag{12.1}$$

be the rate (i.e., mass per unit time) of the fuel consumption. The incremental mass $\mathrm{d}M$ is the change of the rocket's mass over a small time interval $\mathrm{d}t$ (since the rocket's mass is decreasing in the process, this change is, of course, negative). The "−" sign here is owing to the fact that ejecting a mass $\mathrm{d}M$ *decreases* the rocket's rest mass. Thus, in an instantly comoving inertial frame we can write

$$M\mathrm{d}v = -V\mathrm{d}M. \tag{12.2}$$

Special Relativity and How it Works. Moses Fayngold

ISBN: 978-3-527-40607-4

Here, $M(t)$ is the rocket's rest mass at a given moment t and dv is the incremental change in its speed caused by the ejection of mass dM. Considering (12.2) as the differential equation and integrating it, we obtain

$$v(t) = V \ln \frac{M_0}{M(t)}. \tag{12.3}$$

This is known as mass ratio formula or Tsiolkovsky's rocket equation. It relates the final velocity of the rocket $v(t)$ to its initial and final mass.[1] The difference between these masses is determined by the amount of the ejected fuel. Since the equation has been obtained for the nonrelativistic case, it only holds for $v \ll c$. On the other hand, there is no reference in the final expression (12.3) to the incremental masses of the ejected material, even though we might have explicitly considered it as consisting of particles of mass dm. Now, in nonrelativistic mechanics the mass is considered as constant, whereas the actual relativistic mass is velocity-dependent. One might therefore think that Equation (12.3) must hold only for $v \ll c$ and $V \ll c$. But this is true only for the first inequality, not for the second; once the incremental mass does not appear explicitly in Equation (12.3), this equation is applicable to an *arbitrary* jet-stream nozzle velocity V.

Let us consider a few special cases.

(1) $v = 0$ (the initial moment). In this case, Equation (12.3) gives the expected answer $M(t) = M_0$ (the rocket has its initial mass – not a single drop of fuel has yet been spent).

(2) $v(t) = V$ (the rocket has gathered up speed relative to the launching pad equal to the speed of the jet stream relative to the rocket). Equation (12.3) says that at this moment $M(t) = e^{-1} M_0$, where e is the base of the natural logarithm, $e \approx 2.71$. If $V = 2000$ m/s, which is a typical speed of hot gas ejected from the nozzle of a rocket, then in order for the rocket to acquire the same speed 2000 m/s relative to the launching pad, its initial mass, which includes the necessary fuel supply, must be 2.71 times its final mass (the useful load). In other words, at the moment of start, the fuel must constitute 1.71 of the final mass of the rocket $M_f = M(t)$. For $M_f = 1$ ton, the mass of the fuel must be 1.71 ton.

(3) Suppose now that we want to launch a satellite around the Earth, which requires the final speed of the rocket to be roughly 8 km/s. According to (12.3), this would require $M_0 = M_f e^4 \approx 60\, M_f$. If the satellite together with the necessary equipment has a mass of 1 T, then the minimal mass of the fuel must be roughly 60 T (For a primitive rocket with only one "stage" (container), its actual amount must be significantly greater, because it has to accelerate the body of the rocket, container, engines, etc., as well.).

We see from these examples that the necessary mass of the fuel increases much faster than the required velocity of the last stage of the rocket – the spacecraft itself.

1) To illustrate the basic principles, we are using here a simplified description of a rocket. A modern multistage rocket has a few containers with fuel and each consecutive container detaches from the rest of the rocket immediately after having been emptied. For more details, see, for example, Ref. [119].

The reason for this is that, from the start, each consumed kilogram of the fuel has to accelerate not only the useful load (spaceship or satellite with equipment, etc.) but also all the rest of the loaded fuel. Roughly speaking, the fuel has to accelerate itself together with the load. Therefore, each new kilogram of the fuel that we need to add to achieve a greater final velocity is being spent less and less efficiently. To compensate for this, we must add even more fuel, thus decreasing its efficiency even more. This vicious cycle can produce a monster rocket consisting of only the fuel with almost nothing left for the useful load. As a result, there is no hope of reaching distant objects like stars within reasonable time, using conventional chemical jet engines. Is there a way around this hurdle?

The answer is yes. As Equation (12.3) suggests, it would be better, instead of increasing the ratio $M_0/M(t)$, to try to increase the nozzle velocity of the jet stream. Imagine, for instance, that in the above example (3) the jet velocity V is increased from 2000 m/s to, say, 80 000 000 m/s (10^4 the speed of the satellite). In this case Equation (12.3) would, for $v(t) = 8\,\mathrm{km/s}$, give

$$\ln\frac{M_0}{M(t)} = \frac{v(t)}{V} = 10^{-4},$$

which corresponds to $M_0 = 1.0004\,M$. In other words, the necessary amount of ejected fuel would in this case be only $M_0 - M = 0.0004\,M_0$ – quite minuscule with respect to the rocket's mass.

With this kind of fuel one could already think about interstellar missions (with the *ship's* final velocities comparable with c) in practical terms. But this requires, together with new technology, a far more accurate description of the rocket's motion. Equation (12.3), being nonrelativistic (and yet holding for an arbitrary V), becomes inadequate at high speeds of the ship. Thus, we come to the problem of finding a *relativistic* equation describing jet propulsion.

... So imagine now that you are a spaceship commander in a *Star Trek* serial. Your ship is a technological breakthrough, with huge energy resources, powerful lasers, and so on. But when you want to turn back to your home planet upon completing your mission, you realize that the ship's engine stopped dead. What would you do?

Well, you could consider an interesting option of using your lasers as the ship's engines. The same laser beam that vaporized the target ("an alien cruiser") in Ref. [1] could accelerate your cruiser! Each laser photon could play the role of a fish thrown away in your boat experiment, but this time with the maximal possible speed – the speed of light!

With this in mind, we can try to write down an equation similar to (12.1). We will again use an inertial reference frame (RF) that is instantaneously comoving with the rocket at each stage of acceleration. Considering the acceleration in a comoving inertial frame is now especially convenient since this allows us not to bother about the velocity dependence of the rocket mass. But here we have to keep in mind that the mass $M(t)$ at each stage, including the final moment, will be the *rest mass* of the rocket. In the end we can, if we wish, express it in terms of the corresponding relativistic mass as observed from the launching pad.

Now, even though we have originally assumed the jet stream to be that of photons, we will, for greater generality, write here the equation for a stream of "bullets" of arbitrary mass (and accordingly, arbitrary nozzle speed V). In the end, we may set $V=c$ to consider the efficiency of the "laser jet engine" as a special case.

We have

$$M\mathrm{d}v' = -V\mathrm{d}M. \tag{12.4}$$

This equation looks quite similar to nonrelativistic Equation (12.2) (after all, it is nothing else but the universal statement of momentum conservation). But the meaning of the quantities involved is different from that of (12.2). Namely, here, in contrast to (12.2), the incremental mass $\mathrm{d}M$ of the ejected "bullet" is *different* from its rest mass $\mathrm{d}M_0$:

$$\mathrm{d}M = \gamma(V)\mathrm{d}M_0 = \frac{\mathrm{d}\mathcal{E}}{c^2} \tag{12.5}$$

(recall Section 5.3, Equations (5.14) and (5.21)). In the case when the bullet is a photon, we have an extreme example of this difference; in this case, $V=c$ and the Lorentz factor is infinite, so $\mathrm{d}M/\mathrm{d}M_0=\infty$; and for any finite $\mathrm{d}M$, we have $\mathrm{d}M_0=0$ (a photon has the zero rest mass!). The relativistic mass of the emitted photon is determined by its energy $\mathrm{d}\mathcal{E}$, which, in turn, depends on the photon's frequency ω:

$$\mathrm{d}M = \frac{\mathrm{d}\mathcal{E}}{c^2}\left(=\frac{\hbar\omega}{c^2}\right) \tag{12.6}$$

(see Equation (10.1)). This shows that, in any event, we should not worry too much about the incremental *rest* mass $\mathrm{d}M_0$. To accelerate this mass up to the nozzle speed V, the energy source does work equivalent to the energy $(\mathrm{d}M(V) - \mathrm{d}M_0)\,c^2$. This energy, *together* with the rest energy of the ejected mass $\mathrm{d}M_0$ is equal to $\mathrm{d}\mathcal{E}$ in (12.5) and this net amount of incremental energy transferred from the rocket source to the ejected mass is what really matters. (Strictly speaking, the energy $\mathrm{d}\mathcal{E}$ of the source is distributed between the ejected particle and the rocket, which is receiving a slight boost $\mathrm{d}v$ from its initial zero speed. But the rocket's share of this distribution is negligible so far as $\mathrm{d}v \ll V$, so in the limit $(\mathrm{d}v/V) \to 0$ all the released energy $\mathrm{d}\mathcal{E}$ (in the instantaneous rest frame of the rocket!) goes to the jet stream.).

Further, as already mentioned, $M=\mathcal{E}/c^2$ on the left of (12.4) has the meaning of the whole spaceship's rest mass corresponding to its rest energy $\mathcal{E}$ at a given moment t. Finally, the incremental speed $\mathrm{d}v'$ acquired by the ship in the instantaneously comoving inertial frame K′ after the ejection of the "bullet" is not the same as the corresponding increment measured in the initial reference frame K ("the launching pad"). The relative speed between these two frames is the speed v acquired by the ship over the preceding acceleration time. After the ejection of the next "bullet," the ship is moving at a small speed $\mathrm{d}v'$ relative to K′. In the frame K, its speed becomes $v+\mathrm{d}v$. According to Equation (3.22) of Section 3.2, the two speeds $v+\mathrm{d}v$ and $\mathrm{d}v'$ are related as

$$v+\mathrm{d}v = \frac{v+\mathrm{d}v'}{1+(v\mathrm{d}v'/c^2)}. \tag{12.7}$$

Now, because $dv' \ll c$, the right-hand side of (12.7) can be written as

$$(v + dv')\left(1 - \frac{v dv'}{c^2}\right) \approx v + \left(1 - \frac{v^2}{c^2}\right) dv'.$$

It follows

$$dv = \gamma^{-2}(v) dv'. \tag{12.8}$$

Therefore, Equation (12.4), expressed in terms of dv, takes the form

$$M \frac{dv}{1 - (v^2/c^2)} = -V dM \tag{12.9}$$

or, denoting $v/c = \beta$:

$$\frac{d\beta}{1 - \beta^2} = -\frac{V}{c}\frac{dM}{M}. \tag{12.10}$$

Integrating this equation, we obtain on the left

$$\int \frac{d\beta}{1 - \beta^2} = \int \frac{1}{2}\left(\frac{1}{1+\beta} + \frac{1}{1-\beta}\right) d\beta = \frac{1}{2}(\ln|1+\beta| - \ln|1-\beta|) = \frac{1}{2}\ln\left|\frac{1+\beta}{1-\beta}\right|.$$

Therefore, the integration of (12.10) results in

$$\frac{1}{2}\ln\left|\frac{1+\beta}{1-\beta}\right| = \ln\left(\frac{M}{M_0}\right)^{-(V/c)} \tag{12.11}$$

or

$$\frac{1+\beta}{1-\beta} = \left(\frac{M_0}{M}\right)^{2V/c}, \quad \text{that is,} \quad v = c\frac{(M_0/M)^{2V/c} - 1}{(M_0/M)^{2V/c} + 1}. \tag{12.12}$$

This formula was first derived by Eno Peltri in 1930.

Thus, while Equation (12.3) is only true for $v \ll c$, the sought-for relativistic Equation (12.12) must hold for all physically possible v, that is, $0 \leq v < c$.

In the special case when we use a laser as the engine, we will have $V = c$ and

$$\frac{1+\beta}{1-\beta} = \left(\frac{M_0}{M}\right)^2, \quad \text{or} \quad M(t) = M_0\sqrt{\frac{1-\beta}{1+\beta}}. \tag{12.13}$$

It may be interesting to recast these results in a different form, involving the proper time of the rocket needed for its acceleration to a given speed. To this end, we only have to introduce explicitly the fuel consumption rate according to Equation (12.1).

(According to our comments on Equations (12.5) and (12.6), the quantity dM there can be considered as the incremental rest mass of the system "rocket + fuel" converted into the relativistic mass of the ejected fuel in the instantaneous rest frame of the rocket). Assuming $\mu = \text{const}$, we will have simply

$$\mu = \frac{M_0 - M(t)}{t}, \quad \text{so that} \quad M(t) = M_0 - \mu t. \tag{12.14}$$

Here, the acceleration time ranges within the region

$$0 \le t \le t_f \equiv \frac{M_0 - M_f}{\mu}, \tag{12.15}$$

where M_f is the final mass of whatever remains after the acceleration period, when the corresponding amount of the source's energy has been consumed and the corresponding amount of fuel ejected. Putting (12.14) into (12.12) and solving for $v(t)$ yields

$$v(t) = c\frac{1-(1-(\mu/M_0)t)^{2V/c}}{1+(1-(\mu/M_0)t)^{2V/c}}. \tag{12.16}$$

Let us play a little with these equations. First, we want to make sure that in the nonrelativistic limit $v \ll c$, Equation (12.12) reduces to Tsiolkovsky's formula (12.3). This can be seen immediately from (12.11), which at $\beta \to 0$ takes the form

$$\ln\frac{1+\beta}{1-\beta} \underset{\beta \ll 1}{\cong} \ln(1+2\beta) \cong 2\beta = 2\frac{V}{c}\ln\frac{M_0}{M} \tag{12.17}$$

or

$$v(t) = V\ln\frac{M_0}{M(t)}. \tag{12.17a}$$

This is Equation (12.3).

(Note again that the nonrelativistic case does *not* require that $V \ll c$. Suppose that you have the above-described fancy spaceship equipped with the laser gun, so you can use a beam of emitted light as a jet stream. Even then you could still describe the ship's acceleration nonrelativistically so far as its speed relative to your frame of reference remains much less than c.)

Consider now a situation when you, to reach a distant celestial object within a certain specified time t_f, need to accelerate the spacecraft up to $v(t_f) = 0.8\,c$. Using the laser guns ($V = c$) and applying Equation (12.13) gives $M_f = M(t_f) = (1/3)M_0$. Whatever is left after the accelerating process is one third of the original mass of the system. If the useful load (spacecraft with equipment) is 10 T, then the minimal necessary amount of "fuel" must be 20 T (the "fuel" in this case may be an unstable medium; for example, the one with inverse population (see Sections 7.3 and 7.4 of Ref. [1]), whose huge internal energy accounts for the additional mass of 20 T. The best-known candidate for such a medium is a system of particles and equal amount of antiparticles.

It is instructive to compare this result with the case of a more conventional source producing a jet stream, say, with $V = 0.01\,c$. In this case ($v = 0.8\,c$), our more general Equation (12.12) would give

$$\left(\frac{M_0}{M(t)}\right)^{0.02} = 9, \quad \text{that is,} \quad \frac{M_0}{M(t)} = 9^{50}.$$

This means that the mass of the fuel must be roughly 10^{45} of the spacecraft's mass. So forget it! No known conventional source can in this respect compete with a source of photons.

Thus, the laser gun (or whatever else produces the beam of light) can, potentially, be the most efficient jet propulsion engine. Can we take advantage of this fact to accelerate the spacecraft to the speed of light? Look at Equation (12.12). Its answer is NO. If you set in it $v(t) = c$, that is, $\beta = 1$, the ratio $M_0/M(t)$ becomes infinite. No object with finite rest mass can be accelerated to the speed of light.

How about Equation (12.16)? It tells us that by the moment of time $t_c = M_0/\mu$ we must have $\beta = 1$. This corresponds to the final speed $v(t_c) = c$ even if the *jet* speed is less than c. Alas, there is nothing to celebrate. Equation (12.16) is defined only within the range (13.15) of the acceleration time. The moment t_c is later than t_f, it is *beyond* this range for any finite remaining mass M_f. The only way you can stretch the range $(0, t_f)$ up to $(0, t_c)$ is to set the final mass $M_f = 0$. If you hoped that Equation (12.16) allows "something" to achieve exactly the speed of light, this something turns out to be nothing. At best you could claim that since $M(t_f)$ is the *rest* mass, $M_f = 0$ means the zero *rest* mass of the remainder and this could be a photon, which is a little more than nothing. But even so, this photon would be the last thing that is left from the former rocket – you have burned up everything, converting the whole spacecraft to electromagnetic radiation. This *is*, in principle, allowed by Nature, but this is *not* what you wanted! You wanted to get a spacecraft of a *finite* rest mass, zipping as fast as light, – well, but this, as we have learned before, and now found again from our analysis of the above equations, is impossible.

Whatever approach we try to reach the light barrier for a nonzero rest mass, relativity leads us to a dead end.

Having thus shown the consistency of the theory, I now want to show its another twist.

According to relativistic Equation (12.12) or (12.16), we haven't yet exhausted *all* possibilities to reach the forbidden! What about using the tachyons as a fuel? Once we have not ruled out the theoretical possibility of their existence, we may inquire into this subject here as well. This would be an example when an attempt to solve a technical problem in applied physics can raise fundamental questions.

So let us assume again for a moment that tachyons do exist, and think how we could harness them to thrust our spacecraft up to the light barrier.

Well, since the speed V in Equations (12.12) or (12.16) is not restricted by anything, the tachyons used instead of "introductory" fishes can materialize the possibility $V > c$! And once we have trodden onto this path, let us consider the extreme case $V = \infty$. It looks all the more attractive that, as we know from Section 11.2, the tachyons with infinite speed carry no energy, so we can take an infinite number of them, while adding nothing to the start mass M_0 of our rocket!

So imagine a bunch of tachyons rushing back and forth between two parallel mirrors like photons in a conventional laser. But, unlike photons, our tachyons are all moving with infinite speed. Then, at the initial moment, one of the mirrors is made partially transparent, letting the tachyons leak out. Immediately, our system becomes the jet propulsion engine. We may expect astounding efficiency, since the ejected material has an infinite speed! To see the results, one only has to look at Equations (12.12) and (12.16).

Let us start with (12.12). Our hope was that the ratio M_0/M, being greater than 1 and raised into an infinite power, would give an infinite number. Then the fraction in the right-hand side of (12.12) would yield just 1, giving $v = c$.

Alas! Precisely in the case of an infinite V the ratio of the initial and final masses of the rocket is *not* greater than 1, but exactly equal to one, since, as we have already mentioned, the ejected fuel weighs nothing in this case. Therefore, the above reasoning cannot be applied.

Consider now Equation (12.16). Here the expression in the parentheses on the right is less than 1. If raised into an infinite power, it would produce zero. Therefore, the whole fraction would be reduced to 1 and the equation would give $v = c$.

But . . . Alas again! As we already know, precisely in the case of an infinite V, the product μt giving the net ejected mass is zero, so the expression in the parenthesis is *not* less than 1, but exactly 1. Again, the reasoning leading to the conclusion that we can reach $v = c$ by this path cannot be applied.

If we want to find the exact answer for this case, we need to start from the general case with arbitrary V, express the net ejected mass in terms of V, and then consider the limit $V \to \infty$.

The relation between the ejected mass and V depends on jet propulsion mechanism. We will consider the above model of tachyons zipping back and forth between the two mirrors, one of which is partially transparent. With V finite, it takes finite time t_f for all the tachyons to leak out of the system. We assume (without discussing how it could be done) that special care is taken to maintain both V and the consumption rate μ constant during all this time. Under these assumptions, using Equations (11.40) and (11.41), we can write

$$\mu t = M_0 - M = m_0 N \hat{\gamma}(V), \tag{12.18}$$

where N is the total number of the tachyons.

Put this into either Equations (12.12) or (12.16). We will get

$$v(t) = c\frac{1 - (1 - (m_0 N\hat{\gamma}(V)/M_0))^{2V/c}}{1 + (1 - (m_0 N\hat{\gamma}(V)/M_0))^{2V/c}}. \tag{12.19}$$

When $V \gg c$, we have $\hat{\gamma}(V) \approx c/V$, and (12.19) takes the form

$$v(t) \cong c\frac{1 - (1 - (m_0 Nc/M_0 V))^{2V/c}}{1 + (1 - (m_0 Nc/M_0 V))^{2V/c}} \cong c\frac{1 - e^{-\alpha}}{1 + e^{-\alpha}}, \tag{12.20}$$

where

$$\alpha \equiv 2\frac{m_0 N}{M_0}. \tag{12.21}$$

In the limit $V \to \infty$, Equation (12.20) becomes exact. And it tells us that even in this limit, the final speed of the spacecraft, irrespective of the value of α, always remains less than c. We can even estimate by how much less for each possible combination of numbers m_0, N, and M_0. And we can say what to do in order for the spacecraft to get as close as possible to the speed of light. The greater is α the closer to 1 is the fraction on the right of (12.20). Therefore, we get arbitrarily close to the speed of light by increasing, say, the total initial number of tachyons. For a sufficiently large number N, the final speed of the spacecraft approaches the speed of light as

$$v(t) \cong (1 - 2e^{-\alpha})c. \tag{12.22}$$

As we make N greater and greater, the difference between v and c becomes *exponentially* small. And the beauty of this is that, *at infinite* V, the production of the huge number of needed tachyons does not cost us anything in terms of energy, since, in this case, the *relativistic* energy (and accordingly, the relativistic mass) of the whole bunch of tachyons is zero.

As you start thinking of it, this may at first seem to be nonsense. Indeed, how is this consistent with the fact that a spaceship accelerated to nearly the speed of light will, in the rest frame of the launching pad, have an overwhelmingly greater energy than its initial energy $\mathcal{E}_0 = M_0c^2$? Where will all this additional huge energy come from?

The answer is: from the tachyons, of course! Recall the transformational properties of tachyons discussed in Sections 11.2–4. According to these rules, a tachyon emitted with positive energy and moving into the future in one reference frame, may have negative energy and move into the past (backward in time!) in another reference frame. But, as has been pointed out in Refs [105,106], such an object is physically equivalent to an antitachyon with positive energy moving from the past into the future (forward in time). According to the reinterpretation principle, we will get an adequate description of this phenomenon if we simultaneously change signs of the transformed tachyon's time and energy. They become positive again and what is recorded as tachyon emission in the first reference frame is recorded as tachyon absorption in the second reference frame.

Let us apply this to our situation here.

We have a tachyon with the zero energy $\tilde{\mathcal{E}} = 0$ and momentum $\tilde{p} = m_0c$ in the instantaneous rest frame of the spaceship. In the reference frame of the launching pad, the spaceship is moving at a speed v. Therefore, the tachyon's energy in this frame is

$$\tilde{\mathcal{E}}' = \gamma(v)(\tilde{\mathcal{E}} - v\tilde{p}) = -m_0cv\gamma(v). \tag{12.23}$$

A moment t of time along the tachyon's path x in the ship's frame corresponds to the moment

$$t' = \gamma(v)\left(t - \frac{v}{c^2}x\right) = \gamma(v)x\left(\frac{t}{x} - \frac{v}{c^2}\right) \tag{12.24}$$

in the frame of the launching pad. But $t/x = 1/V = 0$. Therefore,

$$t' = -\gamma(v)\frac{v}{c^2}x. \tag{12.25}$$

Thus, both tachyon's time and its energy are negative in the frame of the launching pad.

But, according to the reinterpretation principle, this is physically equivalent to the tachyon with a *finite* and *positive* energy moving forward in time. Changing simultaneously the signs of time and energy in (12.23) and (12.25), we will obtain just what we would *actually observe* if such an effect really took place: positive tachyons coming from infinity and whacking into the cruiser's nozzle. Being positive, they carry finite momentum in the direction of their motion and transfer it to the ship. The ship accelerates, gaining energy from the powerful flux of tachyons spontaneously born in the distant past in the infinite depths of space and miraculously converging all on the same object. For the observers at the launching pad, the spaceship would accelerate in the desired direction entirely at the cost of a spontaneous natural process.

We have started with situations in applied physics and wound up with a theoretically possible phenomenon no less spectacular than anything one could watch in a *Star Trek* serial. Speculative as it may be, it looks like our intuitive expectations that tachyons could be an extremely efficient propellant may have a sound theoretical basis. If these particles are discovered in some future and put to use, they could, in principle, push a spacecraft closer to the light barrier than any other known source or mechanism. But not even they can make the spacecraft *attain* the barrier. In addition, according to Section 11.4 their irreducible length increases with their speed, thus effectively slowing them down. And, as shown in Refs [10,11], they cannot be used at their best as described, because they cannot move in a straight line.

12.2
Principles of Design and Functioning of Particle Accelerators

Some of the existing machines that could, *in principle*, be used for accelerating future spaceships to relativistic velocities are still a way below the level necessary to achieve this goal, but at the same time they are very efficient for studying the microworld. Accordingly, in this section we will redirect our focus from astronomical objects and distances to the subatomic systems. This is the subject of high energy and particle physics, and much of what we have learned in this field was discovered using particle accelerators. Particles accelerated by these big machines play the same role in studying the subatomic world as light does in the study of small objects under a microscope. The relatively small wavelength of visible light ($0.4-0.8\,\mu$m) allows one to see small organisms (e.g., bacteria) and their details, structure of a cell, and so on. The smaller details we want to resolve the smaller wavelengths we need.

As we know, any particle has its corresponding de-Broglie wave. The basic properties of waves are the same regardless of their physical origin. One important property common for waves of any nature is that shorter waves can carry more information about an illuminated object than longer waves. Therefore, just as in the case of optical microscopy, to see the world on a smaller scale, we need particles with shorter de-Broglie wavelengths (recall Section 10.1). And the shorter these waves are the greater the particle's energy, according to the de-Broglie relations (Equation (10.1)). We conclude that what we need is high-energy particles.

Let us estimate the typical energies necessary to travel into the depths of the microworld.

We will use here the energy units universally accepted in high-energy physics: "electron volt" (eV). This is the energy acquired by an electron (or any other particle with equal charge) across the potential difference of 1 V ($1\,\text{eV} = 1.6\times10^{-19}\,\text{J}$).

The de-Broglie wavelength of an electron in a hydrogen atom is about 10^{-10} m. Such a wave, "wrapped around" the nucleus along an imaginary loop and reiterating itself, determines the size of the loop (e.g., Bohr's radius). According to (10.1), this wavelength corresponds to the momentum $p = 2\pi\hbar/\lambda \approx 6 \times 10^{-24}$ J s/m and kinetic energy $K = p^2/2m_e \approx 6.5$ eV. This is quite negligible with respect to the electron's rest energy, equal to $\approx 5\times10^5$ eV. This is why we can safely use the nonrelativistic equation for the electron's kinetic energy in a hydrogen atom.

In Rutherford's famous experiments that lead to the discovery of the atomic nucleus, the wavelength of the "probing" electrons must have been significantly shorter. Furthermore, to resolve the details of a nucleus, the wavelength must be less than the nucleus size (10^{-15} m). In this case, the corresponding kinetic energy of an electron may be comparable with or even much greater than its rest energy. This means that we have left the domains of Newtonian mechanics and must instead use relativistic equations.

Relativistic energy of a particle with a rest mass m_0 and momentum p is given by Equation (5.24):

$$\mathcal{E} = \sqrt{m_0^2 c^4 + p^2 c^2}.$$

To find the energy input necessary to "build up" this energy, we must subtract the intrinsic rest energy of the bombarding particle:

$$K = \mathcal{E} - m_0 c^2 = \sqrt{m_0^2 c^4 + p^2 c^2} - m_0 c^2 = m_0 c^2 \left(\sqrt{1 + \frac{p^2}{m_0^2 c^2}} - 1 \right). \tag{12.26}$$

Combining this with de-Broglie's relation (10.2), we express the necessary kinetic energy directly in terms of the de-Broglie wavelength:

$$K = m_0 c^2 \left(\sqrt{1 + \frac{\hbar^2}{4\pi^2 m_0^2 c^2 \lambda^2}} - 1 \right). \tag{12.27}$$

Since the rest energy of a relativistic particle is only a small fraction of its kinetic energy, the difference between kinetic energy and total relativistic energy is usually small:

$$\mathcal{E} - K \ll \mathcal{E}, \quad \text{so that} \quad K \approx \mathcal{E}. \tag{12.28}$$

In the above case of relativistic electrons with the wavelength of about 10^{-15} m, we have the ratio in the square root of Equation (12.27) close to 1.2×10^3, which means that the energy of such an electron is more than $\sqrt{1200} \approx 35$ times greater than its rest energy. In such cases, we can just use the term "energy" without specifying whether it is kinetic energy or total relativistic energy – the difference is small. In these domains, it is convenient to measure the energy in the units of the rest energy of

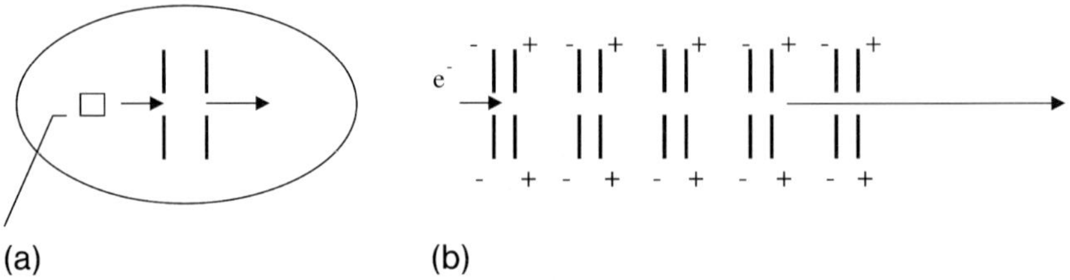

Figure 12.1 Schematic representation of a possible linear accelerator: (a) an electron from a hot cathode (the square inside the oval) enters the spacing between the plates of a capacitor, where it is accelerated by the electric field; (b) the same electron in a linear array of consecutive capacitors. The arrows represent initial and final momentum of the electron.

the corresponding particle. The reader can easily check that in these units, the relativistic energy is a number equal to the corresponding Lorentz factor γ.

Now we can turn to the basic technical question – how can we pump a big amount of energy into a particle?

The first and most practical option that comes to mind is to accelerate charged particles in an external electric field. This idea was used in the construction of the first working particle accelerators. Its basic principle can be described as follows. Imagine first a charged parallel-plate capacitor in a vacuum chamber. Both plates of the capacitor are perforated and an electron released from a hot cathode gets into the capacitor through the hole in the corresponding plate (Figure 12.1a)

Once inside the capacitor, the electron is accelerated by the electric field and exits with energy higher than its initial energy. Then, we can reiterate the process by putting another similar capacitor down the electron's path, then another one, and so on (Figure 12.1b). In this way, we can, in principle, arbitrarily increase the electron's energy by passing it through a sufficient number of capacitors. The electron's energy increases at the cost of the energy of the source (say, a battery) charging the capacitors.

This principle was utilized in the first accelerators. In essence, such an accelerator is a linear succession of capacitors within a long vacuum tube. Accordingly, it has been called the linear accelerator.

To obtain sufficiently high energies, we need a large number of accelerating units (capacitors) and, accordingly, a great size of the machine. The typical length of a linear accelerator of electrons is a few kilometers. Naturally, the question arises whether it is possible to achieve the desired result with smaller machines. For example, is it possible to bend the particle's linear path into a circle to force them to return back and use repeatedly one and the same capacitor?

The answer is yes – by using magnetic field.

Recall that magnetic field, even though it does not change the kinetic energy of a charged particle, curves its trajectory. In a uniform magnetic field perpendicular to the particle's velocity, this trajectory becomes a circle. The new type of accelerator based on this principle was first designed by the American physicist Lawrence in 1931.

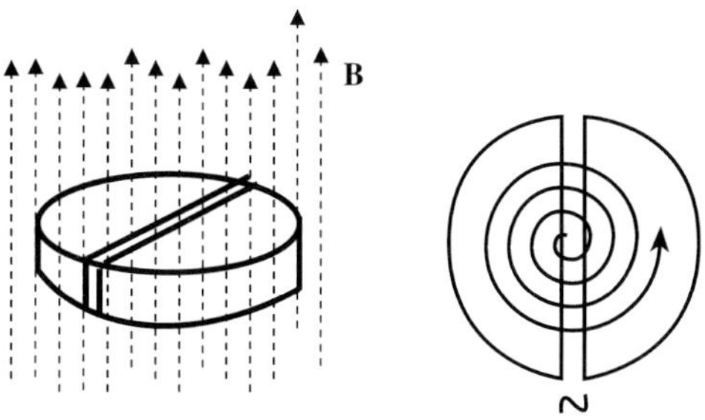

Figure 12.2 (a) Schematic representation of the basic units of a cyclotron; (b) trajectory of an accelerated particle in the cyclotron (plane view).

Imagine a flat metallic container shaped as a round "pillbox." The "pillbox" is cut along its diameter in two equal parts and put into a vacuum chamber between the poles of a big electromagnet producing a uniform magnetic field. This field is directed vertically, perpendicular to the pillbox, which is lying flat. The two halves of the "pillbox" are separated by a narrow gap and can be used as two opposite elements of a capacitor. They are called the "duants" (Figure 12.2).

Now, suppose we apply a voltage V to the duants.

This will produce an electric field in the gap between the edges of the pillbox. Suppose further that, just as we have applied voltage, we put the charged particles, say, protons, at the center of the gap. The protons will be immediately accelerated by the applied electric field directed from one duant to the other and enter the other duant with some initial kinetic energy $K = qV$. Once inside the duant, they will start moving in a circle perpendicular to the magnetic field with constant speed. In essence, they will materialize the situation discussed in the beginning of Section 5.3. Therefore, half a period later the protons will reenter the gap, having traced out a semicircle. Now, suppose that precisely by that moment we reverse the applied voltage. The proton will find itself again in the electric field, but this time pointing in the opposite direction. Since its velocity now is also opposite to the initial one, the field will boost it up. The proton will leave the gap and enter the second duant with a greater speed than it had when entering the first one. Once within the duant, it will again move in a semicircle and return to the gap. By that moment, we again reverse the applied voltage! As a result, the proton will reenter the first duant with a greater speed than before. If we keep on reversing the voltage at the right moments, the process will reiterate with successive incremental increases of the proton's speed.

A crude but figurative mechanical model of this process might be a horse running around the master on a circus' stage. A leash held by the master ensures that the horse stays on a circular path. Imagine that the master thrashes the horse with the long whip each time when the horse completes a cycle and each such thrash instigates the horse to run faster. The result will be the successive incremental increases of the horse's speed, until the nearly mad horse will rush at the natural limit of its agility.

In this model, the leash plays the role of magnetic field keeping the particle in its circular orbit; the whip plays the role of the electric field signaling the particle to accelerate. (The analogy ends here: unlike the particle, the poor animal gains in kinetic energy at the cost of its own resources – its internal chemical energy.).

Now, what is the trajectory of the described motion?

The radius of a circle traced out by a charged particle in a magnetic field stands in proportion to its momentum (Equation (5.17)). Therefore, the radius of the proton's orbit will increase with each cycle, along with its energy. The protons will move along a curve of the type of unwinding spiral.

But the most interesting thing is that the period of each cycle, despite the increasing size of its corresponding spiral loop, remains constant. Recall that according to Equation (5.17), the period of orbital motion of a charged particle in a uniform magnetic field does not depend on the size of the orbit. It is determined only by the magnetic field strength and the mass of the particle. As long as the mass remains constant, the period does not change. This allows successive incremental acceleration of protons by periodic reversal of the voltage U across the gap at the moments when the particles pass through the gap. This can be done using a high-frequency generator, which recharges the duants every half-period, working with the frequency determined from Equation (5.17) (the cyclotron frequency). A special device at the periphery of the camera either leads the accelerated protons out or retains them inside the camera for smashing into an inserted target.

Realization of this idea allows one to accelerate protons up to the energies $K = 20–25$ MeV using the maximal voltage of about 100 KV. The reader can easily estimate that under these conditions, the complete acceleration would require about 100 cycles. In other words, the cyclotron, compared to a linear accelerator, can replace 100 or more accelerating units with only one such unit.

In 1939, Lawrence was awarded the Nobel Prize in physics for developing and practical implementation of his elegant idea.

However, the cyclotron also has its limitations. They are imposed by the relativistic mass increase with velocity. If you remember, Equation (5.17), according to which the cyclotron frequency does not depend on the particle's speed and the orbit's radius, was based on the assumption that the particle's mass is its invariable characteristic. But, as we know, this assumption holds only at relatively low kinetic energies. For instance, if we estimate the mass of the proton with kinetic energy $K = 100$ MeV, using the relativistic formula that follows from (5.20) to (5.21)

$$m = \frac{m_0 c^2 + K}{c^2}, \tag{12.29}$$

we will obtain $M \approx 1.02 M_0$, that is, relativistic mass exceeds the rest mass by only 2%. The corresponding speed of the proton is about 60 000 km/s, that is, 20% of the speed of light. Further acceleration will result in greater increase of the relativistic mass and accordingly ever-increasing departure from Equation (5.17).

The correct relation between the cyclotron's frequency and the particle's energy (or, which is the same, its speed) is given by Equation (5.16a) with the relativistic mass of a particle instead of its rest mass. This equation tells us that the period of the particle's motion in the cyclotron is not constant, but increases with the particle's speed (energy). If we keep the generator's frequency fixed, then, as the particle's energy increases, there will eventually appear an increasing mismatch between the generator's frequency and the actual cyclotron's frequency. The particle will successively cross the gap between the duants with ever increasing retardation relative to the pulses of accelerating voltage. Eventually, it will fall out of synchrony with the generator and its motion will no longer be accelerated. This imposes the limit on the amount of energy that can be pumped into the particle in the cyclotron. On the other hand, the existence of this limit, for example, the fact that one cannot accelerate protons in a cyclotron above the kinetic energy $K = 20\,\text{MeV}$ (corresponding to the Lorentz factor $\gamma \approx 1.02$), is another experimental evidence of the correctness of the relativistic Equation (5.14).

The maximal energy that is allowed for an accelerated particle in a cyclotron depends on the particle's charge and mass. For instance, if we take an electron instead of the proton, the same Lorentz factor $\gamma \approx 1.02$ will correspond to a much smaller kinetic energy of only $K = 0.01$ MeV. This is because the Lorentz factor is the ratio of the relativistic energy to the rest energy. As the latter is about 2000 times smaller for the electron than for the proton, so is the corresponding total energy and, accordingly, the kinetic energy, for the same Lorentz factor. The cyclotron turns out to be of no use for acceleration of electrons! The experiment confirms this negative conclusion as well, giving thereby an additional confirmation of correctness of the relativistic dynamics.

However, the above analysis gives us a hint at how to get around this difficulty. We can extend the acceleration process to much higher energies if we compensate the change of the cyclotron frequency due to the relativistic mass increase, by a corresponding change of the generator's frequency $\omega_0 \to \omega(t)$, using Equation (5.16a). Alternatively, we can change the magnetic field $B \to B\,(t)$, increasing it with time at the same rate as the relativistic mass increase. Then the ratio $B(t)/m(t)$, and thereby the frequency of a cycle, can be kept fixed at the original frequency ω_0. This method of keeping the two frequencies tuned (in phase) by changing either ω at fixed B, or the magnetic field B at fixed $\omega = \omega_0$ is called autophasing. It was suggested almost at the same time independently by two physicists – the Russian V. Veksler (1944) and the American E. McMillan (1945).

And there is yet another possibility: instead of changing B with time, we can design a system with constant but nonuniform magnetic field, whose strength increases with distance s from the symmetry axis. As an example, consider magnetic field of a circular current loop.

In the plane of the loop, the magnetic field lines are "packed" tighter at the periphery than at the center, indicating that the magnetic field gets stronger as you go from the center of the loop to its periphery (where the current flows). It is possible to adjust the distribution of currents in the loop so as to produce the magnetic field of a specific configuration, for which the increase of the field with the distance s exactly matches the corresponding relativistic mass increase of the proton. As is seen in Figure 12.2b, an accelerated proton, while getting heavier, finds itself on larger and larger orbits and, accordingly, in regions with stronger magnetic field.

In this way, the condition of resonance can be maintained for a large group of accelerated particles despite the relativistic mass increase. The process of acceleration can last up to the energies $\mathcal{E}/\mathcal{E}_0 = \gamma \approx 1000$ and more. This already corresponds to particle speeds that are very close to the speed of light!

This improvement has another important advantage: the above-described dephasing, so pronounced for the electrons in a cyclotron, is no longer a factor in the new machines; therefore, they can be used not only to reach much higher proton energies than in the original cyclotron scheme, but also to accelerate electrons.

The corresponding machines – the next generation after cyclotrons – have various names: synchrotrons (for electrons) and phasotrones and synchrophasotrones (for protons).

As an example of some big accelerators working on these principles consider the Serpukhov proton accelerator, which was built in the late 1960s [120]. It consists of a few basic units: foreinjector, linear accelerator, and electromagnetic ring.

In the foreinjector, hydrogen atoms get stripped of their electrons and the naked protons are accelerated up to the energy of 760 keV by an impulse electric field. The further acceleration takes place in the linear accelerator. Here they gain energy up to 100 MeV. Then a special system injects them into the main ring.

The radius of the main ring of the Serpukhov proton accelerator is 235 m. If you would want to make a "circumnavigation" along the ring's tunnel (a huge thorus), it would take you about 20 min of brisk walk. The protons at the final stages of the acceleration make one such circumnavigation 100 000 000 times faster – in about 6×10^{-6} s. The protons, of course, travel in another, much more narrow tunnel with air pumped out. This tunnel is all within the bigger one, its walls are made of stainless steel, and its cross section is an ellipse with major and minor axes of 19.5 and 11.2 cm, respectively. There are 120 electromagnets along the tunnel, whose job is to keep the bunch of accelerated protons within the camera and prevent them from dispersing. The protons moving along this narrow circular corridor without touching its walls travel a net distance of about 600 000 km. This corresponds to 400 000 cycles of acceleration and results in a proton energy of 76 000 MeV (75 times its rest energy). The proton beam intensity (the amount of protons per one hypercycle of complete acceleration) is 10^{12}.

Efficient as these accelerators are, they are not free of limitations either. These limitations are caused by centripetal acceleration. Since any accelerated charged particle radiates electromagnetic waves, a particle on a synchrotron orbit also becomes a source of radiation. We can get the idea of the basic properties of this radiation in the following way. Consider a small element of an arch of the circle traced out by the particle. The particle's instantaneous acceleration is perpendicular to its instantaneous velocity and is pointing toward the center of the orbit. We can think about it as an instant in the life of a microscopic "antenna" (more accurately, an oscillator) with its axis pointing along the acceleration – toward the center of the circle. Acceleration is equivalent to an alternating current along this axis. An antenna with alternating current radiates electromagnetic waves. For an inertial observer moving together with the particle at this instant, the waves are spreading out symmetrically about the axis, with maximum intensity over the equatorial plane.

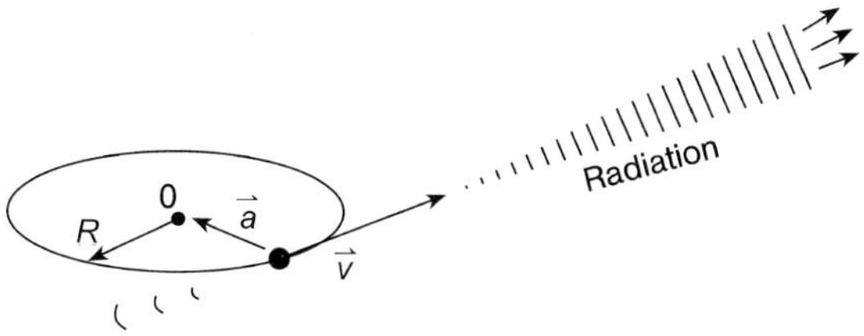

Figure 12.3

This plane can be visualized as perpendicular to the radius and tangential to the circle. For the lab observers, the oscillator is rushing along the circle almost with the speed of light. We see a fast-moving "antenna" oriented perpendicularly to the instant direction of its motion (Figure 12.3).

The fact that this "antenna" is a relativistic object moving as indicated in Figure 12.3 has dramatic consequences. Frequency of the waves from a fast-moving source undergoes changes due to the relativistic Doppler effect (Sections 9.4 and 10.5):

$$\omega(\theta) = \frac{\omega_0}{\gamma(v)}\left(1 - \frac{v}{c}\cos\theta\right)^{-1}. \qquad (12.30)$$

Here, ω_0 is the proper frequency, $\mathbf{v}$ is thc instantaneous velocity of the source (the particle), and θ is the angle between $\mathbf{v}$ and the direction of radiation. At $\theta = 0$ (forward radiation) Equation (12.5) yields

$$\omega = \omega_0\sqrt{\frac{1+\beta}{1-\beta}}, \quad \beta \equiv \frac{v}{c}. \qquad (12.30a)$$

At $\theta = \pi$ (backward radiation) we have

$$\omega = \omega_0\sqrt{\frac{1-\beta}{1+\beta}}. \qquad (12.30b)$$

For an ultrarelativistic particle, $\beta \to 1$, and we have $\omega(0) \to \infty$, $\omega(\pi) \to 0$.

Since radiation can be considered as a flux of photons, and the photon's energy is proportional to its frequency, Equations (12.30a and b) tell us that most of the energy will go in the forward direction in the form of hard X-ray and gamma radiation and only a very small fraction will go in the backward direction as IR or radio waves.

Also, as is seen in Figure 12.3, an open solid angle around the direction of particle's velocity in the comoving frame transforms, due to aberration of light, into a narrow solid angle in the laboratory frame. Thus, most of the radiation emitted by a fast-moving and accelerated particle goes in the forward direction. It is called the *synchrotron radiation.*

We can thus visualize an accelerated particle in a synchrotron as a powerful projector and the synchrotron radiation as a very narrow and intense projector beam instantly tangential to the projector's path. The spectrum of synchrotron radiation is fixed and uniquely determined by the speed of the particle and the radius of its orbit. In some respects, it is similar to a laser beam. And just like the beam from Captain Fletcher's laser in book [1], it also sweeps out all the exterior of the orbital plane. The difference is that, unlike Captain Fletcher's laser, which was only spinning around itself with its barrel always *along* a radial line, the fictitious laser representing the radiating particle in a cyclotron is orbiting together with it and has its "barrel" *perpendicular* to the instant radial line (parallel to the particle's instant velocity). It shoots a huge amount of high-energy photons in the forward direction tangential to the orbit and a small amount of low-energy photons in the backward direction. In the first approximation, we can neglect the latter kind of radiation and only consider the forward-directed beam.

Now, since the energy and momentum of the beam can only come from the kinetic energy and momentum of the radiating particle, you can immediately see the consequences. An ever-increasing part of the energy pumped by the accelerator into the particle leaks out into the space as synchrotron radiation.

You can also describe this process in terms of laser jets of the previous section. The difference is that in a "Star Trek" cruiser, the laser jet shoots the photons in the backward direction, thus accelerating itself and the cruiser. Here the laser jet shoots the photons in the forward direction, thus decelerating itself. You can think of it as the "Star Trek" cruiser attempting to land at its destination and accordingly having its laser jets flipped over to slow down.

No matter which model you use to describe the process, the outcome is clear: beyond a certain point, the particle acceleration process in a synchrotron becomes ineffective.[2)] This is the limitation we were talking about.

We could reduce energy losses by decreasing the centripetal acceleration a. Since $a = v^2/R$, the only way to do it is to increase the radius R of the orbit and thereby the size of the system. But the very idea of synchrotrons was to replace linear accelerators so that the size could be reduced. Thus, we came to the point where the further enhancement of synchrotrons' performance begins to conflict with the justification of their existence.

The maximum possible energies that can be achieved in synchrotrons depend on the mass of the particle and the maximal size of the orbit that we can afford. Thus, in the large electron–positron collider (LEPC) at the Centre Europeenne pour la Recherche Nucleare (CERN) near Zeneva the electrons can be propelled up to 99.999 999 999% of c, which corresponds to $\gamma(v) \sim 10^6$, that is, an electron's relativistic energy about million times of its rest energy .

Is there a way to circumvent this limitation? Again, the answer is yes! How? Very simple: if we cannot hurl the particles fast enough onto the target, we can hurl the target toward the coming particles! Consider, for instance, an experiment where we study the structure of a proton by bombarding it with other protons and then observing the debris of their collision. Such processes, when a great part of the initial kinetic energy of the colliding particles goes into the internal energy of the system and changes its structure, are called deep inelastic collisions. The greater the initial kinetic energy the more dramatic are the possible changes in the particles' internal state in the case of collision

2) As mentioned above, synchrotron radiation is in many respects similar to laser radiation. It is therefore a very useful by-product of the described process and has found many important applications, especially in the high frequency range. But this may be a topic for another book.

and, accordingly, the greater the amount of information about their structure that can be extracted. If, instead of having the target protons idly waiting for the coming of the "intruders," we subject them to the same acceleration procedure in another synchrotron and then direct them with the same maximum achievable energy to meet the "intruders" on the head-on collision course. Now we cannot say who are the intruders and who are not, but the point is that the inelastic collision energy will increase. By how much? The first answer may be: by a factor of 2, of course!

The correct answer is: by a factor of 4 in the domain of Newtonian mechanics and much more in relativistic mechanics!

Let us first start with the classical (nonrelativistic) analogy. Imagine a bunch of cars moving along a straight highway all in one direction. The "collision" between any two cars in the bunch is harmless under these conditions. The relative velocity being zero, no damage is done.

The situation becomes much more serious if one of the cars is stationary (the usual perception of the target!) on the highway. In this case, we can have a more or less dramatic car crash. The crash occurs because part of the initial kinetic energy goes into the "restructuring" of the system. From our human perspective it may result in a disaster.

But an optimist could still say that we are relatively lucky, because this is not yet the worst possible scenario. Indeed, imagine what could happen if the second car were moving toward you rather than just being stationed on the road!

Now, the effect that is so destructive in a car collision may become constructive in high-energy physics, with colliding particles instead of cars. In this case, instead of (or in addition to!) the initial particles, there may appear new particles after the collision – and this is precisely what the physicists are frequently looking for in such experiments. What is most interesting is that the net mass of the newly born particles can by far exceed the sum of the rest masses of the initial particles before the collision. Using again the analogy with a traffic accident, this looks as if, say, two motorcycles crashing into each other produced a few military trucks, and, to make things even more exciting, sometimes emerged themselves intact from the crash. The only loss they suffer in this case is the reduction of their initial relativistic energy (or mass, for that matter): it went into the mass (and its respective energy) of the trucks. Think of it as another illustration of the equivalence of mass and energy discussed in Section 4.3.

These results can be explained formally from the velocity-dependence of kinetic energy: quadratic in Newtonian mechanics and through the Lorentz factor in relativistic mechanics.

Let us describe this process quantitatively for a collision between two identical particles with the rest mass m_0. We will consider a deep inelastic collision with the maximal possible loss of the initial kinetic energy of the system. The loss is maximal when the colliding particles stick together like two pieces of putty.[3] We will compare

3) The resulting object can later decay with more than one possible outcome. There may be a chance of decay into two original particles with the same net kinetic energy. In such cases, we are dealing with collisions that are effectively elastic via an "inelastic channel." In the macroworld, this would correspond to a car crash resulting in reemergence of both cars in their initial internal states out of the debris, although, maybe, with changed velocities. The probability of such an event is zero in the macroworld but may be nonzero in the subatomic world.

two cases: one with a stationary target particle and one with both particles moving on a head-on collision course with equal speeds. Also, first we will consider the process in the framework of Newtonian mechanics and then in the framework of relativistic mechanics.

Case 1. Only one particle is moving, so the initial net momentum and kinetic energy of the system are

$$P = m_0 v, \qquad K = \frac{1}{2} m_0 v^2 = \frac{P^2}{2m_0}. \tag{12.31}$$

After the collision we have the combined particle with the double mass moving with the same momentum $P' = P$; therefore, the final kinetic energy of the system is

$$K' = \frac{P'^2}{2m'} = \frac{P^2}{4m_0} = \frac{1}{2} K. \tag{12.32}$$

The lost kinetic energy is

$$\Delta K = K - K' = \frac{1}{2} K. \tag{12.33}$$

Case 2. Both particles are now moving toward each other with the same speed v. The initial (and, accordingly, the final) net momentum of the system is zero. The initial net energy is twice the initial energy of case 1:

$$\bar{K} = 2\left(\frac{1}{2} m_0 v^2\right) = m_0 v^2 = 2K. \tag{12.34}$$

After the collision, we have one stationary object with zero momentum and zero kinetic energy. All the initial kinetic energy has been lost to internal energy, so that

$$\Delta \bar{K} = \bar{K} - \bar{K}' = \bar{K} = 2K. \tag{12.35}$$

Comparison with (12.32) shows that the amount of energy transferred into the internal energy of the system has increased by a factor of 4 in the head-on collision case.

The reason for the quadruple, rather than double, gain is that in the case 1, half the initial kinetic energy goes to the kinetic energy of the debris rather than to restructuring of the internal state of the system. In the case 2 we have, first, twice the initial kinetic energy and second, *all* of it goes to the restructuring of the system.

Now, consider the same two cases in the framework of relativistic mechanics.

Case 1. Only one particle is moving, so the initial net momentum and kinetic energy of the system are

$$P = \frac{1}{c}\sqrt{\mathcal{E}^2 - m_0^2 c^4} = \sqrt{\gamma^2 - 1}\; m_0 c, \qquad K = \mathcal{E} - m_0 c^2 = (\gamma - 1) m_0 c^2, \tag{12.36}$$

where $\mathcal{E}$ is the initial relativistic energy of the bombarding particle. The total energy of the system is the sum of $\mathcal{E}$ and the rest energy of the target particle

$$\mathcal{E}_t = \mathcal{E} + m_0 c^2. \tag{12.37}$$

After the collision we have one combined particle moving with the same momentum $P' = P$ and the net energy $\mathcal{E}' = \mathcal{E}_t$, but we cannot say that its rest mass is $2m_0$. According to the "mass–energy" equivalence, a certain part of the rest mass of the new particle may come from the initial kinetic energy of the system. Therefore, we denote it as μ_0 and consider as yet unknown quantity to be determined from the conservation laws.

Thus, the final momentum and kinetic energy of the system can be written as

$$P' = \frac{1}{c}\sqrt{\mathcal{E}'^2 - \mu_0^2 c^4} = \sqrt{\gamma'^2 - 1}\,\mu_0 c, \qquad K' = \mathcal{E}' - \mu_0 c^2 = (\gamma' - 1)\mu_0 c^2. \tag{12.38}$$

Here, γ' is the Lorentz factor corresponding to the energy $\mathcal{E}'$. According to the conservation laws, the total energy $\mathcal{E}'$ after the collision equals the total energy $\mathcal{E} + m_0 c^2$ before the collision, that is,

$$\mu_0 \gamma' = m_0 (1 + \gamma). \tag{12.39}$$

The momentum P' of the new particle is equal to the initial momentum P of the system. Using expressions (12.36) and (12.38) for the corresponding momenta, we have

$$\mu_0 \sqrt{\gamma'^2 - 1} = m_0 \sqrt{\gamma^2 - 1}. \tag{12.40}$$

Now we have two simultaneous Equations (12.38) and (12.39) for the two unknowns μ_0 and γ'. Solving them gives

$$\mu_0 = m_0 \sqrt{2(1+\gamma)}, \qquad \gamma' = \sqrt{\frac{1}{2}(1+\gamma)}. \tag{12.41}$$

It is easy to see that μ_0 is greater than $2m_0$ and γ' is less than γ. I leave it to the reader to find the asymptotic behavior of both quantities in the nonrelativistic and ultrarelativistic limits (Problem 12.12).

The lost kinetic energy is

$$\Delta K = K - K' = [(\gamma - 1)m_0 - (\gamma' - 1)\mu_0]c^2. \tag{12.42}$$

Using Equation (12.41), we obtain

$$\Delta K = \left[\sqrt{2(1+\gamma)} - 2\right] m_0 c^2. \tag{12.43}$$

It is easy to show that for $v \ll c$ this expression reduces to $(1/2)K$, as it should for the nonrelativistic case (Problem 12.13). When $v \to c$, Equation (12.43) can be approximated by

$$\Delta K \cong \sqrt{2\gamma}\, m_0 c^2. \tag{12.43a}$$

As we can see from these equations, the increase in the rest mass of the system (equivalent to the lost kinetic energy) can by far exceed the sum of the initial rest masses of the particles; the same, however, and even to a greater extent, can be said about the *kinetic* energy of the debris (in this case, the combined particle): in the ultrarelativistic limit, according to Equations (12.38) and (12.41), it is by a factor of about $\gamma^{1/2}$ greater than their *rest* energy and by a factor of about γ greater than the rest energy of the initial particles. All this energy is lost – instead of smashing the particles against one another, it went into moving the debris. Spectacular as the results may be in terms of possible new discoveries, the process itself becomes increasingly inefficient as we go to higher energies, in terms of energy consumption.

Case 2. Both particles are now moving toward each other with the same speed v. The initial (and, accordingly, the final) net momentum of the system is zero. The initial net energy is twice the initial energy in the case 1:

$$\bar{\mathcal{E}} = 2\mathcal{E} = 2\gamma m_0 c^2. \tag{12.44}$$

The corresponding net kinetic energy before the collision is

$$\bar{K} = 2(\gamma - 1)m_0 c^2. \tag{12.45}$$

After the collision we have one stationary object with zero momentum and zero kinetic energy. Now *all* the initial kinetic energy has been transferred to internal energy, so that

$$\Delta\bar{K} = \bar{K} - \bar{K}' = \bar{K} = 2(\gamma - 1)m_0 c^2. \tag{12.46}$$

In this case, with only twice the expenses, we can expect the process to be about $\gamma^{1/2}$ times more effective than in the case 1.

There is another way to see it: look at the process from the viewpoint of an imaginary observer Tom sitting on one of the particles. According to the addition rule for velocities, Tom would see the other particle approaching with the speed

$$v' = \frac{2v}{1 + (v^2/c^2)} = \frac{2v}{2 - \gamma^{-2}}. \tag{12.47}$$

At v close to c, this gives

$$v' = \frac{2v}{2 - \gamma^{-2}} \approx v\left(1 + \frac{1}{2}\gamma^{-2}\right). \tag{12.48}$$

Tom's first thought might be that there is little to celebrate – practically no gain in speed! And, as we know, this is quite natural in view of the impossibility to attain the light barrier. Close to the light barrier, relativity almost entirely cancels any gain in speed that is promised by nonrelativistic mechanics. But we also know that very close to the barrier, even tiniest change in speed will cause huge change in kinetic energy. Therefore, Tom should express the expected effect of collision in terms of the *energy* of relative motion of the approaching particle, rather than in terms of its speed. Then he will find that the amount of energy going to the *inelastic* process as defined above will be the same as in our analysis in the center-of-mass reference frame.

Thus, in the relativistic mechanics, the idea to use oncoming beams, while so unappealing in terms of speed, shows all its beauty when considered in terms of energy.

However, there is still a price to pay for the energy gain. It is in the probability of close collisions, which are necessary if we want to study deep inelastic effects. This probability is proportional to the particle concentration (number of particles per unit volume) in both the incident beam and the target. Therefore, it is proportional to the product of the corresponding concentrations. Now, the particle concentration in a stationary target may be up to about $10^{24}\,\mathrm{cm}^{-3}$ – just the amount of atoms squeezed within one cubic centimeter in a solid-state body. The particle concentration in the beam is, of course, many orders of magnitude less. When we convert the target into a similar beam, the corresponding concentration accordingly drops down to the same value as in the first beam. The productivity – the number of the informative collisions – drops in the same proportion! This is the downside of the oncoming beams experiments. But even so, the use of oncoming beams has proved extremely effective in our study of the microworld.

Problems

12.1 We said in the text (Section 12.1) that at each moment all the energy of the source goes to the jet stream, rather than to the rocket. How then does the rocket end up accelerating at all (this requires energy!)?

12.2 Suppose you have a laser jet propulsion vehicle in space. You can tune it to the emission of the continuous infrared radiation with the wavelength $1\,\mu\mathrm{m}$, or, instead, use all the available energy to emit only one γ-photon of huge frequency and, accordingly, huge momentum. What is more efficient in terms of
(a) acceleration time?
(b) the final speed of the spacecraft?

12.3 For a system with $M_0 = 1000$ Tand the final speed $v = 0.9\,c$ (using conventional fuel), find
(a) the *maximal* possible rest mass M of the spacecraft;
(b) its corresponding relativistic mass;
(c) its momentum;
(d) mass and momentum of the ejected fuel in the rest frame of the launching pad.

12.4 The rocket on the launching pad has a mass $M_0 = 106\,\mathrm{kg}$. Its engines eject the fuel with the nozzle speed of $0.1\,c$ at a rate $10\,\mathrm{kg/s}$. For the final velocity of the spacecraft to be $v = 0.95\,c$,
(a) what must be its maximal rest mass?
(b) how long will it take to reach the final speed by the clock of the rocket?

12.5 In the previous problem, what time will it take in the rest frame of the launching pad for a spacecraft to reach its final velocity?

12.6 We showed in the end of this section that using tachyons for jet propulsion can accelerate a spaceship very close to the speed of light and it may be free of energy-cost if we manage to produce tachyons with infinite speed. Describe the motion of the spaceship as observed in the reference frame of the launching pad. Find the ship's final energy as a function of its final velocity.

(*Hint*: Consider incremental energy gains described by Equation (13.23) and sum them up for all speeds ranging from zero to the final speed.)

12.7 Instead of a rocket with constant fuel consumption rate, consider a rocket with constant proper acceleration, similar to the spacecraft in the end of Section 5.6. However, in contrast to the case of that spacecraft, *do not* assume now that the system's rest mass remains constant. This means that now we need an internal source of energy necessary to accelerate the rocket, rather than an external source.
For such a rocket, find its fuel consumption rate $\mu(\tau)$ as a function of its proper time τ.

12.8 Consider a project for a travel to the center of our Milky Way Galaxy (about 30 000 LY from us). The rocket is designed so as to maintain constant acceleration $a = g$ in the rest frame of the spaceship. The trip starts on Earth from rest and acceleration program works all the time until the spaceship reaches its destination.
(a) How long will the trip last by clock of the Earth-based observer?
(b) How long will it last for the ship's crew?

12.9 In the previous problem, assuming the ideal rocket (all its rest mass M_0 converts into radiation which is all channeled in the backward direction) find the M_0/m ratio.

12.10 For the project in Problem 12.8, find the travel time elapsed on Earth and on the spaceship if the rocket accelerates the first half of the way and decelerates the second half of the way.

12.11 We observe two identical colliding particles from our inertial reference frame K and see Tom sitting on one of them. In Tom's reference frame, find
(a) the kinetic energy of the approaching particle in terms of its rest mass and of its Lorentz factor in *our* frame;
(b) the amount of kinetic energy that would go to the rest energy of the combined particle after the totally inelastic collision (in terms of the same parameters);
(c) how does the result (b) compare with (12.44)?

12.12 Prove that in Equations (12.39) and (12.40), μ_0 is always greater than $2m_0$ and γ' is always less than γ. Obtain the asymptotic expressions for both quantities in the limits $\gamma \to 1$ $(v \ll c)$ and $\gamma \gg 1$ $(v \to c)$.

12.13 Find the asymptotic behavior for the lost kinetic energy (12.43) in the above two limits of the previous problem.

12.14 For the case 1, show that the lost kinetic energy is equal to the increase in the rest mass of the system, that is, $\Delta K = (\mu_0 - 2m_0)\, c^2$.

12.15 For a charged particle in a cyclotron with the electric field E between the duants, find the ratio of centripetal to linear (along the velocity) acceleration as a function of v. Using the relativistic equation for the radiated power, evaluate the energy losses for each acceleration and find their ratio.

13
A Bit of General Relativity

This chapter does not present the general relativity (GR). It only outlines very briefly the underlying fundamental ideas and basic concepts forming the structure of this theory and preliminary steps necessary for its formulation.

13.1
Basic Ideas of GR

The general relativity, as follows from its name, is a generalization of special relativity. It removes the restriction according to which only the *inertial* reference frames (RFs) are equivalent. The basic idea of GR is that the equivalence of the *inertial* reference frames is only a part of a much more general principle of equivalence of *all possible* reference frames. According to GR, *all* RFs are equivalent because the laws of Nature are the same in all of them. But this generalization implies totally new view of gravitation. According to this view, the arbitrarily accelerated RF turned out to be intimately related to a certain kind of gravitational fields. This, in turn, leads to a conclusion that gravity, which appears to us as a field of physical force, is actually manifestation of curvature of space-time. According to Newton, any mass produces a field of force in the 3-space around itself – the gravitational field, but leaves the geometry intact. According to Einstein, any mass (more accurately – any 4-momentum) curves space-time around itself. This curvature changes geometry and is therefore manifest in motion of physical bodies in a given locality. The change in motion of a physical body appears to us as the action of a physical force on the body.

After creating the special theory of relativity (STR), Einstein started asking himself an apparently childish question: "What happens in a freely falling elevator?" The question, however, was not casual. Einstein thought that his principle establishing the equivalence of all *inertial* reference frames could be generalized onto systems moving with acceleration. An example of such a system is a breaking car where all the passengers regardless of their mass experience a common acceleration forward. This imitates the behavior of objects in a gravitational field, where objects also have common acceleration regardless of their mass or chemical composition. It is just because of this fundamental fact that weightlessness

Special Relativity and How it Works. Moses Fayngold

ISBN: 978-3-527-40607-4

appears in a freely falling elevator: the Earth pulls the elevator down from under the feet of a passenger with the same acceleration, with which the passenger himself is falling. As a result, the passenger does not press onto the elevator's floor and hovers freely in the cabin. Which means that the gravity is locally eliminated within a freely falling system!

Thus, a free fall is the motion with acceleration under the action of gravitational force from the viewpoint of the Earth-based observer, but is a motion without acceleration and without gravity for a falling observer. And vice versa, as we had seen in an example with the breaking car, the accelerated system in a space without gravity can be substituted with an inertial system in which there is gravity. In particular, one can consider a breaking car as stationary, but there emerges horizontal gravitational field.

To illustrate this point more specifically, let us elaborate our example with a spaceship in Chapter 1. You fall asleep in that spaceship again and the engines are turned on while you are asleep. The spaceship is accelerated relative to distant stars. But this time you wake up *during this acceleration process*. You did not know that the engines had been turned on. But, in the framework of special relativity (SR) you can immediately learn that you are now in an accelerated reference frame, just as Tom in Chapter 1 could see from the experiment with water in the tank, deflected chandelier, and other "weird" effects that he was in an accelerated RF. Similarly, you now know that you are in an accelerated RF because you experience the inertial force permeating the whole space around you.

But this is a definite knowledge only within the framework of special relativity and if you do not consider gravity.

A crucial step to GR is that the inertial force is in principle indistinguishable from gravitational force. All the objects in your accelerating ship behave in exactly the same way as they do in a nonaccelerating ship standing still on a launching pad on the surface of a planet with an appropriate gravitation. Therefore you cannot distinguish between these two possibilities. You cannot be sure whether you are in an accelerating ship in free space or you are in the stationary ship on the surface of a planet. As an observer in Chapter 1 could not distinguish between the two inertial RFs, you now cannot distinguish between the two different RFs – one stationary in a space with gravitational field and one accelerated in a free space without gravitational field. Both cases are equivalent.

Similarly, all phenomena in a freely falling elevator are exactly the same as in an inertial RF in a free space without gravitational field. A spaceship freely falling in the field of gravity is equivalent to an inertial spaceship far away from all sources of gravity. You remember, we described various experiments in such a ship to illustrate the basic laws of mechanics. Now we see that all these experiments can as well be performed within a spaceship in a state of a free fall.

This is a famous equivalence principle of Einstein.

Another aspect of the equivalence principle is the equivalence between the inertial and gravitational mass. A body's response to a nongravitational force is determined by its *inertial* mass; on the other hand, if we consider gravity as a field of force, then the same body's response to it must be determined by its *gravitational* mass, just as a body's response to an electric force is determined by its electric charge. The fact that the resulting acceleration is in both cases the same means that both kinds of masses are the same. Numerous experiments demonstrate the independence of acceleration of free fall on mass or composition of the falling body; two bodies with equal inertial mass but different compositions have equal acceleration of free fall. This means that their gravitational masses are also equal, which implies that gravitational mass of each body stands in strict proportion to its inertial mass.

For a comparison, consider an electrically charged capsule falling onto an object with an opposite charge; we cannot eliminate even locally the electric field within this capsule. We can always find or prepare an item with the charge/mass ratio other than that of the capsule, so that the item might have an acceleration different from that of the capsule not only in magnitude but also in direction. This would indicate the presence of an electric field in the capsule.

In contrast, it is impossible to change the (inertial/gravitational mass) ratio of any item within a capsule freely falling in a *gravitational* field. This ratio is the same for all objects of Nature and this statement is known as the equivalence of both kinds of mass. As you think of it, it seems only natural – as another confirmation of the mass–energy equivalence. If inertia is an inalienable (intrinsic) attribute of energy, on the one hand, and any energy is the source *and* the detector of gravity, on the other hand, then the equivalence of the inertial and gravitational mass follows as an inevitable consequence. It would not be very meaningful to ask what actually generates gravitational field around an object – its mass or its energy.

But energy is only the temporal component of a 4-vector. As we know, according to Lorentz transformations, the energy of a system in one RF is a linear combination of the same system's energy *and* momentum measured in another RF. Therefore, the gravity must originate from both – energy and momentum. If you wonder whether a photon – a particle with the zero rest mass – can produce and respond to gravity, the answer following from the mass–energy equivalence of the special relativity is unequivocal "yes." According to (5.13) and (5.18), the 4-momentum is produced by motion of *relativistic* mass, which is nonzero even for photons and is, in fact, the measure of flow of energy. In this respect it provides an analogy between gravitation and the magnetic field, which is produced by flow of electrical charges. Just as electric current responds to an external magnetic field and generates one of its own, so a photon responds to gravity and generates it.

Thus, the equivalence of *all* RF is another side of the equivalence between the inertial and gravitational mass. It is because of this latter equivalence that a gravitational field can be created by switching to a noninertial RF with an appropriate acceleration or it can be eliminated by a transition to a RF freely falling in such a field. These simple statements form a foundation of the GR.

In developing his program of "geometrization of gravity" Einstein had to resolve a few apparent contradictions, whose resolution gave us insight into the mystery of gravitation. Here is one of them: we have come to conclusion that in a freely falling elevator there is no gravity. Hence, in the absence of other forces, all the objects in it should either stay at rest or move in straight lines. In reality, however, the two specks of dust in the elevator, which were initially at rest relative to one another, will soon

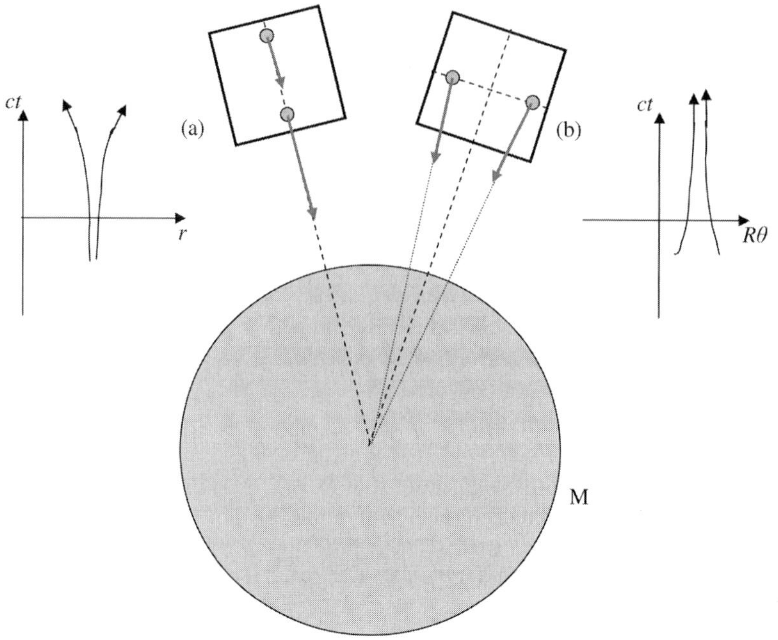

Figure 13.1 The tidal effect. Straight arrows indicate accelerations. (a) Two items on one *radial* line within a free-falling elevator. Their world lines are diverging; (b) two items on one *meridian* line within a free-falling elevator. Their world lines are converging. Since the items are subject to no forces (there is no gravity in *free-falling* RF!), they move along straight world lines (geodesics), according to the least-action principle (or the principle of extremal aging) applied to free particles. The fact that the geodesics here are diverging or converging indicates the non-Euclidean geometry of space-time around a massive object *M*.

start moving with respect to each other (Figure 13.1). They will either get closer if they were arranged horizontally or move farther apart if they were along the vertical line. The reason for this is the difference in acceleration of free fall at different points in space – the so-called tidal effect.

From the viewpoint of the observer in the elevator, there is no gravity around him. And yet he sees one of the specks accelerating toward the other or away from it. Which means that they trace out a curved trajectory in the absence of all forces! This seems to contradict Nature. Einstein had, after extensive thoughts, found an unexpected explanation: since the curvature of trajectory is the same for all bodies – specks of dust, leaden balls, and so on, it can be considered as an intrinsic property of the space-time itself. In other words, all objects move, as they should in the absence of forces, in the straightest possible paths ("geodesics"), but the paths appear to be curved because the space-time itself around massive bodies is curved.

This conclusion can be illustrated by a couple of simple examples.

As a first example illustrating the basic idea consider a heavy marble on an elastic film (a rubber band) (Figure 13.2). Without the marble the film is flat, its surface is described by the Euclidean geometry. A small particle bounced across the film will

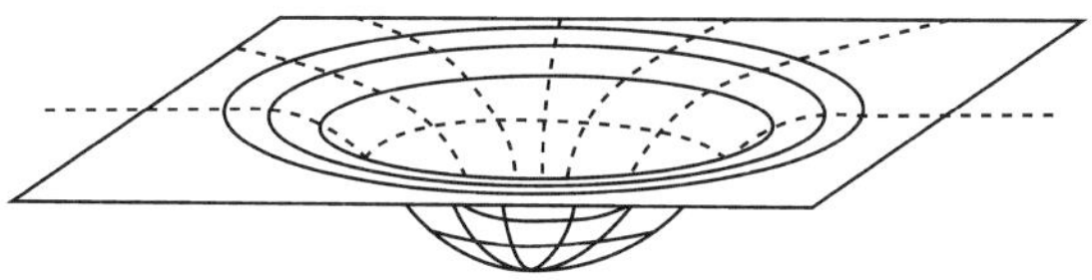

Figure 13.2

move in a straight line. If we include time, we will get three-dimensional space-time (one temporal dimension and two spatial dimensions). The interval will have the form $ds^2 = c^2dt^2 - dx^2 - dy^2$.

Now, put the marble on the rubber band. The marble will deform the band, reshaping it in a form of a funnel. This will result in two changes. First, the geometry of space-time will no longer be pseudo-Euclidean because the surface of the film is no longer flat. Second, a small body launched across the film will no longer move in a straight line (for an external 3D-observer), but will rather curve toward the center of the funnel.

This example still involves gravity (it is necessary here to explain the formation of the funnel under the marble and resulting bending of the trajectories close to the funnel). Thus, gravity is explained through gravity. The reader can find a better example in the remarkable book [3] on gravitation. It illustrates how what we perceive as gravitational force can arise from pure geometry.

Take an apple with a few ants crawling over its surface and put a drop of honey on it. The ants, attracted by flavor, will set out to reach it. Each ant will try to run along the shortest possible path that plays for the apple the role of the straight line. Indeed, if we cut out a thin path along which one of the ants made its shortest way to the honey and then spread this thin strip on the table, we will see that the obtained strip of the peel is straight as a laser beam. However, these straight lines, when on the apple's surface close to the graft, diverge and then converge again, so the trajectories of two ants seem to be curved relative to one another. If the ants are imparted with reason, then one of them can describe the curvature of all the trajectories near the graft as the result of a force emanating from the graft. This could work perfectly well for the ants for a while, even though we see what the ants do not – that there are no forces arising from the graft, but the surface of the apple is curved. However, there may be a certain genius among the ants that could come up with the same explanation as we do, even though it does not see what is obvious to us, but can only reach the same description by an effort of the imagination. This ant would come up with a theory stating that the attractive "force" so long taught to us in school is an illusion here and the real cause of the curvature of the trajectories is the curvature of the surface on which the ants live, or better still, of the space-time itself. This analogy (the best I know of) is still not complete. The model with the funnel does not include the flow of time into the picture. If we include the time, we could notice a remarkable thing about these curved trajectories that they do not depend on physical structure or chemical composition or mass of a small test particle, but only on direction and magnitude of its velocity. In this respect it would be similar to a motion in a real gravitational field. If it does not depend on individual properties of moving objects, then it must be the properties of space-time itself. The "gravitation" produced by 4-momentum vector of matter is actually the manifestation of the fact that the 4-momentum of matter curves all space-time around itself. In the words of the authors of book [3]: "The matter tells space-time how to curve. The space-time tells the matter how to move."

In the above illustration with an apple, it is easy for us to see the curvature of the two-dimensional surface because we are, together with the surface, in a three-dimensional space and can see the third dimension into which the 2-surface is curved. But to discern the curvature of the 3-space itself, and even more – of the whole space-time, is an extraordinary intellectual achievement. But Einstein has gone much farther than just formulating the idea. He developed a consistent and complex theory (GTR) that connected physics with geometry and allowed to determine exactly the type and the amount of curvature of space-time as a function of distribution and motion of matter. The math of this theory falls beyond the scope of this book. Nevertheless the essence of the theory can be expressed in a few words: matter curves the space-time and what appears to us as gravity is just manifestation of this curvature. The free particles move along the shortest "straight" lines in space-time and the apparent curvature of these lines near the massive objects is the manifestation of the curvature of the space-time around these objects. Thus, GTR establishes deep connections between space, time, and matter.

Remember, we described in Chapter 1 the principle of relativity as the equivalence between different inertial reference frames. Now we can, following Einstein's famous example with the elevator, describe the generalized principle of relativity as the equivalence between any arbitrarily moving reference frames.

However – and this is another important aspect – we should remember that this equivalence is exact only locally. The experiments in a falling ship would show the same results as in a uniformly moving ship in space without gravity only within a sufficiently short time intervals (e.g., those intervals should definitely be less than the time it takes to crash into the planet or whatever else producing the field), and within sufficiently small regions of space (small enough to neglect the tidal effects).

This also shows that the gravitational fields, emerging within and around accelerated systems, while being real thing are different from the gravitational fields produced by matter. The former are not associated with any curvature of space-time; the latter come together with it as the north pole of a magnet bar comes together with the south pole. In other words, the gravitational force, while being intimately connected with curvature of space-time, is at the same time a more "flexible" entity – you can change the local gravitational force by changing RF, but you cannot change the local curvature. This is the prerogative of matter only. In this respect, the curvature is more fundamental than gravity. The latter is relative; the former is absolute – it is an invariant under transformations between any reference frames.

From these simple observations there follows a possibility to generalize the principle of relativity to accelerated systems: we just consider such a system as stationary and inertial, but say that there is a gravity field in it. As a magnetic field can be generated by boarding a train (recall Section 6.2), so a gravitational field can be generated by boarding an accelerated system. Similarly, as some configurations of magnetic field can be eliminated in the rest frame of a charge, so the gravity field can be locally eliminated in the rest frame of a freely falling body.

There is another line of reasoning showing that Newton's description of gravity as a field of a 3-vector force is in conflict with relativity.

Consider two point masses M_0 and m_0, which are stationary in system K and separated by distance r. Newtonian gravitational force between them is

$$f_0 = G\frac{M_0 m_0}{r^2}, \tag{13.1}$$

where G is the gravitational constant. According to Newtonian physics, this force is independent of masses' state of motion. According to relativity, it must depend on motion in two ways: through retardation effect similar to that of EM interaction and through velocity dependence of mass itself. The best way to treat these dependencies is to use the relativistic transformation rules for the force, which are universal and take care of all the accompanying effects automatically. Therefore, we can use here the same approach as in Section 6.2, where we unveiled the relativistic nature of magnetic field. But this time we will play with masses instead of charges. This will be a little more involved because mass is not an operational invariant and depends on velocity via Equation (5.14).

Let us consider the force on M in two different cases involving motion: first, when only m is moving and second, when both masses are moving together. As in Section 6.2, in both cases the direction of motion is perpendicular to separation vector $\mathbf{r}$ between the masses, so we can write for velocity $\mathbf{V} \perp \mathbf{r}$. Imagine an observer Howard (system K′) sitting on the moving mass m. For this observer m is stationary, so $m = m_0$, but M is moving with a speed V in the static gravitational field produced by m_0. Therefore $M = M_0\gamma(V)$ and we can assume that the force f on M as measured by Howard in K′ is given by

$$f = f_0\gamma(V), \tag{13.2}$$

where the Lorentz factor takes account of the relativistic mass increase due to motion of M. Once we get information of this result, we can determine the force on M in *our* frame K (where $M = M_0$!) from the last of Equations (5.84):

$$f_{M_0} = \frac{f}{\gamma(V)\left[1-(Vv_{\parallel}/c^2)\right]}. \tag{13.3}$$

(We have dropped the labels indicating components since there is only one component of velocity in both frames.) In the case under consideration we have $v_{\parallel} = V$ and (13.3) reduces to

$$f_{M_0} = f\gamma(V) = f_0\gamma^2(V). \tag{13.4}$$

In the second case both masses are moving in K but stationary in K′ and Howard measures usual Newtonian force f_0 on M. Accordingly, we now set in (13.3) $f=f_0$ and $v_{||}=0$ and obtain

$$f_M = \frac{f_0}{\gamma(V)} = \frac{f_{M_0}}{\gamma^3(V)}. \tag{13.5}$$

Comparing Equations (13.4) and (13.5)z, we see that gravitational force on an object depends on its state of motion and this dependence cannot be explained only by the velocity dependence of the object's mass. Indeed, the mass M is greater than M_0, therefore our naïve expectation would be for the force on it when in motion in K to be greater than when at rest in K.

Instead, the relativistic rules tell us that, in the considered situation, the actual gravitational force on the object *decreases* when the object is moving and is accordingly more massive.

The only possible explanation of this discrepancy is that there has to be an *additional* effect due to motion of a mass, which is manifest independently of the mass increase with velocity. In a way, this additional effect is similar to the appearance of magnetic field due to motion of the electric charges. There has to be an additional component of gravitational field appearing with motion of its source as such, just as the magnetic component of the EM field appears with motion of the source of electric field. We can think of it as "magnetic" analog (magnetic face) of the field seen from a RF where the source of the field is moving.

Thus, the analogy between some properties of the gravitational field and EM field is pretty deep. It emphasizes the fundamental feature of both fields: they are relative characteristics and both transform at the change of a reference frame.

On the other hand, this analogy is far from complete. First, the magnetic field can be "created" by switching between two *inertial* reference frames, whereas the gravitational field can be created only by switching from an inertial to an *accelerated* reference frame. Second, in order to "generate" the magnetic field by transition to another reference frame, we need to have a very specified configuration of matter like a system of electric charges. This is not required for generation of the gravitational field – all you need is two frames, one inertial and the other accelerated.

Third, the analogy with respect to "creating the fields" does not work the same way in reverse: while *any* stationary gravitational field can be locally eliminated by boarding an appropriate system (e.g., a freely falling capsule), there are stationary magnetic fields that *cannot* be eliminated even locally by any transition to another reference frame. For instance, if I have an electrically neutral loop of wire flown around by a stationary current, no change of reference frame can eliminate even locally the magnetic field due to this current.

Thus, the similarity between the gravitational and magnetic phenomena is far from being simple and straightforward. Mathematically, one of the manifestations of the difference between them is that magnetic field, together with its electric counterpart, is a vector field in space (first-rank tensor under 3-rotations). In space-time, as we know (Section 6.10), it can be represented as a second-rank tensor F_{jk}. This tensor is antisymmetric, which makes it transform as a vector under

3-rotations. The gravitational field, on the other hand, turns out to be a higher rank tensor field describing the curvature of space-time. The curvature tensor has a different structure and symmetry. In other words, the EM field is a field of physical force, whereas gravitational field must be a geometrical effect of the space-time curvature due to the *energy–momentum tensor.* The equations of the EM field are linear, whereas Einstein's equations describing the space-time geometry are nonlinear.

13.2 Kinematics of GR

Already in STR the time interval dt between two events does not necessarily represent the proper time $d\tau$ between them. When the events are separated by a displacement $d\mathbf{r} = (dx^1, dx^2, dx^3)$, the proper time between them is found by choosing a reference frame where $d\mathbf{r} = 0$, so that the corresponding timelike interval is reduced to a pure time interval:

$$ds^2 = c^2 dt^2 - dr^2 = c^2 dt^2 - v^2 dt^2 = (c^2 - v^2) dt^2 = c^2 \frac{dt^2}{\gamma^2(v)} = c^2 d\tau^2. \quad (13.6)$$

Here v is the speed of a new RF relative to the initial RF. In the new RF both events occur at the same point in space and the time interval $d\tau$, which is a computational invariant, gives the proper time between these events. This proper time is related to the time interval dt read by the clocks in the initial RF by

$$d\tau = \gamma(v) dt. \quad (13.7)$$

Thus, a necessary condition for measuring the proper time between the two events directly is the choice of a RF where these events occur at one point in space (recall Section 9.1).

The same is true in GR. In order for us to proceed, we must only introduce a new concept resulting from the possibility to use arbitrary RF in the presence of arbitrary gravitational field (and thereby arbitrary space-time geometry). In this broader domain the interval (13.6) between the two events remains invariant under transformations between different coordinate systems. However, instead of Pythagorean theorem (13.6) in pseudo-Euclidean space-time, its mathematical expression takes a more general form for two reasons. First, even in SR we are not necessarily bound to Cartesian coordinates only. We could use arbitrary coordinate systems (nonrectangular, spherical, cylindrical, etc.). Our preference throughout the book has been purely pragmatic: we had used the coordinate system in which the treatment of a problem was most simple. Second, in GR with its arbitrarily moving RF and nonzero curvature of space-time, the use of Cartesian coordinates in many cases is not possible in principle (try to use the Cartesian x, y, to map the surface of the globe and you will see why).

Therefore, when formulating the GR, we must from the very beginning use an arbitrary coordinate system with coordinates that may be both – curved and nonrectangular. We have already encountered a case with nonrectangular coordinates in Section 4.7. Look again at Equation (4.101) for the space interval between the two points in these coordinates. It shows, first, that there are paired products of the *different* coordinates in the expression and second, the coefficients before the squares of coordinates and their products are numbers different from 1 or −1. Those familiar with curved coordinates (see also Section 4.7) know that the most general expression for the square of the interval is a quadratic form of the type

$$ds^2 = g_{ik}dx^i dx^k. \tag{13.8}$$

Here either of the indexes i and k ranges through 0, 1, 2, 3. And, since either index enters expression (13.8) twice, there is summation over them. So we have double sum here. In contrast to SR in Cartesian coordinates, where g_{ik} are given by the diagonal matrix (4.98), the coefficients g_{ik} here are, generally, themselves functions of all four coordinates. They are the elements of a 4×4 matrix (4.97) and form what we call the second-rank covariant metric tensor. The second-rank covariant tensor is a matrix whose elements transform as the products of covariant coordinates (the components of a contravariant tensor would transform as products of contravariant coordinates). Mathematically, the components g_{ik} of the metric tensor in (13.8) must be covariant in order for them, multiplied by the products or squares of the contravariant coordinates in (13.8), to produce a *scalar* on the left. This is why their indexes are subscripts.

In the explicit form, the 16 elements g_{ik} can be written as

$$g_{ik} = \begin{pmatrix} g_{00} & g_{01} & g_{02} & g_{03} \\ g_{10} & g_{11} & g_{12} & g_{13} \\ g_{20} & g_{21} & g_{22} & g_{23} \\ g_{30} & g_{31} & g_{32} & g_{33} \end{pmatrix}. \tag{13.9}$$

The components g_{ik} here can be considered as symmetrical in indexes i, k, that is, $g_{ik} = g_{ki}$, because they are determined from the symmetric form (13.8) which they enter multiplied by the same product $dx^i dx^k$. Therefore there are generally only 10 independent components g_{ik}, 4 with $i = k$, and 6 with $i \neq k$. In a special case of the pseudo-Euclidean geometry of the SR this reduces to the familiar form

$$g_{ik} = \begin{pmatrix} 1 & 0 & 0 & 0 \\ 0 & -1 & 0 & 0 \\ 0 & 0 & -1 & 0 \\ 0 & 0 & 0 & -1 \end{pmatrix}. \tag{13.9a}$$

We see from (13.9a) that the determinant (denoted as g) of the matrix g_{ik} is always negative:

$$g < 0. \tag{13.10}$$

This is another manifestation of the specific (negative) signature of the space-time's metric mentioned in Section 4.7.

Note that if the elements g_{ik} depend on coordinate x^0, which in most cases has the meaning of time, then the metric is a function of time. This may happen just because of use of an arbitrarily accelerated RF (e.g., a nonuniformly rotating merry-go-round). In more "serious" cases the source of a gravitational field may be nonstationary (in particular, it may produce gravitational waves). In these cases the curvature of space-time is itself a function of time.

Now we have the necessary tools for determining the true time between the events in GR. We use the same principle: consider a RF where these events happen at the same point in space. Thus, setting $dx^1 = dx^2 = dx^3 = 0$ in (13.8) gives [29]

$$ds^2 = c^2 d\tau^2 = g_{00}(dx^0)^2, \tag{13.11}$$

so that

$$d\tau = \frac{1}{c}\sqrt{g_{00}}\, dx^0. \tag{13.12}$$

To find the total proper time between *any* two events M and N happening at the same place, we must integrate the above expression:

$$\tau = \frac{1}{c}\int_M^N \sqrt{g_{00}}\, dx^0. \tag{13.13}$$

The integration here is carried out between the initial and final event; if the element g_{00} depends on the spatial coordinates as well, then the integration is taken for the fixed values x^α specifying the point where the process in question takes place.

As we see from (13.10), the determinant of the metric tensor has to be negative for any possible structure of real space-time. The component g_{00}, on the contrary, has to be positive for real processes of the world, as is clearly seen from Equation (13.13).

What if the component g_{00} of the metric tensor is negative? The negative sign of g_{00} would only mean that the corresponding RF where this occurs cannot be materialized by real objects. We have already dealt with a similar situation in Section 6.13, where we considered the mechanism of radiation of light from a sphere that was supposed to trap it. We have realized that such a system, if rigidly extended to $r \to \infty$, would require an impossible thing – the superluminal motion of its remote parts. Now we can add that this would also lead to a nonphysical expression for physical time (we are not considering the tachyons here). Let us simplify the problem by considering again a very large rotating disk of Section 8.4. Suppose we are interested in two events at some point on the disk, separated by the interval dt of our platform time. To visualize the situation better, suppose that our old friend, the disk resident Paul, is sitting there and smoking a cigar and the two events are the moments when Paul first lights the cigar and then finishes smoking and throws the cigar butt away. The 4-interval between the two events in our coordinates is given by expression (13.6), where v is Paul's speed relative to us. If Paul's radial coordinate (distance from the center) is r and the angular velocity of disk is Ω, then

$$v = \Omega r \equiv v_r. \tag{13.14}$$

According to our definition, the *proper* time $d\tau$ between the events is *Paul's time* it takes him to smoke out his cigar. We could find it directly from the time dilation effect (Equation (2.26)) of Chapter 2, but here we will find the same from expression (13.7) of GR. We need to introduce the primed coordinates to be used by disk inhabitants: $x^1 = x'$, $x^2 = y'$, $x^3 = z' = 0$. Since in the given case both events happen in the same place of the disk, $dx' = dy' = dz' = 0$ and expression (13.6) reduces to (13.11). Because the interval is invariant, we can equate it to (13.6) written in the platform coordinates (therefore I did not bother to prime it):

$$ds^2 = \frac{c^2 dt^2}{\gamma^2(v_r)} = g_{00}(dx^0)^2. \tag{13.15}$$

Suppose that Paul uses *our* time as the time coordinate on his disk. Then, $dx^0 = c dt$ and we find from (13.15) that

$$g_{00} = \gamma^{-2}(v_r), \qquad \sqrt{g_{00}} = \gamma^{-1}(v_r) = \sqrt{1 - \frac{\Omega^2 r^2}{c^2}}. \tag{13.16}$$

On the platform, our time coordinate t is the real time read by the clocks. On the disk, it is merely the coordinate time as defined in Sections 8.2 and 8.3. Paul's *proper* time can be determined from

$$d\tau = \sqrt{g_{00}} c dt = \sqrt{1 - \frac{\Omega^2 r^2}{c^2}} c dt. \tag{13.17}$$

Even in this simple case, the component g_{00} of the metric tensor turns out to be the function of position. We can make a few conclusions from this fundamental fact.

First, we see that sufficiently far from the center, g_{00} does indeed become negative. This happens for all $r > r_c$, where

$$r_c \equiv \frac{c}{\Omega}. \tag{13.18}$$

In this domain, two things happen: the speed of the disk and all the objects on it becomes superluminal and the proper time becomes imaginary – Equation (13.17) gives a nonphysical solution for time. The first of these things has been discussed briefly in Chapter 11. Already there we concluded that in the corresponding domain the considered system just cannot be materialized by known (tardyonic) physical objects. Our choice of the rotating reference frame cannot be supported by Nature in those remote regions. It does not at all mean that the physical laws are different there or something horrible is happening there. It means only that we have chosen inappropriate reference frame, which becomes unrealistic beyond $r = r_c$. We can even use corresponding abstract coordinates (not material bodies!) for that domain, but the results obtained there would require a very specific interpretation. The real world exists beyond $r = r_c$, but it cannot be described in a simple way, let alone measured or observed locally, in a rotating RF. We can only do this by returning to x, y, z, or r, θ, φ. In these coordinates, the description of the world becomes uniform and $r > r_c$ does not represent anything different from $r < r_c$. In Paul's rotating world, on the other hand, there is an absolute boundary imposed first by special and now by the general relativity and determined by Equation (13.18).

Second, within the allowed domain ($r < r_c$) Paul's time flows differently depending on his distance from the center. Only when he is exactly at the center, his time, as we have noted above, is identical with ours (this might be a good reason for his choice of our time as his x^0-coordinate). As he recedes from the center, his time runs slower. As he gets closer to the border r_c of his world, his time almost stops. We recognize something familiar in this effect. From the viewpoint of the stationary observer, it is the time dilation because Paul, while idling on his disk, is moving relative to us and the closer to r_c is his location, the closer to c is his speed. Also, if we have ensconced somewhere close to the rim to see Paul periodically each time he passes by, we can observe already familiar twin paradox. If the disk is large enough, it may take quite a while (months, years, centuries) of our time for Paul to make one rotation and pass by again, but we can notice that he is still smoking the same cigar – a year interval between our meetings will take only minutes of his time. At the critical distance from the center ($r = r_c$), Paul would freeze in time, as far as we are concerned.

Third, the GR brings in a totally new dimension to this effect. We know that when the two events happen in the same place in *both* reference frames (we sit still on the platform waiting for Paul's next appearance and so does he on his disk), the discrepancy between his and our time is absolute. Therefore, Paul also observes the slowing down of all the processes on his disk, *as compared with our time*, or with time at the disk's center. What can account for this effect in *his* RF? Here we have come to a point we tried to avoid in Section 9.4. But now we know that there is a gravitational field in Paul's frame and since Paul is stationary in that frame, this field is the only possible physical reason for slowing down of all the processes there. Moreover, we are now in the position to determine slowing quantitatively!

Let us consider a particle of mass m_0 stationary in Paul's frame. The force on it, from our perspective, is the centripetal force; from Paul's viewpoint, it is a force balancing the inertial (centrifugal) force, which, according to the equivalence principle, is identical to the gravitational force diverging from the rotational axis. Unusual as it may appear to Paul, this force is as good in its actions as a gravitation force produced by matter.

In our RF this force is

$$f = m_0\gamma(v_r)\frac{v_r^2}{r} = m_0\gamma(v_r)\Omega^2 r. \tag{13.19}$$

The Lorentz factor here accounts for the relativistic mass increase. Together with this factor, the force f becomes infinite as $r \to r_c$.

In Paul's frame (more accurately, in an inertial RF instantly comoving with Paul) the magnitude of this force can be found from (13.19) by using the second Equation (5.84), where we must set $v_x = V = v_r$. We obtain

$$f' = f\gamma(v_r) = m_0\gamma^2(v_r)\Omega^2 r, \tag{13.20}$$

or

$$f' = m_0 \Omega^2 \frac{r}{1 - (\Omega^2 r^2 / c^2)} = m_0 \Omega^2 \frac{r}{1 - (r^2 / r_c{}^2)}. \tag{13.20a}$$

Since in Paul's frame the force $\mathbf{f}' = f'\hat{\mathbf{r}}$ is indistinguishable from the gravity force (and Ω is assumed to be constant), it can be conventionally described as the negative gradient of the corresponding static gravitational potential $\Phi_g(r)$:

$$\mathbf{f}' - \vec{\nabla} \Phi_g(\mathbf{r}), \tag{13.21}$$

where

$$\Phi_g(r) = -\frac{1}{m_0} \int_0^r f'(r') \mathrm{d}r' = -\Omega^2 \int_0^r \frac{r' \mathrm{d}r'}{1 - (r'^2 / r_c{}^2)}. \tag{13.22}$$

This integral can be easily taken and yields

$$\Phi_g(r) = \frac{1}{2} c^2 \ln\left(1 - \frac{r^2}{r_c{}^2}\right) = c^2 \ln \sqrt{1 - \frac{r^2}{r_c{}^2}} \quad (r < r_c). \tag{13.23}$$

Comparing this with (13.16) we find a simple relation between the component g_{00} of the metric tensor and the gravitational potential in this system:

$$\Phi_g(r) = \frac{1}{2} c^2 \ln g_{00}(r), \quad \text{or} \quad g_{00}(r) = e^{2\Phi_g(r)/c^2} \quad (\Phi_g(r) \leq 0). \tag{13.24}$$

The potential is zero at the center and negative elsewhere, increasing in absolute value toward the periphery. As the radial distance approaches the critical value $r \rightarrow r_c$, the potential $\Omega_g(r) \rightarrow -\infty$, thus forming an infinitely deep and narrow circular potential well around the center. Note that this potential has its reference point at the origin, as opposed to potential of a localized mass, whose reference point is usually taken at infinity. The different choice of reference points for these two cases is based on difference in "global" characteristics of the corresponding potential fields. As mentioned in the previous section, the gravitational field caused by acceleration permeates the whole space, while the field of a localized mass vanishes at infinity. This difference in the boundary conditions reflects the difference in the origin of the field. While the origin of the "regular" gravitational field of mass is mass itself, the origin of the field of the type (13.23) is acceleration of the corresponding RF.[1)]

Combining our results for proper time and for gravitational potential, we conclude that out of two clocks, the one at a lower potential ticks slower than the one at higher potential.

This can be used for prediction of one of the crucial experimental tests of GR – the gravitational redshift. Consider a cylindrical wave converging to the rotational axis of

1) The concept of gravitational potential as defined by (13.22) and (13.23) in an accelerated frame is restricted to static fields and stationary masses. A moving mass is subject to additional forces (the Coriolis force), which requires the corresponding generalization of the concept of potential.

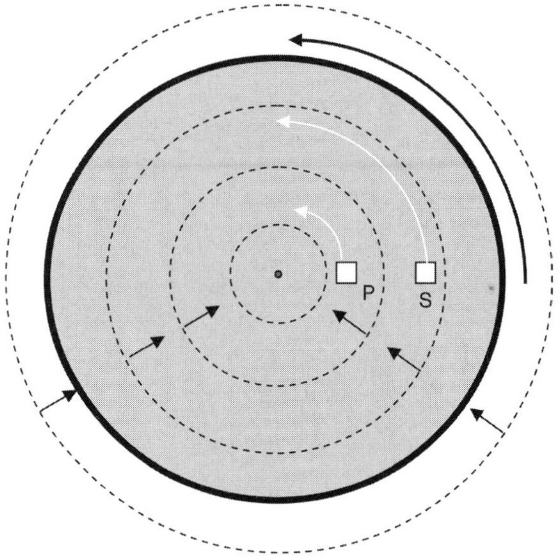

Figure 13.3 Paul and Sam's frequency-recording experiment on rotating disk. White arrows indicate Paul's and Sam's rotational velocities. The disk inhabitants live in a gravitational field generated by rotation of their RF. For them, this is an experiment on *gravitational redshift*. Dashed circles with radial arrows represent the converging cylindrical radio wave. In case of radio broadcast *from* the center, the observed effect would be the *blueshift*.

our spinning disk (Figure 13.3). This process is observed from the disk by Paul and his friend Sam who happens to be there on a visit. Both friends are positioned on the same radial line, at points P and S, respectively. Let T be the period of this wave measured in inertial frame K attached to the center of the disk. Consider two events – the passing of the two consecutive wave crests by Paul. The time interval in K between these two passages is T. For Paul, both events happen at the same point and therefore the time he measures between these events is the proper time between them (physically, this is the proper period τ_P of the passing wave at Paul's location). Using (13.17), we can write

$$\tau_P = \sqrt{g_{00}(r_P)}\, cT = \sqrt{1 - \frac{r_P^2}{r_c{}^2}}\, cT. \tag{13.25a}$$

Similarly, the proper period measured by Sam is

$$\tau_S = \sqrt{g_{00}(r_S)}\, cdt = \sqrt{1 - \frac{r_S^2}{r_c{}^2}}\, cT. \tag{13.25b}$$

Introduce the corresponding proper frequencies and take their ratio:

$$\frac{\omega_S}{\omega_P} = \frac{\tau_P}{\tau_S} = \sqrt{\frac{g_{00}(r_P)}{g_{00}(r_S)}}. \tag{13.26}$$

We see that the measured period (or frequency) of the same wave is different at different locations in an accelerated frame. Since there always is a gravitational field in such a frame, we can attribute this effect to the influence of this field. Using (13.24), we can write

$$\frac{\omega_S}{\omega_P} = e^{\Delta\Phi_g/c^2}, \tag{13.27}$$

where

$$\Delta\Phi_g \equiv \Phi_g(r_P) - \Phi_g(r_S) \tag{13.28}$$

is the potential difference between two locations.

In case when $\Delta\Phi_g \ll c^2$, we can expand the exponent into the Taylor series up to the first power and write Equation (13.27) in the form

$$\frac{\omega_S}{\omega_P} \cong 1 + \frac{\Delta\Phi_g}{c^2}, \tag{13.29}$$

or, denoting $\omega_S - \omega_P \equiv \Delta\omega$ and dropping the labels

$$\frac{\Delta\omega}{\omega} \cong \frac{\Delta\Phi_g}{c^2}. \tag{13.30}$$

Since gravitational field associated with a noninertial RF is locally indistinguishable from a field due to gravitating masses, Equations (13.26–13.30) must hold for *any* gravitational field. When applied to light propagating in such a field, they describe a famous "gravitational redshift" – the decrease in frequency of light propagating from region with lower gravitational potential to a region with higher potential. This shift was one of the first predictions of GR (the other two are the precession of planetary orbits and deflection of light rays near a massive body). This prediction (as well as two others) has been experimentally confirmed in numerous astronomical observations of spectral lines of known chemical elements in the atmospheres of massive stars. In 1959 it was confirmed in spectacular experiments by Pound and Rebka [126], who compared the proper frequency of the same spectral line at two different levels – in the basement and at the top of the Jefferson tower at Harvard University, using a phenomenally sensitive technique based on the Mossbauer effect [127].

Note that the term "gravitational redshift" is generic for *any* frequency shift of light in gravitational field. If the source of light is at a higher potential than the detector, we will observe the blueshift. For instance, imagine that we observe a spaceship from a *Star Trek* serial, hovering over a surface of a neutron star or may be even of a black hole. If the ship's crew signal to us using light beams, we will find this light redshifted. If we answer them by sending similar light signals, they will find the coming light blueshifted.

Sometimes the gravitational redshift is explained in the framework of quantum mechanics, using the concept of photons. It is considered as the actual frequency shift of a photon as it travels from one place to another. Suppose a photon is propagating in a static gravitational field of a planet of mass M_0. It experiences the gravitational force $f_g = G(M_0 m/r^2)$, where

$$m = \frac{\mathcal{E}}{c^2} = \frac{\hbar\omega}{c^2} \tag{13.31}$$

is the relativistic mass of the photon and r is its distance from the center of the planet. Suppose the photon is moving up. As its altitude increases, so does its potential energy. Since the rest mass of the photon is zero, all its energy in free space consists of kinetic energy. Thus, we have $\mathcal{E} = K = \hbar\omega$. The kinetic energy is the only source for the photon to do work against the gravity force. This work is equal to the change in the photon's potential energy,

$$\Delta W = -\Delta K = m\Delta\Phi_g = \hbar\Delta\omega, \tag{13.32}$$

and since the latter increases for the rising photon, the kinetic energy and accordingly, the photon's frequency decreases. Quantitatively, this follows from (13.32). In view of (13.31), we obtain from (13.32) the fractional change of frequency

$$\frac{\Delta K}{\hbar\omega} = \frac{\Delta\omega}{\omega} = \frac{\Delta\Phi_g}{c^2}, \tag{13.33}$$

which is identical with the limit (13.30).

This explanation is very appealing to our intuition, but one could argue that it is misleading [128]. First, according to QM, the quantity $\hbar\omega$ is an eigenvalue of the photon's Hamiltonian, which includes its *total* energy. Equation $\hbar\omega = K$ holds only in free space; in a potential field, we have $\hbar\omega = \mathcal{E} = K + U$, which implies that the photon's frequency must remain constant in the process of its propagation.

Second, the statement that the photon's frequency depends on detector's location would only make sense if the detector's clock has the same rate in all locations. But this contradicts the basic result (13.25a) and (13.25b), according to which it is the proper time of the detector that depends on location. So the redshift must result from difference in ticking rate of identical clocks at two different points, rather than from different frequency of the same photon at these points.

It would, therefore, be better to formulate the gravitational redshift using the Kirchoff's law, according to which if an atom emits a certain frequency, it absorbs the same frequency. In other words, two identical atoms are ideally tuned to one another. According to the Doppler effect, this law is modified when the atoms are in relative motion. According to GR, this law is modified in a gravitational field. If a source and detector, containing identical atoms, are stationary, but held at points with different gravitational potentials, they are detuned. Suppose that the detector is at a higher potential. Then, in order to tune them again, we must move either detector or emitter, or both, toward one another. If the emitter is at a higher potential, then in order to get them tuned we must move them away from one another.

This formulation avoids the explicit reference to any specific "culprit" in detuning, but correctly describes the effect in the operational terms.

Whatever the interpretation, we have now the answer to the question asked in Section 9.4 about how the disk resident Paul explains the difference in frequency of the emitted and detected light, if it cannot be explained in terms of the Doppler effect. The explanation is that the identical measuring clocks, although at rest relative to one another, are detuned (have different ticking rates) because they are located at points with different gravitational potentials.

Now we need an expression for the true distance between the two different points in space. Let us call it dl. To get the right expression for dl, we must make sure that both points are considered at the same moment of true time. This requires two clocks – one at each point and both synchronized so that they read the same proper time. But we cannot set the same proper time for two spatially separated clocks by just setting $dx^0 = 0$, because due to dependence of component g_{00} on position, the proper time is obtained from time coordinate x^0 differently for different locations. Indeed, applying (13.12) to two processes at two different locations A and B, we have

$$d\tau_A = \frac{1}{c}\sqrt{g_{00}(x^0, x_A^\alpha)}dx^0, \qquad d\tau_B = \frac{1}{c}\sqrt{g_{00}(x^0, x_B^\alpha)}dx^0. \tag{13.34}$$

As we have found, even if the coordinate x^0 is the same in the two expressions, the intervals of the true time are generally different for different points A and B. In order to determine dl, we can do the following. Consider the light signal starting from A to B and then right back to A along the same way. Then the round-trip time multiplied by c will give us dl – the separation between A and B. Let us write down expression (13.7) for the interval, separating the time and space coordinates

$$ds^2 = g_{00}(dx^0)^2 + 2g_{0\alpha}dx^0dx^\alpha + g_{\alpha\beta}dx^\alpha dx^\beta \tag{13.35}$$

(recall that the metric tensor is symmetrical: $g_{ik} = g_{ki}$). The interval between the departure of the light signal from A and its arrival at B is zero (it is a lightlike interval!). Setting expression (13.35) to zero, we obtain the quadratic equation for dx^0:

$$g_{00}(dx^0)^2 + 2g_{0\alpha}dx^0dx^\alpha + g_{\alpha\beta}dx^\alpha dx^\beta = 0. \tag{13.36}$$

Its solution gives the two values for dx^0:

$$dx^0 = \frac{1}{g_{00}}\left\{\begin{array}{l} -g_{0\alpha}dx^\alpha + \sqrt{(g_{0\alpha}g_{0\beta} - g_{00}g_{\alpha\beta})dx^\alpha dx^\beta} \\ -g_{0\alpha}dx^\alpha - \sqrt{(g_{0\alpha}g_{0\beta} - g_{00}g_{\alpha\beta})dx^\alpha dx^\beta} \end{array}\right. . \tag{13.37}$$

One corresponds to travel time from A to B, the other gives the travel time from B to A.

We can represent the situation graphically as two world lines for the particles A and B and the broken world line of the photon traveling between them (Figure 13.4). Let x^0 be the moment of the photon arrival at B. Then $x^0 + dx^0_{(1)}$ is the moment of its departure from A and $x^0 + dx^0_{(2)}$ is the moment of its return to A. The time between these two moments is

$$dx^0_{12} = (x^0 + dx^0_2) - (x^0 + dx^0_1) = dx^0_2 - dx^0_1 = \frac{2}{g_{00}}\sqrt{(g_{0\alpha}g_{0\beta} - g_{00}g_{\alpha\beta})dx^\alpha dx^\beta}. \tag{13.38}$$

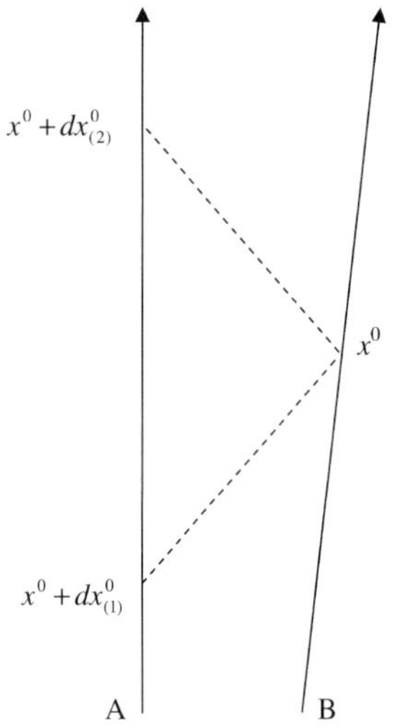

Figure 13.4

Corresponding interval of the proper time between these events is, according to (13.12)

$$d\tau_{12} = \frac{2}{c\sqrt{g_{00}}}\sqrt{(g_{0\alpha}g_{0\beta} - g_{00}g_{\alpha\beta})dx^{\alpha}dx^{\beta}} \tag{13.39}$$

and the true distance between A and B is

$$dl_{12} = \frac{1}{2}c d\tau_{12} = \sqrt{\left(-g_{\alpha\beta} + \frac{g_{0\alpha}g_{0\beta}}{g_{00}}\right)dx^{\alpha}dx\beta}. \tag{13.40}$$

The same process can be used for synchronization of clocks A and B. We just apply Einstein's definition of simultaneity to this situation because both special and general relativity are based on the postulate of the invariance of local speed of light. By definition, the moment at A that is simultaneous with the moment x^0 at B is the reading of A clock halfway between the moments of the signal departure and arrival

$$x^0_A = x^0_B + \frac{1}{2}\left(dx^0_{(1)} + dx^0_{(2)}\right). \tag{13.41}$$

The moment x^0_A at A, simultaneous with the moment x^0_B at B, is generally not equal to x^0_B.

According to (13.41) and (13.36), the difference between them is

$$\delta x^0 = x^0_B - x^0_A = -\frac{g_{0\alpha}}{g_{00}}dx^{\alpha}. \tag{13.42}$$

The farther apart are the world lines A and B (the larger dx^{α}), the greater δx^0.

Let us consider a special case when the gravitational field does not depend on time. Then we can extend the synchronization procedure between A and B considering B being farther and farther apart from A. As a result, we will obtain an expression for the difference in local times of the two simultaneous events at the end points of an arbitrary curve AB:

$$\Delta x^0_{\mathrm{AB}} = -\int_{\mathrm{A}}^{\mathrm{B}} \frac{g_{0\alpha}}{g_{00}} \mathrm{d}x^\alpha. \tag{13.43}$$

We are facing an apparently crazy conclusion: two simultaneous events may be characterized by the different local times. However, terrifying, as this may seem, it is (at least partially) a familiar situation. When two persons – one in New York and the other in Moscow, view a broadcast from a TV station happening to be midway between these geographic points, they see each episode simultaneously because they are equidistant from the station. Nevertheless, their local times are different.

We can go farther and close the line, so that its end points merge together, say, at A. Then, returning to the starting point at the same moment (integration is not the real motion along the curve, it is an instantaneous mathematical operation), we nevertheless realize that there is time discrepancy at this point equal to

$$\Delta x^0_{\mathrm{AA}} = -\oint \frac{g_{0\alpha}}{g_{00}} \mathrm{d}x^\alpha. \tag{13.44}$$

This generalizes expressions (8.26) and (8.27) for the time lag for a point A on a circle centered at the origin of a rotating disk.

Appendix A
State Function and the Continuity Equation

A certain quadratic form of a state function satisfies the continuity equation similar to that of the mass and flow density in fluid dynamics, or charge and current density in EM. Accordingly, such a form can be interpreted as a density flow type characteristic of a system described by this state function.

Consider the Schrödinger equation for a state function Ψ

$$i\hbar\frac{\partial\Psi}{\partial t} = \hat{H}\Psi, \tag{A1}$$

where

$$\hat{H} = -\frac{\hbar^2}{2m}\nabla^2 + U \tag{A2}$$

is the Hamiltonian of the particle, with a scalar function $U(\mathbf{r}, t)$ representing particle's potential energy in an external potential field.

Take a complex conjugate of Equation (A1). Since the Hamiltonian is a real function, the result will be

$$-i\hbar\frac{\partial\Psi^*}{\partial t} = \hat{H}\Psi^*. \tag{A3}$$

Now multiply (A1) through from the left by Ψ^* and (A3) by Ψ. We will obtain

$$i\hbar\Psi^*\frac{\partial\Psi}{\partial t} = \Psi^*\hat{H}\Psi, \tag{A4}$$

$$-i\hbar\Psi\frac{\partial\Psi^*}{\partial t} = \Psi\hat{H}\Psi^*. \tag{A5}$$

Take the difference of the last two equations:

$$i\hbar\Psi^*\frac{\partial\Psi}{\partial t}\left(\Psi^*\frac{\partial\Psi}{\partial t} + \Psi\frac{\partial\Psi^*}{\partial t}\right) = \Psi^*\hat{H}\Psi - \Psi\hat{H}\Psi^*. \tag{A6}$$

Special Relativity and How it Works. Moses Fayngold

ISBN: 978-3-527-40607-4

Since U is a real function, the right-hand side of (A6) reduces to

$$-\frac{\hbar^2}{2m}(\Psi^*\nabla^2\Psi - \Psi\nabla^2\Psi^*) = -\frac{\hbar^2}{2m}\vec{\nabla}(\Psi\vec{\nabla}\Psi^* - \Psi^*\vec{\nabla}\Psi) \tag{A7}$$

and consequently, Equation (A6) takes the form

$$\frac{\partial(\Psi^*\Psi)}{\partial t} = \frac{i\hbar}{2m}\vec{\nabla}(\Psi\vec{\nabla}\Psi^* - \Psi^*\vec{\nabla}\Psi^*). \tag{A8}$$

Denote

$$\rho \equiv \Psi^*\Psi \quad \text{and} \quad \mathbf{j} \equiv \frac{i\hbar}{2m}(\Psi\vec{\nabla}\Psi^* - \Psi^*\vec{\nabla}\Psi) = \frac{\hbar}{m}\mathrm{Im}(\Psi\vec{\nabla}\Psi^*). \tag{A9}$$

Then (A8) can be written as

$$\frac{\partial\rho}{\partial t} = -\vec{\nabla}\cdot\mathbf{j}. \tag{A10}$$

This is a continuity equation for a flowing entity with a local density ρ and the corresponding flow density $\mathbf{j}$. We can therefore interpret ρ and $\mathbf{j}$ as defined by (A9) as the probability and probability flux density, respectively.

Appendix B
Representation of Observables by Operators

Consider first an arbitrary state $\Psi(\mathbf{r})$ in $\mathbf{r}$-representation (we drop here the possible t-dependence). According to the probabilistic interpretation, $|\Psi(\mathbf{r})|^2$ is the probability density to find the particle (originally prepared in the state $\Psi(\mathbf{r})$) in the vicinity of point $\mathbf{r}$. Since the Nature, according to Born's interpretation, is intrinsically probabilistic, another measurement on an identical particle originally in the same state $\Psi(\mathbf{r})$ may find it at another location $\mathbf{r}'$. Because of the variability of the outcomes in the measurement of the same characteristics in the same state, we generally characterize the state by the *mean* value of this characteristic. We can thus introduce the mean position (an average $\langle \mathbf{r} \rangle$) in a given state:

$$\langle \mathbf{r} \rangle = \int \mathbf{r}\mathcal{P}(\mathbf{r})d\mathbf{r} = \int \Psi^*(\mathbf{r})\mathbf{r}\Psi(\mathbf{r})d\mathbf{r}. \tag{B1}$$

Consider now the *momentum* measurement in the same state. We know that the state $\Psi(\mathbf{r})$ can be represented as a superposition of the monochromatic plane waves (wave packet):

$$\Psi(\mathbf{r}) = \int \Phi(\mathbf{p})e^{i(\mathbf{p}/\hbar)r}\, d\mathbf{p}. \tag{B2}$$

From the mathematical viewpoint, this is a Fourier expansion of $\Psi(\mathbf{r})$. The expansion coefficient $\Phi(\mathbf{p})$ has a physical interpretation as the probability amplitude to find the particle's *momentum* in the vicinity of $\mathbf{p}$. The set of values $\Phi(\mathbf{p})$ (or just the function $\Phi(\mathbf{p})$) completely specifies the physical state $\Psi(\mathbf{r})$, and vice versa. Once we know $\Psi(\mathbf{r})$, we can uniquely determine $\Phi(\mathbf{p})$

$$\Phi(\mathbf{p}) = \int \Psi(\mathbf{r})e^{-i(\mathbf{p}/\hbar)\mathbf{r}}\, d\mathbf{r}. \tag{B3}$$

Since $\Phi(\mathbf{p})$ is explicitly expressed in terms of momentum $\mathbf{p}$, it can be called the state function in $\mathbf{p}$-representation. Note that it describes *the same* physical state as $\Psi(\mathbf{r})$ but in a different representation. In this respect, $\Psi(\mathbf{r})$ and $\Phi(\mathbf{p})$ (when related to one another by (B2) and (B3)) are like the two linguistic versions of the *same* story, for example, one in English and one in Russian.

Special Relativity and How it Works. Moses Fayngold

ISBN: 978-3-527-40607-4

Now, we may be interested in another characteristic of the system – its average momentum $\langle \mathbf{p} \rangle$. Then the **p**-representation of the given state (the state function in the form $\Phi(\mathbf{p})$) provides us with the most natural tool for its calculation. In complete analogy with (B1), we can write

$$\langle \mathbf{p} \rangle = \int \Phi^*(\mathbf{p})\mathbf{p}\Phi(\mathbf{p})d\mathbf{p}. \tag{B4}$$

But, on the other hand, if both $\Phi(\mathbf{p})$ and $\Psi(\mathbf{r})$ are only different but equivalent representations of the same state, then either of them stores exactly the same (and maximal possible) information about the state. In this case, the rules of quantum mechanics must have a procedure for calculation of the mean *momentum* from $\Psi(\mathbf{r})$ as well. Such a procedure does exist. Indeed, in view of (B3), the previous equation can be written as

$$\begin{aligned}\Phi(\mathbf{p}) &= \int\left(\int \Psi^*(\mathbf{r}')e^{i(\mathbf{p}/\hbar)\mathbf{r}'}d\mathbf{r}'\mathbf{p}\int \Psi(\mathbf{r})e^{-i(\mathbf{p}/\hbar)\mathbf{r}}d\mathbf{r}\right)d\mathbf{p}\\ &= -i\hbar\int\left(\int \Psi^*(\mathbf{r}')e^{i(\mathbf{p}/\hbar)\mathbf{r}'}d\mathbf{r}\int \Psi(\mathbf{r})\nabla e^{-i(\mathbf{p}/\hbar)\mathbf{r}}d\mathbf{r}\right)d\mathbf{p}.\end{aligned} \tag{B5}$$

and then rearranged as

$$\begin{aligned}\langle p \rangle &= i\hbar\int d\mathbf{p}\iint d\mathbf{r}d\mathbf{r}'\Psi^*(\mathbf{r}')\Psi(\mathbf{r})\nabla e^{i\mathbf{k}(\mathbf{r}'-\mathbf{r})}\\ &= i\hbar\iint \Psi(\mathbf{r}')\Psi(\mathbf{r})d\mathbf{r}d\mathbf{r}'\nabla\int e^{i\mathbf{k}(\mathbf{r}'-\mathbf{r})}d\mathbf{p}.\end{aligned} \tag{B6}$$

But the last integral in the expression on the right is one of the definitions of δ-function. Therefore, integrating by parts and using the properties of δ-function, we finally obtain

$$\begin{aligned}\langle \mathbf{p} \rangle &= i\hbar\iint \Psi^*(\mathbf{r}')\Psi(\mathbf{r})d\mathbf{r}d\mathbf{r}'\,\vec{\nabla}\,\delta(\mathbf{r}-\mathbf{r}')\\ &= -i\hbar\iint \Psi^*(\mathbf{r}')\,\vec{\nabla}\,\Psi(\mathbf{r})\delta(\mathbf{r}-\mathbf{r}')d\mathbf{r}d\mathbf{r}'\\ &= \int \Psi^*(\mathbf{r})\hat{\mathbf{p}}\,\Psi(\mathbf{r})d\mathbf{r},\end{aligned} \tag{B7}$$

where $\hat{\mathbf{p}}$ is precisely the operator determined by (10.37).

Appendix C
The Pauli Matrices

As any other observable, spin can be represented by the corresponding operator. For the particles with spin $(1/2)\,\hbar$, the spin projection onto any direction can have only two values. Accordingly, the spin operator can be represented by 2×2 matrices. Since spin is a 3-vector, we must have three such matrices – one for each out of three spin components. It is convenient to separate the factor $(1/2)\,\hbar$, writing

$$\hat{s} = \frac{1}{2}\hbar\hat{\sigma}. \tag{C1}$$

In this case, the problem reduces to finding three matrices $\hat{\sigma}_\alpha$, $\alpha = 1, 2, 3$. The eigenvalues of these matrices, in view of the normalization (C1), must be $+1$ for the positive spin projection onto the corresponding axis and -1 for the negative projection along this axis. It follows that for any α the eigenvalues of the *squares* $\hat{\sigma}_\alpha^2$ are all equal to $+1$. Thus, we can write

$$\hat{\sigma}_\alpha^2 = \begin{pmatrix} 1 & 0 \\ 0 & 1 \end{pmatrix}, \quad \alpha = 1, 2, 3. \tag{C2}$$

Let us now choose a representation in which the measurement results will be obtained for the z-component of the spin (i.e., we are going to measure the spin component along the z-direction). This corresponds to $\alpha = 3$, and so the matrix $\hat{\sigma}_3$ will be in its own representation. A matrix operator in its own representation is diagonal, with diagonal elements being its eigenvalues. Therefore, in this representation we have

$$\hat{\sigma}_3 = \begin{pmatrix} 1 & 0 \\ 0 & -1 \end{pmatrix}. \tag{C3}$$

We need to find the explicit form of the remaining two matrices.

As was mentioned in Section 10.1, the different components of the angular momentum operator do not commute. This is a general property of the angular momentum, be it orbital or intrinsic. Although the latter is not a function of coordinates, its components satisfy the same commutation relations that hold for

Special Relativity and How it Works. Moses Fayngold

ISBN: 978-3-527-40607-4

the angular momentum:

$$\begin{aligned}
\hat{\sigma}_1\hat{\sigma}_2 - \hat{\sigma}_2\hat{\sigma}_1 &= 2i\hat{\sigma}_3,\\
\hat{\sigma}_2\hat{\sigma}_3 - \hat{\sigma}_3\hat{\sigma}_2 &= 2i\hat{\sigma}_1,\\
\hat{\sigma}_3\hat{\sigma}_1 - \hat{\sigma}_1\hat{\sigma}_3 &= 2i\hat{\sigma}_2.
\end{aligned} \tag{C4}$$

As we see, these relations are obtained from one another by cyclic permutation.

Since the commutators (C4) are not zeros, there are indeterminacy relations between the variances of the different components. Therefore, in a state in which the z-component of spin has a definite value (be it $\sigma_3 = 1$ or $\sigma_3 = -1$), the other two components are indetermined. This means that in the σ_3-representation the other two matrices are nondiagonal:

$$\hat{\sigma}_1 = \begin{pmatrix} a_{11} & a_{12} \\ a_{21} & a_{22} \end{pmatrix}, \qquad \hat{\sigma}_2 = \begin{pmatrix} b_{11} & b_{12} \\ b_{21} & b_{22} \end{pmatrix}. \tag{C5}$$

Put this into (C2), which holds in any representation since the unitary matrix is the invariant. After simple algebra we will obtain the relations

$$a_{12}(a_{11} + a_{22}) = 0, \qquad a_{11}^2 + a_{12}a_{21} = a_{22}^2 + a_{12}a_{21} = 1 \tag{C6}$$

and similar relations for the elements of matrix $\hat{\sigma}_2$.

Since the nondiagonal elements of matrices (C5) are nonzero, conditions (C2) require that the traces of these matrices be zero. Denoting $a_{11} = -a_{22} \equiv a$ and $b_{11} = -b_{22} \equiv b$, we can write

$$\hat{\sigma}_1 = \begin{pmatrix} a & a_{12} \\ a_{21} & -a \end{pmatrix}, \qquad \hat{\sigma}_2 = \begin{pmatrix} b & b_{12} \\ b_{21} & -b \end{pmatrix}. \tag{C7}$$

Now put this result into the last two equations (C4). After the same simple algebra we will obtain

$$a = b = 0, \qquad b_{12} = -ia_{12}, \qquad b_{21} = ia_{21}. \tag{C8}$$

It follows then from the second of Equations (C6)

$$a_{12}a_{21} = 1. \tag{C9}$$

From the symmetry considerations we can assume that the symmetric nondiagonal elements must have the same magnitude, in which case it follows $|a_{12}| = |a_{21}| = 1$, and we can write

$$a_{12} = e^{i\alpha}, \qquad a_{21} = e^{-i\alpha}, \tag{C10}$$

where α is an arbitrary real number. Without any loss of generality, we can set $\alpha = 0$, so that $a_{12} = a_{21} = 1$. Combining this with (C8), we can finally write

$$\hat{\sigma}_1 = \begin{pmatrix} 0 & 1 \\ 1 & 0 \end{pmatrix}, \qquad \hat{\sigma}_2 = i\begin{pmatrix} 0 & -1 \\ 1 & 0 \end{pmatrix}, \qquad \hat{\sigma}_3 = \begin{pmatrix} 1 & 0 \\ 0 & -1 \end{pmatrix}. \tag{C11}$$

Appendix D
Dimensionality of Dirac's Matrices

We need to determine the number N for matrices in Dirac's equation. To this end, forget for a while about physics and recall some linear algebra. Consider determinants of matrices (10.124):

$$D(\hat{\gamma}^{(j)}) = \begin{bmatrix} \gamma_{11}^{(j)} & \gamma_{12}^{(j)} & \cdots & \gamma_{1N}^{(j)} \\ \gamma_{21}^{(j)} & \gamma_{22}^{(j)} & \cdots & \gamma_{2N}^{(j)} \\ \cdots & \cdots & \cdots & \cdots \\ \gamma_{N1}^{(j)} & \gamma_{N1}^{(j)} & \cdots & \gamma_{NN}^{(j)} \end{bmatrix}, \qquad j = 0, 1, 2, 3. \tag{D1}$$

Note that the determinant of the product of two matrices equals the product of determinants of the individual matrices. In our case,

$$D(\hat{\gamma}^{(0)}\hat{\gamma}^{(\alpha)}) = D(\hat{\gamma}^{(0)})D(\hat{\gamma}^{(\alpha)}). \tag{D2}$$

But determinants are just numbers, so the product on the right-hand side in (D2) can be written in reverse order and therefore,

$$D(\hat{\gamma}^{(0)}\hat{\gamma}^{(\alpha)}) = D(\hat{\gamma}^{(\alpha)}\hat{\gamma}^{(0)}). \tag{D3}$$

In other words, while the matrix products $(\hat{\gamma}^{(0)}\hat{\gamma}^{(\alpha)})$ and $(\hat{\gamma}^{(\alpha)}\hat{\gamma}^{(0)})$ are generally different matrices, they have equal determinants.

Now, it follows from the anticommutation conditions (10.128) that in our case

$$\hat{\gamma}^{(0)}\hat{\gamma}^{(\alpha)} = -\hat{\gamma}^{(\alpha)}\hat{\gamma}^{(0)} = -\hat{I}\hat{\gamma}^{(\alpha)}\hat{\gamma}^{(0)}, \tag{D4}$$

where $\hat{I}$ is the identity matrix. Therefore,

$$D(\hat{\gamma}^{(0)}\hat{\gamma}^{(\alpha)}) = D(\hat{\gamma}^{(\alpha)}\hat{\gamma}^{(0)}) = D(-\hat{I}\hat{\gamma}^{(0)}\hat{\gamma}^{(\alpha)}) = D(-\hat{I})D(\hat{\gamma}^{(0)}\hat{\gamma}^{(\alpha)}) \tag{D5}$$

(note that $D(-\hat{I})$ is *not* the same as $-D(\hat{I})$).
It follows

$$D(-\hat{I}) = 1, \quad \text{that is,} \quad (-1)^N = 1. \tag{D6}$$

Thus, N must be an even number. The simplest case is $N = 2$, but there are only four linear-independent 2×2 matrices, of which one *commutes* with other three and thus

Special Relativity and How it Works. Moses Fayngold

ISBN: 978-3-527-40607-4

does not satisfy the anticommutation condition (10.128). In contrast, for $N=4$ there exist at least four linear-independent matrices, all satisfying these conditions. Specifically, the reader can verify by inspection that four matrices

$$\hat{\gamma}^{(0)}=\begin{pmatrix}1&0&0&0\\0&1&0&0\\0&0&-1&0\\0&0&0&-1\end{pmatrix},\quad \hat{\gamma}^{(1)}=\begin{pmatrix}0&0&0&1\\0&0&1&0\\0&1&0&0\\1&0&0&0\end{pmatrix},$$
$$\hat{\gamma}^{(2)}=\begin{pmatrix}0&0&0&-i\\0&0&i&0\\0&-i&0&0\\i&0&0&0\end{pmatrix},\quad \hat{\gamma}^{(3)}=\begin{pmatrix}0&0&1&0\\0&0&0&-1\\1&0&0&0\\0&-1&0&0\end{pmatrix}\tag{D7}$$

satisfy all the requirements and are Hermitian, so they can represent physically measurable quantities, that is, be their operators. They are called the *gamma matrices* or the Dirac's matrices. But if you take a closer look at them, you recognize that they are composed of the Pauli matrices! Indeed, we can rewrite (D7) in a more compact form using the notations (C11) for the Pauli matrices:

$$\hat{\gamma}^{(0)}=\begin{pmatrix}\mathrm{I}&0\\0&-\mathrm{I}\end{pmatrix},\quad \hat{\gamma}^{(1)}=\begin{pmatrix}0&\sigma_1\\\sigma_1&0\end{pmatrix},$$
$$\hat{\gamma}^{(2)}=\begin{pmatrix}0&\sigma_2\\\sigma_2&0\end{pmatrix},\quad \hat{\gamma}^{(3)}=\begin{pmatrix}0&\sigma_3\\\sigma_3&0\end{pmatrix}.\tag{D8}$$

Here each element of the matrices in (D8) is itself a 2×2 matrix.

Appendix E
Optical Barriers for a Photon

For a medium such as water or glass ($n > 1$), our first impulse is to relate positive potential U to it. One reason for this is a false analogy with the bounce of a tennis ball from a concrete wall. Both the wall and the glass plate are barriers for the ball and its recoil from the barrier is quite natural behavior for a classical particle. And since the phase velocity of light in glass is less than in vacuum, the glass plate impedes the propagation of a light pulse, as the potential barrier with $U > 0$ would do. Therefore, it seems natural to associate potential $U > 0$ with a transparent medium having $n > 1$.

The facts we have used for this analogy are correct. But the conclusion is wrong.

In terms of potential energy, a slab of glass plays the role of the *potential well* attracting light particles (Section 6.10). In other words, to a material with $n > 1$ we must assign a potential with $U < 0$. This is a well-known fact widely used in fiber optics. The notion of a dielectric substance as a medium that attracts light can be traced back to Newton, who explained the bending of a refracted ray toward the normal to the interface as a result of attracting forces exerted on light particles as they pass from air into dielectric medium.

Reversing the argument, we can say that by the same token the passing of light from glass into air can be described as it entering the region with a potential higher than in glass. In other words, if we set the potential in glass to be zero, we must attribute to air a potential $U > 0$. Accordingly, an air layer between the two glass plates is a potential barrier for a photon. Similarly, a layer of plasma, for frequencies greater than the corresponding plasma frequencies, also plays the role of potential barrier.

But within the framework of this argument there arise two questions.

First, if glass plays the role of a potential well, then (assuming that all incident particles are identical) there appears to be no room left for the light reflection from glass under given conditions. No classical particle can bounce off a boundary of a well that attracts it.

The answer to this question lies in realizing the wave nature of light. As mentioned above, the wave can reflect from a well. While tunneling is universally recognized as a specific wavelike behavior, the reflection of light from a glass plate as a specific wavelike behavior is much less appreciated. But, in the language of potential, it is the reflection from a potential well.

Special Relativity and How it Works. Moses Fayngold

ISBN: 978-3-527-40607-4

The second confusing issue is the above-mentioned fact that the speed of light in glass is smaller than in air, whereas one would expect it to increase as a result of the attraction toward the glass.

But this inconsistency is illusory, because generally it is the *momentum* that must, under conditions, increase that is just what happens to light in a medium with refraction index $n > 1$. The increase of the momentum with simultaneous decrease of speed is due to the fact that a photon in transparent dielectric acquires an effective *rest* mass! The rest mass emerges here through the photon's entanglement with the medium, just as an electron in a crystalline lattice is characterized by an effective mass that is different from its mass in a free space [112].

Considering the photon in a medium with negligible absorption as a relativistic particle with the effective rest mass m_0 in an effective potential field U, we can find the values of m_0 and U from the equations for relativistic momentum and energy for such particle:

$$Pn = \frac{m_0 v}{\sqrt{1-(v^2/c^2)}} = \frac{m_0 c}{\sqrt{n^2-1}} \tag{E1}$$

and

$$\mathcal{E} = \sqrt{(Pn)^2 c^2 + m_0^2 c^4} + U. \tag{E2}$$

Here, $P = \mathcal{E}/c$ is the momentum in vacuum of a light pulse with energy $\mathcal{E}$ (for a single-photon state, $\mathcal{E} = \hbar\omega$), Pn is its momentum in the medium, and $v = c/n$ is its group velocity in the medium (for a nonabsorbing and hence nondispersive medium, the group velocity equals the phase velocity). If we put these expressions into the above equations and solve them for m_0 and U, we will have

$$m_0 = \frac{\mathcal{E}}{c^2} n\sqrt{n^2-1}, \qquad n > 1, \tag{E3}$$

and

$$U = \mathcal{E}(1-n^2). \tag{E4}$$

The rest mass disappears, as it should, in a free space. On the other hand, for sufficiently large n, the rest mass becomes greater than the total mass $\mathcal{E}/c^2$ of the free photon. This "contradiction" is explained by the above-mentioned fact that, according to (E4), a photon in the medium with refractive index $n > 1$ is characterized by negative potential energy. This negative contribution compensates for the rest mass increase.

(For high frequencies, n is less than 1 and is determined by $n = \sqrt{1-(\omega_p^2/\omega^2)}$, where ω_p is the plasma frequency [17]. In this case, the group velocity

$$v = \frac{d\omega}{dk} = \frac{d\omega}{d((\omega/c)n)} = \frac{c}{n + \omega(dn/d\omega)} = cn, \qquad n < 1, \tag{E5}$$

and solving Equation (E1) for m would give $m = (\mathcal{E}/c^2)\sqrt{1-n^2}$. We cannot use Equation (E3) in this case because plasma is an essentially dispersive medium and the definition of electromagnetic energy in a dispersive medium is more complicated [17].)

Thus, we have found that passing of light into glass can be described in terms of a particle that is attracted to the region with negative potential; similarly, passing of light into a medium with $n < 1$ (say, plasma) can be described in terms of particle that is repelled from the region with positive potential.

In the work [124], a single photon tunnels through an air gap between two prisms. The gap serves as a potential barrier. At a certain width of the barrier (approximately 0.1 of the photon's wavelength), there is a 50-50 chance for a photon either to tunnel through or to be deflected by 90°. In the "normal" situation (indefinite number of photons), the system works as a beam splitter. In a single-photon state, however, with two detectors placed each in one of its two respective paths, the photon turns out to have chosen only one option. The one corresponding to tunneling through the splitter reveals the wavelike nature of the photon and the fact that only one detector clicks (the photon does not split but is detected as one single object) reveals its particlelike nature in the same experiment. Here a single-photon state turns out to be neither wave nor particle by displaying both wavelike and particlelike features in one single trial.

There is another way to produce the effective repelling (or attracting) potential for a particle – by producing the alternating regions of different potentials (Figure 11.13a). Such potential field is known as the Kronig–Penni's potential. The corresponding systems have been manufactured and widely used in modern technology. They are known as periodic structures. The energy spectrum of a particle in such a structure consists of alternating regions of allowed and forbidden energy bands (Figure 11.13b). A particle with energy within an allowed band can propagate freely (dissipation neglected) through the structure. If the particle's energy falls within a forbidden band of the spectrum ("energy gap"), it is not allowed to live within the structure. However, if the structure fills only semispace, then a particle entering it from the free space with energy that is forbidden in the structure can still penetrate a certain distance into the structure. This distance, called penetration depth, is a characteristic depending on the parameters of the structure and particle's energy. The wave function of the particle in such state consists of the two running waves (one incident and one reflected) in free space and one exponentially decaying wave within the structure (Figure 11.14a). The latter wave is of the type mentioned above, with an imaginary wavenumber. The other wave of this type, exponentially increasing, while being mathematically possible, is not physically allowed, since it would increase with the distance from the interface. As a result, the wave entering from the left can only survive within a certain region (penetration depth) near the interface. Since this is not a running wave, there is no energy flux associated with it. It can be thought of as the incident energy (particle) just peeping in and turning back. Accordingly, the reflected wave has the same amplitude as the incident one – the structure acts as an ideally reflecting barrier (you can call it the mirror) for certain energies.

If the structure has a finite width, then the exponentially increasing wave inside is also allowed physically, since its amplitude does not have a chance to build up to infinity. Due to continuity of the wave function, this results in the emergence of a nonzero running wave in free space on the opposite side of the structure (Figure 11.14b). In the "particle and barrier" language, there appears a nonzero chance of particle's transmission (a particle that peeped inside the barrier may be lucky to reach its opposite side before it knows that it is forbidden to be within).

If the particles in question are the photons, the corresponding layered periodic structure playing the role of the optical barrier can be made of the alternating transparent glass films with different indexes of refraction.

Appendix F
Cause and Effect

Suppose a certain factor $F(t)$ (cause) that may last for a long time produces a certain effect $\epsilon(t)$ at a moment of time t. We assume that the process leading to this effect is linear: for each small time interval dt' around a moment t', there is an incremental contribution $d\epsilon(t)$ to $\epsilon(t)$ proportional to $F(t')$ at this moment:

$$d\epsilon(t) = G(t, t')F(t')dt'. \tag{F1}$$

The proportionality coefficient $G(t, t')$ determining the response $d\epsilon(t)$ is, generally, a function of moments t and t'. It is a special case of an extremely important concept of the Green's function [122].

Our world is time uniform: any process repeated under the same conditions an arbitrary time T later looks the same as before. The only possible time dependence of the Green's function on moments t and t', which would satisfy this requirement, is dependence on difference $t - t'$: the shift by T will not change this difference and thereby the response. Thus,

$$G(t, t') = G(t - t'), \tag{F2}$$

so that

$$d\epsilon(t) = G(t - t')F(t'). \tag{F3}$$

The net effect at time t will be the sum of the incremental effects (F3) from all preceding moments t':

$$\epsilon(t) = \int_{-\infty}^{\infty} d\epsilon(t) = \int_{-\infty}^{\infty} G(t - t')F(t')dt'. \tag{F4}$$

Consider an example: a motion of a particle in a straight line along a certain direction $\hat{\mathbf{r}}$ under a force $f(t)\hat{\mathbf{r}}$. In this case, the force is a cause and the resulting displacement of the particle is the effect, that is,

$$F(t) = f(t), \qquad \epsilon(t) = r(t). \tag{F5}$$

What will be the displacement by moment t resulting from the force f applied all the time?

Special Relativity and How it Works. Moses Fayngold

ISBN: 978-3-527-40607-4

Suppose the force is small enough so that the particle's speed accumulated even over a long time remains much less than c. In this case, we can describe the motion by nonrelativistic equations. We can find the answer in two different ways. First by performing straightforward integration of the equation of motion

$$m_0\ddot{r} = f(t). \tag{F6}$$

Since we are interested only in the displacement caused by this force, we must look for a special solution satisfying the conditions that the initial displacement and velocity of the particle be zero:

$$r(t_0) = 0, \qquad \dot{r}(t_0) = 0. \tag{F7}$$

Here the word "initial" denotes the moment when the force started acting. The solution can be written as

$$r(t) = \int_{t_0}^{t} \dot{r}(t')\mathrm{d}t' = \int_{t_0}^{t}\left(\int_{t_0}^{t'} \ddot{r}(t'')\mathrm{d}t''\right)\mathrm{d}t' = \int_{t_0}^{t}\left(\int_{t_0}^{t'} m_0^{-1} f(t'')\mathrm{d}t''\right)\mathrm{d}t'. \tag{F8}$$

In case when the force and, accordingly, acceleration, are constant, $\ddot{x}(t) = \mathrm{const} = a$, we get the familiar result $r(t) = (1/2)a(t - t_0)^2$.

In general case, however, when the force may be an arbitrary function of time, the double integration, while being straightforward, may not be very efficient; all the more so in the case of equations involving derivatives of order higher than 2. In this case, a better approach would be to chop the acting agent $F(t)$ into a succession of actions lasting over infinitesimally small segments of time, find the effect of each such short action as described in the beginning of the section, and then add them up together. In our case, when the acting agent is a force f, we think of it as composed of very short impulses $\mathrm{d}p' = f(t')\mathrm{d}t'$. Each such impulse produces the corresponding incremental change in velocity $\Delta v' = \Delta p'/m_0$; were this short impulse the only actor in the play, then such velocity would, in turn, produce the displacement by the moment t:

$$\mathrm{d}r(t) = \frac{\Delta p'}{m_0}(t - t') = \frac{f(t')}{m_0}(t - t')\mathrm{d}t'. \tag{F9}$$

This simple argument also explains how an impulse at a moment t' can cause an effect at a much later moment t: in this case it acts through inertia – the incremental velocity $\Delta v'$ acquired from this single impulse would last forever, accordingly producing ever increasing displacement. Adding up all the displacements corresponding to their respective impulses yields

$$r(t) = \int_{-\infty}^{t} G(t, t') f(t')\mathrm{d}t'. \tag{F10}$$

Comparing with (F9) shows that

$$G(t, t') = \frac{1}{m_0}\begin{cases} (t - t'), & t > t', \\ 0, & t < t'. \end{cases} \tag{F11}$$

This is the Green function of the operator d^2/dt^2. We see that $G(t, t')$ has a very simple physical meaning: it is a response at a moment t to an instant impulse of normalized force at an earlier moment t'. Therefore, we can also call it the response function. We see that it satisfies much harsher initial conditions than the displacement itself. The latter is only required to have the zero values at a specified moment $t = t_0$. The Green's function, on the contrary, is required to be zero at all moments before $t = t'$. This requirement reflects a simple but fundamentally important physical condition: there is no effect before its cause. No impulse acting at $t = t'$ can produce any effect at an earlier time $t < t'$. All physically observed effects at a given location happen either simultaneously with or after the corresponding cause at this location. Applied to our situation, there is no displacement before the corresponding impulse of force causing it.

This condition allows us to rewrite solution (F10) in a more simple and convenient form, retaining the time t dependence only in the integrand, but not in the upper limit of the integral:

$$r(t) = \int_{-\infty}^{\infty} G(t, t') f(t') \mathrm{d}t' \tag{F12}$$

(all the impulses acting on the particle *after* the moment of interest t cannot contribute to the displacement at time t due to condition (F11)).

The Green function satisfying this condition says that any effect observed in the real world happens with some retardation relative to the corresponding acting cause. Accordingly, this type of Green's function is called the retarded Green function. (It is mathematically possible to introduce an advanced Green's function – the opposite to the retarded one. We will discuss such a possibility later.)

Since a temporal Green's function turns out to reflect one of the most fundamental properties of time, and these properties are also studied in relativity, it is worthwhile to devote some attention to general properties of the Green function of an arbitrary linear differential operator. We will describe this in the spirit of quantum mechanics, in terms of the eigenfunctions and eigenvalues of the corresponding operator.

So consider a linear operator $\hat{L}$ with the spectrum of its eigenfunctions Ψ_s and eigenvalues L_s:

$$\hat{L}\Psi_s = L_s \Psi_s. \tag{F13}$$

Consider now the inverse operator, that is, an operator which, when applied together with $\hat{L}$ will produce a unit operator:

$$\hat{L}^{-1}\hat{L}\Psi = \Psi. \tag{F14}$$

Applying $\hat{L}^{-1}$ to both parts of Equation (F13) yields

$$\Psi_s = L_s \hat{L}^{-1} \Psi_s \tag{F15}$$

or

$$\hat{L}^{-1}\Psi_s = L_s^{-1}\Psi_s. \tag{F15a}$$

From the purely mathematical viewpoint, (F15) is the Fredholm's homogeneous integral equation of the first kind with a characteristic number L_s^{-1}. We see that the set L_s^{-1} forms the spectrum of eigenvalues of the inverse operator $\hat{L}^{-1}$. The direct and inverse operators have the same set of eigenfunctions and eigenvalues L_s or L_s^{-1}, respectively.

Consider now an arbitrary function Ψ from the appropriate class of functions (square-integrable, etc.) and expand it over the eigenfunctions Ψ_s:

$$\Psi = S\, b_s \Psi_s. \tag{F16}$$

Symbol S here denotes the summation over discrete and integration over continuous ranges of the L_s-spectrum.

The expansion (F16) allows us to find a special solution of the inhomogeneous equation

$$\hat{L}\Psi = \rho \tag{F17}$$

with the given right-hand side (the source), $\rho(x)$. Indeed, such a solution can be written immediately using the inverse operator $\hat{L}^{-1}$:

$$\Psi = \hat{L}^{-1}\rho. \tag{F18}$$

Thus, our problem now reduces to finding the explicit expression for the inverse operator.

Assuming that $\rho(x)$ belongs to the same class $\Psi(x)$, we expand it over the same eigenfunctions:

$$\rho = S\, a_s \Psi_s. \tag{F19}$$

Here,

$$a_s = \int \rho \Psi_s dx. \tag{F20}$$

Now, find the relation (F17) between ρ and Ψ in the $\hat{L}$-representation. Using expansions (F16) and (F19) for Ψ and ρ respectively and keeping in mind that Ψ_s are the eigenfunctions of $\hat{L}$, we obtain

$$S\, b_s L_s \Psi_s = S\, a_s \Psi_s. \tag{F21}$$

It follows

$$a_s = L_s b_s; \quad b_s = L_s^{-1} a_s. \tag{F22}$$

That's it! The differential relation (F17) (or its integral counterpart (F18)) in the x-representation converts into a simple algebraic relation (F22) in the $\hat{L}$-representation.

Using this result, we obtain the explicit expression for solution of Equation (F18) as a series

$$\Psi = S\, a_s L_s^{-1} \Psi_s. \tag{F23}$$

The same can be obtained by putting (F16) and (F19) directly into (F18) (Problem 11.4).

We have expressed the sought-for solution in terms of a_s, that is, through the $\hat{L}$-representation of the source ρ. In view of (F20), we can also express it directly in terms of the source in x-representation:

$$\Psi(x) = S\, L_s^{-1} \left\{ \int \rho(x')\Psi_s^*(x')\mathrm{d}x' \right\} \Psi_s(x) = \int S \frac{\Psi_s(x)\Psi_s^*(x')}{L_s} \rho(x')\mathrm{d}x'. \tag{F24}$$

The sum

$$S \frac{\Psi_s(x)\Psi_s^*(x')}{L_s} \equiv G(x, x') \tag{F25}$$

is the Green function of the $\hat{L}$-operator. Comparing this with (F18), one can see immediately that the Green's function is the kernel of the inverse operator $\hat{L}^{-1}$, which turns out to be, as one could have suspected from the very beginning, an integral operator:

$$\hat{L}^{-1} = \int \mathrm{d}x' G(x, x') \ldots \tag{F26}$$

Just as a differential operator is "poised" in wait to act on a function, so is the integral operator. The function (in our case $\rho(x')$) is to be put in the dotted place in (F26). Doing this, we finally get what we wanted – the explicit expression for solution in terms of the source in the x-representation:

$$\Psi(x) = \int G(x, x')\rho(x')\mathrm{d}x'. \tag{F27}$$

The sum (F25) can be considered as the expansion of the kernel of the $\hat{L}^{-1}$-operator over its eigenfunctions. If we expand over $\Psi(x)$, then we get $g(s, x') = L_s^{-1}\Psi(x')$ as the expansion coefficients. If we expand over $\Psi^*(x')$, we get for the expansion coefficients the expression $g(s, x) = L_s^{-1}\Psi^*(x')$. Finally, we can consider the expanding $G(x, x')$ as a function of two variables, over the set of products $\tilde{\Psi}_{s,s'}(x, x') \equiv \Psi_s(x)\Psi_{s'}^*(x')$ in the two-dimensional space (x, x'), since such set also forms an orthogonal system of functions in (x, x'). In this case, we get for the expansion coefficients $g(s, s') = L_s^{-1}$. Which means they have a remarkable property

$$g(s, s') = L_s^{-1} \begin{cases} \delta_{ss'}, & s \text{ discrete}, \\ \delta(s - s'), & s \text{ continuous}. \end{cases} \tag{F28}$$

Let us consider the Green function of a unit operator $\hat{L} = \hat{L}^{-1} = 1$. All its eigenvalues are equal to 1. By definition of $\hat{L}$, any eigenfunction of any operator will also be the eigenfunction of $\hat{L}$ with the unit eigenvalue. Therefore, the Green's function $G_U(x, x')$ of the unit operator $\hat{L}$ can be equally composed out of any complete set of eigenfunctions. This means that all spectra of all linear operators obey one common fundamental law:

$$\left.\begin{array}{l} S\psi_s(x)\psi_s^*(x') \\ S\phi_s(x)\phi_s^*(x') \\ \dots \\ S\varphi_s(x)\varphi_s^*(x') \end{array}\right\} = G_U(x, x'). \tag{F29}$$

It is easy to find the explicit expression for the universal Green function $G_U(x, x')$ in a closed form. We have from the definition of the unit operator,

$$\hat{L}\Phi = \hat{L}^{-1}\Phi = \Phi. \tag{F30}$$

According to this definition, the solution of operator equation with the $\hat{L}$-operator and with any source is always identical to this source. Expressing this solution in terms of $G_U(x, x')$ by using (F27) gives

$$\Phi(x) = \int G_U(x, x')\Phi(x')\mathrm{d}x'. \tag{F31}$$

Since this must hold for *any* function $\Phi(x)$, we see that $G_U(x, x')$ is nothing else but Dirac's delta function:

$$G_U(x, x') = \delta(x - x').$$

And vice versa, the delta function is, from this viewpoint, the Green function of the unit operator.

Combining Equations (F29) and (F31) gives the result that for a complete set of eigenfunctions of any linear operator one has:

$$\left.\begin{array}{l} S\psi_s(x)\psi_s^*(x') \\ S\phi_s(x)\phi_s^*(x') \\ \dots \\ S\varphi_s(x)\varphi_s^*(x') \end{array}\right\} = \delta(x - x'). \tag{F32}$$

This can be considered as x-representation of the fundamental property of eigenfunctions

$$\int \psi_s(x)\psi_{s'}^*(x)\mathrm{d}x = \begin{cases} \delta_{ss'}, & s \text{ discrete;} \\ \delta(s - s'), & s \text{ continuous.} \end{cases} \tag{F33}$$

Now we are prepared for finding the equation for the Green function of an arbitrary linear operator. To this end, express the unit operator as a product of the operator in question $\hat{L}$ and its inverse and then use the integral representation (F26) of the inverse operator:

$$\hat{L} = \hat{L}\hat{L}^{-1} = \int \hat{L}G(x - x')\dots\mathrm{d}x' = \int \delta(x - x')\dots\mathrm{d}x'. \tag{F34}$$

It follows

$$\hat{L}G(x,x') = \delta(x-x'). \tag{F35}$$

We can get the same result applying $\hat{L}$ directly to (F25) (Problem 11.5).

Thus, the Green function of a linear operator is the solution of an inhomogeneous equation (F35) with a singular point – precisely as defined in the beginning of this section from purely physical considerations.

The Green function (F11) is one of the simplest possible. The initial operator generating this function is

$$\hat{L} = \frac{d^2}{dt^2}. \tag{F36}$$

It has the eigenvalues $L(\omega) = -\omega^2$ and the corresponding normalized eigenfunctions are $\psi_\omega(t) = (2\pi)^{-1/2}e^{-i\omega t}$. Using (F25), we find the Green function (F35) in the ω-representation (which is just its Fourier transform):

$$G(t,t') = -\frac{1}{2\pi}\int_{-\infty}^{\infty} \frac{e^{i\omega(t-t')}}{\omega^2} d\omega. \tag{F37}$$

The integration can be carried out by writing $\omega \to \omega - i\alpha$, then using Cauchy's theorem

$$\frac{1}{2\pi i}\oint \frac{f(z)}{(z-a)^2} dz = f'(a), \tag{F38}$$

and then taking the limit $a = i\alpha \to 0$. The result coincides with (F11). This Green function is somewhat exotic in the sense that it does not satisfy the standard boundary conditions.

As another example consider an operator

$$\hat{L} = \frac{d^2}{dt^2} + \gamma\frac{d}{dt} + \omega_0^2. \tag{F39}$$

You can easily recognize this as the differential operator of the equation describing a harmonic oscillator under an external force $f(t)$:

$$\hat{L}y(t) = \left(\frac{d^2}{dt^2} + \gamma\frac{d}{dt} + \omega_0^2\right)y(t) = f(t). \tag{F40}$$

We assume here that the cause (force $f(t)$) is acting only along the y-direction.

The corresponding Green function can be found in two different ways. We can apply the rule (F35) and find a solution of the equation

$$\left(\frac{d^2}{dt^2} + \gamma\frac{d}{dt} + \omega_0^2\right)G(t,t') = \delta(t-t') \tag{F41}$$

satisfying the condition

$$G(t,t') = 0, \quad t < t' \tag{F42}$$

(no effect before the cause) (Problem 11.7). Or, we can first find the Green function G (ω) in the ω-representation and then determine its Fourier transform:

$$G(t,t') = \frac{1}{2\pi}\int G(\omega)e^{i\omega(t-t')}d\omega \tag{F42a}$$

(from now on, the absence of the integration limits will mean integration from minus to plus infinity).

We will use the second way. The eigenvalue $L(\omega)$ of the operator (F40) is

$$L(\omega) = \omega_0^2 - \omega^2 + i\gamma\omega, \tag{F43}$$

so that according to the general approach outlined in (F22), we have

$$G(\omega) = L^{-1}(\omega) = \frac{1}{\omega_0^2 - \omega^2 + i\gamma\omega}. \tag{F44}$$

Solving the equation $L(\omega) = 0$, we find that $G(\omega)$, considered as a function of complex variable, has two poles at

$$\omega_1 = \omega_\gamma + i\frac{\gamma}{2} \quad \text{and} \quad \omega_2 = -\omega_\gamma + i\frac{\gamma}{2}, \qquad \omega_\gamma \equiv \sqrt{\omega_0^2 - \frac{1}{4}\gamma^2}. \tag{F45}$$

At $\gamma > 0$, both poles are in the upper half of the complex ω-plane (Figure F1) and, consequently, the function $G(\omega)$ is holomorphous in the lower part of this plane (situation would be the opposite at $\gamma < 0$). At $\omega_\gamma = 0$, both poles merge into one pole of second order. Since $e^{i\omega(t-t')}$ is holomorphous in all planes, the whole integrand in (F42a) has the same properties as $G(\omega)$. This allows us to find the integral (F42a). Denote

$$\tau \equiv t - t', \tag{F46}$$

and consider separately $\tau < 0$ and $\tau > 0$. In the first case, the integral (F42a) can be completed by the integral over infinitely remote semicircle in lower semiplane. The resulting integral over the closed curve is zero in view of the analiticity of the integrand in the lower semiplane.

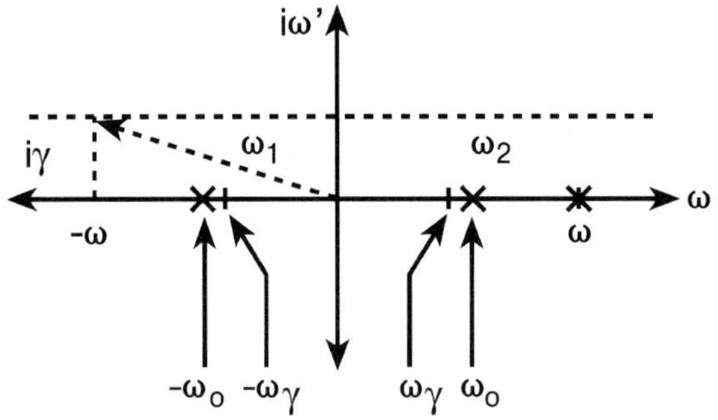

Figure F1

In the second case, (F42a) is also equivalent to analogous integration over a closed curve, but now in the upper semiplane. Excluding the case $\gamma = 2\omega_0$, this integral is the sum of residues of the function $G(\omega)e^{i\omega\tau}$:

$$G(\tau) = -\frac{i}{2\omega_\gamma}\left(e^{i\omega_\gamma\tau} - e^{-i\omega_\gamma\tau}\right), \quad \tau > 0. \tag{F47}$$

If $\gamma > 2\omega_0$, then ω_γ is imaginary and (F47) becomes the sum of two real terms – exponential functions of time. Mathematically, both terms enjoy equal rights, but physically, we retain only the one with "descent" asymptotic behavior, that is, decreasing with time.

Combining all these results, we have the following expression for the Green's function in the τ-representation:

$$G(\tau) = \begin{cases} 0, & \tau < 0, \\ \left.\begin{matrix} \omega_\gamma^{-1} e^{-(1/2)\gamma\tau}\sin\omega_\gamma\tau, & 0<\gamma<2\omega_0 \\ -\dfrac{1}{2|\omega_\gamma|}e^{-(|\omega_\gamma|-(1/2)\gamma)\tau}, & \gamma>2\omega_0 \end{matrix}\right\}, & \tau > 0. \end{cases} \tag{F48}$$

The graph of $G(\tau)$ is shown in Figure F.2. We recognize the familiar behavior of a damped oscillator after the action of instant force. The oscillator is stationary before the zero moment $\tau = 0$, then starts oscillating after being subject to instant hit at the zero moment; these oscillations gradually die out due to the damping factor γ. If the medium is so viscous that $\gamma > 2\omega_0$, the oscillator, after being displaced by an instant infinite force (the finite instant force would not produce a displacement), gradually returns to its equilibrium position without vibrations.

The Green function (F48) describes natural behavior: the system responds after the action of an external agent. The effect follows the cause. Such Green function is called the retarded Green function.

What if the damping factor is negative? Then it would be more convenient to call it the "pumping" factor. The oscillator would be pumped up with energy converging at it from the medium. The corresponding Green function in the ω-representation would be holomorphic in the upper semiplane and would have poles in the lower semiplane. The same calculation as before would show its τ-representation to be zero at $\tau > 0$ and nonzero at $\tau < 0$. If we, as before, consider the instant force at $\tau = 0$ as the cause, then the system's behavior would appear acausal: it gradually starts vibrating long before the hit, the oscillation amplitude is increasing, then there is a "hit" at the zero moment and the system stops. This is equivalent to running the previous

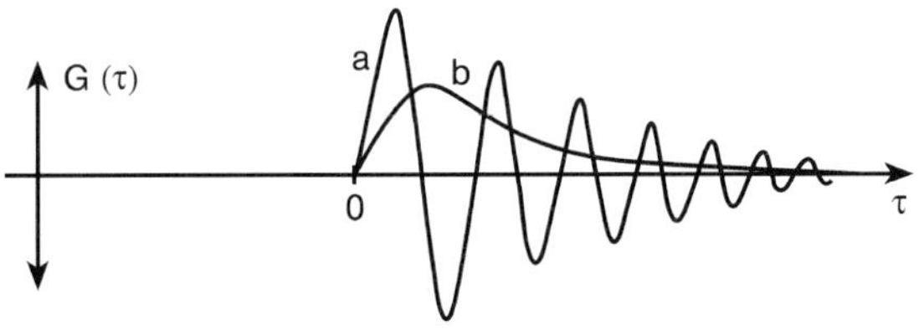

Figure F2

situation in reverse. From this viewpoint, the causality, in a sense, is not violated. The hit that provided the system with energy in the previous situation is now replaced by "antihit" taking the energy from the system. A dissipative medium, draining the energy out of the system, is now replaced by an "antidissipative" medium pumping the energy into the system. It seems that the cause and effect here just exchange places. The cause of the seemingly unprovoked excitation of the system in the second situation is the antientropic action of the medium pumping the energy into it (in a very coordinated way to keep and even increase the system's oscillations); the effect is the recoil of the "antihitting" mechanism, to which the system transfers all the energy accumulated from the medium, and stops. Unusual as it looks, it seems logically consistent. Except for one thing: if the medium was the cause of the system's excitation all the time before the zero moment, $-\infty < \tau < 0$, then why does not the same repeat after the moment $\tau = 0$? We do not have this difficulty in the previous situation. There are two possibilities to solve this puzzle in the second situation.

1. We can consider the whole process as a gigantic statistical fluctuation. But this would be already beyond the scope of mechanics describing the medium by a phenomenological parameter γ. And, worse, this would be an implicit appeal to the previous (direct) process, as it is this process that creates absolutely improbable conditions leading to such a coordinated and lasting fluctuation after the reverse.

2. We can postulate that in the second situation the cause of system's excitation during all the time $\tau < 0$ is the "hit" (actually, "antihit") that has to happen at $\tau = 0$, rather than coincidental help from the medium. This would be close to a kind of "predestination" known in some religious traditions.

Since in the second case the system responds in advance to the action of the cause, the corresponding Green function is called the advanced Green function.

In both cases, we come to a fundamental conclusion:

> *The retarded or advanced character of the Green function of a linear differential operator is intimately linked to analyticity of its Fourier transform* $G(\omega)$ in one of the semiplanes ω, where ω is considered as a complex variable.

These results have very important physical applications in the description of the EM wave propagation in a medium.

Appendix G
Permittivity and Refractive Index

We will now apply the above results to describe the collective response of many harmonic oscillators on light propagation in a medium. We consider the medium as an ensemble of uniformly distributed atoms or molecules that can be polarized by the electric field of a passing wave. This polarization, in turn, produces the secondary field (the medium's response). For each monochromatic component of the field, this response consists of two contributions. The first contribution exactly cancels the incident wave (Ewald–Ozeen's extinction theorem [123]) and the remaining residue propagates with phase velocity u different from c and depending on frequency ω:

$$u(\omega) = \frac{c}{n_r(\omega)}. \tag{G1}$$

The quantity $n_r(\omega)$ is the real part of the refraction index of a given medium:

$$n(\omega) = n_r(\omega) + i n_i(\omega). \tag{G2}$$

Its imaginary part $n_i(\omega)$ determines the absorption rate for a given frequency.

The refraction index of a medium arises as the medium's response to applied external field. It is a collective effect of the individual responses, and we want to trace out its origin.

As we just mentioned, the applied field polarizes each atom of the medium, that is, induces in each atom the corresponding dipole moment. So let us first focus on this individual effect, making a few simplifying assumptions. We will represent each atomic electron as the classical point charge held in its equilibrium position by a restoring force

$$\mathbf{f}_r = -\kappa\delta\mathbf{r} = -m\omega_0^2\delta\mathbf{r}. \tag{G3}$$

Here, κ is the effective "spring constant" of this force, m is the electron's mass, $\delta\mathbf{r}$ is a displacement from the equilibrium position, and ω_0 is the corresponding proper frequency. The electron is also subject to collisions with neighboring particles, which produce the damping force on it:

$$\mathbf{f}_d = -m\gamma\frac{d\delta\mathbf{r}}{dt}. \tag{G4}$$

Special Relativity and How it Works. Moses Fayngold

ISBN: 978-3-527-40607-4

We also neglect the effect on the electron from the secondary field due to polarization of all other atoms. The corresponding approximation is very accurate for a medium like a dilute gas.

If the field is not outrageously strong, the resulting motion is accurately described by the equation of motion in the nonrelativistic limit:

$$\hat{L}(t)\delta\mathbf{r}(t) \equiv \left(\frac{d^2}{dt^2} + \gamma\frac{d}{dt} + \omega_0^2\right)\delta\mathbf{r}(t) = \frac{q}{m}\mathbf{E}(t). \tag{G5}$$

We are interested not so much in displacement, as in the resulting dipole moment. Writing $\delta\mathbf{p} = q\delta\mathbf{r}$ for an induced dipole moment due to one electron, we see that we must simply multiply the solution of the differential equation (G5) by the electron charge q.

The differential operator $\hat{L}(t)$ in this equation exactly coincides with that in (F40). Therefore, we can immediately write the expression for the induced elementary dipole moment $\delta\mathbf{p}$ in terms of the corresponding response function:

$$\delta\mathbf{p}(t) = \int \alpha(t,t')\mathbf{E}(t')dt', \qquad \alpha(t,t') = \frac{q^2}{m}G(t,t'), \tag{G6}$$

where $G(t,t')$ is the Green function of the operator $\hat{L}(t)$. The result (G6) shows that the instant displacement of the electron at a moment t is determined by the local values of the applied field at all preceding moments $t' < t$. But we can express the result in a much more simple form if we consider, instead of an arbitrary function of time $\mathbf{E}(t)$, the result of action of each monochromatic component of this field. In other words, we can rewrite (G6) in Fourier representation. According to general result (F44) obtained in the previous section, we have

$$\delta\mathbf{p} = \alpha(\omega)\mathbf{E}, \tag{G7}$$

where

$$\alpha(\omega) = \frac{q^2}{m}G(\omega) = \frac{q^2/m}{\omega_0^2 - \omega^2 + i\gamma\omega}. \tag{G8}$$

Here, the quantities $\delta\mathbf{p}$ and $\mathbf{E}$ mean the Fourier amplitudes of the dipole moment and field, respectively. We see that an induced instant polarization is determined by the field at the same instant only in a monochromatic or in a stationary field.

Expression (G8) constitutes the embryo out of which the refraction index forms. Although we have used the classical model for its derivation, the obtained result for $\delta\mathbf{p}$ is actually quantum-mechanical expectation value of the induced dipole moment.

As a next stage, we must take into account that even in a one-electron atom there are many characteristic frequencies associated with the corresponding optical transitions between different stationary states. Therefore, we replace in (G8) $\alpha \to \alpha_j$ (where j labels an initial state of the illuminated atom) and $\omega_0 \to \omega_{jk} = (\mathcal{E}_j - \mathcal{E}_k)/\hbar$. According to QM, each such transition contributes to the response function α_j, so that

$$\alpha_j(\omega) = \frac{q^2}{m}\sum_k \frac{f_{jk}}{\omega_{jk}^2 - \omega^2 + i\gamma\omega}. \tag{G9}$$

Here, f_{jk} is the weight of the corresponding contribution called the *oscillator strength*

$$f_{jk} \equiv \frac{2}{3}\frac{m\omega_{kj}|\mathbf{d}_{kj}|^2}{\hbar}, \tag{G10}$$

and $\mathbf{d}_{kj}$ is the displacement matrix element between states j and k.

The constant $\alpha_j(\omega)$ is called the atomic polarizability for a given type of atom, due to one electron in a state j.

Now we can evaluate the collective effect of the medium's response. To this end, we consider a volume element, which is large enough to contain many atoms but small enough to consider the field amplitude as constant within this volume. In other words, the linear size of the volume must be much less than the characteristic scale of spatial variation of the monochromatic field. In the case of a plane monochromatic wave, such scale is determined by the corresponding wavelength λ in the medium, and our condition then means that

$$a \ll l \ll \lambda, \tag{G11}$$

where a is the interatomic distance and l is the linear size of the chosen volume element.

If only one state j is populated, we can write the expression for resulting polarization (the induced dipole moment in unit volume) as

$$\mathbf{P} = N\alpha\mathbf{E} \equiv \varepsilon_0\chi_e\mathbf{E}, \tag{G12}$$

where N is the concentration of atoms. The new proportionality constant

$$\chi_e \equiv \frac{N\alpha}{\varepsilon_0} \tag{G13}$$

is called the electric susceptibility of the given medium.

The condition (G11) then allows us to consider $\mathbf{P}$ as a function of position. It is also a function of time, since the induced dipoles oscillate with the frequency of the field. Thus, we have $\mathbf{P} = \mathbf{P}(\mathbf{r}, t)$.

As we know, in a real situation the electrons are distributed between different states j with certain probabilities, depending, among other things, on temperature. Denote as N_j the number density of atoms in energy eigenstate j, so that $\sum_j N_j = N$. Then the polarizability averaged over all the states is given by

$$\alpha(\omega) = \frac{q^2}{m}\sum_j\sum_k \frac{N_j f_{jk}}{\omega_{jk}^2 - \omega^2 + i\gamma\omega}. \tag{G9a}$$

Consider now a volume element within a polarized material. Since polarization of the material can be different at different locations, it may result in the emergence of an effective nonzero charge within such volume, even though the medium as a whole is electrically neutral. We can visualize this effect if we imagine each dipole as a small stretched out snakelike creature with positive head and negative tail. It may so happen that because of, say, inhomogeneity of polarization function, the dipole concentration or dipole strength (the length of the creatures) will be different at different points. As a result, there may be more positive heads peeping into the volume on the one side, then peeping out on the other side. This will result in the appearance of the net nonzero charge within the volume. This charge is owing entirely to polarization of the material and as such is a bound charge: each electron can only move within the bounds of its respective atom. The resulting amount of the bound charge within the chosen volume is proportional to the net flux of polarization vector through the surface enclosing the volume. Introducing the bound charge density ρ_b, we can express this statement in differential form in terms of ρ_b:

$$\rho_b = -\vec{\nabla} \cdot \mathbf{P}. \tag{G14}$$

There will also appear the bound polarization currents owing to the oscillating motions of the constituting dipoles that form the bound charge:

$$\mathbf{j}_p = \frac{\partial \mathbf{P}}{\partial t}. \tag{G15}$$

Similarly, we can introduce magnetization **M** of the medium as the magnetic moment per unit volume and the corresponding bound magnetization current

$$\mathbf{j}_m = \vec{\nabla} \times \mathbf{M}. \tag{G16}$$

Now consider the implications. Our fundamental equations (6.104), being very general, hold in this case, too, but we have to rewrite them to include explicitly the bound charges and currents. Specifically, the divergence of **E** is determined by the *net* charge density, which is the sum of the free and bound densities, and curl of **B** is determined by the net current density in given location, which includes contribution from free current as well as polarization and magnetization bound currents in given location:

$$\begin{aligned} \varepsilon_0 \vec{\nabla} \cdot \mathbf{E} &= \rho_f + \rho_b, \\ \mu_0^{-1} \vec{\nabla} \times \mathbf{B} &= \mathbf{j}_f + \mathbf{j}_p + \mathbf{j}_m. \end{aligned} \tag{G17}$$

Using (G14)–(G16) yields

$$\vec{\nabla} \cdot \varepsilon_0 \mathbf{E} = \rho - \vec{\nabla} \cdot \mathbf{P} \quad \text{and} \quad \vec{\nabla} \times \frac{\mathbf{B}}{\mu_0} = \mathbf{j}_f + \nabla \times \mathbf{M} + \frac{\partial \mathbf{P}}{\partial t}.$$

It seems natural to collect together terms with $\vec{\nabla}$-operators. Then we will have

$$\vec{\nabla} \cdot \mathbf{D} = \rho_f \quad \text{and} \quad \vec{\nabla} \times \mathbf{H} = \mathbf{j}_f + \frac{\partial \mathbf{P}}{\partial t}.$$

Here we use the notations

$$\mathbf{D} \equiv \varepsilon_0 \mathbf{E} + \mathbf{P} \quad \text{(electrical displacement)} \tag{G18}$$

and

$$\mathbf{H} \equiv \frac{\mathbf{B}}{\mu_0} - \mathbf{M} \quad (\text{the auxiliary field } \mathbf{H}). \tag{G19}$$

Just as polarization, being a linear response to electric field, is a linear function (G12), the magnetization vector is in most cases a linear function of **H**:

$$\mathbf{M} = \chi_m \mathbf{H}, \tag{G20}$$

with frequency-dependent proportionality constant called the *magnetic susceptibility*. Using (G12) and (G20), Equations (G18) and (G19) can be written in a more simple form

$$\mathbf{D} = \varepsilon \mathbf{E}, \qquad \mathbf{B} = \mu \mathbf{H}. \tag{G21}$$

Here the coefficients

$$\varepsilon \equiv \varepsilon_0(1+\chi_e) \quad \text{and} \quad \mu \equiv \mu_0(1+\chi_m) \tag{G22}$$

are, respectively, the *permittivity* and *permeability* of the medium. Together with χ_e and χ_m, they are functions of ω. Equations (G21) are called *constitutive relations*.

The original Maxwell equations for linear media can now be written as

$$\begin{aligned} &\vec{\nabla} \cdot \mathbf{D} = \rho_f, \qquad \vec{\nabla} \times \mathbf{E} = -\frac{\partial \mathbf{B}}{\partial t}, \\ &\vec{\nabla} \cdot \mathbf{B} = 0, \qquad \vec{\nabla} \times \mathbf{H} = \mathbf{j}_f + \frac{\partial \mathbf{D}}{\partial t}. \end{aligned} \tag{G23}$$

We are interested in EM propagation through a homogeneous medium with no free charges or currents. Setting here $\rho_f = 0$, $\mathbf{j}_f = 0$ and using (G21) yields

$$\begin{aligned} &\vec{\nabla} \cdot \mathbf{E} = 0, \qquad \vec{\nabla} \times \mathbf{E} = -\frac{\partial \mathbf{B}}{\partial t}, \\ &\vec{\nabla} \cdot \mathbf{B} = 0, \qquad \vec{\nabla} \times \mathbf{B} = \mu\varepsilon \frac{\partial \mathbf{E}}{\partial t}. \end{aligned} \tag{G24}$$

This system of coupled equations is different from the corresponding equations (6.104) in vacuum by changing $\mu_0 \to \mu$, $\varepsilon_0 \to \varepsilon$. This change reflects the effect of a medium on wave propagation.

Now we are getting closer to the description of the effect itself. Decoupling the system gives

$$\left(\nabla^2 - \frac{1}{\mu\varepsilon}\frac{\partial^2}{\partial t^2}\right)\mathbf{E} = 0, \tag{G25}$$

and the similar equation for **B**. This can be written as

$$\left(\nabla^2 - \frac{1}{u^2}\frac{\partial^2}{\partial t^2}\right)\mathbf{E} = 0, \qquad u(\omega) = \frac{c}{n(\omega)}, \qquad n^2(\omega) \equiv \frac{\mu(\omega)\varepsilon(\omega)}{\mu_0\varepsilon_0}. \tag{G26}$$

Equation (G26) describes wave motion with a speed $u(\omega) = c/n(\omega)$. In contrast to the vacuum case, in which all the waves move with the same invariant speed, in the medium the speed u is different for different frequencies. Moreover, according to Equations (G22), the index of refraction is a complex number and so is $u(\omega)$.

Indeed, write the harmonic solution as

$$E = E_0 \mathrm{e}^{i(kx - \omega t)}. \tag{G27}$$

Putting it into (G25) gives the dispersion equation for ω and k:

$$k^2 - \frac{\omega^2}{c^2} n^2 = 0. \tag{G28}$$

Introduce explicitly the real and imaginary components:

$$k = k_\mathrm{r} + ik_\mathrm{i}, \qquad n = n_\mathrm{r} + in_\mathrm{i}. \tag{G29}$$

Then (G28) takes the form

$$k_\mathrm{r} + ik_\mathrm{i} = \frac{\omega}{c}(n_\mathrm{r} + in_\mathrm{i}), \tag{G30}$$

or

$$k_\mathrm{r} = \frac{\omega}{c} n_\mathrm{r}, \qquad k_\mathrm{i} = \frac{\omega}{c} n_\mathrm{i}. \tag{G31}$$

It follows

$$E = E_0 \mathrm{e}^{i(k_\mathrm{r} + ik_\mathrm{i})x - i\omega t} = E_0 \mathrm{e}^{-(\omega/c)n_\mathrm{i}x} \mathrm{e}^{i((\omega/c)n_\mathrm{r}x - \omega t)} = E_0 \mathrm{e}^{-(\omega/c)n_\mathrm{i}x} \mathrm{e}^{ik_\mathrm{r}(x - u_\mathrm{r}t)}, \tag{G32}$$

where

$$u_\mathrm{r} = \frac{\omega}{k_\mathrm{r}} = \frac{c}{n_\mathrm{r}}. \tag{G33}$$

Combining Equations (G9a), (G13), (G22) and (G26) we can write for a sufficiently diluted medium

$$n(\omega) = 1 + \frac{1}{2\varepsilon_0} \sum_j N_j \alpha_j(\omega) = 1 + \frac{q^2}{2m\varepsilon_0} \sum_j \sum_k \frac{N_j f_{jk}}{\omega_{jk}^2 - \omega^2 + i\gamma\omega}. \tag{G34}$$

References

1 Fayngold, M. (2002) *Special Relativity and Motions Faster than Light*, Wiley-VCH Verlag GmbH, Weinheim, Germany.
2 Taylor, E.F. and Wheeler, J.A. (1966) *Spacetime Physics*, Freeman & Co., San Francisco, CA.
3 Misner, C.W., Thorne, K.S. and Wheeler, J.A. (1973) *Gravitation*, Freeman & Co., San Francisco, CA.
4 Herbert, N. (1982) *Foundations of Physics*, **12**, 1171.
5 Herbert, N. (1988) *Faster than Light: Superluminal Loopholes in Physics*, New American Library, New York.
6 Eberhard, P. (1978) *Nuovo Cimento*, **46B**, 392.
7 Wooters, W.K. and Zurek, W.H. (1982) *Nature*, **299**, 802.
8 Dieks, D. (1982) *Physics Letters A*, **92** (6), 271.
9 Cramer, J.G. (1997) Quantum nonlocality and the possibility of superluminal effects. Proceedings of the NASA Breakthrough Propulsion Physics Workshop, August 12–14, Cleveland, OH.
10 Fayngold, M. (1981) *Soviet Physics – Teoreticheskaya: Matematicheskaya Fizika*, **46** (6), 395.
11 Fayngold, M. (1982) *Soviet Physics – Ukrainskiy Fizicheskiy Zhurnal*, **27** (3), 440.
12 Chiao, R.Y., Kwiat, P.G. and Steinberg, A.M. (2000) *Scientific American*, 98–106.
13 Steinberg, A.M., Kwiat, P.G. and Chiao, R.Y. (2001) The single-photon tunneling time. Proceedings of the 28th Rencontre de Moriond, Editions Frontieres, France.
14 Kuzmich, A., Dogarin, A., Wang, L.G., Milonni, P.W. and Chiao, R.Y. (2001) *Physical Review Letters*, **86** (18), 3925–3929.
15 Brillouin, L. and Zommerfeld, A. (1960) *Wave Propagation and Group Velocity*, Academic Press, New York.
16 Leontovich, M. (1972) The wave front propagation, in *The Seminar on Optics, Relativity, and Quantum Mechanics* (ed. L.I. Mandelstamm), Science, Moscow (in Russian).
17 Landau, L.D. and Lifshits, E.M. (1998) *Electrodynamics of Continuous Media*, Butterworth-Heinemann, Oxford.
18 Milonni, P.W. (2005) *Fast Light, Slow Light, and Left-Handed Light*, IOP Publishing.
19 Hafele, J.C. (1972) *American Journal of Physics*, **40**, 81–85.
20 Hafele, J.C. and Keating, R.E. (1972) *Science*, **177**, 166–170.
21 Tipler, P. (1992) *Elementary Modern Physics*, Worth Publishers, New York.
22 Krane, K. (1992) *Modern Physics*, John Wiley & Sons, Inc., New York.
23 Hewitt, P.G. (1992) *Conceptual Physics*, Harper Collins College Publishers, New York.
24 Bell, J. (1987) How to teach special relativity, in *Speakable and Unspeakable in Quantum Mechanics*, Cambridge University Press.

Special Relativity and How it Works. Moses Fayngold

ISBN: 978-3-527-40607-4

25 Fock, V.A. (1958) *The Theory of Space, Time, and Gravitation*, Pergamon Press, Oxford.

26 Woodhouse, N.M.J. (2003) Special Relativity, Springer.

27 Mandelstam, L.I. (1972) *Lectures on Optics, Relativity, and Quantum Mechanics*, Nauka (in Russian).

28 Fix, J.D. (1997) *Astronomy. Journey to the Cosmic Frontier*, WCB/McGraw-Hill, New York.

29 Landau, L.D. and Lifshits, E.M. (1998) *The Classical Theory of Fields*, Butterworth-Heinemann, Oxford.

30 Bergmann, P.G. (1942) *Introduction to the Theory of Relativity*, Prentice-Hall, Englewood Cliffs, NJ, Chapters 4 and 5.

31 Moeller, C. (1962) *The Theory of Relativity*, Oxford University Press.

32 Tonnelat, M.-A. (1959) Les Principes de la Theorie Electromagnetique ae de la Relativite, Massonet C12, Editeurs (in French).

33 Gilde, W. (1979) *Gespiegelte Welt*, Veb Fachbuch Verlag, Leipzig (in German).

34 Feynman, R.P., Leighton, R.B. and Sands, M. (1964) *The Feynman Lectures on Physics: Electrodynamics*, Addison-Wesley.

35 Wheeler, J.A. and Feynman, R.P. (1945) *Reviews of Modern Physics*, **17**, 157; Wheeler, J.A. and Feynman, R.P. (1949) *Reviews of Modern Physics*, **21**, 425.

36 Cramer, J.G. (1983) The arrow of electromagnetic time and generalized absorber theory. *Foundations of Physics*, **13**, 887.

37 Greene, B. (2000) *The Elegant Universe*. Vintage.

38 Wilczek, F. (1999) Mass without mass. *Physics Today*, Nov. 1999, p. 11–12.

39 Landau, L.D. and Lifshits, E.M. (1998) *Mechanics*, Butterworth-Heinemann, Oxford.

40 Ford, K.W., Hill, D.L., Wakano, M. and Wheeler, J.A. (1959) *Annals of Physics*, **7**, 239.

41 Fayngold, M. (1968) Some features of the spiral scattering, Reports of Uzbek Acad. of Sci., **6**, p. 50–54 (Russian).

42 Fayngold, M. (1977) *Astrometry and Astrophysics*, **33**, 35–40.

43 Fayngold, M. (1979) *Astronomical Journal*, **56** (3), 541–548.

44 Griffiths, D. (2006) *Introductory Electrodynamics*, Prentice-Hall.

45 Jackson, J.D. (1999) *Classical Electrodynamics*, 3rd edn, John Wiley & Sons, Inc., New York.

46 Housdorff, F. (1957) Set Theory, Chelsea Publ. Co.

47 Vilenkin, N. (1965) *Stories About Sets*, Nauka (in Russian).

48 Lem, S. (1985) *The Star Diaries*, Harvest/HBJ.

49 Einstein, A. and Laub, J. (1908) Über die Electromagnetischen Grundgleichungen für Bewegte Körper. *Annals of Physics (Leipzig)*, **26**, 532–540.

50 Hestenes, D. (2003) *American Journal of Physics*, **71** (2), 104–121; Hestenes, D. (2003) *American Journal of Physics*, **71** (7), 691–714.

51 Hecht, E. (2002) *Optics*, 4th edn, Addison-Wesley, San Francisco, CA.

52 Kerker, M. (1969) *The Scattering of Light and Other Electromagnetic Radiation*, Academic Press.

53 Nussenzveig, H.M. (2006) *Diffraction Effects in Semiclassical Scattering*, Cambridge University Press.

54 Arnold, S. (2001) Microspheres, Photonic Atoms and the Physics of Nothing. American Scientist, **89** (5), 414.

55 Aharonov, Y. and Rorlich, D. (2005) *Quantum Paradoxes*, Wiley-VCH Verlag GmbH, Weinheim, Germany.

56 Bolotowsky, B.M. (1985) On visible shape of fast moving bodies, in *Einstein Collection*, Science, Moscow. pp. 142–168; Bolotowsky, B.M. (1990) On visible shape of fast moving bodies, in *Einstein Collection*, Science, Moscow. pp. 279–328.

57 Bolotowsky, B.M. (1989) What do the superluminal objects look like? in *Problems of Theoretical Physics and Astrophysics*, Science, Moscow, pp. 24–56.

58 Landau, L. (1953) On multiple particle generation in the ultra-fast collisions.

Reports of the Academy of Sciences USSR, **17**, 51.

59 Fermi, E. (1924) Uber die Theorie des Stosses zwishen Atomen und electrish geladenen Teilchen. *Zeitschrift fur Physik*, **29**, 315–327.

60 Terrell, J. (1959) Invisibility of the Lorentz contraction. *Physical Review*, **116**, 1041–1045.

61 Feinberg, F.L. (1975) *Uspekhi Fizicheskikh Nauk*, **116** (4), 709–730 (Russian).

62 Dewan, E.M. (1963) *American Journal of Physics*, **31**, 383–386.

63 Crøn, Ø. (1977) *American Journal of Physics*, **45** (1), 65–70.

64 Nicolic, H. (1999) *American Journal of Physics*, **67** (11), 1007–1012.

65 Jefimenko, O.D. (1994) *American Journal of Physics*, **62** (1), 79–85.

66 Sorensen, R.A. (1995) *American Journal of Physics*, **63**, 413–45.

67 Pauli, W. (1958) *Theory of Relativity*, Pergamon Press, Oxford.

68 Zeldovich, Ya.B. and Novikov, I.D. (1971) *Relativistic Astrophysics*, vol. 1, University of Chicago Press, Chicago, IL, pp. 9–12.

69 Marder, L. (1971) *Time and Space Traveler*, Allen & Unwin, Chapter 3.

70 David Green, private communication.

71 Smorodinskii, Ya. (1972) *JETP Letters*, **16** (8), 356–357 (English translation).

72 Alessi, J.S., private communication.

73 Ehrenfest, P. (1909) *Physikalische Zeitschrift*, **10**, 918.

74 Rizzi, G. and Ruggiero, M.L. (eds) (2004) *Relativity in Rotating Frames*, Kluwer.

75 Wucknitz, O. (2004) Sagnac effect, twin paradox, and space–time topology – time and length in rotating systems and closed Minkowski space–times, arXiv:gr-qc/0403111v1.

76 Fayngold, M. Dynamics of Relativistic length contraction and the Ehrenfest paradox, arXiv: 0712.3891v1.

77 Sagnac, M.G. (1914) *Journal of Physics*, **4** (5), 177.

78 http://www.mathpages.com/rr/s2-11/211.htm.

79 Taylor, E.F. and Wheeler, J.A. (2000) *Exploring the Black Holes: Introduction to General Relativity*, Benjamin Cummings.

80 Einstein, A. (1905) On the electrodynamics of moving bodies. *Annalen der Physik*, **XVII**, 891–921.

81 Einstein, A. and Infeld, L. (1938) *Evolution of Physics*, Simon & Shuster.

82 Tolman, R.C. (1962) *Relativity, Thermodynamics, and Cosmology*, Clarendon Press, Oxford.

83 Planck, M. (1901) On the law of distribution of energy in the normal spectrum. *Annalen der Physik*, **4**, 553.

84 Einstein, A. (1905) On a heuristic viewpoint concerning the production and transformation of light. *Annalen der Physik*, **17**, 132–148.

85 Bohr, N. (1913) On the constitution of atoms and molecules, Part 1. *Philosophical Magazine*, **26**, 1–25.

86 de Broglie, L. (1926) *Waves and Motions*, Gauthier-Villars, Paris.

87 Landau, L. and Lifshits, E. (1981) *Quantum Mechanics: Non-Relativistic Theory*, Butterworth-Heinemann, Oxford.

88 Born, M. (1963) *Atomic Physics*, Blackie & Sons, London.

89 Dirac, P. (1964) *The Principles of Quantum Mechanics*, Yeshiva University, New York.

90 Blokhintsev, D.I. (1964) *Principles of Quantum Mechanics*, Allyn & Bacon, Boston, MA.

91 von Neumann, J. (1996) *Mathematical Foundations of Quantum Mechanics*, Princeton University Press.

92 Landau, L.D. and Peierls, R. (1931) *Zeitschrift fur Physik*, **69**, 56.

93 Berestetskiy, V.B., Lifshits, E.M. and Pitaevskiy, L.P. (1968) *Relativistic Quantum Theory*, Nauka, p. 1.

94 Davydov, A.S. (1965) *Quantum Mechanics*, Pergamon Press.

95 Klein, O. (1926) *Zeitschrift fur Physik*, **37**, 895.

96 Fock, V.A. (1926) *Zeitschrift fur Physik*, **38**, 242; Fock, V.A. (1926) *Zeitschrift fur Physik*, **39**, 226.

97 Gordon, W. (1926) *Zeitschrift fur Physik*, **40**, 117.

98 Born, M. (1951) *The Restless Universe*, Dover Publications, New York.

99 Bohm, D. (1952) *Quantum Theory*, Prentice-Hall.

100 Fixsen, D.J., Gheng, E.S., Gales, J.M., Mather, J.C., Shafer, R.A. and Wright, E.L. (1996) The cosmic microwave background spectrum from the full COBE FIRAS data set. *Astrophysical Journal*, **473**, 576–587.

101 Shore, S.N. (2003) *The Tapestry of Modern Astrophysics*, John Wiley & Sons, Inc., New York.

102 Thorne, K. (1994) *Black Holes and Time Warps: Einstein's Outrageous Legacy*, Norton & Co.

103 Branco, G.C., Lavoura, L. and Silva, J.P. (1999) *CP Violation*, Clarendon Press, Oxford.

104 Streater, R.F. and Wightman, A.S. (1964) *PCT, Spin and Statistics, and All That*, W.A. Benjamin, Inc., New York.

105 Bilaniuk, O.M., Deshpande, V.K. and Sudarshan, E.C.G. (1962) *American Journal of Physics*, **30** (10), 718–723.

106 Bilaniuk, O.M. and Sudarshan, E.C.G. (1969) *Physics Today*, **22** (5), 43–51.

107 Feinberg, J. (1967) *Physical Review*, **159** (5), 1089–1105.

108 Jones, F.C. (1972) *Physical Review D*, **6** (10), 2727–2735.

109 Dirac, P. (1931) Quantized singularities in the electromagnetic field. *Proceedings of the Royal Society A*, **133**, 60.

110 Jeon, H. and Longo, M.J. (1995) *Physical Review Letters*, **75**, 1443–1446.

111 Carroll, L. (1899) *Through the Looking Glass*, Mansfield & Wessels, New York.

112 Kittel, C. (1961) *Introduction to Solid State Physics*, John Wiley & Sons, Inc.

113 Greenberger, D.M., Horne, M.A. and Zeilinger, A. (1993) *Physics Today*, **46** (8), 22.

114 Nimtz, G. (2003) *Progress in Quantitative Electronics*, **27**, 417.

115 Goodman, J.W. (1968) *Introduction to Fourier Optics*, McGraw-Hill Book Co.

116 Wimmel, H.K. (1972) *Nature: Physical Science*, **236** (66), 79–80.

117 Aspect, A., Grangier, P. and Roger, G. (1982) *Physical Review Letters*, **49**, 91–94.

118 van Enk, S.J., arXiv: quant-ph/9803030 v.1, 13 Mar 1998.

119 Turner, M.J.L. (2004) *Rocket and Spacecraft Propulsion: Principles, Practice and New Developments*, Springer, Berlin.

120 Mukhin, K. (1972) Zanimatelnaya Yadernaya Fizika, Atomizdat (in Russian).

121 Davies, P. (2002) *How to Build a Time Machine*, Viking Penguin, New York.

122 Challis, L. and Sheard, F. (2003) The Green of Green's functions. *Physics Today*, **56** (12), 41–46.

123 Born, M. and Wolf, E. (1968) *Principles of Optics*, 4th edn, Pergamon Press.

124 Ghose, P., Home, D. and Agarval, G.S. (1991) *Physics Letters*, **A153**, 403–406.

125 Mizobuchi, Yu. and Ohtake, Yo. (1992) *Physics Letters*, **A168**, 1–5.

126 Pound, R.V. and Rebka, G.A. (1959) *Physical Review Letters*, **3** (9), 439–441.

127 Mossbauer, R.L. (1958) *Zeitschrift fur Physik*, **151**, 124.

128 Okun, L.B., Selivanov, K.G. and Telegdi, V.L. (2000), *Am. J. Phys.*, **68** (2), 115–119.

Index

Special Relativity and How it Works. Moses Fayngold

ISBN: 978-3-527-40607-4